BIOGEOGRAPHY AND ECOLOGY IN NEW ZEALAND

MONOGRAPHIAE BIOLOGICAE

Editor

J. ILLIES

Schlitz

VOLUME 27

DR. W. JUNK b.v. PUBLISHERS THE HAGUE 1975

BIOGEOGRAPHY AND ECOLOGY IN NEW ZEALAND

Edited by

G. Kuschel

Auckland

DR. W. JUNK b.v. PUBLISHERS THE HAGUE 1975

ISBN 90 6193 079 0

Cover Design: Max Velthuijs, The Hague
Printed in the Netherlands
Zuid-Nederlandsche Drukkerij N.V., 's-Hertogenbosch

CONTENTS

AUTHORS' ADDRESSES

P. C. Bull, Ecology Division, Department of Scientific and Industrial Research, Lower Hutt, New Zealand

F. M. Climo, National Museum, Wellington, New Zealand

J. D. Coulter, New Zealand Meteorological Service, Wellington, New Zealand

I. G. Crook, New Zealand Wildlife Service, Department of Internal Affairs, Wellington, New Zealand

J. S. Dugdale, Entomology Division, Department of Scientific and Industrial Research, Auckland, New Zealand

C. A. Fleming, New Zealand Geological Survey, Department of Scientific and Industrial Research, Lower Hutt, New Zealand

R. R. Forster, Otago Museum, Dunedin, New Zealand

E. J. Godley, Botany Division, Department of Scientific and Industrial Research, Christchurch, New Zealand

R. C. Green, Department of Anthropology, University of Auckland, Auckland, New Zealand

G. A. Knox, Department of Zoology, University of Canterbury, Christchurch, New Zealand

R. M. McDowall, Fisheries Research Division, Ministry of Agriculture and Fisheries, Wellington, New Zealand

I. D. McLellan, P.O. Box 95, Westport, New Zealand

B. Reid, New Zealand Wildlife Service, Department of Internal Affairs, Wellington, New Zealand

Q. W. Ruscoe, Soil Bureau, Department of Scientific and Industrial Research, Lower Hutt, New Zealand

J. T. Salmon, Zoology Department, Victoria University, Wellington, New Zealand

V. M. Stout, Department of Zoology, University of Canterbury, Christchurch, New Zealand

J. C. Watt, Entomology Division, Department of Scientific and Industrial Research, Auckland, New Zealand

A. H. Whitaker, Ecology Division, Department of Scientific and Industrial Research, Lower Hutt, New Zealand

G. R. Williams, New Zealand Wildlife Service, Department of Internal Affairs, Wellington, New Zealand

CHAPTERS' CONTENTS

complex, consists of thicker crust (20–25 km), and is perhaps older than the narrow simple Macquarie Ridge (> 6 km).

The other system of rises and ridges, including the Campbell Plateau, is believed to be composed of more ancient rocks – parts of the foreland (Precambrian and Lower Palaeozoic) and sediments (Carboniferous to Jurassic) of the New Zealand Geosyncline, disposed in a dominantly north-westerly direction but curving sinuously into a goose-necked orocline in the south, and broken up by narrow depressions (Norfolk, New Caledonia and Bounty basins) of such depth as to imply attenuation if not absence of sialic crust. The rises are underlain by sub-continental crust (15–25 km) thickening to continental dimensions (> 30 km) below New Zealand and its shallower culminations (CULLEN, 1970). Whereas the Kermadec-Macquarie ridge systems appear to be youthful thickenings of an originally oceanic crust (< 10 km), the more ancient rises are probably relics of an area of fully continental crust, subsequently thinned.

New Zealand and Gondwanaland

To geologists who believe, following CAREY (1955), that the curvature of oroclines is the result of later torsion of an originally straight or gently curved structure, the New Zealand Geosyncline, extending from New Caledonia to the Chatham Islands, seems to have been distorted and ruptured during the Early Cretaceous climax of the Rangitata Orogeny, which coincided in time with the main dispersal movements of component continental fragments of Gondwanaland. In any construction of Gondwanaland, therefore, the remnants of the foreland, now constituting the Lord Howe Rise, the westernmost parts of the South Island and the Campbell Plateau, must be treated as part of Gondwanaland itself. The New Zealand Geosyncline then falls into place as a linear trough marginal to the continent (Fig. 2), facing the ancient Pacific – forming an asymmetric active region comparable (in later Paleozoic and early Mesozoic time) with the asymmetric active regions of modern continental margins and island arcs (EVISON, 1968).

In west Antarctica the Jurassic at least is represented by essentially identical geosynclinal rocks (LAUDON, *et al.*, 1969) and the geosyncline may have extended into the Andean Geosyncline of western South America, before it was disrupted by formation of the swan's-neck orocline of the Scotia Arc. These geosynclinal fragments locally had the same marginal relationships to the Antarctic and South American fragments of Gondwanaland as the New Zealand Geosyncline had to the Antarctic and Australian fragments of Gondwanaland. Prior to the disruption of Gondwanaland, therefore, the New Zealand Geosyncline was perhaps continuous with the West Antarctic and Andean Geosynclines, although they are now distorted by oroclines and separated by an extensive gap formed by the subsequent production of new ocean floor generated from

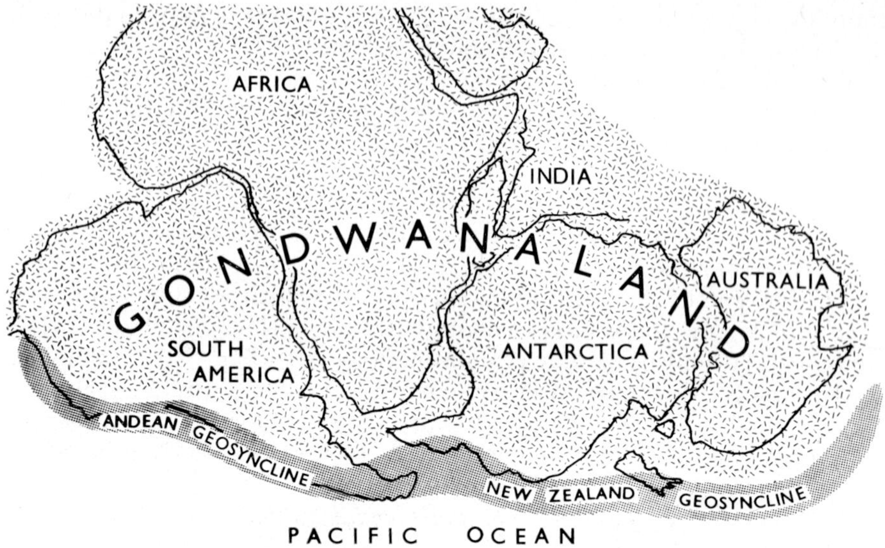

Fig. 2. A reconstruction of Gondwanaland and its marginal geosyncline in the Permian. (After FLEMING, 1971a.)

the South Pacific Rise. The position of New Zealand at the beginning of its geological history is thus part of Gondwanaland as close to the Pacific coast of Antarctica (Marie Byrd Land) as to the east coast of Australia, which lay adjacent to the Wilkes Land Coast (SPROLL & DIETZ, 1969). New data, for instance from the Deep Sea Drilling Project (Scientific Staff, Glomar Challenger, 1973b) and new interpretations (e.g. HAYES & RINGIS, 1973; GRIFFITHS, 1973 in FRASER, 1973), which have appeared since this article was written make it unwise to be dogmatic.

Early Paleozoic

The history of New Zealand during Precambrian and early Paleozoic time is neither well enough known nor sufficiently relevant to the biogeography of the present fauna and flora to justify elaboration in this article. Continental crust, and thus probably at least intermittent land, has been present in the region since the Precambrian, and Upper Cambrian, Ordovician and Silurian sediments were laid down in early geosynclinal phases, presumably marginal to Gondwanaland. The Lower Devonian sediments that followed include orthoquartzites, shale and limestones, suggesting a halt in tectonic activity and a period of dominantly chemical weathering, if not peneplanation, of adjacent land. During the early Devonian, shallow seas that transgressed on the basement rocks of Antarctica, at the base of the Beacon Group, supported a brachiopod genus common to New Zealand, South Africa and Bolivia *(Pleurothyrella)*, supporting the concept of a Malvinocaffric Province, marginal to Gondwanaland, at this time, but with the New Zealand faunas less intimately related to those of the other regions than they were to each other (BOUCOT *et al.*, 1963). Relationships are less clear between New Zealand and East Australia, where marine deposition was widespread.

During the Middle Devonian eastern Australia suffered the disruption of the Tabberabberan Orogeny while in Antarctica non-marine formations of the Beacon Group accumulated, preserving the rich remains of fishes, Antarctica's first recorded vertebrates (WHITE, 1968; RITCHIE, 1971), but in New Zealand no sedimentary record of Middle to Upper Devonian time is preserved. The orogenic movements during and after this period, setting the stage for the New Zealand Geosyncline, have been called the Tuhua Orogeny.

The New Zealand Geosyncline

A fundamental reorganisation of New Zealand structure and the pattern of sedimentation followed the Devonian. The next sediments are extremely thick deposits accumulated east of the foreland of older rocks (of which Fiordland, coastal Westland and north-west Nelson are the emergent parts) in a rapidly subsiding geosyncline embracing the remaining parts of the South and all the North Island. The name New Zealand Geosyncline has generally been applied to this structure, but some geologists extend the term to include the earlier rocks dealt with in the previous section.

The oldest dated rocks in the New Zealand Geosyncline are Upper Carboniferous marble (metamorphosed limestone) containing conodonts, exposed near Kakahu in South Canterbury (JENKINS & JENKINS, 1971). Much of the sedimentary content of the geosyncline is now altered to schist, which may also contain Carboniferous rocks, but the known occurrences of fossils in metamorphosed rocks are Triassic. The Carboniferous Kakahu Marble is not typical of the geosynclinal rocks, in which limestone is rare, and it may represent sediments deposited prior to the main development of the geosyncline.

The Permian rocks crop out on the west and south-west margins of the geosyncline in the South Island, where they are exposed on both limbs of a great marginal syncline (WELLMAN, 1956), and to the east, in the region considered to be more axial, they appear in structural 'highs' – in South Canterbury, in the Marlborough Sounds and in Northland. In addition, considerable parts of the schist, representing the more deeply buried rocks of the geosyncline, are probably Permian. The pre-Triassic rocks of the marginal syncline span most of Permian time and include thick marine volcanic tuffs, sandstone, thin-bedded slates, and limestones, much better differentiated in lithology than the more monotonous Mesozoic beds that overlie them (WATERHOUSE, 1964).

Already, in the Permian, the main geographical elements in the New Zealand Geosyncline, which lasted until Jurassic, can be recognised. A western foreland, represented by Precambrian (?) and older Paleozoic rocks of southern Stewart Island and western South Island, and their hypothetical extensions offshore, is bounded by a line (Median Tectonic Line; LANDIS & COOMBS, 1967) representing the continental margin of the geosyncline, perhaps the late Palaeozoic boundary of the Pacific, the hinge line between regions of dominant uplift and subsidence and between contrasting geological processes. The third element, bordering the foreland to the east, consists of thick clastic moderately fossiliferous sediments in a broad western syncline, attributed to shelf deposition, but with some freshwater formations, now ruptured 450 km by subsequent horizontal movement of the Alpine Fault. The Pacific margin of these 'shelf' sediments is separated by a discontinuous zone of ultramafic and related intrusives, marking a 'facies junction', from a belt, up to 170 km wide, of thick uniform complexly-deformed indurated sediments (greywacke suite), poor in fossils, that have been named Torlesse Group and interpreted as eugeosynclinal deposits of the axial region of the geosyncline, together with metamorphosed rocks of the same original composition (Haast Schist Group). Still further east, presumably, lay the ocean floor of the ancient Pacific.

Permian Biogeography

Apart from indeterminable fragments, the oldest New Zealand plants, from Southland, are members of the *Glossopteris* flora characteristic of Gondwanaland – *Equisetites*, *Cladophlebis*, *Sphenopteris*, ?*Linguifolium*, *Noeggerathiopsis* (MCQUEEN, 1954) and *Glossopteris* itself (MILDENHALL, 1970), confirming, if need be, New Zealand's association with other southern lands during the Permian. Among the diverse Permian formations, no evidence has yet been found of the glacial episodes that characterise contemporary deposits of Australia, Antarctica, South America and South Africa, but WATERHOUSE (1967, 1970) has interpreted the succession of marine faunas as indicating alternations of dominance by cool (Austral) and warm (Tethyan) elements, compatible with a succession of glacial and interglacial climates on the adjacent lands of the Gondwanaland continent; in detail he recognises three episodes of Permian refrigeration in New Zealand and Tasmania.

The Austral element among marine Permian invertebrates that were dominant in cool assemblages includes solitary corals of the *Cyathaxonia* fauna and brachiopods of the genera *Horridonia*, *Plekonella*, *Ingelarella* and *Terrakea*. The Tethyan elements suggestive of warm water include reef corals in Northland, Fusulinid Foraminifera, and other brachiopods such as *Aulosteges*.

In the Permian, therefore, we see clearly for the first time two biogeographic elements that recur with different representatives and to varying extent in later times: (1) a cool, Southern Hemisphere, more or less circum-polar *Austral* element distributed perhaps with the aid of westerly winds and currents round the south polar coasts of Gondwanaland, and (2) a tropical or subtropical *Tethyan* or Indo-Pacific element distributed from equatorial regions. Much of the later biogeographic history of New Zealand concerns the interplay of these two elements.

It is theoretically possible that ancestors of some New Zealand organisms, such as the southern conifers and older groups of terrestrial invertebrates, may have already been in the region in Permian times, but for this there is as yet no fossil evidence.

Triassic

The New Zealand Geosyncline continued to sink and fill during the Triassic at an increasing rate, so that limestones (which depend on lack of detritus from land) are almost absent. The main sediments are the altered and strongly folded greywackes that now form the backbone of New Zealand, including the Southern Alps. Some geologists believe that much of the sediment was deposited rapidly near the centre of the geosynclinal trough by turbidity currents from its steep sides, in an environment unfavourable for benthic organisms. Others, as the result of detailed studies, attribute the rocks to rapid shallow-water deposition but agree that sedimentation was rapid and organic populations low. Most of the axial greywackes contain few traces of life: tubeworms (*Torlessia*, *Titahia*), plant fragments, and odd fossils transported from marginal shelves or dropped from the pelagic zone. In the western marginal facies, however, bottom-dwelling brachiopods and molluscs abounded. The western foreland rose rapidly to supply abundant fresh sediment, and at times deltas advanced on to the shelves and even towards the axis so that plant remains are preserved at Mt Potts and conglomerates elsewhere in Canterbury. Granitic boulders in Triassic conglomerates near Marakopa (west coast of South Auckland) are so large that they must have come from a steep rocky coast not far west of the present coast. Volcanoes were active on the western land, supplying ash to the geosyncline. Some of the younger granites of west Nelson were perhaps emplaced at this time.

The Triassic rocks and fossils of New Caledonia are so like those of New Zealand that the geosyncline and its western border ridge probably extended at least to there, whether or not the ridge was continuous land. In Eastern Australia, on the other hand, the Triassic sediments are very different: non-marine quartz sandstones from a stable continent which give little evidence of a Tasman Sea, but also speak against the close proximity of the actively rising ridge that supplied sediment to form the New Zealand greywackes.

Many Triassic bottom-dwelling brachiopod and mollusc genera are confined to New Zealand, or occur elsewhere only in New Caledonia, and the two countries formed a marine faunal province in Triassic and Lower Jurassic time – the Maorian Province, characterised by its endemic archaic brachiopods (Fig. 3). The Maorian Province was apparently somewhat isolated from the Tethyan Province, and this is indirect evidence against Triassic land connection towards Indonesia. On the other hand, ammonites, which probably lived in the surface waters, and pterioid bivalves, with pelagic larvae, migrated freely. There are no distinct faunal provinces among Triassic and Lower Jurassic ammonites but local faunas vary in richness, New Zealand being relatively poor in species and genera. The pterioids, *Daonella*, *Halobia*, *Monotis* (Fig. 4), and *Rhaetavicula*, spread dramatically at successive stages in the Mid and Upper Triassic so that similar forms occur in distant parts of the globe. Other groups, perhaps less easily dispersed, suggest that the Pacific and Arctic oceans supported similar faunas (the so-called Pacific-Arctic Realm, of which the Maorian Province is a part). On the other hand, the Tethys, from South-East Asia to the Mediterranean, had a somewhat different fauna, and the shallow seas that flooded northern Eurasia supported the impoverished Muschelkalk fauna related to the Tethyan.

The concept of a Pacific-Arctic Realm, contrasted with those of the Tethys and Europe, may be illustrated by the history of the Trigoniidae, a group of bivalves that flourished in the Jurassic and Cretaceous, derived from the Upper Paleozoic and Triassic family Myophoriidae. The first Trigoniidae occur in the middle Triassic of Chile and New Zealand, and they became diverse (at the generic level) in the Upper Triassic of circum-Pacific lands (Fig. 3): Peru, western North America, Alaska, the Arctic, Japan, Timor, New Caledonia and New Zealand. During this time Myophoriidae took their place in the Tethys and Eurasia, overlapping with Pacific trigoniids in Japan, New Guinea and North America. Only near the end of the Triassic did an odd Trigoniid reach Europe, heralding their subsequent dramatic replacement of Myophoriids in the Jurassic.

An even more striking distribution within the Pacific-Arctic realm is that of *Otapiria* (Fig. 3) a pterioid bivalve first named in New Zealand (Rhaetic and Lower Lias) and now recognised in Siberia, Alaska and Chile.

Under the continental drift hypothesis, such a 'Pacific-Arctic Realm' would occupy the peripheral coasts of a restored Pangaea continent, surrounding the Tethys and Muschelkalk realms that extended irregularly across Pangaea. A faunal province entirely surrounded by another faunal province seems unlikely; a more reasonable conceptual model

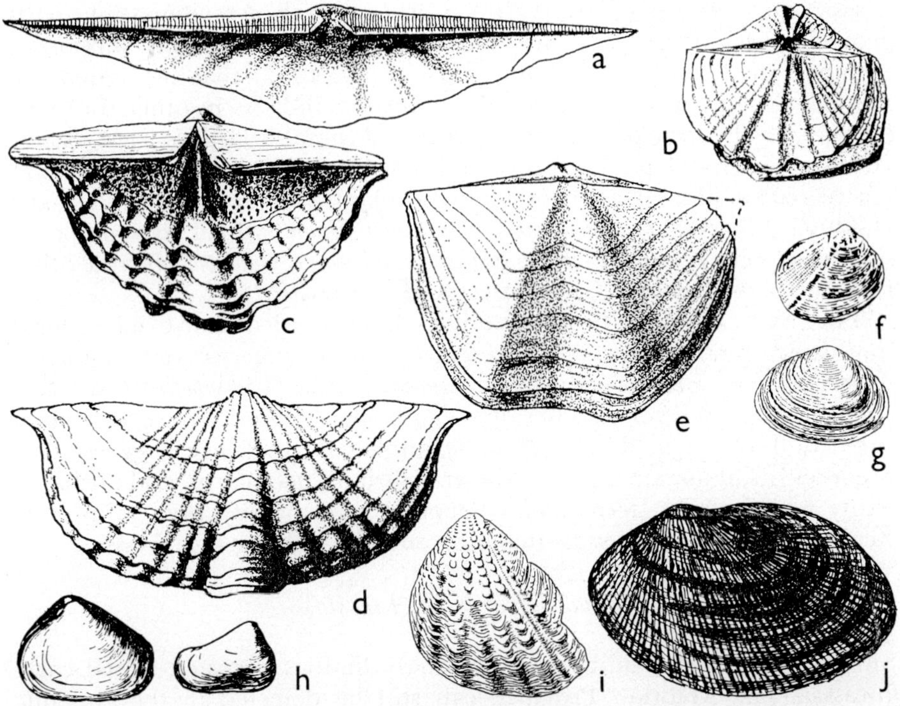

Fig. 3. Endemic marine Triassic and Lias invertebrates of the Maorian Province and Pacific-Arctic Realm: a, *Rastelligera* (U. Trias); b, *Mentzeliopsis* (M. Trias); c, d, '*Spiriferina*' (n. gen.) *kaihikuana* (M. Trias); e, *Clavigera* (U. Trias–L. Lias); f, *Praegonia* and g, *Agonisca* (M. Trias Trigoniidae); h, *Pseudaucella* (U. Lias); i, *Maoritrigonia* (U. Trias); j, *Otapiria* (U. Trias–Lias, Pacific–Arctic). (Figs h and j by J. MARWICK.)

would be to interpret the Arctic-North Pacific and the South Pacific parts of the 'realm' as Boreal and anti-Boreal provinces, with certain bipolar elements in common.

When we look elsewhere to find representatives of a Triassic Austral or Anti-boreal province in southern regions other than New Zealand, we are handicapped by the absence of fossiliferous Triassic in southern Australia, South America, South Africa and Antarctica. Conceivably, then, the Maorian Province of New Zealand and New Caledonia may be all that remains of a Triassic Austral realm; the complete lack of corals is compatible with such a model.

The only New Zealand Triassic vertebrate fossils are marine ichthyosaurs, capable of long-distance dispersal.

No Triassic land fauna is known, but vegetation of Triassic to Middle Jurassic time is quite well represented by macroscopic remains from several localities. Unfortunately their preservation is poor, their taxonomy in need of revision, and their stratigraphic position arguable, the Rhaetic-

Liassic age determined by ARBER (1917) not always compatible with geological relationships. They indicate that New Zealand supported members of the *Dicroidium* flora (TOWNROW, 1957) that succeeded the *Glossopteris* flora in the Triassic of Gondwanaland. As in other southern lands, early Podocarpaceae are represented, notably *Mataia podocarpoides* (Ettingshausen) (Fig. 5), recorded from Mt Potts (TOWNROW, 1967) where recent observations indicate the plant-beds are Middle Triassic (Ladinian) or older. Associated are *Linguifolium* (which SEWARD considered closely related to *Glossopteris*), *Equisitales*, *Cladophlebis* and other fern-like plants, *Baiera* (Ginkgoales) and *Araucarites*.

Two New Zealand Triassic samples have yielded sparse microfloras that supplement macrofossil evidence, spores of ferns and lycopods, *Osmundacidites*, *Ginkgocycadaphytus*, *Araucariacidites*, *Classopollis* (an early conifer), *Vitreisporites* and other disaccate pollen grains that are attributed to early Podocarps. Thus the palaeontological evidence suggests that predecessors, if not ancestors, of the araucarians, podocarps, and probably many lower plants (ferns and lycopods) were already present in New Zealand, as in other Gondwanaland countries, during the Triassic.

Lower Jurassic (Liassic)

Similar geological conditions and faunal affinities persisted in the early Jurassic. The Maorian Province can still be detected in the endemic benthos of New Zealand and New Caledonia, represented by the brachiopod *Clavigera* and the bivalve *Pseudaucella* (Fig. 3) whereas some of the pelagic ammonites are world-wide in distribution *(Psiloceras, Dactylioceras)*. Isolation of the benthos is further indicated by the complete absence from the Liassic of New Zealand of belemnites and trigoniids, abundant in other parts of the world.

While the New Zealand Geosyncline's history continued through the Jurassic, there were some premonitions of change attributed to the beginning of the movements that were later to lead to its evagination to form an elevated area. A Liassic dating for the Bounty Island granite is compatible with this. At both ends of the western syncline shallowing and filling during the Middle Jurassic, a complete absence of marine Oxfordian sediment, and widespread occurrence of fresh water ('continental') deposits are symptoms of these changes, which may account for the complete lack of dated Bathonian-Oxfordian rocks in the eastern eugeosynclinal trough, in which rapid sedimentation was resumed in the late Jurassic (Kimmeridgian). In the eastern eugeosyncline this break in marine deposition may be represented by plant beds of uncertain Rhaetic-Liassic age.

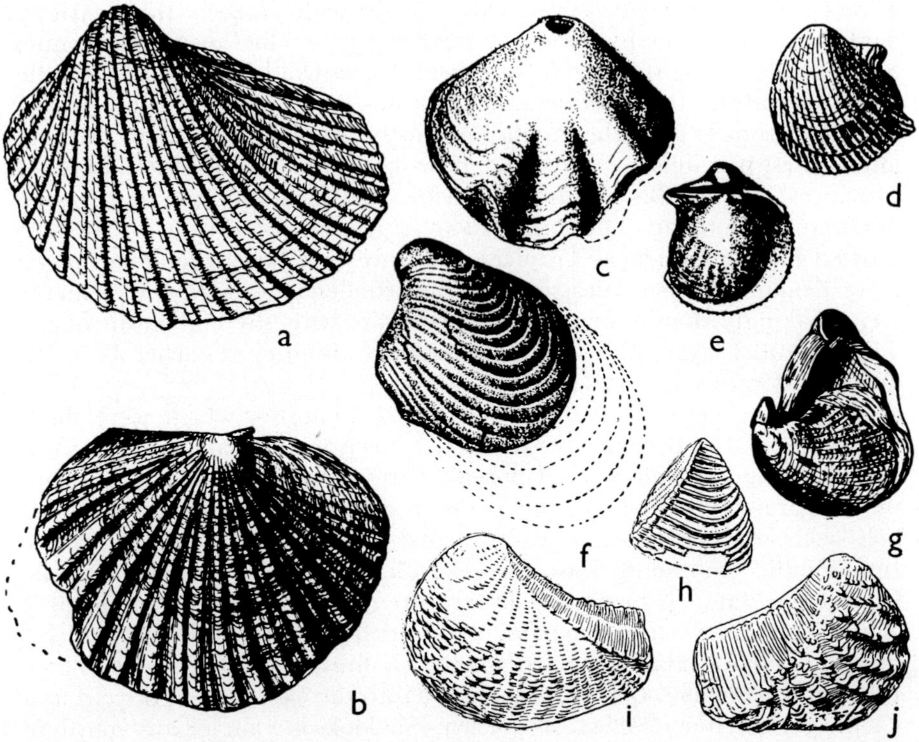

Fig. 4. Tethyan and cosmopolitan Upper Triassic and Later Jurassic marine invertebrates: a, b, *Monotis* (U. Trias); c, *Kutchithyris* (U. Jurassic); d, e, *Meleagrinella* (M. Jurassic); f, *Buchia;* g, *Malayomaorica;* h, *Trigonia;* i, *Orthotrigonia;* j, *Myophorella* (last five Upper Jurassic). (Figures by J. MARWICK, R. C. BRAZIER & C. A. FLEMING.)

Middle and Upper Jurassic

MARINE FAUNA

In the shallow water faunas a dramatic change took place at the end of the Liassic. Middle and Upper Jurassic benthic faunas have strong affinities with those of the Tethys (Fig. 4). Some species (*Inoceramus galoi*, the genus *Malayomaorica*) range to the East Indies (hence some palaeontologists have suggested that there was a Malayo-Maorian Faunal Province), others have relationships as far away as India (*Buchia* species, belemnites, *Kutchithyris*), and a few are even European species. A few endemic lineages linger from the Triassic as a minor faunal element but most have disappeared, their places taken by immigrants (Fig. 8). For instance, instead of the endemic or Pacific Trigoniidae of the Triassic, there are representatives of six European Jurassic trigoniid genera. Some

Jurassic elements were practically world-wide rather than strictly Tethyan, but the Tethys seems to have been the chief dispersal avenue. Some palaeontologists suggest that there was now a barrier in the Middle East separating the Mediterranean from the Tethys proper, which extended from Iran to the Pacific. According to ARKELL (1956), pelagic ammonites, previously world-wide, now show three rather distinct faunal provinces: Pacific, Boreal, and Tethyan. New Zealand ammonites were dominantly Tethyan, but some (such as *Idoceras*, occurring in Mexico) showed Pacific influence. Tropical reef-corals are not represented in the New Zealand Upper Jurassic, but nevertheless the Tethyan elements were strikingly dominant, in strong contrast with the endemism of the Triassic and Liassic, and with the southern affinities of earlier Permian and later Cretaceous faunas.

Among the upper Mesozoic squid-like belemnites, which were shelf-dwelling nekton perhaps restricted by deep-water barriers, STEVENS (1963) recognises well-defined Boreal, Tethyan and Indo-Pacific realms, the Boreal north temperate, the Tethyan extending in a quasi-equatorial belt from southern and central Europe towards Australasia, and the Indo-Pacific embracing East Africa, India, Indonesia and Australasia. During the later Jurassic the latter province was merely an offshoot of the Tethyan with a somewhat reduced fauna. As Tethyan elements present in New Zealand (in particular a belemnite, *Conodicoelites* (STEVENS, 1967) and a bivalve, *Orthotrigonia*) range into the extension of the circum-Pacific geosyncline in West Antarctica, one looks in vain for any southern hemisphere 'anti-boreal' or Austral Realm during the Upper Jurassic. For the complete obliteration of an anti-boreal province in the Southern Hemisphere, during part of the interval between the Permian, when there is good evidence of the existence of such a province, and the Cretaceous when a new Austral fauna appeared, it seems necessary to postulate climatic as well as geographic controls, but as there was a well-defined Boreal Province in the Northern Hemisphere at the time, the situation cannot be explained by symmetrical expansion of climatic belts unless the whole of the potential south temperate zone was remote from the areas where Jurassic fossils are now preserved.

JURASSIC VEGETATION

No terrestrial Jurassic animals are known from New Zealand, but the flora is quite well represented, by Rhaetian to Liassic assemblages of uncertain date in the eastern geosynclinal trough (Clent Hills, Black Jacks, etc.), by Middle Jurassic in the Southland Syncline, including the well-known fossil forest of Curio Bay, Waikawa, and by Tithonian plant beds at Waikato South Head (from which angiosperm fossils were once recorded in error).

The similarity of the Jurassic flora to that of the Trias is indicated by

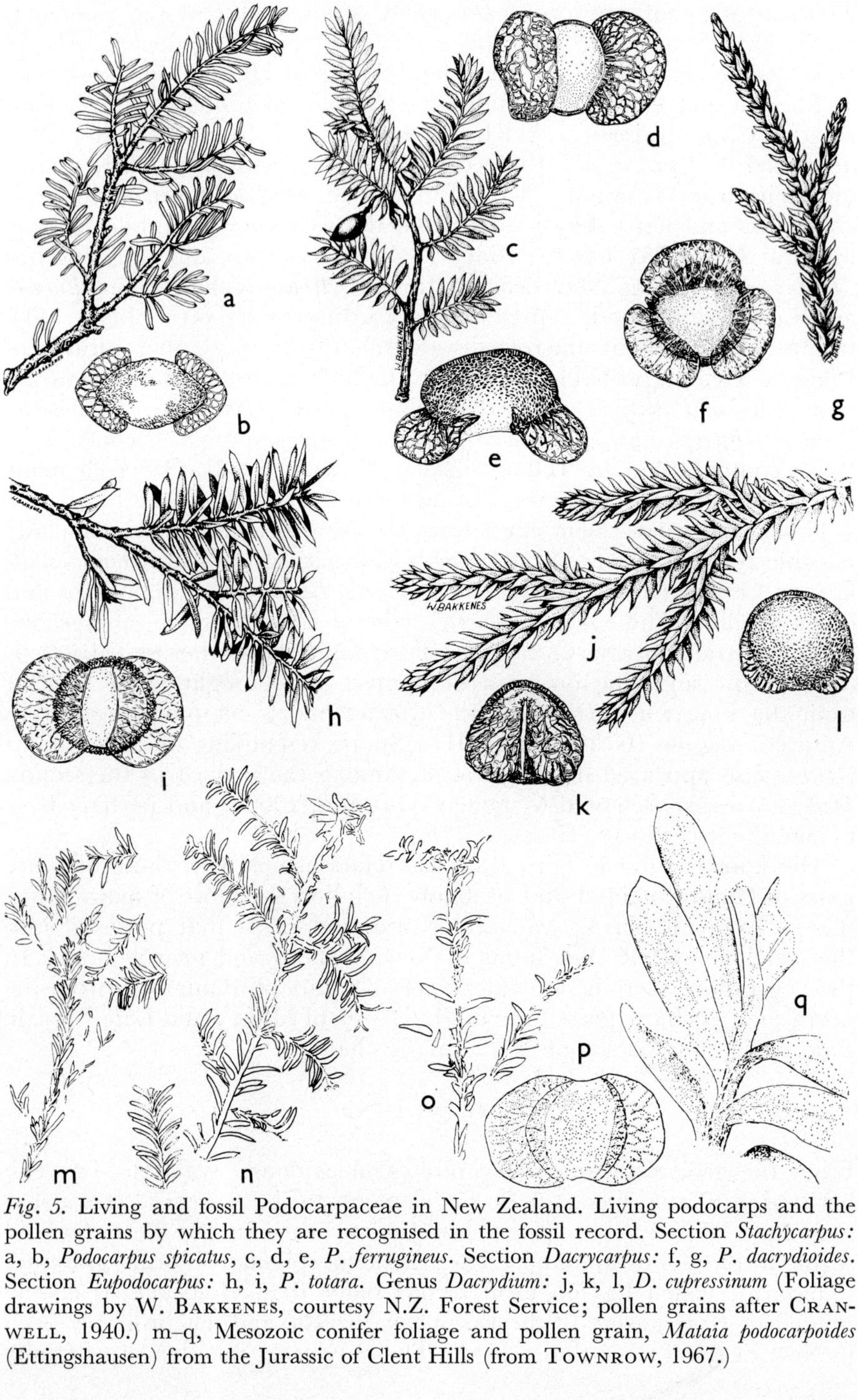

Fig. 5. Living and fossil Podocarpaceae in New Zealand. Living podocarps and the pollen grains by which they are recognised in the fossil record. Section *Stachycarpus:* a, b, *Podocarpus spicatus*, c, d, e, *P. ferrugineus*. Section *Dacrycarpus:* f, g, *P. dacrydioides*. Section *Eupodocarpus:* h, i, *P. totara*. Genus *Dacrydium:* j, k, l, *D. cupressinum* (Foliage drawings by W. Bakkenes, courtesy N.Z. Forest Service; pollen grains after Cranwell, 1940.) m–q, Mesozoic conifer foliage and pollen grain, *Mataia podocarpoides* (Ettingshausen) from the Jurassic of Clent Hills (from Townrow, 1967.)

TOWNROW's confirmation of *Dicroidium odontopteroides* in the Callovian of Owaka, Southland, and the occurrence of the podocarp *Mataia podocarpoides* (Fig. 5) in the Clent and Malvern Hills (date uncertain), at Mokoia and Curio Bay (middle Jurassic), and probably at Waikato Heads (Upper Jurassic). A Rhaetic-Liassic flora from Waitaki Valley described by BELL *et al.* (1956) contains *Ginkgo* and *Linguifolium* similar to species from Tasmania (J. A. TOWNROW, *pers. comm.*) in addition to *Dicroidium* and ferns. Fossil wood from the well-known Middle Jurassic forest at Curio Bay has been identified as *Osmundites* and *Araucarioxylon (Dadoxylon)*. Among New Zealand Jurassic *Araucaria* the sections *Eutacta* and *Colymbea* are both represented according to FLORIN (1940). The twenty or more plant macrofossils identified from the Upper Jurassic of Waikato Heads (probably Tithonian) include an araucarian cone-scale and cones of *Palissya* close to Australian species, *Coniopteris*, *Taeniopteris spatulata*, *Pterophyllum*, *Elatocladus plana*, and the pentoxylean cone *Carnoconites* comparable with Indian species (HARRIS, 1962). Other elements (such as *Cladophlebis antarctica*) are recorded from West Antarctica.

Pollen and spore assemblages from the New Zealand Jurassic include possible representatives of *Sphagnum*, *Lycopodium* of the *fastigiatum-volubile* Group, Osmundaceae, Equisitales, *Pilularia*, *Leptolepia*, of podocarps and other conifers. The spore *Cyathidites australis* is very similar on the one hand to certain Jurassic *Cladophlebis* spores and on the other to *Thyrsopteris elegans*, now surviving on Juan Fernandez and thought to be a relict from the time when *Dicksonia* and *Cyathea* had a common ancestor in Antarctic regions (COPELAND, 1947). Spores resembling those of Recent *Cyathea* also appeared in the Jurassic. Among the podocarps the section *Dacrycarpus* is represented by pollen (COUPER, 1960b) and perhaps by a foliage shoot (FLORIN, 1940).

The known Jurassic flora thus has relationships with those of other parts of Gondwanaland and probably included ancestors of many lower plants, lycopods, ferns, araucarians and podocarps that persisted into the Tertiary and Recent floras of New Zealand and nearby lands. In the vegetation, there is no evidence of an influx of immigrants from the north such as characterised the marine fauna of Middle and Late Jurassic time, nor of any corresponding climatic change.

LIFE ON LAND

From the geological and palaeontological evidence available, one can perhaps justifiably speculate on the elements in the terrestrial fauna that may have already been present in Jurassic time in the New Zealand segment of Gondwanaland. The rhynchocephalian reptile *Sphenodon* (Tuatara), which has no fossil record older than Holocene, is closely related to *Homoeosaurus* of the European Jurassic and has probably been in New Zealand since that period. (The only other member of its family so far known from southern continents is the Triassic *Palacrodon* from South Africa.) The New Zealand frog, *Leiopelma*, likewise has an Upper Jurassic relative classed in the same family in the primitive anuran suborder Archaeobatrachia. For archaic invertebrates with far-flung relationships in southern lands and with limited ability to cross the sea, late Jurassic or early Cretaceous are likely times for dispersal in southern continents, before the disruption of Gondwanaland had proceeded to any considerable degree.

According to SIMPSON (1961) 'ancestors of the monotremes, still near the nominal reptile-mammal line, probably reached Australia in the Jurassic or late Triassic and there evolved into the Monotremata'. Such an early date of entry, whether it was from the north (South-east Asia) or from other segments of Gondwanaland, implies early isolation of Australia from the rest of Gondwanaland since the available evidence of late Mesozoic climate in Antarctica and other southern lands does not point to ecological barriers to monotreme dispersal. If they were more widely distributed (in Antarctica, South America and New Zealand), their subsequent extinction would be understandable where higher mammals penetrated (South America) or when climate deteriorated (Antarctica) but not in New Zealand unless purely by chance since other colonists from the same period *(Sphenodon, Leiopelma)* have survived. The absence of monotremes from New Zealand thus suggests (but by no means establishes) an earlier isolation of the segments of Gondwanaland than does other evidence.

The Rangitata Orogeny and the Disruption of Gondwanaland

Towards the end of the Jurassic, sediment had accumulated in the New Zealand Geosyncline to depths comparable with or greater than the thickness of the earth's crust in continental areas, perhaps locally at an increasing rate. Tectonic movement of foreland and geosyncline had been continuous, but the orogenic movement that followed, known as the Rangitata Orogeny, was more rapid and it replaced the relatively simple structural and palaeogeographic arrangement by a more complex pattern of discrepant and smaller scale structures.

The New Zealand Geosyncline, persistent over most of New Zealand from pre-Permian to Jurassic times, became greatly restricted in the early Cretaceous, if, indeed, it persisted at all as a locus of marine deposition in this region. Most of its contained sediments, indurated by burial and deformed by folding, were elevated, and underwent erosion as land. During the ten million years or so of Upper Tithonian and Neocomian times, the sea may have retreated beyond our shores, for no fossils of these ages have been found. By Aptian time, when marine fossils take up the record once more, the area of geosynclinal deposition was restricted to part of Northland and the east coast, from Raukumara

Peninsula south to Marlborough. The early Cretaceous has always been considered the period of maximum land extension, but all through Permian-Jurassic time the geosyncline was fed by a persistently rising ridge of land to the west, so that the Lower Cretaceous land area was not necessarily greatly enlarged and there is no geological evidence to show how far the New Zealand ridge extended north or south. The land was at first mountainous, but was worn down to low relief before the end of the Cretaceous, when renewed transgressions indicate restriction of the New Zealand land area.

If as has been supposed, the geosyncline was originally a straight or gently curved belt parallel to the coast of Gondwanaland, then the distortion of the belt during folding, into its present S-shaped pattern, and its rupture at the Alpine Fault probably took place during the Rangitata Orogeny. The north-east trend, parallel to the Alpine Fault, dominated later New Zealand history, breaking across the old geosynclinal axis where the latter is not flexed parallel, as it is on the central limb of the S (FLEMING, 1970a). The orogeny took place at a time when the pattern of sea floor spreading was changing, from directions normal to the geosyncline (NE-SW in terms of modern coordinates) to the present direction, normal to the South-East Pacific Rise, which has generated new ocean floor in a north-easterly direction since the Upper Cretaceous. The contortion of the geosynclinal rocks into an orocline entailed rotation and drifting of the adjacent crustal segments that had formed the geosynclinal foreland. Thus the Rangitata Orogeny, dated by geologists as reaching its climax in the Lower Cretaceous, probably represents the start of the disruption of Gondwanaland in the New Zealand sector.

The chronology of the break-up and dispersion of other parts of Gondwanaland is by no means certain; it has been discussed by HEIRTZLER (1968) and by SMITH & HALLAM (1970) who arrive at somewhat different dates for the main separations but agree in general on the late Mesozoic initiation of most of the breaks (apart from HEIRTZLER's late date for the separation of Australia from Antarctica). The sequence of events illustrated in Fig. 6. (from SUGGATE, 1972) follows a somewhat different chronology, stressing the early formation of the Tasman Sea and the late separation of Australia from Antarctica. The data provided by marine geophysical studies and by the Deep Sea Drilling Project on the Glomar Challenger since this article was first drafted, have apparently narrowed the range of options still open. Thus the Tasman Sea is thought to have opened in a phase of sea floor spreading between 80 and 60 million years ago, i.e. in the Late Cretaceous (HAYES & RINGIS, 1973), much earlier than the separation of Australia from Antarctica that began about 56 million years ago, i.e. in the Eocene (Scientific Staff, Glomar Challenger, 1973b). Such a timetable is quite compatible with the Upper Cretaceous dispersal of *Nothofagus* and other early angiosperms. On the other hand such late separations demand special explanations for the

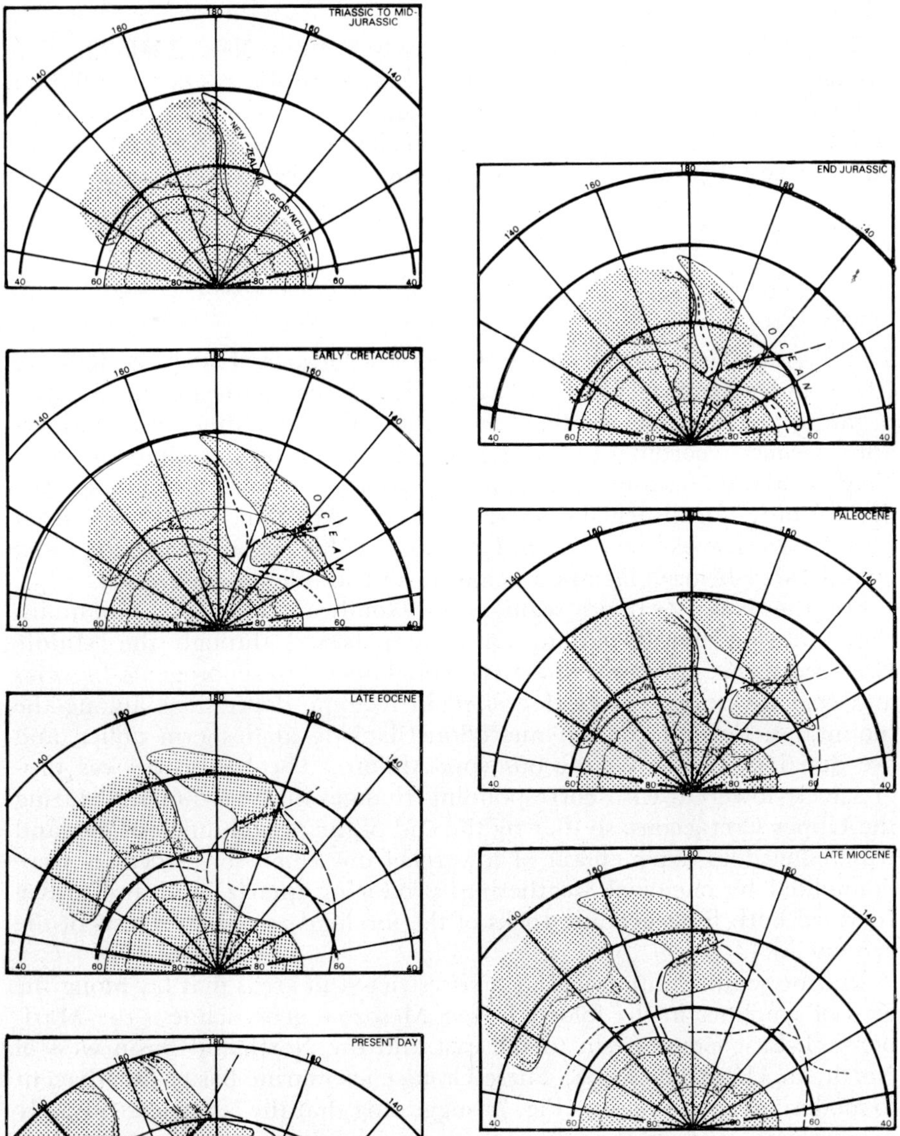

Fig. 6. Progressive development of the New Zealand region from mid-Jurassic to the present day. Active axes of spreading are shown in short dashes and major fracture zones and faults in long dashes. The continental areas are bounded approximately by the present 4000 m contour which however is not a shoreline. (From SUGGATE, 1972.)

absence of snakes and marsupial mammals from New Zealand. The Cretaceous was such a critical period in biogeographic history, in relation to the development and dispersal of angiosperms, mammals, ratite birds and snakes that data from geophysics, geology, paleontology and biology must all be taken into account in arriving at a preferred hypothesis.

Cretaceous

Paleogeography

The Tasman Sea opened in the Upper Cretaceous (Hayes & Ringis, 1973) and according to Heirtzler (1968) the southeast Pacific Rise began to produce new sea-floor crust about the same time, but considerably earlier according to Suggate (1972). From this time onwards New Zealand became progressively more isolated, decreasing the opportunities for dispersal of land organisms. Australia's drift north from East Antarctica was delayed until the Eocene (Fig. 6), and had progressed considerably before glaciation began to restrict life on Antarctica.

On the New Zealand segment of Gondwanaland, the mountains raised during the Rangitata Orogeny persisted through the Middle Cretaceous. Freshwater sediments – conglomerates, synorogenic breccias and coal measures – were deposited in tectonic depressions among the mountains. Several contain microfloras lacking angiosperm pollen and are dated as Lower Cretaceous (pre-Albian). The landscape was progressively lowered, with corresponding changes in the sediments, during the Upper Cretaceous, so that by the end of the period much of the land was reduced to a peneplain of low relief on which soil processes were dominated by chemical weathering, producing quartz sand and gravel (derived both from granitic rocks of the foreland and from schists of the geosyncline).

Marine sedimentation was at first restricted to areas that lay along the axes of synclines in the folded Lower Mesozoic geosyncline – the Marlborough-East Coast Basin to the east and the Northland Basin west of North Auckland Peninsula. These Cretaceous marine basins lie adjacent to modern deep-sea basins (Fig. 7) suggesting that the latter were already in existence, and in the two named basins, the occurrence of spilitic basalts suggests the presence of oceanic crust. In Northland, this indicates that the New Caledonia Basin, a window of ocean crust separating the foreland from the New Zealand Geosyncline, had already been formed. By late Cretaceous time marine conditions had extended south along the east coast of the South Island and began to spread west of the South Island before the Maastrichtian; a west coast for the New Zealand land was thus delineated, implying for the first time the presence of a Tasman Sea. To the east, the sea reached to the Chatham Islands by Middle Cretaceous time, and to Campbell Island by the beginning of the

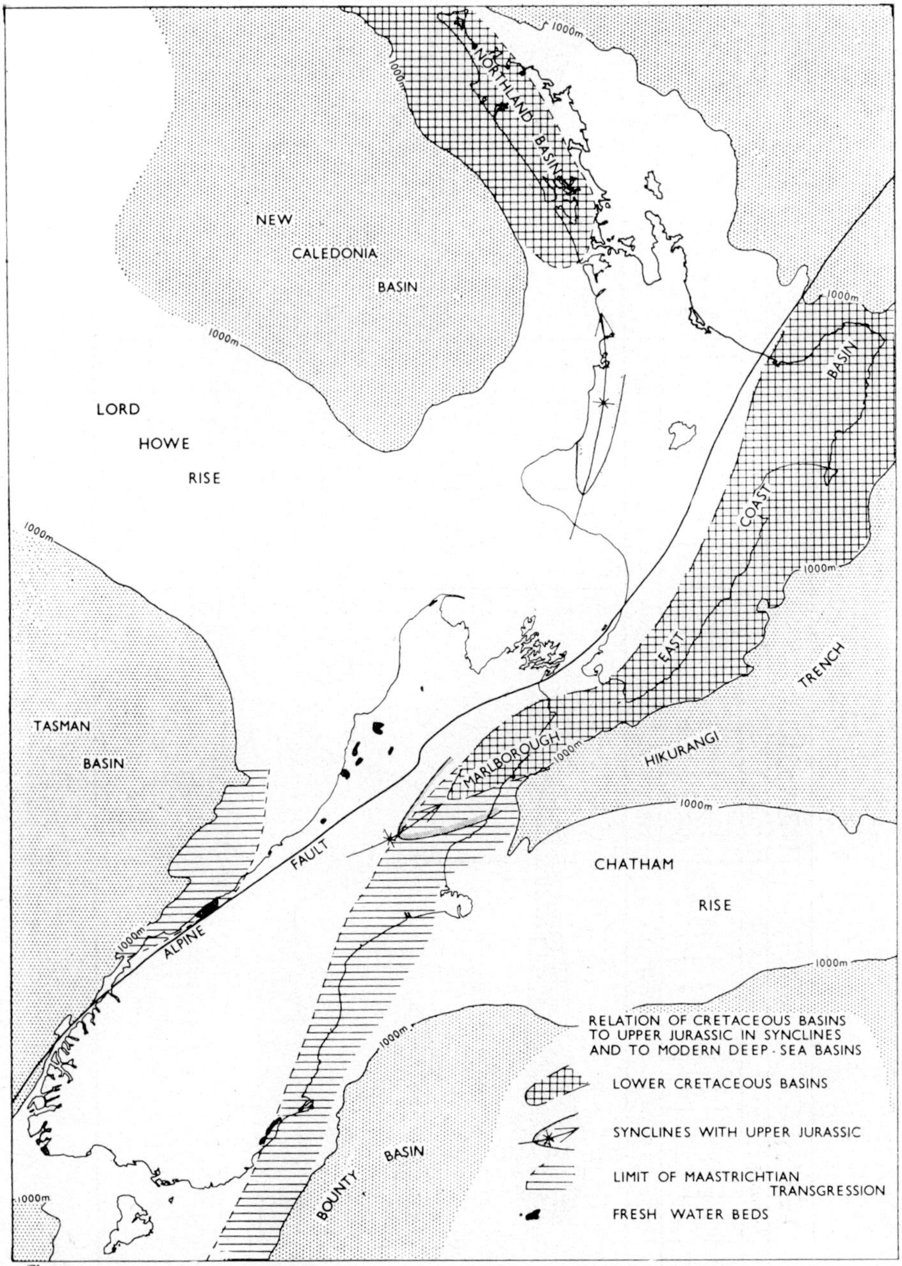

Fig. 7. Relation of Cretaceous basins of deposition to modern deep-sea basins and to synclines occupied by Upper Jurassic rocks. The Northland Basin probably represents the extent of Upper Cretaceous transgression, Lower Cretaceous being restricted to its northern half. (From FLEMING, 1970, Fig. 20.)

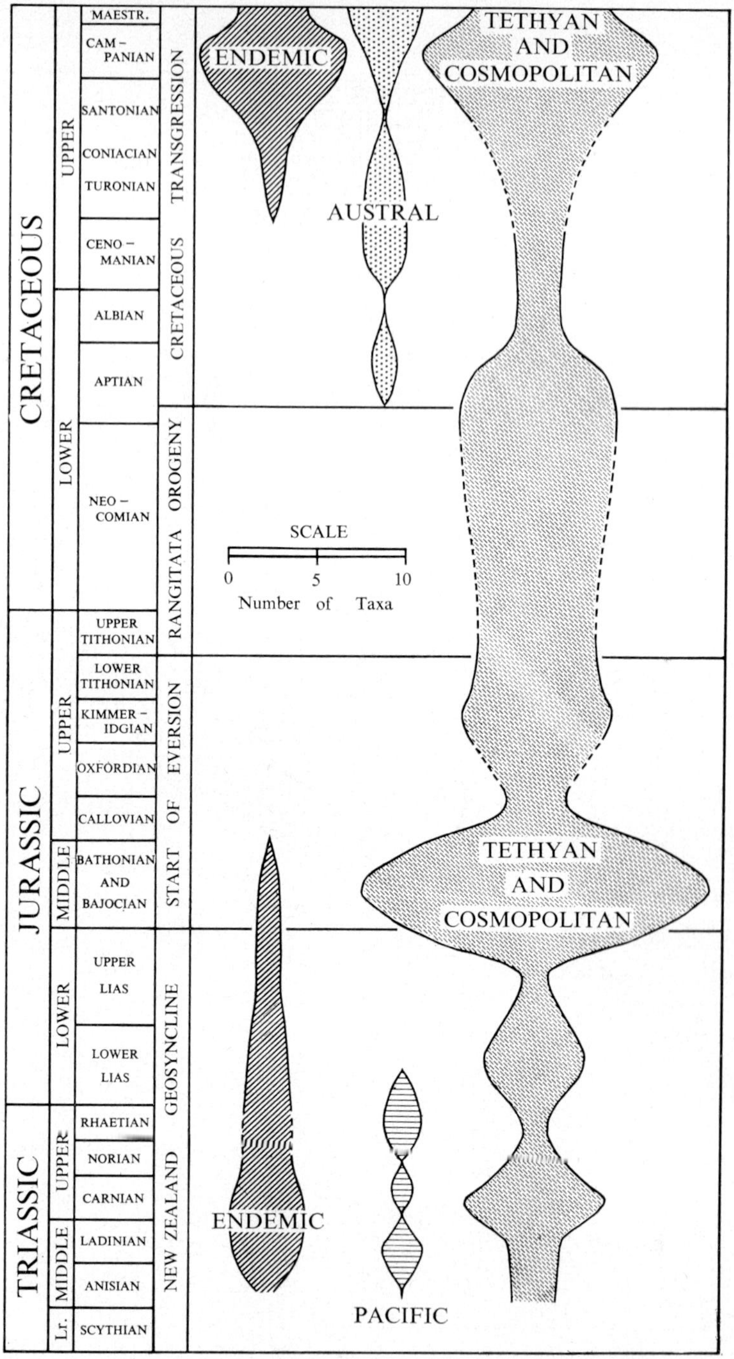

Fig. 8. Faunal changes in the Mesozoic of New Zealand. The diagram shows the relative abundance of the several biogeographic elements represented among incoming taxa of bivalve and gastropod molluscs in each stage in the Mesozoic. (From Fleming, 1967b.)

Paleocene. There is no evidence of marine conditions on Lord Howe Rise to the north-west before late Cretaceous – in fact latest Cretaceous (Maastrichtian) sediment overlies rhyolite flows, perhaps subaerial, about half-way between New Zealand and Lord Howe Island (Burns *et al.*, 1972). Land may have extended in this direction and also south on Campbell Plateau, but probably to a decreasing extent as the sea transgressed over lowland swamps and alluvial plains fringing a land of low relief. Flints and other siliceous sediments were deposited in the offshore waters to the east. This phase of peneplanation and quartzose sedimentation, represented by widespread coal measures, locally lasted until the Oligocene and represents the only really stable phase in New Zealand history since the Devonian.

Cretaceous marine faunas

The Korangan (probably Upper Aptian) marine fauna, still inadequately recorded, includes many widely distributed forms of Tethyan or cosmopolitan affinity, together with species of the bivalve genus *Maccoyella*, otherwise known from Australia and South America, and an *Aucellina* related to the South American and Antarctic *A. radiatostriatus*, these marking the return of a recognisable Austral element in the fauna for the first time since the Permian (Fig. 8). The succeeding mid-Cretaceous faunas include ammonites and benthic bivalves that presumably came by larval dispersal via the Tethys, for instance the European *Hyphantoceras*, *Inoceramus concentricus* and *Exogyra*, others with Austral relationships (*Iotrigonia* spp., *Eselaevitrigonia* and belemnites of the family Dimitobelidae).

Early Senonian marine faunas are similar in affinities to those of the mid Cretaceous. Campanian and Maastrichtian faunas include even more Austral elements – pelagic ammonites (e.g. Kossmaticeratidae) and benthonic lamellibranchs (e.g. *Lahillea*, *Pacitrigonia*) and gastropods *(Struthioptera)* with relations in Chile and Seymour Island (Antarctica); other forms are more widespread, and several genera endemic. Some characteristic Tertiary genera appear for the first time. Vertebrates are represented by several sharks, a Chimaerid and an elephant fish *(Callorhynchus)* and by quite an impressive list of marine reptiles, including Ichthyosaurs (Albian), giant Mosasaurs and Plesiosaurs and a reputed Crocodile in the Maastrichtian.

By the end of the Cretaceous the marine fauna included some ancestors of Tertiary and Recent New Zealand mollusca (e.g. Struthiolariidae, *Glycymerita*, *Cucullaea*). Perhaps we may one day find ancestors of Tertiary Penguins in the sediments of this period, their likely time of origin.

Cretaceous vegetation

Lower Cretaceous vegetation is known by spores and pollen from the

Puysegur Formation of Fiordland and from the Ohika, Hawkes Crag and parts of the Topfer formations of southern Nelson. The first New Zealand representatives of the ferns *Gleichenia* and '*Blechnum*' (the latter perhaps not a reliable identification), the oldest pollen known related to that of the podocarp *Microcachrys* (now restricted to Tasmania) and other podocarps of less precise affinity appear in the Lower Cretaceous.

The greatest biogeographic event in the mid Cretaceous was the dispersal of angiosperms, which apparently originated in the Lower Cretaceous.

The first angiosperm pollen is reported from Albian in Clarence Valley (COUPER, 1960b) and East Wellington (HASKELL, 1971), thus approximately from the same horizon as in Australia and North America. The vegetation was still dominated by Gymnosperms, including podocarps, of which the pollen representing *Microcachrys* (known also as a fossil from Australia and Kerguelen) is characteristic, together with tree ferns and other ferns. *Nothofagus* pollen first appears in the Jay Coal measures of Greymouth about Coniacian in age and thus slightly later than its uppermost Turonian appearance in Australia (EVANS & STOVER, 1971), occurring with a distinctive assemblage of coniferous pollen species, including pollens closely similar to those of *Agathis australis* (the kauri), of the *Dacrydium franklinii* group (now persisting in Tasmania), *Microcachrys* (see above), *Podocarpus* (section *Dacrycarpus*) and other podocarps. The wide southern-hemisphere distribution of these plants in the Upper Mesozoic and Tertiary was discussed by COUPER (1960a); some of them have been recorded from Antarctica by CRANWELL (1959), CRANWELL *et al.* (1960), and by MCINTYRE & WILSON (1966). In New Zealand, the first *Nothofagus* are of the *brassi* group in the Lower Senonian, followed by the *menziesii* group in the Campanian (rare) and by the *fusca* group in the Maastrichtian.

The podocarps and southern beeches have always been considered the most distinctive Austral elements, but many other disjunct distribution patterns have been explained in terms of the emerging geological history of Australasia (RAVEN & AXELROD, 1972).

Another characteristic Austral angiosperm family, the Proteacea, has a similar history, its pollen first appearing in the Middle (?) Senonian (Mid to Upper Turonian in Australia) and becoming more diverse in the Maastrichtian, by which time pollen closely similar to that of *Beauprea* and *Knightia* had joined half a dozen undifferentiated Proteaceous pollen species. Proteacea continued as prominent members of the New Zealand flora throughout the Tertiary but were much reduced in the Pliocene and Early Pleistocene. *Knightia* persists in the Recent flora and was joined by the Australian genus *Persoonia*, apparently at a very late date (as its characteristic pollen is not known fossil) and thus, by implication, by transoceanic dispersal. The pollen *Haloragacidites harissii*, believed to represent *Casuarina*, appeared in the Maastrichtian and remained characteristic throughout the Tertiary (MILDENHALL & HARRIS, 1971).

As the angiosperms increased, many characteristic Cretaceous lower plants and gymnosperms became extinct, mostly at the end of the Maastrichtian, judged by their spore and pollen record, but others appeared for the first time; thus pollen closely related to that of *Dacrydium cupressinum* appeared in the Maastrichtian. By the end of the Cretaceous substantial generic elements of the living flora had appeared, judged by their pollen record: *Lycopodium* (*fastigiatum-volubile* group), *Hymenophyllum*, *Blechnum*, *Pilularia*, *Dicksonia*, *Cyathea*, *Adiantum*, *Gleichenia* (aff. *circinata*), *Pteris*, Araucarians including *Agathis*, *Podocarpus*, *Dacrydium* (aff. *cupressinum*), *Knightia*, *Nothofagus*, Cruciferae, together with elements that did not survive to Recent time (*Dacrydium* aff. *franklinii*, *Microcachrys*, *Ephedra*, *Beauprea* and other Proteacea, *Anacolsia* and *Casuarina*). Macrofossil remains support some of these identifications (e.g. *Pteris* related to *P. tremula*, *Blechnum* related to *B. lanceolatum*, *Dacrydium* related to *cupressinum* from Pakawau, *Agathis* wood from Kaipara) and odd others, not confirmed by pollen (*Sticherus*, *Podocarpus* (sections *Dacrycarpus* and *Eupodocarpus*) from Pakawau, *Beilschmiedia* identified by Dr. W. R. B. OLIVER from Shag Point). Additional taxa identified from macroscopic remains only are *Athrotaxus* (Shag Point), *Cinnamomum* (Senonian Paparoa beds and Maastrichtian of Shag Point), *Dryandra* (Trelissic Basin; but its distinctive pollen has not been found), *Ripogonum* (Ohai, identified by OLIVER, but horizon uncertain), and *Acer* and *Betula* (from leaves, identified by OLIVER). The *Athrotaxus* is confidently recognised by FLORIN (1940), but without further evidence the remainder must be considered uncertain. The podocarp-*Nothofagus* forests extended to Campbell Island in the Maastrichtian (D. G. MCINTYRE, pers. comm.).

Students of recent representatives of the genus *Nothofagus* and of podocarps insist that their dispersal requires a continuous land surface or, in the case of podocarps with fleshy fruit, a close approximation of land to permit bird transport (PREEST, 1963). Dispersal southwards across the tropics from the northern hemisphere to southern lands is no longer a tenable hypothesis, in view of their fossil record (COUPER, 1960a). It would be an attractive hypothesis to attribute these plants to migration on a Lower Cretaceous land extension, with the possibility of land connections with other lands, but their fossil record suggests that they came after the Rangitata mountains were reduced, when the seas were spreading over a subdued land and reducing its extent once more. Distribution among the segments of Gondwanaland prior to its complete disruption is much more probable. The fossil record of their pollen grains suggests that such dispersal continued until late Cretaceous or even early Tertiary time. Thus the podocarp *Phyllocladus* (formerly present in Australia, New Zealand, South America and West Antarctica) is not so far known before the early Tertiary (first appearance, in New Zealand, Upper Eocene).

Such a late persistence of dispersal pathways between the several segments of Gondwanaland as is implied (i.e. Eocene) runs counter to the datings derived from sea-floor spreading, from geological evidence, and from the presence and absence of certain groups of animals which might be expected to take advantage of them.

Terrestrial fauna

If the palaeognathous birds (the ratites) are monophyletic and result from an early avian stock that became flightless in the Cretaceous, their dispersal in the various segments of Gondwanaland before its fragmentation is plausible – indeed, the distinction of the two-toed ostriches from the three-toed ratites is compatible with the early separation of Africa suggested by independent evidence. The oldest fossil ratite is Eocene, but birds have a notoriously fragmentary fossil record. For discussion of other bird families for which a Gondwanaland dispersal has been suggested, see Serventy (1972), Cracraft (1973), Mayr (1972) and Keast (1971).

The primitive Booid snakes (Ophidia) are first known as fossils in the Upper Cretaceous, and had already reached South America before the end of the Cretaceous. Their absence from New Zealand implies that isolation was achieved before that date.

No terrestrial Mesozoic reptilia have been found in New Zealand, although they occurred in other segments of Gondwanaland. If we accept the general thesis that land connections with Antarctica and adjacent parts of Gondwanaland persisted long enough for access by late Cretaceous angiosperms, we are forced to adopt special explanations for the absence of terrestrial reptiles and, indeed, the ancestors of egg-laying monotremes which Simpson (1961) considers reached Australia in the Late Triassic or in the Jurassic. *Sphenodon*, as we have concluded above, was almost certainly present. If others occurred, but have not been found as fossils (which is likely, as appropriate sediments are either marine or strongly acid, unfavourable for preservation of bone), their subsequent extinction must be explained. World-wide extinction was indeed the fate of most Mesozoic vertebrates, with Australia's monotremes and New Zealand *Sphenodon* as exceptions nominated by chance. But at the time the marsupials reached Australia, probably in the late Mesozoic or earliest Cenozoic (Simpson, 1961), New Zealand's isolation must have been sufficient to prevent their colonisation and, as I have already noted, the drift from Gondwanaland was well under way by this time although it may not have proceeded far enough to prevent dispersal of some Austral plant groups, especially those with bird-transported fruits. Thus the date of marsupial dispersion to Australia sets a plausible limit on the duration of New Zealand's close connection with Gondwanaland. This date is unknown, and both the dispersal abilities and fossil records of many plants are uncertain.

The difficulties discussed in the preceding sections must be faced both by supporters of the Gondwanaland hypothesis and by those who believe, like G. G. Simpson, that Australasia gained its vertebrate faunas from south-east Asia. On the whole, data do not seriously conflict if we assume that New Zealand's isolation became effective for land vertebrates before the end of the Cretaceous, a date compatible with the more than 80 million years of sea-floor spreading on both sides of New Zealand indicated by the correlation of sea-floor magnetic anomalies.

Cretaceous climate

Lower Cretaceous marine invertebrates (such as *Spondylus*) suggest warm-temperate seas, but there is no evidence of really tropical conditions.

Upper Cretaceous climate was probably moist warm-temperate, warmer than the present judged by the presence in the South Island of the tropical family Olacaceae (represented by the pollen *Anacolosidites*) and large marine reptiles, but the seas were not warm enough to support reef rudistids and corals. Isotope temperatures show a steady decline of marine temperatures in the late Cretaceous to a mean figure of 14.3 °C in the eastern South Island, that is roughly present temperature (Clayton & Stevens, 1968). Increasing peneplanation probably caused deep podsolisation and slow replenishment of soil nutrients, and perhaps rainfall was lowered through reduction of topographic relief. It is most unlikely that there were any subalpine or even montane biotopes in New Zealand during the period from Maastrichtian to Oligocene.

Cenozoic

In New Zealand, it is inconvenient to separate the Tertiary and Quaternary because deposition continued without interruption in many areas. Compared with many countries, New Zealand's record of life in shallow seas is remarkably continuous throughout the Cenozoic and the record of terrestrial plant life is also very complete though not yet fully studied. New Zealand geologists have used a local system of Cenozoic Stages (Fig. 9) to classify the rocks and fossils of this era. But in this article ages are mainly cited in terms of the standard international epochs (Paleocene, Eocene, Oligocene, Miocene and Pliocene).

Judged by the correlation of sea-floor magnetic anomalies to the south east, the New Zealand segment of Gondwanaland had already drifted some 1500 km from the shelf of West Antarctica by the beginning of the Tertiary, and movement continued during the Tertiary (Fig. 6). The sea lapped over the margins of this segment of continental crust, leaving a much smaller area as emergent land.

Series	Stage		Correlation	Typical Sediments
HAWERA	Post-glacial	H	Holocene	Ash showers, dune-sands, river flats
	Several glacial and interglacial stages		U. Pleistocene	Terrace deposits, moraine, volcanic breccias
WANGANUI	Castlecliffian	Wc	M. Pleistocene	Papa and shellbeds, Wanganui, Ohope, Cape Kidnappers
	Okehuan	Wk		
	Nukumaruan	Wn	L. Pleistocene	Papa and shellbeds, Wanganui, Hawke's Bay
	Hautawan	Wh		Shell limestones, Wairarapa, Hawke's Bay; Moutere Gravels
	Waitotaran	Ww	Pliocene	
	Waipipian	Wp		Papa, Hawera-Taihape, Wairoa, Westland
	Opoitian	Wo		Otahuhu well, Awatere
TARANAKI	Kapitean	Tk	U. Miocene	Papa, shellbeds, East Cape-Hawke's Bay
	Tongaporutuan	Tt		Taranaki, Marlborough, Westland, Te Waewae Bay
SOUTHLAND	Waiauan	Sw	M. Miocene	Sandstones, Taranaki, North Island east coast, Westland, Waipara, Southland
	Lillburnian	Sl		
	Clifdenian	Sc		Siltstones (as above); limestone, Southland
	Altonian	Sa		
PAREORA	Awamoan	Pa	L. Miocene	Sandy shellbeds, greensand, Oamaru, Waipara
	Hutchinsonian	Ph		Sandstone, Northland; mudstone, King Country, Canterbury
	Otaian	Po		
LANDON	Waitakian	Lw	U. Oligocene	Mudstone, limestone, most districts
	Duntroonian	Ld		Limestone, greensand, Chatton sand
	Whaingaroan	Lwh	L. Oligocene	Calc. mudstone, Cobden, Amberley Limestone
ARNOLD	Runangan	Ar	U. Eocene	Dark siltstones, Reefton, Greymouth
	Kaiatan	Ak		
	Bortonian	Ab	M. Eocene	Greensand, Northland, Otago
DANNEVIRKE	Porangan	Dp		Amuri Limestone; offshore mudstones, etc.
	Heretaungan	Dh	L. Eocene	
	Mangaorapan	Dm		
	Waipawan	Dw	Paleocene	Wangaloa shellbed; Waipara greensand
	Teurian	Dt		

Fig. 9. Table of divisions of the New Zealand Cenozoic. The succession of series and stages has been established for New Zealand sediments and fossils and mapping symbols (third column) are convenient abbreviations. Recently the Hutchinsonian and Awamoan have been merged in the Altonian. A few examples of sediments are given in the fourth column (see also FLEMING (ed.), 1959.)

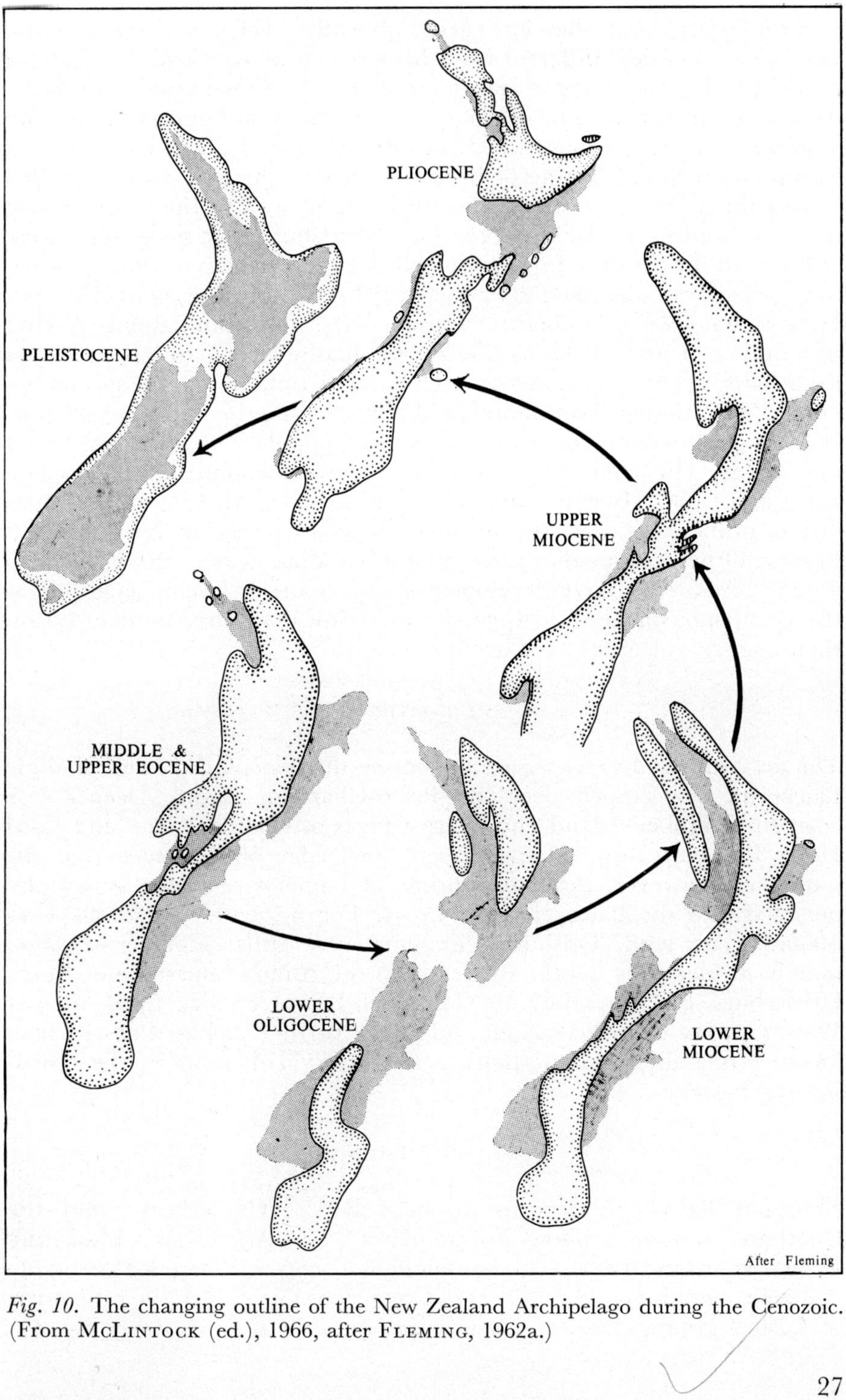

Fig. 10. The changing outline of the New Zealand Archipelago during the Cenozoic. (From McLintock (ed.), 1966, after Fleming, 1962a.)

Structurally, and therefore geographically, Tertiary New Zealand became a good deal different from Mesozoic New Zealand. The pattern of folds, welts, and troughs that developed was on finer scale than in the Mesozoic; instead of a broad geosyncline some two hundred kilometres wide and thousands long, the land moved up and down as a series of narrow short interfingering or branching folds. The welts, which tended to be submarine ridges or land, were small, so we can think of Tertiary New Zealand as an archipelago (Fig. 10). Changes in geography were frequent. Troughs sank rapidly but filled with sediment as they subsided so were seldom submerged to bathyal depths. Welts rose in complementary fashion, but owing to constant erosion were not mountainous. A kind of writhing of part of the mobile Pacific margin seems to have gone on in the later Tertiary. A changing archipelago would encourage speciation, i.e. the formation of two or more species by successive invasions, geographic isolation, re-invasions or back-invasions of populations from one island to another. This could account for genera with a multitude of species. Long ago F. W. Hutton suggested that the diversification of the Moas (Dinornithiformes) was due to an archipelagic stage in New Zealand history. Other groups that show adaptive radiation (e.g. the Callaeidae among birds) may have developed on the changing archipelago just as the Galapagos finches (Geospizinae) and Hawaiian Drepaniidae did on their ancient volcanic island groups.

Paleocene

The sea continued to transgress further on the peneplaned land. Shallow Paleocene seas lapped over the Proto-Chatham Island. Deeper seas covered Campbell Island and at least parts of Chatham Rise and Lord Howe Rise north-west of the present shelf edge off Taranaki. For the most part, however, the main outline of Upper Cretaceous geography persisted into the Paleocene and Lower Eocene, with increasing peneplanation on land. Offshore sediments were either globigerina ooze, mainly in moderate depths (typically in the Amuri Limestone of North Canterbury but extending to Campbell Island and at times to east Wairarapa) or fine siliceous silt and mud (Northland, East Coast) which includes turbidites and perhaps locally bathyal deposits in the down-sagging basins.

Marine fauna

Marginal Paleocene deposits at Eyre River (Canterbury) and the Chathams contain orbitoid Foraminifera (*Asterocyclina*, etc.) indicating at least subtropical seas. The benthic fauna (e.g. at Wangaloa) is highly endemic, containing descendants of Cretaceous forms, including members of Austral groups (*Lahillea*, *Perissodonta*, ancestor of *Struthiolaria*, Fig. 11)

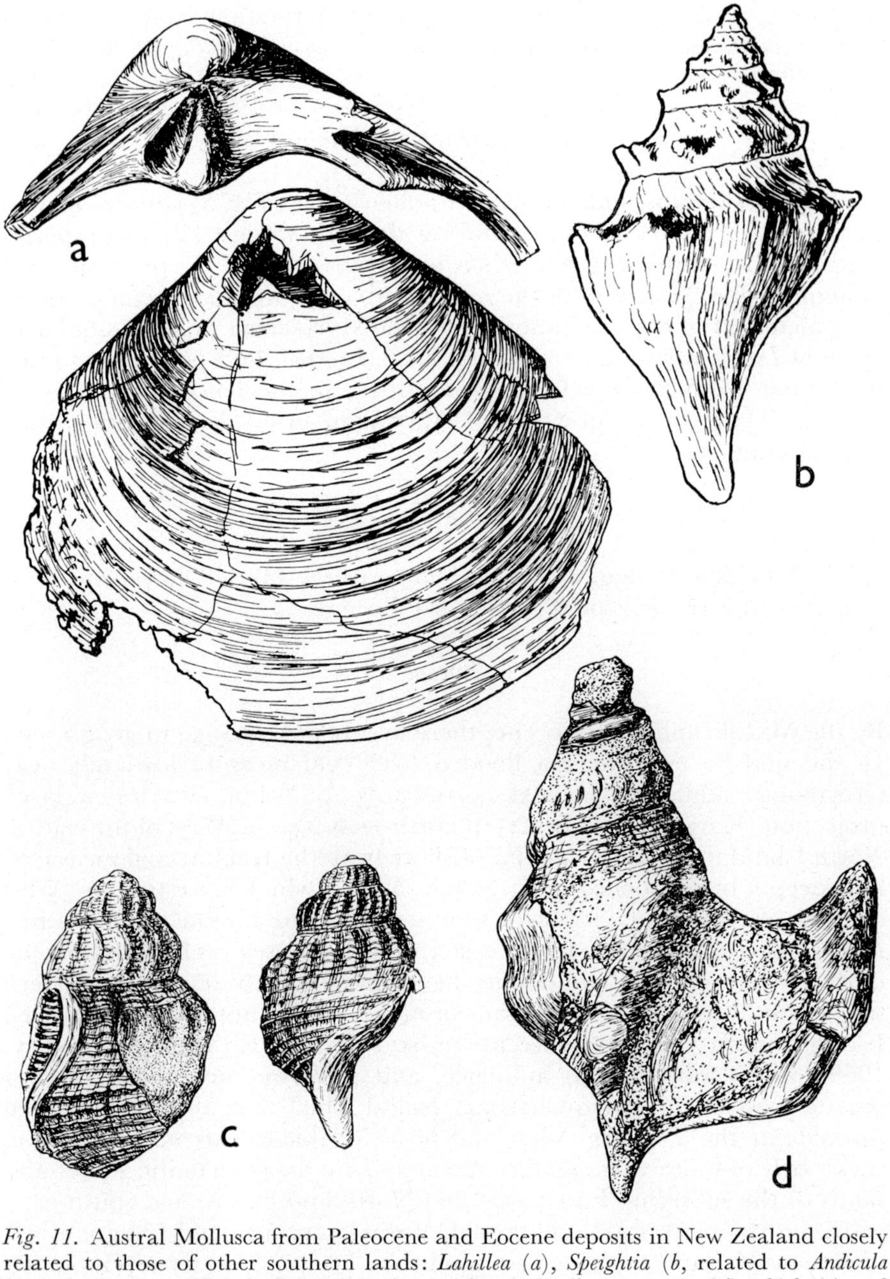

Fig. 11. Austral Mollusca from Paleocene and Eocene deposits in New Zealand closely related to those of other southern lands: *Lahillea* (*a*), *Speightia* (*b*, related to *Andicula* from the Eocene of Peru), *Perissodonta* (*c*, syn. *Struthiolarella*, ancestor of *Struthiolaria*) and *d*, *Struthioptera*. (Figs. a, c, d by J. Marwick.)

but some of the newcomers in it suggest derivation from the north (e.g. *Costacallista*, *Polinices*, *Sigaretotrema*, *Priscoficus*), as do the additions to the flora.

Terrestrial vegetation

On land, podocarps (judged by their pollen grains, Fig. 5) still dominated the earliest Tertiary forests as during the Cretaceous, but angiosperms became more important. The *Nothofagus brassi* group retreated from New Zealand while *Nothofagus* of the *fusca* group and *Casuarina* became significant members of the vegetation. The first Myrtaceous pollens, similar to those of *Leptospermum* and *Metrosideros*, of *Ascarina*, and pollen attributed to the pan-tropical tribe Cupanieae appear in the Paleocene or Lower Eocene. These newcomers are of Australian or Malayo-Pacific distribution today.

Paleocene climate

Limited evidence suggests rising temperatures in the Paleocene, culminating in a tropical peak during the Eocene.

Eocene

By the Middle and Late Eocene, there had been a change in geography. In the mid-Eocene, the sea flooded over coal measure lowlands near Greymouth and later pushed northwards to Nelson, leaving a large projection (Karamea Peninsula) in north-west Nelson. West of the central North Island the coast lay not far offshore from the Kapuni coal measures, now deeply buried beneath Taranaki, some of which are estuarine. West of Karamea Peninsula, sea-floor samples of marine Eocene sediment give rather uncertain control. In the eastern Bortonian sea (as from Paleocene onward) foraminiferal limestone, like globigerina ooze of the modern sea bottom, was deposited in Marlborough, North Canterbury, Campbell Island and on Chatham Rise at bathyal depths (Kennett & Casey, 1969), away from coastal influence, and shallower limestones covered part at least of the Proto-Chatham Island which was an oceanic island throughout the Tertiary. West and north of this calcareous sea bottom was a belt of more variable fine calcareous muds, glauconitic shales and sands in the subsiding East Coast and Northland basins, and south-eastward, on the more stable shallow shelf of Canterbury and Otago, a belt of glauconitic sand merging into mudstone in deeper water at Hampden (North Otago). Glauconite formation required special biochemical conditions in seas where terrigenous detritus was minimal. On land, quartz gravels and sands formed wide swampy plains through which flowed sluggish meandering rivers, draining a lowland of deep leached podsols.

Late Eocene seas locally transgressed further over this peneplain. Its stability was probably only relative, however. In the newly-formed Kaiata Gulf inside 'Karamea Peninsula' dark carbonaceous silt was deposited above coal measures, bordered by land that rose quickly enough locally to form steep cliffed coasts in late Eocene time. In the Waikato Basin, too, there was appreciable relief in the coal-measure landscape.

Marine fauna

The Bortonian marine shelf fauna was one of the most distinctive in New Zealand's Tertiary history. It contained several restricted endemic genera and the first members of many common Tertiary genera.

The new gastropod arrivals include *Arca* (s. str.), *Astrea*, *Xenophora*, members of the Cypraeidae, *Architectonica*, *Ficus*, Mitridae, *Eocithara* (a harp-shell), Conidae, *Gemmula*, suggesting dominance of Malayo-Pacific dispersal avenues and subtropical climates. *Neilo* and *Speightia* (aff. *Andicula* of Peru, Fig. 11) are Austral immigrants at this time. A notable invasion of coral genera took place but they do not indicate sea temperatures higher than 10 °C to 16 °C (Squires, 1958). The most important of many Foraminifera which entered in the Middle and Upper Eocene is *Hantkenina*, world-wide in distribution. In the Upper Eocene a marked warming almost to tropical conditions (Hornibrook, 1953) is suggested by the return of the Orbitoid *Asterocyclina*, the appearance of the cidarid *Eucidaris* (at the Chatham Islands; Fell, 1954), the regular echinoid *Brochopleurus*, the lamellibranch *Hinnites* and the gastropod *Colubraria*. Some of the warm-water genera mentioned in this paragraph are illustrated in Fig. 13. Although the chief source of immigrants was the Indo-West Pacific, some Middle and Upper Eocene Foraminifera, corals (*Discotrochus*, *Asterosmilia*), crabs (*Laeviranina*, *Portunites;* Glaessner, 1960) and the gastropod *Personella* show rather notable American relationships.

The earliest fossil penguins (*Palaeeudyptes marplesi* and *Pachydyptes ponderosus*) are late Eocene (Simpson, 1971).

Terrestrial vegetation

The podocarps that had been abundant since the Cretaceous at first remained dominant but in the middle Eocene *Nothofagus* of the *fusca* group increased at their expense in the South Island, suggesting cooling climate; ferns remained abundant. Pollen questionably related to that of the Australian-New Zealand genus *Plagianthus* first appears in the Lower Eocene.

The main vegetational change in the Upper Eocene was the return of *brassi*-group beeches to the record as dominant trees for the first time. The Austral podocarp *Phyllocladus* (confirmed by macrofossils), the mistletoes *Elytranthe* (Malayo-Pacific), and the grasses appeared for the first time. The only records of the pollen *Banksiaeidites* (representing Proteaceae, *Banksia* or *Dryandra*) are mid-upper Eocene. *Podocarpus (Dacrycarpus)* pollen (as distinct from macrofossils) is recognised for the first time. Couper recorded the earliest pollen of Araliaceae, *Dysoxylum*, the palm *Rhopalostylis* (all of Malayo-Pacific affinity) and the endemic (or New Zealand-based) *Hoheria*, *Phormium* and *Dracophyllum* in the Upper Eocene, but the botanical relationships of the pollens are not always unambiguous. Macroscopic remains of two kinds of palm have been recognised. Leaves of the liane *Rhipogonum* (Australia-New Guinea) were recognised by the late Dr. W. R. B. Oliver from Ohai (buts its pollen is not recorded till Miocene). Leaves identified as *Cinnamomum* and *Acer* persist, together with a *Beilschmiedia* related to *B. taraire* by Oliver (but pollen of *Beilschmiedia* does not enter the record until Miocene).

Eocene climate

Hornibrook (1971) summarised evidence for tropical-subtropical marine climates during the early Eocene, rising to a peak in the middle-Upper Eocene when isotope data (Fig. 12) suggest mean temperatures of about 22 °C (Devereux, 1967). On land, the increase of *fusca*-beeches in middle Eocene, suggesting cool temperate conditions, is discrepant with the marine evidence of at least subtropical seas, but the sudden dominance of *brassi*-beeches (now confined to the montane tropics) in the Upper Eocene, apparently accompanied by warmth-loving plants such as Cupanieae and palms, agrees with the indications of marginal-tropical coastal conditions. The late Eocene warm period was an outstanding event.

Tertiary erratics of probably Eocene age at McMurdo Sound, Antarctica (Cranwell *et al.*, 1960; McIntyre & Wilson, 1966), contain pollen indicating nearby temperate forests in which *fusca*-beeches were dominant over *brassi*-beeches, with Podocarps, Araucarians, Myrtaceae, Proteaceae and *Casuarina* present as minor constituents – all prominent members of New Zealand early Tertiary floras. How long such vegetation persisted, as a legacy of a common inheritance shared with other southern lands, is speculative, but it may be assumed to have lasted until the Tertiary climatic deterioration that culminated in the Pleistocene Ice Ages (see Fig. 20); thus Antarctica, though at an increasing distance from New Zealand, probably continued to function as a Tertiary avenue for Austral plant dispersal and as a source for organisms sufficiently vagile to overtake the southern continents as they continued to drift away (Fig. 6).

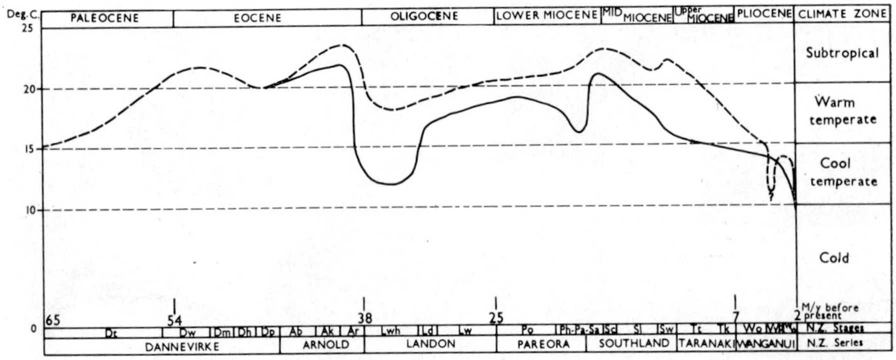

Fig. 12. New Zealand Cenozoic climates shown on an absolute time-scale. Temperatures from oxygen isotope analyses by DEVEREUX (1967) normalised to the latitude of Wellington (41 °S) are shown by a solid line, those estimated from paleontological evidence, at the same latitude, by a dashed line. The Awamoan (Pa) downward fluctuation is based on isotope measurements from North Otago where paleontological evidence suggests a local cool current. (From HORNIBROOK, 1971.)

Oligocene

The transgression that began in the Upper Cretaceous reached its climax in the Oligocene. Proto-Chatham Island remained emergent. Early Oligocene seas flooded over the Late Eocene coal swamps of Northland, the Waikato, King Country and Taranaki, eastward at least to National Park, and persisted in the Eastern Geosyncline. 'Karamea Peninsula' was submerged, the South Island land almost obliterated in its northern parts (Fig. 10). The lithologic and faunal facies of many Oligocene marls, limestones and chalks (impure 'globigerina ooze') confirms the impression that land was distant or very low-lying. Geologists are rather uncertain about the whereabouts of land at the height of the Landon transgression, but glauconitic sandstones west of National Park and near the west coast, flanking the main development of limestone (Te Kuiti Group), suggest low islands on either side. In central New Zealand, Lower Oligocene calcareous mudstones and limestones are widely distributed (including remnants infaulted by later earth movements), so that the sea extended widely across the axis of New Zealand in this area. In Otago, shallow seas and estuaries lapped inland but this maximum transgression was temporary and probably did not submerge the whole land. The Fiordland region was probably submerged, and thick sediments filled a trough in the Waiau-Eglinton region. In most districts the same conditions persisted into the Upper Oligocene, with local signs, however, of incipient earth movements. Thus in Auckland Province, limestone gave way to Mahoenui mudstone and graded-bedded sandstone suggesting an increase of sediment from rising land, and sea

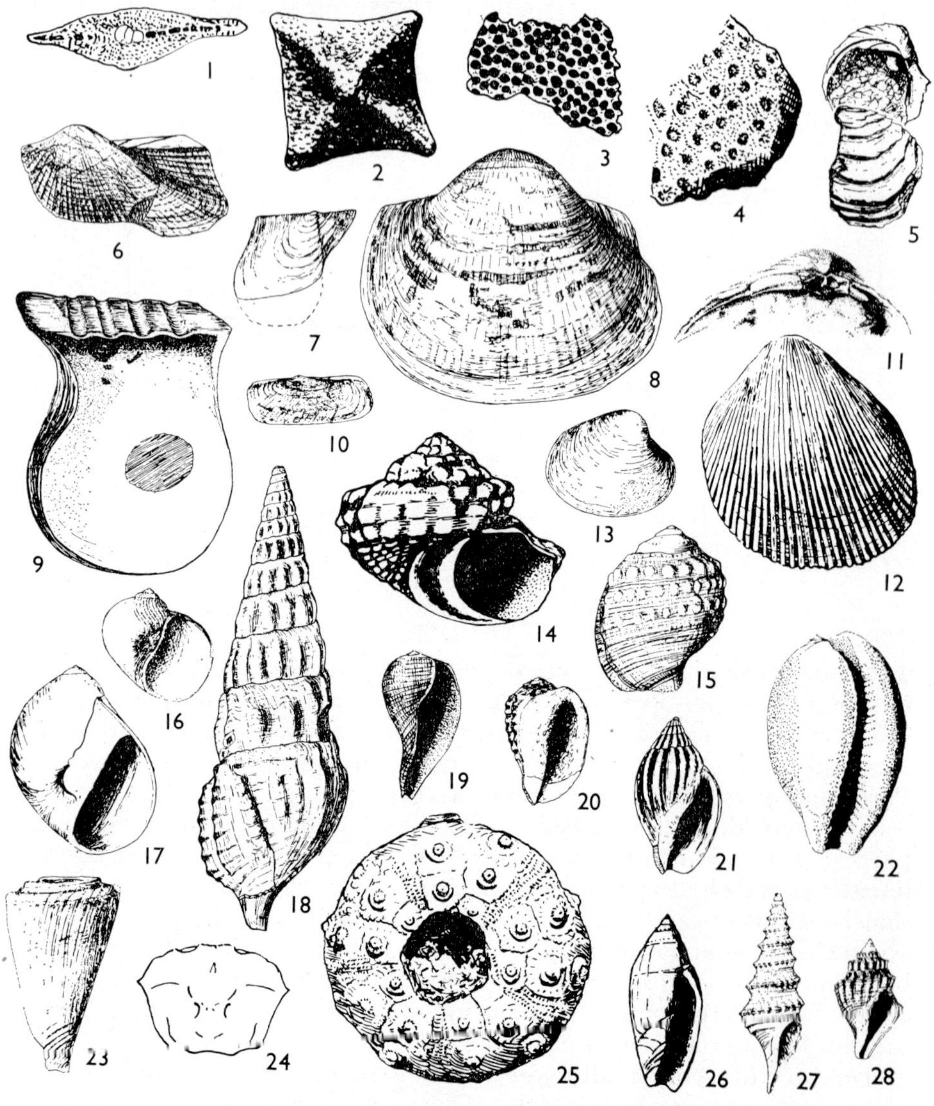

Fig. 13. Warm-water marine invertebrates of Malayo-Pacific origin from the New Zealand Tertiary, now locally extinct. 1, *Lepidocyclina* (Mio.); 2, *Asterocyclina* (Eo.); 3, 4, *Alveopora*, *Turbinaria* (reef corals); 5, *Chama;* 6, *Arca;* 7, *Electroma;* 8, *Cucullaea;* 9, *Isognomon;* 10, *Solecurtus;* 11, *Maoricardium;* 12, *Trachycardium;* 13, *Pitar (Hyphantosoma);* 14, *Sarmaturbo;* 15, *Echinophoria;* 16, *Korovina;* 17, *Polinices;* 18, *Pyrazus;* 19, *Ficus;* 20, *Morum;* 21, *Lyria;* 22, *Cypraea;* 23, *Conus;* 24, *Galene* (Brachiura); 25, *Histocidaris;* 26, Olividae *(Lamprodomina);* 27, *Gemmula;* 28, *Marshallena.* (Figures by J. Marwick, M. F. Glaessner (24) and C. A. Fleming (1–4, 14, 19, 22, 26).)

retreated from Central Otago, where deposition of quartzose alluvium was resumed.

Marine fauna

Many mollusc genera appear for the first time as fossils in the Oligocene, which is locally richly fossiliferous. Newcomers include elements of Australian affinity *(Bassina, Eucrassatella)* and many endemic forms of unknown origin, but the majority are Malayo-Pacific genera of warm-waters: *Solecurtus*, *Maoricardium* (living in the Indian Ocean), *Pyrazus*, *Isognomon*, *Bathytoma* (see Fig. 13). The characteristic New Zealand *Struthiolaria* appeared (Duntroonian Stage), a descendant of the middle Eocene *Monalaria*, but intermediate stages apparently lived beyond our shores. The cidarids *Phyllacanthus* and *Eucidaris*, the regular echinoid *Grammechinus*, irregulars such as *Echinolampas* and *Planilampas*, abundant Alcyonarians *(Graphularia, Isis, Moltkia)* and Scleractinian solitary corals (species of *Notocyathus*, *Stephanocyathus*, *Conocyathus*, etc.), emphasise the Indo-Pacific relationships of the invertebrate fauna.

Most of the fossil penguins described from New Zealand (Marples, 1952; Simpson, 1971) are Oligocene, including *Palaeeudyptes antarcticus* Huxley, species of *Platydyptes* (2), *Korora*, *Archaeospheniscus* (2) and *Duntroonornis*. The Landon Series, particularly in North Otago, is also the preeminent horizon for fossil whales in New Zealand. *Kekendon onamata* from the Lower Oligocene of Wharekuri, Waitaki Valley, is a large Archaeocete with characteristic teeth, classed by Romer (1966) in the family Doradontidae. Odontoceti are represented by the squalodonts *Microcetus*, *Prosqualodon*, and *Squalodon*, and Mysticeti (whalebone whales) by four species of the Cetothere genus *Mauicetus* Benham (Marples, 1956). Marples (1946) also described the furcula of a flying bird *(Manu antiquus)* of uncertain affinity, perhaps an albatross, and indeterminable femora.

Oligocene vegetation

Nothofagus of the *brassi* group continued to dominate the Oligocene forests, with podocarps, *Casuarina*, and *fusca*-beeches as only minor elements and fern spores, abundant until mid-Eocene, unusually rare.

Additions to the flora in the early Oligocene (judged by pollen) were *Lycopodium* of the *billiardieri-varium* group, *Microsorium*, *Bombax*, (tropical, but rare), *Myrsine*, rushes (Juncaceae) and the banded sedges of the Restionaceae, a predominantly Austral family. In the late Oligocene (Waitakian Stage) several genera prominent in the living flora appeared – *Pseudowintera* (Austral), *Epilobium*, *Laurelia* (Austral according to Scottsberg), *Elytranthe* related to living species, *Coprosma* (Malayo-Pacific but headquarters in New Zealand), *Astelia* (Austral), and *Typha* (cosmopolitan). Lower Oligocene plant macrofossils have been attributed (mostly by OLIVER, MS) to *Agathis australis*, *Athrotaxis*, *Coprosma*, and *Euphorbia*, the Australian-New Zealand genus *Pomaderis* and the dominantly Malayo-Pacific *Pittosporum*, *Carpodetus*, *Heimerliodendron*, and *Geniostoma*, identifications from leaf-impressions with all their shortcomings.

OLIGOCENE CLIMATE

Despite the dominance of Malayo-Pacific elements in the biota, the absence of orbitoids, nummulites and reef-corals and the disappearance of certain nannoplankton suggest that conditions were not fully tropical and the abundance of penguins and whales gives an impression of coolness supported by oxygen-isotope measurements that indicate an abrupt drop in temperature after the late Eocene warm period. Shallow-water Foraminifera (*Amphistegina*, *Operculina*, etc.) and the rich cidarid fauna indicate, however, that marine temperatures were warmer than at present, as were air temperatures, to judge by the vegetation. The reality and extent of an Oligocene cooling are thus somewhat controversial (Fig. 12; HORNIBROOK, 1971).

Lower Miocene

Increased topographic relief, noticeable in the Oligocene of the King Country, continued and extended to the rest of the country in the Lower Miocene, here taken to include the Otaian to Clifdenian Stages of New Zealand stratigraphy.* Short gaps in the record suggest local withdrawals and readvances as the restless land began to stir. The Cretaceous to Oligocene peneplain was now a thing of the past, yet there is little evidence of truly mountainous relief. In many areas sandstones and unweathered mudstones derived from rising land overlie Oligocene limestones. In Northland, the Waitemata Sandstone was deposited in shallow water, east of a rising ridge composed of Paleozoic granitic rocks (or perhaps of Triassic conglomerates) soon to be the site of a chain of andesite volcanoes from Auckland to Hokianga, source of the Manukau Breccias. To the east, sea transgressed over greywacke land that had been emergent long enough to lose nearly all trace of the early Tertiary peneplain and its covering Oligocene sediments. Later in the Miocene, volcanoes were active on this eastern ridge as well, producing the rocks now forming picturesque peaks at Coromandel, Great Barrier, Whangarei and Whangaroa. The Waitemata sea penetrated into the Waikato Basin from the west coast south of Port Waikato, perhaps linking southwards with the deeper Mahoenui sea, which occupied the southern King Country

* A reduction in the number of formal stages in the Lower Miocene is anticipated in the light of G. H. SCOTT's conclusions from microfossil studies.

and Taranaki. The sea extended from north-west Nelson south to beyond Punakaike, depositing calcareous polyzoal sands and silts. In the south, limestones and glauconitic sediments at first persisted. Sea occupied the Waiau Syncline in Southland, whence shallow-water limestones extend east to Forest Hill. In eastern Otago and Canterbury, a shallow shelf of sand, locally glauconitic, bordered fossiliferous silts (Blue Cliffs). The Lower Miocene sea probably occupied the East Coast basins continuously, but some of the microfossil zones are absent, perhaps owing to the unsuitable offshore facies of the deposits.

During the successive stages of the Lower Miocene, rising land, local volcanic outbursts and the consequent increase in sedimentation caused the sea to retreat from most of Northland and parts of the King Country (where the Mokau coals formed in swampy deltas), and perhaps also from parts of Southland. Apart from volcanoes, however, the land was not yet mountainous.

Marine faunas

Two strongly contrasting climatic provinces are apparent in the shallow-water Lower Miocene marine fauna. In the eastern South Island, the molluscan fauna does not greatly differ from that of the preceding late Oligocene; in Northland, in contrast, a marginal tropical fauna is marked by the first appearance of many new genera, including Australian (e.g. *Cratis*, *Herpetapoma*, *Murexsul*) and especially Indo-Pacific elements, including Conidae, *Murex* (s.str.), *Chicoreus* (gastropods), *Ctenoides*, *Septifer*, *Cardita* (bivalves), the only reef corals known in the Tertiary record (for instance *Lobophyllia*, *Turbinaria* and *Alveopora*) and orbitoid Foraminifera (*Lepidocyclina*, etc.). Some characteristic mid-Tertiary warm-water invertebrates are shown in Fig. 13. All told, the Lower Miocene marks a high-point of mid-Tertiary warmth and of the Indo-Pacific faunal influence that it brought (Fleming, 1962a; 1967a; Beu & Maxwell, 1968); it may even have exceeded the Eocene temperature maximum. Additional genera of Australian and Malayo-Pacific origin reached the eastern South Island in the succeeding stages: *Eumarcia*, *Cleidothaerus* (Australian), *Lutraria*, *Cymatiella*, *Aphera*, *Cronia*. As illustrated in Fig. 14, about 175 mollusc genera appeared for the first time in the New Zealand record during the first three Lower Miocene stages, most of Indo-Pacific affinity, but some distinctly Australian, and a few of Austral distribution, now characteristically Subantarctic (*Hochstetteria*, *Kidderia*). The marked Lower Miocene influx, spread over the Otaian to Altonian Stages, is also shown by corals, Foraminifera and Ostracoda.

Even during this Lower Miocene climax of Indo-Pacific influence, when reef corals equivalent to those of the Capricorn Group off Queensland were living in northern New Zealand (Squires, 1962a, b), the number of Indo-Pacific genera amounted to less than a quarter of the

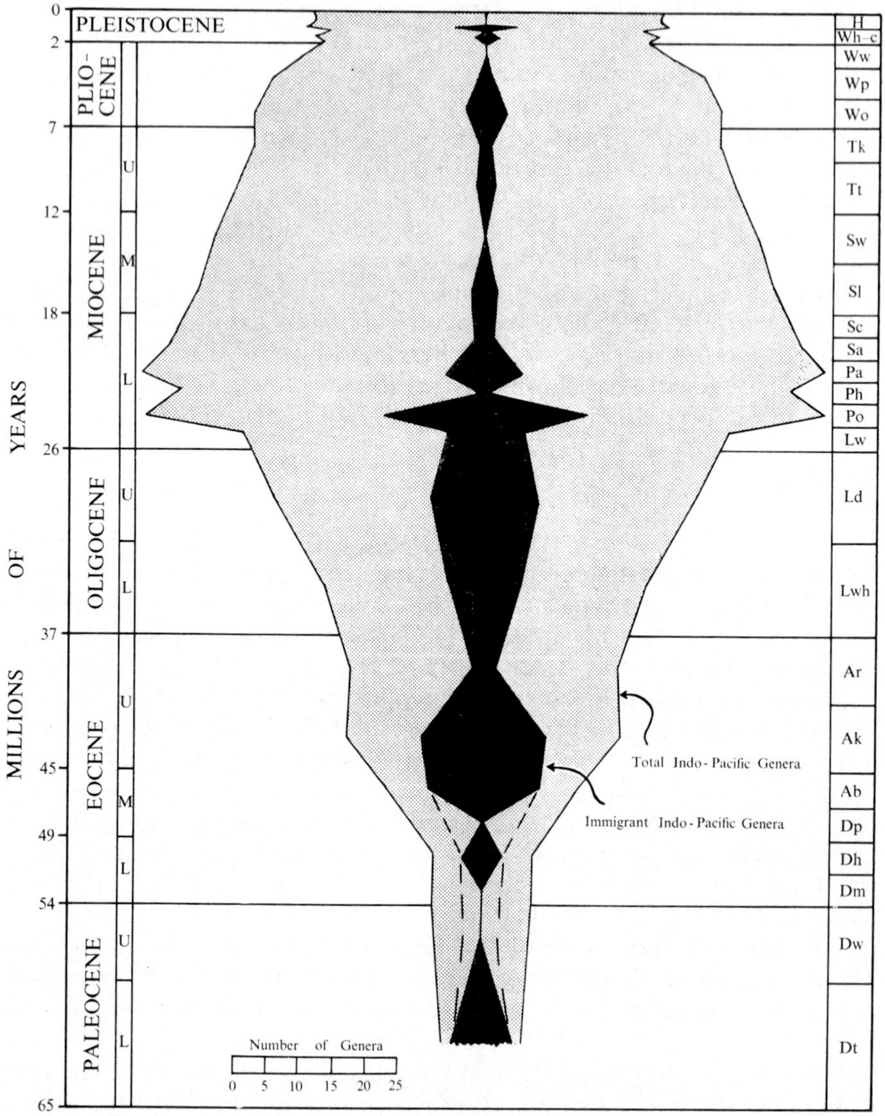

Fig. 14. Diagram illustrating fluctuating total numbers of Indo-Pacific warm-water mollusc genera and of Indo-Pacific immigrant genera during the Cenozoic in New Zealand. (From FLEMING, 1967a.)

fauna and many of the Indo-Pacific immigrants were short-lived, apparently unable to maintain themselves in New Zealand environments. Many characteristic Indo-Pacific genera of Mollusca were not present (e.g. *Malleus*, *Tridacna*, *Fimbria*, *Strombis*, *Lambis*, *Gafrarium*, *Oliva*, *Terebra*), suggesting that New Zealand always lay outside the Indo-Pacific Realm, even when Indo-Pacific elements were at their maximum (FLEMING, 1967a).

TERRESTRIAL VEGETATION

In the early Miocene forest vegetation, judged by the pollen-rain, *brassi*-beeches generally remained dominant, but *fusca*-beeches and podocarps became important sub-dominant trees and at some times and localities one or other of these latter groups attained dominance, mostly to the south in Otago (COUPER, 1960b, fig. 3). These changes were accompanied by a recovery of the tree ferns and other ferns that had become rare in the Upper Oligocene, and by abundant palms among which *Cocos* is confirmed by fossil nuts (Fig. 15), which occur with a fruit *(Phymatocaryon)* of proteaceous affinity.

A considerable number of groups have their first pollen or macrofossil record in the Lower Miocene: the ferns *Schizaea* cf. *fistulosa*, *Trichomanes*, *Arthropteris tenella*, *Pteridium aquilinum*, *Histiopteris incisa*, podocarps of the *Dacrydium kirkii-bidwilli* group, *Podocarpus spicatus* and *P. totara*, and the following flowering plants: a. *Beilschmiedia* (but note earlier macrofossil identifications of this genus, the pollen of which is not particularly distinctive), *Elaeocarpus*, *Geniostoma*, *Macropiper*, *Alectryon*, ?*Avicennia*, *Rhipogonum* (earlier macrofossils), *Cordyline* (all Malayo-Pacific); b. *Muehlenbeckia*, *Fuchsia*, *Weinmannia*, *Griselinia* (the last also as leaf impressions) (Austral); c. *Coriaria* (Eurasia, New Zealand, South America, with Austral (trans-Pacific) affinities at species level) and *Aristotelia* (Australia, New Hebrides, South America); d. *Myriophyllum*, Compositae, *Hydrocotyle* and *Potamogeton* (Cosmopolitan); e. *Phormium* related to Recent species (endemic) and *Melicytus* cf. *ramiflorus* (virtually endemic but on some Pacific islands).

Phormium of Recent type presumably evolved from extinct predecessors *(P. recticulatum)* which had been in New Zealand since the Eocene. Among presumptive immigrants, Malayo-Pacific forms dominate, as with marine immigrants at this time. *Coriaria* and *Aristotelia* both occur to the north but have outlying representatives in South America, so may perhaps be thought of as immigrants by the Malayo-Pacific route that later attained a partially Austral distribution along with more strictly Austral immigrants of this period *(Muehlenbeckia*, *Weinmannia*, *Fuchsia*, and *Griselinia)*.

By the Miocene, New Zealand had drifted sufficiently far north to be under dominantly subtropical influence and was a considerable distance from Antarctica although that continent still, presumably, supported vegetation and was available as a stepping stone for organisms that attained an Austral distribution. Submarine drilling (Scientific staff, Glomar Challenger, 1973b) has now indicated that extensive

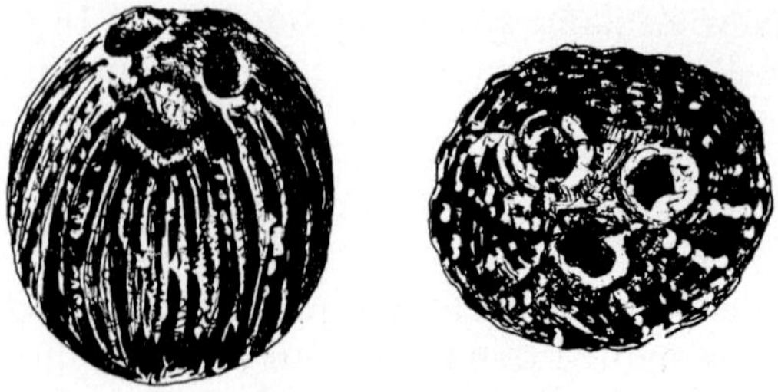

Fig. 15. The New Zealand Miocene coconut *Cocos zeylanica* Berry (natural size).

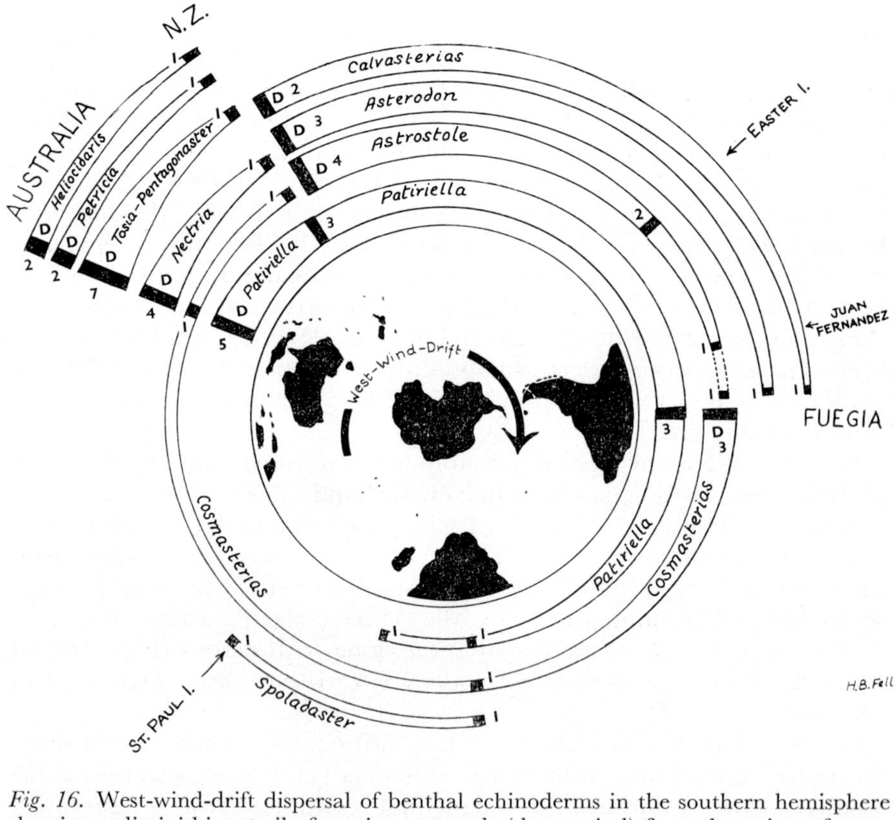

Fig. 16. West-wind-drift dispersal of benthal echinoderms in the southern hemisphere showing a diminishing trail of species eastwards (down-wind) from the points of entry in Australia, New Zealand, South America and South Africa. Australasia is inferred to have been a major point of entry of Indo-West-Pacific elements into the southern hemisphere. (From FELL, 1962.)

glaciation in Antarctica had begun by early Miocene, but this is not incompatible with vegetation persisting at sea-level. H. B. FELL (1962) has developed the view (previously advanced by OLIVER, 1925, in respect to land plants) that many organisms of northern origin which penetrated to the southern continents were able to extend their range eastward, carried on the West Wind Drift of surface water (and air) to produce an attenuating stream of derivative species to the east of their point of origin (Fig. 16), for instance 'a regular trail of Fuegian plants eastwards to the Falklands, South Georgia, and other subantarctic islands' from South America (OLIVER, 1925). Thus *Coriaria*, *Aristotelia* and *Sophora* (the last without any fossil record) may be primarily Malayo-Pacific in origin and only secondarily Austral. Such a dispersal mechanism implies the Tertiary existence of the West Wind Drift, at least in high southern latitudes, as a more plausible hypothesis than a land connection for Austral organisms of Tertiary age.

EARLY MIOCENE CLIMATE

Isotopic analyses indicate peaks of 19° and 20 °C for seas in the latitude of Wellington (thus less than the Eocene figures), while the Kaipara Harbour reef corals suggest a seasonal range of 19° to 28 °C, temperatures compatible with the known tolerances of other marine benthic climatic indicators in the Miocene. Larger Foraminifera, at first confined to the north, reached S. Lat. 46° in the west of the South Island before the end of the Lower Miocene (HORNIBROOK, 1968), thus suggesting both north-south and east-west temperature gradients, as at present; nanno-plankton elements also support a late Lower Miocene climatic peak (EDWARDS, 1968). Fig. 12 summarises the evidence for the whole Cenozoic.

While the climate on land was generally subtropical (on the evidence of *Bombax*, Cupanieae, *Cocos*), dominance of *fusca*-beeches in Otago suggests latitudinal (or perhaps partly edaphic) gradients.

Middle and Upper Miocene

Middle Miocene includes the Lillburnian and Waiauan stages of the New Zealand classification, Upper Miocene the Tongaporutuan and Kapitean stages. The pattern of Lower Miocene sedimentation at first continued in the Middle Miocene but the sea had left the central trough of Northland and most of Auckland, where the late Tertiary earth movements began at a relatively early (late Miocene) date. Basalts were erupted at Ti Point and on the Tutamoe Plateau. Sediments became coarser on the East Coast (Tutamoe Beds), in Taranaki, Marlborough, and the Waiau Syncline of Southland and the sea retreated locally from Canterbury and Otago. North Taranaki received ash from the Mohakatino volcanoes on the south end of the western ridge of Northland and andesitic volcanoes were active on the east (Miranda-Maramarua). Graded beds were deposited locally in sinking troughs, especially on the East Coast.

Some Early Tertiary geosynclinal troughs had become 'evaginated' to form rising anticlines in the Late Tertiary, and some formed short-lived islands. Synclinal or fault-angle troughs between anticlinal shoals or islands subsided rapidly as tectonic relief increased, sometimes reaching even bathyal depths (e.g. in Wairarapa district), but generally they were filled as they subsided with thick sediments derived from the 'highs', deposited in only moderate shelf depths. Most Miocene sediments do not suggest that adjacent lands were truly mountainous but the Putangirua Conglomerate, the Great Marlborough Conglomerate, and the thick Longford Formation (Murchison) were accumulated near rising mountains, mainly in fresh water. Short-lived peneplains of late Tertiary age – plain-like surfaces of low relief cut by erosion across older rocks – have been recognised in Otago, Fiordland and the King Country, so 'mountains' were local and not of alpine height.

South of the shallow Mid Miocene (Lillburnian) seas in Marlborough, sediments of deeper facies were deposited both near the present coast at Kekerengu, Amuri Bluff, and Gore Bay (North Canterbury), and inland near Mandamus. About this time, the sea extended to within twenty miles of the present divide near the head of the Esk River (WILSON, 1956) but had retreated from the Weka Pass District, from Otago-South Canterbury, and from some of the West Coast basins. It persisted in the Waiau Syncline of Southland.

During the following stage (Waiauan), the sea remained in most of the districts of Lillburnian sedimentation and transgressed over some areas previously emergent – for instance at Weka Pass, Fox River, Cape Foulwind, and on the margin of the stable schist platform at Otago Peninsula where volcanic activity was initiated at this time. Tongaporutuan beds followed conformably in the main basins, but locally the sea transgressed further, on the west flank of the East Coast basins towards Dannevirke in the north and Awatere Valley in the south. In Hawke's Bay, rapidly deposited graded sandstones filled deep depressions attributed to active transcurrent faulting during phases of accelerated earth movement (KINGMA, 1958). Some have suggested that rhyolitic tuffs near Mahia and Porangahau came from volcanoes off the east coast, but a source in the Taupo Zone seems more likely. The youngest Miocene stage (Kapitean) is represented by marine beds in the same areas, but its seas spread further, towards Manawatu Gorge, and from Taranaki towards the west Taupo mountains, heralding the Pliocene climax of late Tertiary transgression. An acute spasm of mountain-building in Marlborough led to thick conglomerate deposits in Awatere district.

Marine fauna

In the middle and late Miocene successive marine bottom faunas record a progressive cooling of the seas around New Zealand, by the extinction of subtropical genera that had marked the Mid Tertiary thermal maximum. Among the organisms that became extinct were the corals *Platyhelia*, *Dendrophyllia*, *Flabellum pavoninum distinctum* and *Stephanocyathus ixine* (still living in the Indo-Pacific); *Schizaster* and other echinoids; the mollusca *Morum*, *Conospirus*, *Hinnites*, *Solecurtus*, *Aturia* and others of longer residence including endemic genera (such as *Hedecardium*, *Serripecten*, *Magnatica*, *Neocola*); the brachiopods *Lingula*, *Magadina*, *Rhizothyris;* and the commonest Tertiary crab (*Tumidocarcinus*, endemic to Australasia). Temperatures may have fluctuated but remained higher than now, as shown by the persistence of *Cucullaea* and the cidarid *Phyllacanthus* into the Kapitean.

Despite cooling conditions, Malayo-Pacific and Australian immigrants continued to colonise: the bivalves *Pitar* and *Katelysia* in the Middle and *Trachycardium* in the Upper Miocene (which had a short-lived existence in New Zealand), *Atrina* and *Mactra* (Mid Miocene), *Striacallista*, *Zeacumantus* (Upper Miocene) and the spider-crab *Leptomithrax* (Upper Miocene) which remain characteristic members of the Recent fauna. Immigrant Foraminifera also came from the Indo-Pacific (e.g. *Bolivinita quadrilatera* in the Upper Miocene) and from Australia, so that the assemblages on opposite sides of the Tasman look more alike than previously. The gastropod *Lamprodomina*, a member of the Indo-Pacific subfamily Olivinae, is common to New Zealand and Chile in the Upper Miocene-Pliocene and is perhaps a genus of tropical origins that achieved a partially Austral distribution using the West Wind Drift, like the plants *Coriaria* and *Aristotelia* earlier in the Miocene. Some mollusc incomers are southern or endemic forms which had perhaps begun to move north to New Zealand latitudes as climate cooled: *Cratis*, *Buccinulum*, and *Aeneator*.

Terrestrial vegetation

The forests continued to be *brassi*-beech forests, with *fusca*-beech, podocarps and ferns becoming rare towards the end of the Middle Miocene but recovering in relative abundance in the Upper Miocene. Figure 17 illustrates the three species-groups of *Nothofagus*, that played leading roles in the Upper Cretaceous and Cenozoic vegetation.

No additions to the flora are recorded in the Middle Miocene; in fact, a number of mid-Tertiary pollen species and *Cocos zeylanica* are recorded for the last time. In the Upper Miocene a number of Recent spore and pollen types appear. Thus spores closely similar to those of the ferns *Lindsaea cuneata*, *Nephrolepis cordifolia*, *Rumohra adiantiformis* are first recorded from localities correlated with the Taranaki Series. *Podocarpus*

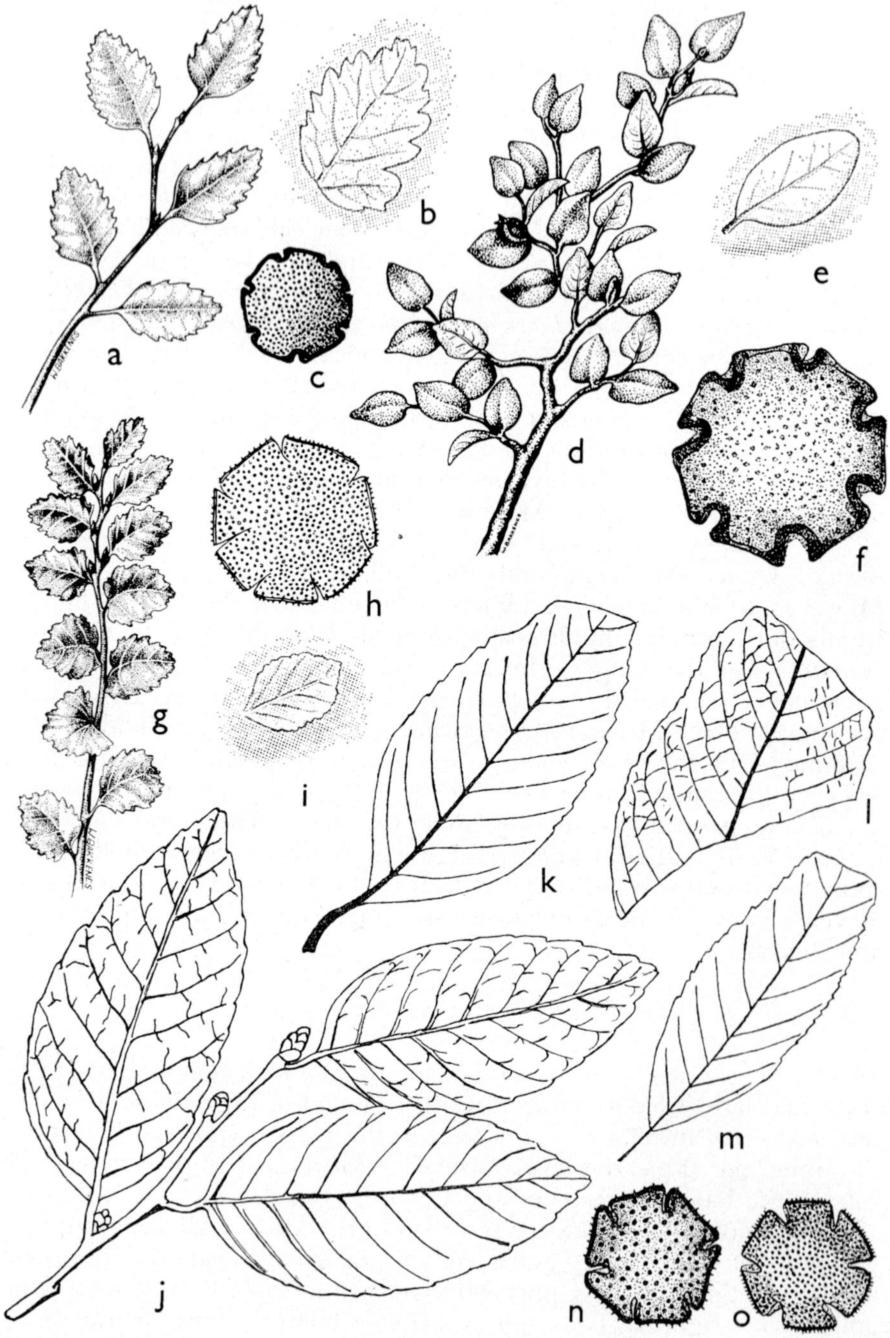

Fig. 17. The Southern Beeches *(Nothofagus)*; characteristic foliage and pollen grains of the three species groups. I. The *fusca*-group, represented by foliage and fossil leaves

pollen recognisable as related to *P. ferrugineus* (sect. *Stachycarpus*, see Fig. 5) first appears at this time; it probably does not indicate an immigrant but rather evolution from within the large assemblage of podocarps present since the Cretaceous.

Two notable Upper Miocene plant fossil localities (at Great Barrier Island and at Kaikorai, Dunedin) have yielded identifications by OLIVER of leaves mostly confirming spore and pollen evidence of the Miocene flora. Exceptions are *Libocedrus plumosa* (pollen not recorded until Lower Pleistocene), *Melicytus, Heimerliodendron, Pittosporum, Neopanax, Geniostoma, Senecio* (no pollen record) and an extinct genus *Parafagus* (believed by OLIVER to have been related to the oaks).

By the end of the Miocene a number of characteristic Tertiary plants (mostly of uncertain botanical affinity) had become extinct.

TERRESTRIAL FAUNA

Moa bones (Dinornithiformes) recorded by OLIVER (1949) from an area of Uppermost Miocene marine sediments at Maungapurua, Wanganui Valley, would be the oldest known if really derived from those rocks. As they are identified as the Recent species *Pachyornis mappini* Archey they suggest either that speciation in the order was well advanced before the Pliocene, or that the bones really came from Quaternary deposits.

UPPER MIOCENE CLIMATE

Isotope temperatures of 16° in the Middle and Upper Miocene appear too low for the paleontological evidence of sea temperatures above those of today, such as Lepidocyclinidae as far south as Southland, and *Globorotalia menardii*, now restricted by a minimum temperature of 20 °C. The persistence of the bivalve *Cucullaea*, large *Conus* (gastropoda) and the Cidarid *Phyllacanthus* until latest Miocene (Kapitean Stage) and of Mid Miocene warm-temperate indicators south to 51° S Lat. at the Auckland Islands (FLEMING, 1968) are evidence that sea temperatures remained relatively high till the end of the Miocene. On the other hand, progressive extinctions among corals and molluscs, the fleeting appearance of tropical pelagic Foraminifera, the appearance of cool-water

---

of *N. fusca* (a, b), *N. solandri* var. *cliffortioides* (d, e) and pollen grains of *N.* aff. *fusca* (c) and *N. flemingii* (f, Eocene). II. The *menziesii*-group, represented by foliage (g), pollen grain (h), fossil leaf (i, Lower Pleistocene Ohuka Creek) of *N. menziesii*. III. The *brassi*-group, represented by foliage of *N. perryi* from New Guinea (j), Oligocene leaf impression of *N. ninnisiana* (k, l, m) from New Zealand, and pollen grains of *N. mataura-ensis* (n, Eocene-Pliocene) and *N. kaitangata* (o, Upper Cretaceous). (a, d, g by W. BAKKENES, by courtesy of N.Z. Forest Service; pollen grains from COUPER, 1953.)

genera (e.g. the corals *Oulangia* and *Lophelia*) and foraminiferal evidence from the 99% sinistral coiling in *Globorotalia pachyderma* (KENNETT, 1968) of a late Miocene cooling of catastrophic scale (when the Antarctic Convergence reached the Mid North Island) suggest that there were marked fluctuations of temperature. Certainly there was a temperature gradient between the subtropical Waikuku Limestone of Northland (LEITCH *et al.*, 1969) and the more or less contemporary Mid Miocene of the Auckland Islands.

On land, the plant fossils suggest warm-temperate to subtropical conditions in the Middle Miocene but some cool-temperate vegetation on uplands in the Upper Miocene.

Pliocene

TECTONICS AND GEOGRAPHY

During the Pliocene earth movements became even more intense, approaching the climax of the Kaikoura Orogeny to which New Zealand owes its present geography. Differential movements were most acute in a mobile belt trending from East Cape obliquely south-westward through the Southern Alps to Fiordland, and decreased towards the more stable districts of Northland and Otago-Stewart Island on either side. The last major marine transgression, begun in upper Miocene times, reached its maximum in the Pliocene. In stable Northland the sea flooded the Manukau-Lower Waikato depression, as far south as Huntly and Meremere. Sea occupied Wharekahika Graben at East Cape. The southern third of the North Island was widely inundated, between the persistently subsiding Wanganui and Hawkes Bay basins, with extensions north to Wairoa and south to Palliser Bay. Deep-water Opoitian mudstones (continental slope according to P. V. VELLA) suggest that coastal ranges east of Eketahuna were not elevated as continuous land till late Wanganuian time. Between these areas marine straits crossed sags separating the rising axial ranges, which were sometimes emergent, supplying coarse sediment, at other times reduced to islands and shoals. It is uncertain whether the Wellington area (flanked by sea in the Wairarapa and at Makara) formed another island or was linked to the emergent areas of Marlborough Sounds and Nelson. On the east side of the South Island, marine Pliocene beds are present in the Awatere and in Canterbury south to Timaru (here buried by later deposits) and have been identified from the sea bed off Oamaru. On the West Coast marine Pliocene deposits occur from Karamea to Hokitika and at Resolution Island. A shallow sea still penetrated the Waiau syncline. The lands were not as mountainous as today but relief increased at the end of the epoch when schist in the alpine backbone was exposed to erosion and gravels from the rising South Island Mountains drove back the coasts.

Straits and islands

The Far North, beyond Ahipara, consists of an ancient archipelago now linked together by sand bars that were almost certainly sea during the Pliocene and some of the interglacial ages (Fig. 18). At these times, too, the sea flowed across the Auckland Isthmus from Hauraki Gulf to the Manukau. The chronology of junctions and separations has yet to be worked out. The influence of such former straits on speciation is, however, obvious in land mollusca (see Powell, 1949; Climo, in this volume). Thus *Rhytida duplicata* ('Aupouri Island') is isolated from *R. dunniae* (Northland) by the 'Ahipara Strait', *R. dunniae* is now sympatric with *R. greenwoodi* in the general area of the 'Manukau Straits', which was a more effective barrier than the present Colville Channel and Tamaki Strait. *Paryphanta watti* (Far Northland) and the *ambagiosus* group of *Placostylus* in 'Aupouri Island' are isolated from *P. busbyi* and *Placostylus hongii* (Northland) by the Ahipara Strait. The same isolating barriers seem to have been responsible for speciation in the stag beetle family Lucanidae (Holloway, 1961). In both groups an additional species in northern Northland suggests speciation originating from a further island, probably on Kerikeri Peninsula. These speciation patterns probably date from the Pliocene.

Volcanoes

Isotope datings (Stipp, 1968; Stipp & McDougall, 1968; Stipp & Thompson, 1971) now permit greater precision than previously in the chronology of volcanic eruptions. During the Pliocene, volcanoes, mainly andesites, erupted at Whangaroa, Northland and at the Coromandel Range, south to Kaimai. On the western side of the North Island, the ancient andesite volcanoes of Orangiwharo and Whareorino were active. In the central volcanic region further south, however, none of the diverse eruptive rocks that now cover thousands of square kilometres have proved to be as old as Pliocene. At Lyttelton, the latest eruptions (Diamond Harbour Group) were Pliocene. The Chatham and Campbell Islands, emergent during at least part of the Miocene, were lapped by Pliocene seas and were the sites of volcanoes at least partly Pliocene.

Marine fauna

Pliocene marine faunas (Opoitian-Waipipian) retained many long-lived Tertiary genera, some warm-water ones such as the corals *Lophelia*, *Carophyllia*, the bivalves *Arca*, *Eucrassatella*, *Miltha*, *Trachycardium*, and the gastropods *Polinices*, *Galeoocorys*, and even gained additional forms from the north such as the Foraminifera *Bulimina echinata*, *Globorotalia inflata*, and *Globigerinoides rubra* and the Mollusca *Glycymerula*, *Patro* and *Zethalia*.

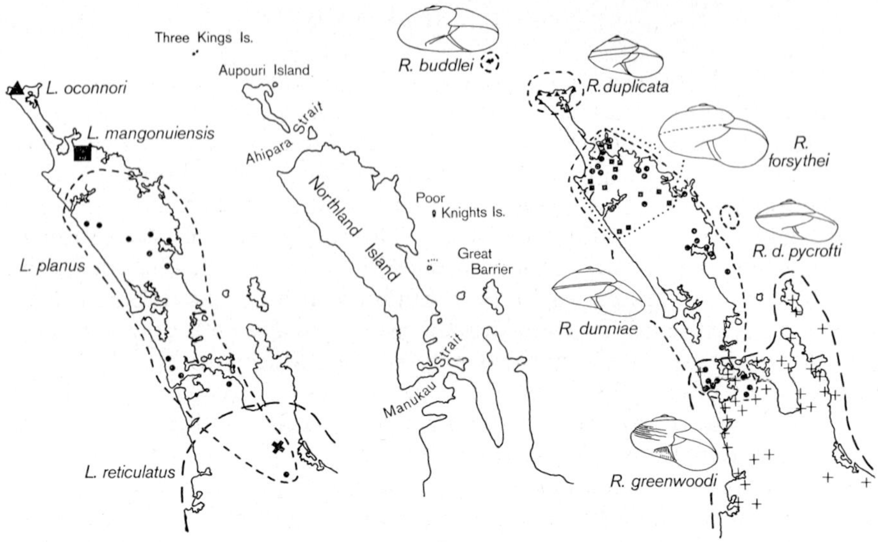

Fig. 18. Speciation in pulmonate land snails *(Rhytida)* and stag beetles *(Lissotes)* on the Pliocene archipelago in North Auckland. Data mainly from POWELL (1949) and HOLLOWAY (1961).

Pliocene sea temperatures in central New Zealand were evidently nowhere cooler than those of modern Northland. But the most dramatic faunal event of Pliocene time was the progressive extinction of Miocene warm-water groups: the *Haeuslerella* lineage, the *Globorotalia miozea* lineage and other Foraminifera; *Gemmula*, *Maoricardium*, *Mauicassis* and many other Mollusca; ostracods such as *Loxoconcha australis* and *Cytherelloidea auricula*. Other characteristic mid-Tertiary marine invertebrates, known or inferred to have been stenothermal in warm waters, had disappeared before the beginning of the Pliocene. Gaps in the fauna were only partly filled by newcomers to the fossil record. Ancestors of many groups still living appeared, some of them dominant organisms of our shores: for instance *Flabellum rubrum* (Scleractinia); *Mesodesma*, *Cyclomactra*, *Notirus* (Bivalves); *Nerita*, some groups of *Buccinulum* and *Cominella*, *Penion* of the *adusta* group, *Paratrophon*, *Zeatrophon*, *Amphibola* (Gastropoda); *Coronula* (whale-barnacle); *Cancer novaezealandiae*, *Ovalipes*, *Hemiplax* (Crabs); *Fellaster* (Echinoidea) became abundant. Some of the Pliocene immigrants are of Australian or northern origin but others are clearly Austral groups of cool water, distributed by the West Wind Drift (e.g. *Lissarca*, *Aulacomya*), and others 'cryptogenetic' – endemic genera of unknown ancestry. Some molluscan genera with a tranquil history of mid Tertiary stability showed a minor burst of speciation in the Pliocene (e.g. *Bassina*). At about this time the echinoid *Pseudechinus* arose in Australasia from *Brochopleurus*, and became distributed in a southern circumpolar zone by epi-planktonic dispersal in the West Wind Drift (FELL, 1962).

FLEMING (1962b) has suggested that the unknown Tertiary bird fauna probably included many unsuspected endemic forms, in addition to the endemic families of New Zealand Wattle-birds, Wrens and Thrushes that survived the Pleistocene, as the result of colonisation from across the Tasman (such as still continues) throughout the long epochs of Tertiary time. That the fauna included unsuspected elements is shown by the discovery of a bony-toothed seabird, *Pseudodontornis*, in a Pliocene concretion from Motunau (HOWARD & WARTER, 1969). Its nearest relatives are from the Miocene of North America, so it may already have been a 'living fossil' in the Pliocene of New Zealand.

VEGETATION

Pliocene plant fossils also give evidence of cooling climate. Indeed, plant microfossils from one horizon near Waverley (MILDENHALL & HARRIS, 1970) indicate climate cooler than the present at sea level (whereas associated marine faunas are of warm facies). This is anomalous in relation to other evidence of Pliocene warmer temperatures such as the dominance of *brassi*-beeches at Dunedin, and persistence of *Dysoxylum*, *Knightia* and *Agathis*, south of their present limits, together with the last *Ephedra* and Cupanieae, neither of which survived the Pliocene. The pollen attributed to *Casuarina* is so rare that its source tree must have been reduced to a rarity. Newcomers to the flora, judged by pollen, are *Cyathea medullaris*, *Quintinia*, *Salicornia*, *Eugenia* (and "*Myrtus*' s. lat.), and Compositae of *Taraxacum* type.

CLIMATE

COUPER & MCQUEEN (1954) found evidence of geographical differentiation of Pliocene vegetation not evident in Miocene samples, and a later assessment (MCQUEEN *et al.*, 1968) mentions a mixed cool and warm flora, suggesting a relief of at least 500 m, with *brassi*- and *fusca*-beeches occurring together in samples from near the head of the Rangitikei River. The evidence suggests regionally and altitudinally diverse climates, fluctuating in time from warm temperate to somewhat cooler than now, on a land of increasing topographic relief, but with sea-temperatures still subtropical, at least north of Lat. 43 °S.

Early Pleistocene

GEOGRAPHY

In New Zealand, early Pleistocene deposits, consisting of four stages grouped as Upper Wanganui Series (Fig. 9), follow directly on the Pliocene in the Wanganui and Hawke's Bay basins, and in Westland, North Canterbury and Marlborough. Elsewhere the land was emergent and eroding, or river gravels were built up as alluvial terraces filling intermontane depressions (Otago-Southland) or fans sloping out from rising mountains. By analogy with later glaciations, sea-level was probably relatively low during the cold phases, when the chief deposits were alluvium built up by aggrading rivers, whereas the higher seas of warmer phases have left elevated marine deposits. The 'low' sea levels of early glaciations may, however, be above the present sea level in areas of predominant tectonic uplift. Only in subsiding geosynclinal basins was a relatively complete record deposited. The first glacial stage, following on the Pliocene, has been called the Hautawan, followed by the Nukumaruan, at least in part covering the first interglacial. Some writers have rejected Hautawan as a stage because it is difficult to define faunally but agree that the base of the Pleistocene falls between Waitotaran and Nukumaruan stages. The second period of cooling (perhaps multiple) spans latest Nukumaruan and early Okehuan time, and is followed by the second interglacial culminating in the Castlecliffian (= Putikian) Stage. The date of the Plio-Pleistocene boundary is uncertain but probably between $1\frac{3}{4}$ and $2\frac{1}{2}$ million years BP.

Some of the Northland basalts, the extrusive Franklin Basalts of the Waikato, andesites in the Bay of Plenty, the volcanoes of Pirongia and Karioi on the North Island West Coast, Paritutu and the Sugar Loaves off New Plymouth, the andesites of Titiraupenga and several ignimbrite sheets were erupted in the early Pleistocene.

Hautawan

Marine Hautawan deposits are largely confined to areas flanking the axial ranges of the central part of the mobile belt – the Wanganui Basin on the west and the complex of synclines from Hawke's Bay to North Canterbury to the east. The Pliocene Manawatu Strait linking these areas was soon temporarily bridged by a gravel spit fed by coarse detritus from the rising Ruahine-Tararua Ranges. At about this date extensive and thick aggradation gravels from young mountains were deposited in the Reikorangi and Kaitoke districts of Wellington, in Nelson (Moutere Gravels) and on both sides of the Southern Alps (Kowai and Old Man Gravels). Locally they contain cold climate pollen assemblages, the Old Man Gravel, in Westland, also including moraine and varve-silts of the first or Ross Glaciation (GAGE, 1961). Other non-marine deposits correlated with the Ross Glaciation have been recognised by cool-climate plant fossils in the Waikato, on the West Auckland coast, at Timaru, and Invercargill. The extent of cooling is shown by the presence of *Nothofagus solandri* var. *cliffortiodes* (Mountain Beech) and *N. menziesii*

(Silver Beech) near Port Waikato at the coast (Kear, 1957) and widespread subalpine grassland and scrub at low altitudes in the South Island.

Marine animals now confined to subantarctic waters came north to the southern North Island: the fan shell *Chlamys delicatula*, the cockle *Tawera subsulcata*, the gastropods *Zeacolpus (Stiracolpus) symmetricus*, *Maurea couperi* (aff. *spectabile*), *Cominella (Eucominia)* aff. *nassoides*, *Trophon* s. str., the crab *Jacquinotia*, an early relation of the Ross Seal *(Ommatophoca)*, and the echinoid *Notocidaris vellai* Fell. These immigrations suggest climatic zones moved at least 5° Lat. (600 km) north of their present positions. Despite these indications of cold climate, a number of warmth-loving organisms survived the Ross Glaciation.

Nukumaruan

Seas of this interglacial age occupied the flanking basins of Wanganui and Hawke's Bay-Wairarapa and were again linked through Manawatu Strait but left no certain traces in Auckland and South Island districts. Rhyolite volcanism, producing pumice and ignimbrite, began near Taupo, judged by the appearance of pumice silt in the Wanganui Basin. Ninkovich (1968) concluded the earliest pumice ash on the sea bottom east of New Zealand (and therefore the earliest ignimbrite eruptions on land) was 0.8 m.y. BP but most local geologists would date them as over a million years old.

Several warm-water bivalve genera thrived for the last time: *Patro*, *Isognomon*, *Glycymeris* (s. str.), *Lutraria* and *Eumarcia*, and a few immigrant gastropods from Australia or the north trickled in: *Pterynotus angasi*, *Ellatrivia*, *Leucotina ambigua*.

Pollen assemblages attributed to the Nukumaruan contain few extinct species and indicate climates either similar to the present or cooler, those of Wanganui (Maxwell Formation) showing fluctuating climate. Several plants have their first records in the Nukumaruan: *Libertia*, *Plantago*, *Cardiomanes* and several other ferns.

In the late Nukumaruan of Wanganui district, two shallow-water marine incursions, perhaps representing interstadials or interglacials, interrupted alluvial deposition of delta sediments deposited when vegetation and climate resembled that of coastal south-east Otago (equivalent to a lowering of vegetation zones by up to 650 m), probably during glacial stages.

At the end of the Nukumaruan the sea retreated from Wanganui Basin, perhaps a eustatic withdrawal during a glacial phase.

Okehuan

Evidence for cold climate is weaker than in the Hautawan or in the later part of Nukumaruan time and the type Okehuan may represent the transgression at the beginning of an interglacial stage, rather than the glacial stage itself. On the east flank of the Tararua Range, however, periglacial vegetation is shown by a pollen sample from the Mangahou Formation, which may be Okehuan judged by its pumice content (G. NEEF, Thesis, 1967).

The Australian genus *Acacia* appeared briefly in New Zealand during the Okehuan (MILDENHALL, 1971). The marine faunas are impoverished, even stunted locally, and lack some of the warm-water elements that persisted through the Nukumaruan, but include no distinctive cool-water forms.

Sea had retreated permanently from most of the Hawke's Bay – Palliser Bay area and temporarily from Wanganui Basin as alluvial aprons spread out from rising ranges and from the central volcanic zone of the North Island, where ignimbrite eruptions produced vast quantities of pumiceous sediments. Submarine pumice layers in the Pacific east of New Zealand dated by paleomagnetic stratigraphy as 0.86 and 0.73 million years BP are probably Okehuan. Fur seal vertebrae have been identified from Wanganui.

Castlecliffian

The marine fauna suggests temperatures warmer than now. The sea lapped back on to the land in several regions, perhaps in the Manukau depression, certainly in parts of the Bay of Plenty from Matata to Cape Runaway, at Gisborne, Mahia Peninsula and Cape Kidnappers, and occupied restricted parts of Wanganui Basin, where Castlecliffian fossils are best known, and of North Canterbury. Otherwise, Castlecliffian geography did not greatly differ from the present. Earth movements were accelerated. Ignimbrite eruptions continued and the Tongariro andesite volcanoes began to erupt. Immigrant mollusca, mainly from Australia, but also from the north, included *Xenophalium*, *Pecten* (s. str.), *Anadara trapezia*, *Bembicium melanostomum*, *Eunaticina*, *Agnewia*, *Pterynotus zelandicus*, *Zelippistes* (now northern Northland). On land, kauri lived near Wanganui, and the *brassi* group of beeches flourished for the last time.

Late Pleistocene

Late Pleistocene stratigraphy and chronology are active fields of research. Owing to earth movements during the Kaikoura Orogeny, late Pleistocene sediments (Hawera Series) in central New Zealand are clearly separated from those of the early Pleistocene. Most of them are either in coastal terraces formed during the interglacial stages of high sea levels or lahars and alluvial deposits of river beds aggraded during glacial stages of low sealevel and of rapid erosion, caused by periodic deforestation and periglacial processes in the highlands. At least four glaciations and three

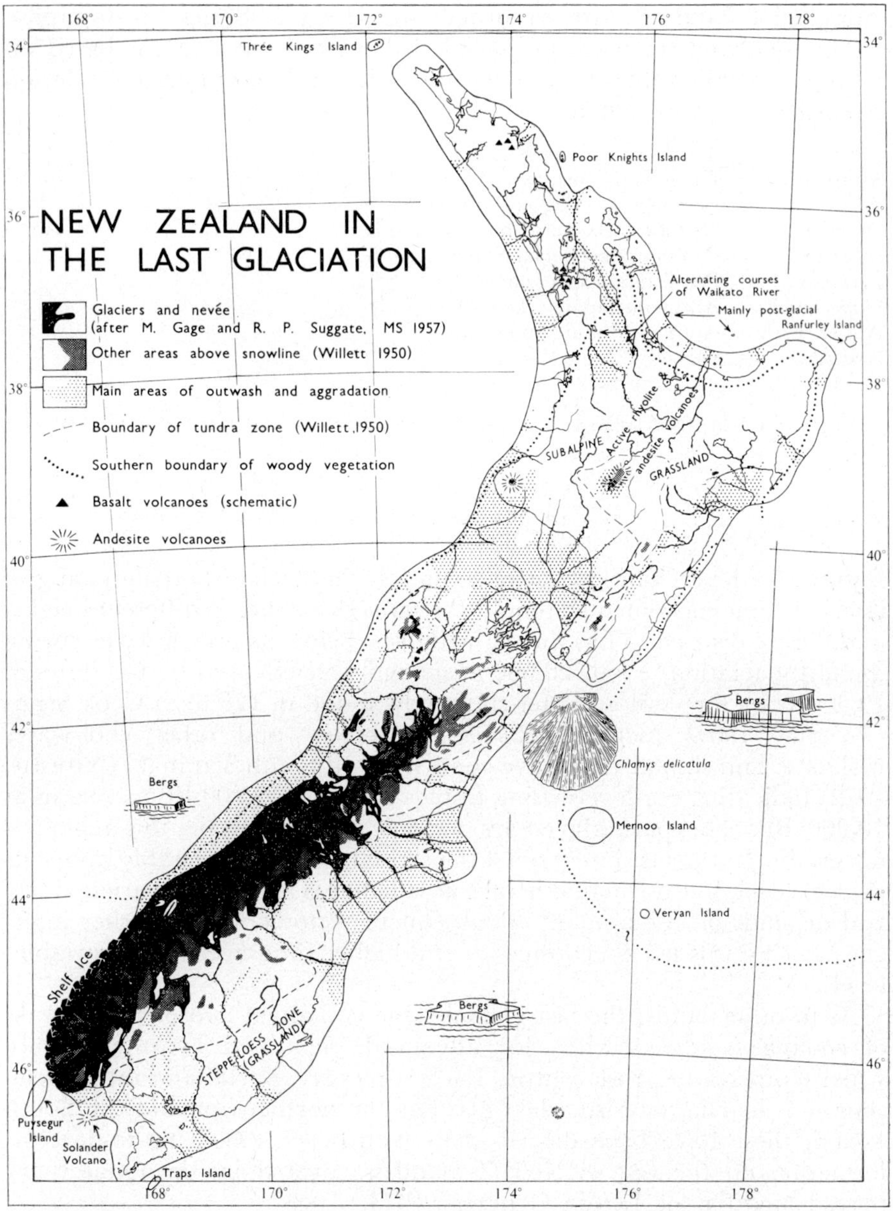

Fig. 19. New Zealand during the Last (Otiran) Glaciation. Distribution of ice based on a map by M. GAGE and R. P. SUGGATE, MS, 1957. Judged by the evidence for refugia of scrub in the Foveaux Strait-Otago region, the boundary of woody vegetation is too far north. (FLEMING, 1962a.)

interglacial intervals are recorded by Hawera Series stratigraphy.

The following sequence of glacial and interglacial stages is based on SUGGATE (1965); there are some difficulties in determining the relation of Porikan to Castlecliffian.

Aranuian:	Post-glacial (begins c. 1400 yr B.P.)
Otiran:	Last Glacial Stage (3 advances)
Oturian:	Last Interglacial (2 high sea-levels)
Waimean:	Penultimate Glacial Stage
Terangian:	Penultimate Interglacial
Waimaungan:	Antipenultimate Glacial
Waiwheran:	Antipenultimate Interglacial
Porikan:	Glacial

Figure 19 is an attempt to show New Zealand during the last of the glacial ages.

SEA LEVEL AND GEOGRAPHY

Eustatic sea-level changes controlled the late Pleistocene shoreline changes; later, tectonic movements brought the same shorelines to different heights in different districts. Thus the sea withdrew below its present level during the last glaciation, about 20,000 years ago, as evidenced by C^{14} dates of 19,500 years for shallow water shells now found in 128 m in Cook Strait (PANTIN, 1957) including *Chlamys delicatulus* and other cool-water mollusca, and similar data have come from off South Taranaki (NORRIS, 1972) indicating sea-levels rising from – 167 m at 17,200 BP to – 90 m at 10,000 BP. The part played by tectonic movements in this region is uncertain. In the last interglacial the shoreline lay considerably above its present level, but the actual height at which it is now found varies widely and depends on the amount of subsequent tectonic uplift. Earlier interglacial sea levels left their traces at still higher (and even more variable) levels.

As in other lands, the last glacial stage contained two or more peaks of cooling (stadials) when ice advanced, separated by interstadials, when temperatures rose a little. Each successive glacial and interglacial stage left its characteristic deposits. For the northern part of the South Island, these have been described by SUGGATE (1965); up-to-date information on the rest of New Zealand is covered by the Quaternary maps issued for the INQUA Congress, 1973.

ISLANDS

Important biogeographical results followed from the linking of the main islands of New Zealand to each other and to near-shore islands; the persistent separation of others led to some degree of endemism, especially in the north, where climatic deterioration at each glacial stage was not sufficiently severe to eliminate the fauna and flora.

The Three Kings Islands 60 km northwest of Cape Reinga presumably originated before the Pliocene. (Pliocene marine deposits have been identified offshore). They have remained as islands ever since, separated from the mainland by deep water. Both the invertebrate fauna and the flora are highly endemic.

Most of the islands off the Northland coast were linked with the mainland during the Last Glacial Stage, but the Poor Knights, perched on the shelf edge at a depth of about 125 m, has had a longer history of isolation judged by its endemic snails *(Rhytida, Allodiscus)* and wingless Orthoptera *(Deinacrida)*.

Further south, the Solanders and Snares were probably persistent islands, but they suffered severe climates during the last glaciation, so that the only organisms that could have inhabited them long enough to become strongly endemic must have tolerated tussock and herb field communities.

To an even more marked extent, life at the Subantarctic Islands was severely reduced during the glacial phases of the Pleistocene. Campbell Island arose from volcanism, at first submarine, on a domelike uplift of the sea-bottom in latest Miocene or earliest Pliocene time, and built up a shield volcano that was never, so far as known, linked to other lands. Auckland and Antipodes Islands probably had similar histories. Whatever fauna and flora had developed by the late Pliocene must have been decimated in the Pleistocene, when permanent snow reached below 350 m so that woody vegetation was eliminated. Most, if not all, strongly endemic invertebrates at Auckland Islands are members of tussock grassland or herb field communities (e.g. the endemic land snail genus *Ranfurlya*) whereas in the strictly forest community of the Aucklands the flora and fauna lack endemic species or genera.

The Chatham Islands, east of New Zealand, have a long history as land but their Pleistocene climatic history is as yet uncertain. That the vegetation was reduced to scrubland is suggested by the virtual lack of endemic plants and land snails in the coastal forest. Most bush birds too are only endemic at subspecies level and the exceptions (a warbler and several rails) could tolerate scrub vegetation.

Judged by bottom contours, additional islands, most of them now completely submerged, were exposed above sea-level during glacial stages. In addition to those shown on Fig. 19 (Ranfurley, Mernoo, Veryan, Traps and Puysegur), a large subantarctic island formerly occupied the site of Pukaki Bank and appears to have been almost completely truncated by marine erosion during one of the later glacial ages, so that its flattish surface now lies at a depth of 146 m and the uneroded summit is submerged to a depth of 60 m. Such lost islands, hundreds of square kilometers in area, may have played an important rôle in speciation.

It has been customary to consider Cook Strait a comparatively young feature in relation to the older 'Manawatu Strait' although the evidence is not unequivocal. In the later Pleistocene Cook Strait was bridged at each ice age by a broad corridor now represented by sea bottom shallower than −100 m between South Taranaki and Northwest Nelson. When the sea rose rapidly after the last glaciation it reached −100 m about 10,500 years ago (judged by local dates, Norris, 1972), and flowed across the barrier to form a strait at least 50 km wide within the next 500 years. Thus its age of most recent formation as a barrier to most land animals is about 10,000 years.

Foveaux Strait is much shallower than Cook Strait and its last formation is therefore younger, but not much younger because the post-glacial rise of sea level was rapid (Cullen, 1967). The sea apparently flowed across the Foveaux Strait land bridge in a depth of about 30 m at about 9500 years BP.

Just as there are lost islands, now submerged, so there are lost straits, now bridged by sediment. The straits separating Northland from Aupouri Island to the north and from the main North Island to the south were re-formed at each interglacial, reinforcing the effects of these earlier (Pliocene) barriers. Presumably the straits now isolating Waiheke, Ponui, Cuvier, and Great Barrier in Hauraki Gulf arose during the sea-level changes of the Pleistocene. Further south, during at least some of the Interglacial transgressions, a broad strait separated Banks Island (as Captain Cook called it in error) from the outwash gravels of the preceding glaciation, so that 'insular endemics' restricted to Banks Peninsula occur in cicadas (Fig. 24) and some other invertebrate groups.

Conditions during Glacial stages

During the first Early Quaternary Ross Glaciation (Gage, 1961), glacial advance was contemporary with rapid uplift of the Southern Alps. The coastline moved seawards partly through emergence and partly by progradation due to abundant supply of waste, but elevation was not the sole cause of glacierisation since cooling near sea level can be demonstrated by pollen analysis before the ice reached the site of deposition which was still not far above sea level. Wellman (cited by Gage, 1961) visualised an early stage of mountain building when relief and scenery were less strongly alpine than in the late Quaternary, with extensive plateaus on which ice-caps could have accumulated. The paleobotanical evidence suggests considerable devegetation during the Ross Glaciation in the South Island, and forests of silver beech *(Nothofagus menziesii)* and probably Mountain beech *(N. cliffortioides)* growing near sea level in southwest Auckland (Couper, 1960b). Thus, although little detailed information on any one of the earlier glacierisations is as yet available, they seem to have been roughly comparable in severity with later ones.

For the 'Last Glaciation', locally known as the Otira Glaciation, there is considerably more data. WILLETT (1950) estimated an average lowering of snow-line by about 1000 m, which GAGE (1965) considered satisfactory for the last glaciation maximum; outside the glaciated mountains vegetation zones were lowered 800 m (COUPER & MCQUEEN, 1954; WARDLE, 1963). Precipitation was controlled by westerly to north-westerly winds, so the snowline descended westward and rain shadows were developed east of the main ranges. Large areas, not all peripheral to areas of glacial snow and ice, suffered periglacial climate, defined by SOONS (1962, following TRICART) as a regime where frost action replaced erosion by running water. Solifluxion (COTTON & TE PUNGA, 1955) was widespread in middle and southern New Zealand and the process was repeated several times in most districts. Soliflual deposits enclose fossil floras showing alpine scrub and herbs near sea level. Permafrost, however, was probably restricted to Central Otago and Marlborough mountains. GAGE (1965, 1966) considers New Zealand moraines to be small, but the associated fluvioglacial accumulations to be bulky, suggesting continuous action of abundant unfrozen running water, so that the 'winters were not unbroken seasons of frost' but characterised by frequent freeze-and-thaw; he confirmed WILLETT's estimate of an average of 6 °C depression of temperature in the Last Glaciation.

The extent of de-forestation during the last and preceding stages of glacierisation has been argued. WILLETT (1950) concluded that forest was almost exterminated in the South Island, except for possible coastal refugia. Early pollen analyses of sea-level samples either without (or with very rare) forest pollen grains (e.g. COUPER & MCQUEEN, 1954) led FLEMING (1962a) to show forest largely excluded from the South Island and much of the North Island (Fig. 19), a conclusion debated by WARDLE (1963) who assumed that areas with many endemic taxa or parts of the ranges of discontinuously distributed taxa are areas where considerable floras survived glaciation. Such areas are Otago-Southland and Nelson-Marlborough on either side of the harshly-glaciated 'waist' of the central South Island, separated from the North Auckland 'refugium' by a second area of severe Pleistocene climates in the southern North Island. None of the South Island endemics cited by Wardle are obligate forest dwellers, so that they could have survived even if true forests were unable to persist.

The empirical data for judging post-glacial climates have recently been increased by post-glacial pollen diagrams from Canterbury, Nelson and North Westland. Although MOAR (1971) concluded that forest was not established in the South Island before 10,000 years ago, he did not find evidence to contradict suggestions that stands of *Nothofagus* survived the Otira Glaciation in sheltered positions outside the ice limit. On the other hand, Otiran samples within the range 15,100 to 22,330 years BP by carbon-fourteen dating from Lake Hawea outlet, from the Grey Valley (several), from Greta (North Canterbury) and from Orepuki, Southland (not precisely dated) imply a barren treeless landscape, with a mosaic of grass, herb and scrub vegetation, the absence or extremely low percentage of anemophilous tree pollen suggesting that there were no refugia near these South Island districts (McKellar, 1960; Suggate & Moar, 1970; Couper, 1960b). Several Southland sequences ranging from cold grassland to mild climate *Podocarpus* forest are almost completely lacking in *Nothofagus* pollen grains. In the southern North Island, at 19,200 years BP, *Nothofagus* and other forest trees were close enough to distribute a few pollen grains to west-facing periglacial fans on which vegetation some 100 m above the ice-age shoreline was reduced to grassland with Compositae, and *Nothofagus menziesii* was close enough to invade the area rapidly during a brief warming (Fleming, 1970b). Glacial climate was cold enough to deforest coastal Taranaki, so that during two earlier Otiran stadials (represented by Stratford and Opunake formations, the latter 32–38 thousand years BP) vegetation at present sea-level was similar to present vegetation at 4,000–5,000 ft, grassland and subalpine bogs with *Dacrydium bidwillii* or *D. biforme* scrub (Harris in Grant-Taylor & Rafter, 1962). Pollen samples from the Huka Falls Formation give evidence of cold climate near Taupo in the centre of the North Island (Vucetich & Pullar, 1969). Even as far north as the Hauraki Plains, at the foot of the Coromandel Range, the post-glacial section overlies samples with *Hebe* scrub dominant in a more or less treeless assemblage (Harris & Norris, 1972). Still further north, in Northland, the only Pleistocene climatic evidence is alternation of *Nothofagus*-dominant and *Agathis*-dominant forest pollen assemblages near Kaipara Harbour.

In summary, there is much evidence for bleak treeless landscape in southern New Zealand during periods of glacierisation, with woody vegetation reduced to scrub, but there is as yet no proof that there were no forest refugia in the South Island.

Origin of Alpine Biota

The evidence of warm Tertiary climates and of minimal relief of the land prior to the Kaikoura Orogeny suggests that the alpine biota is geologically young. The positions of the vegetation zones in relation to the hypothetical orographic relief in former times have been reconstructed in Fig. 20(a)–(d), using the differences between fossil and present distributions of *Nothofagus* beeches, *Agathis*, *Dysoxylum*, Nautiloids, *Cocos*, and reef corals.

During the mid-Tertiary (Fig. 20a) the zones are displaced 15° Lat. southwards, this being the latitudinal difference between Miocene reef-corals in Auckland and comparable living coral-reef assemblages (Squires, 1962a, b). Still greater displacement would result from choice of other climatic indices, e.g. Nautiloids or *Cocos*, so the diagram probably does not represent the period of maximum warmth when podocarps and *Nothofagus* lived at Seymour Island (Lat. 64°15'S) and shed pollen near McMurdo Sound (Cranwell *et al.*, 1960; Wilson, 1967). The Alpine Zone was most unlikely to have been present on the low-relief land of mid-Tertiary New Zealand. Even in Pliocene, when there is geological evidence of rising mountains, the Alpine Zone, widespread in coastal Antarctica (Fig. 20b) may have failed to reach New Zealand. The Alpine Zone was probably first established in the Pleistocene, when the rising mountains first intersected the theoretical timber-line in its slow northward movement towards its fluctuating Ice Age position (Fig. 20c).

Working from these assumptions, Fleming (1963b) suggested that New Zealand plants and animals now endemic to the Alpine Zone have had only a short existence as alpines in New Zealand. He suggested that the alpine biota can be classed in several groups: 1. Old alpine plants from Antarctica that colonised New Zealand when the zone was first established in the Pleistocene (including such strongly endemic genera as *Hectorella* (? aff. *Lyallia*, Kerguelen). 2. Old alpines with a long history in the Holarctic (Boreal) region that were able to cross the equator when the alpine zone was first established on tropical mountains. 3. Facultative alpine organisms with a longer history in New Zealand with wide enough tolerance and plasticity to colonise the alpine zone, some of which have speciated to produce alpine-restricted forms. This category will include both open country, coastal and rock-dwelling plants and organisms from the forests that covered most of pre-Quaternary New Zealand.

The view that the alpine flora is young has been debated by Wardle (1968) who found evidence of much greater antiquity than Quaternary in small taxonomically isolated genera or subgeneric groups restricted to cool habitats, which he considers already occupied cool habitats in the Southern Hemisphere during the Tertiary. While these could be elements of category (1) above, Wardle prefers the hypothesis that they occupied 'infertile soils on peneplained uplands drenched by persistent mist and rain'. Such an explanation may well apply to plants like *Nothofagus menziesii* that are known to have been present throughout the Tertiary.

Dawson (1971) considered origins in categories (1) or (3) above to be equally possible for two characteristic Umbelliferous genera, *Aciphylla* and *Anisotome* (also in Australian alpine sites).

The invasion of the alpine zone by forest organisms may be illustrated by speciation in the *Nestor* parrots (Fig. 21). During the Early Quaternary Ross Glaciation, a marine Manawatu Strait crossed the southern North Island so that southern populations were forced to adapt to subalpine scrub and herbfield as the climate cooled and forests retreated. When the

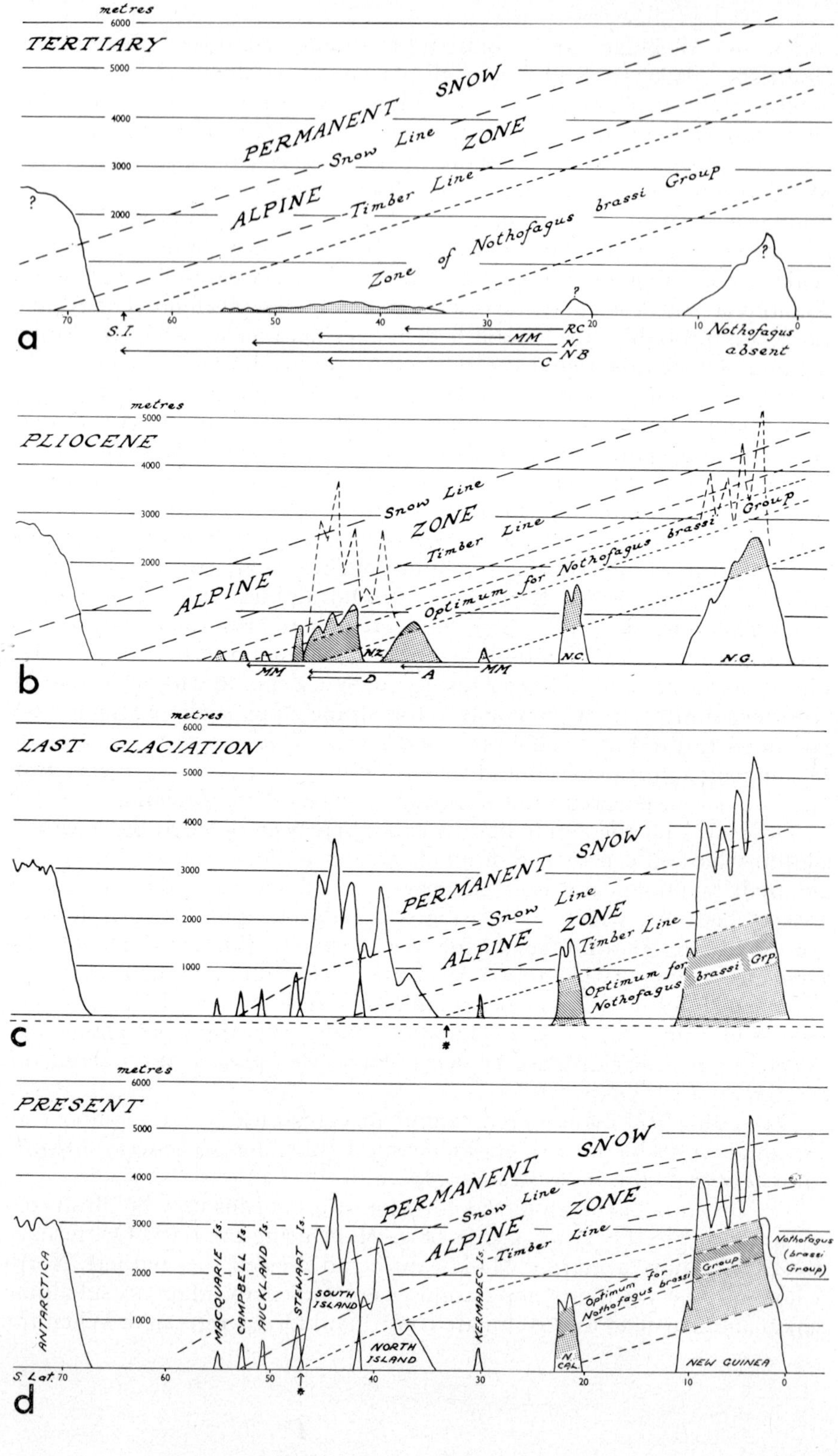
a
TERTIARY
metres
6000
5000
4000
3000
2000
PERMANENT SNOW
Snow Line
ALPINE ZONE
Timber Line
Zone of Nothofagus brassi Group
S.I.
RC
MM
N
NB
C
Nothofagus absent
70
60
50
40
30
20
10
0
b
PLIOCENE
metres
5000
4000
3000
2000
Snow Line
ALPINE ZONE
Timber Line
Optimum for Nothofagus brassi Group
NZ
N.C.
N.G.
MM
D
A
MM
c
LAST GLACIATION
metres
6000
5000
4000
3000
2000
1000
PERMANENT SNOW
Snow Line
ALPINE ZONE
Timber Line
Optimum for Nothofagus brassi Grp.
d
PRESENT
metres
6000
5000
4000
3000
2000
1000
PERMANENT SNOW
Snow Line
ALPINE ZONE
Timber Line
Optimum for Nothofagus brassi Group
Nothofagus (brassi Group)
ANTARCTICA
MACQUARIE Is.
CAMPBELL Is.
AUCKLAND Is.
STEWART Is.
SOUTH ISLAND
NORTH ISLAND
KERMADEC Is.
N CAL
NEW GUINEA
S. Lat. 70
60
50
40
30
20
10
0

daughter stocks subsequently came together they co-existed as sympatric species ecologically differentiated in contrasting environments. In exactly the same way, in the genus *Xenicus* (New Zealand Wrens) an alpine derivative (Rock Wren) of the forest-dwelling Bush Wren was formed. The same type of speciation pattern, resulting in one or more alpine or montane species in the South Island derived from forest ancestors persisting at least in the North Island, has been invoked (FLEMING, 1971b) to explain speciation in cicadas (genus *Maoricicada*) and is a plausible mechanism for the occupation of the Alpine Zone by many plants and insects (e.g. *Hemideina*). In some groups repetition of the process seems to have led to several or many Alpine species sharing the available niches in different altitudinal zones.

A multiplicity of species, many of them hybridising, is characteristic of many alpine plant genera (*Aciphylla, Anisotome, Hebe, Celmisia, Ranunculus, Dracophyllum, Coprosma*, etc.), so that whatever their ultimate origin, they indicate recent, rapid and continuing evolution, while reported inter-generic hybridisation suggests to WARDLE (1963) that 'intense selection pressures . . . led to morphological evolution outstripping the evolution of genetic barriers.'

Distribution patterns of many alpine organisms reflect former barriers of severe glaciation. Thus the central 'waist' of the South Island between Canterbury and Westland, where glaciers extended to the lowlands and produced barren aprons of outwash gravels, has gained many of its alpine plants in post-glacial time as sparse invasions from north and south (BURROWS, 1969); similar patterns can be seen in some alpine cicadas (Fig. 22). Although many New Zealand mountain ranges stand isolated from their fellows by low passes or tectonic lowlands, there has seldom been sufficient time for evolution of endemic populations since their biotas merged in the intervening lowlands during the last glaciation some 20,000 years ago. An exception is in the genus *Paryphanta*, where many distinct geographic races occupy ridge-tops in the northwestern South Island (Fig. 23, from POWELL, 1936).

Isolation in refugia, separated by ice, or by barren outwash fans of gravel, allowed evolution of local endemics, both in the alpine grassland

---

Fig. 20a–d. Cross-sections with greatly exaggerated vertical scale from Antarctica through New Zealand and New Caledonia to New Guinea, showing altitudinal vegetation zones in relation to topography. (From FLEMING, 1963b.) (a) Mid-Tertiary, with the zones displaced 15° Lat. southward at sea-level, the minimum displacement to conform with the occurrence of reef corals in Northland (RC), warm-water marine Mollusca (MM), Nautiloids (N), *brassi*-group beeches (NB) and *Cocos* (C). (b) Pliocene, with zones displaced to conform with occurrence of certain marine Mollusca (MM), *Agathis* (A) and *Dysoxylum* (D). Orogenic heightening of New Zealand is inferred. (c) Last Glaciation, with glacial sea-level shown and topography fully developed. Asterisk at southern potential limits of *brassi*-beeches, which were exterminated in New Zealand. (d) Present, with asterisk at potential southern limit of *brassi*-beeches.

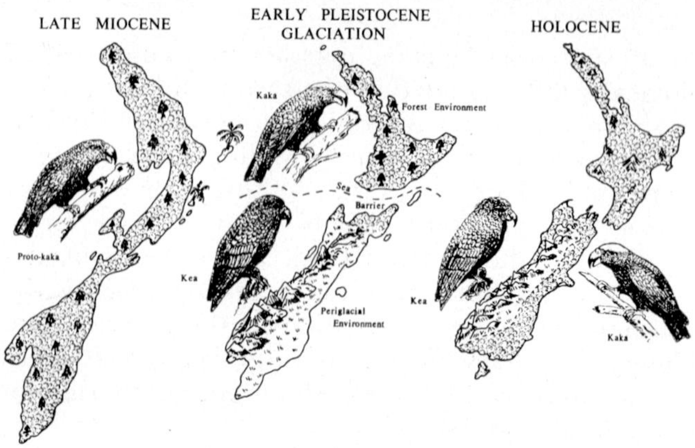

Fig. 21. Speciation in *Nestor* parrots. A Tertiary forest-dwelling parrot (Proto-Kaka) is inferred to have subspeciated during the Early Pleistocene on either side of the Manawatu Strait, with strong ecological differentiation at the onset of the Ross Glaciation. The daughter stocks (Kaka and Kea) subsequently became sympatric in contrasting forest and alpine scrub environments in the South Island. The Kaka subsequently subspeciated on either side of Cook Strait.

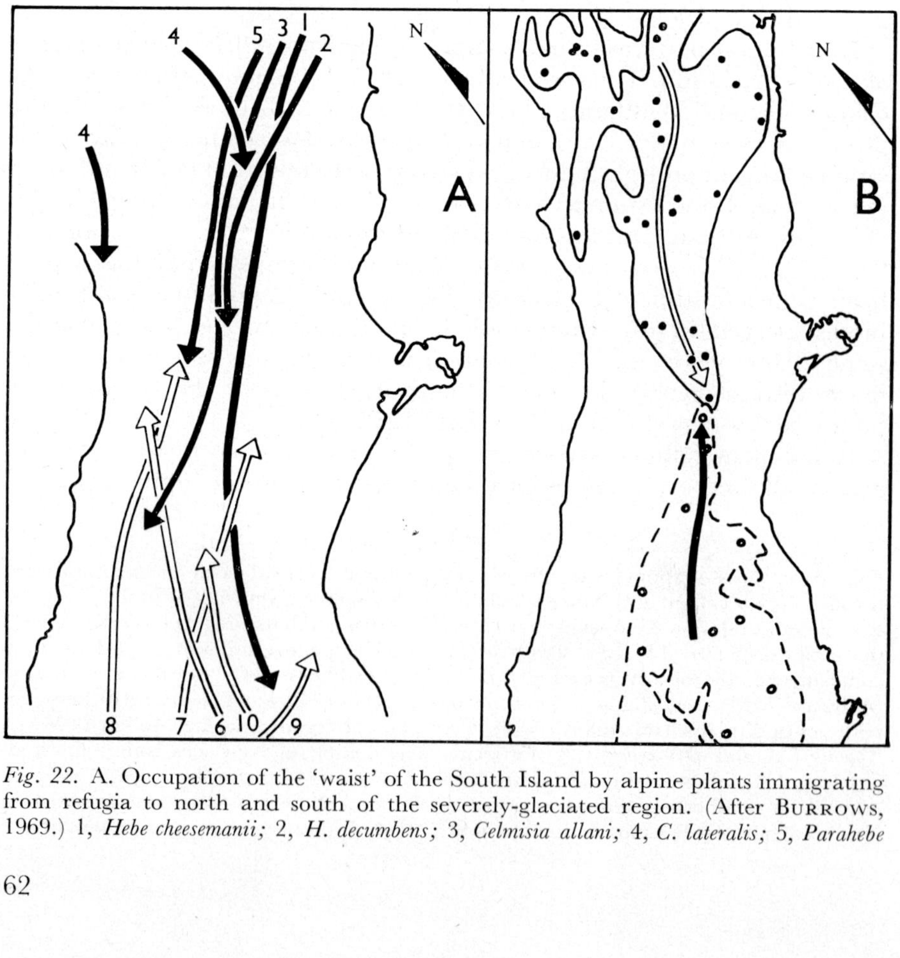

Fig. 22. A. Occupation of the 'waist' of the South Island by alpine plants immigrating from refugia to north and south of the severely-glaciated region. (After Burrows, 1969.) 1, *Hebe cheesemanii;* 2, *H. decumbens;* 3, *Celmisia allani;* 4, *C. lateralis;* 5, *Parahebe*

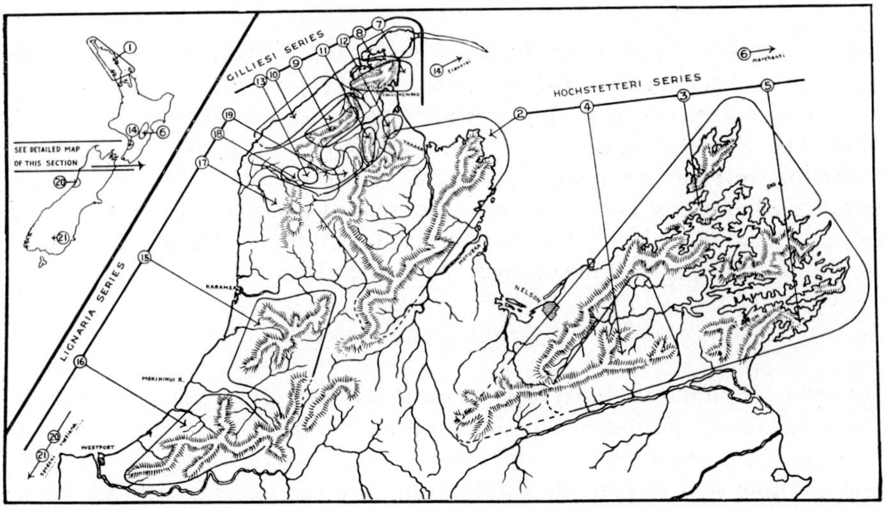

Fig. 23. Subspeciation in the land snail *Paryphanta.* (After POWELL, 1936.) Each numbered locality represents a named allopatric form in a polytypic species, due to post-glacial evolution mainly on uplands isolated from each other by river valleys.

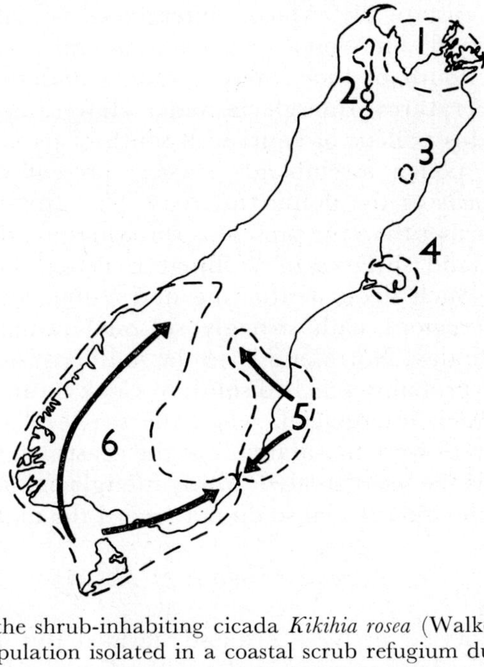

Fig. 24. Races of the shrub-inhabiting cicada *Kikihia rosea* (Walker), each apparently derived from a population isolated in a coastal scrub refugium during the last glacial stadial.

---

cheesemanii; 6, *Lobelia linnaeoides;* 7, *Raoulia hectori;* 8, *Celmisia hectori;* 9, *C. ramulosa;* 10, *Aciphylla dobsonii.* B. Locality records for two species of mountain cicada *(Maoricicada)* inhabiting subalpine scrub with *Dracophyllum*, suggesting post-glacial immigration into the severely glaciated 'waist' of the South Island from refugia to north *(Maoricicada cassiope)* and south (*Maoricicada* n. sp. cl.).

flora (BURROWS, 1969) and in inhabitants of scrub vegetation. Thus the scrub-inhabiting cicada *Kikihia rosea* (Walker) is now broken into six allopatric subspecies differing in morphology and/or acoustic behaviour (Fig. 24). Each probably arose in a scrub refugium near the glacial age shoreline and has since spread inland, in post-glacial time, but only in one area have any two races met. Some still occupy lowlands, one (6) extends from coastal scrub to alpine scrub, while another (2) is confined to alpine scrub, and is not known from low altitudes.

Interglacial Climates

Interglacial climate is hard to determine. For one thing, each interglacial followed a cold period in which stenothermal warmth-loving organisms had been exterminated owing to the limited northward extent of land and shallow seas to provide a refugium. Many interglacial fossil assemblages seem to indicate present climatic conditions, rather than the warmer conditions we would expect by analogy with interglacial climates of other lands and by inference from the amount of de-glaciation necessary to raise sea-level eustatically. Marine interglacial faunas of the Hawera Series contain very few mollusca (e.g. *Leucotina* and *Marginella mustellina* near Wanganui) south of their present ranges and thus suggestive of warmer sea temperatures. Interglacial older Hawera Series lignite near Wellington contains pollens of kauri well south of its present range but most interglacial pollen assemblages suggest present conditions (with *Dacrydium cupressinum* as the dominant tree). The strongest evidence for an interglacial warmer than the present is the widespread red-weathering of old soils and land surfaces in Wellington Peninsula, apparently at several horizons. Such deep weathering and 'rubification' are characteristic of warm regions with strongly seasonal rainfall and savanna vegetation. Even today, Northland is on the southern fringe of the Trade Wind belt, which probably reached south to Cook Strait in several interglacial ages. In each interglacial, when the sea level was higher than now, many low areas were inundated, but the coasts later prograded and built dune fields as the sea retreated. Many interglacial lignites date from this later part of the interglacial so do not record the maximum warmth.

Post-Glacial (Holocene) History

The post-glacial period in New Zealand, termed the Aranuian Stage, saw the fluctuating climatic recovery from the Kumara 3 glacial episode about 14,000 years ago, the rapid rise of sea level to its present level (or somewhat above it) in what has widely been called the Flandrian transgression, and the re-vegetation of the areas that had been deforested. Aranuian is not an exact equivalent of Holocene as it begins with the first warming after the Kumara 3 glacial advance.

Pioneer pollen diagrams from Southland and Otago (CRANWELL & VON POST, 1936; see also MCINTYRE & MCKELLAR, 1970) have been supplemented by a regional study (with C^{14} dates) in Canterbury, Nelson and north Westland by MOAR (1971), while some North Island post-glacial pollen diagrams from Wallaceville (Wellington), Hamilton and the Hauraki Plains, have been studied by HARRIS (1951, 1963).

Whatever the vegetation was at the peak of glaciation, the succession in different South Island districts was through grassland-shrubland to shrubland and then to forest. The particular shrubland community in Westland *(Coprosma-Myrsine)* differed from that of inland Canterbury (*Dacrydium bidwillii* or *biforme* plus *Phyllocladus*) and the particular forest community established also differed – first *Weinmannia* and then Rimu *(Dacrydium cupressinum)* in north Westland but *Podocarpus* from East Otago (MCINTYRE & MCKELLAR, 1970) to central Canterbury, whereas in north Canterbury an initial *Podocarpus-Nothofagus* mosaic locally gave way to *Nothofagus*. Forest was not re-established in the South Island before 10,000 years ago and its spread was not necessarily synchronous. Nor was the later spread of *Nothofagus* synchronous, affecting Nelson 7,800 years ago but not reaching north Westland until about 2400 years ago. MOAR concludes that the slow-spreading vegetation was still recovering from the disruption of the Otira Glaciation and changes do not necessarily reflect climate changes. There is in fact no definite evidence of a post-glacial thermal maximum and claims that minor climatic changes were responsible for changes in the vegetation (RAESIDE, 1948; HOLLOWAY, 1954) have been contested by CUMBERLAND (1962). There have certainly been fluctuations in the terminal faces of the glaciers, perhaps corresponding with historic advances of glaciers in Europe.

The retreat of South Island glaciers did not really begin until after the Kumara 3 advances at 16,500 and 14,250 years BP (corresponding with the Poulter in Canterbury and correlative events, e.g. Hinuera aggradation in the Waikato Valley), but after 14,000 BP warming was uninterrupted and sea-level rose rapidly and was complete at about 5000 BP. Valleys that had been excavated while sea-level was low were drowned to form harbours and sounds. Shoreline processes, as in other parts of the world, began to rectify coastlines by building spits and retrograding headlands, and the drowned estuaries were soon shallowed by formation of tidal flats seaward of filled valley heads in which post-glacial alluvium penetrates a considerable distance inland. On the open coast, the transgressing sea generally cut cliffs, but as the burden of longshore drift fed from major river-estuaries increased, the shores shoaled, and dunes were nourished from prograding high-tide berms. Progradation and dune advance began soon after sea level stabilised (say 4000 to 5000 BP) and has locally continued, building dunefields up to about 10 km wide, made up of successive transverse dune-belts overlying beach deposits and frequently ponding lakes and swamps.

Inland, post-glacial sediments include river alluvium, loess (locally persistent after the end of glacierisation), peats and bog deposits. Active volcanism continued in restricted districts: the Bay of Islands and Auckland Isthmus (basalt cones), Mt Egmont (andesite cones producing ash showers) and the Taupo Zone, where a variety of volcanic events, mainly pumice eruptions but also andesites from the Tongariro volcanoes mantled the central North Island with ash-fall deposits as follows (VUCETICH & PULLAR, 1964):

	ASH SHOWER	AGE IN YEARS
	Rotomahana Mud	1886 A.D. (historic)
	Tarawera Ash and lapilli	
	Kaharoa Ash	930 ± 70 B.P.
	Taupo Pumice (with several members)	1824 ± 24 B.P.
	Waimihia Lapilli	3270 ± 200 B.P.
Holocene	Rotokawau Ash	
	Whakatane Ash	
	Mamaku Ash	
	Rotoma Ash	
	Unnamed Brown Ash	
	Waiohau Ash	
	Rotorua Ash	
Pleistocene	Rerewhakaaitu Ash	14,700 ± 400 B.P.

Although forest vegetation had developed over much of the area buried by the Taupo Pumice before the arrival of man, the repeated history of catastrophic destruction during the post-glacial period had almost certainly eliminated some of the more sensitive organisms from the central North Island. The distinctive physical and chemical properties of pumice soils may also be responsible for the presence or absence of some or other species of plants and animals. Thus the pulmonate snails *Paryphanta* are curiously absent from the central North Island between Waitotara River and Te Kuiti.

The majority of Aranuian sediments date from the late Holocene, after sea level stabilised at about 5000 BP. The deposits formed during the late Holocene phase of progradation, dune advance, ponding of streams to produce peat swamps, and alluviation of inland valley-bottoms, after an interval of early Holocene down-cutting that followed the last Otiran glaciation, are the main sources of bones of the moa (Dinornithiformes), of extinct birds of other groups, and shells of locally extinct land mollusca (*Placostylus*, *Paryphanta*, etc.). It seems doubtful if any major source of subfossil bird bones so far known in New Zealand is as old as early Holocene (Fig. 25). Among the considerable number of birds represented in the late

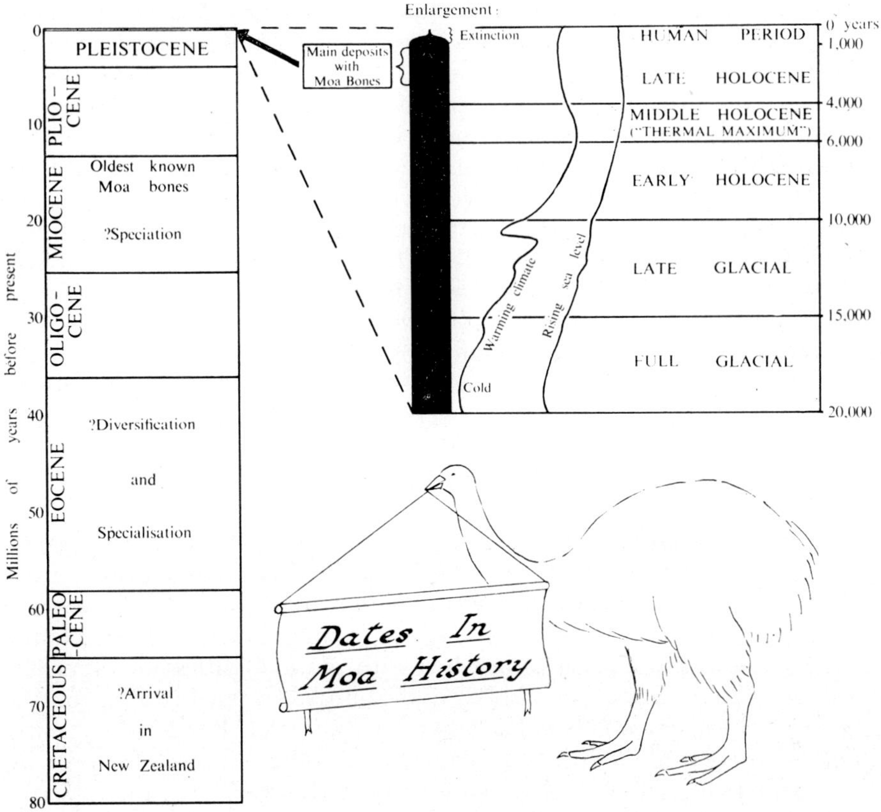

Fig. 25. Although the Dinornithiformes undoubtedly had a long history in New Zealand, the oldest fossil bones are not older than Upper Miocene and the main sources of bones are late Holocene. The enlarged time scale (right) shows the extinction of moas in the human period, thousands of years after the climatic changes of the Pleistocene to which it has sometimes been attributed. (From FLEMING, 1962c.)

Holocene deposits are 24 species (7 genera) of Dinornithiformes (Moa), *Apteryx* and many other birds that still survive, and some 20 extinct carinates including pelican, swan, goose, four ducks (2 endemic genera), harriers, two eagles (one endemic genus), four endemic flightless genera of rails, an endemic owlet–nightjar genus, and an endemic crow; many of these occur associated with middens or campsites of Polynesian man (KINSKY *et al.*, 1970). Extinctions during the Holocene can mostly be attributed to the profound ecological changes brought about by the arrival of man with fire, rats, and dogs (FLEMING, 1962c).

Judged by their lack of fossil record, many Foraminifera, Echinoderms and Mollusca colonised New Zealand in post-glacial time. Most of the latter are of Australian origin *(Saccostrea, Agnewia tritoniformis, Notocochlis migratoria, Tonna tetracotula)* and it is likely that many Australian species in groups without fossil record are post-glacial immigrants. Unassisted immigrants among insects (e.g. the Monarch Butterfly) and birds (e.g. White-eye, White-faced Heron, Spur-winged Plover) have colonised in the past century, and many others became established earlier, but perhaps also in post-glacial time (e.g. Pukeko, Crested Grebe, Pied Stilt, Grey Duck, Kingfisher). If, however, we subtract from our list of species of land and freshwater birds the species of known and inferred late arrival from Australia, we are left with rather few groups of probable Tertiary history in this country – a result of Pleistocene decimation.

The only Holocene extinction in the marine invertebrate fauna is the 'Sydney Cockle', *Anadara*, which came to New Zealand during the Castlecliffian interglacial and returned to Northland for a brief interval of post-glacial time, judged by its occurrence at Hokianga.

The coming of Man

The arrival and spread of Polynesian man in New Zealand a thousand years ago (greater precision is scarcely justified) initiated a long train of biological events that continued at more rapid pace after the visits of Tasman (1642) and Cook (1769–77). Early man found a diverse fauna of Moas (Dinornithiformes), together with giant flightless or otherwise sensitive carinate birds (see above) and abundant seals breeding on the coasts at least as far north as Coromandel Peninsula. For some centuries the newcomers exploited these relatively easily-hunted sources of protein, thus earning the name of Moa Hunters recently applied to them. Large areas of forest, on the east of the South Island divide and in the North Island, were destroyed by fire and replaced by tussock-grassland, bracken or scrub, producing disjunct distributions in *Nothofagus* forests. Dunes were activated or rejuvenated and overwhelmed forests.

The Polynesian rat (*Rattus exulans* Peale or *Kiore*) and a domestic dog accompanied early human colonists (if not indeed the first of them). The rat is known to have undergone 'population explosions' and its long-term effects on plant and animal communities had probably not been fully achieved before it was joined by European rats, pigs and mustellids. Modern ecologists find vast differences between plant and invertebrate communities on islands with kiore and those lacking all rat species; moreover *R. exulans* is capable of attacking large ground-nesting birds and causing their death (Kepler, 1967). Undoubtedly the kiore can be blamed for many subtle ecological changes in the human period. The Maori dog may have gone wild and hunted in packs, thus contributing to the reduction of flightless birds, but if so it had become extinct as a wild animal and left no traditional record of such behaviour.

Man, with fire, dog and rat had produced considerable effects by the time European contacts began in 1769 (Fleming, 1962c). The moas had become extinct, or, if *Megalapteryx* alone remained, it was reduced to a rare relict in Fiordland, on the verge of the extinction that followed soon after. A score of carinate birds as well are known to have disappeared. Many other bird species suffered partial extinction, some in Northland (e.g. *Strigops*), others in the North Island *(Apteryx oweni)*, or everywhere except in Fiordland *(Notornis)* or on outlying islands *(Coenocorypha, Mergus)*. Fur seals and sea elephants no longer bred in the north. The Tuatara *(Sphenodon)* had become greatly restricted until it too now survives only on outlying islands.

There is some evidence for extinction of land and fresh-water invertebrates during the same period as the more conspicuous extinction or restriction of vertebrates. Two fresh-water Ostracoda and a tendipedid midge from the Pyramid Valley lake deposits have not yet been found living. Disjunct distribution of giant wetas *(Deinacrida)*, the snail *Paryphanta*, and other invertebrates probably dates from the early human period.

Even the marine invertebrata changed to a minor degree; the bivalve *Paphies australis* is appreciably larger in moa-hunter middens and mid-Holocene deposits than in nearby harbours today, perhaps as the result of silting due to forest clearance, and subfossil toheroa *(P. ventricosa)*, 1030 years old, are larger than contemporary shells; the rock-dwelling gastropods in middens at Coromandel Peninsula are also larger than those that can be collected today.

Biogeographic Elements

Botanists and zoologists on the whole have agreed on the main elements in the biota, classified according to their geographic relationships and probable origin, but have used a variety of names for them. The names used by the author (FLEMING, 1963a) in an attempt to reconcile different nomenclatures (Fig. 26) have been used throughout this paper.

MALAYO-PACFIC
(Malayan, Malaysian, Palaeotropic, Indo-Pacific etc.).

This element includes plants and animals of tropical and subtropical derivation (other than those that came via Australia). Many organisms without an adequate fossil record are attributed to this element: *Placostylus*, plants like *Meryta* and *Cordyline*, sea-birds such as *Pterodroma hypoleuca*. It is impossible to tell whether some colonists came direct from the north via the islands and shallows of the submarine ridges (perhaps more emergent in the past than now) or via Australia. In groups that were possibly once cosmopolitan but are now restricted to southern lands, it may be difficult to tell whether immigrants to New Zealand came by the Malayo-Pacific or Austral route and this led FORSTER (1961) to

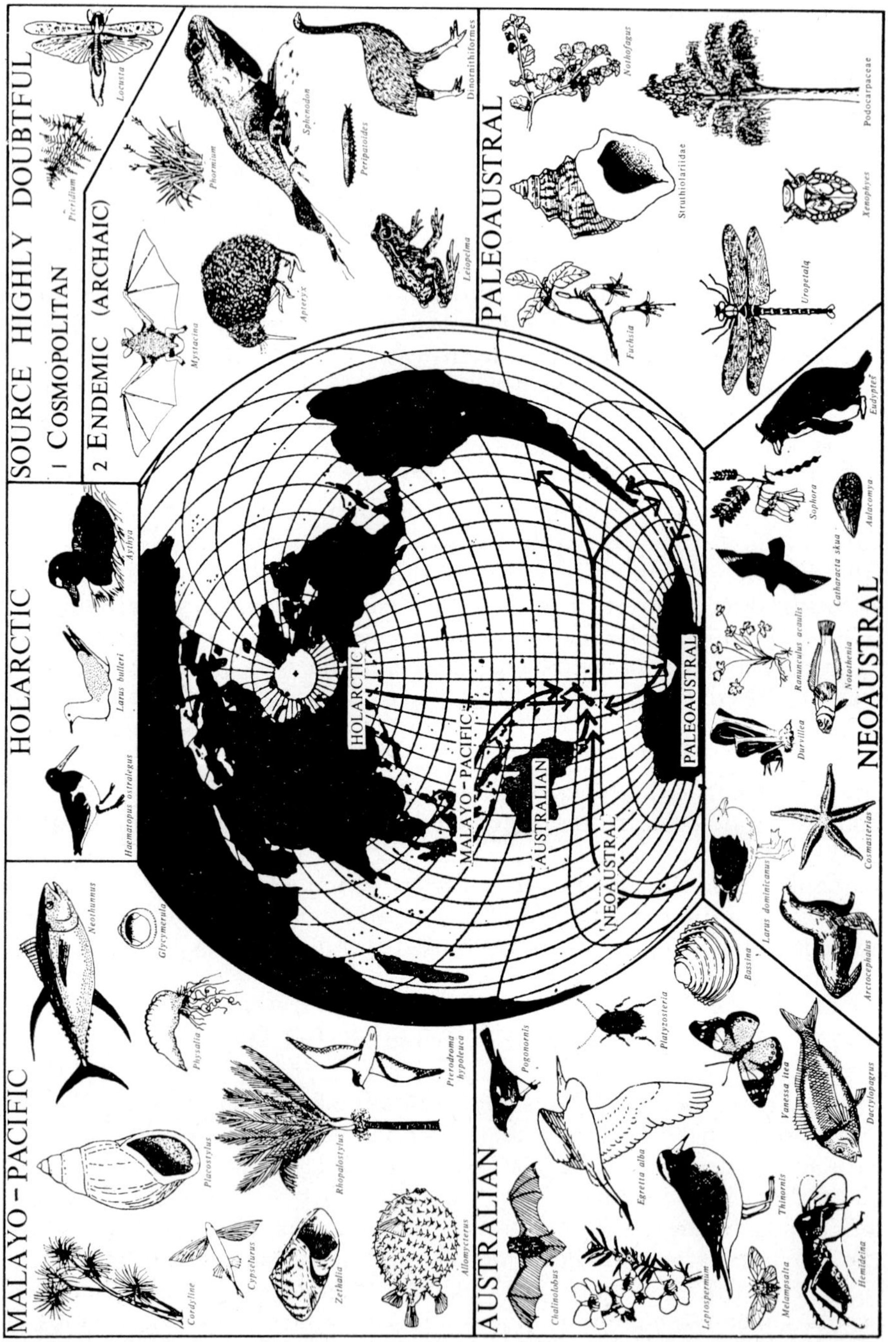

Fig. 26. New Zealand Biogeographic Elements, with some examples. (From FLEMING, 1962a.)

abandon the attempt to determine their route of origin and to group all older elements in an 'Archaic' category.

Australian element

Australian relationships have always been strong in the fauna, and were much stronger in the flora before the rise of the xerophilous plants to dominance on the other side of the Tasman, a comparatively late development. Australia has been the source of the majority of New Zealand birds (Falla, 1953), both in the recent past and in more distant times (Fleming, 1962b). In other groups, particularly of ancient dispersal, it is sometimes hard to distinguish Australian from Malayo-Pacific elements. For instance, according to Professor J. T. Salmon *(pers. comm.)* the wetas (*Hemideina*, etc.), though related to the King Crickets of Australia, may be Malayo-Pacific elements that reached both sides of the Tasman independently.

Austral
(Antarctic, Subantarctic, Fuegian, Antarcto-Tertiary in part).

Many organisms have a circumpolar, or partially circumpolar distribution in cool-temperate latitudes but are absent from the warmer climatic zones to the north. Some have colonised, in post-glacial time, islands severely glaciated in the Pleistocene (Dawson, 1958), most have adaptations for dispersal, and some suggest by their limited fossil record that dispersal has occurred late in geologic time (Pliocene and later). This element, evidently distributed by the West Wind Drift of air and sea, under present geographic conditions, has been called *Neoaustral*. Some of its marine members can be derived from originally tropical or subtropical stocks that came to the Southern Ocean at any one of the three major points of entry (Australasia, South Africa, South America). Oliver (1925) pointed out that each southern land seems to have supplied colonists to the west wind zone, so that a decreasing number of species is found as a genus is traced down-wind (eastwards) from its headquarters in one of the southern entry points. Other Neoaustral species (the mussel *Mytilus edulis*, the King-crab *Lithodes*, the skua *Catharacta*) are bipolar (antitropical) in cold temperate seas of both hemispheres, and neither their hemisphere of origin nor their longitude of contact across the tropics is certain.

The older and more puzzling division of the Austral Element is the *Paleoaustral*, consisting of plants and animals more distantly related to those of other southern lands, some with a fossil record going back to Tertiary or Mesozoic times, which are no longer being dispersed (judged by their systematic distinctness) and which have such poor dispersal abilities that many biologists have considered they were distributed along land connections linking southern lands with Antarctica. They include the podocarps, Proteaceae, *Nothofagus* and other angiosperm genera (*Aristotelia, Laurelia, Astelia, Fuchsia, Weinmannia*). The mollusc *Struthiolaria*, descended from Paleocene *Perissodonta* which ranged to South America and Graham Land (and now survives in South Georgia and Kerguelen), is an example from the littoral fauna, and there are many among the terrestrial invertebrates, including those of the forest floor litter. Their fossil record suggests that the Paleoaustral elements entered New Zealand over a long period. Some of them, originally Malayo-Pacific, may have come south through New Zealand and subsequently spread in the Austral region.

Endemic element
(Aboriginal, Palaeozelandic in part, Archaic).

Groups that have no close relations in other countries to indicate their place of origin, either because they developed here (primary endemics) or because they became extinct elsewhere (secondary endemics). The endemic element, shorn of forms that can be attributed to other elements, includes the moas, kiwis, short-tailed bat, and tuatara among vertebrates, and many invertebrates (e.g. *Peripatoides*).

Holarctic element

A few organisms related to north-temperate forms, absent in the intervening tropics, especially certain birds (Scaup, Merganser, Black-billed gull, South Island Pied Oystercatcher), plants such as *Euphrasia* and *Festuca novaezelandiae* (aff. *ovina*) and others in the alpine flora. The alpine or montane habitats of such Holarctic elements suggest that they joined the Alpine Biota under category 2; there may also be a link with Neoaustral elements of bipolar distribution.

Cosmopolitan element

Organisms so widespread that their particular route of colonisation cannot be determined on available evidence.

The Fossil record and its Defects

The data of neontology cannot tell us the extent and duration of past distributions; but they supply the warp into which the woof of paleontology can be woven to give a pattern in time and space still full of imperfections but with hope of probability.

The fossil record includes remains of many living plant and animal taxa identifiable with varying degrees of confidence (Figs. 26, 27). The time-spans covered by the recognisable predecessors (some certainly ancestors) of living forms vary greatly. We obviously cannot accept the short fossil record of *Sphenodon*, of the snail *Rhytida*, or of the endemic fern genus *Cardiomanes* (for example) as significant of anything but the inadequacy of the record. The interpretation of such a fallible record must always be subject to revision as new information comes to hand.

Incomings in the fossil record (Figs. 27, 28) may be due to chance preservation or identification (as with *Jasus* and *Notothenia*, represented by single records). They may be due to evolution from an ancestral form, as with *Struthiolaria*, which appeared in the Oligocene but is a descendant of *Monalaria* (Eocene) and *Perissodonta* (Paleocene). In the diagrams, living forms are shown at the point in the record where their presumptive ancestors first appeared, even if they differ generically or specifically from the first fossil record of the taxon they represent. (Some ranges have altered from new discoveries since the diagrams were compiled but the general picture remains essentially correct).

In many families and genera, the earliest fossil representatives are not the direct ancestors of the living species, which are the result of re-invasion of new stock from overseas. Thus the living *Architectonica reevei* and *Chlamys (Mesopeplum) convexum* are not direct descendants of the first New Zealand members of these genera known in the Eocene and Oligocene but are due to re-invasion at much later dates. Immigration has been important throughout Tertiary time, and we no longer believe that the whole New Zealand marine fauna has been isolated here since the Cretaceous. In the marine fauna, therefore, the first appearance of a taxon that has no recorded ancestors but occurs regularly later may be taken (as a working hypothesis) to indicate an immigration. As such 'first appearances' are important to the stratigrapher, they are being constantly tested and amended.

Study of pollen and spores has yielded much data on the fossil history of living plant taxa. In some plants, pollen is so distinctive and abundant that its absence from older rocks gives confidence in the hypothesis that its appearance in the record is due to immigration. Other cases are ambiguous. Thus pollen of the miro-matai type (*Podocarpus spicatus* and *ferrugineus*) ranges back to Upper Miocene but the section *Stachycarpus* to which these species belong probably ranges back at least to Upper Cretaceous (Shag Point) judged by macrofossils. Similarly, pollen resembling that of *Podocarpus dacrydioides* first appears in the Upper Eocene, but macrofossils suggesting section *Dacrycarpus* range back into the Mesozoic, perhaps to Jurassic. Podocarps have a long history since the Triassic in the Southern Hemisphere, so the appearance of distinctive pollen in the Tertiary is probably due to evolution (at least of the pollen) in New Zealand, but the possibility of immigrations at late dates cannot be excluded.

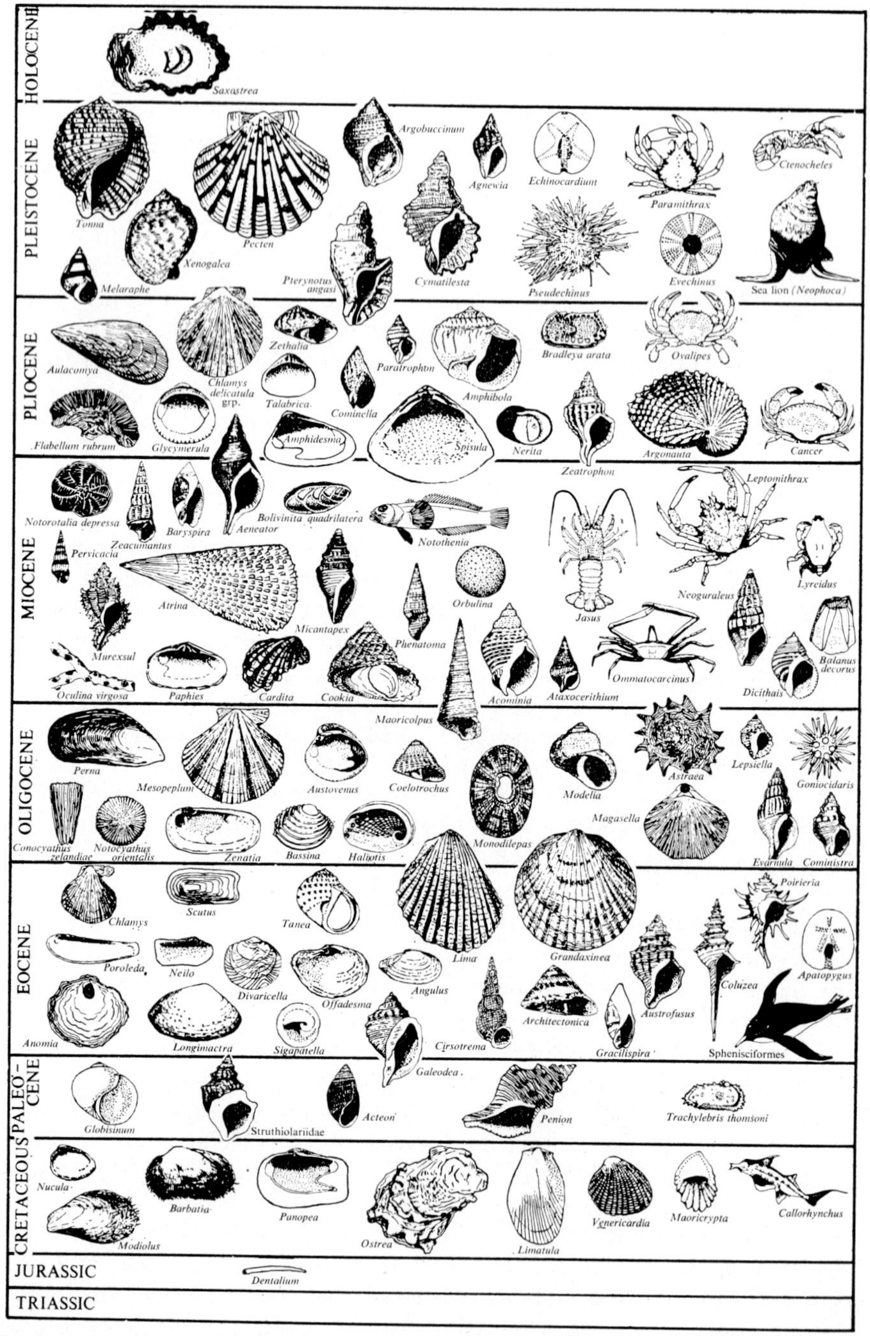

Fig. 27. First appearance in the New Zealand fossil record of members of the marine fauna, illustrated by living species shown at the date in the geological time-scale when the taxa to which they belong are first recorded by fossil predecessors in New Zealand. Time-scale distorted. (From FLEMING, 1962a.)

Fig. 28. First appearance in the New Zealand fossil record of members of the land and fresh-water fauna and flora, illustrated by living (or recently extinct) species shown at the dates in the geological time-scale when the taxa to which they belong are first recorded by fossil predecessors in New Zealand. Time-scale distorted. (From FLEMING, 1962a.)

The classification of New Zealand organisms into biotic elements from their geographic affinities (Fig. 26) can be combined with paleontological data on their probable age in the biota (Figs. 27, 28) to give a picture of changing biogeographic influences in later Mesozoic and Cenozoic time (Figs. 29A, 29B). Endemic, cosmopolitan, and Holarctic elements are not shown as they either add nothing to the picture or have an inadequate fossil record. Extinct organisms of known geographic affinities are included as they contribute greatly to the understanding of past changes in the biota.

The alternation of Tethyan (tropical) and Austral (southern) influences in Paleozoic and Mesozoic faunal history was emphasised in an earlier section. The Middle and Upper Jurassic period of Tethyan dominance left little trace in later marine faunas. In the Tertiary, Tethyan, Austral and Australian influences seem to have persisted, but were of unequal and fluctuating importance.

Whatever their ultimate origin, Paleoaustral conifers (Podocarps and Araucarians) were already prominent in New Zealand vegetation in the Jurassic and remained so in the Cretaceous, when the Paleoaustral influence was strengthened by the appearance of the southern angiosperms *Nothofagus* and the Proteaceae, and of marine groups (Fig. 11) such as ancestors of the Struthiolariidae (Gastropoda), *Lahillea* and *Pacitrigonia* (Bivalvia). The Paleoaustral influence remained strong in the early Tertiary and contributed plant immigrants (*Fuchsia*, etc.) as late as Miocene. By this time, however, the Malayo-Pacific influence (successor to the Tethyan) had become dominant, and it seems possible that some Austral plant genera that first appeared in New Zealand in the mid Tertiary (*Astelia*, *Aristotelia* and *Weinmannia*, for instance) are in fact Malayo-Pacific elements that entered the Austral region at this time. It is most unlikely that land connections with Antarctica persisted through the long period of Paleoaustral invasion indicated by the paleobotanical record, and all we can be sure of is that dispersal was easier in the past than it is now. Antarctica was not heavily glaciated and was at least partly vegetated in the earlier Tertiary with Paleoaustral plants, as pollen studies have shown (Cranwell, 1959; *et al.*, 1960; McIntyre & Wilson, 1966), and this undoubtedly contributed (irrespective of land extensions) to the dispersal of the Paleoaustral elements, most of which had probably entered the area from the north down one of the southern land masses. Some Paleoaustral organisms (e.g. *Fuchsia* and the extinct penguins) were certainly distributed across the sea like their Neoaustral successors. (*Fuchsia* reached the oceanic volcanoes of the Society islands).

Australian elements can be detected as early as Lower Cretaceous (*Maccoyella*) and continued to contribute a steady trickle, often hard to disentangle from the Malayo-Pacific, perhaps fluctuating in importance.

Australia shared with New Zealand the Paleoaustral elements, so that it is questionable whether Proteacea should be classed as Australian or Paleoaustral.

The dominant Malayo-Pacific influence (Fig. 29B) in the middle Tertiary is most obvious in the marine fauna, as illustrated by Fig. 14 for the Mollusca, but it also brought plant immigrants *(Metrosideros, Dysoxylum, Freycinetia, Rhopalostylis, Cocos, Elytranthe, Macropiper, Avicennia, Dodonaea)*. The Malayo-Pacific influence died away after a climax in the early Miocene and many of its immigrants became extinct. The most important of all Malayo-Pacific immigrants – Polynesian man – came in the Late Holocene to lay the foundation-stones of New Zealand human culture and to sound the death-knell for an ecology of isolation.

In the Pliocene and Pleistocene an Austral influence again became important, marked by the appearance of marine animals classed as Neoaustral, distributed in the West Wind zone. The plants that (by inference) accompanied such animals are not well represented as fossils. I would judge that *Ranunculus acaulis*, *Sophora* and *Hebe* gained their wide southern distribution by easterly dispersal at this time. The Neoaustral is not a direct continuation of the Paleoaustral – it is much more restricted. There has been no further dispersal of beeches, Proteaceae and podocarps between southern lands, so that the appeal to land extension or continental drift to explain the difference is not unreasonable.

Epilogue

This is the author's third attempt to describe the history of life in New Zealand. It still remains necessary to emphasize that tentative conclusions are hypotheses for testing and that the gulf between raw data and paleogeographic conclusions has often been crossed by a rather flimsy bridge of projection and extrapolation. The principle of correlation of organs with organisms enables a specialist to identify a mammal from one tooth, a gastropod from its operculum or a tree from a pollen grain. The principle of economy of hypothesis (Occam's razor) requires us to conclude that identical things represent identical things until evidence shows they represent different things, even if this road has pitfalls. Unrelated plants can indeed produce indistinguishable pollen grains, but in tracing back the history of a vegetation nothing is gained by assuming that pollen indistinguishable from that of a living taxon (whether it be species, genus or family) was produced by quite a different taxon. The accumulating data commonly force us to complicate our hypotheses but there is no justification for doing so in advance of such evidence.

Apart from the deletion of a seal that was wrongly dated as Pliocene, the illustrations published as Figures 26 to 29 have not been revised since they were published in 1962 and some of the time-ranges have since been modified by new data; but the largely subjective interpretation they

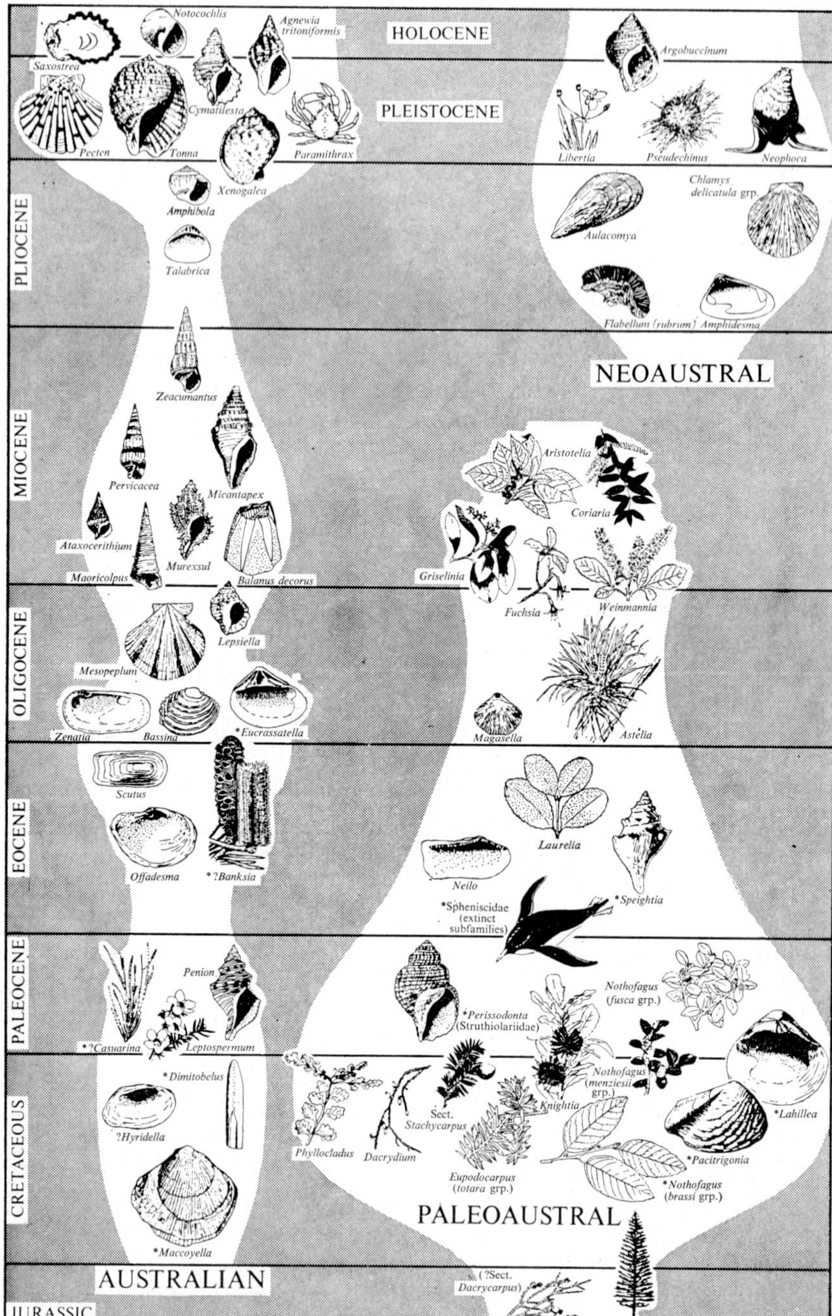

Fig. 29A, 29B. Interpretation of the history of biogeographic elements. Immigrant taxa, living and extinct (the latter marked by asterisks) classified under biogeographic elements from their geographic relationships, are placed near the geological date of their first appearance as fossils to give a picture of changing geographic influences through Mesozoic and Cenozoic time. (From FLEMING, 1962a.)

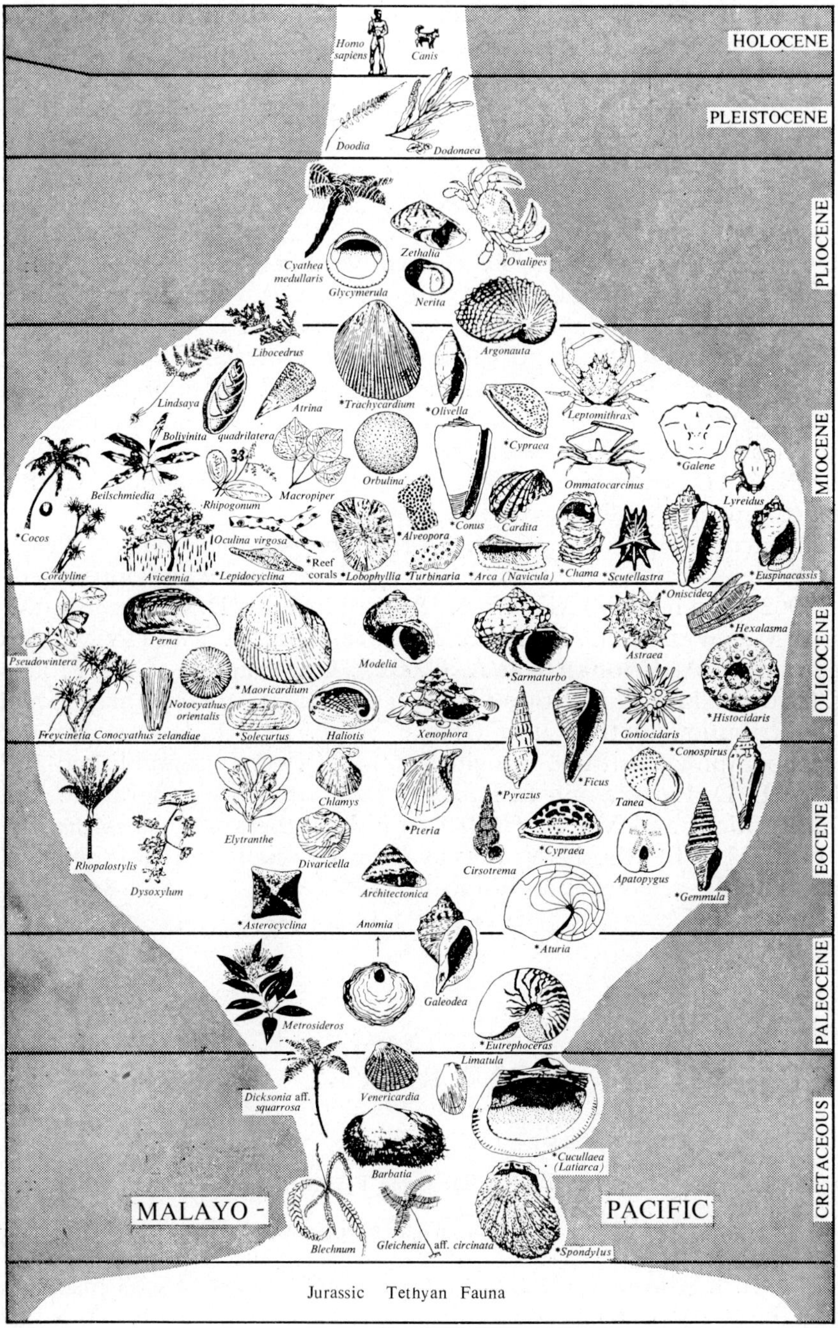

Fig. 29B. For caption see facing page.

picture still remains (in the writer's opinion) a valid one. In the very inexact science of biogeography, hypotheses can seldom be directly proved, though they are tested by new data at every step, and in CHARLES DARWIN's words 'the doctrine must sink or swim according as it groups and explains phenomena'.

Acknowledgements

This article is a considerably revised version of 'New Zealand Biogeography. A Paleontologist's Approach' (1962), published in *Tuatara*, the journal of the Biological Society, Victoria University of Wellington, which has kindly agreed to the use of the material in this way. The cartographic section of the Department of Scientific & Industrial Research provided technical help with some illustrations. Others are based on drawings by Dr. J. MARWICK, Dr. A. W. B. POWELL and the writer. The N.Z. Forest Service kindly gave permission to republish W. BAKKENES' drawings of forest vegetation.

The manuscript owes a great deal to cooperation of my colleagues at the New Zealand Geological Survey, both those who contributed to the 1962 version and to Miss M. I. DICKSON, Messrs D. C. MILDENHALL, P. A. MAXWELL, Drs. N. DE B. HORNIBROOK, G. R. STEVENS, I. G. SPEDEN, R. A. COOPER and A. G. BEU. Once more I thank Professor M. GAGE and Dr. R. P. SUGGATE for permission to use their map showing ice distribution in the Otira Glacial Stage. Dr. D. J. CULLEN, N.Z. Oceanographic Institute, provided some welcome unpublished data. Messrs L. D. HOMER and L. R. LARONDE supplied photographic copies of illustrations and Mr. R. C. BRAZIER drew the fossil *Nothofagus* leaves in Figure 17. To Mr. I. W. KEYES I am especially grateful for help in compiling the references and drafting maps.

REFERENCES

ARBER, E. A. N. 1917. The Earlier Mesozoic Floras of New Zealand. *Paleont. Bull. N.Z. geol. Surv.* 6. 80 pp. 14 pls.

ARKELL, J. W. 1956. Jurassic Geology of the World. Oliver & Boyd Ltd., Edinburgh and London. 806 pp.

BELL, S., H. J. HARRINGTON & I. C. MCKELLAR. 1956. Lower Mesozoic plant fossils from Black Jacks, Waitaki River, South Canterbury. *Trans. R. Soc. N.Z.* 83(4): 663–72.

BEU, A. G. & P. A. MAXWELL, 1968. Molluscan evidence for Tertiary sea temperatures in New Zealand: a reconsideration. *Tuatara* 16(1): 68–74.

Boucot, A. J., K. E. Caster, D. Ives & J. A. Talent, 1963. Relationships of a new Lower Devonian Terebratuloid (Brachiopoda) from Antarctica. *Bull. Amer. Paleont.* 46(207): 81–151.

Burns, R. E., J. E. Andrews *et al.* 1972. *Glomar Challenger* down under. *Geotimes* 17(5): 14–16.

Burrows, C. J. 1969. Alpine Grasslands. Pp. 133–66 in Knox, G. A. (Ed.) 'The Natural History of Canterbury'. A. H. & A. W. Reed Ltd. Wellington. 620 pp.

Carey, S. W. 1955. The orocline concept in geotectonics. *Proc. R. Soc. Tasmania* 89: 255–88.

Clayton, R. N. & G. R. Stevens. 1968. Paleotemperatures of the New Zealand Jurassic and Cretaceous. *Tuatara* 16(1): 3–7.

Copeland, E. B. 1947. Genera Filicum; the genera of ferns. Chronica Botanica, Waltham, Mass., U.S.A. 247 pp. 10 pls.

Cotton, C. A. & M. T. te Punga, 1955. Solifluxion and periglacially modified landforms at Wellington, New Zealand. *Trans. R. Soc. N.Z.* 82(5): 101–31.

Couper, R. A. 1953. Upper Mesozoic and Cainozoic Spores and Pollen Grains. *Paleont. Bull. N.Z. geol. Surv.* 22. 77 pp. 9 pls.

Couper, R. A. 1960a. Southern Hemisphere Mesozoic and Tertiary Podocarpaceae and Fagaceae and their Palaeogeographic significance. *Proc. R. Soc. Lond. B*, 152(949): 491–500.

Couper, R. A. 1960b. New Zealand Mesozoic and Cainozoic plant Microfossils. *Paleont. Bull. N.Z. geol. Surv.* 32. 87 pp. 12 pls.

Couper, R. A. & D. R. McQueen. 1954. Pliocene and Pleistocene Plant Fossils of New Zealand and their climatic interpretation. *N.Z. Jl. Sci. Technol.* B35(5): 398–420.

Cranwell, L. M. 1940. Pollen grains of the New Zealand Conifers. *N.Z. Jl Sci. Technol.* B22(1): 1–17.

Cranwell, L. M. 1959. Fossil pollen from Seymour Island, Antarctica. *Nature* 184 (4701): 1782–85.

Cranwell, L. M., H. J. Harrington & I. G. Speden. 1960. Lower Tertiary microfossils from McMurdo Sound, Antarctica. *Nature* 186(4726): 700–2.

Cranwell, L. M. & L. von Post. 1936. Post Pleistocene pollen diagrams from the Southern Hemisphere. *Geogr. Ann. Stockh.* 18(3–4): 308–47.

Cracraft, J. 1973. Continental drift, paleoclimatology, and the evolution and biogeography of birds. *J. Zool. Lond.* 169: 455–545.

Cullen, D. J. 1967. Submarine evidence from New Zealand of a rapid rise in sea level about 11,000 B.P. *Palaeogeography Palaeoclimatol. Palaeoecol.* 3: 289–98.

Cullen, D. J. 1970. A tectonic analysis of the South-West Pacific. *N.Z. Jl Geol. Geophys.* 13(1): 5–20.

Cumberland, K. E. 1962. 'Climatic change' or cultural interference? New Zealand in Moa-hunter times. Pp. 87–142 in McCaskill, M. (Ed.) 'Land and Livelihood'. N.Z. Geographical Soc. Special Publ. Miscel. Series No. 4. Christchurch. 280 pp.

Dawson, J. W. 1958. Interrelationships of the Australasian and South American floras. *Tuatara* 7(1): 1–6.

Dawson, J. W. 1971. Relationships of the New Zealand Umbelliferae. Pp. 43–62 in 'The Biology and Chemistry of the Umbelliferae' Supplement 1, *Bot. Jl Linn. Soc.* 64.

Devereux, I. 1967. Oxygen isotope paleotemperature measurements on New Zealand Tertiary fossils. *N.Z.Jl Sci.* 10(4): 988–1011.

Edwards, A. R. 1968. The calcareous nannoplankton evidence for New Zealand Tertiary marine climate. *Tuatara* 16(1): 26–31.

Evans, P. R. & L. E. Stover. 1971. Late Cretaceous to Eocene stratigraphic palynology of the offshore Gippsland Basin, Southeastern Australia. Aust. N.Z. Assoc. Adv. Sci. 43rd Congress (Brisbane). Abstracts Sect. 3 (Geol.): 14–15.

Evison, F. F. 1968. Active regions of the south-west Pacific. *Can. Jl Earth Sci.* 5(4): 1045–9.

Falla, R. A. 1953. The Australian element in the avifauna of New Zealand. *Emu* 53: 36–46.

Fell, H. B. 1954. Tertiary and Recent Echinoidea of New Zealand: Cidaridae. *Paleont. Bull. N.Z. geol. Surv.* 23. 62 pp. 16 pls.

Fell, H. B. 1962. West-wind-drift dispersal of Echinoderms in the Southern Hemisphere. *Nature* 193(4817): 759–61.

Fleming, C. A. (Ed.). 1959. New Zealand. Lexique Stratigraphique International Vol. 6 (Océanie) Fasc. (4). Paris. 527 pp.

Fleming, C. A. 1962a. New Zealand Biogeography. A paleontologist's approach. *Tuatara* 10(2): 53–108.

Fleming, C. A. 1962b. History of the New Zealand land bird fauna. *Notornis* 9(8): 270–4.

Fleming, C. A. 1962c. The extinction of moas and other animals during the Holocene period. *Notornis* 10(3): 113–7.

Fleming, C. A. 1963a. The nomenclature of biogeographic elements in the New Zealand biota. *Trans. R. Soc. N.Z. (General)* 1(2): 13–22.

Fleming, C. A. 1963b. Age of the alpine biota. *Proc. N.Z. Ecological Soc.* 10: 15–18.

Fleming, C. A. 1967a. Cenozoic history of Indo-Pacific and other warmwater elements in the marine mollusca of New Zealand. *Venus* 25(3–4): 107–17.

Fleming, C. A. 1967b. Biogeographic change related to Mesozoic orogenic history in the South-west Pacific. *Tectonophysics* 4(4–6): 419–27.

Fleming, C. A. 1968. Tertiary fossils from the Auckland Islands. *Trans. R. Soc. N.Z. (Geol.)* 5(11): 245–52.

Fleming, C. A. 1970. The Mesozoic of New Zealand: Chapters in the history of the circum-Pacific mobile belt. *Quart. Jl Geol. Soc. Lond.* 125(2): 125–70.

Fleming, C. A. 1970b. Radiocarbon dating and pollen analyses from Otiran Periglacial fans in Western Wellington. *Trans. R. Soc. N.Z. (Earth Sci.)* 7(11): 197–208.

Fleming, C. A. 1971a. Continents in dispersion. Pp. 12–21 in Knox, R. (Ed.) 'New Zealand's Heritage', Vol. 1, Pt. 1. Paul Hamlyn Ltd., Wellington.

Fleming, C. A. 1971b. A new species of cicada from rock fans in southern Wellington, with a review of three species with similar songs and habitat. *N.Z. Jl Sci.* 14(3): 443–79.

Florin, R. 1940. The Tertiary fossil conifers of south Chile and their phytogeographical significance; with a review of the fossil conifers of southern lands. *K. Svensk. vet. Akad. Handl.* 19(2): 1–107.

Forster, R. R. 1961. The New Zealand fauna and its origins (Rep. 9th Sci. Congr. 1960) *Proc. R. Soc. N.Z.* 89(1): 51–5.

Fraser, R. (Compiler) 1973. Oceanography of the South Pacific 1972 New Zealand National Commission for UNESCO. 524 pp.

Gage, M. 1961. On the definition, date and character of the Ross Glaciation, Early Pleistocene, New Zealand. *Trans. R. Soc. N.Z.* 88(4): 631–8.

Gage, M. 1965. Some characteristics of Pleistocene cold climates in New Zealand. *Trans. R. Soc. N.Z. (Geol.)* 3(2): 11–21.

Gage, M. 1966. The climate of New Zealand during cool phases of the Pleistocene. Pp. 83–94 in Blumenstock, D. I. (Ed.), 'Pleistocene and Post-Pleistocene Climatic Variations in the Pacific Area'. Bishop Museum Press, Honolulu.

Gaskin, D. E. 1970. The origins of the New Zealand Fauna and Flora; A Review. *Geographical Review* 60(3): 414–434.

Glaessner, M. F. 1960. The fossil Decapod Crustacea of New Zealand and the evolution of the Order Decapoda. *Paleont. Bull. N.Z. geol. Surv.* 31. 63 pp. 7 pls.

Grant-Taylor, T. L. & T. A. Rafter. 1962. New Zealand Radiocarbon age measurements – 5. *N.Z. Jl Geol. Geophys.* 5(2): 331–59.

Griffiths, J. R. 1973. Plate tectonics and the evolution of New Zealand. Pp. 263–9 in Fraser, R. 1973.

HARRIS, T. A. 1962. The occurrence of the fructification *Carnoconites* in New Zealand. *Trans. R. Soc. N.Z. (Geol.)* 1(4): 17–27.
HARRIS, W. F. 1951. New Zealand plants and their story: Clues to the past. *N.Z. Sci. Review* 9(1–2): 3–7.
HARRIS, W. F. 1963. Paleo-ecological evidence from pollen and spores. *Proc. N.Z. Ecol. Soc.* 10: 38–44.
HARRIS, W. F. & G. NORRIS. 1972. Ecologic significance of recurrent groups of pollen and spores in Quaternary sequences from New Zealand. *Palaeogeography, Palaeoclimatol. Palaeoecol.* 11(2): 107–24.
HASKELL, T. R. 1971. Motuan (Late Albian) microplankton floras from the Tinui area, Wairarapa, New Zealand. Aust. N.Z. Assoc. Adv. Sci. 43rd Congress (Brisbane) Abstracts, Sect. 3 (Geol.): 14.
HAYES, D. E. & J. RINGIS. 1973. The early opening of the Tasman Sea. P. 338 in FRASER, R. 1973.
HEIRTZLER, J. R. 1968. Sea floor spreading. *Scientific American* 219(6): 60–70.
HOLLOWAY, B. A. 1961. A systematic revision of the New Zealand Lucanidae (Insecta: Coleoptera). *Bull. Dominion Mus. (Wellington)* 20. 139 pp.
HOLLOWAY, J. T. 1954. Forests and climates in the South Island of New Zealand. *Trans. R. Soc. N.Z.* 82(2): 329–410.
HORNIBROOK, N. DE B. 1953. Immigration of Foraminifera to New Zealand in the Upper Cretaceous and Tertiary. *N.Z. Jl Sci. Technol.* B34(6): 436–44.
HORNIBROOK, N. DE B. 1968. Distribution of some warm water Foraminifera in the New Zealand Tertiary. *Tuatara* 16(1): 11–15.
HORNIBROOK, N. DE B. 1971. New Zealand Tertiary climate. *N.Z. geol. Surv. Report* 47. 19 pp.
HOWARD, H. & S. L. WARTER, 1969. A new species of bony-toothed bird (Family Pseudodonthornithidae) from the Tertiary of New Zealand. *Rec. Canterbury Mus.* 8(4): 345–57.
JENKINS, D. G. & T. B. H. JENKINS. 1971. First diagnostic Carboniferous fossils from New Zealand. *Nature* 233(5315): 117–18.
KEAR, D. 1957. Stratigraphy of the Kaawa-Ohuka coastal area, west Auckland. *N.Z. Jl Sci. Technol.* B38(8): 826–42.
KEAST, A. 1971. Continental drift and the evolution of the biota of southern continents *Q. Rev. Biol.* 46: 335–78.
KENNETT, J. P. 1968. Paleo-oceanographic aspects of the foraminiferal zonation in the Upper Miocene – Lower Pliocene of New Zealand. *Giornl. Geol. Ser.* 2a, 35(3): 143–56.
KENNETT, J. P. & R. E. CASEY. 1969. Foraminiferal evidence for a pre-Middle Eocene age of the Chatham Rise, New Zealand. *N.Z. Jl mar. Freshwat. Res.* 3(1): 20–28.
KEPLER, C. B. 1967. Polynesian rat predation on nesting Laysan Albatrosses and other Pacific seabirds. *Auk* 84: 426–30.
KINGMA, J. T. 1958. Possible origin of piercement structures, local unconformities and secondary basins in the Eastern Geosyncline, New Zealand. *N.Z. Jl Geol. Geophys* 1(2): 269–74.
KINSKY, F. C. (Ed.) 1970. Annotated checklist of the birds of New Zealand including the birds of the Ross Dependency. Ornithological Soc. N.Z. A.H. & A.W. Reed Ltd. 96 pp.
LANDIS, C. A. & D. S. COOMBS. 1967. Metamorphic belts and orogenesis in southern New Zealand. *Tectonophysics* 4(4–6): 501–18.
LAUDON, T. S., L. L. LACKEY, P. G. QUILTY & P. M. OTWAY. 1969. Geology of eastern Ellsworth Land. Geol. Map of Antarctica, Sheet 3, American Geographical Soc.
LAURENCE, P. 1967. New Zealand Region: Bathymetry, 1:6,000,000. Chart, Miscellaneous Series No. 15, N.Z. Oceanographic Institute.
LEITCH, E. C., J. A. GRANT-MACKIE & N. DE B. HORNIBROOK. 1969. Contributions to the geology of northernmost New Zealand: I – The mid-Miocene Waikuku Limestone. *Trans. R. Soc. N.Z. (Earth Sci.)* 7(2): 21–32.

MARPLES, B. J. 1932. Early Tertiary Penguins of New Zealand. *Paleont. Bull. N.Z. geol. Surv.* 20. 66 pp. 8 pls.
MARPLES, B. J. 1946. Notes on some neognathous bird bones from the early Tertiary of New Zealand. *Trans. R. Soc. N.Z.* 76(2): 132–4.
MARPLES, B. J. 1956. Cetotheres (Cetacea) from the Oligocene of New Zealand. *Proc. zool. Soc. Lond.* 126(4): 565–80.
MAYR, E. 1972. Continental drift and the history of the Australian bird fauna *Emu* 72: 26–28.
MCINTYRE, D. J. & I. C. MCKELLAR. 1970. A radiocarbon dated post glacial pollen profile from Swampy Hill, Dunedin, New Zealand. *N.Z. Jl Geol. Geophys.* 13(2): 346–9.
MCINTYRE, D. J. & G. J. WILSON. 1966. Preliminary palynology of some Antarctic Tertiary erratics. *N.Z. Jl Bot.* 4(3): 315–21.
MCKELLAR, I. C. 1960. Pleistocene deposits of the Upper Clutha Valley, Otago, New Zealand. *N.Z. Jl Geol. Geophys.* 3(3): 432–60.
MCLINTOCK, A. H. (Ed.) 1966. An Encyclopaedia of New Zealand. 3 Vols. Government Printer, Wellington.
MCQUEEN, D. R. 1954. Palaeozoic plants from New Zealand. *Nature* 173: 88.
MCQUEEN, D. R., D. C. MILDENHALL & C. J. E. BELL. 1968. Paleobotanical evidence for changes in the Tertiary climates of New Zealand. *Tuatara* 16(1): 49–56.
MILDENHALL, D. C. 1970. Discovery of a New Zealand member of the Permian *Glossopteris* flora. *Aust. Jl Sci.* 32(12): 474.
MILDENHALL, D. C. & W. F. HARRIS. 1970. A cool climate pollen assemblage from the type Waipipian (Middle Pliocene) of New Zealand. *N.Z. Jl Geol. Geophys* 13(3): 586–91.
MILDENHALL, D. C. & W. F. HARRIS. 1971. Status of *Haloragacidites* (al. *Triorites*) *harrisii* (Couper) Harris comb. nov. and *Haloragacidites trioratus* Couper, 1953. *N.Z. Jl Botany* 9(2): 297–306.
MOAR, N. T. 1971. Contribution to the Quaternary history of the New Zealand flora. 6: Aranuian pollen diagrams from Canterbury, Nelson and north Westland, South Island. *N.Z. Jl Bot.* 9(1): 80–145.
NEEF, G. 1967. The geology of Eketahuna. Unpublished Ph.D. Thesis lodged in Victoria University of Wellington Library.
NINKOVICH, D. 1968. Pleistocene volcanic eruptions in New Zealand, recorded in deep-sea sediments. *Earth Planet. Sci. Letters* 4: 89–102.
NORRIS, R. M. 1973. Shell and gravel layers, western continental shelf, New Zealand. *N.Z. Jl Geol. Geophys.* 15(4): 572–89.
OLIVER, W. R. B. 1925. Biogeographical relations of the New Zealand region. *Jl Linn. Soc. Bot.* 47: 99–140.
OLIVER, W. R. B. 1949. The Moas of New Zealand and Australia. *Bull. Dominion Mus. (Wellington)* 15. 204 pp.
OLIVER, W. R. B. 1950. The fossil flora of New Zealand. *Tuatara* 3(1): 1–11.
PANTIN, H. M. 1957. Fossiliferous concretions from the shelf south-east of Cape Campbell, New Zealand. *N.Z. Jl Sci. Technol.* B38(7): 781–91.
POWELL, A. W. B. 1936. The Paryphantidae of New Zealand. III. Further new species of *Paryphanta* and *Wainuia*. *Rec. Auckland Inst. Mus.* 2(1): 29–41.
POWELL, A. W. B. 1949. The species problem in New Zealand land snails. Rep. 6th Sci. Congr. 1947, *Trans. R. Soc. N.Z.* 77(5): 202–8.
PREEST, D. S. 1963. A note on the dispersal characteristics of the seed of the New Zealand podocarps and beeches and their biogeographical significance. Pp. 415–24 in GRESSITT, J. L. (Ed.) 'Pacific Basin Biogeography', Bishop Museum Press, Honolulu.
RAVEN, P. H. & AXELROD, D. I. 1972. Plate tectonics and Australasian paleobiogeography. *Science*, 176: 1379–86.
RAESIDE, J. D. 1948. Some post-glacial climatic changes in Canterbury and their effect on soil formation. *Trans. R. Soc. N.Z.* 77(1): 153–71.

RITCHIE, A. 1971. New fossil discoveries in Antarctica. *Modern World*, June 1971, p. 10–14.

ROMER, A. S. 1966. Vertebrate Paleontology (3rd Edit.). University of Chicago Press. 468 pp.

SCIENTIFIC STAFF, *GLOMAR CHALLENGER*, 1973a. Leg 28 Deep-sea drilling in the Southern ocean. *Geotimes* 18(6): 19–24.

SCIENTIFIC STAFF, 1973b. Southwest Pacific structure, Leg 30: Deep Sea Drilling Project. *Geotimes* 18(9): 18–21.

SERVENTY, D. L. 1972. Causal zoogeography; Australia. Proc. 15 internat. ornith. Congr.

SIMPSON, G. G. 1961. Historical zoogeography of Australian mammals. *Evolution* 15: 431–46.

SIMPSON, G. G. 1971. A review of the pre-Pliocene penguins of New Zealand. *Bull. Amer. Mus. Nat. Hist.* 144(5): 321–78.

SMITH, A. G. & A. HALLAM. 1970. The fit of the southern continents. *Nature* 225 (5228): 139–44.

SOONS, J. M. 1962. A survey of periglacial features in New Zealand Pp. 74–87 in MCCASKILL (Ed.) 'Land and Livelihood'. N.Z. Geographical Soc. Special Publ. Miscel. Series No. 4. Christchurch. 280 pp.

SPROLL, W. P. & R. S. DIETZ. 1969. Morphological continental drift fit of Australia and Antarctica. *Nature* 222(5191): 345–48.

SQUIRES, D. F. 1958. The Cretaceous and Tertiary Corals of New Zealand. *Paleont. Bull. N.Z. geol. Surv.* 29. 107 pp. 16 pls.

SQUIRES, D. F. 1962a. Additional Cretaceous and Tertiary Corals from New Zealand. *Trans. R. Soc. N.Z. (Geol.)* 1(9): 133–50.

SQUIRES, D. F. 1962b. A Scleractinian Coral faunule from Cape Rodney. *N.Z. Jl Geol. Geophys* 5(3): 508–14.

STEVENS, G. R. 1963. Faunal realms in Jurassic and Cretaceous belemnites. *Geol. Mag.* 100(6): 481–97.

STEVENS, G. R. 1967. Upper Jurassic fossils from Ellsworth Land, west Antarctica, and notes on Upper Jurassic biogeography in the South Pacific region. *N.Z. Jl Geol. Geophys.* 10(2): 345–93.

STIPP, J. J. 1968. The geochronology and petrogenesis of the Cenozoic volcanics of North Island, New Zealand. Ph.D. Thesis, Australian National University, Canberra, A.C.T.

STIPP, J. J. & I. MCDOUGALL. 1968. Geochronology of the Banks Peninsula volcanoes, New Zealand. *N.Z. Jl Geol. Geophys.* 11(5): 1239–60.

STIPP, J. J. & B. N. THOMPSON. 1971. K/Ar ages from the volcanics of Northland. *N.Z. Jl Geol. Geophys.* 14(2): 403–13.

SUGGATE, R. P. 1965. Late Pleistocene geology of the northern part of the South Island, New Zealand. *Bull. N.Z. geol. Surv.* 77. 91 pp.

SUGGATE, R. P. 1972. Mesozoic-Cenozoic development of the New Zealand region. *Pacific Geology* 4: 113–20.

SUGGATE, R. P. & N. T. MOAR. 1970 Revision of the chronology of the Late Otira Glacial. *N.Z. Jl Geol. Geophys.* 13(3): 742–6.

TOWNROW, J. A. 1957. On *Dicroidium*, probably a pteridospermous leaf, and other leaves now removed from this genus. *Trans. geol. Soc. S. Africa* 60: 21–56.

TOWNROW, J. A. 1967. On *Rissikia* and *Mataia* podocarpaceous conifers from the Lower Mesozoic of Southern Lands. *Proc. R. Soc. Tasmania* 101: 103–36.

VUCETICH, C. G. & W. A. PULLAR. 1964. Stratigraphy of Holocene ash in the Rotorua and Gisborne Districts. Pt. 2 (p. 43–78) in HEALY, J, C. G. VUCETICH & W. A. PULLAR, Stratigraphy and Chronology of Late Quaternary Volcanic Ash in Taupo, Rotorua and Gisborne Districts. *Bull. N.Z. geol. Surv.* 73: 88 pp.

VUCETICH, C. G. & W. A. PULLAR. 1969. Stratigraphy and Chronology of Late Pleisto-

cene Volcanic Ash Beds in Central North Island, New Zealand. *N.Z. Jl Geol. Geophys.* 12(4): 784–837.
WARDLE, P. 1963. Evolution and distribution of the New Zealand flora as affected by Quaternary climates. *N.Z. Jl Bot.* 1(1): 3–17.
WARDLE, P. 1968. Evidence for an indigenous pre-Quaternary element in the mountain flora of New Zealand. *N.Z. Jl Bot.* 6(1): 120–5.
WATERHOUSE, J. B. 1964. Permian stratigraphy and faunas of New Zealand. *Bull. N.Z. geol. Surv.* 72. 101 pp.
WATERHOUSE, J. B. 1967. Proposal of series and stages for the Permian in New Zealand. *Trans. R. Soc. N.Z. (Geol.)* 5(6): 161–80.
WATERHOUSE, J. B. 1970. The world significance of New Zealand Permian stages. *Trans. R. Soc. N.Z. (Earth Sci.)* 7(7): 97–109.
WELLMAN, H. W. 1956. Structural outline of New Zealand. *Bull. N.Z. Dep. scient. ind. Res.* 121. 36 pp.
WHITE, E. I. 1968. Devonian fishes of the Mawson-Mulock area, Victoria Land, Antarctica. *Trans. Antarctic Exped.* 1955–58, *Scintific Rep.* 16: 1–26. 3 pls.
WILLETT, R. W. 1950. The New Zealand Pleistocene snow-line; climatic conditions and suggested biological effects. *N.Z. Jl Sci. Technol.* B32(1): 18–48.
WILSON, D. D. 1956. The late Cretaceous and early Tertiary transgression in South Island, New Zealand. *N.Z. Jl Sci. Technol.* B37(5): 610–22.
WILSON, G. J. 1967. Some new species of lower Tertiary Dinoflagellates from McMurdo Sound, Antarctica. *N.Z. Jl Bot.* 5(1): 57–83.

POSTSCRIPT

The reader's attention is drawn to the following information that became available after the foregoing article was prepared.

At the Geological Society of New Zealand conference in November 1974, G. W. GRINDLEY presented paleomagnetic evidence from the Chatham Islands for post-Jurassic rotation to form the New Zealand orocline (pp. 3, 16). Many recent authorities have settled for an age of c. 0.8 m.y. for the Plio-Pleistocene boundary (p. 50). MOAR & SUGGATE (*N.Z. Jl. Geol. Geophys.* 16(3): 333 found evidence that *Nothofagus* forest persisted during the Otira Glacial near Hokitika (p. 57–8). Identification of kauri *(Agathis)* pollen near Wellington and its interglacial age (pp. 57–8) are now seriously questioned. In the light of C_{14} dates, it is doubtful if any New Zealand *Anadara* occurrences are as young as Holocene (p. 68) and all may well be interglacial. Several generic names have been changed (e.g. *Pterotyphis* for *Pterynotus*, pp. 51, 52).

II. THE CLIMATE

by

J. D. COULTER

New Zealand's place in world climates is determined by its location and its geographical form. It is a long, narrow, mountainous land, the main islands extending about 1900 km from latitude 34 °S to 48 °S and mostly between 100 to 300 km wide from east to west. No place is more than 130 km from the sea, and the ocean extends at least 1600 km to the next large land mass.

The main mountain axis lies northeast – southwest from near East Cape to the southwest of the South Island and is high enough to have a dominating influence on weather in the South Island and the southern half of the North Island. Less continuous and generally lower hills in the west and north of the North Island have a smaller effect.

The Broad-Scale Atmospheric Environment of New Zealand

Energy balance

If the mean intensity of the solar beam outside the earth's atmosphere is taken as 1.36 kW m^{-2} (1.95 cal cm^{-2} min^{-1} or 1.95 ly min^{-1}) the extra-atmospheric solar energy input at latitude 40 °S ranges from 44.6 MJ m^{-2} (1060 ly) per day in midsummer to 12.5 MJ m^{-2} (300 ly) per day in midwinter. Averaged over the whole earth, the proportion reflected by the earth atmosphere system (albedo) is close to 30%. It is about 25 percent in the area around northern New Zealand, and about 35 percent in the vicinity of the southern end of the country. The average effective solar radiation (absorbed in the atmosphere or at the earth's surface) at 40 °S is about 30 MJ m^{-2} (700 ly) per day in January and 8.5 MJ m^{-2} (200 ly) per day in July, seasonal changes in albedo contributing to the difference to a small extent.

Long wave 'thermal' radiation lost to space averages 20 MJ m^{-2} (480 ly) per day in the New Zealand area over the year as a whole. It is about 10 percent greater in January-February than in July-August, and about 5 percent greater in the north of New Zealand than in the south. Hence the net radiation budget over the year as a whole is quite small. There is a net gain in summer, the average for December-January-February being approximately 8 MJ m^{-2} (190 ly) per day at 40 °S, and in winter a negative value of approximately the same magnitude. There is a marked poleward decrease in net radiation in latitudes 30–50 °S, especially in autumn and winter.

The poleward energy transport required by the annual mean net radiation gradient reaches a maximum around latitude 35°S (in terms of total energy crossing a given parallel of latitude). How the transport varies with time of year depends on the amounts of energy held in storage as latent and sensible heat in the atmosphere, oceans and land masses concerning which reliable information is still lacking. It is known, however, that the transport is at a relative maximum from May to October and is low in summer. At 40 °S, the latitude of central New Zealand, ocean currents account for about a fifth of the total transport. The atmosphere carries the rest, about as much in the form of latent heat of water vapour as by way of sensible heat plus mechanical energy*.

The above values are mostly hemispherical averages. Recent satellite data show some variations with longitude, resulting from variations in albedo and in effective radiating temperature. The data are not yet sufficient to provide valid climatological information in any detail, but New Zealand appears to lie in a region where net radiation is predominantly zonally distributed. Land-sea difference in the terms of the energy budget resulting from the greater thermal inertia of the oceans have an important influence on the climate in maritime areas. Ship data, while insufficient to provide accurate values, give an indication of the way in which transfers of sensible and latent heat from ocean to atmosphere contribute to the total heat transfer in the area west and north of New Zealand. Some results from the Tasman Sea are as follows: 1. Total heat transfer is greatest just off the Australian coast and smallest just west of New Zealand in all seasons. 2. The seasonal means in winter and autumn are greater than those in spring and summer. 3. In individual seasons large variations from the mean can occur. 4. The proportion of sensible heat to latent heat ranges from around .05 in summer to around .20 in winter, i.e., most heat transfer is by evaporation. 5. Average values in areas bounded by 5 degree lines of latitude and longitude range from approximately 14.6 to 16.8 MJ m^{-2} (350 to 400 ly) per day in autumn and winter in latitude 25–30 °S to 6.3 to 8.4 MJ m^{-2} (150 to 200 ly) per day in latitude 35–45 °S in spring and summer (DE LISLE, 1970).

WIND CIRCULATION AND BAROMETRIC PRESSURE PATTERNS

The southern hemisphere atmospheric circulation patterns are simpler than those of the northern hemisphere because of the smaller influence of continental land masses. Mean wind flow tends to be zonal. Though there are important day-to-day changes resulting from moving disturbances these are largely lost in the average circulation patterns.

* For detailed data and discussion concerning the southern hemisphere energy balance see, e.g., VONDER HAAR & SUOMI (1971), GABITES (1960), NEWELL, VINCENT, DOPPLICK, FERRUZZA & KIDSON (1970).

In the free air over New Zealand, and at most places at the surface as well, the mean wind is westerly, and above 1 to 2 km the instantaneous winds in the troposphere are usually from some westerly direction. The preponderance of westerlies is especially marked in the south of New Zealand. In winter the upper air westerlies commonly have a 'jet stream' core of high velocity winds in a narrow band around latitude 25 to 30 °S at a height around 10 to 12 km where the average speed is nearly 60 m/s and speeds sometimes exceed 75 m/s. In summer the main maximum of the mean westerly flow is found to the south of 50 °S, with mean speed greater than 30 m/s, at about 12 km. The subtropical jet stream is still identifiable in the mean flow in about the same latitude and height as in winter but with mean speed only about 30 m/s.

Over New Zealand itself, between 1 and 10 km, the average velocities for the most part are between 5 and 20 m/s, with weak maxima below about 6 km around June and November, the spring (November) maximum being slightly more pronounced over central and southern New Zealand than in the north.

The direction of the mean wind vector is generally within about 15 degrees of 270°. Only at relatively low levels in the stations near Cook Strait where there is orographic deflection of wind flow are large S-N components found in the mean winds.

In summer and autumn months (January-April), maps of mean atmospheric pressure at sea level show the high pressure belt between sub-tropical easterlies and mid- and high-latitude westerlies in latitude 33–37 °S in the New Zealand region, bringing the easterly wind zone just on to the far north of New Zealand. The high pressure belt is interrupted and displaced to the south in the Australian area. In winter (May, June, July) the highest pressure belt is about 10 degrees further north (25–27 °S). The isobars (and resulting general wind flow) are then oriented from about 230–240 °, between the winter mean anticyclone over Australia and a low pressure centre to the northeast of the Ross Sea, and the pressure gradient over New Zealand is somewhat greater than in summer. By October the mean isobars are nearly zonal (from 290 ° around northern New Zealand) and the mean pressure gradient over the country is at its maximum.

Temperature of the air at sea level

Isotherms of average air temperature at sea level in the region around New Zealand are oriented broadly from east to west, apart from local distortions introduced by the land itself. However, over the southern ocean at higher latitudes the Indian Ocean sector is colder than the Pacific sector, the 0 °C isotherm in July being at about 53 °S at longitude 90 °E and at 59 °S in longitude 180 °W. The air over the Tasman Sea however is about 1–2 °C warmer than at corresponding latitudes to the

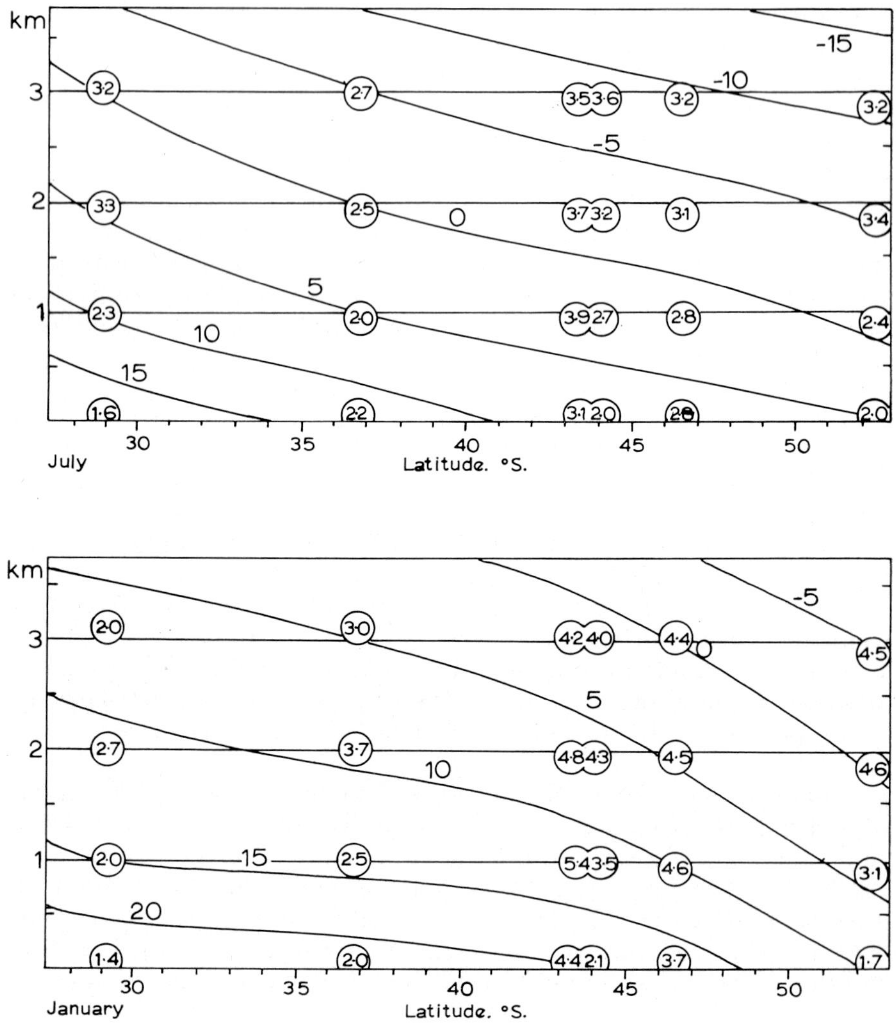

Fig. 1. Average temperature of the free air over New Zealand at about 12 hours NZST in July and January – °C. The numbers in circles are standard deviations of daily temperatures at constant pressure levels.

east of New Zealand in July and August. In summer (January, February) the 10 °C isotherm is at about 45–46 °S at 90 °E and at about 52 °S at 180 °W, but again the temperatures over the Tasman Sea are about 1 ° higher than those to the east of New Zealand (Taljaard *et al.*, 1969).

Temperatures in the upper air

Figure 1 shows the mean temperatures at mid-day in the free air over the New Zealand area in January and July – the warmest and coldest months.

In January the 0 °C isotherm rises from about 1850 m at Campbell I. (52.6 °S) to 2950 m at Invercargill (46.4 °S) to 3900 m at Auckland (36.8 °S) and to 4530 m at Raoul I. (29.2 °S), while in July it rises from 800 m at Campbell I. to 1500 m at Invercargill, 2100 m at Auckland and 3125 m at Raoul I.

There are strong gradients of mean temperature in the vertical (lapse rates) in the lowest kilometre or so over land near mid-day when surface temperatures are on average 1° to 3 °C above the mean for the day. Above this boundary layer, mean lapse rates in the troposphere are around 4° to 5 °C per km in all seasons at all of the New Zealand upper air stations. The temperatures in the free air have only slight diurnal variation.

Temperatures are most variable near the tropopause level south of 30 °S (standard deviation 5–6 °C in daily values) and at low levels at stations in the South Island.

Large scale transient features of the circulation

At any given time the tropospheric wind flow over a sector of the earth such as that around New Zealand may be predominantly zonal, broadly resembling the mean flow pattern described above. However, the westerlies undergo large scale wave-like oscillations hundreds and thousands of kilometres in length. The jet stream axis assumes a meandering sinuous form, with variations of intensity along its length and it occasionally divides, branches or loops as the amplitude of the perturbations increases.

On conventional surface weather maps, New Zealand lies in the path of an irregular succession of migrating anticyclones that typically lie in latitudes 33 to 40 °S, extend some 1000 km from east to west, and pass eastward at three to seven day intervals. Individual centres may pass to north or to south of the country but the more northerly tracks are most common in spring. Anticyclone centres occasionally appear well to the south of New Zealand, often associated with a slowly-changing 'blocking' situation. The most intense anticyclones are commonly found in winter. Cyclonic circulations (clockwise in the southern hemisphere) around low pressure centres are usually present in high latitudes south of Australia, the Tasman Sea and New Zealand where they travel rapidly eastward in a relatively unbroken succession. Associated with these depressions, and extending to the north of them, are low pressure 'troughs' which usually contain one or more cold fronts. They frequently pass over New Zealand between pairs of anticyclones. These meridionally or NW-SE oriented cold fronts comprise one of the commonest types of weather disturbance in New Zealand. They characteristically produce a period of rain of several hours' duration at their passage, which is preceded by northerly winds, increasing cloud, and varying amounts of orographic rain in windward areas. The passage is followed by colder southerly to westerly winds and showery weather in those areas exposed to the new southwesterly air-stream.

Sometimes cyclonic centres form in the vicinity of New Zealand and pass over or close to the country, occasionally developing into large and vigorous depressions which can give several days of rain in much of the country. Some of the most severe storms have originated as tropical cyclones in the sea areas between Fiji and New Guinea from December to April. Though they take on extra-tropical storm characteristics before reaching New Zealand, nevertheless they can still produce very heavy and widespread rain and destructive gales over large areas.

'Cold-' or 'cut off'-lows are occasionally experienced, usually lying to the north or northwest of a 'blocking' anticyclone of relatively high latitude. As the name implies, they are regions with colder air than in the surroundings, in a deep layer. They may be of only weak intensity at sea level, but the cyclonic circulation increases aloft. They are usually slow moving, and are characterised by extensive cloud sheets and widespread rain which may be little influenced in its distribution by orography.

Small scale disturbances, weather distribution

At the other end of the scale, tornadoes in New Zealand characteristically cause destructive winds in belts 30–100 m wide and 1 to 2 km long. There are perhaps 30 per year, on average, in New Zealand. Westland, North Taranaki, Waikato, Auckland and Bay of Plenty appear to be the most susceptible areas. They tend to occur during periods of unstable windy conditions within a generally cyclonic regime, are usually associated with thunderstorms and nearly all are associated with an active front.

Thunderstorms are most frequent in the north and west (15 to 20 days per year). Most are associated with fronts but some originate within unstable air masses, e.g. in moist air of tropical origin in summer in the north of the North Island.

Hail is most frequent in the southwest (30 days per year) also mostly associated with fronts. Most of the hailstones are small, but occasionally large stones cause local damage, chiefly in horticultural areas in Canterbury and Hawkes Bay.

The distribution of precipitation and weather phenomena are markedly influenced by the topography, the local effects depending on whether the prevailing air-stream has an onshore trajectory, or has crossed high land. In the latter case, except near the centre of a vigorous depression or a pressure trough or front, skies will be clear or partly cloudy (often with parallel bands of wave cloud at levels above 3 km) with low humidity and relatively mild temperatures. Such 'föhn' conditions are most common in the eastern districts of the South Island where they produce the well-known 'Canterbury nor'wester'. They are also found east of the higher ranges of the North Island in westerly winds and in west coast districts of the South Island in less frequent periods of southerly or easterly wind flow. By contrast, in coastal areas winds from the sea tend to bring low cloud, showers, and occasionally fog and drizzle, especially on higher ground.

An important consequence is the distribution of rain fall. Orographically induced or augmented rain produces the highest rainfall amounts where the mountains are exposed to the direct sweep of the westerly and north-westerly winds. By contrast areas in the lee of the mountains receive little precipitation.

Air-streams originating from such extreme environments as the arid hot centre of Australia or the ice of Antarctica reach New Zealand modified by the long ocean paths. Thus there is an absence of extreme variations of temperature. In winter or spring exceptionally severe southerly outbreaks will bring winds with sea level temperatures just above freezing point (and dew point one or two degrees below) to south-eastern parts of the South Island.

In mid and late summer northerly air streams carry air of tropical origin over northern New Zealand which may attain temperatures of about 25 °C and dew point of up to 23 °C persisting for 24 hours or more. With humidity near 100 percent in a deep layer of the troposphere, such airstreams contain the limiting amount of precipitable water (65 mm approx.) found in New Zealand – a necessary condition for maximum precipitation events.

In many instances the changes of air mass properties at fronts or front-like weather systems are rather small. Nevertheless, about eastern coasts and Cook Strait in particular, it is not uncommon for temperature to change suddenly by as much as 10 °C or 15 °C when air which has been heated over land by sunshine or the föhn effect (or both) is replaced by a southerly outbreak. Because of the mountains, the southerlies flow directly from at least the latitude of the southern end of the country. These very sharp discontinuities often arrive with a conspicuous squall. They resemble the well known 'southerly buster' of the eastern Australian coast.

Meteorological situations

Some characteristic meteorological situations and the weather distribution in New Zealand associated with them are illustrated in figure 2. The examples include some weather events of unusual severity. Figure 2a shows a period of essentially zonal circulation, with disturbances of relatively small amplitude moving rapidly eastward. In these 'westerly' conditions there may be rather persistent precipitation in the western mountains of the South Island, but elsewhere only brief periods of mainly

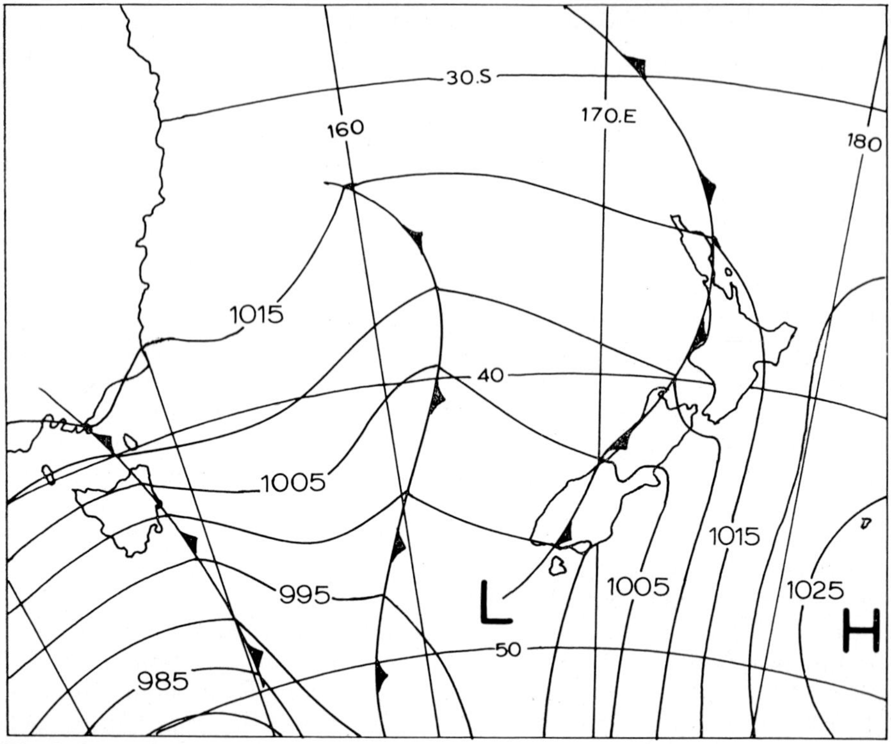
30.S
160
170.E
180
1015
40
1005
995
985
50
1015
1005
1025
L
H

Fig. 2a.

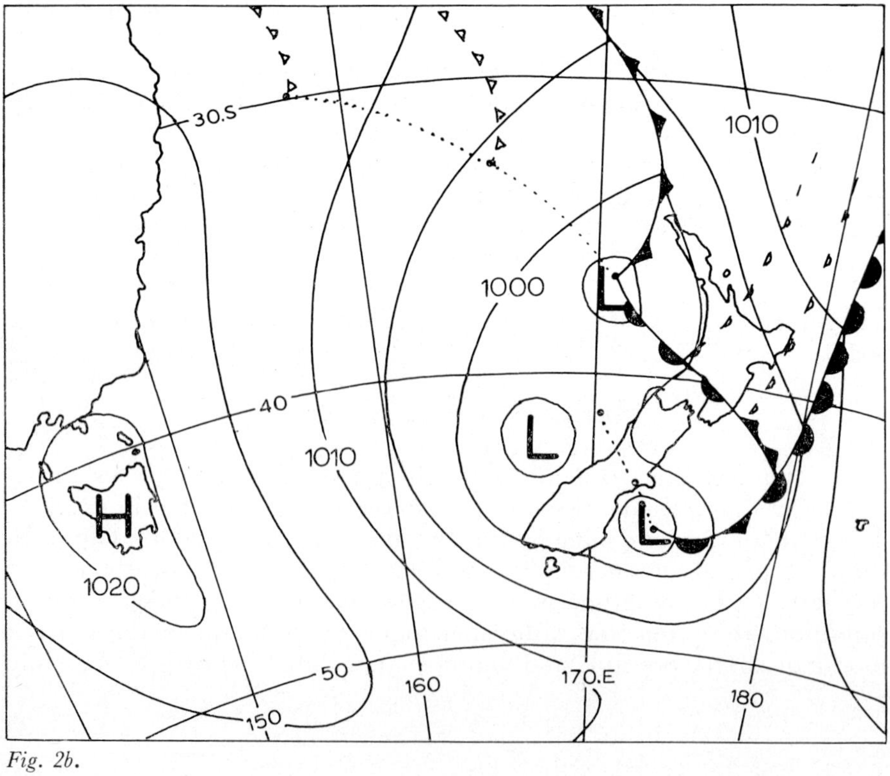
30.S
1010
1000
L
40
L
1010
H
L
1020
50
160
170.E
180
150

Fig. 2b.

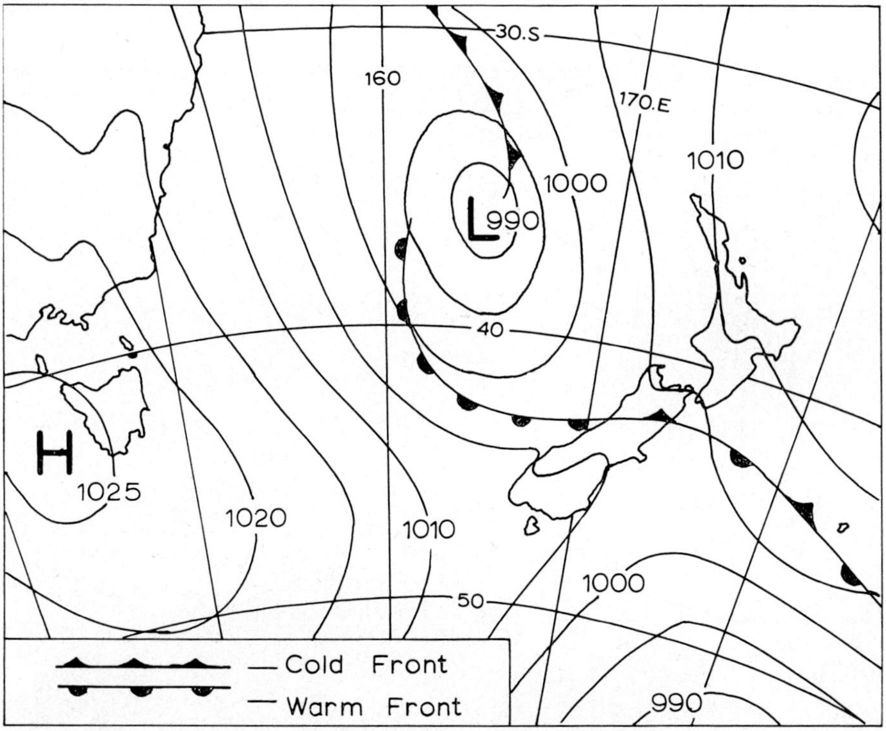
30.S
160
170.E
1010
1000
L990
40
H
1025
1020
1010
1000
50
990
Cold Front
Warm Front
Fig. 2c.

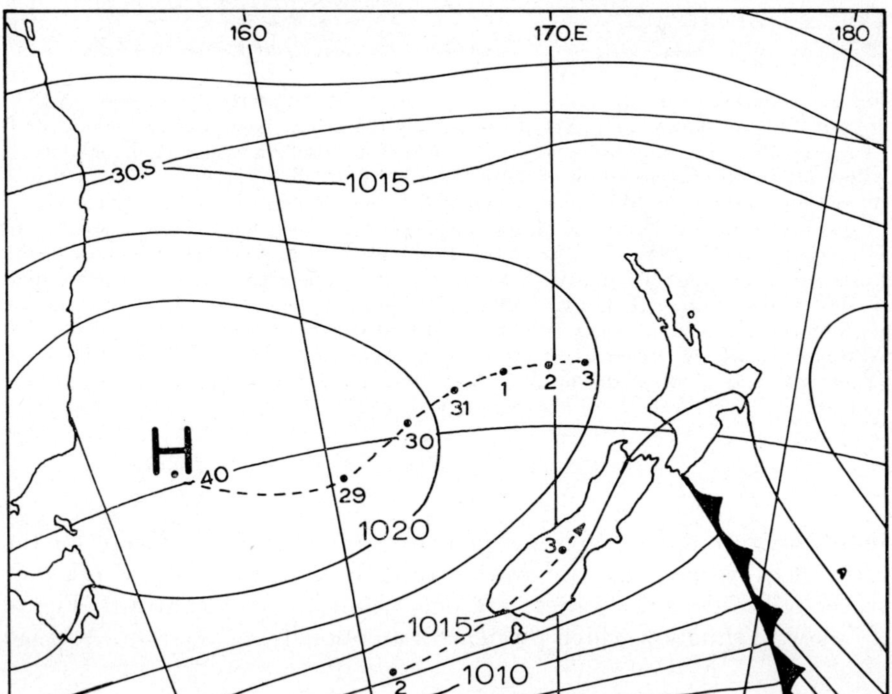
160
170.E
180
30.S
1015
H
40
29
30
31
1
2
3
1020
3
1015
1010
2
Fig. 2d.

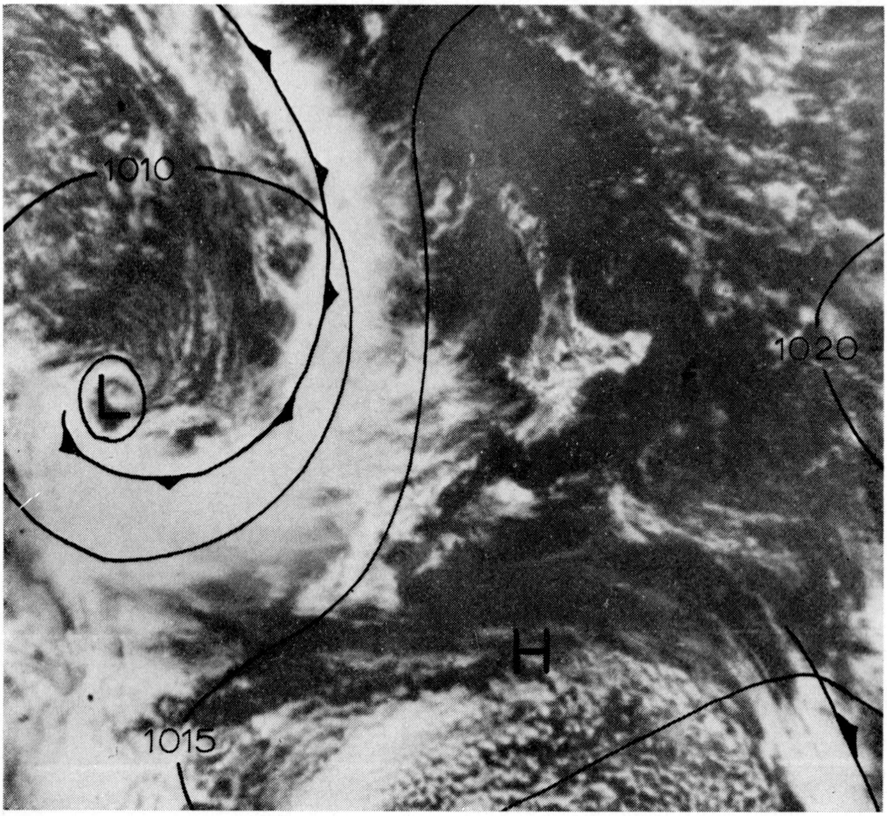

Fig. 2. Meteorological situations: a. (Top on p. 94) westerly type; 0000 NZST, 1 April 1971. b. (Bottom on p. 94) heavy rainfall in northern New Zealand; 0000 NZST 11 March 1964. Positions of low pressure centres and fronts shown at 12 hour intervals before map time. Precipitation of more than 300 mm fell in the estern Bay of Plenty district and more than 100 mm in much of the northwestern half of the North Island in the 3 days ending 9 a.m. 12 March. c. (Top on p. 95) extensive heavy snowfall in Canterbury; 0600 NZST, 16 November 1967. There was very cold air south of the front across the South Island. d. (Bottom on p. 95) anticyclonic situation with settled weather; 0600 NZST, 28 January 1965, with positions of anticyclones shown for the six following days. e. (Above) satellite picture of clouds in New Zealand area; 1228 NZST, 27 February 1970, with meteorological situation for 1200 NZST superimposed. (Figures 2b and 2d adapted from *N.Z. Geogr.* Vol. 22, 1965: 22–34, 2c from *N.Z. Geogr.* Vol. 26: 20–35, 2a from *N.Z. Entomologist* Vol. 5: 253–268.)

light rain accompany each low pressure trough or front. Except for the differences between the air flowing directly from the sea and föhn air in the lee of the ranges there are only slight air-mass contrasts. Figure 2b shows a situation which produced exceptionally heavy rain. A broad stream of moist warm air originating in low latitudes and ascending in a convergent trajectory into the rain area in the northern New Zealand is the source of water vapour. The pattern persisted without much change for more than 24 hours. Figure 2c is a typical but severe snow situation affecting the eastern areas of the South Island. It was followed by damaging frosts when clearing skies and a drop in winds allowed radiative cooling of the already cold air. Figure 2d shows one of a series of anticyclones which moved rather slowly across the country giving a long period of generally fine dry weather. Figure 2e is a satellite photograph of the South Pacific region with the corresponding meteorological situation superimposed. It shows, for example, mainly clear skies in the anticyclonic areas, and a large cloudy area near an intense low pressure centre and the front which extends northwards from it.

At times there are persistent departures from the mean pattern of wind and weather. For example, in December 1960 there was recurrent cyclone activity east of New Zealand, and anticyclones over the Tasman Sea. Southerly winds predominated with average temperatures for the month about 1 °C below normal, and with heavy rains in eastern districts from Banks Peninsula to East Cape and very dry conditions in Taranaki and Westland. By contrast in December 1961 there was persistent anticyclonic activity over and east of New Zealand, resulting in a northerly wind flow over the country. Temperatures were generally 1.5 °C to 2 °C above normal, and apart from Bay of Plenty most of the country was much drier than normal.

Such long lasting anomalies of weather and circulation constitute an ecologically important aspect of climate through the long term variability that they introduce. Their effects are not generally uniform, as one area may be experiencing settled anticyclonic conditions while others further east and west may have predominantly cyclonic weather. Changes in the frequency and more especially of preferred positions of such anomalies would account for some of the climate fluctuations which have occurred in the recent past (and which presumably also occurred at earlier times in New Zealand, as elsewhere). Other climate changes – smaller in degree and on a longer time scale which appear to have been more or less contemporaneous over the earth have presumably resulted from, or been associated with other changes in the general circulation. (See section on climate fluctuations, p. 127).

Recently acquired information from satellite photographs has given a new perspective on the regional pattern of cyclogenesis and storminess and associated cloud cover, though not yet available over a sufficient period to establish long term climatological averages. The photographs show that the weather systems tend to pass rather steadily through the New Zealand area. However, developments in low latitudes north of New Zealand tend to move into a broad zone which extends southeastward from the New Guinea – Fiji area towards the Southern Ocean

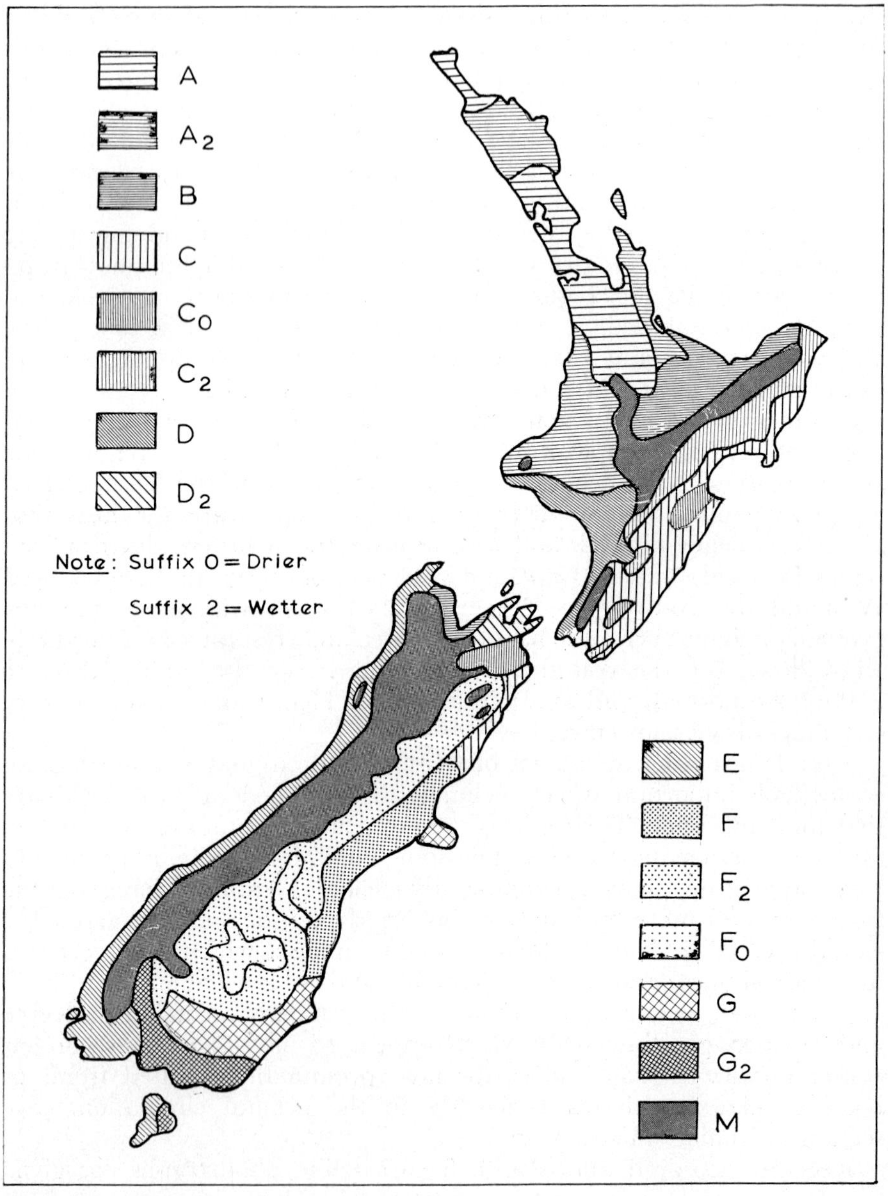

Fig. 4. Major climate types in New Zealand (after ROBERTSON, 1951, with slight simplifications). A. Warm humid summers, mild winters. Annual rainfall 1100–1500 mm with maximum in winter. Prevailing wind S.W., but occasional strong gales and heavy rains from E. or N.E. from Auckland northwards and about Coromandel. A2. Similar to type A but much wetter. Rainfall 1500–2500 mm. B. Sunny rather sheltered areas which receive rains of very high intensity at times from N.E. and N. Very warm summers and mild winters. Annual rainfall 1000–1500 mm with winter

and Antarctica in longitudes far to the east of New Zealand to produce there a more consistently cloudy zone than is found elsewhere in the Pacific hemisphere.

Elements of the Surface Climate of New Zealand

General

Locations of climate stations and other places named in the text are shown in figure 3.

On Koppen's classification all New Zealand (except for small areas in Central Otago with B type climate and the high mountains) is classed Cfb.

In Thornthwaite's 1948 scheme, again excluding the mountain tops, New Zealand falls within moisture categories ranging from 'super-humid' to 'semi-arid' and four thermal categories – 'microthermal' and 'mesothermal', each with two subdivisions. (Garnier, 1951, Cox, 1968).

Figure 4 gives a broad subdivision of New Zealand into major climatic districts according to a descriptive scheme designed to emphasise the more significant climatic variations found within New Zealand. In most instances the transition from one type to the next is gradual, but climatic contrasts are generally much sharper from west to east than from south to north. In some inland areas of the South Island, and to a lesser degree in the east of both islands, the climate is distinctly continental in character.

---

maximum. C. Very warm summers; day temperatures occasionally rise above 32 °C with dry fohn N.W. wind blowing. Rainfall 1000–1500 mm per annum; marked decrease in amount and reliability of rain in spring and summer; moderate winter temperatures with maximum rainfall in this season. C0. Drier than type C; rainfall 650–900 mm. Very sunny. C2. Cooler and wetter hill climates. Very heavy rains at times from S. or S.E.; annual rainfall mainly 1500–2000 mm. D. W. to N.W. winds prevail with relatively frequent gales. Mean annual rainfall 900–1250 mm. Rainfall reliable and evenly distributed throughout the year. Warm summers, mild winters. D2. Wetter than type D; rainfall 1250–2000 mm. E. Mild temperatures, high rainfall increasing rapidly inland with height; minimum rainfall in winter especially in the south. Prevailing winds S.W. but gales not frequent at low levels in spite of exposed coastline. F. Low rainfall; 600–750 mm; in the south slightly more in summer than in other seasons. Warm summers with occasional hot föhn northwesterlies giving temperatures above 32 °C. Cool winters with frequent frosts and occasional light snowfalls. Prevailing winds N.E. near the coast; N.W. inland. F2. Cooler and wetter hill climates; rainfall 750–1500 mm. N.W. winds prevail with occasional very strong gales, especially along river courses. Snow may lie for weeks in winter. F0. Semi arid areas; rainfall 330–500 mm. Very warm dry summers; cold winters. G. Warm summers, cool winters. Rainfall 650–900 mm, evenly distributed except for slight falling off in winter. G2. Wetter and slightly cooler than G climate; rainfall 900–1250 mm; in coastal districts cloudy windy conditions and frequent showers. M. High rainfall; mountain climate.

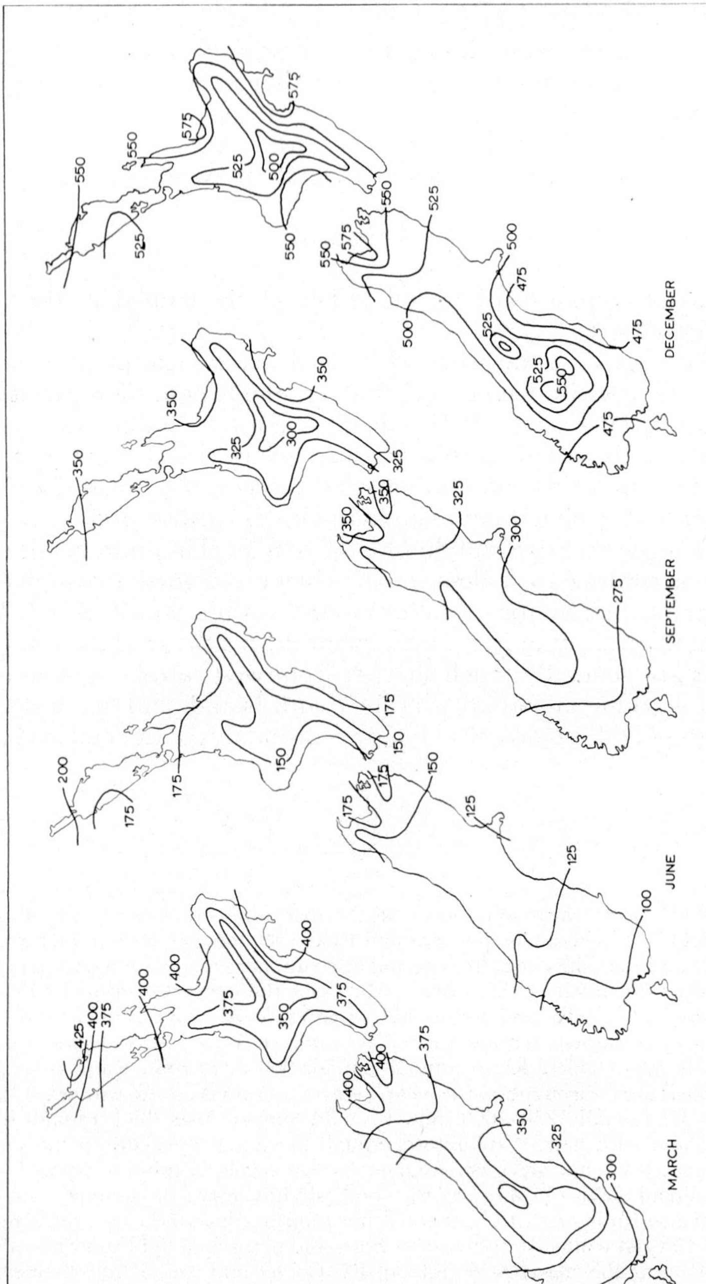

Fig. 5. Average global radiation at surface – Cal cm^{-2} day^{-1} (ly/day). a. March, b. June, c. September, d. December. (After DE LISLE, 1966.)

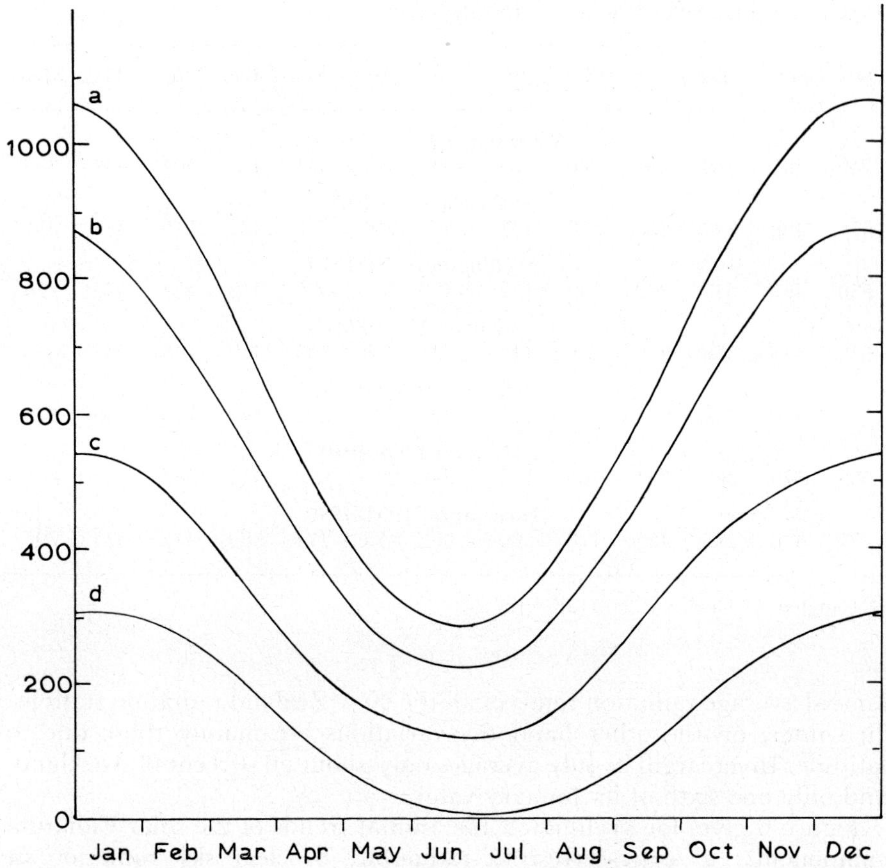

Fig. 6. Annual variation of solar radiation on a horizontal surface at Wellington – ly/day. a. extra-atmospheric radiation, b. maximum values at ground level with clear sky, c. average radiation at ground level, 1954–70, d. average net radiation at Taita (near Wellington) 1965–68.

Solar radiation and sunshine

The average distribution of incoming solar radiation at the surface ('global radiation') is shown for December, March, June and September in figure 5, based on data from pyranometers supplemented by sunshine records. The summer values do not differ much from north to south, the major differences being because of differences of cloudiness rather than of extra-atmospheric radiation. In January for example, the average at Invercargill in the south is about 95 percent of that at Auckland in the north, and 90 percent of that at Mt John. This high level station in an intermountain basin in the South Island has the highest summer and

Table 1. Average Solar Radiation (Langleys/day).

Jan.	Feb.	Mar.	Apr.	May	Jun.	Jul.	Aug.	Sep.	Oct.	Nov.	Dec.	Mean
					Whenuapai 1954–1969							
540	487	399	295	200	165	187	240	334	427	501	537	359
					Ohakea 1954–1970							
571	488	372	268	178	139	150	216	322	422	519	563	351
					Wellington 1954–1970							
536	467	347	242	149	122	126	189	300	410	484	523	325
					Taita 1965–1970							
522	475	356	229	149	114	128	178	303	419	443	503	318
					Christchurch 1960–1970							
517	460	324	235	144	117	130	200	292	424	514	549	326
					Mt John 1966–1970							
583	529	366	254	165	133	154	222	320	472	568	625	366
					Invercargill 1954–1970							
489	428	295	187	118	89	108	175	274	389	472	525	296

(1 Langley = 1 cal/cm^2 = 41.86 kJm^{-2}).

annual average radiation totals of all the New Zealand radiation stations. In winter, on the other hand the variations are mainly those due to latitude. Invercargill in July averages only about 50 percent of Auckland, and only one sixth of its January value.

Figure 6 gives for Wellington the annual trends of the solar radiation components: a. extra-terrestrial radiation; b. clear sky radiation at ground level; c. average global radiation. Average net radiation as measured at Taita (near Wellington) is also included. Table 1 gives average incoming solar radiation at stations equipped with standard 'Eppley' thermocouple pyranometers.

Sunshine duration

Average duration of bright sunshine ranges from over 2400 hours annually near Nelson and Blenheim in two areas in the north of the South Island well protected by mountain ranges, to less than 1800 hours in the south of the South Island and in the mountains of both islands. Other sunny areas (more than 2200 hours) are in the Bay of Plenty and Hawkes Bay in the North Island, and inland Canterbury. Westland, despite its high rainfall, has 1900 hours. Canterbury has mostly between 1800 and 2000 hours. In Southland it falls to near 1600 hours. Campbell Island (53 °S) has only 650 hours. There is a marked increase in cloudiness in winter in the North Island and in Southland as shown in Table 2.

Table 2. Duration of sunshine as percentage of the possible sunshine.

	Summer	Winter		Summer	Winter
Auckland	51	44	Nelson	58	57
Hamilton	52	44	Hokitika	43	44
New Plymouth	53	47	Christchurch	46	45
Napier	57	52	Dunedin	41	43
Wellington	52	42	Invercargill	42	35

Reflectivity, net radiation

Apart from global radiation and sunshine duration measurements, few data concerning radiation components are available from New Zealand. Net radiation above a short grass surface was measured at Invercargill and Wellington during the IGY and between 1965 and 1968 at Taita, a valley site near Wellington (Fig. 6).

A series of total shortwave and spectral reflectivity measurements have been made for a number of crops and pastures in New Zealand. (Scott *et al.*, 1968). Total reflectivity varied from 19 percent to 29 percent. It is not possible therefore to present any general climatological data for the radiation components (other than global radiation) on the basis of network measurements. From the nature of the plant cover of the country, expected values for reflectivity for the drier areas of the country, i.e. eastern districts and much of the inland area in the South Island would be 30–40 percent. For the more humid improved grasslands and cultivated areas 20–30 percent, and for forest and scrub and exotic tree plantations 10–20 percent.

Precipitation

The mean annual precipitation (Fig. 7) ranges from as little as 350 mm in a small area in Central Otago to over 8000 mm in the Southern Alps. Much of the country receives between 800 and 1600 mm, a range regarded as favourable for plant growth in the temperate zone. There are substantial areas in the east of the South Island and small areas in the east of the North Island with less than 800 mm, and these experience rather frequent dry spells and droughts which necessitate special dryland farming methods, or supplementary irrigation. Some areas of productive farmland have upwards of 1600 mm, especially in the northern half of the North Island. Areas with over 3200 mm are mainly mountainous and unoccupied.

The rainfall is spread rather evenly over the year, on average. In the north, winter has almost twice as much rain as summer and winter rainfall predominates in most of the country. In much of the southern half of the South Island however winter is the season of least rainfall and

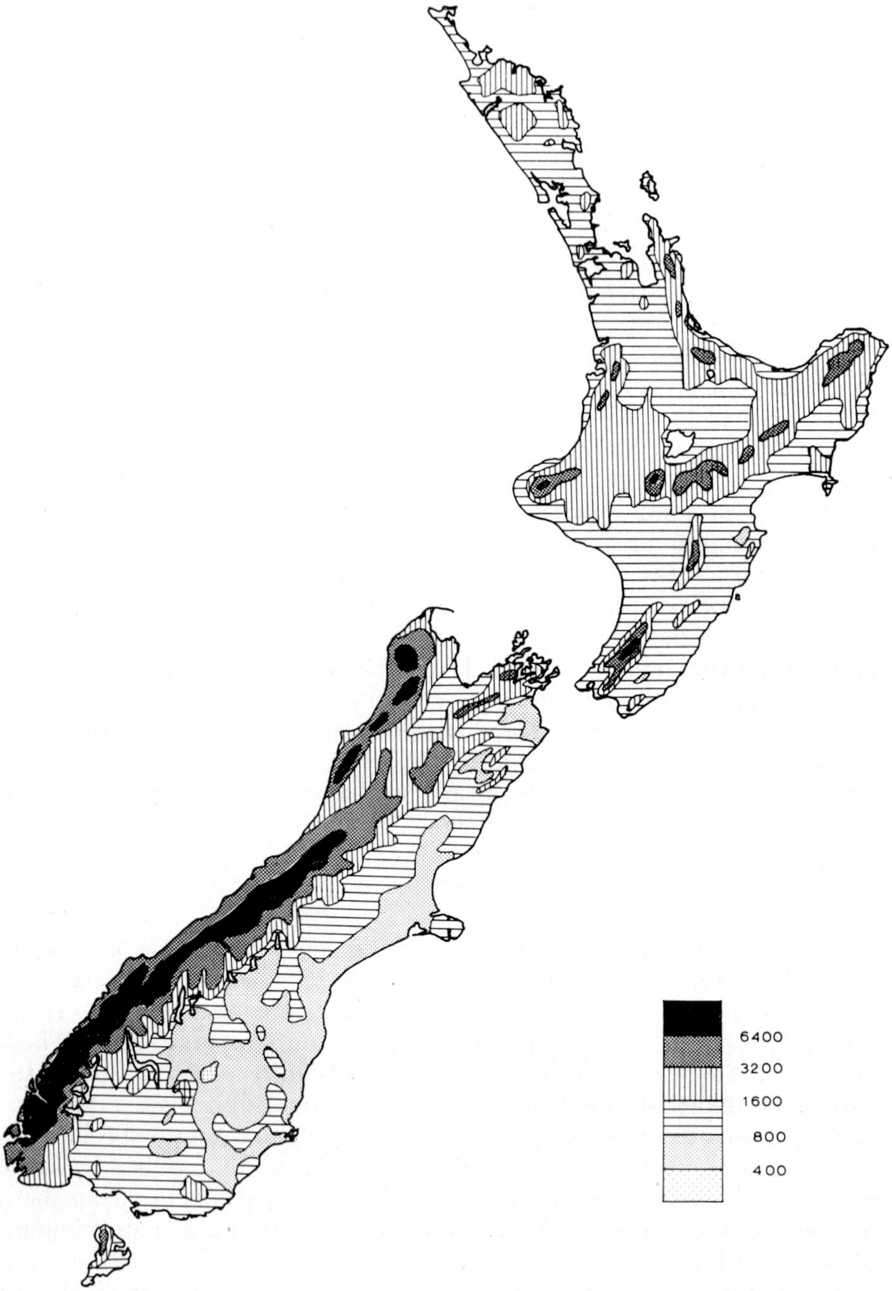

Fig. 7. Average annual precipitation, 1921–50 – mm.

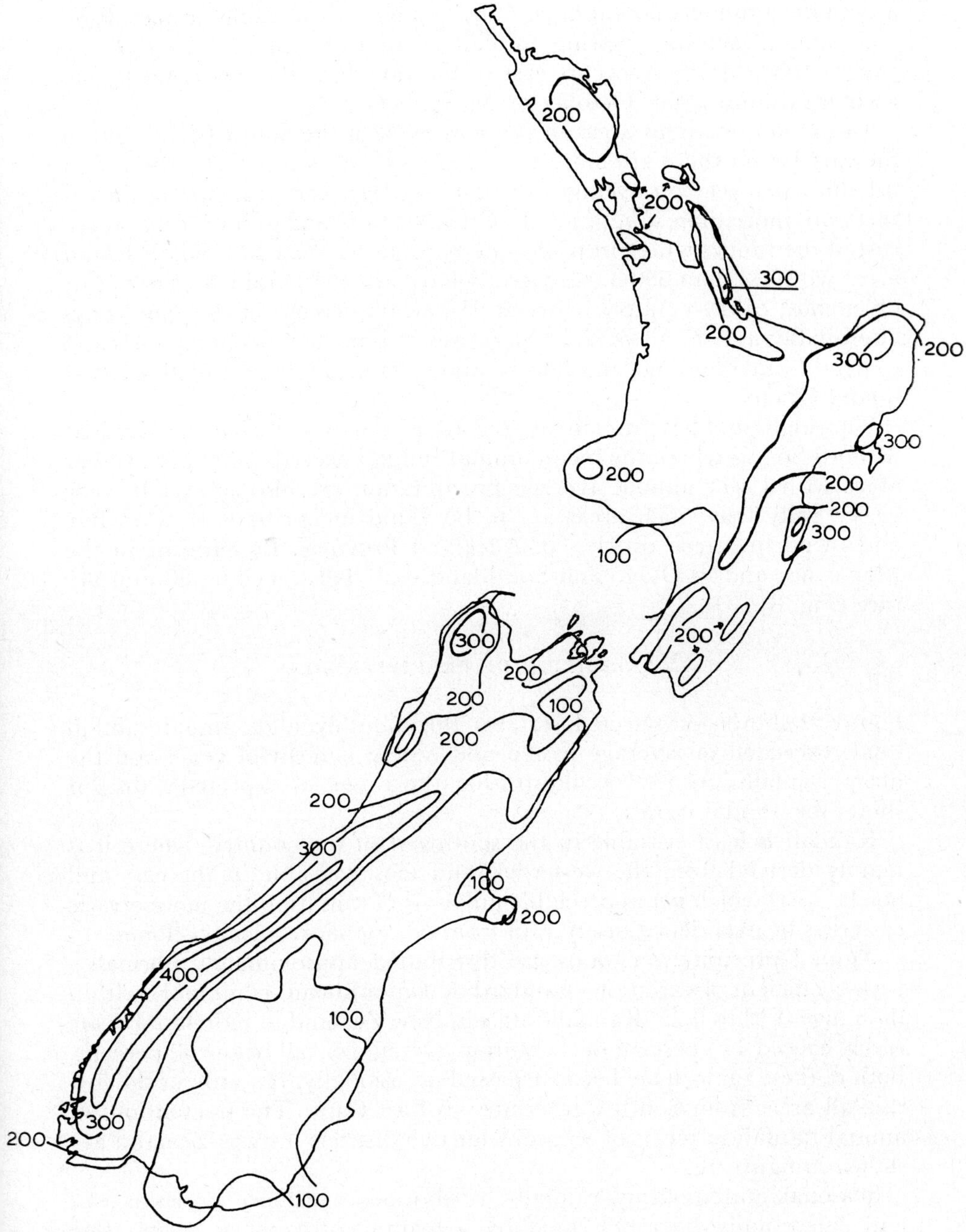

Fig. 8. Maximum daily precipitation likely to be exceeded once in 20 years – mm.

a definite summer maximum is found inland which results in part from convectional showers. Spring rainfall is increased in and west of the ranges and a decrease occurs east of the ranges as the westerlies rise to their maximum about October or November.

In the mountainous areas in the southwest of the South Island and in Stewart Island there are more than 200 precipitation days (with at least 0.1 mm) per year on average. In other western and southwestern areas there are more than 175, in much of the North Island at least 150, except east of the ranges where in places there are fewer than 125. South Island areas with less than 600 mm generally have about 100 rain days per year. Over most of the country between 55 and 65 percent of the 'rain' days also qualify as 'wet' days (2.5 mm or more) but the percentage increases to over 70 in Westland and falls to about 40 in the low rainfall areas of inland Otago.

The highest day's rainfall on record is 560 mm which occurred at Milford Sound where the mean annual rainfall exceeds 6000 mm. Other areas with lower annual average precipitation are also subject to very heavy daily falls; such areas are to be found in northern Hawkes Bay and in northeastern districts of Auckland Province. By contrast in the Manawatu and in Otago and Southland daily falls reaching 80 mm are rare (Fig. 8).

Variability of precipitation

Figure 9 shows for selected stations the monthly and annual rainfall totals exceeded on average in ten and ninety percent of years and the mean rainfalls. As well as illustrating the range of expected values it shows the annual trends.

Rainfall is least variable in the southwest of the country, where it is mainly derived from the westerlies, and most variable in the east and north, areas which get appreciable amounts of rain from the more erratic easterlies or experience heavy rain from occasional tropical cyclones.

Annual precipitation totals are distributed approximately normally, with coefficient of variation (standard deviation/mean) commonly within the range 0.12 to 0.25. Rainfall totals in New Zealand in individual years rarely exceed 175 percent of the station average, or fall below 50 percent. Both of these limits have been surpassed occasionally, but only in the low rainfall areas from South Canterbury to East Cape. The fluctuations of annual rainfall in terms of an index for two districts in New Zealand are shown in figure 18.

Seasonal and monthly rainfall distributions are more or less skew, and can usually be represented by a 'gamma' frequency curve. The coefficient of variation of seasonal totals in general ranges from 0.20–0.30 in spring and winter in western areas to around 0.5 in summer in areas in the north and east of the North Island. For monthly rainfalls similar

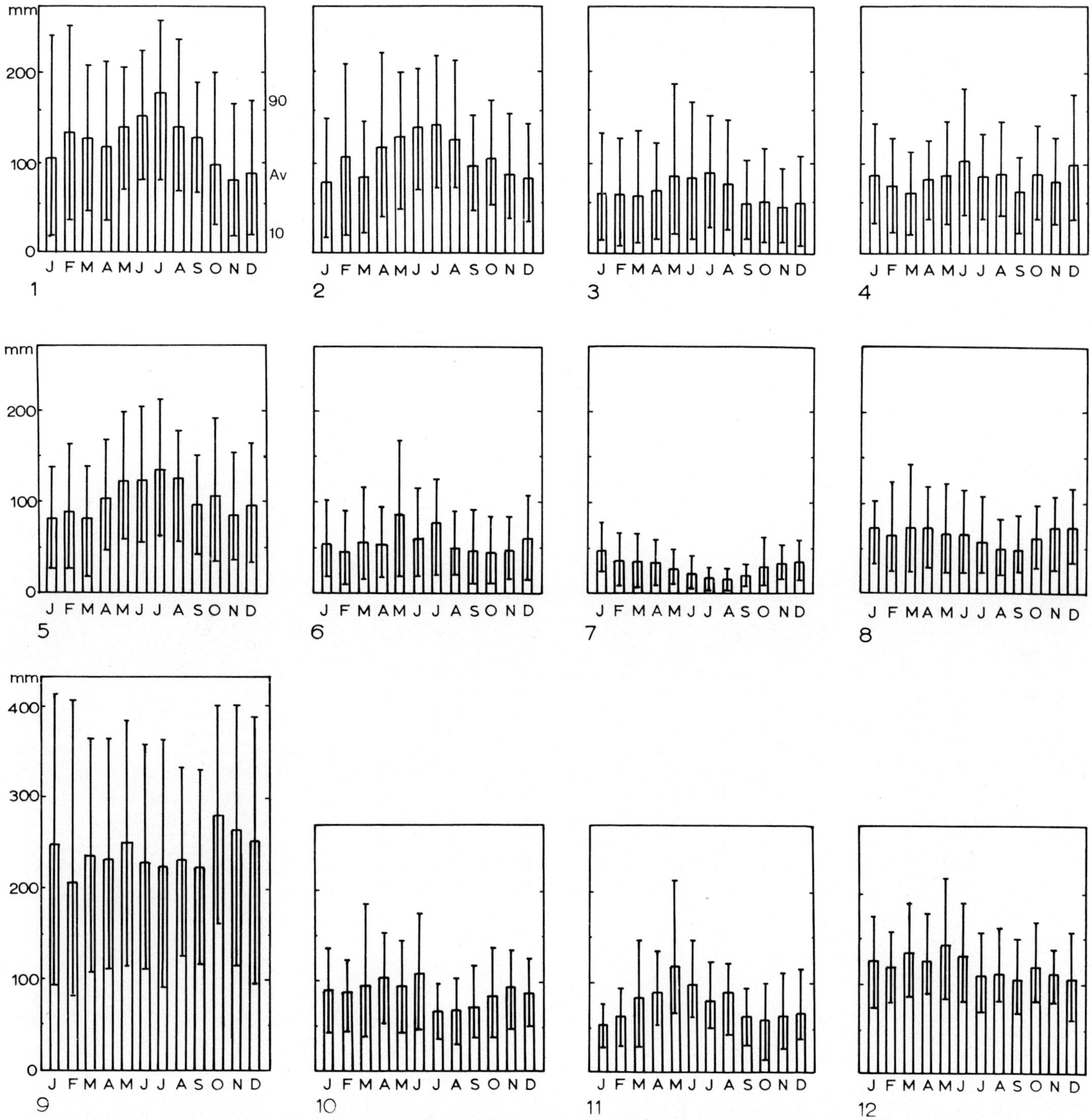

Fig. 9. Average precipitation and amount not exceeded in 10 percent and ninety percent of years. (1) Raoul Is. 1951–70, (2) Auckland 1910–69, (3) Hastings 1892–1966, (4) Palmerston North 1928–69, (5) Wellington 1928–69, (6) Christchurch 1894–1969, (7) Alexandra 1922–69, (8) Dunedin 1918–69, (9) Hokitika 1866–1965, (10) Invercargill 1939–69, (11) Chatham Is. 1951–70, (12) Campbell Is. 1951–70.

relationships are found, coefficients of variation from 0.8 to 0.9 have been found in stations in the north and east of North Island in late summer months and between 0.35 and 0.5 in western and southern districts in winter and spring months. In the North Island variability tends to be greatest in summer and least in winter months; in most of the South Island greatest in summer and autumn (but in July in Canterbury), and least in spring.

There is a slight persistence in monthly rainfalls, such that (for a typical New Zealand station) while the probability of any month having rainfall below average is about 0.55 if the preceding month was 'wet', it is nearly 0.60 if the preceding month was 'dry'. Runs of 'dry' months are appreciably more common than corresponding runs of 'wet' months.

There are few areas in which there are not periods without rain long enough to cause a reduction of transpiration from most vegetation and to dry out the soil and vegetation to such an extent that there is a risk of fire. The possible exceptions are the mountains, particularly the Fiordland area and the west of the Southern Alps.

On average, there is not a very large difference between rainfall at different times of day and night. The north has generally a maximum during the afternoon resulting from convective activity (e.g. Ruakura, in the Waikato Valley, gets only about 42 percent of its summer rainfall at night – 9 p.m. to 9 a.m.). Districts round Cook St. and the west and south of the South Island tend to have a rainfall maximum at night (e.g. Wellington, Hokitika and Invercargill get between 54 and 57 percent at night in spring and summer).

Evaporation and evapotranspiration, water balance and agricultural drought

Tank evaporation has been measured at a number of places in New Zealand. Annual evaporation from open water, derived from tank data by applying empirical factors is shown schematically in figure 10a. Monthly average values of evaporation from the standard raised pans (U.S.W.B. Class A type) range from approximately 150 mm to above 200 mm in January and from about 10 to 50 mm in June and July.

Evaporation from a well watered continuous short vegetation cover (potential evapotranspiration) varies from about 100–150 mm per month in summer to about 30 mm in winter months in the north of the country and to negligible amounts inland in the South Island in mid winter. (Fig. 10b, c).

Evaporation is to a large extent a function of the radiation energy available, and hence is great in the areas which have high global radiation. (These are also areas of high temperature maxima and much sunshine.) Evaporation is also influenced by the wind and relative humidity of the air however, and for this reason evaporation is augmented in some of the

Fig. 10. a. Average annual open water evaporation – mm, b. Average potential evapotranspiration, January – mm, c. Average potential evapotranspiration, July – mm. (Open water evaporation estimated from tank evaporation data; potential evapotranspiration calculated by means of 'Penman' formula.)

eastern areas which are windy as well as sunny. Lake Grassmere, near the northeast extremity of the South Island for example has been chosen as the site for evaporation of sea water to produce salt. Similarly the foothill ranges to the lee of the main divide frequently experience strong föhn winds with low humidity and much sunshine, and in these areas at these times the potential evaporation is likely to be unusually great. The annual totals however do not appear to be excessive. Evaporation is probably low on average in the higher mountain areas.

For the most part only indirect estimates of potential and actual evapotranspiration are available. Estimates of potential evapotranspiration made from mean temperature data by means of the well known 'Thornthwaite' formula have been widely used. They give a fairly satisfactory representation of average seasonal trends but they are not satisfactory for detailed comparisons between different periods or places. More reliable estimates by means of 'combination' type formulae (e.g., PENMAN's) which make use of more extensive climatological data (radiation, as measured or as calculated from sunshine duration; wind; humidity; temperature) are available for a limited number of places. The 'Penman' estimates for summer months are as much as 50 percent greater than 'Thornthwaite' estimates for some stations in the dry eastern and central areas. (COULTER, 1973).

When averaged over periods of a month or even of one or two weeks, evaporation from water surfaces and potential evaporation from vegetation do not generally vary greatly, and it is thus possible to use average evaporation values in water balance calculations as the major changes result from the variability of the rainfall. Hourly or daily rates of evaporation of course vary widely depending on prevailing weather.

Comparisons of average monthly values (of precipitation and 'Thornthwaite' potential evapotranspiration) which show periods of water surplus, soil moisture withdrawal, water deficiency (i.e. rainfall plus available soil moisture less than potential evapotranspiration), and soil moisture recharge for individual stations have been widely used to characterise climates. An arbitrary allowance of 100 mm for soil moisture storage capacity has generally been adopted, and on this basis limited areas in New Zealand show water deficits. These are small areas near Gisborne, Napier, Palmerston North and Masterton in the North Island and near Nelson and Blenheim in the South Island, and a large area in the South Island which includes the Canterbury Plains and most of lowland Otago except the far west and south (GARNIER, 1951).

This method of analysis has been found especially useful for correlation with soil types. In New Zealand the areas with the driest climatic types, ('Semi-arid Mesothermal') are characterised by the 'brown-grey earths' zonal soil group; yellow-grey earths are found in the slightly moister 'Sub-humid Mesothermal' areas; and at the other extreme, podzols are found in the 'Superhumid' regions. In the intermediate 'Sub-humid' and

'Humid Mesothermal' climatic classes, mainly yellow-brown earths are represented. Thus the climatic classes are for the most part related to the main zonal soil groups. For some classes however the relationship is rather broad, and pedologists have found it necessary to modify the limiting values of the thermal and moisture indices used to define several of the climate classes to reach a satisfactory correlation with the New Zealand soil groups (Cox, 1968).

Besides representing an index of leaching in the soil forming process, the amount of any water surplus provides a rough measure of the average 'runoff' available for long term ground water recharge and stream flow. Most areas of New Zealand show a surplus at some season of the year, and most of the country ('Humid B' and 'Superhumid' areas) a surplus in most months.

In much of New Zealand while the average monthly rainfalls indicate little or no moisture deficiency, the variable incidence of rainfall is such as to bring about more or less serious deficits in some years and surpluses in others, or even at different times in the same season. These short term changes are important in agriculture and probably in some natural eco-systems also, but are not revealed in the mean climatic data. They may be examined for broad comparative purposes by means of a simplified water balance model to estimate soil moisture changes. Assuming a constant soil moisture capacity, a standard rate of withdrawal of soil moisture (potential evapotranspiration) so long as any soil moisture remains, and water input as given by measured daily precipitation amounts, 'agricultural drought' is defined as commencing whenever calculated soil moisture is exhausted and ending when a day's rainfall in excess of the daily potential evapotranspiration is recorded.

For broad comparisons, taking 75 mm as a more generally appropriate value for soil moisture capacity and using 'Penman' estimates of average monthly potential evapotranspiration, it is found that in most parts of New Zealand the incidence of daily rainfall is such that periods of agricultural drought occur, but not in all years. Figure 11a gives schematically the average number of days per year. It occurs most frequently of course in places with lowest annual rainfall – in Central Otago on up to 200 days in a year, average 150 – but even areas which support a strong dairying industry are subject to such dry periods, e.g., Ruakura has 45 such days per year on average and has had 94 in one season. The 1969–70 season was one of widespread drought. There were 60 days of agricultural drought at Ruakura, and dairy production in the district was reduced to about 90 percent of average.

On the same basis, figure 11b shows the average amount of water deficit or irrigation need, figure 11c the frequency percent of years in which deficits occurred, and figure 11d the average annual water surplus (rainfall minus calculated evapotranspiration) available for deep percolation and runoff.

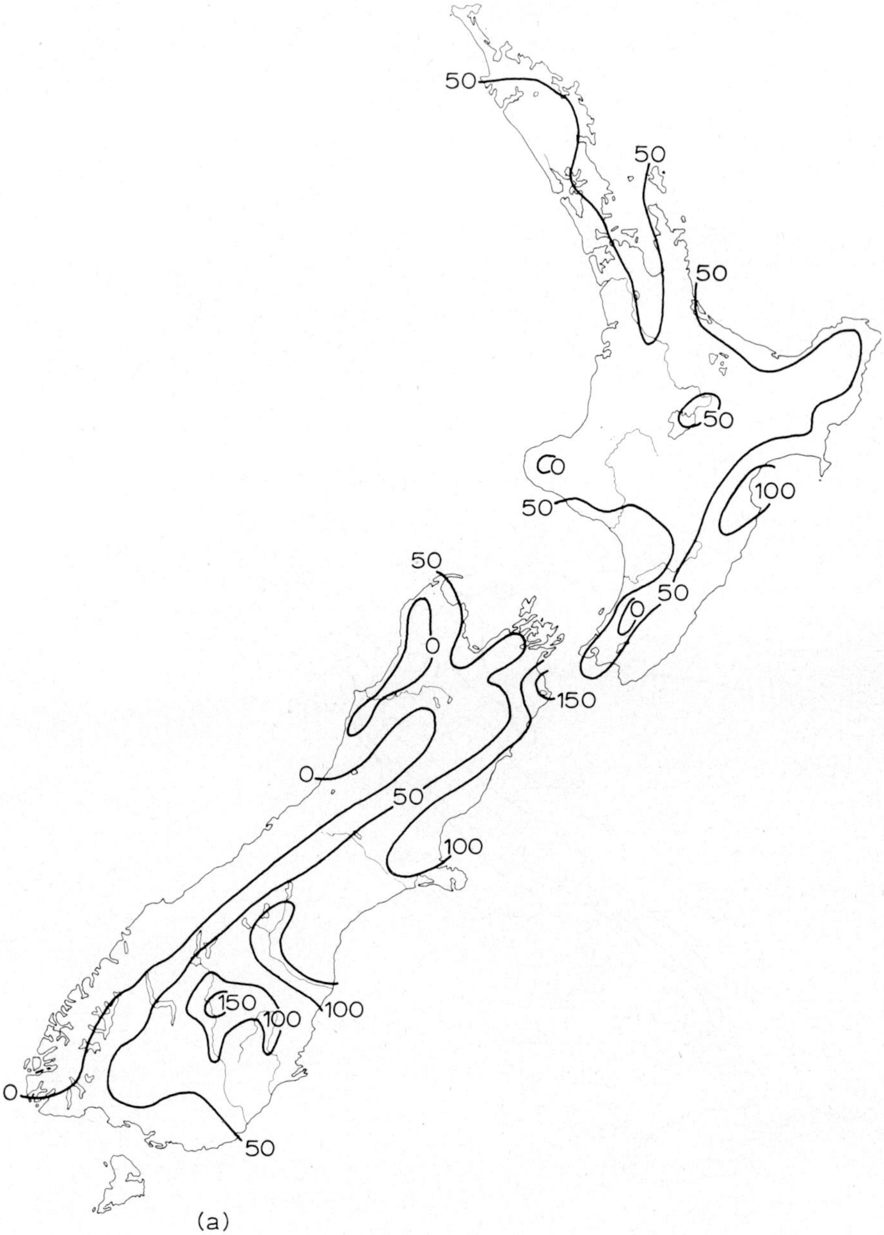

Fig. 11.

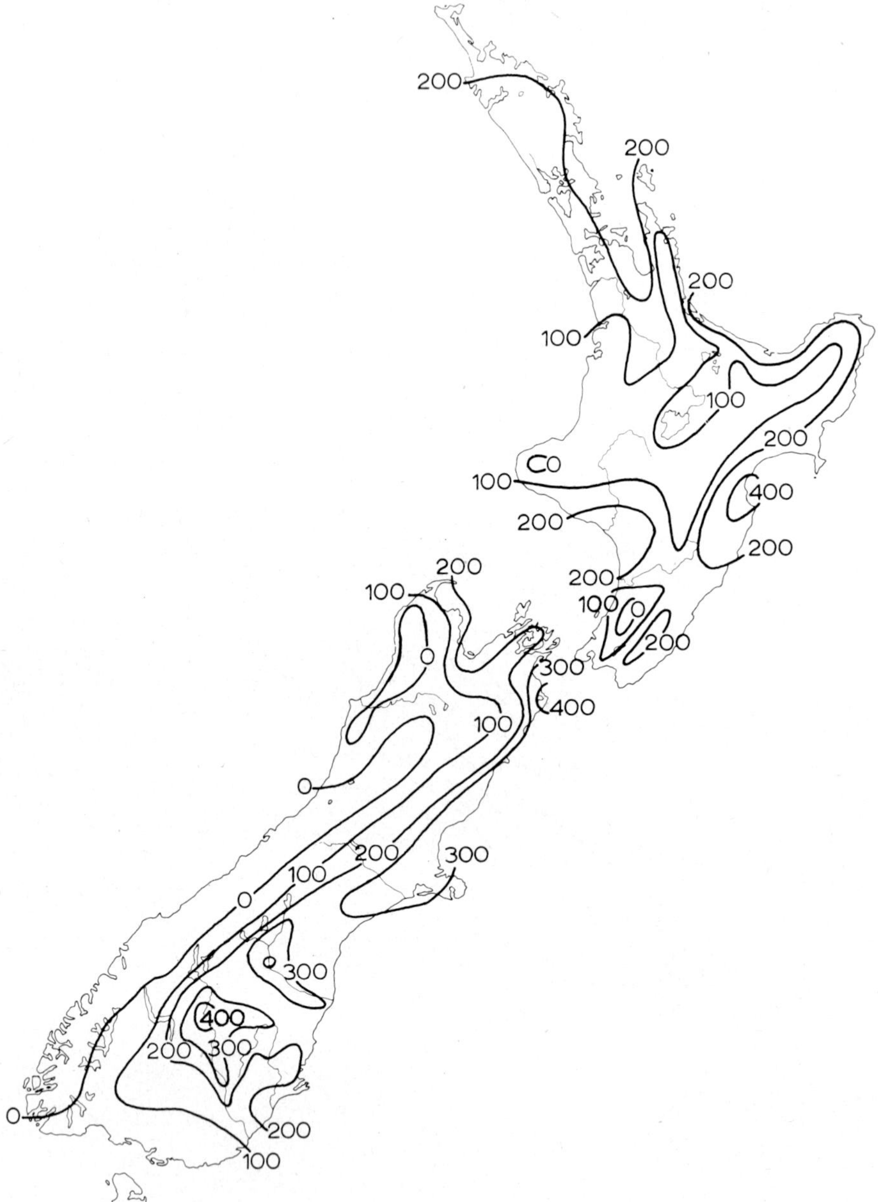
200
200
200
100
100
200
C0
100
400
200
200
200
100
200
100
0
200
0
300
400
100
0
300
200
100
0
300
400
200
300
0
200
100

(b)

Fig. 11.

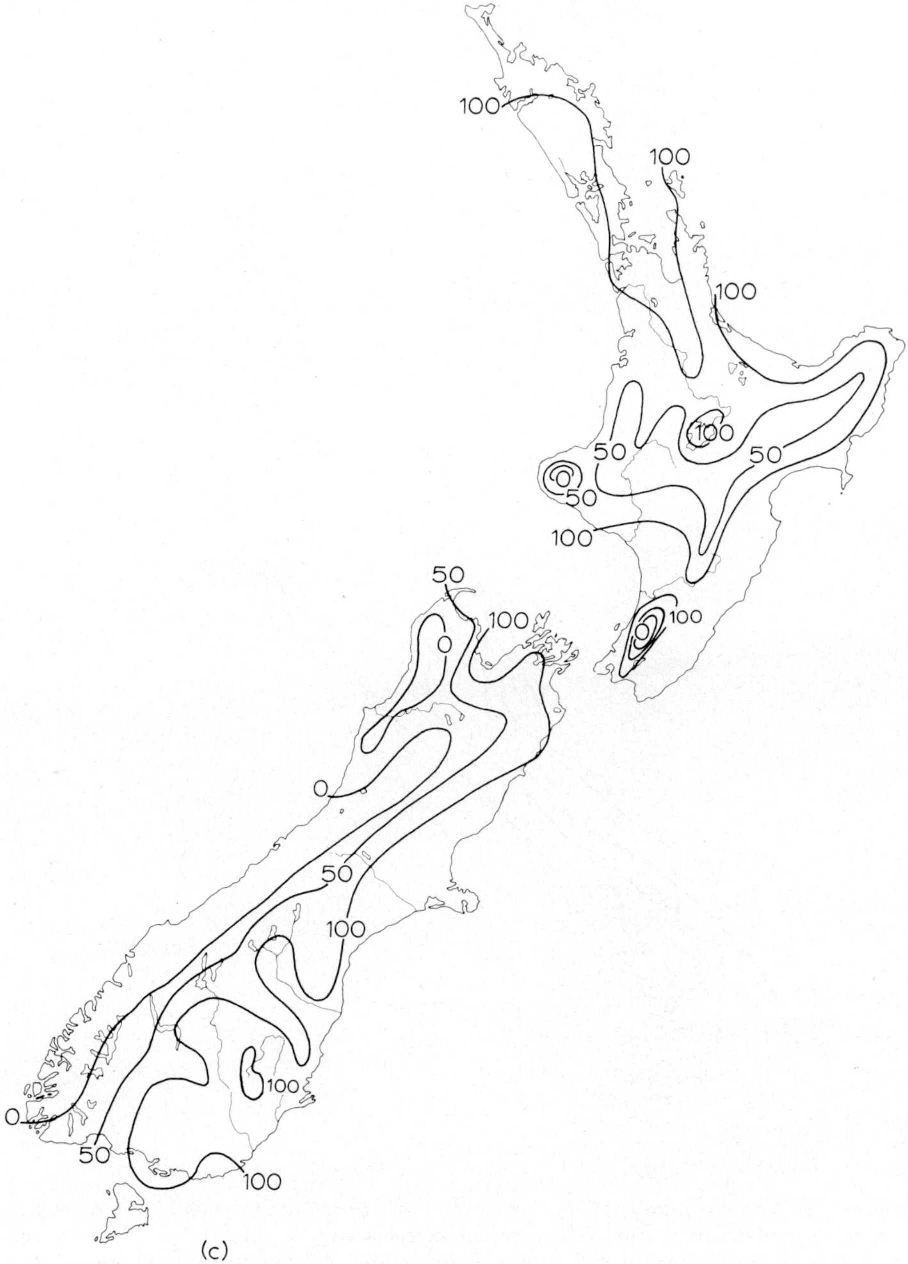
100
100
100
100
50
50
50
100
100
50
100
0
100
0
0
50
100
100
0
50
100
(c)

Fig. 11.

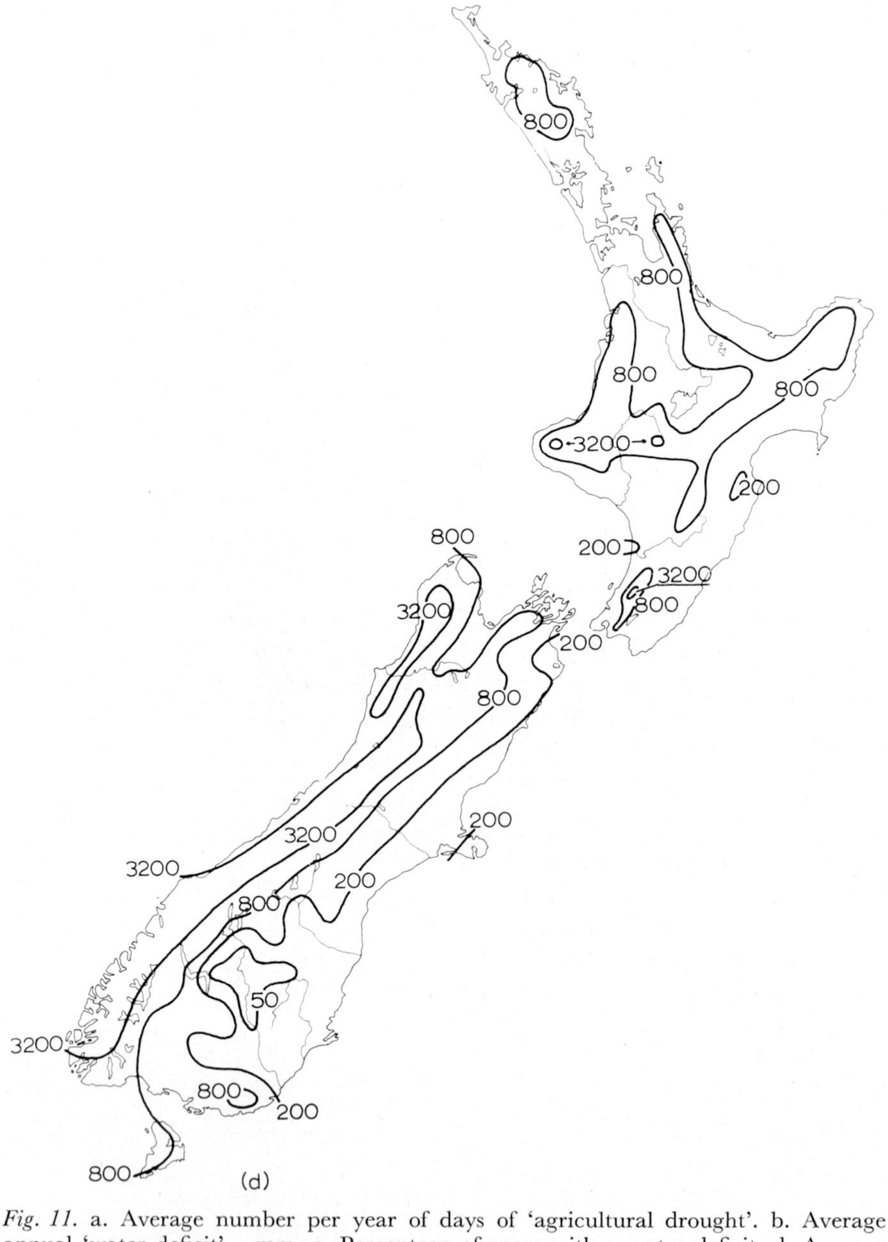

Fig. 11. a. Average number per year of days of 'agricultural drought'. b. Average annual 'water deficit' – mm. c. Percentage of years with a water deficit. d. Average annual water surplus or 'run-off' – mm. (Based on data calculated from a simple water balance model using daily precipitation data. Evapotranspiration as calculated (AE) takes place at the average monthly 'Penman' potential evapotranspiration rate (PE) until 75 mm of soil moisture are used; it then ceases and 'agricultural drought' and 'water deficit' (PE-AE) begin. 'Run-off' is calculated as the surplus of precipitation in excess of that used to provide for AE and recharge of soil moisture.)

The choice of parameters was intended to provide an agriculturally useful drought index. A smaller deficiency of rainfall – or shorter period without rain – may be important for small and shallow rooting pasture or crop plants. On the other hand trees and most native vegetation survive more extreme water shortages without much apparent effect.

Snow

At sea level in the North Island and in the north and west of the South Island snow falls only very rarely – once in one to three years on average in the south and virtually never in the far north. In east and south coast areas of the South Island, it falls on a few days in the year, and in some years lies for a few days even at sea level. The North Island has a small permanent snow field above about 2500 m on the central plateau, but snow rarely descends below 500 m even for brief periods in winter. In the South Island the snow line is around 2000 m in summer, being somewhat higher in the north and east, and lower in the south and on the western side of the Southern Alps. In winter and early spring there is persistent snow cover above about 1500 m in the North Island or 1000–1300 m in the South Island. In some seasons there may be fairly lengthy periods of snow cover at heights as much as 700 m lower than these.

Many observers have suggested that snow cover in the mountains was greater some 50 to 100 years ago but there are no detailed records to substantiate this. The fact that glaciers have receded markedly throughout New Zealand since the late nineteenth century lends some support to the suggestion. The frequency and amount of snow varies greatly from winter to winter. Wellington city does not appear to have had another fall as great as those recorded in 1849, soon after the foundation of the settlement, when there were several inches of snow lying on the hills around the town. The winter of 1895 was a disastrous one for many of the pastoralists of the South Island high country from Otago to Marlborough; widespread deep and persistent snow cover caused extremely heavy losses of sheep. On the other hand some recent snow falls such as that of November 1967 in Canterbury illustrated in figure 2c appear to have been as heavy and extensive for the time of year as any previous ones on record (Tomlinson, 1970).

Relative humidity, dew and fog

Relative humidity by day is commonly between 70 and 80 percent in coastal areas and about 10 percent lower in inland areas. (Table 3). Very low humidities – below 30 percent – occur at times in the lee of the ranges and, less frequently, at high mountain sites. In the lee of the Southern Alps the föhn effect is often very marked and in summer the hot dry Canterbury nor'wester is generally an unpleasant wind which

Table 3. Average Relative Humidity – Percent.

	January		July	
	3 a.m.	3 p.m.	3 a.m.	3 p.m.
Auckland (Mechanics Bay)	85	63	90	74
Gisborne	89	62	90	72
Ohakea Aerodrome	87	62	89	72
Wellington	89	71	87	77
Christchurch	87	61	89	73
Hokitika	95	78	93	75
Invercargill	90	72	91	78

causes a rapid drying out of vegetation and soil. On mountain tops, low humidity occurs with descent of air from high in the tropopause during periods of broad scale atmospheric subsidence. In the north of the North Island humid midsummer weather is occasionally oppressive though temperatures rarely reach 30 °C. Dull cool humid spells are generally not prolonged anywhere except in the hills exposed to ocean winds, but their frequency shows a marked increase in the south.

Western mountains in particular are subject to long periods of high humidity and mist which contribute towards an environment favourable for epiphytic mosses and lichens in upland forests and the development of bogs. In the lowlands the duration of conditions of saturated or nearly saturated air, or of the presence of liquid water (fog, dew) affects in particular the spread and development of fungi including pathogens. Detailed knowledge of these conditions which may often be needed on a micro-climatic scale, is not readily obtainable. The presence of dew at fixed times – usually at 9 a.m. – is recorded at some climate stations, but there have been few attempts to record the duration of dew or leaf-wetness.

Fog is often local, occurring principally at night and in the early morning as radiation fog, but sea fogs sometimes spread on to coastal areas, especially about Cook Strait and the east coast of the South Island. Morning fogs, sometimes persisting till mid-day, are a prominent feature of the winter climate of some sheltered lowland areas, e.g., parts of the Waikato and Wanganui Valleys in the North Island and in some of the west coast valleys in the South Island and east and south coastal plains. (e.g., Taumarunui, upper Wanganui River, has 100 days of fog per year, 15 in July, on average; Murchison, Buller Valley, South Island, probably has as many; and stations in Auckland, Hamilton, Christchurch and Invercargill urban areas average more than 45 days per annum). Some of the exposed mountain ranges (e.g., western Tararua Range, Wellington Province) probably exceed 200 days per year of fog or drizzle (at heights above 1000 m) but few detailed observations are available. Eastern ranges in general are less subject to fog, e.g., at Makahu Saddle at 1006 m in the Hawkes Bay ranges, there are only 24 days of fog, (but at Black Birch Range, Marlborough 175 days at 1396 m). Many stations in the lowlands have fewer than 15 days with fog per year.

Surface wind

Over the oceans around New Zealand and in the free air over it winds from a westerly quarter prevail at all seasons. Over the land surface, considerable local modifications to the general flow occur during its passage across the mountainous terrain, but nevertheless in most places westerlies still predominate. On approaching the main ranges the westerly wind flow turns towards the northeast, and southwesterlies are more frequent west of the ranges. On crossing the ranges and passing through Cook Strait the wind swings towards the southeast, giving rise to a preponderance of northwesterlies in inland districts of the South Island and about Cook Strait, and strong gales from this quarter are often experienced in these areas, especially in late spring and early summer. On the east coast, however, easterlies or northeasterlies are almost as common as the prevailing southwesterlies mainly as a result of daytime sea breezes during the summer, augmented by the deflected northerly flow through Cook Strait and by the low pressure trough which develops over the plains in the lee of the mountains. Cook Strait, the only substantial gap in the mountain chain, acts as a natural funnel for the air flow and is thus a particularly windy area with occasional gales from the southeast as well as from northwest. A similar funnel effect is found in Foveaux Strait (south of South Island) where the general westerly airflow is increased, and on a smaller scale also in many mountain gaps and valleys. North of Taranaki the general air flow is more southwesterly, and there is a noticeable reduction in windiness in summer.

As might be expected, the wind speeds are generally much less in low lying inland areas than about the coasts (but stronger on open hill and mountain tops). The proportion of calms is much greater at night and in winter than at other times, especially in inland areas.

The mountainous terrain of many areas results in gusty turbulent winds, rather than a particularly great average wind speed. Wellington Airport for example has over 41 days annually with gusts exceeding 27 m/s, and an overall average wind speed of 7.7 m/s (cf. Invercargill 10 days and 4.5 m/s respectively).

In most places there is a simple diurnal variation of wind speed with maximum in early afternoon and minimum after midnight. (Table 4).

Wind frequency diagrams are illustrated in figure 12 for selected stations. Kaitaia in the far north gets an appreciable proportion of easterlies – it is north of the mean high pressure axis in summer. Rotorua,

Table 4. Average surface wind speed – m/s: (a) 03 hrs, (b) 15 hrs, (c) average of 24 hours daily. (based on hourly data from anemometers about 10 m above ground).

	Jan.	Feb.	Mar.	Apr.	May	Jun.	Jul.	Aug.	Sep.	Oct.	Nov.	Dec.	Year or Mean
Kaitaia 1962–1969													
a.	3.1	3.5	3.6	4.0	3.5	4.0	4.4	4.1	3.8	3.5	3.9	3.3	3.7
b.	6.6	6.6	6.3	6.4	5.8	6.0	6.4	6.5	7.0	6.7	7.4	6.7	6.5
c.	4.5	4.5	4.5	4.6	4.1	4.6	4.9	4.9	4.8	4.6	5.3	4.7	4.7
New Plymouth 1962–1969													
a.	3.1	2.9	3.3	3.9	4.2	4.6	4.5	4.9	4.2	4.0	4.1	3.2	3.9
b.	5.9	5.3	5.0	5.3	4.9	5.7	5.6	6.1	5.8	6.1	6.7	5.8	5.7
c.	4.2	3.7	3.8	4.3	4.3	4.8	4.8	5.3	4.7	4.6	5.4	4.4	4.5
Rotorua 1964–1969													
a.	2.1	2.0	1.8	2.1	2.3	2.2	2.4	3.0	2.6	2.3	2.9	2.3	2.3
b.	5.2	4.9	4.6	4.6	4.4	4.1	4.6	4.7	5.2	5.3	6.0	5.5	4.9
c.	3.5	3.3	3.0	3.1	3.0	2.9	3.0	3.5	3.7	3.7	4.3	3.7	3.4
Wellington 1963–1970													
a.	7.1	6.5	6.0	7.1	6.8	7.1	6.6	7.4	7.0	7.5	7.6	6.5	7.0
b.	9.4	9.2	8.4	8.4	8.0	8.3	8.0	8.5	8.5	9.5	10.2	9.3	8.8
c.	8.0	7.5	7.1	7.5	7.3	7.5	7.1	7.7	7.5	8.2	8.8	7.7	7.7
Nelson 1962–1969													
a.	2.2	2.1	2.0	1.7	1.9	1.4	1.5	2.0	2.0	2.5	2.8	2.5	2.1
b.	6.7	6.4	5.4	4.8	3.8	3.1	3.3	4.4	5.4	6.3	7.4	6.8	5.3
c.	4.0	3.7	3.0	2.7	2.4	1.9	1.9	2.7	3.0	3.7	4.6	4.0	3.1
Hokitika 1964–1968													
a.	2.5	1.4	2.2	2.2	2.5	2.4	2.3	2.6	2.3	2.4	3.2	1.8	2.3
b.	5.0	4.6	4.2	4.0	3.0	2.6	2.6	3.8	4.5	5.2	5.6	4.9	4.7
c.	3.5	2.8	2.8	2.7	2.5	2.5	2.4	2.8	3.0	3.6	4.0	3.0	3.0
Christchurch 1960–1968													
a.	3.2	3.0	2.9	2.7	2.6	2.4	2.7	2.7	2.9	2.8	2.9	3.1	2.8
b.	6.8	6.6	5.9	5.3	4.3	3.8	4.5	4.8	6.0	6.4	7.1	6.8	5.7
c.	4.7	4.3	4.0	3.5	3.2	2.7	3.2	3.3	4.0	4.2	4.6	4.6	3.9
Invercargill 1960–1968													
a.	3.9	3.6	3.4	3.5	3.7	3.8	2.9	2.9	3.5	4.5	4.5	3.5	3.6
b.	7.7	6.9	6.2	6.0	4.8	4.9	4.0	5.0	6.4	7.8	8.1	7.3	6.2
c.	5.4	4.9	4.4	4.3	3.9	4.1	3.1	3.7	4.4	5.8	6.1	5.0	4.6

an inland station, has a large proportion of calms and few strong winds. At Ohakea westerly air flow is locally diverted to northwesterly. At Wellington the influence of Cook Strait is predominant – winds almost invariably blow through the strait from the north or from the south – any other winds are light. Christchurch on the east coast of the South Island has many easterly sea breezes. Invercargill has a preponderance of westerlies caused by funnelling through Foveaux Strait.

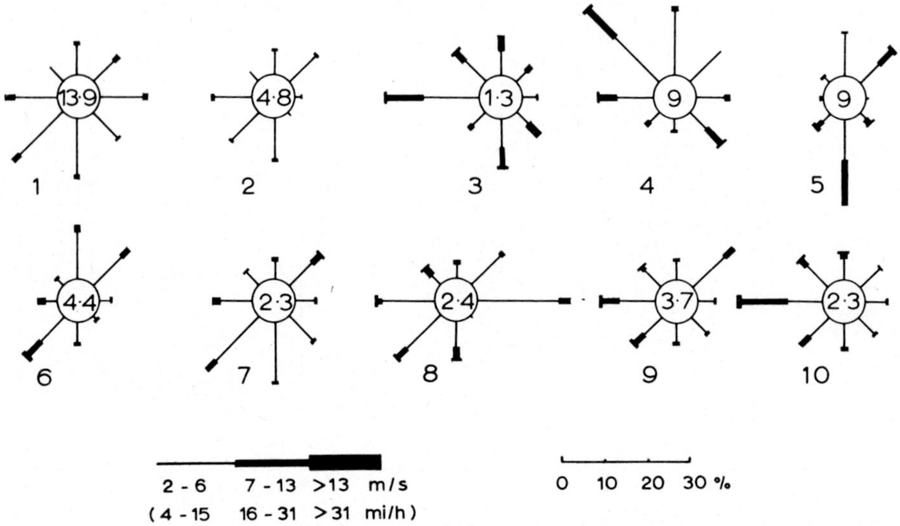

Fig. 12. Average frequency (percent) of surface wind speed and direction and of calms (in circle). (1) Kaitaia, (2) Rotorua, (3) New Plymouth, (4) Ohakea, (5) Wellington (Moa Point), (6) Nelson, (7) Westport, (8) Christchurch (Harewood), (9) Taieri (near Dunedin), (10) Invercargill.

Air temperature

Mean annual temperature at sea level ranges from over 15 °C about and north of Auckland to less than 10 °C in Southland. At mountain stations at about 1000 m in the northern half of the North Island it is about 9 °C, and in the south of the South Island about 6 °C. This represents an overall lapse rate of approximately 5 °C per 1000 m.

Mean monthly temperatures (Fig. 13, 14) are averages of mean daily maxima and mean daily minima measured in a standard Stevenson screen at 1.5 m above ground at 9 a.m. local time. The major influence on temperature is altitude, and the mountains make representation of actual temperatures difficult. The isopleths on the maps refer to the level of the stations for which data are available, and they are mainly on the coastal lowlands or in valleys, the great majority being below 500 metres above sea level. The isotherms as drawn are generalised and refer to altitudes up to about 1000 m.

July is the coldest month and January or February which are nearly equal, the warmest. Coastal districts exposed to the west or southwest have a relatively small range of temperature between winter and summer (7–9 °C). Inland and to the east of the mountains the range is much greater (11–14 °C).

Similarly, the mean daily range has lowest values of about 7 °C in

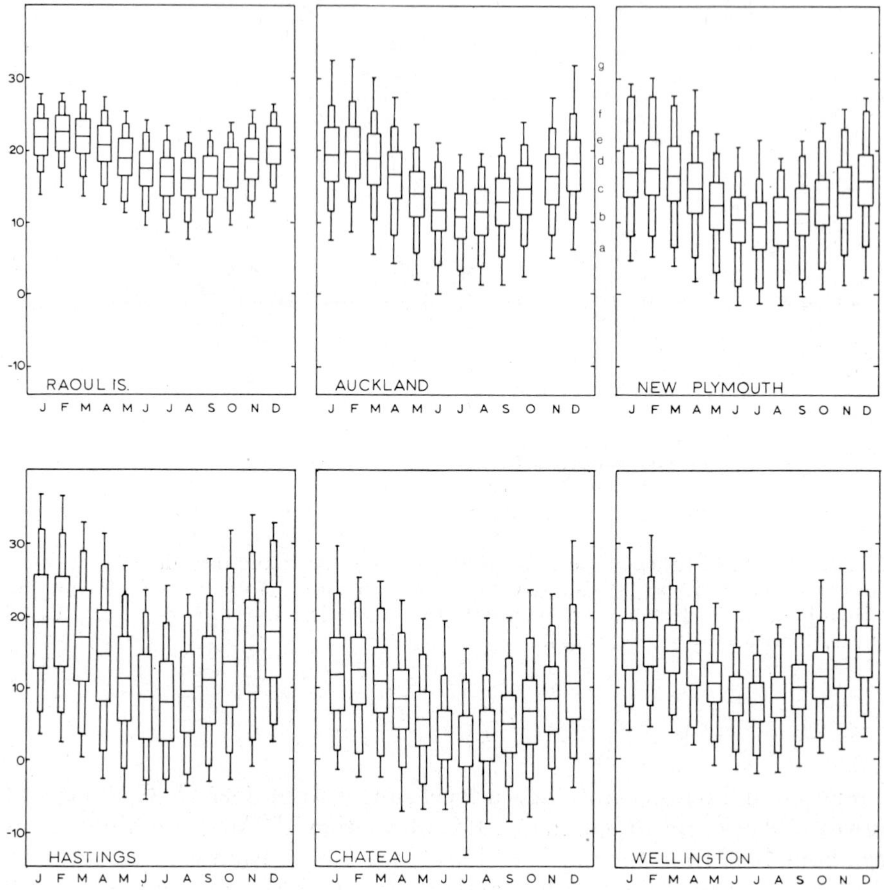

places exposed to wind from the sea, and is greatest (up to 14 °C) in inland and eastern areas (Figs. 13, 15).

Maximum temperatures show little latitudinal variation (Fig. 15). Stations east of the main mountains and in the far north occasionally have maxima in excess of 30 °C. The highest air temperatures recorded in New Zealand until recently were 38 °C, at Ashburton, Lincoln, Roxburgh and Gisborne, but in February 1973 most stations in Canterbury exceeded this and a new record of 42 °C was established at Christchurch, Rangiora in North Canterbury and Jordan (in the Awatere Valley, Marlborough). East of the main ranges and in parts of Auckland Province temperatures exceed 30 °C several times a year on average, the number reaching 10 per year in Central Otago, near Christchurch and near Napier. In some western areas of both islands and at several high level stations a temperature of 30 °C has never been recorded. Figures

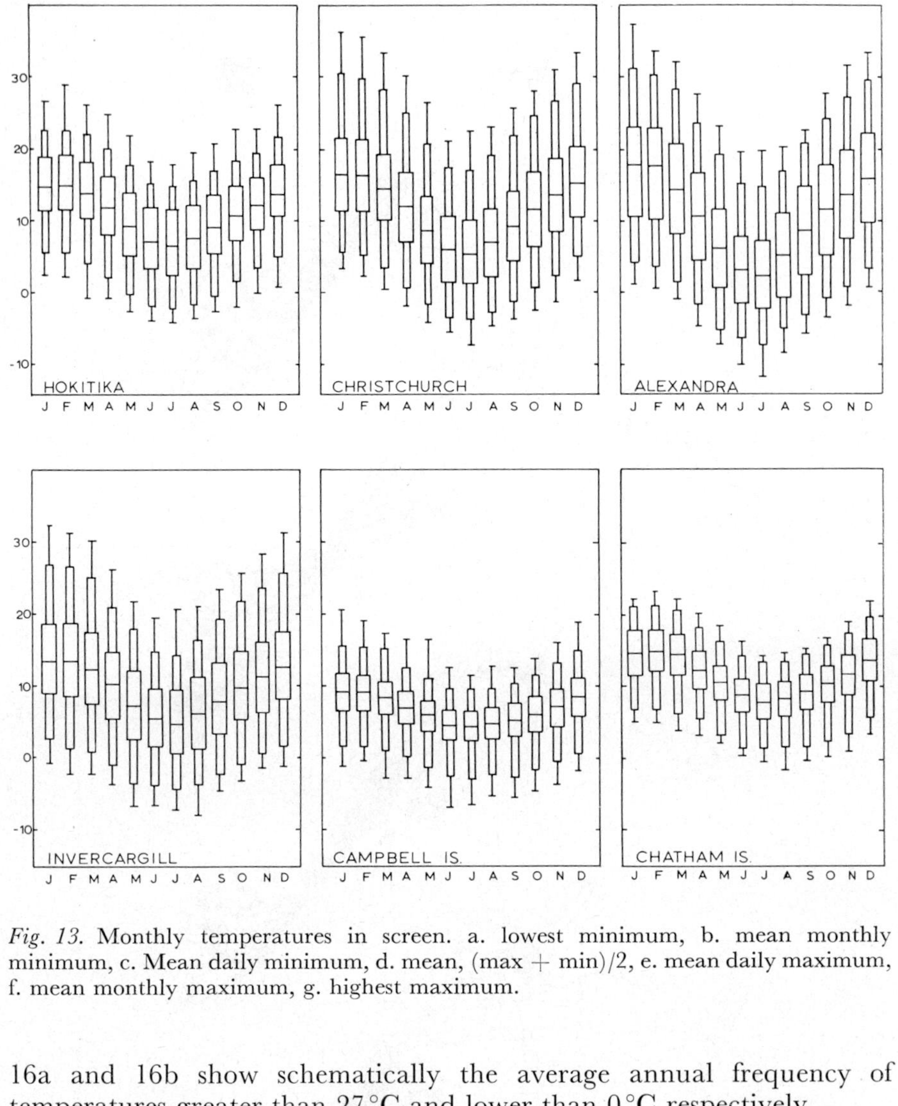

Fig. 13. Monthly temperatures in screen. a. lowest minimum, b. mean monthly minimum, c. Mean daily minimum, d. mean, (max + min)/2, e. mean daily maximum, f. mean monthly maximum, g. highest maximum.

16a and 16b show schematically the average annual frequency of temperatures greater than 27 °C and lower than 0 °C respectively.

The extreme low temperatures on record have been at inland basins in the South Island at moderate heights above sea level. The lowest air temperature so far recorded was —19 °C at Ophir (305 m) in Central Otago. Temperatures below —12 °C are very rarely recorded – in Central Otago, the inland plains and basins of Canterbury and Marlborough and on the central plateau of the North Island. No measurements are available from sites at very high altitudes where still lower temperatures might be expected.

An index of agricultural potential in a temperate climate with respect

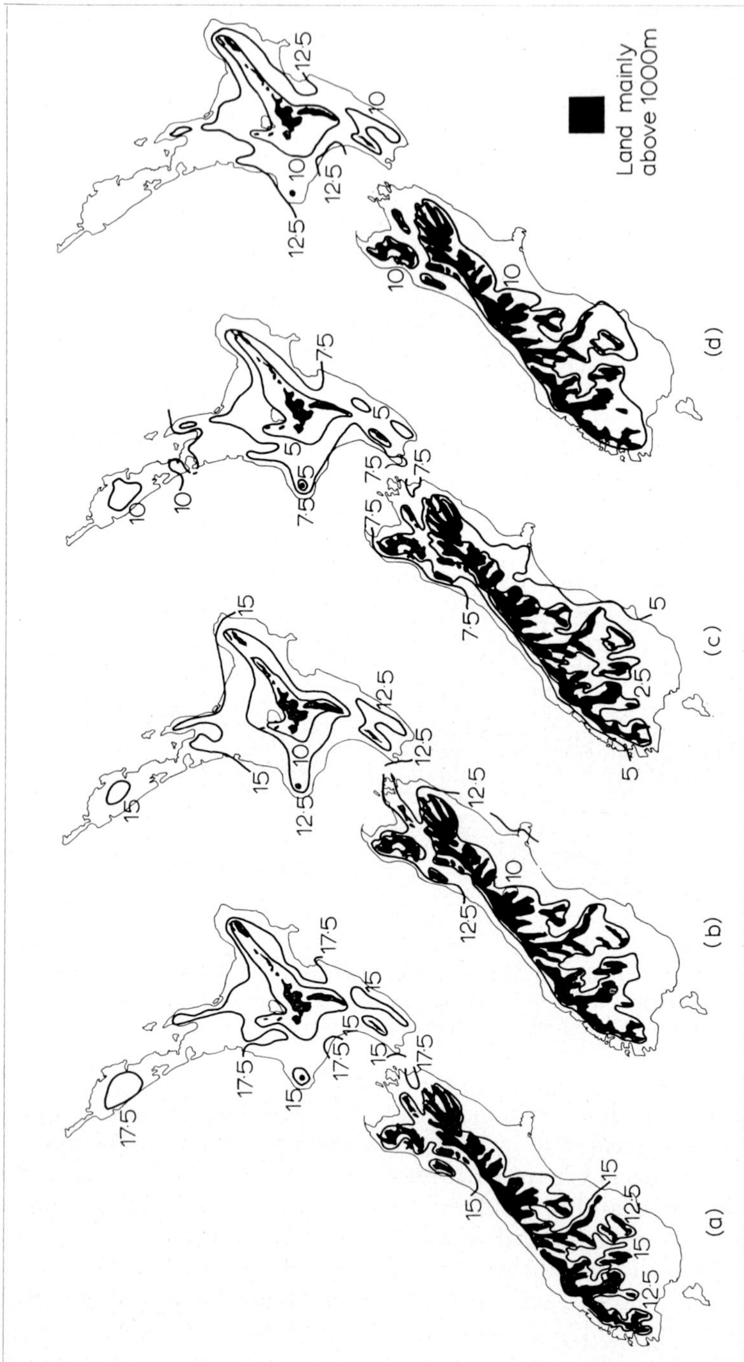

Fig. 14. Mean daily temperatures, (max + min)/2, in screen – °C. a. January, b. April, c. July, d. October.

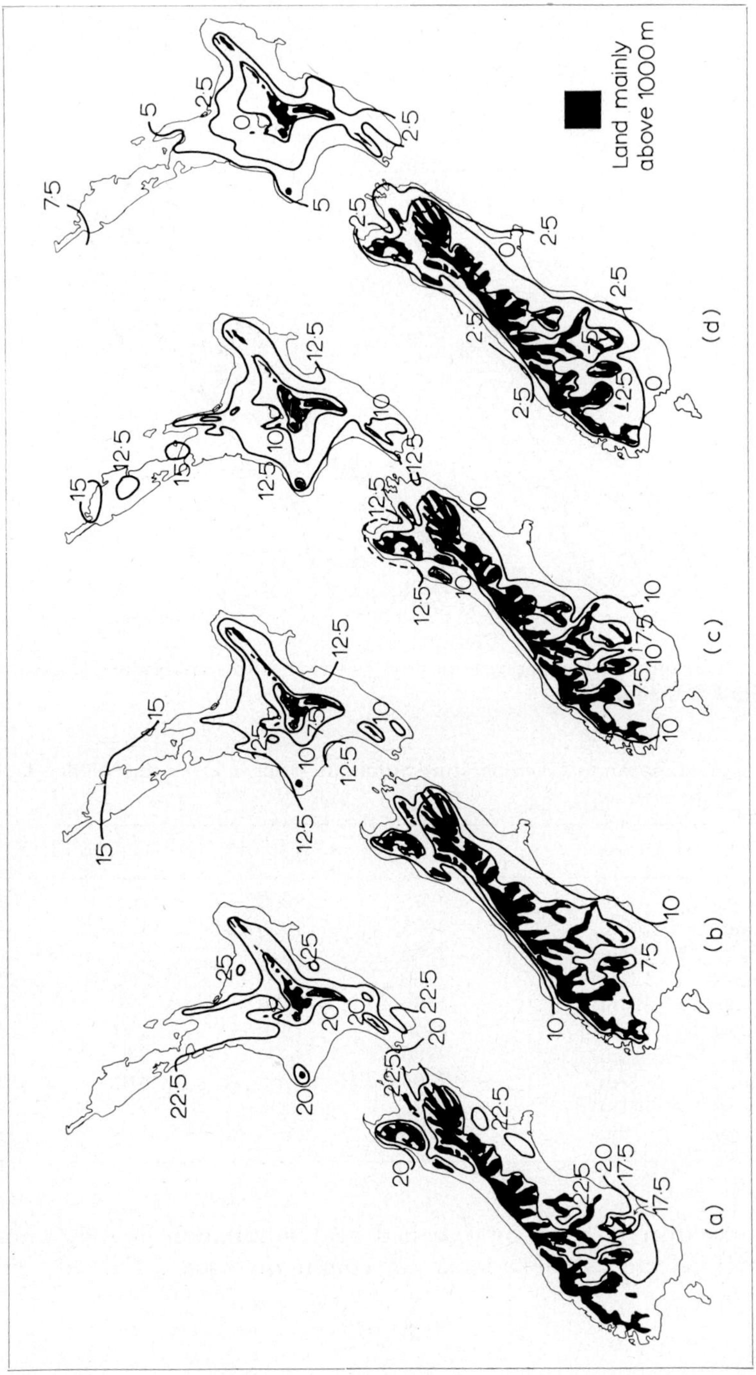

Fig. 15. Mean daily temperatures in screen – °C. a. January maximum, b. July maximum, c. January minimum, d. July minimum.

Fig. 16. Average number per year of days per year with temperature a. greater than 27 °C, b. less than 0 °C.

Table 5. Average Annual Temperature Summation above given thresholds – C degree-days.

Threshold temperature		5 °C (41F)	10 °C (50F)	12.8 °C (55F)	15 °C (59F)
Kaitaia	(80m)	3875	2060	1165	630
Auckland	(49m)	3765	1970	1115	615
Ohakea	(48m)	2965	1310	665	320
Wellington	(127m)	2780	1145	524	220
Molesworth	(893m)	1465	490	191	65
Christchurch	(30m)	2440	1070	550	280
Hokitika	(4m)	2280	790	298	100
Dunedin	(2m)	2205	785	318	125
Alexandra	(141m)	2205	990	519	270
Invercargill	(0m)	1755	515	175	60

to temperature is given by so-called heat summation figures. These are given for selected stations for a few commonly used threshold values in Table 5.

Frost

Frostiness is locally variable, even within small areas. Gently sloping land with a northerly aspect tends to be least affected by frost. At a number of stations, e.g., Albert Park in Auckland City, Kaitaia, Kelburn in Wellington City screen temperatures below 0 °C have been reported only once or twice or not at all in records of up to 50 years, while places a few miles away register several such frosts per year (Whenuapai near Auckland, 7; Wallaceville near Wellington, 35). The average number of days between first and last screen frost each year ranges from less than 50 in some coastal areas north of Auckland, about Hawkes Bay and Wellington and in Taranaki to more than 250 in central valleys and basins and on high plateaus of both islands. In some such localities, e.g. in Central Otago and the Mackenzie Basin of inland South Canterbury severe frosts occur on practically every clear night in winter. The horticultural areas around Tasman Bay, Nelson, are notable for an absence of late spring frosts although there are some 20 to 40 frosts per year.

Visibility, air pollution, atmospheric chemistry

Because of its turbulent winds, frequent rain, small land area and limited industrialisation New Zealand by and large enjoys clean air. This is evidenced by visibility observations, for example at lighthouses, visibility better than 9 km (5 nautical miles) was recorded at Mokohinau (Hauraki Gulf) on 92 percent of 3 hourly observations (90 percent in summer); at Centre Island (Foveaux St) 95 percent (93 in summer), and in the Cook St area, Brothers Island 83 percent, Cape Campbell 99 percent and Stephens Island 87 percent. In inland and urban areas visibility is often less than at coastal stations, but still good compared with typical European cities – Wellington at 6 a.m. had 83 percent of observations with 10 km ($6\frac{1}{2}$ miles) or better, Auckland 89 percent while Kew, England, had only 31 percent (at 7 a.m.). Nevertheless urban air pollution is often noticeable especially in the large cities of Auckland and Christchurch, and most noticeably after still 'inversion' nights in winter. Another source of visible haze is sea salt which gives a soft outline to the landscape, more especially in western districts, when the winds off the sea are strong. By the standards of northern hemisphere continental areas the visibility in these conditions (10–15 km) would still be considered good.

Measurements of air pollutants have been made in some of the cities, especially Auckland and Christchurch where the problem has been most severe, and in connection with industrial plants. With some exceptions levels of pollution are generally low to moderate compared with overseas standards, because source strengths are much lower. In a normal winter, parts of Christchurch experience levels of smoke pollution (150 $\mu g/m^3$ of air, approx.), comparable with parts of London, but sulphur dioxide concentration (150 $\mu g/m^3$ of air, approx.) is substantially less than in most urban areas of England. Pollution (smoke and SO_2) is usually negligible in summer. Auckland pollution is more uniformly spread through the year, but less on average (smoke, 60–70 $\mu g/m^3$; SO_2 50–120

$\mu g/m^3$ in industrial localities). Its source is industrialisation and motor vehicles rather than domestic. If trends continue sulphur dioxide concentration in some areas could soon reach significant levels by accepted standards. Locally, roadside measurements of carbon monoxide from motor vehicles have reached up to 117 ppm by volume with a maximum over eight hours of 18 ppm – comparable to those measured in major cities overseas. (N.Z. Board of Health Committee 1970).

Daly noted that air pollution had modified the distribution of byrophytes and lichens in and around Christchurch; some 45 species being found in the outskirts of the city but less than 20 near the centre of highest pollution. The level of pollution is not sufficient to cause measurable damage to trees and gardens. (DALY, 1970).

A locally important natural source of sulphur compounds in the atmosphere is geothermal activity near Rotorua, Taupo and Kawerau (Bay of Plenty) in the North Island.

Measurements of the salts precipitated with rainwater have been insufficient for general conclusions to be drawn about the various elements in New Zealand precipitation, but do show that the levels are similar to those in other maritime countries. The amount of solids decreases inland; within 10 km of the sea it may be more than 200 kg/ha/year on average, and possibly only 50–100 kg/ha/year further inland. Near the sea, and in strong wind periods, the ionic ratios were similar to those in sea water, and total salt much increased, suggesting that much of the salt comes directly as spray from the sea.

Comparison of recent data with measurements made last century near Christchurch in the 1880's and 1900's shows that no great change has taken place except in the city centre or near industrial plants. At Taita inputs from rainwater provided the net nutrient requirements of beech forest for most elements. MILLER (1961) obtained the following amount of the major elements in rainwater averaged over a 3 year period at Taita, in kg/ha/year: Na 59, Cl 116, S 8.4, N 2.8, Mg 11.2, Ca 7.3, K 6.4. Taita is about 7 km from the nearest sea (Wellington harbour) but the predominantly northwesterly winds have a path of some 16 km over land. Later studies showed lower inputs during a less rainy period, and revealed wide differences between different sampling points (CLARIDGE, 1970). Sulphur in rainwater at Otara, a suburb of Auckland averaged 13 kg/ha/yr from 1954–60, then fell to about half this amount following diversion of trade wastes from a nearby harbour inlet, but values gradually increased to 17 kg/ha/yr in 1968, presumably as a result of increased urban and industrial development (N.Z. Fertiliser Manufacturer's Research Assn.).

An appreciable amount of 'dry' wind blown salt appears to be deposited on vegetation, at least in areas near the coast. While the major input of soluble ions in precipitation is of oceanic origin other particulate matter in the atmosphere and deposited from it is of terrestrial origin. Wind blown dust has built up deep loess deposits or contributed to soil layers in many areas. Wind transport of dust is still very evident near the wide gravel river beds in Canterbury during strong northwest winds and no doubt deposition occurs in places. However destruction of vegetation has allowed accelerated wind erosion in other areas, particularly in the drier mountain and hill areas in the east of the country.

Continuous recording of concentration of carbon-dioxide and carbon monoxide has commenced recently to provide information on global atmospheric pollution. (N.Z.D.S.I.R., 1971).

Volcanic dust from Krakatoa and from South American volcanoes has been seen in New Zealand skies. More recently, dust from Mt Agung, Indonesia, which erupted in March 1963, was detected at about 18–25 km altitude over New Zealand in late April and was again present during the following two winters. It produced optical effects and appears to have caused some reduction in solar radiation, especially in the intensity of the direct beam.

Smoke and dust of Australian origin have often been seen in New Zealand skies, and on one occasion detectable amounts of dust were deposited on snow fields on the Southern Alps, and at other times dust evidently of Australian origin has been brought down by rain in western districts.

Radioactive fallout has been detected in rainwater and regular measurements of its amount have been maintained since 1955. Strontium 90 input increased fourfold from 1960 to reach a maximum about 1964, thereafter decreasing to a minimum in 1968. Smaller increases occurred in the next few years. Levels never reached more than about a fifth of those in the northern hemisphere and in 1973 the inputs decreased to less than one tenth of the 1964 maximum. (N.Z. Dept. of Health, 1974).

Total ozone in the atmospheric column has been measured by means of a Dobson Spectrometer at Wellington from 1951 to 1970 and Invercargill since 1970 (FARKAS, 1973). A series of ozone-sonde measurements have also been made. Monitoring of ozone concentration near the ground commenced at Wellington in 1972. The general level has been about two parts per hundred million by volume and it has not been found to exceed five parts per hundred million. There appear to have been no reliable reports of damage from low level ozone pollution in New Zealand.

LONG TERM CLIMATE FLUCTUATIONS AND TRENDS

Climate records in New Zealand go back more than a century, but data for the earliest years are limited in number and some are of doubtful accuracy. Temperature trends from the most reliable of the long records, adjusted to relate to the present observing site whenever possible are shown in figure 17. They show typical fluctuation of varying periodicity superimposed on a longer term rise of temperature which began towards

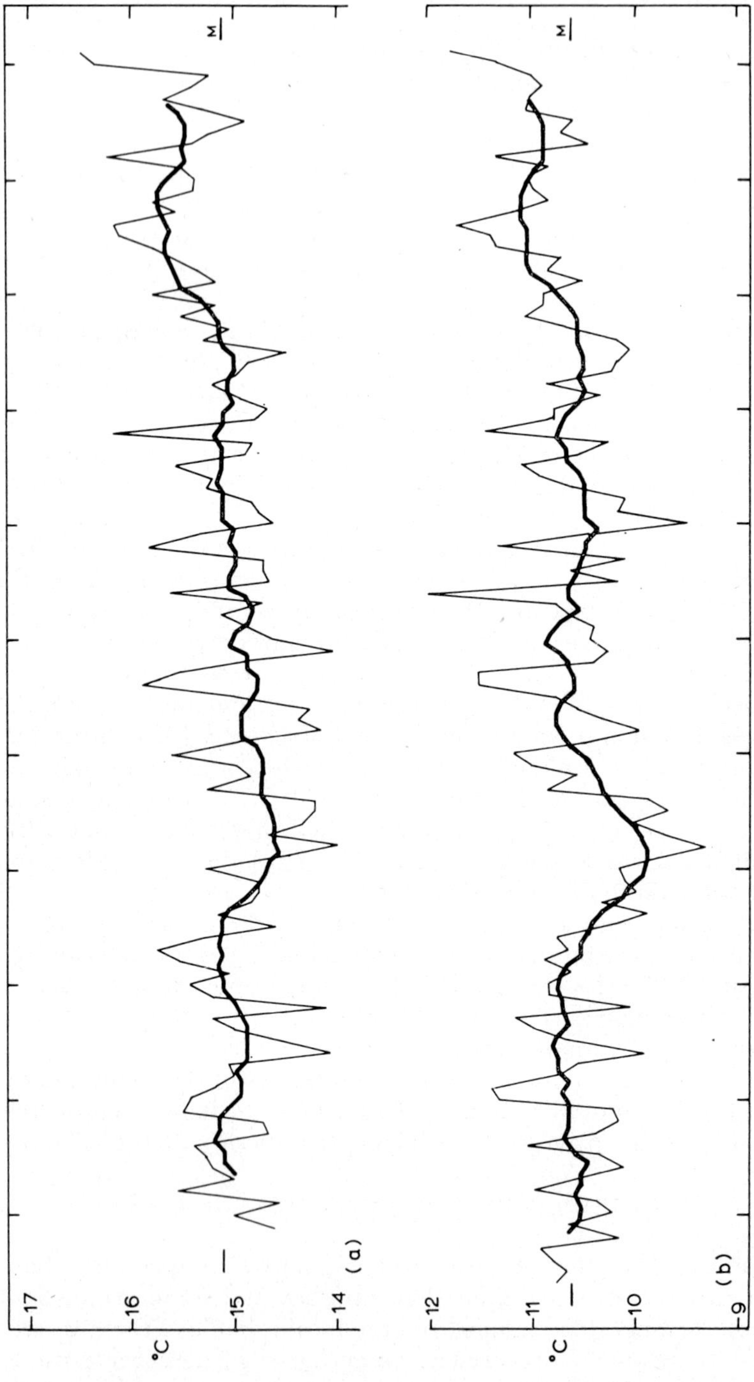

Fig. 17. Temperature trends a. Auckland, b. Dunedin. Annual and 10-year running means, and overall average (m), adjusted to current sites.

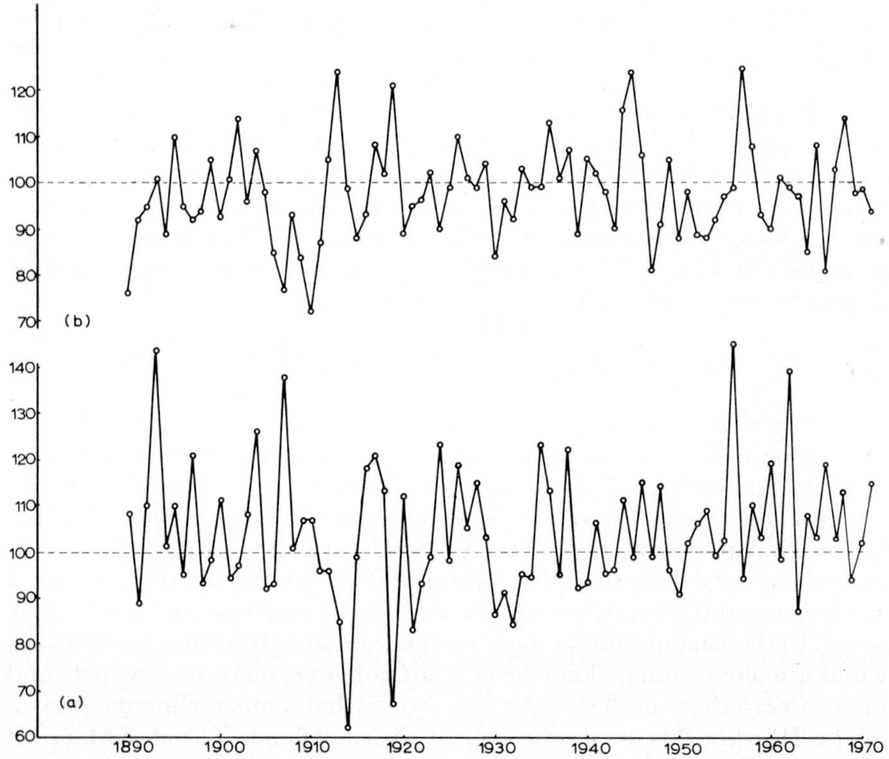

Fig. 18. Precipitation trends, 1890–1971. Annual values of a rainfall index for a. northern area of New Zealand, b. Otago-Southland area. (1911–1940 = 100. The index was based on an average of up to 35 stations in each area. The Northern area is that part of the North Island north of a line roughly from East Cape to Cape Egmont).

the end of the last century and which appears to have continued up to the present (1971). There is little evidence of any decline since 1950 as seen in many northern hemisphere curves. The amplitude of the recent trends (and of annual temperature anomalies) appears to be appreciably less than corresponding values from land areas bordering the North Atlantic.

A set of annual rainfall indices since 1890 has been derived for several areas of New Zealand (Fig. 18). These show short term fluctuations which are largely unrelated from area to area but longer periodicities (period greater than 12 years) show a parallelism in the northern areas probably because of variations in the number of depressions passing over northern New Zealand. There is little evidence of any appreciable long term trend. There is some evidence for a weak relationship with sunspot number.

There has been much speculation about possible secular changes or cyclical fluctuations of climate in post glacial time.* Temperature, rainfall and wind regime changes have been proposed to account for pollen profiles, vegetation patterns, and other biological and geophysical evidence. Temperature curves deduced from isotope ratios in speleothems show a parallelism with some northern hemisphere paleotemperature reconstructions for the past 4000 years. (Hendy & Wilson, 1968). Some uncertainty remains in the paleoclimatic interpretation of these observations, but as they provide quantitative information, isotope methods should help substantially in future towards the understanding of past environments.

Mountain and local climates and microclimates

It is only since the second world war ended that any appreciable investigation of the climate of the mountains has been undertaken. Consequently knowledge of them is still selective and incomplete. Most attention has been given to the ranges on the eastern flanks of the Southern Alps where the mountain lands are more accessible, where large areas have been occupied for grazing sheep, and where problems of accelerated erosion and downstream aggradation and flooding have become acute. In these mountains precipitation ranges from 1000 mm to 4000 mm, winds are often strong, cloud cover is not excessive, there are dry spells and many freeze-thaw cycles of the soil. Somewhat similar climates prevail in the Hawkes Bay mountains of the North Island. The mountains of Otago, further to the south, are somewhat similar, but are colder and cloudier on the tops, have reduced evaporation, and the contrast with adjacent lowlands is greater.

The western ranges of Fiordland and the western flank of the Southern Alps are much wetter (precipitation up to 8000 mm per year), have more persistent cloud cover and experience few drying spells. Snow and glacier levels are considerably lower than in the east coast ranges. Few observations have been made in this area. Western mountains in the North Island are similar but less extreme, with high rainfall and much enveloping cloud especially in the south (Tararuas, southern Ruahines) and Mt Egmont.

On exposed ridges and peaks and on high plateaus such as the summits of the Otago block mountains the surface winds and temperatures resemble those at the same level in the free air, so that free air climate may be taken as an approximation to the surface climate of such places.

The diurnal variation of temperature and wind velocity is negligible in the free air above about 1 km above the surface. On exposed mountains there is often a slight diurnal change in wind speed, for example at Black

* For a recent review see for example Molloy (1969).

Table 6. Air Temperatures at Mountain Climate Stations Mean daily maximum/Mean daily minimum – °C.

January	April	July	October	Year
Chateau (1931–1970) 38°12′S, 175°32′E 1120 m				
17/7	12/4	6/—1	11/2	11/3
Makahu Spur (1969–1971) 39°17′S, 176°25′E 1480 m				
16/7	9/4	4/—2	8/1	9/3
Black Birch Observatory (1961–1963, 1971) 41°45′S, 173°48′E 1395 m				
15/7	10/3	2/—2	9/2	9/3
Camp Stream (1967–1971) 43°08′S, 171°42′E 1430 m				
15/6	9/3	3/—2	10/0	9/2
Ski Basin (1967–1971) 43°08′S, 171°41′E 1550 m				
13/5	8/2	1/—4	6/—1	8/1
Mt John (1963–1971) 43°59′S, 170°28′E 1026 m				
19/8	13/4	4/—2	13/2	13/3
Old Man Ra.* (1963–1968) 45°21′S, 169°13′E 1590 m				
11/0	6/—3	—3/—9	5/—4	5/—3

* Data from MARK & BLISS (1970), other data from N.Z. Meteorological Service records.

Birch Observatory there were two maxima, at about 00 and 15 hours local time, of about 7.5 m/s, the average being 7 m/s. Diurnal changes in air temperature at such sites are small but still appreciable (Mean daily range at Black Birch was 8 °C in January, 4.5 °C in July).

Conditions on mountain slopes and in valleys are very different. Wind then becomes an extremely local phenomenon with some areas generally sheltered and others subject to local increases. The radiation and temperature climate becomes a function of slope, aspect and shading. In sheltered areas the amplitude of diurnal temperature changes is great.

Table 6 summarises data from some of the mountain climate stations in New Zealand.

Local, topographically determined, variations of climate and microclimates in different vegetation environments may be as important to plants and animals as the macroclimate. There have been a number of investigations of these small scale variations of climate in New Zealand. It is not possible to describe them in detail. Nor is it necessary for present purposes as the factors which cause these small scale differences and their characteristics are not peculiar to New Zealand. The scope of some of the more important investigations will however be indicated briefly: 1. The New Zealand Forest Service has maintained a series of basic climate stations between 915 and 1740 m on the Craigieburn Range since 1961. The sites were in mountain forest and at various places above-timber-line in the subalpine and alpine zones. Air temperatures and in some instances precipitation and soil temperature were measured. Local and microclimate variations dependent on slope and aspect and on position within vegetation were investigated, and temperatures of plant and soil surfaces measured. Solar radiation was recorded at two sites and snow surveys and hydrological studies carried out. (E.g., MORRIS, 1965, MORRIS & O'LOUGHLIN, 1965). 2. Air and soil temperature, wind and soil moisture measurements over a range of altitude have been made on the Otago Mountains, especially the Old Man Range. (E.g., MARK, 1965, MARK & BLISS, 1970). 3. Hydrologic studies of 'experimental basins' have included surveys of space variations of precipitation and in some instances of other climate elements (wind, temperature etc.) over small areas in mountain and lowland catchments, e.g., Moutere Hills (near Nelson), Taita and Makara (near Wellington). 4. Alexandra – Clyde area in Otago. A survey of night minimum temperatures was carried out in connection with the frost problem in orchards. 5. Microclimate investigations made in New Zealand in connection with agriculture include the following topics a. Wind profiles near typical shelter belts as used in various farming areas, and their influence on soil moisture and on crop productivity. (STURROCK, 1969, 1970). b. Temperature and light distribution in native grasses (e.g. SCOTT, 1962) and in pasture swards (e.g. MITCHELL & CALDER, 1958, MITCHELL & BIELSKI, 1964, BROUGHAM, 1959) in connection with pasture and crop production. c. Surface climate in high country areas in relation to erosion and revegetation problems, freezing and frost heave (e.g. O'CONNOR & LAMBRECHTSEN, 1967, GRADWELL, 1960).

Bioclimatology in New Zealand

Weather and climate influence, directly or indirectly, many aspects of the growth and survival of plants and animals. Climate usually acts as only one of a number of factors in the environment, and its effects cannot always be readily isolated. In many biological investigations climate records have been taken merely as an adjunct to the study. In others however climate has been a major element. Some bioclimatic relationships have already been noted; a few others are outlined briefly as follows:

CLIMATE AND THE DISTRIBUTION OF PLANTS

Climate is one of the environmental factors which limits the distribution of plant and animal species and communities. Many observations on these relationships have been made by ecologists and agriculturalists during the history of scientific work in New Zealand, (e.g., KIDSON, 1938, WARDLE, 1964). In recent years there have been experiments aimed at elucidating the particular environmental factors involved and their quantitative effects.

As already noted (p. 99) most of New Zealand is included in KOPPEN's cfb, a climate characteristic of forest vegetation. Much of the forest of the wetter areas remains, but much has been cleared to grassland – large areas in the less humid east in pre-European times. The semi-arid areas in the South Island mostly remain as dry tussock grassland, or are used for crops or improved pasture – often with irrigation. The far north of the North Island approaches a Mediterranean type climate but supported a forest vegetation for the most part.

In New Zealand the forest ends at a well defined timber line above which there are woody shrubs, grasses and herbs. Many speculations on the cause of the timber lines have been put forward. A recent explanation, namely that they represent the highest altitude at which the summer growing period is sufficiently long for shoots of woody plants to grow and ripen under the air temperatures that prevail at the height of tree canopies, has been supported by evidence from experiments in the Craigieburn Ra. Reversed timber lines at valley and basin bottoms are often apparently a result of winter frost severity in the open bottoms which kills tree seedlings. In some cases, though long established, these features do not seem to be permanent, as forest may spread from the borders. The alteration of microclimate (and suppression of grasses) at the forest margin allows a gradual encroachment into the grassland. (WARDLE, 1971).

The climate range for optimum production of the common temperate crops is relatively well-known. For example, sub-tropical fruits (citrus, tamarillo, chinese gooseberry) are grown commercially mainly in the Auckland Peninsula and Bay of Plenty, especially in localities with much sunshine and which are relatively frost free. This area corresponds roughly with that of a number of native plants – kauri *(Agathis australis)*, mangrove *(Avicennia resinifera)*, pohutukawa *(Metrosideros exelsa)* and others – and its boundary is roughly the 13 °C mean annual isotherm. On the other hand temperate fruits such as apricots and black currants do better in the south with its colder winter.

CLIMATE AND AGRICULTURAL AND FOREST PRODUCTIVITY

New Zealand's farming industry depends very largely on sheep and cattle, and its climate is well suited to them. The winters in the lowlands being mild by the standards of continental areas in like latitudes, the animals can be kept out in the open all the year, though high country flocks – which graze up to 2000 metres above sea level in summer in places in the South Island east of the main divide – are generally brought to lower levels in winter.

Pasture production continues for most of the year, though some of the summer production needs to be saved, or other crops grown to meet winter feed requirements. If a mean temperature of 10 °C be taken as a rough limit for satisfactory pasture growth there are then on average 5 months with reduced productivity from this cause in the south of the country and one month or none in the north (at sea level). Supplementary water is needed in many areas in dry periods in summer to utilise the production potential given by relatively high global radiation. Annual net productivity (above ground) of improved pasture with dry season irrigation has reached 23,000 kg/ha at Palmerston North. (BROUGHAM, 1959).

Pinus radiata plantations are efficient producers in the climate of most of the country's lowlands, especially in the Bay of Plenty hinterland in the North Island where there are relatively high radiation values, a favourable growing season with regard to temperature, good water availability from deep pumice soils replenished by an adequate average rainfall and adequate nutrients. In this area an annual net production of 36,000 kg/ha was found for a good quality 15 to 20 year old stand near Rotorua (WILL, 1966). The two production values quoted represent photosynthetic energy conversion of approximately 0.8 and 1.3 percent (in terms of total solar radiation received) – relatively high values for temperate latitude crops (EGUNJOBI, 1969).

AERIAL TRANSPORT OF ORGANISMS

Organisms have been carried to New Zealand by the wind, e.g., insects, including the Australian Blue Moon butterfly *Bolina nerina* which breeds in Australia and has been found many times in western areas of New Zealand, spores of wheat rusts, probably from Australia (ALLEN, 1961, CLOSE, 1967) and pollen (MOAR, 1969). Seeds and birds have probably often been blown across the Tasman Sea. Favourable air trajectories across the Tasman are not an every-day occurrence, but occur from time to time. See, for example, fig. 2a which illustrates an occasion when there was a possibility that aphids had been carried from Eastern Australia Canterbury, and plausible trajectories across the Tasman Sea could be drawn. (TOMLINSON, 1973).

FUNGAL DISEASE, ANIMAL HEALTH AND CLIMATE

Fungus dispersal and development are often critically dependent on humidity and temperature. *Dothistroma pini*, a fungus which has caused serious damage to planted pine forests, first appeared in northern New Zealand about 1964. It then spread southward, presumably air-borne, through western and central districts – into areas where humid air streams are likely to carry inoculum from the source areas to regions with a high probability of being wet. (GILMOUR, 1967). Up to 1972 it had not been reported in the east and south of the South Island (BASSETT, 1972).

The poplar leaf rusts *Melampsora medusae*, first noticed in Northland in March 1973, and *M. Larici-populina*, which appeared in Taranaki a little later, are almost certainly cases of trans-Tasman wind-borne dispersal. They spread rapidly over much of the North Island and infected trees were found near Nelson in the South Island by the end of the year. Warm humid conditions are considered to promote growth and spread of the fungi. Implications for future use of poplars in New Zealand, mainly for soil conservation, are being studied. Whether the final distribution of these pathogens will reflect climatic differences to any great extent has yet to be discovered.

Facial eczema in sheep and cattle is caused by a poisonous fungus *Pithomyces chartarum* which proliferates on dead material in pasture where night temperature levels remain for a time above 16 °C and moisture is present. It does not grow in most of the South Island where temperatures are too low or the pastures often too dry. Fairly reliable warning of toxic conditions can be made on the basis of a knowledge of the pre-disposing weather. (MITCHELL, WALSHE & ROBERTSON, 1959). Other animal health problems in which climatic influences may be important and which have come under investigation recently include liver-fluke, nemotode parasites, cattle ticks, and leptospirosis (see, e.g., annual research reports, Ministry of Agriculture and Fisheries).

WEATHER CATASTROPHES

The ecological effects of particularly severe storms have become recognized in recent years, (e.g. MOLLOY, 1969). For example erosion initiated in the 1938 floods in Hawkes Bay will take a long time to heal, and there is evidence of far reaching effects of earlier events of this sort in alluvial deposits and patterns of vegetation in some other parts of the country. Trees have died in some particularly cold winters (e.g. *Eucalyptus* near Queenstown in the late sixties) and after a succession of particularly dry (e.g. 1945 and 1971 in parts of the North Island) or wet years. Pathogens favoured by the unusual weather may of course have been the direct cause of mortality. The effect of climate on heat losses in relation to nutrition and productivity has been studied (e.g., JOYCE, 1968). The most obvious effects are in severe weather, e.g., lamb losses are often serious when cold wet windy conditions (or snow) persist more than a few hours at the critical lambing time in early spring. Effects on animal populations of short term weather extremes and of longer term weather or climate fluctuations have been examined, e.g. the probable decline of numbers of the takahe *(Notornis mantelli)* in recent years (REID, 1967).

In the Canterbury plains in particular, occasional strong winds have caused extensive wind throw in exotic forest plantations, necessitating a shortened rotation period (WENDELKEN, 1966).

Conclusion

Though not subject to the extremes of heat and dryness, New Zealand's climate nevertheless spans a wide range of conditions if the mountains are taken into account. The lowland climate is relatively well known in so far as the distribution of air temperature, humidity, precipitation and broad scale weather events are concerned. The climate of the largely unoccupied mountainous areas is still largely unmeasured. Knowledge of the surface wind climate which is extremely variable from place to place is inadequate for many purposes, as is that of actual and potential evaporation from open water and vegetation, and of the radiation climate. More detailed information on microclimate and small scale local climate variations is often wanted, both as to natural variations and possible changes brought about by land developments and vegetation changes.*

Acknowledgement

Thanks are due to past and present members of the New Zealand Meteorological Service whose work has been freely used without specific acknowledgements.

REFERENCES

Allen, J. D. 1961. Wheat rusts in New Zealand. *N.Z. Wheat Rev.* 8: 27–31.

Bassett, C. 1972. The dothistroma situation, 1972. *Farm forestry*, 14: 47–52.

Brougham, R. W. 1959. The effects of season and weather on the growth rate of a rye-grass and clover pasture. *N.Z.J. Agric. Res.* 2: 283–296.

Claridge, G. G. C. 1970. Studies in element balances in a small catchment at Taita, New Zealand. *IASH Publication* No. 96: 523–540.

Close, R. L. 1967. Stem rust of wheat in Canterbury. *N.Z. Wheat Rev.* 10: 47–52.

Coulter, J. D. 1973. The prediction of evapotranspiration from climatological data. In Proceedings of Soil and Plant Water Symposium, Palmerston North, April 1973. N.Z. DSIR Inf. Ser. in press.

Coulter, J. D. 1973b. Climate. In Natural History of New Zealand, Ed. G. R. Williams, A. W. and A. H. Reed, Wellington.

* For fuller accounts of New Zealand weather and climate see Maunder (1970), Garnier (1958), Kidson (1936). Coulter (1973b) contains an historical review of bioclimatology in New Zealand with numerous references. Sparrow & Healy (1968) provide a comprehensive bibliography.

Cox, J. E. 1968. Evaluation of climate and its correlation with soil groups. In Soils of N.Z. Part 1: 33–44. *N.Z. Soil Bur. Bull.* 26 (1).
Daly, G. T. 1970. Byrophyte and lichen indicators of air pollution in Christchurch, New Zealand. *Proc. N.Z. Ecol. Soc.* 17: 70–79.
De Lisle, J. F. 1966. Mean daily insolation in New Zealand. *N.Z. J. Sci.* 9: 992–1005.
De Lisle, J. F. 1970. Heat transfer from ocean to atmosphere in the Tasman Sea area. *N.Z. J. Sci.* 13: 166–176.
Egunjobi, J. K. 1969. Primary productivity and nutrient cycling in terrestrial ecosystems. *Tuatara* 17: 49–66.
Farkas, E. 1973. Total ozone variations in New Zealand 1951–1972. *N.Z. Met. Service Tech. Note* 219.
Gabites, J. F. 1960. The heat balance of the antarctic through the year. In Antarctic Meteorology: 370–376. Pergamon Press, Oxford.
Garnier, B. J. 1951. Thornthwaite's new (1948) system of climatic classification and its application to New Zealand. *Trans. Roy. Soc. N.Z.* 79: 87–103.
Garnier, B. J. 1958. *The climate of New Zealand.* Arnold, London.
Gilmour, J. W. 1967. Distribution and significance of the needle blight of pines caused by *Dothistroma pini* in New Zealand *Pl. Dis. Reptr.* 51: 727–730.
Hendy, C. H. & Wilson, A. R. 1968. Palaeoclimatic data from speleothems. *Nature* 219: 48–51.
Joyce, J. P. 1968. Effect of climate on sheep production. *Proc. N.Z. Agric. Sci.* 2: 174–179.
Kidson, E. 1936. Climate. In Agricultural organisation in New Zealand: 82–122, ed. H. Belshaw, Melbourne Univ. Press.
Mark, A. F. 1965. Vegetation and mountain climate. In Central Otago: 69–91. N.Z. Geograph. Soc. Misc. Ser. 5.
Mark, A. F. & Bliss, L. C. 1970. The high-alpine vegetation of Central Otago, New Zealand. *N.Z. J. Bot.* 8: 381–451.
Maunder, W. J. 1970. Climate of New Zealand. In World survey of climatology, 13; Climates of Australia and New Zealand: 213–268. Elsevier, Amsterdam.
Miller, R. B. 1961. The chemical composition of rain water at Taita, New Zealand, 1956–58. *N.Z. J. Sci.* 4: 844–853.
Mitchell, K. J. & Bieleski, R. L. 1964. Note on the temperatures of leaf and meristematic tissues of plants of short-rotation ryegrass in summer conditions. *N.Z. J. Agric. Res.* 7: 761–765.
Mitchell, K. J. & Calder, D. M. 1958. The light regime within pastures. *N.Z. J. Agric. Res.* 1: 61–68.
Mitchell, K. J., Walshe, T. O. & Robertson, N. G. 1959. Weather conditions associated with outbreaks of facial eczema. *N.Z. J. Agric. Res.* 2: 584–604.
Moar, N. T. 1969. Possible long-distance transport of pollen to New Zealand. *N.Z. J. Bot.* 7: 424–426.
Molloy, B. P. J. 1969. Evidence for post-glacial climatic changes in New Zealand. *J. Hydrol. (N.Z.)* 8: 56–67.
Morris, J. Y. 1965. Climate investigations in the Craigieburn Range, New Zealand. *N.Z. J. Sci.* 8: 556–582.
Morris, J. Y. & O'Loughlin, C. 1965. Snow investigations at the Craigieburn Ra. *J. Hydrol. (N.Z.)* 4: 2–16.
Newell, R. E., Vincent, D. G., Dopplick, T. B., Ferruzza, D. & Kidson, J. W. 1970. The energy balance of the global atmosphere. In Global circulation of the atmosphere: 42–90. Ed. G. A. Corby Roy. Met. Soc. London.
N.Z. Board of Health, Committee on Air Pollution, 1970: Air Pollution. N.Z. Bd. Health Report Series 15. Wellington.
N.Z. Dept of Health, 1974. Environmental radioactivity annual report, 1973.
N.Z. D.S.I.R. 1971. *Research* 1971. N.Z. D.S.I.R., Wellington.
N.Z. Fertiliser Manufacturers' Res. Assn. Annual Reports 1960–61 *et seq.*
O'Connor, H. F. & Lambrechtsen, N. L. 1967. Some ecological aspects of vegetation

of eroded Kaikouras soil at Black Birch Range, Marlborough. *Proc. N.Z. Ecol. Soc.* 14: 1–7.

RAMSAY, G. W. & ORDISH, R. G. 1966. The Australian blue moon butterfly *Hypolimnas bolina nerina* (F.) in New Zealand. *N.Z. J. Sci.* 9: 719–729.

REID, B. 1967. Some features of recent research on the Takahe *(Notornis mantelli)*. *Proc. N.Z. Ecol. Soc.* 14: 79–87.

ROBERTSON, N. G. 1957. The climatic districts in New Zealand. Proc. *N.Z. Ecol. Soc.* 4: 6 and 22–23.

ROBERTSON, N. G. 1963. The frequency of high intensity rainfall in New Zealand. N.Z. Met. S. Misc. Publ. 118.

SCOTT, D. 1962. Temperature and light microclimate within a tall tussock community. *N.Z. J. Agric. Res.* 5: 179–182.

SCOTT, D., MENALDA, P. H. & BROUGHAM, R. W. 1968. Spectral analysis of radiation transmitted and reflected by different vegetations. *N.Z. J. Bot.* 6: 427–449.

SPARROW, C. J. & HEALY, T. R. 1968. Meteorology and climatology of New Zealand, A. bibliography. Oxford Univ. Press, Auckland.

STURROCK, J. W. 1969. Aerodynamic studies of shelterbelts in New Zealand – 1 Low to medium height shelterbelts in mid-Canterbury. *N.Z. J. Sci.* 12: 754–776.

STURROCK, J. W. 1970. Studies of the effect of wind reduction on soya beans – (1) A preliminary assessment. *N.Z. J. Agric. Res.* 13: 33–44.

TAALJARD, J. J., VAN LOON, H., CRUTCHER, H. L. & JENNE, R. L. 1969. Climate of the upper air. Southern Hemisphere. Vol. 1. U.S. Dept. Commerce, Asheville.

TOMLINSON, A. 1970. The snowstorm of Nov. 1967. *N.Z. Geographer.* 26: 20–35.

TOMLINSON, A. 1973. Meteorological aspects of Trans-Tasman Insect Dispersal. *N.Z. Entomologist* Vol. 5: 253–268.

VONDER HAAR, T. H. & SUOMI, V. E. 1969. Satellite observations of the earth's radiation budget. *Science* 163: 667–669.

WARDLE, P. 1964. Facets of the distribution of forest vegetation in New Zealand. *N.Z. J. Bot.* 2: 352–365.

WARDLE, P. 1971. An explanation of alpine timberline. *N.Z. J. Bot.* 9: 371–402.

WENDELKEN, W. J. 1966. Eyrewell forest: a search for stable management. *N.Z. J. Forestry* 11: 43–65.

WILL, G. M. 1964. Dry matter production and nutrients uptake by *Pinus radiata* in New Zealand. *Commonw. For. Rev.* 43: 57–70.

WILL, G. M. 1966. Root growth and dry matter production in a high producing stand of *Pinus radiata. N.Z. For. Res. Note* 44.

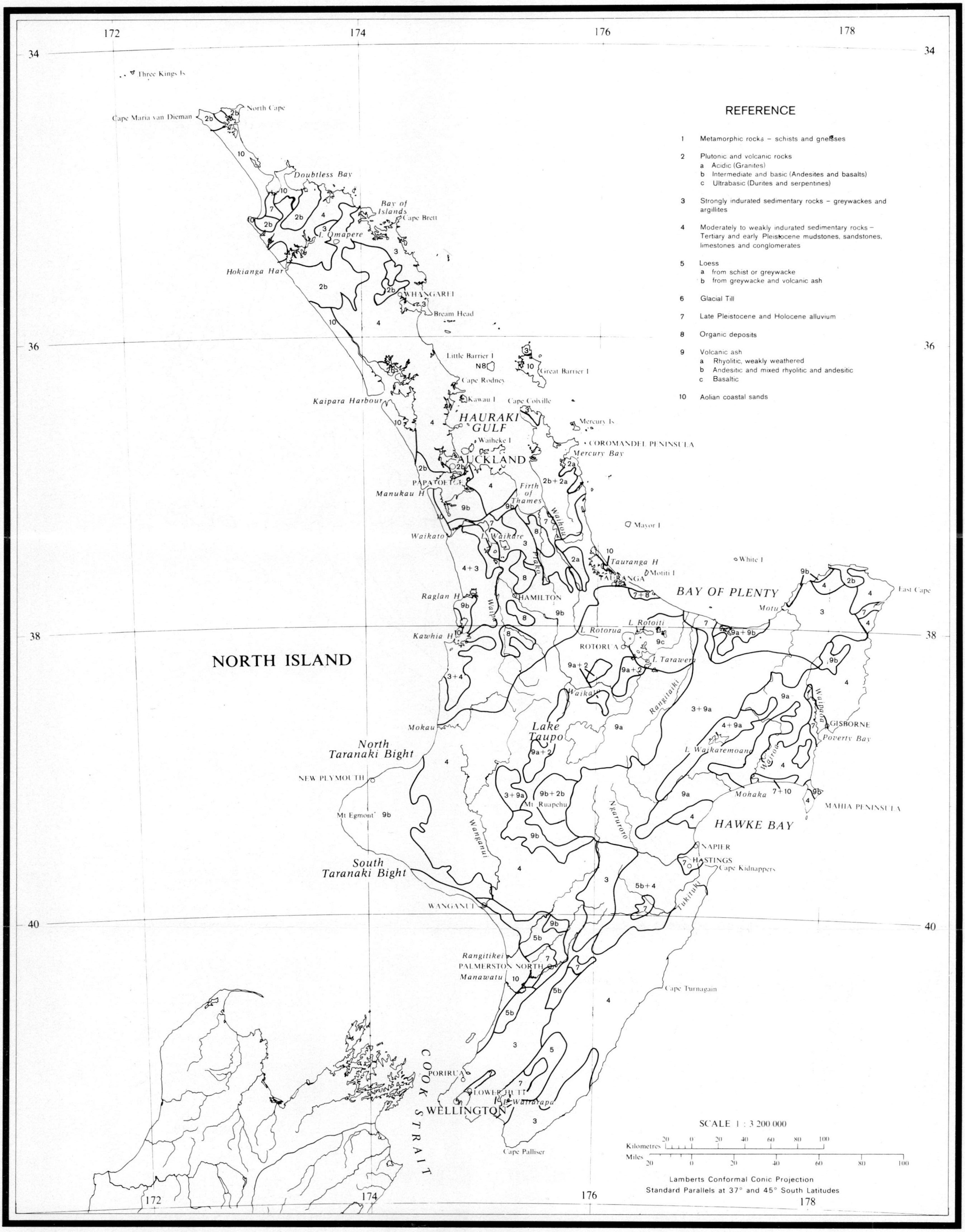

Fig. 1a. Soil parent materials, North Island.

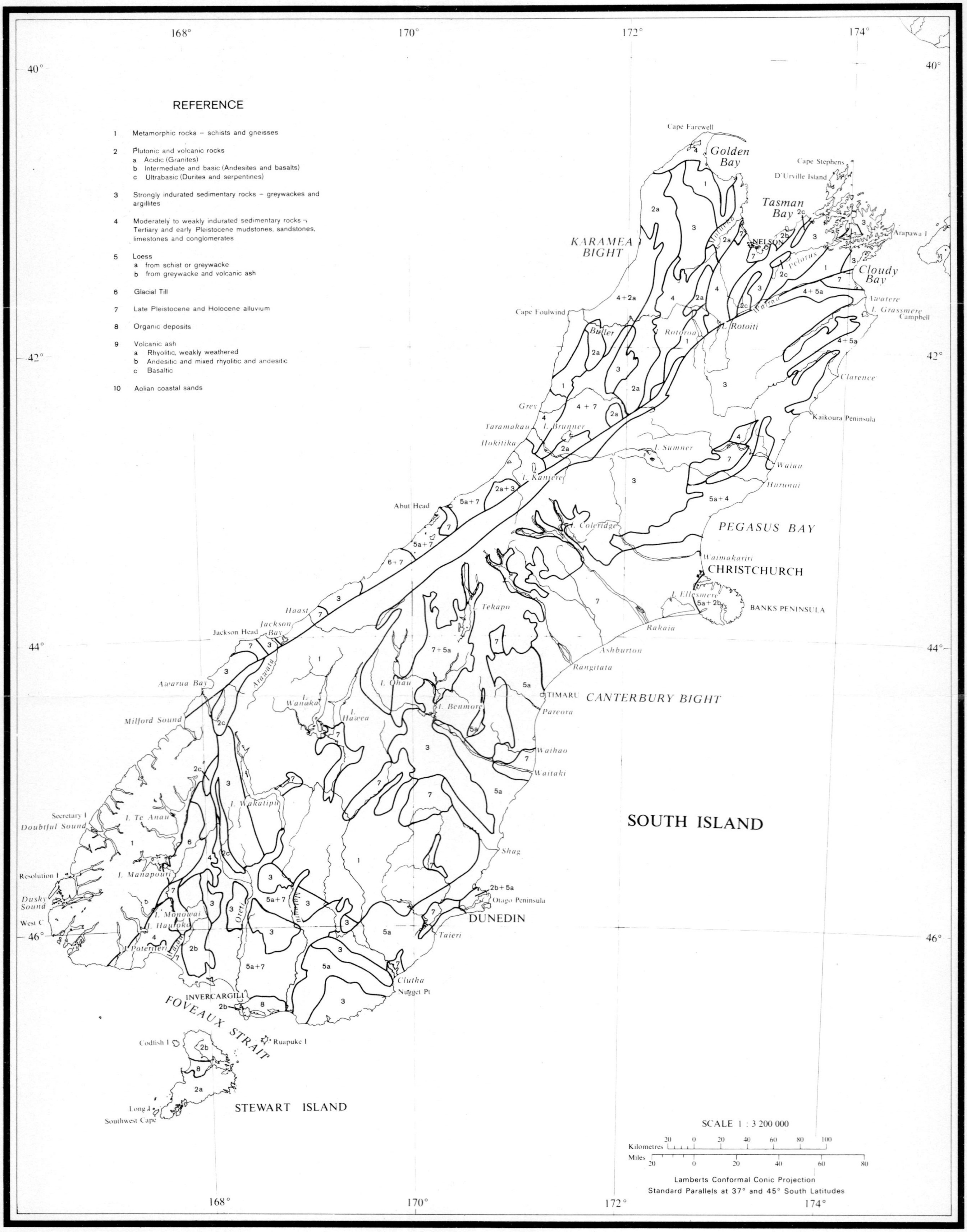

Fig. 1b. Soil parent materials, South Island.

III. THE DISTRIBUTION AND PROPERTIES OF SOILS, AND THEIR BIOTA

by

Q. W. RUSCOE

Introduction

The soil pattern of New Zealand is a complex one. This is the result of a wide range of soil-forming factors acting on parent materials including many kinds of igneous rocks (from ultrabasic to acidic), metamorphic and sedimentary rocks (schists, gneisses, conglomerates, sandstones, mudstones and limestones), loess, and volcanic rocks (Fig. 1). Land forms are varied; about 50 percent of New Zealand is steep, 20 percent is hilly, and only 30 percent is rolling or flat. The climatic types under which the rocks weather vary generally from oceanic to subcontinental, and from cool temperate to subtropic, often with considerable local diversity, as shown for example in the contrast between the semi-arid basins of Central Otago and the adjacent superhumid mountains and lowlands of Westland. Soil formation occurred under a native vegetative cover ranging from lowland tussock grassland or shrubland in the drier areas to podocarp-dicotylous forest and beech forest in the more humid regions. In the warmer northern part of the North Island local areas of kauri forest occurred. Subalpine grasslands, herbfield, moorland and scrubland occurred at altitudes above 900 m. The normal course of pedogenesis has been interrupted in many localities by events such as the flooding of rivers over alluvial plains, the drifting of sand and dust, the fall of ash from erupting volcanoes and by continuing erosion, both normal and accelerated, of steep hillsides. Of more widely disruptive influence were the glacial and periglacial processes of the Pleistocene which removed the former soil mantle, or disturbed it, or buried it under new accumulations. Hence, whereas in the unglaciated subtropical north soils are generally well weathered and clay is the predominant texture, in the temperate remainder few soils are older than 20,000 years and many are younger than 10,000 years, they contain much weatherable mineral material, and the textures are predominantly silt loams and sandy loams. The considerable glaciation, post-Tertiary volcanic activity, and continuing tectonic uplift keeping many slopes steep and rejuvenating, have all had a marked influence on soil development. The resulting soil pattern contrasts strongly with that developed under the stable conditions of the neighbouring Australian continent.

Distribution and Properties of New Zealand Soils

(in collaboration with J. D. Cowie)

An overall evaluation of the soil pattern of New Zealand was obtained through the completion of a general survey of the North Island in 1948 (New Zealand Soil Bureau, 1954) and of the South Island in 1966 (New Zealand Soil Bureau, 1968). In the reconnaissance survey, the whole of New Zealand (265,000 km²) was covered on a scale of about 1:250,000. In addition, district surveys on a scale of 1:63,360 and 1:126,720 have covered almost one third of the land area, and detailed surveys on a scale of 1:15,840 and 1:31,680 have covered small areas of fertile and intensively farmed land. The progress of soil survey and the stages in the evolution of soil science in New Zealand have been chronicled by Pohlen (1957).

The following description of the major soil groups in New Zealand considers the soils recognised by the New Zealand genetic classification of Taylor & Pohlen (1959). This scheme classifies the soil as a dynamic system resulting from three interdependent processes, namely, the wasting, organic and drift regimes (Taylor & Pohlen, 1970).

Differing kinds and intensities of these three regimes acting on the wide range of rocks present gives a complex variety of soils but several basic soil forms can be recognised which are the basis of subdivision into eighteen distinctive soil groups. These groups are listed in Table 1, together with their approximate international equivalents. Their geographic distribution is shown by the soil map (Fig. 2).

The first eight groups are those which were formerly considered as zonal soils and their distinguishing characteristics are controlled by climate and the resulting vegetation acting on rocks of normal siliceous composition. Table 2 shows that the climatic classes (moisture and thermal efficiency) of Thornthwaite (1948) are generally closely related to these groups. The groups are the brown-grey earths of semi-arid and nearly semi-arid areas with mean annual rainfall (m.a.r.) of 300–500 mm, the yellow-grey earths of subhumid areas with m.a.r. of 500–1,000 mm, the yellow-brown earths of humid areas with well distributed rainfall and m.a.r. of 900–2,500 mm, the associated podzolised yellow-brown earths and podzols resulting from excessive leaching under an acidic surface litter of decomposing vegetation, and the gley podzols of superhumid regions.

The next seven groups are those formerly considered as intrazonal soils with characteristics reflecting the effect of some local factor such as extreme of parent rock or poor drainage. The remaining groups are the azonal soils whose characteristics are strongly modified by such causes as instability or shortness of time during which the soils have been developing.

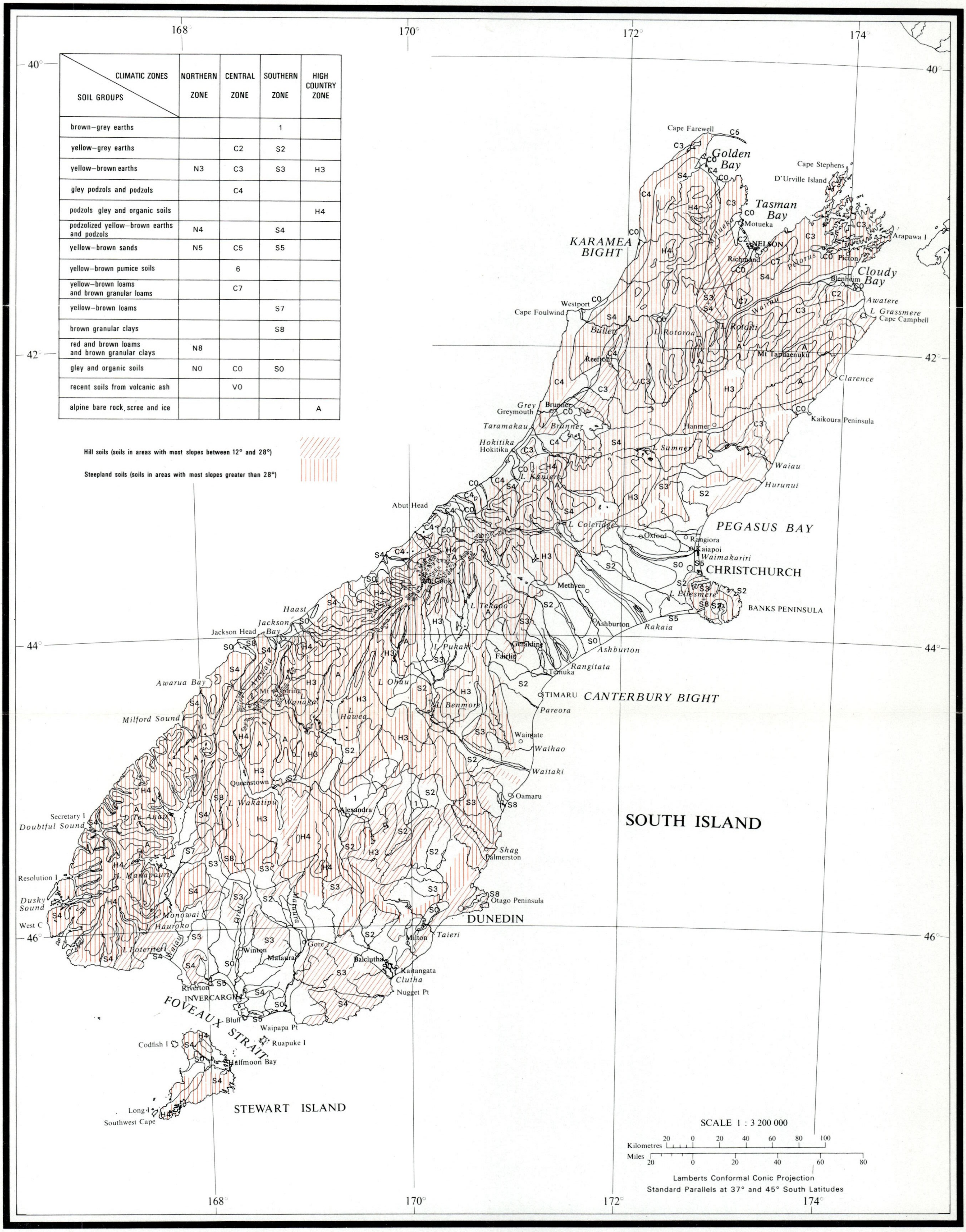

Fig. 2b. Soils of South Island.

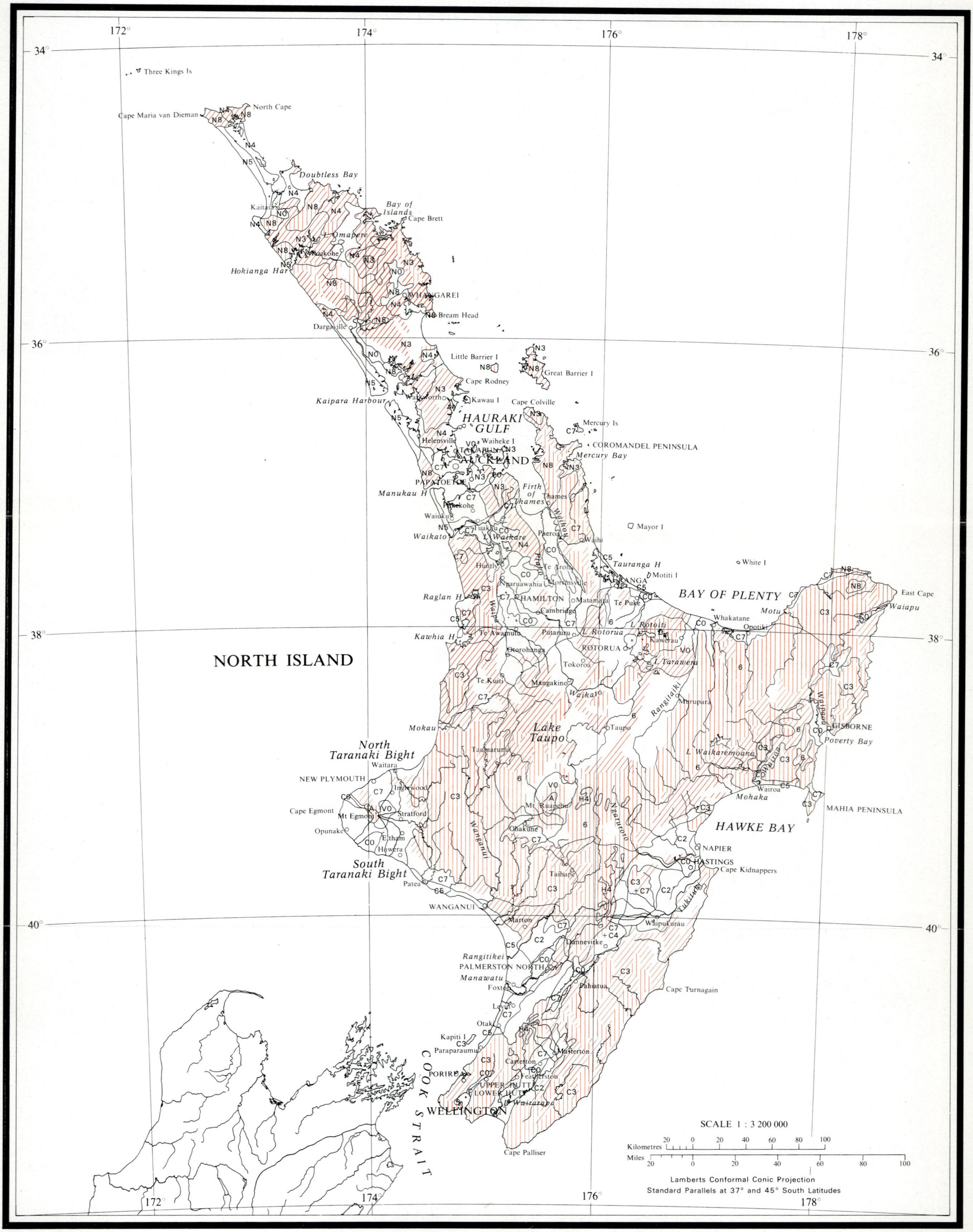

Fig. 2a. Soils of North Island.

Table 1. Approximate Correlation of the Soil Groups (Figure 2) of the New Zealand Genetic Classification with suborders of the U.S. Taxonomic Classification and units of the International Classification.

New Zealand Genetic Classification	U.S. Taxonomic Classification	International Classification
Brown-grey earths	Ustalfs	Calcic Luvisols, Haplic Phaeozems
Yellow-grey earths	Ustalfs, Udalfs, Aqualfs	Gleyic, Orthic, Luvisols, Haplic Phaeozems
High country yellow-brown earths	Ochrepts (Cryic, Dystric)	Dystric Cambisols
Southern and central yellow-brown earths	Ochrepts, Umbrepts	Dystric, Humic Cambisols
Northern yellow-brown earths	Humults, Udults	Humic, Orthic, Acrisols
Podzolised yellow-brown earths and podzols	Udults, Orthods, Aquods	Orthic Acrisols, Placic, Orthic, Gleyic Podzols
Gley Podzols	Aquods	Gleyic Podzols
Subalpine gley soils and gley podzols	Aquepts, Aquods	Humic Gleysols, Gleyic Podzols
Rendzinas and rendzic intergrades	Rendolls, Udolls, Ochrepts	Rendzinas
Yellow-brown sands	Psamments, Ochrepts (Psammic)	Eutric Regosols, Eutric Cambisols
Yellow-brown pumice soils	Andepts (Vitric)	Vitric Andosols
Yellow-brown loams	Andepts	Humic, Ochric Andosols
Red and brown loams and brown granular loams and clays	Andepts, Humults, Udults, Humox, Orthox	Humic Acrisols, Humic Andosols, Orthic Ferralsols, Eutric Nitosols
Organic soils	Histosols	Eutric, Dystric Histosols
Gley soils	Aquepts, Aqualfs, Aquults	Dystric, Humic, Mollic Gleysols
Recent soils from alluvium	Fluvents, Aquents, Psamments	Eutric Fluvisols
Recent soils from volcanic ash	Andepts (Vitric), Psamments	Vitric Andosols
Steepland soils	Ochrepts, Umbrepts, Andepts, Udents	Dystric, Humic, Eutric Cambisols, Vitric Andosols, Dystric Regosols, Lithosols

Table 2. Zonal Soil Groups Related to Climatic Classes.

Climatic Class		
Moisture Class	Thermal Efficiency Class	Zonal Soil Groups Characteristic Of The Climatic Classes
Semi arid	Mesothermal '1'	Brown-grey earths
Semi arid	Microthermal '2'	Intergrades between brown-grey earths and yellow-grey earths (high country)
Subhumid	Mesothermal '2'	Yellow-grey earths
Subhumid	Mesothermal '1'	Yellow-grey earths
Subhumid	Microthermal '2'	High country yellow-brown earths; high country intergrade between yellow-brown earths and yellow-grey earths
Subhumid to humid 'A'	Mesothermal '2'	Intergrades between yellow-grey earths and yellow-brown earths; some yellow-grey earths
Subhumid to humid 'A'	Mesothermal '1'	Some yellow-brown earths
Subhumid to humid 'B'	Mesothermal '2'	Intergrades between yellow-brown earths and yellow-grey earths
Subhumid to humid 'B'	Mesothermal '1'	
Humid 'A'	Mesothermal '2'	Northern yellow-brown earth; some central yellow-brown earths
Humid 'B'	Mesothermal '2'	Northern yellow-brown earth
Humid 'A'	Mesothermal '1'	Central and Southern yellow-brown earths
Humid 'B'	Mesothermal '1'	Central yellow-brown earth
Humid 'A'	Microthermal '2'	Blanket peat (subantarctic)
Superhumid	Mesothermal '1'	Gley podzols, some central and southern yellow-brown earths
Superhumid	Microthermal '2'	Subalpine gley podzols
Superhumid to humid 'A'	Megathermal	Not known
Humid 'B'	Megathermal	Not known
Humid 'A'	Mesothermal '3'	Not known

1. References; New Zealand: TAYLOR, N. H. & POHLEN, I. J. 1959. Soils and land use. Pp. 28–33 in A Descriptive Atlas of New Zealand (ed. A. H. MCLINTOCK). Government Printer, Wellington. U.S.: Soil Survey Staff 1960: Soil classification – A Comprehensive System. 7th Approximation. U.S. Department of Agriculture, Washington. 265 p. International: DUDAL, R. 1968. Definitions of soil units for the soil map of the World, April, 1968. F.A.O. World Soil Resources Reports 33. 72 p.
2. Correlations; The correlations can not be exact as the units are derived in different ways. The U.S. and international units shown are those that most closely relate to the central concept(s) of the New Zealand groups and their main subgroups.

Brown-grey earths

The brown-grey earths and associated solonetzic soils occupy 160,000 hectares of flat and rolling, and 40,000 hectares of hilly country in the semi-arid and nearly semi-arid intermontane basins of Central Otago and the McKenzie Plains in the South Island. They are formed under relatively low temperatures, averaging 8 °C and rainfall of 300 to 600 mm. As a result of the low rainfall, the soils are dry for a considerable part of the year and soil moisture is inadequate for the growth of vegetation for most of the growing season. Brown-grey earths developed under lowland short tussock (bunch grass) grassland in which *Poa colensoi* (blue tussock), *P. laevis* (silver tussock) and *Festuca novae-zelandiae* (hard tussock) were co-dominant. This original vegetative cover has now almost completely disappeared and, except where irrigation is applied or where dryland pastures have been established, the soils are inadequately protected by scattered cushion plants and ephemeral weeds such as *Raoulia* spp. and *Rumex* spp.

Brown-grey earths are characterised by sandy, greyish brown topsoils with a platy structure. Subsoils are generally olive to yellowish-brown in colour with an accumulation of clay as coatings on aggregate faces and through aggregates to give a compact clay pan which is massive to weakly prismatic. This clay pan is best developed in the older and deeper soils. On shallow soils over gravels, the clay pan is represented by a layer of reddish brown to yellowish red clay – and iron-cemented gravels which may be up to 60 cm thick in the older soils. Below the clay pan or cemented gravel layer, lime accumulations generally occur and soluble salts are not uncommon in lower horizons. In low-lying parts of the lower land surfaces soluble salts come to the surface and small scattered patches of solonetzic soils are common in the inland basins of Central Otago. The concentration of salt in the various horizons of the rooting zone tends to vary with the seasons, efflorescences being most conspicuous during long dry periods.

Because of low winter temperatures and lack of soil moisture, weathering of parent schist, greywacke or loess is at an early stage and weathering of the micas is the dominant process of clay formation. Weakly hydrated micas are the dominant clay minerals with minor amounts of chlorite, quartz, felspar, vermiculite and montmorillonite. Both clay and organic matter contents are relatively low and this is accompanied by low cation exchange capacities. However, because of the weak leaching, base saturations are generally high in topsoils and increase with depth generally to 100 percent saturation in the lower subsoil. pH's are slightly acid to near neutral in the topsoil increasing to alkaline with depth. Phosphate soluble in dilute acids is high to very high in the topsoil and phosphate retentions are low. Potassium reserves are high. Available sulphur tends to be low in topsoils, but may be high in deep subsoils. The main limitation for plant growth is the lack of soil moisture during the growing season and irrigation is necessary for intensive farming or horticulture, although some success has been obtained in establishing and utilising dryland pastures in which lucerne (alfalfa) is the chief component. With shallow rooting pastures, good sulphur responses are usually obtained and slight responses to phosphates.

Yellow-grey earths

The yellow-grey earths occupy 1.4 million hectares of flat and rolling, and 400,000 hectares of hilly country. They are most extensive in South Island where they occur on terraces, downlands and hills of coastal Marlborough, Canterbury and Otago. They also occur in North Island in the lowland areas of Hawkes Bay, Wairarapa and Manawatu. The average mean temperature is 10 °C with rainfalls from 500–1,000 mm. Rainfall is not well distributed through the year and these soils dry out for varying periods, which may be up to 5 months in the drier areas. The drier yellow-grey earths such as those of Marlborough and parts of Canterbury and Otago were formed under lowland short tussock grassland whereas the wetter ones of Hawkes Bay, Wairarapa and Manawatu were formed under shrub- or fernland with some areas of dicotylous forest. Parent materials are mainly loess or alluvium derived from schist or greywacke and to a lesser extent weakly to moderately consolidated sedimentary rocks. In the North Island the loess contains varying admixtures of air-fall volcanic ash.

Characteristically the yellow-grey earths have greyish brown fine sandy loam to silt loam topsoils with weak to moderate nut structures overlying pale-coloured olive or olive-grey silt loam to clay loam subsoils which are compact, have a weak to moderate blocky structure, show weak to strong clay illuviation, and contain some brownish mottles. At depths of from 45 to 90 cm a near massive very compact fragipan occurs which is traversed by mainly vertical grey veining, called gammation (Taylor & Pohlen, 1970). However, within the group, the characteristic features of the subsoil and fragipan can vary a great deal according to the rainfall and duration of the dry season, and according to such local factors as the thickness and texture of the parent loess, its age, and by the nearness or otherwise of buried soils.

In lower rainfall areas, with long dry periods, subsoils tend to be brownish in colour with only slight gleying or clay illuviation. The fragipan is close to the surface and gammation is weak or absent. In the moderate rainfall areas subsoils are olive in colour and more strongly mottled with weak to moderate clay illuviation. The fragipan is slightly deeper in the profile and more strongly expressed, and there is moderate development of vertical grey veining. In the higher rainfall areas where the soils are integrading to yellow-brown earths, subsoils are gleyed with olive to olive-grey colours and have abundant yellowish brown mottles, and there is moderate to strong clay illuviation. The fragipan is deeper (up to 90 cm) and is not well developed, but both horizontal and vertical grey veining is strong giving the fragipan a coarse prismatic to blocky structure.

The degree of weathering shown by the yellow-grey earths is slightly more advanced than in the brown-grey earths. Clays are more hydrated and consist predominantly of illite and hydrous micas with some vermiculite. The several stages of micaceous weathering present show that there is active weathering of micaceous minerals.

The organic matter content of topsoils is low to moderate but is higher than in the brown-grey earths. Generally these soils are moderately acid, but in parts of Hawkes Bay and North Canterbury, they are less acid. Percentage base saturations are generally moderate in the upper horizons and stay constant or are higher at depth. Cation exchange capacity is also higher in the subsoil following the increase in clay content. Both exchangeable magnesium and sodium are higher at depth, with magnesium becoming the dominant ion in the deeper horizons. Phosphorus soluble in dilute acids is low to medium in topsoils and phosphate retention is low. Fertiliser requirements for pastures are variable. In the drier areas of Hawkes Bay only slight responses to phosphorus are obtained; in the wetter areas of the Manawatu, responses are obtained to phosphorus, lime and potassium, whereas in the South Island, phosphorus, lime, sulphur and molybdenum responses are recorded.

The yellow-grey earths are used mainly for pastoral farming and for the growing of grain crops in rotation with pastures.

Stony soils associated with the yellow-grey earths occupy 0.3 million hectares mainly on the Canterbury Plains of the South Island. They are excessively draining, shallow soils over gravels with browner and more friable subsoils than in the yellow-grey earths. Some are well suited for irrigation and are used for intensive sheep farming and cereal and pasture seed crops.

Yellow-brown earths

The yellow-brown earths are extensive throughout New Zealand, occurring in humid regions where rainfall is adequate for plant growth and subsoils remain moist for most of the year. They are mostly formed from sedimentary rocks, loess or alluvium but in a few places are from acid igneous rocks. They occur in a wide range of landscapes including the easy and moderately steep slopes of coastal hills, downs, plains, upland plateaus, and mountains and mountain valleys. Considerable variation in morphology and chemistry occurs, however the mode is a moderately leached friable soil with a yellowish brown free-draining subsoil with nutty structure. The yellowish brown colour of the subsoil results from the weathering of iron-bearing minerals under moist conditions to hydrous iron oxides. The trends of micaceous weathering in the brown-grey earths and yellow-grey earths are continued and extended, but a wider range of weathering and leaching occurs because of the wider climatic range. The yellow-brown earths that have been recognised are:

High country yellow-brown earths, which occur in the cold to cool inland regions of the South Island from Marlborough to Southland, where mean temperatures average about 4 °C and m.a.r. ranges from 700 to 1,300 mm.

Southern and central yellow-brown earths which occur in the cool humid lowlands of Southland, Otago and Canterbury and the mild humid lowlands of the northern end of South Island and the southern and central parts of North Island. Average mean temperature is about 11 °C and m.a.r. is 1,000 to 2,500 mm.

Northern yellow-brown earths occur in the warm lowlands and hill country of North Auckland, where the average mean temperature is 14 °C and m.a.r. is 1,250 to 2,500 mm.

High country yellow-brown earths

These occupy 400,000 hectares of rolling, and 400,000 hectares of hilly country, and are derived from greywacke or schist or solifluction material from these. They developed under a vegetative cover of predominantly tussock grasslands at altitudes between 750 and 1,500 m. These soils are mainly used for extensive sheep farming and with overgrazing and burning, the vegetation of some areas has been depleted leading to serious frost and wind erosion. They have dark brown loamy topsoils with yellowish brown friable subsoils and are only weakly weathered with clay minerals mainly illite, weakly hydrated micas and vermiculite. Amorphous clay minerals have been recorded in some soils. The soils are moderately acid with a low to moderate content of plant nutrients. Pastures respond to phosphate, sulphur and molybdenum.

The southern and central yellow-brown earths

These cover 600,000 hectares of flat to rolling land and 1.8 million hectares of hilly country. Most of these soils were formed under podocarp-dicotylous forest, or *Nothofagus* forest (southern beech), on greywacke, schist, loess, sandstones, mudstones or granite. Principal features are loamy, brown to greyish brown topsoils which are high in organic matter and slightly heavier, nutty brownish yellow subsoils. Natural fertility of the soils is strongly conditioned by the composition of the forest cover. Forest containing much *Beilschmiedia tawa* (tawa) and associated fertility-demanding dicotylous trees, was relatively efficient at returning plant nutrients to the soil which became moderately fertile (although low in phosphorus) and only moderately acid. However forests containing much *Dacrydium cupressinum* (rimu) or *Nothofagus* spp., produced more strongly leached and less fertile soils.

Weathering in the southern and central yellow-brown earths is moderate and the main clay minerals are illite and vermiculite with some halloysite. Montmorillonite is present under conditions of weak leaching.

These soils may show a wide range of chemical features dependent on their leaching stage but they are generally moderately to slightly acid with medium percentage base saturations which decrease with depth. Citric-soluble phosphorus is usually low and pasture responds to phosphate, lime and molybdenum. Pastoral farming is the predominant form of land use on the southern and central yellow-brown earths but some areas, especially of the strongly leached and less fertile soils, are in forest or scrub.

Northern yellow-brown earths

These earths occupy 200,000 hectares of flat and rolling and 300,000 hectares of hilly land and were formed on sandstones, claystones and greywacke under podocarp-dicotylous forest. These soils are mainly heavy clays with thin topsoils and subsoils of moderate to low fertility. They are sticky soils which swell and shrink with wetting and drying. The heavy rainfall and high temperatures result in strong weathering of the micas and feldspars to kaolin although where leaching is weak montmorillonite can occur. The degree to which the soils are leached depends on the composition of the forest cover. Where the forest contained much *Beilschmiedia tarairi* (taraire) or *Vitex lucens* (puriri), moderately acid brown granular topsoils and yellowish brown nutty subsoils resulted. Where the forest contained much *Agathis australis* (kauri) and *Dacrydium cupressinum* the soils are less fertile; they have greyish brown topsoils with weakly developed granular structure, and yellowish brown or greyish brown, harsh nutty or blocky subsoils; they are moderately to strongly acid and strongly leached. Much of the land covered by the northern yellow-brown earths has been cleared of the original forest cover and sown to pastures which are used for sheep and cattle grazing. These pastures respond to phosphate, lime and potash and on the more leached members pasture responses to molybdenum and copper have been recorded. Phosphate responses are obtained in exotic plantations of *Pinus radiata* on some strongly leached soils.

Podzolised yellow-brown earths and podzols

These are developed under mor-forming vegetation in humid and superhumid climatic zones (m.a.r. 1,250–2,500 mm). Under the conditions of heavy rainfall, organic acids leached from the mor surface humus further leach out plant nutrients and topsoil clay. The resulting profile shows a thin dark topsoil on a grey loose A_2 horizon over black and reddish brown compacted iron/humus horizons on yellowish brown blocky subsoils. The southern and central podzolised yellow-brown earths and podzols are formed under *Dacrydium cupressinum – Weinmannia racemosa* (rimu – kamahi) forest and cover 56,000 hectares of easy and 80,000 hectares of hilly country mainly in Southland. The northern podzolised yellow-brown earths and podzols occupy 200,000 hectares of easy land and 120,000 hectares of hilly land mainly in North Auckland. Here they form a mosaic with the associated northern yellow-brown earths. Where these soils developed under *Agathis australis* (kauri) they are referred to as 'gumlands'. In these soils, leaching has removed much topsoil clay, producing grey structureless massive topsoils over a grey siliceous horizon which is cemented to a pan in places.

All these soils are strongly acid and low in plant nutrients and requirements for both major and minor plant nutrients are high for grassland farming.

Gley podzols

Gley podzols are the waterlogged counterparts of the southern podzolised yellow-brown earths and podzols, and occur under wetter (m.a.r. 2,000–4,500 mm) and cooler conditions. Moisture is almost always in excess of plant needs. Gley podzols cover 300,000 hectares of flat and rolling country and 40,000 hectares of hilly country mainly in Westland but they also occur in western Nelson, coastal Southland and Fiordland. In parts these are called 'pakihi lands' which are moorland–like areas of *Leptospermum scoparium* (manuka), umbrella ferns and rushes, and also small swampy clearings in forests. A great proportion of pakihi land is waste land resulting from destruction of *Dacrydium cupressinum – D. colensoi* (rimu – silver pine) and *D. cupressinum – Podocarpus ferrugineus – Weinmannia racemosa* (rimu – miro – kamahi) forests. Profiles of gley podzols consist of a dark brown peaty silt loam topsoil overlying a pale grey massive structureless silt loam. Below about 40 to 75 cm this pale grey silt loam changes abruptly to a dark reddish brown iron pan up to 5 cm thick in silty materials or in thin bands over a zone of 15 cm where gravels occur under the silt loam. Profiles are continually waterlogged and the difficulties of drainage coupled with extremely low nutrient levels are major problems for utilisation of these soils.

Steepland soils related to subalpine gley soils and gley podzols

Steepland soils related to subalpine gley soils and gley podzols together with their hilly and rolling counterparts occupy 1.3 million hectares in these same districts. They occur at altitudes above 900 m in the south and 1,500 m in the north, under rainfalls greater than 2,500 mm, and a vegetative cover of subalpine scrub and tussock grassland. In places they have thin iron pans below the gleyed horizons.

Rendzinas and rendzic intergrades

Rendzinas and rendzic intergrades are formed from limestone and other calcareous parent rocks under various climates and vegetation types. They occupy 48,000 hectares in subhumid Hawkes Bay and parts of Canterbury, and 60,000 hectares in more humid areas such as North Auckland. True rendzinas with deep black A horizons resting on the parent rock do occur but they are of small extent and most soils from calcareous rocks in New Zealand are intergrades to zonal soils with brown to reddish brown silt loam to clay subsoils containing lenses of calcium carbonate. These soils are highly fertile and have good structure but swell and shrink severely with seasonal fluctuations of moisture. Their clay fraction consists predominantly of montmorillonite.

Yellow-brown sands

Yellow-brown sands are derived from coastal sand drifts of various ages. Present in most coastal districts they cover about 200,000 hectares with the largest areas bordering the west coast of Manawatu-Wanganui, and in Auckland and North Auckland districts. They occur over a wide range of climates. Soils on young sand drifts bordering the coast usually have poorly developed profiles and are loose and droughty and are very susceptible to wind erosion once the plant cover is weakened or removed. Further inland, soils on older sand drifts have better developed profiles; topsoils are greyish brown to black very friable loamy sands, and subsoils are brownish yellow to strong brown loose structureless loamy sand. These older soils retain moisture better and are less susceptible to wind erosion. The fertility of the yellow-brown sands varies from low to medium, they are moderately to slightly acid, and have low contents of organic matter. The older, more developed soils are used mainly for pastoral farming, whereas the younger soils are planted in trees (chiefly *Pinus radiata*) to protect inland farmed areas from drifting sand.

Yellow-brown pumice soils

Yellow-brown pumice soils are formed from rhyolitic ash and lapilli derived from two series of paroxysmal eruptions that occurred approximately 800 and 1,700 years ago. They occupy 800,000 hectares of flat to rolling country and 800,000 hectares of hilly country, mainly on the volcanic plateau of the central North Island within a radius of 80 km of Lake Taupo but also extending northward to the Bay of Plenty and eastwards to Gisborne. The climate is generally humid with a well distributed rainfall. Pre-European vegetation was largely scrub and fernland, probably perpetuated by Maori fires, but in some areas lowland tall tussock and podocarp – mixed hardwood forest occurred.

The yellow-brown pumice soils are weakly weathered with black to brown sand to sandy loam topsoils and yellowish brown sandy subsoils containing pumice lapilli. Generally subsoils are rather structureless and friable but where the soils are formed from pumice breccias (nuées ardentes) subsoils are compact and penetration by plant roots is difficult.

Because of the weak weathering these soils have undergone and the small amounts of clay, the impact of the other soil factors, especially vegetation, can markedly affect soil morphology and soils under high rainfalls and mor-forming trees show podzol morphology. The clay is dominantly allophane, amorphous hydrous oxides and silicates of iron and aluminium. These give high phosphate retention properties to the soil and form stable complexes with organic matter so this tends to be high in topsoils. The yellow-brown pumice soils are slightly acid and generally medium to high in citric-soluble phosphorus. Exchangeable potassium is high but reserves are low so this element becomes depleted with intensive farming. Percentage base saturations are medium to low.

For pastoral farming regular topdressing with phosphorus, sulphur and potassium is required and cobalt and selenium are necessary for good animal health. Large areas of these soils have been planted with exotic trees (chiefly *Pinus radiata*) and these plantations are the basis of a large pulp and paper industry.

Yellow-brown loams

Yellow-brown loams are formed from fine-textured andesitic or rhyolitic ash beds or from water-sorted materials derived from these ashes. These ash beds were erupted from volcanoes in central North Island and from Mt. Egmont in Taranaki district between 5,000 and 20,000 years ago. They are extensive in the Waikato, Bay of Plenty, Ohakune and Taranaki districts occupying a total area of 1.2 million hectares of flat and rolling land and 400,000 hectares of hilly land. Small areas of soils in Southland formed on alluvium derived from basic rocks are also included in this group. The yellow-brown loams usually occur under humid climates with mean annual rainfalls ranging from 1,000 to over 2,500 mm. The original vegetation included forest, scrub and fernland, with small areas under native tussock.

Characteristically the yellow-brown loams have low bulk densities (about 0.6 g/cm^3) with black, greyish brown, or brown friable silt loam or sandy loam topsoils with a moderate to strong soft granular or crumb structure. Subsoils are yellowish brown very friable silt loams to heavy silt loams with weak blocky structures which break down readily to a fine crumb structure. The soils developed under high rainfalls and mor-forming trees show signs of podzolisation.

The clay mineral complex of the yellow-brown loams is dominated by allophane, an amorphous clay. Allophane has a highly specific surface and is able to absorb large quantities of humic compounds leading to soils with deep, humic topsoils with high carbon and nitrogen contents. The build-up of organic matter which occurs under intensive pastoral farming also locks up phosphorus in a form unavailable to plants. The high phosphate retention capacity of allophane makes regular and large additions of phosphate fertilisers necessary for the growth of high-producing pastures. Exchangeable potassium is generally high, but due to the lack of micaceous clay minerals, potassium reserves are low and become depleted with intensive use. pH's range from around 6.4 to 5.0, decreasing with increasing stage of maturity. Percentage base saturations are medium to low and these decrease with depth and with increased stage of maturity.

The yellow-brown loams, although not particularly fertile soils in their natural state, have good physical properties and high moisture holding capacities and with regular topdressing of phosphorus and potassium can be converted to high producing soils for grassland farming, cropping or horticulture.

Red and brown loams, and brown granular loams and clays

Red and brown loams, and brown granular loams and clays or 'volcanic soils', are derived from basaltic and ultrabasic rocks, and from andesitic ash and andesites respectively. They cover 500,000 hectares, mainly in North Auckland, Waikato and Coromandel districts but they also occur scattered throughout both islands. Red and brown loams are formed from basic and ultrabasic rocks and occur mainly in the North Auckland district although small areas also occur throughout the South Island. They include two sub-groups, the red loams formed from scoria and scoriaceous rocks and the brown loams from denser rocks of lava flows. Topsoils are dark brown to dark greyish brown in colour with a very friable fine granular structure. Subsoils are red to brown, usually friable breaking to fine granular structure. Subsoil consistence is characterised by low plasticity and lack of stickiness. In some of the soils with poorer drainage on basalt flows, subsoils are compact.

Their clays are variable, depending on the climate the soils are formed under and the length of time they have been developing. With weak weathering, the dominant clay minerals are amorphous oxides of iron and aluminium. With increased weathering these oxides become more crystalline and gibbsite and kaolin develop. In the strongly weathered and leached soils iron concretions together with gibbsite nodules are common. These soils are known as 'ironstone' soils and are very droughty.

The weakly weathered and leached soils are fertile, well supplied with plant nutrients but have high phosphate fixing capacity due to amorphous oxides. The more weathered and leached members have much lower levels of nutrients but have low phosphate retention due to the crystalline nature of the oxides. An ecologically interesting sub-group within the red and brown loams is the group of soils derived from ultrabasic rocks such as serpentine (a magnesium iron silicate formed by metamorphosis from peridotite). As in other countries the flora of the ultrabasic soils shows a predominance of xeromorphic species compared with that of the normal basic soils characteristic of the climatic region. Thus, in the Peridotite Belt of Nelson the normal luxuriant *Nothofagus* forests of the neighbouring land-surface is replaced by three main plant-associations. Shrubland comprises species such as *Griselinia littoralis* and *Nothofagus* sp. found in the neighbouring forests but which are considerably dwarfed in the belt, associated with such non-reduced shrubs as *Cassinia vauvilliersii*, *Coprosma propinqua*, *Dracophyllum longifolium* and *Leptospermum scoparium*. Open scrubland contains *Cassinia vauvilliersii*, *Dracophyllum rosmarinifolium*, *Exocarpus bidwillii*, *Hymenanthera alpina*, *Hebe odora*, *H. divaricata*, *Pimelea suteri* and *Muehlenbeckia axillaris*. The third association is tussock grassland dominated by *Chionochloa* sp., with *Phormium cookianum* and *Astelia nervosa* as sub-dominants. The reasons for such floristic changes are not known but ultrabasic soils are highly alkaline, have high levels of magnesium, chromium and nickel, as well as low levels of calcium and other major nutrient elements.

Brown granular loams and clays are formed from old volcanic ash showers or from andesitic rocks. The brown granular loams are extensive on rolling land in the district between Hamilton and Auckland where they are formed from volcanic ashes estimated at between 20,000 and 100,000 years old. They are friable to firm well drained silt loams to clay loams with black to brown nut structured topsoils over brown to reddish brown blocky subsoils. Many have illuviated clay in the B horizon. They differ from the yellow-brown loams in being firmer, having larger and better developed subsoil aggregates, having shallower topsoils and having wider fluctuations in moisture content. In part, these differences are due to the greater weathering these soils have undergone with a resulting change in clay minerals from allophane to kaolin-type clays. As with the red and brown loams, colloidal oxides of iron and aluminium are present and cause high phosphate retention. Brown granular loams are moderately acid and percentage base saturations range from medium to low depending on the leaching stage.

Land use on these soils is fairly intensive and includes dairying, fattening stock, market gardening, and viticulture.

Brown granular clays are similar to the brown granular loams but are formed from andesitic rocks and tend to be more compact in the subsoil. They occur mainly in North Auckland under rainfalls of over 1,250 mm but localised areas occur in South Island under lower rainfalls. The less leached soils under lower rainfalls have expanding layer-type clays and have high cation exchange capacities and percentage base saturations. The more strongly leached soils of North Auckland contain kaolinitic clays and have lower cation exchange capacities and base saturations. In strongly leached soils under mor-forming trees, podzolisation occurs. The weakly leached soils have moderate to high levels of nutrients but the strongly leached soils are deficient in phosphorus, potassium and calcium, and in trace elements such as molybdenum.

Organic soils

The organic soils are peaty and are formed in hollows and low flats where the water table is permanently high, and conditions of acidity and anaerobiosis result in only partial decomposition of plant remains. About 200,000 hectares of organic soils are scattered throughout New Zealand. However, the most extensive and deepest peat deposit is in the Hamilton basin of North Island, where drainage has been hindered by the levees of the Waikato River. The peat-forming flora of the country is more varied than that of Europe, and most peat occurs under medium rainfall conditions indicating that topography is of prime importance in the distribution of intrazonal peats. Organic soils range in colour from light brown, through dark brown, to almost black, and may be classified into 3 subdivisions. Mellow organic soils were derived from a vegetative cover of mainly *Typha muelleri* (raupo), *Podocarpus dacrydioides* (kahikatea) and *Phormium tenax* (New Zealand flax). Alluvium or volcanic ash also accumulated with the peat to produce peaty loams. Many of these soils are less than 120 cm deep, but some are underlain by deep organic peat. They occur especially in Manawatu, Bay of Plenty, Hawkes Bay and North Auckland. Fertility of these soils is moderate to high and with drainage they can be intensively used for market gardening, dairying or fattening. Intermediate organic soils are mainly covered today with *Leptospermum scoparium*, *Gleichenia* (umbrella fern) and rushes, but the original cover is uncertain. They are mainly deep loamy peats, but include some peaty loams and peaty sands. The organic material is in a less humified condition than in mellow organic soils. Loamy peats occur around large swamps in the Waikato and Hauraki lowlands, as well as in small swamps in several districts. The third subdivision, acid organic soils have a similar vegetative cover today to the mellow organic soils, but there are indications that the main original cover was a *Sporadanthus – Calorophus* association. These are mainly deep, highly acid peats of raised and flat bogs in the Waikato and Hauraki lowlands, North Auckland and central North Island and southern South Island. In many of the raised bogs the plant residues are well decomposed but the peat is less humified than in the mellow and intermediate organic soils. The flat bogs close to sea level are at an earlier phase of bog formation. The intermediate and acid organic soils have lower levels of plant nutrients than the mellow organic soils and development for pastoral farming or for horticulture is difficult.

Gley soils

Gley soils occupy 300,000 hectares throughout New Zealand but are most extensive in the Hauraki Plains south-east of Auckland. They occur widely on valley floors and on low-lying land where the watertable is periodically high and also on sloping land under conditions of very high rainfall (2,500–5,000 mm) or very low evaporation (cloudy mountains). Thus topography and climate are the main factors responsible for gleying. Gley soils are formed where frequent waterlogging decreases the rate of decomposition and retards the formation of mull humus in the soil and the development of numerous small granular aggregates. The result is the formation of a grey-coloured subsoil commonly mottled with rust colours. Poor drainage is the main factor limiting the utilisation of the gley soils but where they have been drained they support high-producing pastures used for dairying and fattening.

Recent soils

Recent soils are young soils which lack the differentiation into the normal horizons of older soils. Such differentiation is limited to a melanisation of the surface layer and in the more developed ones to a colour B horizon. Parent materials are ash from active and dormant volcanoes, alluvium from flooding rivers and colluvium from erosion of slopes. Soil formation is slight where deposition is frequent or the deposits too coarse, but slowly-accumulating and finely-divided deposits usually contain sufficient nutrients and water for good plant growth.

Recent soils from alluvium

Recent soils from alluvium cover about 300,000 hectares on modern flood plains mainly associated with coastal lowlands in both islands. They are mainly deep, brown to olive-grey loams with little differentiation of topsoil where accumulation is rapid, or with a deep dark granular topsoil where accumulation is slow. Where drainage is poor as on low-lying flats, subsoils show gleying. These soils where accumulation is not too rapid and where the alluvium is not too coarse, are highly fertile, well supplied with most plant nutrients and are intensively farmed. Where accumulation is rapid and of coarse textured material, the soils are excessively drained and have a lower natural fertility.

Recent soils from volcanic ash

Recent soils from volcanic ash include soils derived from ash (Ngauruhoe Ash, andesitic) accumulating around the active volcanoes of Ngauruhoe and Ruapehu, from material ejected from the eruption of Mt. Tarawera and Lake Rotomahana in 1886 (Tarawera Ash and Lapilli (basalt), and Rotomahana Mud), and from material erupted from other small recent eruptions. These recent soils occupy about 200,000 hectares and apart from small areas of very coarse textured soils and those at high altitudes, are used for pastoral farming or for forestry.

Steepland soils

Steepland soils are important components of the soil pattern of New Zealand as they occupy about half of the country's area, being most extensive on the axial ranges, although large areas also occur in north-west Nelson and east Taranaki. Because of their steep slopes these soils are periodically rejuvenated by downhill movements, and consequently on most slopes there is little differentiation of horizons (other than melanised topsoils and organic surface layers). The dominant formative processes are removal by erosion, accumulation as colluvium, and mixing by soil creep, all of which tend to increase the amount of soil skeleton exposed to weathering and to reduce the thickness of the weathered mantle, and reduce base leaching, podzolisation and clay illuviation. Thus, the fertility of many of these soils is closely related to that of the parent greywackes, argillites, sandstones, mudstones or volcanic ash. These weakly developed soils of the steeplands are associated with the more strongly developed zonal or intrazonal soils on more stable slopes, and share some properties with them as they occur under similar climatic conditions and weather in a similar direction.

Steepland soils associated with brown-grey earths occupy 100,000 hectares in the semi-arid zone. Steepland soils associated with yellow-grey earths occupy 1 million hectares in the sub-humid zone, and lie to the east of the steepland soils associated with high country yellow-brown earths which cover 1.8 million hectares of cool, humid, high country on the eastern side of the Southern Alps. All these soils were originally covered in tussock grassland but they suffered depletion of this cover soon after European occupation through the introduction of grazing animals such as sheep, rabbits, deer and goats, and the continual firing of the tussock lands to produce more palatable feed for sheep. This depletion was most severe on the steepland soils associated with brown-grey earths and led to severe erosion. With continuous reduction in noxious animal numbers, limitation of firing, more lenient grazing of sheep and cattle, and aerial oversowing of clovers and of sulphur-fortified superphosphate, a denser and more erosion-resistant plant cover is being established in many areas.

Steepland soils associated with yellow-brown earths of the milder and warmer areas of New Zealand were originally covered with forest and scrub. They cover a total of 5 million hectares of which 4.2 million are occupied by soils derived from rocks low in plant nutrients, and the remainder by soils well supplied with plant nutrients. Some areas of the low-fertility soils were cleared of the original forest and grassed but farming is difficult and large areas have reverted to scrub and second growth. Pastures for sheep and cattle grazing have been established on most of the higher fertility steepland soils and these have been maintained and improved by aerial oversowing of clovers and regular aerial topdressing of phosphates. Severe slip, slump and gully erosion occurs in many areas but this is being controlled by the maintenance of a dense pasture cover, space planting of poplars on slopes and on the sides of gullies, and the building of debris dams on the floors of gullies. Where erosion is too severe establishment of exotic tree plantations is carried out.

Steepland soils related to podzolised soils are forest covered, very low in plant nutrients, and occupy 2.7 million hectares mainly in west Nelson, Westland and west Southland.

Finally, alpine steepland soils occur at high altitudes (1,500 m in the south and 1,800 m in the north). These are not well developed soils but delineate the area of bare rock, scree and waste rock above the zone of continuous plant and soil cover. They occupy about 1.5 million hectares.

Plant Nutrients in New Zealand Soils

Most New Zealand soils are of low natural fertility. However, many have been improved by cultivation, drainage, regular topdressing with superphosphate, the application of lime and potash and other fertilisers where needed, sub-division, and all-the-year-round grazing. Of the total land area of 26.6 million hectares, 7.4 million are in improved grassland, 0.6 million in field crops, gardens and orchards and 0.4 million are planted with exotic trees. Detailed information based on experimental work on the nutrient levels and fertiliser requirements of farmed lands is given by DURING (1967).

The organic matter levels of most New Zealand soils are in the range 7–17% organic matter in topsoils; there is vigorous production of organic matter, with no serious limitations to its decomposition. High levels of topsoil organic matter occur in the yellow-brown loams, in which stable complexes form between the organic matter and the amorphous, mineral colloids, and in organic soils, gley soils and podzolised soils, in which the rates of decomposition are low. Low levels, generally resulting from low productivity, occur in yellow-brown sands, recent soils from volcanic ash, and drier yellow-grey earths. Very low levels occur in the brown-grey earths of the South Island.

Levels of the macro-nutrient elements of soils, calcium, magnesium, potassium, nitrogen, and phosphorus have been summarised in Single Factor Maps published by the New Zealand Soil Bureau (METSON, 1962). Levels of sulphur have also been extensively studied. Deficiencies of phosphorus, sulphur, calcium, potassium and magnesium occur in various soils and are corrected by the application of fertilisers. Nitrogenous fertiliser is applied in small amounts to crops, but it should be remembered that the New Zealand farming system is predominantly pastoral, based on the grass-clover association with sheep and cattle, and that nitrogen is almost entirely supplied by biological fixation. Approximately 2 million tons of fertiliser are used annually, of which about 90% is based on superphosphate, 9% on potash, and 1% on nitrogen. About 1 million tons of lime are also used. Sulphur is, of course, added with superphosphate; the other elements are mainly applied as mixtures.

Most New Zealand soils require fertilisation with phosphate for intensive production. The amount required depends to a considerable extent on the 'phosphate retention' of the soil, the reduction in solubility and hence availability of applied phosphate, which varies greatly with soil type. Phosphate retention tends to be low in soils such as the brown-grey earths and yellow-grey earths, in which weathering has not been strong enough to release sesquioxides. It tends to be medium in yellow-brown earths where the weathering is stronger, and high in soils from volcanic parent materials (e.g., yellow-brown loams) which weather rapidly to release large amounts of amorphous mineral colloids. Plants native to areas with highly retentive soils are very slow growing, and the faster growing introduced European species will only survive if supplied with additional phosphate. The work of BAYLIS (1967) suggests that the phycomycetous endotrophic mycorrhizas of the dominant forest trees of New Zealand are normally essential for the uptake of phosphate from forest soils. Absorption of phosphorus from soils by mycorrhizal plants has also been studied by MORRISON (1962). He suggested that the apparent solubilising effect of the ectotrophic mycorrhizas of *Pinus radiata* on soil phosphorus may simply be a consequence of the rapid withdrawal of labile phosphorus by the mycorrhizal sheath.

Micro-nutrient levels in New Zealand soils correlate reasonably well with soil types, and levels of some elements such as copper, molybdenum and cobalt, have been shown to decline in developmental sequences of weathering and leaching (TAYLOR *et al.*, 1956). Molybdenum is the most important micro-element limiting plant growth in New Zealand. It is extensively applied to pastures on yellow-grey earths and yellow-brown earths where natural levels are insufficient to maintain the nitrogen-fixing capacity of the clover nodule rhizobia. Boron, copper, manganese, iron and zinc are also required, singly or in combinations, for specialised crops on a few soils.

Micro-elements are also used on certain soils to improve stock growth. In the 1930's severe cobalt deficiency was found to be responsible for 'bush sickness' in sheep and cattle on the yellow-brown pumice soils of the central North Island. This discovery enabled the present highly efficient grassland farming to develop there. Selenium deficiency is also quite widespread, particularly in the South Island, and is the cause of white muscle disease in lambs and unthriftiness in lambs and calves. Copper is required for both plants and animals in a few areas, particularly those with peaty, sandy, or podzolised soils. Levels of copper in soils have to be carefully controlled, particularly in relation to those of molybdenum, as imbalances of molybdenum and copper affect stock in several parts of New Zealand. Naturally occurring molybdenum-copper imbalance occurs where molybdenum concentration in pastures is high, and where copper levels are insufficient for protection. Induced molybdenum-copper imbalance occurs through excessive or unnecessarily frequent application of molybdenum. Iodine levels in soil and pasture appear to be maintained by wind-blown iodine from the sea. Goitre in lambs, common on the yellow-grey earths in Canterbury, may result more from goitrogens in some white clover strains than from low iodine content of the feed. In addition to these mainly limited areas with well defined plant and animal micro-element deficiencies, there are often much larger areas in which the deficiencies are marginal and difficult to identify. It is likely that as more intensive farming develops in New Zealand, it will be necessary to use micro-elements more extensively.

A number of studies of soil-plant-animal relationships have been discussed by HEALY (1968). These include the effect of soils, especially in regard to micro-elements, on the composition of milk, wool, bone and various organs. Under New Zealand conditions grazing sheep and cattle can also ingest substantial amounts of soil, and HEALY (1970) and HEALY *et al.* (1970) have investigated the effects of ingested soils on macro- and micro-element intake by the animal.

With regard to human health, goitre occurred extensively in New Zealand until iodised salt was introduced in the 1920's, and is now a rare condition. The introduction of fluoride into water supplies in various parts of the country has resulted in considerable reductions in the incidence of dental caries.

The Changing Soil

INFLUENCE OF THE MAORI

Until the arrival of the Maoris before the tenth century, the soils of New Zealand developed under a generally closed vegetation cover, free from the influence of man. The only large grazing and browsing animals were the flightless moas, such as the common three-metre-tall grassland moa, *Dinornis maximus*. The early Maoris were nomadic hunters of moas, and they burnt the vegetation in the course of this and possibly also to encourage growth of *Pteridium aquilinum* for its starch-rich rhizome, but they made little impact on soil formation. However, increased interference with the soils followed from the arrival of further migrants between 1200 and 1400 AD. These Maoris settled in villages and practiced agriculture, certainly as far south as Banks Peninsula in South Island. The root crops they cultivated were the yam (*Dioscorea* sp.), taro *(Colocasia antiquorum)*, small gourd (*Cucurbita* sp.) and particularly the kumara *(Ipomoea batatas)*, all of Polynesian origin. TAYLOR (1958) has described the soil problems that faced the Maori in attempting to grow his tropical food crops in generally infertile soils under a temperate climate. Fire was used to clear tracts of forest, after which the land was carefully prepared by removing large stones, cutting surface drains, and using ashes of the original burn or ashes of the fern and brushwood of a fallowed burn as fertiliser. Blackened Maori soils have been found that appear to have been treated more intensively by burning material (manuka) carried from neighbouring hills – a distinct improvement on primitive shifting agriculture. Some of these blackened soils remained more fertile than surrounding soils. Attempts were also made to improve soil texture by mixing sea sand or river gravel with the soils. In producing these blackened gravelly soils the Maori provided an excellent medium for the growth of the kumara. The coming of the European altered this system of Maori agriculture by replacing the wooden digging stick (ko) with iron implements, and the kumara by the potato, which grew well without the careful soil preparation.

INFLUENCE OF THE EUROPEAN

European settlement, which commenced around 1820, first initiated an exploitive stage of soil use. In this large areas of forest and scrub were felled and burnt, but the soil nutrients derived from these practises were soon depleted by cropping, and to a lesser degree by grazing. Cropped soils were then converted to pasture which became colonised by scrub and fern. Increased run-off and flooding resulting from the replacement of the forest litter by slowly permeable topsoils. Irrigation attempts often caused flooding, and seepage caused water logging or toxic accumulation of soluble salts.

Soil erosion was accelerated on forest land sown in pasture, tussock grassland burnt to graze the regenerating young growth, and rolling and plains lands ploughed for repeated crop monoculture. The soils formed under previous regimes became shallower and more skeletal. Unfortunately as accelerated soil erosion was naturally most severe in areas where normal (geological) erosion was formerly most active, the resulting bare rock was often falsely considered to be a result of normal erosion. In fact much of the accelerated erosion was the consequence of applying farming procedures that may have been successful overseas but were unsuited to soil and vegetative conditions in New Zealand.

Serious erosion problems also stemmed from the introduction of certain exotic mammals for sport, food or their skins. Populations of rabbits, pigs, goats, various deer, wallabies, opossums, goats, chamois and thar became major pests in areas where food was abundant, and there was a relative freedom from predators, interspecific competition, diseases of their native lands, and heavy hunting pressure. The effect of their 'uncontrolled' browsing and grazing has been to accelerate normal erosion under the weakened native vegetation, and to slow the revegetation of areas bared by erosion, thus increasing the area of bared country and consequently the amount of run-off and debris fed to the rivers.

This general exploitation of the country's soils continued until the increased demand for food and woollen clothing by the expanding New Zealand population and particularly the development of the European export market, gave farmers the incentive to improve their soils and increase their yields. This they did by buying fertilisers, weedkillers, certified seeds and modern machinery. A concise account of the development of farming in New Zealand can be found in McLintock (1960), and more comprehensive accounts of the development of grassland farming are given by Smallfield (1970) and Levy (1970). However, it is worth noting here that one of the most important factors in the establishment of the meat export trade and the expansion of the sheep industry, was the introduction of refrigerated ships and the building of freezing works, enabling lamb and mutton to reach the British market in prime condition. Also, one of the most significant developments in the upgrading of fertility in the soils was the commencement in the 1920's of topdressing, or spreading fertiliser on existing pastures, especially by the use of aircraft later.

The greatest changes in the New Zealand soil system have thus resulted from the conversion of vast areas of native forest to a predominantly high producing pastoral system of farming. (Native forests originally covered over 66% of the land area, and have now been reduced to about 22% of the land area.) Miller (1962) followed the rapid chemical changes in a New Zealand hill soil that accompanied a change from scrub (mainly manuka) to fertilised grass. The topsoil, originally infertile, became fertile with large increases in phosphate and bases. Increases occurred in all the phosphorus fractions that were separated.

Many improvements occur in the soil system following the establishment of a good grass-clover pasture. The quality and quantity of soil organic matter increases in most soils, there being either an increase in the level of mull organic matter or a gradual conversion of mor into mull organic matter. The C/N ratio is gradually lowered mainly due to increasing levels of available soil nitrogen from biological fixation by the clover-rhizobia association. With the increases in soil nitrogen, increases in organic sulphur and phosphorus and, in some acid soils, adsorbed sulphate and inorganic phosphorus also occur. Moisture-holding capacity and resistance to drought are also improved. The grazing stock have an obvious influence on the soil by returning plant nutrients in their dung and urine. These nutrient returns, which are not equally divided between dung and urine, are mainly phosphorus, nitrogen, potassium and calcium. However, the chemical changes which can be attributed to the return of dung and urine, are apparently rather small and are evident mainly in the top few centimetres of soil (METSON & HURST, 1953).

A consequence of pasture improvements has been that although it allows very high stock densities, with the absence of a springy matted turf it makes the topsoil more susceptible to hoof damage. The treading of stock on wet soils of winter pastures results in deep hoof penetration, accompanied often by sliding actions, which remoulds the soil and damages the grasses and clovers by tearing or burying them. Such treading produces an increase in the density of the surface soil, the slowing of infiltration, and the slowing of drainage of the soil pores following saturation of the 'pugged' soil by rain (GRADWELL, 1956). Reductions in pasture growth under treading have been recorded, due either to diminished soil aeration or mechanical plant damage (GRADWELL, 1968). The compaction effect of treading varies considerably with soil type, young soils from basalt or andesitic ash being the most resistant and zonal soils the least resistant.

Cropping practice has been responsible for deterioration of soil structure in some areas. In South Canterbury poor yields of wheat were obtained in a wheat/green feed and grass/turnips/rape crop rotation on a yellow-grey earth. Measurements of degree of aggregation and stability factor revealed that a marked deterioration in soil structure had occurred as a result of an unsuitable crop rotation system (PACKARD & RAESIDE, 1952). Ploughing was found to increase the total porosity and volume of large pores, but this effect was considered to be short-lived, since the poor stability of the aggregates was not improved by such treatment. Deterioration of soil structure has also resulted from bad practices followed during continuous cropping of the brown granular loam of the Pukekohe (south of Auckland) market gardening district. Aggregates larger than 2 mm in diameter have almost disappeared by breaking down to smaller aggregates, but crop yields have not been seriously reduced (GRADWELL & ARLIDGE, 1971).

At the micromorphological level, BARRATT (1968) studied the effects of land use differences on a range of New Zealand soils. When native forest, tussock grassland, fernland and scrubland topsoils were converted to introduced pasture, three general trends occurred: a decrease in the concentration of organic matter near the surface, a decrease in granular structure and a decrease in interpedal porosity. These trends were towards a weakly granular mull with a weak mull microfabric. When pasture is topdressed these trends are reversed but with an additional thickening of the topsoil, so that they are towards strongly granular mull with a strong mull microfabric.

Exotic European lumbricid earthworms have had a considerable beneficial influence on the fertility and structure of New Zealand pasture soils, and on pasture productivity (STOCKDILL, 1959). Native species of earthworms were replaced by exotic species when native vegetation was replaced by exotic grassland or cultivated, and lumbricids now comprise most of the earthworm populations of the country's soils. In addition, lumbricids have been deliberately introduced into some developed pastures to improve pasture growth. The three common lumbricid species that inhabit pastures are *Allolobophora caliginosa*, *A. terrestris* and *Lumbricus rubellus*. Earthworm activity removes organic debris from the surface and incorporates it into the topsoil, aiding the distribution of nutrients and improving the latter's structure. In addition their activity improves the infiltration rate and the water absorptive capacity of the soil, reducing run-off and the risk of erosion and conserving water for pasture growth.

Exotic afforestation of the yellow-brown pumice soils of central North Island has induced soil changes. Pumice soils may be more rapidly modified by change of vegetation unlike more strongly weathered soils which are in greater harmony with their environment (N.Z. Soil Bureau, 1960). On these soils a crop of *Pinus radiata* grows to maturity in about 30 years, probably the fastest growth rate of any coniferous forest in the world. Following clear felling of the first crop the slash and litter left on the ground rapidly decompose to produce a topsoil in which earthworms are very active. Consequently, the second crop starts off in topsoil that is more fertile than when the original crop was planted, and nitrogen and other elements are released for new growth. WILL (1959) has studied nutrient return in litter and rainfall under exotic conifers. Annual rates of litter fall are higher than those in other countries, and because of this greater weight of litter fall, the total addition of nutrients to the soil is greater. The amounts of potassium and phosphorus reaching the soil in rainwater under *Pinus radiata* and *Pseudotsuga taxifolia* are considerable; at least two thirds of the potassium reaching the soil under *Pinus radiata* and *Pseudotsuga taxifolia* and half the phosphorus under *Pinus radiata* do so in rainwater. It is presumed that these soluble nutrients are then readily available for uptake by the trees.

The main apparent change to pumice soils under *Pinus radiata* is a bleaching of the soil, which was first thought to result from podzolisation. Under conditions of rapid litter decomposition the bleached topsoil can be rapidly 'remelanised', although the bleached subsoil is more slowly reconditioned. The bleached areas are associated with the mycorrhizal development of feeding roots, and may be caused by the efficient transference of nutrients to the tree by the mycorrhizas.

Consideration of changes in the soil biota occurring as a result of changing patterns of land use are included in the following section.

Distribution and Ecology of the Soil Biota

The current trend of soil biological research in New Zealand (tabulated by LUXTON, 1968) is towards a more ecological approach emphasising the role of the soil biota in soil processes. Previously there was a period of accumulation of basic data on systematics and distribution. However, much basic information is still required on the taxonomy of such important groups as Nematoda, Enchytraeidae, and Acari, and various groups of micro-organisms. Soil biological surveys concerned especially with establishing qualitative differences in soil populations, have been carried out mainly by the Soil Bureau of the Department of Scientific and Industrial Research. New Zealand workers have always considered that a soil type possesses distinctive intrinsic biological properties. The following account of the distribution and ecology of major groups of New Zealand's soil biota is necessarily superficial and selective, and STOUT (in press) should be consulted for further information.

The Microflora

FUNGI

The general saprophytic soil mycoflora has been examined from indigenous tussock grassland, indigenous broadleaved-podocarp forest and beech forest, and pastures of introduced grasses and clovers. Isolates from these habitats are generally referable to species of the cosmopolitan soil mycoflora. Thornton has shown that the type of vegetation cover plays a key role in determining the density and species composition of the soil mycoflora. First studies (THORNTON, 1958, 1960a) showed that the predominant fungi in low-tussock grassland soils were non-sporing forms, dominated by *Rhizoctonia* sp., and that the mycoflora did not alter appreciably when the tussock cover was replaced by root crops or introduced pastures. The lack of change was considered to be due to lack of appreciable change in the type and supply of organic substrates being added to the soil. However, when soils under broadleaved-podocarp forest and soils formerly under such forest but now under pasture, were examined, a marked change in the mycoflora occurred associated with the change in vegetative cover, and presumably change in added organic substrates and microclimate (THORNTON, 1960b). In soils under this forest type *Cylindrocarpon destructans (C. radicicola)* was predominant and Fusaria were absent, whereas in soils under pasture Fusaria were predominant, particularly *Fusarium oxysporum*. The fungal patterns for most of the pasture soils were very similar, suggesting that a comparable environment was being developed in the various soils as a result of the

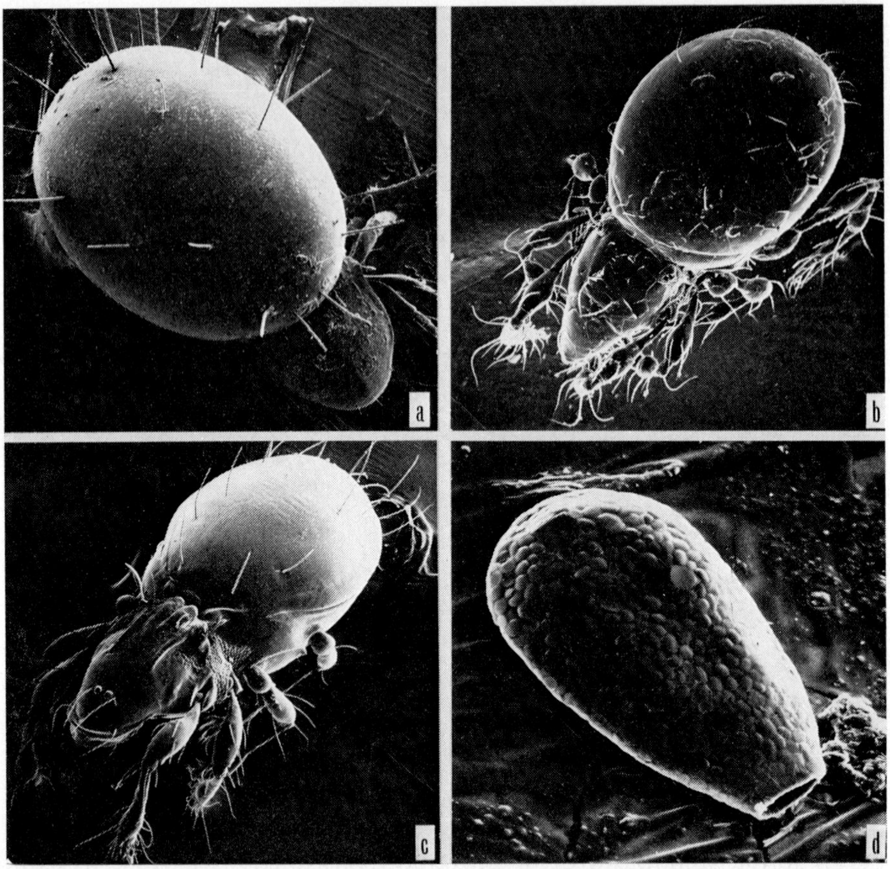

Fig. 3a–d. Common soil animals in *Nothofagus* forest litter. Oribatid mites: a. Phthiracaridae; b. Oppiidae; c. Otocepheidae (*Pseudotocepheus punctatus* Hammer). Testaceous protozoan: d.

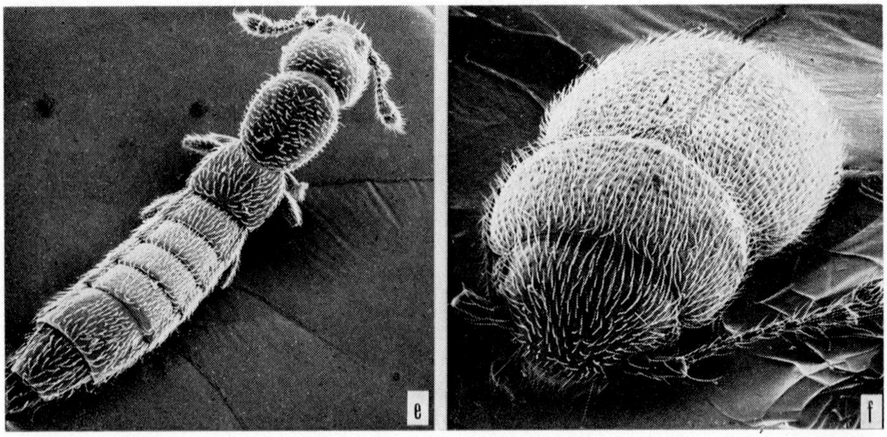

Fig. 3e–f. Common soil beetles in *Nothofagus* forest litter: e. Staphylinidae (*Stenasthetus* sp.); f. Ptiliidae.

establishment of the same vegetative cover, the application of fertilisers and by the grazing of sheep. Soil fungal patterns obtained for hard beech *(Nothofagus truncata)* forest by RUSCOE (1971a) contrast with those obtained by THORNTON (1960b) for podocarp forest. In the hard beech forest soil, *Cylindrocarpon* spp. *(C. didymum* and *C. orthosporum*) were of low frequency. Also fungi recorded by THORNTON (1960b) as contributing more than 5% of his isolates were not recorded from the hard beech forest soil. An exception was *Trichoderma viride*, but THORNTON (1960b) did not isolate it at the high frequency levels with which it occurred in the latter. With the hard beech forest the mycoflora of the surface humus layers and mineral soil horizons were found to be similar, and the most prevalent species were Penicillia, *T. viride*, *Aspergillus versicolor*, *Acremonium* sp., *Tubercularia* sp., *Mucor hiemalis*, *Absidia glauca* and *A. spinosa*.

More intensive studies of fungi in pasture soils were carried out by THORNTON (1965) and JACKSON (1965). The mycoflora colonising live roots of *Lolium perenne* (perennial ryegrass) and *Trifolium repens* (white clover) was dominated by sterile hyaline and dark mycelia. Of the identifiable isolates, a few species such as *Fusarium oxysporum*, *Cylindrocarpon destructans*, *Gliocladium roseum*, *F. culmorum*, *Absidia spinosa*, *Mortierella alpina* and *Penicillium* spp., showed a high frequency of occurrence in particular soils. In warm, moist, moderately acid soils the predominant identifiable fungus was *F. oxysporum*, but this species was rare or absent in cool, moist, less acid soils. In the latter soils *C. destructans* occurred with a high degree of frequency. *C. destructans* was not recorded from soils which had a seasonal drought. As with the live roots, a large proportion of the fungi colonising plant debris in the pasture soils were sterile hyaline and dark forms. Three fungi, *Paecilomyces carneus*, *G. roseum*, and *Cladosporium* sp. occurred on debris in all the soils examined; others occurred in some but not all soils.

Considerable attention has been paid to the mycoflora of plant litter and the leaf surface. RUSCOE (1971b) studied the succession of fungal colonists of the leaf surface and internal tissues of living and dead leaves of *Nothofagus truncata*. He found that young leaves acquired a mycoflora of host-specific internal parasites, and discrete surface colonies soon after unfolding. By the time dead leaves reached the surface litter of the forest floor they were substantially colonised by a variety of Fungi Imperfecti and Ascomycetes. Although the primary saprophytes *Aureobasidium pullulans*, *Cladosporium herbarum*, and *Alternaria tenuis* were common components of successions on leaves of other trees, they apparently colonised the leaves of hard beech at an earlier stage than in the other successions.

Much work has been done on the saprophytic mycoflora of pasture plant materials. Impetus to this was given by the finding that *Pithomyces chartarum*, growing on dead material at the base of pasture, may sporulate profusely enough in warm, humid weather, to make the sward toxic to sheep and cattle. Ingestion of the spores during grazing introduces a hepatotoxin into the animal, which produces liver damage and photosensitisation of the face, symptomatic of the disease called facial eczema (e.g., THORNTON & SINCLAIR, 1960). Fungi on perennial ryegrass, white clover and litter from ryegrass-dominant pasture were investigated by DI MENNA & PARLE (1970); numbers were highest on litter, lower on ryegrass leaves and lowest on clover leaves. The most commonly isolated groups were in approximate order of frequency, pycnidial forms, *Cephalosporium* spp., *Cladosporium herbarum*, *Fusarium* spp. (especially *F. nivale*), *Colletotrichum* spp., *Rhyncosporium* spp., *Verticillium* spp., *Myrothecium* spp., *Pithomyces chartarum*, and *Metarrhizium anisopliae*. Few of the isolates could be placed confidently in the leaf surface mycoflora. MCKENZIE (1971) followed seasonal changes of fungal spore numbers in ryegrass-white clover pasture, by examining both aerial spora and herbage spora. *Puccinia coronata* (from crown rust infection of ryegrass) and *Cladosporium herbarum* spores occurred in greatest numbers, but comparisons of species listed by MCKENZIE (1971) and DI MENNA & PARLE (1970) are difficult due to differences in sampling methods. Finally, DI MENNA's (1959) study of the yeast population in the leaf surface of pasture plants, showed that these yeasts were taxonomically distinct from soil yeasts, and varied with season but not with the localities sampled. An interesting finding was that the yeast content of pasture leaves eaten by herbivores may vary from almost nothing in spring, to about 1% by weight in the early autumn (mainly red pigmented species). Thus grass should not be regarded as grass alone, but as a microcosm.

The soil yeasts, of which there are one or more species in most New Zealand soils, have been particularly intensively studied. Numbers of soil yeasts vary little with season, but considerably with soil type. From a survey of representatives of the major soil groups, DI MENNA (1956) found that numbers of yeasts showed some relation to the amount of organic matter reaching the soil and to the degree of soil aeration. The distribution of species could best be related to soil moisture and temperature. The most common isolates were *Cryptococcus albidus*, a variant of *C. albidus*, *Cryptococcus terreus*, and *Candida curvata*. *C. curvata* was the dominant species of moist soils: those with good drainage and rainfall of over about 1,300 mm p.a., or a lower rainfall and impeded drainage. *Cryptococcus albidus* variant was common only in the semi-arid soils of South Island. *C. albidus* was dominant in the cooler of the well drained soils with a rainfall of less than about 1,300 mm, at or near sea level in the south of South Island, or at higher altitudes in the rest of the country. *C. terreus* was dominant in the well drained soils of warmer districts. The distribution pattern of the genus *Lipomyces* was interesting (DI MENNA, 1966). It occurred in all soil groups examined, but numbers were usually less than those of other yeasts, the genus being dominant only in soils where other yeasts were virtually absent. Unlike other yeasts, a widely fluctuating moisture regime appeared not to be unfavourable to *Lipomyces*. Cultivation of the soil increased the numbers of *Lipomyces*. In a further study (DI MENNA, 1960), two kinds of change were found when broadleaf forest was replaced by introduced ryegrass-white clover pasture. A yellow-brown loam under forest contained small numbers of *Candida curvata;* under pasture yeast numbers were greater and dominated by *Cryptococcus terreus*. In a recent gley soil under forest *Candida curvata* was dominant but *Hansenula* spp. were also present in considerable numbers; under pasture yeast numbers were less, *Hansenula* spp. were absent and *C. curvata* was the sole dominant.

A survey of predacious fungi, mainly nematode-trapping Hypomycetes, showed them to be more numerous in agricultural than virgin soils, in samples containing animal dung than in dung-free samples, and in less acid than acid soils (FOWLER, 1970). It was considered that their presence in soils was probably a result of post-European mechanical introduction.

It is not possible to discuss the ectotrophic and endotrophic mycorrhizae of native and introduced plants here. However, in view of the widespread occurrence of vesicular arbuscular mycorrhizae through the Bryophytes, Pteridophytes, Gymnosperms and Angiosperms, and the abundance of spores of mycorrhizal *Endogone* species in most soils in the world, the survey of MOSSE & BOWEN (1968) is of interest. Soil samples collected from under native forest, grassland and scrub, and under arable, pasture and garden usage, were examined for the presence of spores of *Endogone*. Live spores were more abundant in cultivated soils than under natural vegetation. The forest soils characteristically lacked spores, or contained small, dark-brown, unidentified, dead spores. There also appeared to be few spores in regions of high rainfall or permanent wetness. The survey suggested that the distribution of *Endogone* spores is correlated with land usage and soil moisture rather than with soil type or particular host plants. JACKSON (1965) also found that vesicular-arbuscular mycorrhizal fungi are an important component of the mycoflora of pasture soils.

ACTINOMYCETES

Actinomycetes form an important component of the microflora of New Zealand soils (VERNON, 1968). Numbers of actinomycetes were highest in soils under good pasture, of near neutral pH, in sub-humid areas. Numbers were low in wet, acid soils, and in coarse textured soils of the central volcanic plateau of North Island. They were also low in acid soils, below pH 5.6, and in soils under poor pasture, scrub, fern, tussock grassland, podocarp-dicotylous forest or exotic pines.

Most of the actinomycetes isolated belong to the genus *Streptomyces*. Of the more widely distributed species, *S. venezuelae*, *S. violaceoniger* and *S. viridogenes* are present in, and *S. griseoflavus* absent from, each soil group derived from recent volcanic ash. Also, the numbers of species of *Streptomyces*, as with total number of actinomycetes, tend to be very low in podzolised and organic soils.

STOUT (1961) recognised four distinct bacterial floras associated with the four main phases of the organic cycle, namely floras of living leaves, dead leaves and litter, mineral soil, and peat.

The leaf surface populations can be extremely prolific, representing the highest non-parasitic bacterial populations in nature; e.g., up to 100,000,000 per g wet weight on leaves of ryegrass – clover pasture. Although some components of this flora inevitably enter the litter and soil, they do not persist extensively in time or space. The leaf surface flora is characterised by pigmented bacteria, whose carotenoids appear to have a protective function against the effects of light and photo-oxidation (STOUT, 1960a, b). Taxa include the pseudomonad *Xanthomonas*, the peritrichous *Flavobacterium*, *Micrococcus*, and *Sarcina*. Two common soil genera, *Nocardia* and *Bacillus*, are very poorly represented. The aerogenic fermenter *Enterobacter* may occur, and become dominant on pasture leaves during short periods of rapid pasture growth in spring.

The litter populations tend to bear the impress of the original leaf surface flora. Pasture litter may have large populations, approaching the size of those on the leaf surface. The pasture litter population resembles that of the latter habitat, except that pigmented bacteria are less dominant and there is a higher incidence of soil bacteria like *Pseudomonas* and *Bacillus*. Populations in forest litter are generally lower in magnitude, between 10,000 and 10,000,000 per g wet weight. The composition of the forest litter floras varies with the litter type. Thus, mull litter under *Vitex lucens* (puriri) and *Dysoxylum spectabile* (kohekohe) has a flora related to both leaf and soil floras and without any distinctive litter components. With mor litter under *Leptospermum scoparium* (manuka) and *Agathis australis* (kauri) a similar pattern occurs, except that the leaf flora tends to predominate in the upper litter layers and the soil flora (e.g., *Bacillus*) in the lower layers. Aerogenic fermenters are important in the well-decomposed layers of manuka, beech, and kauri litter, and the well-humified kamahi *(Weinmannia racemosa)* litter of Stewart Island.

Bacterial populations of mineral soil have been studied more extensively for grassland soils using enriched media (STOUT, 1958a, 1960a). Populations may vary greatly with depth and season (e.g., from 100,000 to 10,000,000 per g wet weight), and are greater in the topsoil than in the subsoil. The grassland soil flora consists predominantly of aerobic spore-forming bacteria, and Gram-negative bacteria, especially pseudomonads. Pigmented bacteria, especially flavobacteria, from the leaf flora, may also be present in the uppermost layer of soil. The soil flora may consist very largely of spore-forming bacilli (*Bacillus cereus*, *B. mycoides* and *B. megaterium*), non-pigmented pseudomonads, and smaller numbers of the aerogenic fermenter *Enterobacter*. The latter, however, tend to be dominant in soils rich in organic matter. In the well established grassland soils high in organic matter, pseudomonads are characteristically dominant and the *Pseudomonas fluorescens* group is common. There is also an increase in the proportion and number of taxa of aerogenic fermenters (*Aeromonas*, *Aerobacter*, and especially *Serratia*) in the organically richer soils. STOUT (1961) considered that the *Pseudomonas fluorescens* – *Serratia* flora is an edaphic climax comparable with those found in higher plant ecology. It has been found in soils of widely different origin under native and exotic grasslands, and over a wide range of latitude. The greatest numbers of soil bacteria occur in the rhizosphere of plant roots, and *Aerobacter* and *Serratia* are conspicuous components of this flora.

However, the results of ROBINSON & MACDONALD (1964), and LOUTIT & LOUTIT (1966) differ from those of STOUT (1958a, 1960a). Unlike the latter, these workers isolated soil bacteria using mainly soil extract agars. ROBINSON & MACDONALD (1964) studied the bacterial flora of the same tussock grassland soil as did STOUT (1960), and characterisation of their isolates revealed that *Arthrobacter*, *Agrobacterium* and *Nocardia* accounted for about one-half of the isolates, but pseudomonads comprised less than one-fifth of the isolates. This was in sharp contrast to STOUT's (1960) conclusion that the aerobic bacterial flora was predominantly a pseudomonad one. The work of LOUTIT & LOUTIT (1966) confirmed the results of ROBINSON & MACDONALD (1964) as the former authors found their wasteland and pasture soils to be characterised by species of *Arthrobacter*, *Nocardia* and *Mycobacterium*. Both ROBINSON & MACDONALD (1964), and LOUTIT & LOUTIT (1966) considered that the bacterial floras of the soils that they examined, are similar in general character to those of soils elsewhere in the world.

The peat flora is physiologically less active than the other floras and is less diverse. Except for the surface layers where pigmented bacteria are sometimes present, the flora is characterised by the absence of the pseudomonads of forest litter or grassland soil. Aerobic spore-forming bacteria are less common than in soil, and *Enterobacter* is sometimes present.

Nitrogen-fixing bacteria isolated from New Zealand soils include *Azotobacter*, *Beijerinckia*, Gram-negative facultative anaerobes (e.g., *Klebsiella aerogenes*), spore-forming bacteria (e.g., *Bacillus polymyxa*) and *Clostridium* (STOUT, in press). *Azotobacter*, of which two species *A. chroococcum* and *A. beijerinckia*, occur has been isolated from relatively few soils. These are all soil with near neutral pH's; three high fertility soils, several low fertility soils on sand dunes, and the very young soil formed after the retreat of the Franz Josef Glacier. Some factors influencing the distribution of *Azotobacter* in soils of the South Island are discussed by LINE & LOUTIT (1969). *Beijerinckia* has been isolated from mountain soils of one locality, which are similar to the soils of sand dunes and soils from Franz Josef. The most widely distributed nitrogen fixer is the anaerobic *Clostridium* which is present in most soils. Numbers of this bacterium in soils of uniform texture (e.g., silt loams) are related to rainfall, the lower the rainfall the lower their numbers.

Algae

Algae have been found in New Zealand soils at sea level and up to at least 3,500 m, and they are present in every soil that has been examined. Flint (1968) examined a range of soils under forest, scrub, and pasture, and found that the distribution of the algal classes and some of the species was similar to that observed on soils in other parts of the world. The dominant class in these mainly acid soils of low base saturation, was the Chlorophyceae. Bacillariophyceae were absent from the soils with the lowest base saturation, and generally rare in the other soils. Myxophyceae were prolific only on the more alkaline and base-rich soils where the sparse vegetation cover had bared the soil. Three species of nitrogen-fixing Myxophyceae, *Scytonema hofmanni*, *Nostoc muscorum* and *N. commune*, were found in intensively studied soils. Xanthophyceae were recorded in few soils, with the exception of *Heterothrix* which was widespread. Chryosphyceae were found in two soils. The other classes were represented by only *Porphyridium* (Rhodophyceae) in a salt slick, and *Euglena* sp. (Euglenineae) in certain acid soils and a soil from ultra-basic rocks.

Flint (1968) found that the algae varied qualitatively and quantitatively with the pH, the concentration of bases, and the nature of the vegetation cover. The litter-covered surface of soils under forest and scrub, seemed to be unfavourable to the growth of Bacillariophyceae. When pasture replaced scrub or forest, algae became more numerous and the populations contained more species.

The Microfauna

Protozoa

The protozoa are the most widespread of the microfauna. Several hundred species of shelled and naked amoeboid protozoa and ciliate protozoa are found in various terrestrial habitats, and almost all are known from other parts of the world.

Stout (1958b) studied the microfauna of three tussock-grassland soils, and in each case the ciliate fauna was larger than the rhizopod fauna. The average incidence of the protozoan species was characteristic for each soil type. When such soils were cultivated or had pasture established on them, changes in the protozoan fauna ensued only when these agricultural changes resulted in increased organic activity and a higher bacterial population (Stout, 1960c). The protozoan changes included an increase in the number of species recorded and a more frequent occurrence of species.

Changes in the incidence of protozoa and other microfauna following scrub burning on a hill soil, have been reported by STOUT (1961b). Prior to the burn, the protozoan faunas of the upper and lower litter layers under scrub were very similar. The burn resulted in partial sterilisation of the litter, which was depleted of its microfauna, rhizopods being more severely affected than the ciliates. However, the topsoil fauna was enriched, probably by litter and its associated fauna being added to the topsoil. Thus, many species of rhizopods and ciliates not initially recorded from the topsoil were added to its microfauna. Finally, the newly established grassland soil with its new topsoil niches and new ecological equilibrium produced a microfauna consisting predominantly of ciliates.

NEMATODA

As yet the free-living nematode fauna of New Zealand is poorly known, but clearly contains a very large number of undescribed endemic species. There is even less distribution data than taxonomic data.

YEATES (1967) made a quantitative comparison of the nematode faunas of six dune sand sites under *Ammophila arenaria* (marram grass) or *Desmoschoenus spiralis*. Fifty six, mainly 'terrestrial' species, were found, the majority of the identified species being new taxa. The faunas under the two types of vegetative cover were essentially similar. However, there were more bacterial feeders in the fauna under *A. arenaria* than under *D. spiralis*, apparently connected with the greater amounts of available organic 'residue' in the *A. arenaria* soils. In addition to bacterial feeders, the faunas contained plant feeding, predacious and microherbivorous types. Soil moisture and food sources seemed to be important in determining species distribution. There was also an indication of a north-south differentiation of the fauna, *Aporcelaimellus maitai*, *Labronema rikia*, *Discolaimium sabuli*, *Scutellonema magna* and *Alaimus himatangiensis* being known only from North Island localities, whilst *Stegelleta tuarua*, *Trissonchulus littoralis*, *Helicotylenchus depressus* etc. were known only from South Island localities. Annual variation of the nematode fauna has also been studied in a dune sand site (YEATES, 1968), and variations of the populations with time and soil depth, correlated with variation in soil moisture, temperature, salinity, pH, 'residue', and enchytraeid fauna.

An ecological study has also been made of the nematode fauna associated with apple trees in a grassed orchard (EGUNJOBI, 1968). Thirty nine genera belonging to five orders were identified, the majority of the nematodes being tylenchids. The vertical distribution of *Tylenchus* spp., total nematodes, *Pratylenchus* spp., and *Helicotylenchus* spp. were influenced by soil moisture levels, although root distribution was more important in influencing the distribution of the plant feeders. The effects of the seasons on the vertical zonation patterns of the total nematodes was considered.

Twenty four genera of plant parasitic nematodes are known to occur in New Zealand. These include ovary parasites of wheat, browntop and chewings fescue, stem and bulb parasites of cereals, cultivated bulbs and potatoes, foliar parasites of cultivated flowers and grasses, sedentary root parasites, and migratory root parasites. The most important animal parasites associated with pasture soils are the trichostrongylid nematodes which cause parasitic gastro-enteritis in sheep and cattle.

The Macrofauna

It is not possible to give an adequate treatment here to the macrofauna of New Zealand soils. For an account of their distribution, LEE's (1968) survey of the fauna of a selected range of 45 soils should be consulted. This indicated that a sequence of changes in the macrofaunal populations paralleled the zonal soil sequence of brown-grey earths, yellow-grey earths, yellow-brown earths, podzols and gley podzols. Further information on the distribution and ecology of the soil macrofauna is given by LEE (1959) for the earthworms, and STOUT (in press) for Crustacea, millipedes, chilopods, Collembola, mites, opiliones, spiders, pseudoscorpions, snails and slugs. Some of these, and other invertebrate groups, have been treated in this volume by CLIMO, FORSTER, WATT and DUGDALE.

Acknowledgements

I wish to thank Dr. J. D. STOUT, Soil Bureau, DSIR and Physics and Engineering Laboratory, DSIR, Lower Hutt for the scanning electron microscope photos of the soil animals illustrating this paper.

REFERENCES

BARRATT, B. C. 1968. Micromorphological observations on the effects of land use differences on some New Zealand soils. *N.Z. Jl agric. Res.* 11: 101–130.

BAYLIS, G. T. S. 1967. Experiments on the ecological significance of the phycomycetous mycorrhizas. *New Phytol.* 66: 231–243.

DURING, C. 1967. Fertilisers and soils in New Zealand farming. Wellington, R. E. Owen, Government Printer.

EGUNJOBI, O. A. 1968. An ecological study of some soil nematodes associated with apple trees in a New Zealand grassed orchard. *N.Z. Jl agric. Res.* 11: 386–406.

Flint, E. A. 1968. Algae on the surface of some New Zealand soils. *N.Z. Soil Bur. Bull.* 26(2): 183–190.
Fowler, M. 1970. New Zealand predacious fungi. *N.Z. Jl Bot.* 8: 283–302.
Gradwell, M. 1956. Effects of animal treading on a wet alluvial soil under pasture. *Proc. N.Z. Soc. Soil Sci.* 2: 37–39.
Gradwell, M. 1968. Compaction of pasture topsoils under winter grazing. 9th Int. Congr. Soil Sci. Trans. 3: 429–435.
Gradwell, M. & Arlidge, E. Z. 1971. Deterioration in soil structure in the market gardens of the Pukekohe district, New Zealand. *N.Z. Jl agric. Res.* 14: 288–306.
Healy, W. B. 1968. Some soil-plant-animal relationships. *N.Z. Soil Bur. Bull.* 26(1): 110–112.
Healy, W. B. 1970. Ingested soil as a possible source of elements for grazing animals. *Proc. N.Z. Soc. Animal Production.* 30: 11–19.
Healy, W. B., McCabe, W. J. & Wilson, G. F. 1970. Ingested soil as a source of microelements for grazing animals. *N.Z. Jl agric. Res.* 13: 503–521.
Jackson, R. M. 1965. Studies of fungi in pasture soils. 2. Fungi associated with plant debris and fungal hyphae in soils. *N.Z. Jl agric. Res.* 8: 865–877.
Lee, K. E. 1959. The earthworm fauna of New Zealand. *Bull. N.Z. Dep. sci. industr. Res.* 130.
Lee, K. E. 1968. A preliminary study of soil animals and their relationships to some New Zealand soils. *N.Z. Soil Bur. Bull.* 26(2): 168–183.
Levy, E. B. 1970. Grasslands of New Zealand. Wellington, A. R. Shearer, Government Printer.
Line, M. A. & Loutit, M. W. 1969. Occurrence of *Azotobacter* in some soils of South Island, New Zealand. *N.Z. Jl agric. Res.* 12: 630–638.
Loutit, M. W. & Loutit, J. S. 1966. The aerobic heterotrophic bacteria of two New Zealand soils. *N.Z. Jl agric. Res.* 9: 84–92.
Luxton, M. 1968: Soil biology in New Zealand. *Pedobiologia.* 8: 265–270.
McKenzie, E. H. C. 1971. Seasonal changes in fungal spore numbers in ryegrass-white clover pasture, and the effects of benomyl on pasture fungi. *N.Z. Jl agric. Res.* 14: 379–392.
Menna, M. E. di. 1959. Yeasts from the leaves of pasture plants. *N.Z. Jl agric. Res.* 2: 394–405.
Menna, M. E. di. 1965. Yeasts in New Zealand soils. *N.Z. Jl Bot.* 3: 194–203.
Menna, M. E. di. 1966. Lipomyces in some New Zealand and Pacific soils. *N.Z. Jl Bot.* 4: 406–417.
Menna, M. E. di & Parle, J. N. 1970. Moulds on leaves of perennial ryegrass and white clover. *N.Z. Jl agric. Res.* 13: 51–68.
Metson, A. J. 1962. Soil Bureau Atlas. N.Z. Soil Bureau.
Metson, A. J. & Hurst, F. B. 1953. Effects of sheep dung and urine on a soil under pasture at Lincoln, Canterbury, with particular reference to potassium and nitrogen equilibria. *N.Z. Jl Sci. Techn.* A. 35: 327–359.
Miller, R. B. 1962. Biological and chemical changes following scrub burning on a New Zealand hill soil. 4. Changes in some chemical properties. *N.Z. Jl Sci.* 5: 259–268.
Morrison, T. M. 1962. Absorption of phosphorus from soils by mycorrhizal plants. *New Phytol.* 61: 10–20.
Mosse, B. & Bowen, D. C. 1968. The distribution of *Endogone* spores in some Australian and New Zealand soils, and in an experimental field soil at Rothamsted. *Trans. Br. mycol. Soc.* 51: 485–492.
New Zealand Soil Bureau. 1954. General survey of the soils of North Island, New Zealand. *N.Z. Soil Bur. Bull.* 5.
New Zealand Soil Bureau. 1960. Soils, forestry and agriculture of the northern part, Kaingaroa State Forest and the Galatea Basin. *N.Z. Soil Bur. Bull.* 18.

New Zealand Soil Bureau. 1968. General survey of the soils of South Island, New Zealand. *N.Z. Soil Bur. Bull.* 27.

Packard, R. Q. & Raeside, J. D. 1952. Deterioration of soil structure in a South Canterbury soil. *N.Z. Jl Sci. Techn.* A. 33: 40–53.

Pohlen, I. J. 1957. The history of soil science. In Science in New Zealand (handbook for 1957 A.N.Z.A.A.S conference), Wellington, A. H. & A. W. Reed.

Robinson, J. D. & MacDonald, G. M. 1964. The aerobic bacterial flora of a New Zealand tussock-grassland soil. *N.Z. Jl agric. Res.* 7: 146–157.

Ruscoe, Q. W. 1971a. The soil mycoflora of a hard beech forest. *N.Z. Jl Sci.* 14: 554–567.

Ruscoe, Q. W. 1971b. Mycoflora of living and dead leaves of *Nothofagus truncata*. *Trans. Br. mycol. Soc.* 56: 463–474.

Smallfield, P. W. 1970. The grasslands revolution in New Zealand. Auckland, Hodder and Stoughton.

Stockdill, S. M. J. 1959. Earthworms improve pasture growth. *N.Z. Jl agric.* 98: 227–233.

Stout, J. D. 1958a. Biological studies of some tussock-grassland soils. 4. Bacteria. *N.Z. Jl agric. Res.* 1: 943–957.

Stout, J. D. 1958b. Biological studies of some tussock-grassland soils. 7. Protozoa. *N.Z. Jl agric. Res.* 1: 974–984.

Stout, J. D. 1960a. Biological studies of some tussock-grassland soils. 13. Bacteria of two cultivated soils. *N.Z. Jl agric. Res.* 3: 214–223.

Stout, J. D. 1960b. Bacteria of soil and pasture leaves at Claudelands Showgrounds. *N.Z. Jl agric. Res.* 3: 413–430.

Stout, J. D. 1960c. Biological studies of some tussock-grassland soils. 16. Protozoa of two cultivated soils. *N.Z. Jl agric. Res.* 3: 237–244.

Stout, J. D. 1961a. A bacterial survey of some New Zealand forest lands, grasslands, and peats. *N.Z. Jl agric. Res.* 4: 1–30.

Stout, J. D. 1961b. Biological and chemical changes following scrub burning on a New Zealand hill soil. 3. Microbiological changes. *N.Z. Jl Sci.* 4: 740–752.

Stout, J. D. 1973. Soil Ecology. In G. R. Williams (ed.) The Natural History of New Zealand. A. H. & A. W. Reed, Wellington.

Taylor, N. H. 1949. Soil survey and classification in New Zealand. Proc. 7th Pacif. Sci. Congr. 6: 103–113.

Taylor, N. H. 1958. Soil science and New Zealand prehistory. *N.Z. Sci. Rev.* 16: 71–79.

Taylor, N. H., Cunningham, I. J. & Davies, E. B. 1956. Soil type in relation to mineral deficiencies. Proc. 7th int. Grassl. Congr. 357–367.

Taylor, N. H. & Pohlen, I. J. 1959. Soils and land use. In A Descriptive Atlas of New Zealand. (ed. A. H. McLintock): 28–33. Maps 12 and 13, Government Printer, Wellington.

Taylor, N. H. & Pohlen, I. J. 1970. Soil survey method. *N.Z. Soil Bur. Bull.* 25.

Thornton, R. H. 1958. Biological studies of some tussock grassland soils. 2. Fungi. *N.Z. Jl agric. Res.* 1: 922–938.

Thornton, R. H. 1960a. Biological studies of some tussock grassland soils. 16. Fungi of two cultivated soils. *N.Z. Jl agric. Res.* 3: 203–206.

Thornton, R. H. 1960b. Fungi of some forest and pasture soils. *N.Z. Jl agric. Res.* 3: 699–711.

Thornton, R. H. 1965. Studies on fungi in pasture soils. 1. Fungi associated with live roots. *N.Z. Jl agric. Res.* 8: 417–449.

Thornton, R. H. & Sinclair, D. P. 1960. Some observations on the occurrence of *Sporidesmium bakeri* Syd. and facial eczema disease in the field. *N.Z. Jl agric. Res.* 3: 300–313.

Thornthwaite, C. W. 1948. An approach towards a rational classification of climate. *Geogr. Rev.* 38: 55–94.

VERNON, T. R. 1968. Actinomycetes. *N.Z. Soil Bur. Bull.* 26(2): 202–206.
WILL, G. M. 1959. Nutrient return in litter and rainfall under some exotic conifer stands in New Zealand. *N.Z. Jl agric. Res.* 2: 719–734.
YEATES, G. W. 1967. Studies on nematodes from dune sands. 9. Quantitative comparison of the nematode faunas of six localities. *N.Z. Jl Sci.* 10: 927–948.
YEATES, G. W. 1968. An analysis of annual variation of the nematode fauna in dune sand, at Himatangi Beach, New Zealand. *Pedobiologia*, 8: 173–207.

IV. FLORA AND VEGETATION

by

E. J. GODLEY

Introduction

In the southern spring of 1769, Lieutenant JAMES COOK, commanding His Majesty's bark Endeavour, rediscovered New Zealand and explored its coasts. The botanists with COOK were DANIEL SOLANDER and JOSEPH BANKS and the latter noted in his journal: 'The face of the countrey is in general Mountainous, especially inland, where probably runs a chain of very high hills parts of which we saw at several times; they were generaly coverd with snow and certainly very high –' (BANKS, 1963). We now know that three quarters of New Zealand lies above an altitude of 200 m and that BANKS was describing the mountainous backbone, which runs N.E.–S.W. through most of the land, and is particularly prominent in South Island where 19 mountains or peaks exceed 3000 m in height. The three main islands, North, South and Stewart, have a total length of some 1600 km. They lie between 34°25′ and 47°20′S and form the core of the New Zealand Botanical Region. This also includes the Kermadec Islands (29°15′–31°24′S), some 960 km N.E. of Auckland; the Chatham Islands (43°35′–44°25′S) some 770 km E. of Christchurch; and the Subantarctic Islands, six widely separated groups between 48° and 55°S and 159° to 179°E. The Kermadec Islands lie in a latitude equivalent to northern Florida or the Canary Islands; the New Zealand mainland is diametrically opposite Spain; Antipodes Island is very roughly antipodal to Greenwich; and Macquarie Island lies at the same distance from the equator as Yorkshire or the Aleutians.

These altitudinal and latitudinal changes provide environments for a rich variety of vegetation; but this is not reflected in any extraordinary number of indigenous species (MILLENER, 1960, GODLEY, 1961). Later exploration has confirmed – at least in the flowering plants – the first impression of BANKS: 'Tho the countrey is generaly covered with an abundant verdure of grass and trees yet I cannot say that it is productive of so great a variety as many countries I have seen.' Certainly the number of species of vascular plants in four temperate countries (Table 1) does not indicate exceptional species numbers in New Zealand.

BANKS went on to write: 'The intire novelty however of the greatest part of what we found recompens'd us as natural historians for the want of variety.' After two centuries of investigation what can we say of this early impression? Table 2 shows that, except for the ferns, there is a high degree of endemism on the specific level amongst the native vascular

Table 1. Total number of indigenous species of ferns, lycopods, gymnosperms, and flowering plants, in four temperate countries in relation to area.

	Indigenous species of vascular plants	Area Km²	Species per 1000 Km²
British Islands[1]	1702	307,702	5.5
New Zealand[2]	1996	267,800	7.4
California[3]	3727	401,408	9.3
Japan[4]	4022	377,152	10.7

1. Clapham, Tutin, Warburg, 1962. (2375 species less 673 adventives).
2. Allan, 1961; Moore & Edgar, 1970; V. D. Zotov (pers. comm.).
3. Jepson, 1951 (4019 species less 292 adventives).
4. Jisaburo Ohwi, 1965. (4263 species less 241 adventives).

Table 2. Number of indigenous genera and species of higher plants in New Zealand. Based on – Flora of New Zealand Vol. 1 (Allan, 1961) and Vol. 2 (Moore & Edgar, 1970) with grasses from V. D. Zotov, (pers. comm.).

	GENERA			SPECIES		
	Total	Endemic	% Endemic	Total	Endemic	% Endemic
Ferns and fern allies	47	1	2%	163	67	41%
Gymnosperms	5	0	0%	20	20	100%
Dicotyledons	235	32	14%	1268	1131	89%
Monocotyledons (except grasses)	75	3	4%	339	214	63%
Grasses	31	3	10%	206	186	90%
Total N.Z.	393	39	10%	1996	1618	81%

plants. On the generic level, however, there is only one endemic genus of ferns *(Loxoma)* and 38 of flowering plants, 24 of the 39 being monotypic. These genera are listed in Table 3 and their authors include many famous names in New Zealand botany. The difference between the percentages of endemic genera and endemic species in a country so isolated, the number of monotypic endemics, the relation between species numbers in typically lowland and alpine genera, and the total number of species, have been discussed by Millener *(Op. cit.)* who considers that all result from a decimation in the glacial periods, particularly of the lowland species which could not move away.

Some of these endemic genera contribute towards the distinctive appearance of the vegetation, but no less important in this respect are

Table 3. Endemic genera of New Zealand (currently recognised) arranged in chronological order of description.*

1776	Brachyglottis	J. R. & G. FORSTER	(Compositae)
1814	Entelea	R. BROWN	(Tiliaceae)
1828	Tupeia	A. VON CHAMISSO & D. F. L. VON SCHLECHTENDAL	(Loranthaceae)
1837	Loxoma	R. BROWN ex A. CUNNINGHAM	(Loxomaceae)
1838	Rhabdothamnus	A. CUNNINGHAM	(Gesneriaceae)
1839	Alseuosmia	A. CUNNINGHAM	(Caprifoliaceae)
1839	Ixerba	A. CUNNINGHAM	(Escalloniaceae)
1839	Hoheria	A. CUNNINGHAM	(Malvaceae)
1844	Pleurophyllum	J. D. HOOKER	(Compositae)
1853	Desmoschoenus	J. D. HOOKER	(Cyperaceae)
1854	Teucridium	J. D. HOOKER	(Verbenaceae)
1854	Stilbocarpa	A. GRAY	(Araliaceae)
1857	Notospartium	J. D. HOOKER	(Papilionaceae)
1859	Dactylanthus	J. D. HOOKER	(Balanophoraceae)
1859	Myosotidium	W. J. HOOKER	(Boraginaceae)
1862	Notothlaspi	J. D. HOOKER	(Cruciferae)
1864	Traversia	J. D. HOOKER	(Compositae)
1864	Haastia	J. D. HOOKER	(Compositae)
1864	Hectorella	J. D. HOOKER	(Hectorellaceae)
1867	Pachycladon	J. D. HOOKER	(Cruciferae)
1874	Sporadanthus	F. MUELLER	(Restionaceae)
1877	Oreostylidium	S. BERGGREN	(Stylidiaceae)
1881	Corallospartium	J. B. ARMSTRONG	(Papilionaceae)
1897	Simplicia	T. KIRK	(Gramineae)
1910	Leucogenes	G. BEAUVERD	(Compositae)
1911	Chordospartium	T. F. CHEESEMAN	(Papilionaceae)
1924	Ischnocarpus	O. E. SCHULZ	(Cruciferae)
1925	Pachystegia	T. F. CHEESEMAN	(Compositae)
1933	Pseudowintera	J. E. DANDY	(Winteraceae)
1941	Lophomyrtus	C. E. M. BURRET	(Myrtaceae)
1941	Neomyrtus	C. E. M. BURRET	(Myrtaceae)
1943	Cockaynea	V. D. ZOTOV	(Gramineae)
1946	Aporostylis	H. M. RUPP & E. D. HATCH	(Orchidaceae)
1951	Elingamita	G. T. S. BAYLIS	(Myrsinaceae)
1961	Kirkianella	H. H. ALLAN	(Compositae)
1963	Pyrrhanthera	V. D. ZOTOV	(Gramineae)
1967	Scandia	J. W. DAWSON	(Umbelliferae)
1967	Lignocarpa	J. W. DAWSON	(Umbelliferae)
1973	Damnamenia	D. R. GIVEN	(Compositae)

* The following genera are not included for reasons given in the references attached: *Coxella* (Umbelliferae) (DAWSON, 1968); *Kirkophytum* (Araliaceae) (PHILIPSON, 1965); *Tetrapathaea* (Passifloraceae) (GREEN, 1972); *Parahebe* (Scrophulariaceae) (ROYEN & EHRENDORFER, 1970); *Pygmaea* (Scrophulariaceae) (RATKOWSKY, 1974).

genera which attain their greatest development in New Zealand, although also found elsewhere, such as *Hebe* (Scrophulariaceae), *Celmisia* (Compositae), and *Coprosma* (Rubiaceae), all with 50 or more species, and *Aciphylla* (Umbelliferae), and *Dracophyllum* (Epacridaceae), with 35–50 species. There are also northern hemisphere genera such as *Myosotis*, *Euphrasia* and *Epilobium* for which New Zealand is a southern centre of diversity.

During their three voyages COOK and his shipmates had either deliberately or accidentally introduced factors into the New Zealand environment, which would drastically modify the flora and vegetation. These were exotic mammals and exotic plants.

In primitive New Zealand the vegetation was browsed by several species of moa – flightless birds, some of large size. There were no large land mammals. Later the Maoris brought dogs and rats, but when COOK introduced pigs a new era had begun with increasing pressures on a wider diversity of plants and probably over a wider range of environments. Escape of domestic stock or liberation of animals for sport, food and the skin trade, has had drastic effects on New Zealand vegetation. Australian opossums, red deer, and goats, the first a canopy feeder, the others browsers, could, in combination, destroy forest if the country is steep and stability of the slope easily disrupted. (POOLE & ADAMS, 1963).

The Polynesians brought food plants such as sweet potato, taro, and gourd to New Zealand but it is difficult to be sure what plants they introduced accidentally, if any. With the coming of COOK however, this process began on an ever-increasing scale. H. GUTHRIE-SMITH (1936) wrote: 'We may be certain, therefore, that about their persons or gear would lurk stowaways from English lanes and gardens, representatives of species whose seeds ran to millions per pound weight and which are parasitic to man the world over. – It is at least possible that they may have dropped within a few minutes of the disembarking of the first discoverers.' This adventive element is now a major feature in the vegetation and flora of New Zealand. Mr. A. J. HEALY estimates that there are now about 560 species more or less firmly established in the wild, with some 240 more or less common throughout the country. So extensive has been the introduction of plants by accident or design, that in many lowland areas the visitor may notice little that is native to the country. The plants of pastures, fields, hedges, plantations, roadsides, parks and gardens may be all from other lands. Perhaps only in the occasional undrained swamp, or steep gully useless for agriculture, will native plants be found. Fortunately, however, in more out of way parts of the country, there are still large areas of vegetation where native plants predominate although few areas are unmodified in some way; and scattered throughout the country are nearly 1000 reserves, some of generous size, others a pathetic reminder of the forests that once clothed the hills. (Figs. 1, 2). Further reserves are continually being sought with emphasis on non-forest plant communities.

There are ten national parks in New Zealand covering $7\frac{1}{2}\%$ of the land and protecting a wide variety of vegetation types. The three in North Island with their years of constitution, and area are: Urewera

Fig. 1. Conifer mixed-hardwood forest bordering the Wanganui River, west-central North Island, with southern beech on the ridges. (Photo: Whites Aviation.)

Fig. 2. Remnants of forest on hills cleared for farming near Hunterville, southern North Island. (Photo: New Zealand Aerial Mapping.)

(1954; 199,499 ha); Tongariro (1894; 66,132 ha); Egmont (1900; 33,312 ha); and in South Island we have: Abel Tasman (1942; 17789 ha); Nelson Lakes (1956; 57136 ha); Arthur's Pass (1929; 98,335 ha); Mount Cook (1953; 70,032 ha); Westland (1960; 85,124 ha); Mount Aspiring (1964; 275,708 ha); Fiordland (1952; 1,224,743 ha). Maritime parks involving extensive areas of coast and many islands have been formed in the Hauraki Gulf (North Island) and Marlborough Sounds (South Island). Similar to national and maritime parks in providing natural areas for recreation, but differing in that they may include normal forestry activities such as timber production, are the forest parks set up by the New Zealand Forest Service. In 1972 these covered 728,500 ha with a further 74,900 ha approved. By 1972 the Forest Service had also set aside eight sanctuaries (with others planned) to provide strict protection for unique or representative forest-types.

Basic texts for any student of New Zealand's flora and vegetation are the two volumes of the 'Flora of New Zealand' (ALLAN, 1961; MOORE & EDGAR, 1970) which deal with indigenous plants; taxonomic accounts of the adventive flora by ALLAN (1940) and HEALY (1970); and 'The Vegetation of New Zealand' by L. COCKAYNE (1958). There is also a general account of the vegetation written in German by SCHWEINFURTH (1966), and bryophytes are discussed in *N.Z. Jl. Bot.* (1971).

Relationships of the Flora

The fern flora as we have seen has a low degree of endemism. BROWNLIE (1962) estimates that 50% of the species are also found in South Eastern Australia suggesting recent overseas dispersal by wind. Of the fern genera occurring in New Zealand 85% are represented in Australia. Amongst seed plants, too, the greatest number of species is shared with Australia, with the presence of several orchids and water plants in both countries again suggesting recent trans-tasman dispersal. Of the genera of New Zealand seed plants, 75% are also represented in the Australian flora, but on the other hand there are many typical Australian genera not found in New Zealand. COCKAYNE (1958) drew attention to the absence from New Zealand of important Australian genera of the Myrtaceae *(Eucalyptus, Callistemmon, Melaleuca)* and Rutaceae *(Boronia, Eriostemon)* as well as to the poor representation of the Proteaceae and absence of such families as Mimosacea, and Casuarinaceae.

With respect to the lowland tropics such important families as Palmae and Pandanaceae, Meliaceae and Sapotaceae are on the margin of their distribution and diversity in New Zealand, where they are only represented by one endemic species a-piece. But whereas a New Zealander might see few familiar genera in the lowlands it is different at higher altitudes. On the highlands of New Guinea, Tahiti and Hawaii a small but important component of the vegetation consists of genera also found in New Zealand. In his account of the vegetation of the central highlands of New Guinea ROBBINS (1961) mentions 73 genera of seed plants and 10 genera of ferns, also found in New Zealand.

Relatively close relationships with Chile are shown by the fact that some 43% of the genera of seed plants, and some 40% of the genera of ferns, found in New Zealand are also found there, as well as many shared species, particularly in coastal situations. GODLEY (1968) has shown that 60% of seed-plant genera on the highlands of the Northern Chiloé (Chile) are also found in New Zealand as well as all the genera of ferns.

Life-forms and leaves

The plants of New Zealand are mainly evergreen and perennial, with deciduous trees, summer-green and winter-green herbs and annuals uncommon. The largest leaves in the flora are found in the ferns and the monocotyledons. The slender frond of *Lygodium articulatum* twines upwards for many metres in the forest, while the more conspicuous fronds of the tree-ferns reach a maximum length of 5–6 m in the mamaku *(Cyathea medullaris)* (Fig. 3). Fronds reaching 1 m or more are found in *Pteridium aquilinum* var. *esculentum*, *Blechnum capense*, *B. discolor* and others. In seven monocotyledon families (Palmae, Agavaceae, Liliaceae, Pandanaceae, Typhaceae, Cyperaceae, and Gramineae) several genera

Fig. 3. Tree ferns *(Cyathea medullaris)* with an exotic poplar bordering a dairy farm in western North Island. (Photo: National Publicity Studios.)

contain species with leaves or leaflets between one and three m in length e.g. the nikau palm *(Rhopalostylis sapida)*, the New Zealand flax *(Phormium tenax)*, the ti-ngahere *(Cordyline banksii)*, the bush flax *(Astelia grandis)*, the kauri grass *(Astelia trinervia)*, the kahakaha *(Collospermum hastatum)*, the kie-kie *(Freycinetia baueriana)*, the raupo *(Typha orientalis)*, as well as sedges such as *Gahnia lacera* and *Machaerina sinclairii*, and the giant toetoes, species of the grass genus *Cortaderia* (Fig. 4). Many of these linear monocotyledonous leaves were used by the Maoris for baskets, mats, and twine, and for roofing and wall materials. The culms of *Cortaderia* and *Typha* were also used in building. But the most important general-purpose leaf was that of the New Zealand flax, still cultivated today (Fig. 5), which was either plaited in its natural state or treated for its fibres.

Net-veined leaves are found in *Ripogonum* (Smilacaceae) a monocotyledonous liane, while the typical monocotyledonous leaf with parallel veins is mimicked in the Epacridaceae, a predominantly southern family of dicotyledons.

The woody dicotyledons of New Zealand have relatively small leaves and Table 4 is an attempt to classify these according to a modification

Fig. 4. Cortaderia splendens, a toe-toe grass, cultivated in the experimental gardens of the Botany Division, Department of Scientific & Industrial Research. (Photo: J. SOMERS COCKS.)

of the size classes proposed by RAUNKIAER (1934). The leptophyll class includes the distinctive 'whipcord' species of *Hebe* (Scrophulariaceae), with minute scale-like green leaves on slender flexible tetragonous branches. These species, which are found at higher altitudes, are often

Fig. 5. Commercial production of New Zealand flax *(Phormium tenax)* near Foxton, southern North Island. In the distance the low northern extremity of the Tararua Ranges. (Photo: A. E. ESLER.)

difficult to discriminate one from another, and there are some which suggest quite unrelated plants, as the expressive names indicate – *Hebe salicornioides*, *Hebe cupressoides*, *Hebe lycopodioides*. The only true macrophyll is considered to be *Meryta sinclairii*, an entire-leaved araliad found on islands off the north coast of New Zealand. With respect to leaf anatomy we could note, from the considerable amount of scattered information in the literature, the survey by WYLIE (1954) of 38 species of evergreen woody dicotyledons from North Island. While 'too limited for generalisation it reveals a wide range and a high average both for blade and tissue thickness. Mean blade thickness (406 μ), spongy mesophyll (193 μ), palisade parenchyma (124 μ), cuticular depth (9.3 μ), vein spacing (199 μ), and the proportion having a hypodermis were greater than any

Table 4. Leaf or leaflet sizes of adult dicotyledonous species of trees and shrubs in POOLE & ADAMS (1963) classified according to the system of RAUNKIAER (1934) modified by WEBB (1959).

		No. of species	Percentage
Leptophyll	<25 mm²	126*	24.7
Nanophyll	25– 225 mm²	144	28.2
Microphyll	225– 2,025 mm²	132	25.9
Notophyll	2,025– 4,500 mm²	83	16.3
Mesophyll	4,500– 18,225 mm²	24	4.7
Macrophyll	18,225–164,025 mm²	1	0.2
Megaphyll	>164,025 mm²	0	0
		510	100.0

* The evanescent leaflets of *Carmichaelia* are classified as leptophylls.

group previously studied by the writer.' However before we can claim any peculiarly New Zealand characteristics for our leaves, studies are needed of species of the same genera from other countries.

In evergreens, distinct seasonal periods of growth and leaf production have been demonstrated in subalpine trees and shrubs (P. WARDLE, 1963a) and in *Nothofagus* (BUSSELL, 1968). Whether year-round growth occurs in lowland species particularly in the north of New Zealand remains to be seen. What is certain is that the method of protecting leaf primordia during the relatively unfavourable season is quite diverse in New Zealand trees. Only a proportion have true bud scales and their presence or absence is not closely correlated with environmental conditions. Bud-scales can be absent in subalpine scrub in the south or present on coastal trees in the far north.

Deciduousness is found in varying degrees in the species of *Plagianthus* and *Hoheria* (Malvaceae), *Fuchsia* (Onagraceae), *Sophora* (Papilionaceae), *Aristotelia* (Elaeocarpaceae), *Hymenanthera* (Violaceae) (2 spp).), *Muehlenbeckia* (Polygonaceae); as well as in *Discaria toumatou* (Rhamnaceae), *Coprosma virescens* (Rubiaceae), and a few species of *Olearia* (Compositae). *Fuchsia excorticata* and *Plagianthus betulinus* are usually completely deciduous, while *Aristotelia serrata* may be only weakly so, and *Sophora microphylla* tends to be more deciduous in some races than in others. The mechanism of leaf fall (RUSSELL, 1936) and seasonal growth (BUSSELL, 1968) have been studied in *Aristotelia*, *Fuchsia* and *Hoheria*.

A striking and common growth form is the divaricating shrub, with branches produced at wide angles to form a dense tangled mass, and leaves much reduced in size. This form is found in about 50 species representing 16 families, and the convergence may be so marked that the

Fig. 6. A seedling of *Schefflera digitata* (Araliaceae) from the Waitakere Ranges in northern North Island, showing juvenile dissected leaves. (Photo: C. J. MILES.)

Fig. 7. Cabbage trees *(Cordyline australis)* near the Clarence River mouth, north-east South Island, with an Australian gum, left rear, and *Pinus radiata* shelter belts, right rear. (Photo: E. J. GODLEY.)

uninitiated observer must often wait for flowers or fruit before determining that two plants, apparently closely related, belong to widely separated families. Common examples are: *Plagianthus divaricatus* (Malvaceae), *Coprosma rhamnoides* (Rubiaceae), *Melicope simplex* (Rutaceae), *Myrsine divaricata* (Myrsinaceae), *Pseudopanax anomalum* (Araliaceae), *Corokia cotoneaster* (Cornaceae), *Pittosporum rigidum* (Pittosporaceae). These convergences have been explained as due to the inherited effect of external stimuli, mainly wind, (COCKAYNE, 1912), as adaptations to still existing, albeit dryish forest environments (WARDLE, 1963b) or as having little adaptational value and due to a particular chromosome segment transmitted sexually from one species of a genus to another, or a-sexually between genera and families (WENT, 1971).

Several trees pass through a divaricating juvenile form, later replaced by a non-divaricating habit with larger leaves. In *Pennantia corymbosa* (Icacinaceae) and *Hoheria angustifolia* (Malvaceae) the divaricating juvenile form persists long after adult foliage is produced. In *Sophora microphylla* (Papilionaceae) divaricating juveniles are only found in some races, and persist for varying lengths suggesting that this part of the life cycle is still actively evolving. In this connection we could note the juvenile form in *Schefflera digitata* (Araliaceae) with attractive deeply lobed leaflets, (Fig. 6), which is confined to the northern part of the species range and here is usually found only at low frequency in natural populations.

Perhaps the most attractive of the juvenile forms is the young rimu, *Dacrydium cupressinum* (Podocarpaceae) with dense deep green weeping foliage. Certainly the most bizarre form is the single slender stem and the long rigid deflexed leaves of the young lancewood, *Pseudopanax crassifolium* (Araliaceae).

Tuft trees are seen in *Cordyline* (Fig. 7) and *Dracophyllum* (Epacridaceae) (Fig. 23). Cushion forming species are found in several families such as Donatiaceae *(Donatia)*, Stylidiaceae *(Phyllachne)*, Thymelaeaceae *(Drapetes)*, Centrolepidaceae *(Centrolepis)*, culminating in the Compositae where the large cushions of *Raoulia eximia* and *Haastia pulvinaris* have been aptly termed 'vegetable sheep' (Fig. 24).

Summer-green and winter-green herbs are represented in the lowland vegetation by small terrestrial tuberous orchids such as *Thelymitra* and *Pterostylis*, by a few ferns such as *Hypolepis*, *Pteridium*, *Histiopteris* and *Cheilanthes*, or by plants of special habitats such as *Typha orientalis* (freshwater swamps), *Scirpus americanus* (salt swamps), *Carex coriacea* (damp places) and *Calystegia soldanella* (coastal dunes). In montane tussock-grasslands this life-form is seen in species of *Coriaria* (Coriariaceae) and in *Bulbinella hookeri* (Liliaceae); in alpine meadows, in *Ranunculus lyallii* and *Senecio scorzoneroides;* and it is characteristic of most of the unique scree species of the Southern Alps e.g. *Anisotome carnosula* (Umbelliferae) and *Ranunculus haastii*. In the subantarctic islands the two physiognomically important species *Bulbinella rossii* and *Pleurophyllum criniferum* are summer-green.

Ranunculus lyallii with its large peltate leaves has been called the largest buttercup in the world. Other herbs with giant leaves are endemic to various outlying islands forming a great arc from the Chathams round to Kerguelen. (Fig. 8). Do we see here, with the inclusion of the woody *Meryta sinclarii* to the north, a remnant of a more large-leaved pre-glacial flora?

The low frequency and unimportance of annuals in the New Zealand vegetation cannot be overstressed. Only some 6% of the flowering plants are classified as such (ALLAN, 1937a) (as compared with the 13% of RAUNKIAER's 'normal' spectrum) and even this appears rather a high figure. None of the annuals is of physiognomic importance. Although the climate of the tussock-grassland regions is one which could have favoured the evolution of annual species in New Zealand, ALLAN points out that the closed nature of these associations was not favourable to the establishment of short-lived plants. However no native annuals have

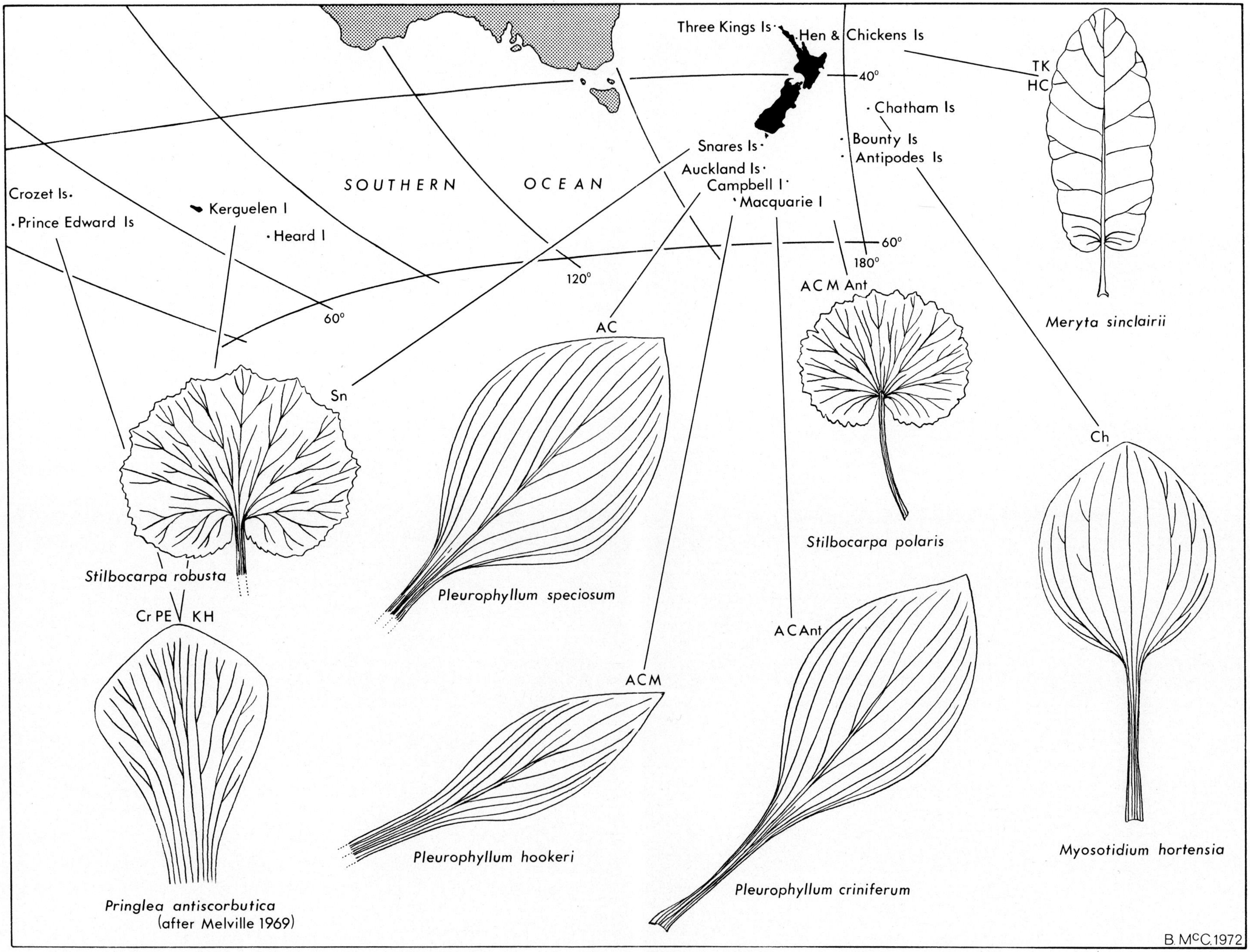

Fig. 8. Large leaved plants on outlying islands. TK (Three Kings); HC (Hen and Chickens); Ch (Chathams); Sn (Snares); Ant. (Antipodes); A (Auckland); C (Campbell); M (Macquarie); K (Kerguelen); H (Heard); Cr (Crozet); PE (Prince Edward). (Figure by Beverley McCulloch.)

evolved to exploit the open spaces of the sand dunes, as do such adventive annuals as the hares-tail grass, *Lagurus ovatus* and the hare's foot trefoil, *Trifolium arvense*. Nor can natives emulate the introduced *Hordeum maritimum* and *Atriplex patula* which thrive annually on bare ground in salt-swamp and salt-meadows.

Flowers and fruits

The seasons in New Zealand are conventionally given as:

Spring	(9–11):	September,	October,	November
Summer	(12– 2):	December,	January,	February
Autumn	(3– 5):	March,	April,	May
Winter	(6– 8):	June,	July,	August

In the lowlands, spring, summer and atumn are important flowering times with some flowering periods extending into or beginning in winter (e.g. *Avicennia resinifera*, *Pseudopanax arboreum*, *Vitex lucens*, *Metrosideros fulgens*, *Dysoxylum spectabile* and races of *Sophora microphylla*). During winter too, flower buds which first appeared in summer gradually enlarge ready for spring flowering (e.g. *Sophora microphylla*, *Elaeocarpus dentatus*, *Pseudowintera colorata*). In alpine situations CLARKE (1968) showed that in northern South Island summer was the main flowering season and further south in Otago, MARK (1970) recorded 81 of 100 alpine species with autumn initiation and successful overwintering of floral primordia. MARK (1965) and CONNOR (1966) have discussed aspects of periodic flowering in *Chionochloa* (snow-grasses) and the latter compares these with beech flowering.

The large white flowers of *Clematis paniculata*, the scarlet brush-flowers of the tree species of *Metrosideros* or the yellow blossoms of the tree *Sophoras* are not characteristic of New Zealand lowland flowers. Many are small and inconspicuous with white and green the predominant colours. (This is not meant to imply that flowers of species of the same genus in other countries are any more conspicuous). The subalpine and alpine meadows and streamsides support many species with conspicuous flowers, particularly if we include the composite heads, but here too, white is an important colour, with yellow also conspicuous. Red, as the main flower colour, is found in some 28 species (1.5%) of the native flowering plants. Predominantly purple or blue flowers occur in 23 species (1.3%). But purple colouration is also found on the standards of many carmichaelias, and certain composites mentioned below have purple disc and white ray-florets. Fully yellow flowers on flower-heads are found in some 111 species (6.1%), the main contributor being *Ranunculus* in which all species have yellow flowers except *Ranunculus lyallii* and *R. buchananii* which have white petals and yellow stamens. Not included in the above total are the many species of *Celmisia* and a few other composites with white ray florets and yellow disc florets.

Many species in the blue-purple flower group occur in the outlying islands and nowhere else. On Chatham Island is the blue-flowered *Myosotidium hortensia* sole member of an endemic genus already noted for its large leaves. Also on Chatham Island is *Olearia semidentata* with all florets purple, while *O. chathamica* has purple disc and white ray florets. On Auckland and Campbell Islands grow the blue-flowered species *Hebe benthamii* and *Myosotis capitata*, while *Myosotis antarctica* of Campbell Island has plants with either blue or white flowers. On Antipodes, Auckland, Campbell and Macquarie islands grow members of the subantarctic endemic genus *Pleurophyllum* with purple disc florets. Purple disc florets are also present in the subantarctic *Celmisia vernicosa* in contrast to the yellow disc florets of the many mainland species. We could also note here that whereas the many mainland species of *Gentiana* have white flowers with rarely a trace of red, the subantarctic species on Auckland, Campbell and Antipodes islands have high proportion of plants with various patterns of red on the petals.

A striking feature of the native species is that some 12–13% are dioecious as compared with the 2% of the British flora. Monoecism is found in some 9% of the flowering plants with *Carex* and *Uncinia* the main contributors. Gynodioecism has been reported in *Fuchsia*, *Cyathodes*, *Pimelea*, *Cortaderia*, *Gentiana*, *Ranunculus* and *Carpodetus* (summary in SHORE, 1969) as well as *Gingidium*, *Lignocarpa* and *Scandia* (DAWSON, 1967a, b). Observations on isolated plants of many species show that hermaphrodite flowers are usually self-fertile and self-fertilised. Self-sterility has been demonstrated in species of *Festuca*, *Poa* and *Trisetum* (CONNOR, 1957, 1960) and in *Pentachondra pumila* (Epacridaceae) (GODLEY, 1966).

Birds known as flower visitors are three species of Honey-eaters – tui, bellbird and the now rare stitchbird – as well as native parrots (kaka and kea) and the introduced wax-eye. There are only 16 species of butterflies in New Zealand, but a large number of moths (including one hawkmoth), a great variety of Diptera and some 40 species of solitary bees – short-tongued Colletidae, and small-tongued Halictidae. There are also two species of native bats.

HEINE (1938) has suggested that the comparatively large percentage of green flowers is due to the predominance of Diptera, and that the lack of reds, blues and purples is due to the absence of native long-tongued bees. She considers that yellow and white flowers are best adapted to short-tongued bees and flies, 'the chief pollinators in New Zealand', and that this adaptation has led to 'a predominance of short tubes and exposed pollen and pistil which in its turn results in reversion to unisexual flowers to avoid self pollination.' However there are many points in this hypothesis which do not stand up to examination. The amount of exposure of pollen and pistil is surely not greatly different between members of the same genera in New Zealand and overseas e.g. there are open flowers in *Clematis*, *Rubus*, and members of the Umbelliferae throughout their

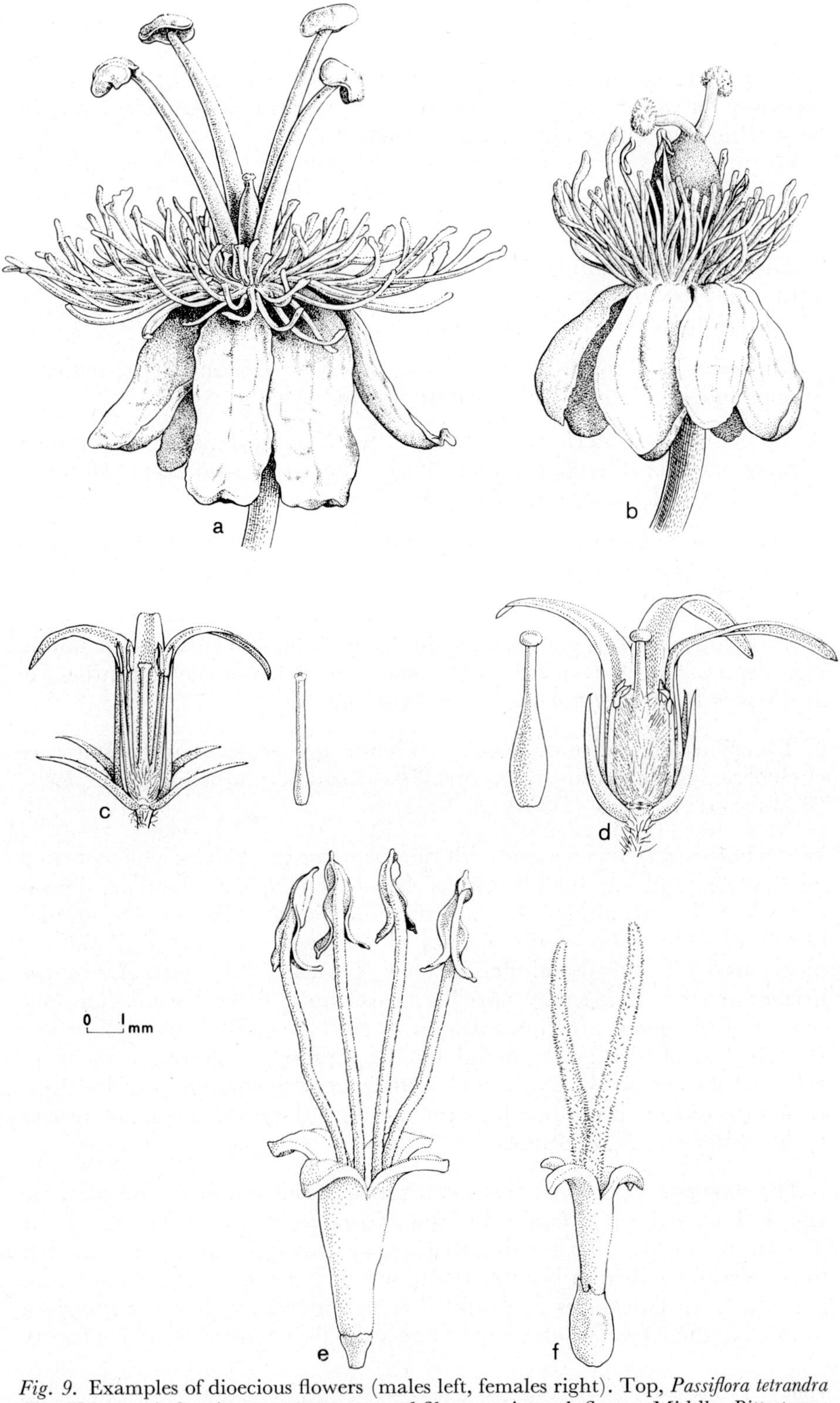

Fig. 9. Examples of dioecious flowers (males left, females right). Top, *Passiflora tetrandra* (Passifloraceae) showing numerous coronal filaments in each flower. Middle, *Pittosporum pimeleoides* (Pittosporaceae). Bottom, *Coprosma robusta* (Rubiaceae). (Drawings by K. R. WEST.)

geographic range. With respect to flower colour we could point out the tendency for purple and blue flowers to be present in subantarctic islands from which bees have not been recorded.

We must also realise that not all the dioecism in the New Zealand flora has necessarily evolved in the New Zealand region. At least seven kinds of dioecism can be recognised.

1. Dioecism as a family character with New Zealand representatives outliers of dioecious groups better represented elsewhere e.g. Pandanaceae *(Freycinetia)*, Restionaceae *(Leptocarpus, Calorophus)*.

2. Dioecism as a family character with a New Zealand representative an endemic genus, e.g. Restionaceae *(Sporadanthus)*.

3. Dioecism as a generic character with the New Zealand representatives outliers of genera better represented elsewhere e.g. *Hedycarya*, (Monimiaceae), *Litsea* (Lauraceae), *Macropiper* (Piperaceae).

4. Dioecism as a generic character with the greatest number of species in the New Zealand region, e.g. *Coprosma* (Rubiaceae) (Fig. 9), *Astelia* Liliaceae).

5. Dioecious endemic genera with dioecism found elsewhere in the family e.g. *Tupeia* (Fig. 9) is not the only dioecious genus in the Loranthaceae and nor is *Passiflora tetrandra* in the Passifloraceae.

6. Dioecious endemic, or almost endemic genera, with dioecism rare elsewhere in the family, e.g. the New Zealand umbellifers, *Aciphylla*, *Coxella*, and *Anisotome* (DAWSON, 1964).

7. Partially dioecious genera with dioecism confined to, or concentrated on New Zealand. e.g. both species of *Aristotelia* in New Zealand are dioecious while the remainder in Australia and South America are hermaphrodite; only some of the species of *Bulbinella* in New Zealand are dioecious while the remainder in New Zealand and South Africa are hermaphrodite. *Rubus* may also be an example in which the dioecious species of the genus are concentrated in New Zealand but the dioecious *R. gunnianus* of Tasmania should not be forgotten. Although DAWSON *(op. cit.)* states that dioecism in *Clematis* is very uncommon outside New Zealand, we should not overlook the fact that there are dioecious species in Australia and New Guinea.

The dioecism in the first three categories could well be of considerable age and have been brought to New Zealand by the ancestors of the present plants. In the fourth category we can only speculate whether the dioecism evolved in New Zealand or whether the ancestral home was elsewhere. In categories 5, 6 and 7 it is very likely that the dioecism evolved in the New Zealand region but it should be noted that in category

5 this breeding system has often evolved elsewhere under presumably quite different conditions.

In the New Zealand flora there are many genera with attractive fleshy fruits which, although often small (e.g. *Coprosma*) are produced in abundance and in a diversity of colours. Examples of larger fruits (2–4 cm long and ovoid) all of which were eaten by the Maoris are: the dark-purple berries of the two species of *Beilschmiedia;* the orange berries of two species of *Solanum;* the purplish 'drupe' of *Podocarpus ferrugineus;* and the orange drupes of the Karaka *(Corynocarpus laevigatus)* which were steamed to remove poison from the seed. The largest is the compound fruit of *Freycinetia baueriana* (Pandanaceae), sometimes eaten by the Maoris, although the fleshy bracts around the inflorescense were more favoured. Another vegetable delicacy of the Maoris was the pollen of the raupo *(Typha orientalis)*, often gathered in bucketsful, which when baked reminded Colenso of London gingerbread.

Coastal plants

The following are some of the salient or peculiar features of the coastal vegetation.

1. A remarkable group of species with tropical affinities is found on the northern offshore islands. These are: *Meryta sinclairii* (Araliaceae), a large leaved tree, growing abundantly on the Three Kings and the Hen and Chickens Islands and representing a genus centred on New Caledonia; *Xeronema callistemon* (Liliaceae), common on Poor Knights and the Hen Islands, and related to *X. moorei* of New Caledonia; *Tecomanthe speciosa* (Bignoniaceae) a woody climber, known in the wild from only a single plant in Great Island, Three Kings, and most closely related to *T. hillii* of Queensland; *Elingamita johnsoni* (Myrsinaceae), a monotypic endemic genus of about a dozen trees on West Island, Three Kings, resembling *Tapeinosperma* of New Caledonia; *Pennantia endlicheri* of Norfolk Island, represented by a single sterile, apparently female tree, on Great Island of the Three Kings group. (This was originally classified as *Plectomirtha baylisiana* of the Anacardiaceae, until corrected by Sleumer (1970)).
2. The mangrove, *Avicennia resinifera* (Fig. 10), is characteristic of muddy tidal estuaries from North Cape to Ohiwa Harbour (38 °S, 10 miles W of Opotiki) on the east coast. On the west coast it does not reach Kawhia and Aotea Harbours, which are about 38 °S.
3. *Salicornia australis* (Chenopodiaceae) is a prominent plant of salt marshes from North Cape to Foveaux Strait.
4. *Juncus maritimus* var. *australiensis* and *Leptocarpus similis* (Restionaceae) either singly or in combination may dominate salt swamps from North Cape to Banks Peninsula and Okarito. Further south *Leptocarpus* alone is found.
5. The sand dunes may be dominated by the sandbinders *Spinifex hirsutus*

Fig. 10. Mangroves *(Avicennia resinifera)* bordering a tidal river near Whangarei, northern New Zealand. Note breathing roots exposed at low-tide on the mud. (Photo: E. J. GODLEY.)

(Gramineae) and *Desmoschoenus spiralis* (Cyperaceae) from North Cape to about Blenheim, south of which only *D. spiralis* is found.

6. Two trees characteristic of the fairly narrow belt of coastal forest are: the pohutukawa, *Metrosideros excelsa* (Myrtaceae) which reaches about 38 °S on each coast; and the karaka, *Corynocarpus laevigata*, (Corynocarpaceae) which crosses into the South Island, and extends down the east coast to Banks Peninsula.

7. On the west and south-east coasts of South Island the lowland forest – in which the southern rata, *Metrosideros umbellata*, is usually an important tree – may reach the edge of beach or cliff, with little modification except the entrance of *Hebe elliptica* or coastal composite shrubs such as *Senecio reinoldii* and *Olearia oporina*. Further south, on eastern Stewart Island and particularly on Auckland Island *Metrosideros umbellata* forms a belt of coastal forest like its relative in the north, *M. excelsa*, but unlike it, extends inland and uphill as a stunted tree or shrub. Coastal composite shrubs on Snares Island are *Senecio stewartiae* and *Olearia lyallii* the latter extending to the north of the Auckland Islands.

8. Some of the more common and widespread coastal plants in New Zealand are also found on the coasts of other continents and islands. Most are small herbs, and examples are: *Ranunculus acaulis* (Ranunculaceae); *Samolus repens* (Primulaceae); *Selliera radicans* (Goodeniaceae); *Calystegia*

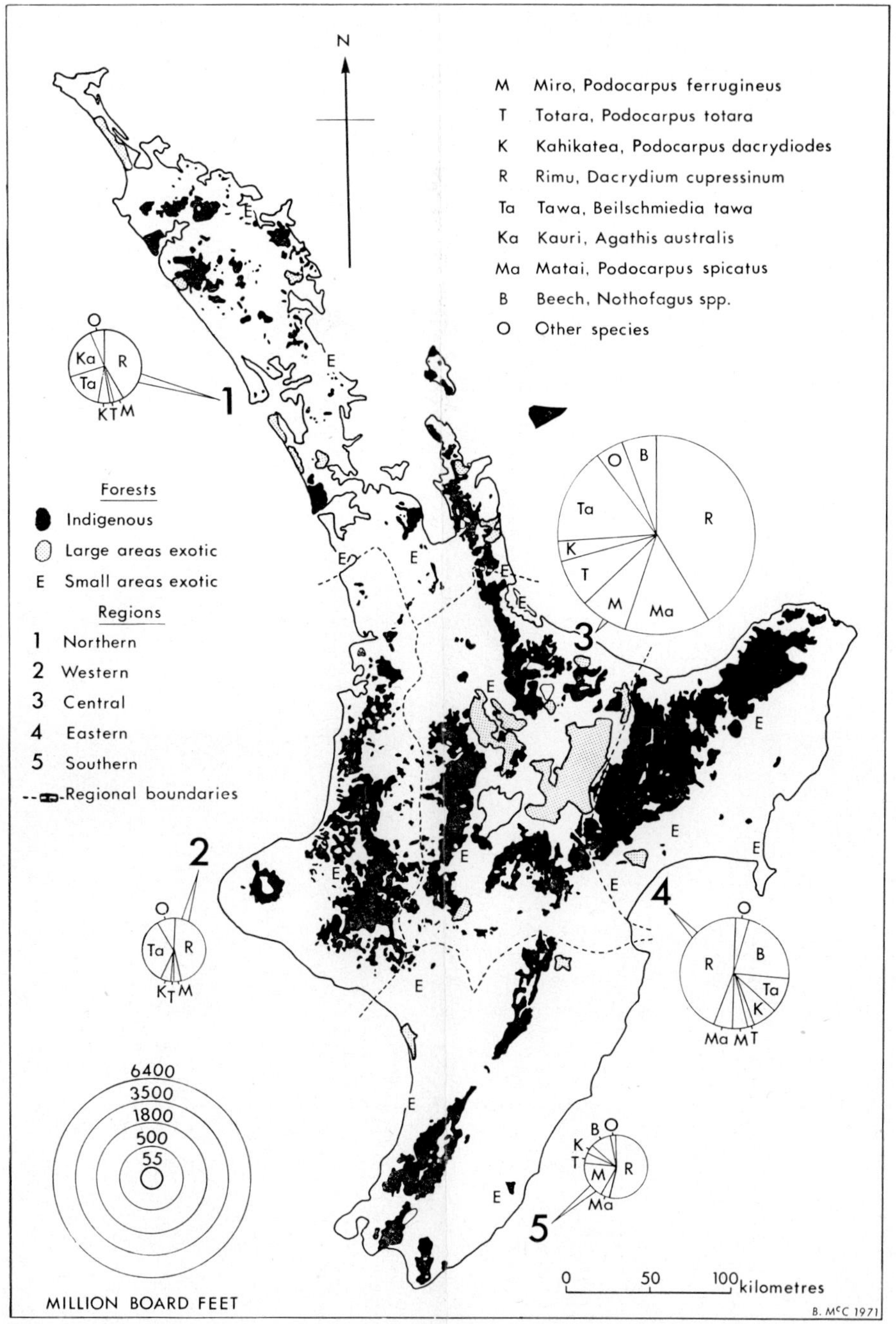

Fig. 11. Distribution of indigenous and exotic forests in North Island (from 'New Zealand Forestry' N.Z. Forest Service, 1964) as well as the indigenous merchantable timber resource by species (adapted from 'Bay of Plenty Natural Resources Survey', Ministry of Works, 1962). (Figure by BEVERLEY McCULLOCH.)

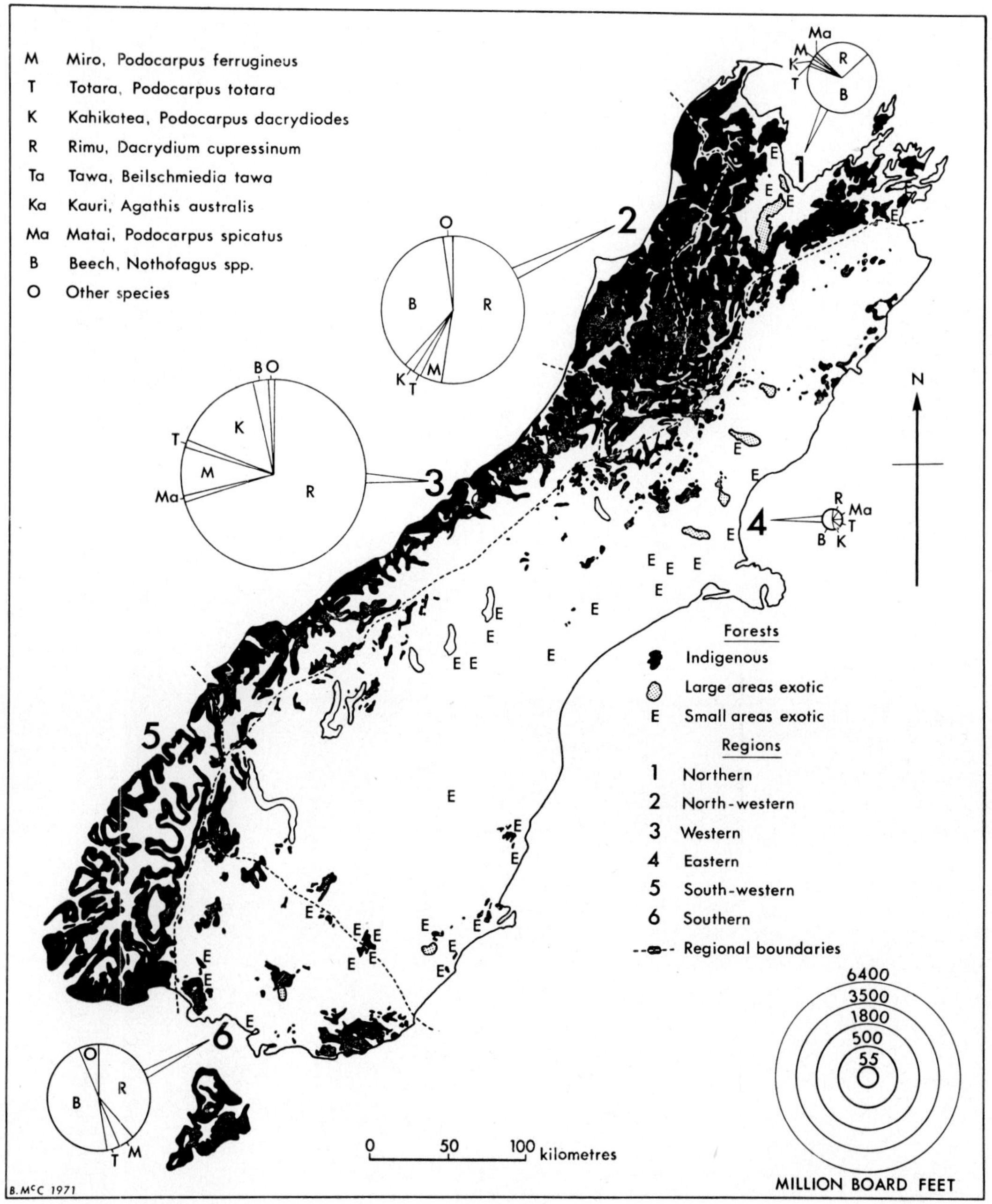

Fig. 12. Distribution of indigenous and exotic forests in South Island (from 'New Zealand Forestry', N.Z. Forest Service, 1964) as well as the indigenous merchantable timber resource by species (adapted from 'Bay of Plenty Natural Resources Survey', Ministry of Works, 1962). (Figure by BEVERLEY McCULLOCH.)

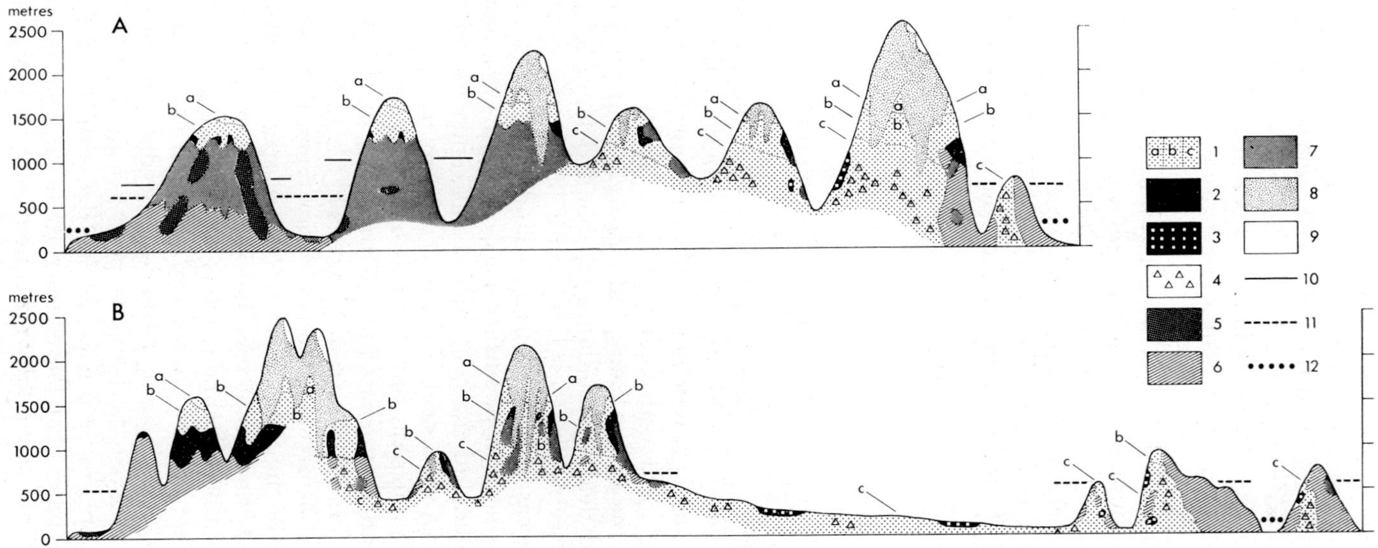

Fig. 12 a. Stylised west-east transects through the South Island of New Zealand at (A) 42°–42°20′ and (B) 43°–43°40′S. Vegetation as at 1840. Drawn by K. R. West, from information supplied by P. Wardle; 1. Grassland and herbaceous vegetation. a) High alpine short grassland and herbfield; b) 'Snow tussock' grassland dominated by *Chionochloa* spp.; c) Montane and lowland grassland, *Pteridium* and swamp; 2. Subalpine scrub; 3. *Leptospermum* woodland; 4. Scrub of *Discaria*, *Cassinia*, etc., scattered and in patches; 5. Stunted forest, scrub and 'pakihi' on wet, sterile soils; 6. Mixed forest (conifers and hardwoods, but *Nothofagus* absent); 7. *Nothofagus* forest; 8. Alpine basins and scree; 9. Permanent snow; 10. Upper limit of *Nothofagus fusca;* 11. Upper limit of tall lowland podocarps; 12. Upper limit of palm *(Rhopalostylis sapida)*.

soldanella (Convolvulaceae); *Carex pumila* and *Scirpus nodosus* (Cyperaceae); *Cotula coronopifolia* (Compositae); *Tillaea moschata* (Crassulaceae); *Tetragonia tetragonioides* (Aizoaceae); and the shrub *Hebe elliptica* (Scrophulariaceae). Common on New Zealand beaches are the yellow seeds of *Sophora microphylla* (Papilionaceae) which definitely float to the Kermadec Islands (Sykes & Godley, 1968). The species also occurs in Southern Chile.

A good general account of coastal plants is given by Moore & Adams (1963). Of the several studies on coastal islands we could mention the Three Kings (Baylis, 1958), the Hen and Chickens (Cranwell & Moore, 1935), the Mercury Group (Atkinson, 1964) (all off the northern coasts) as well as Kapiti (Esler, 1967) off south west North Island and Bird Island (Fineran, 1966) further south in Foveaux Strait.

The Forests

General

A little more than a century ago the early settlers found the greater part of North Island under forest, except for areas recovering from the effects of volcanic activity or where Maori fires or other activities of these people had changed the landscape. Thus in 1835 at the heavily populated Bay of Islands Darwin observed that from a distance the land appeared 'as if clothed with coarse pasture, but this in truth is nothing but fern. On the more distant hills, as well as in parts of the valleys, there is a food deal of woodland.' In South Island forest predominated in the wetter north, west, and south but in the drier east there were extensive areas of indigenous grassland at lower elevations. However evidence in the form of surface logs, wind throw hummocks and hollows, buried wood and charcoal, buried soils, and relics of earlier vegetation showed that a good deal of the lowland east had been under forest in the not too distant past, but that this had been destroyed by a sequence of natural and Polynesian fires (Molloy *et al.*, 1963). The present distribution of forest is shown in Figures 11, 12, and 12A.

Most of the forest can be divided into two major types, the first dominated by one or more of the four native evergreen species of southern beech *(Nothofagus)* and the second dominated by one or more species of the native conifers. But beeches and conifers are not mutually exclusive and podocarps, *Libocedrus*, or even *Agathis* can be found intermixed with *Nothofagus*. P. Wardle (1964) has emphasised that seedlings in *Nothofagus* are produced in great numbers and able to persist in dense shade until gaps appear in the canopy and a quick-growing dense thicket is formed typical of the gregarious habit of all four species. In the large gymnosperms on the other hand seedlings are usually infrequent, and grow slowly and steadily, which results usually in a sparse overstorey to the forest. (Exceptions are seen where special conditions of the soil are unfavourable to the growth of hardwoods, leading for example to dense stands of the Kahikatea, *Podocarpus dacrydioides*, or of rimu, *Dacrydium cupressinum*). *Nothofagus*, too, pioneers more readily than the gymnosperms which usually regenerate under 'nurse' hardwoods.

SCHIMPER (1903) classified the New Zealand forests as temperate rain forests, and grouped them with those of Eastern Australia, South Japan and South Chile. He noted the following general characteristics of such forests. They are composed of evergreen, hygrophilous trees, which lack the large foliage of tropical woody plants. The leaves are mostly small, glabrous, and entire, with firm, leathery consistency, fairly thick cuticle, and glossy upper surface. He points out that plank buttresses are exceptional in such forests, and states that cauliflory and epiphylly are unknown. (In the New Zealand forests some examples of the last two phenomena occur). The undergrowth as SCHIMPER notes may be as dense as in the tropics, but lacks the well developed foliage and feeble branching which characterise tropical forests. Again, lianes and epiphytes may be abundant but show 'weakened development' as compared with their tropical cousins.

L. COCKAYNE (1926) modified the terminology of WARMING (1909) and classified the *Nothofagus* forests as 'sub-antarctic rain forests', and the podocarp dicotylous forests as 'subtropical rain forests'. These terms are not considered particularly appropriate and HOLLOWAY (1946) has pertinently written: 'The use of the words "sub-tropical" and "sub-antarctic" is derived largely by consideration of the affinities of the flora. They belong to the field of plant geography. Used in an ecological sense they imply far more than can be accepted: a greater degree of warmth or a greater degree of cold than normal in a temperate insular climate. The incorporation of the word 'rain' is likewise misleading. Both major forest types can exist in areas where the annual rainfall is no more than sufficient to maintain a forest vegetation. With a decrease in rainfall the forest would be replaced by a vegetation of a completely different type, but not by a drier type of forest. The most that can be said with accuracy is that we are dealing with *Nothofagus* forest or beech forest, and with mixed Podocarp – *Weinmannia* – *Metrosideros* forest or for short, podocarp forest.' And again: 'Both types of forest are temperate forests, the one, more widely developed in the north or at lower altitudes, a warm temperate forest; beech forest, more widely developed in the south or at higher altitudes, a cool temperate forest. Certain types of podocarp forest, with respect to their associated species are, or may be, true rain forest but this is not an essential feature of podocarp forests as such. Similarly, beech forests may be rain forests but are not necessarily so.'

Classifications of the altitudinal sequence of vegetation proposed by ZOTOV (1938) and WARDLE *(op. cit.)* are given in Table 5, and a sequence shown in Fig. 13.

Table 5. Proposed Division of the Altitudinal Sequence of Vegetation, and the Upper Limits of Forest and Scrub Species in Areas Where the Climate is Moist and Equable.

Zotov, 1938	Wardle, 1964	Altitudes in feet at 41°	45°
Upper Warm Subpolar	High Alpine		
		5000 feet	4000
Lower Warm Subpolar	Lower Alpine		
		4000 feet	3000
Upper Cold Temperate	Subalpine		
		3000 feet	2000
Lower Cold Temperate	Cool Temperate or Montane		
Upper Warm Temperate		1500 feet	
Lower Warm Temperate	Warm Temperate		Sea level
		Sea level	

The interaction between past and present climates and the South Island forests has been studied by Holloway (1954a). He has shown that these forests show varying degrees of 'maladjustment' to present climates, and has suggested that this situation was initiated about the 13th century A.D. by a fall in mean or effective temperature. This, he considered, placed the podocarps in general at a disadvantage relative to the beeches and began a series of readjustments between species which still continues. The long-lived podocarps may still persist in situations which have now become unfavourable to them, and the beeches may be seen invading new areas as fast as the wind or water carriage of their seed allows.

'A few types of forest may have achieved a position of near stability but, in general, New Zealand foresters concerned with management of the indigenous forests deal with a peculiarly complex agglomeration of relict types of forest and with a host of new types in which the processes of adjustment and re-adjustment to the new climates are in active progress.' (Holloway, 1954a).

Holloway's hypotheses receive only qualified support from later

Fig. 13. Altitudinal sequence of forest on Mount Fyffe (42°S, north-eastern South Island). At timberline (850–900 m) is *Podocarpus hallii*, passing down into *Podocarpus spicatus* forest and then into conifer-mixed hardwood forest. (New Zealand Forest Service Photo by J. H. G. Johns, A.R.P.S.)

writers. At least in the wetter districts, widespread failure of the podocarps to regenerate indicates that the climatic deterioration became fully effective several centuries later than Holloway suggested (Nicholls, 1956; Wardle, 1963b; Barton, 1972). The palynological evidence suggests that invasion of beech into new areas has been continuing over several thousands of years (MOAR, 1971) while CUMBERLAND (1962) and MOLLOY (1969a) question the whole concept of ecologically significant climatic changes during the last 1,000 years.

CONIFER MIXED-HARDWOOD FORESTS

COCKAYNE (1958) wrote: 'Two features, however, stand out for subtropical rain-forest: that is, it is composed so far as trees go partly of *Podocarpaceae*, *Cupressaceae* or *Araucariaceae*, and partly of broad-leaved dicotylous trees. Unfortunately, for purposes of classification, there is every intermediate stage between a pure podocarp or Kauri community and one made up of dicotylous trees alone.'

North of 38 °S are forests in which the Kauri pine (*Agathis australis*, Araucariaceae) grows in association with podocarps and hardwood trees. South of this latitude the gymnosperms in the forests are podocarps alone *(Podocarpus, Dacrydium, Phyllocladus)* except at higher altitudes where *Libocedrus* (Cupressaceae) may be present. The vegetation in these forests is generally extremely luxuriant with small trees, shrubs, ferns, bryophytes, and epiphytes plentiful. However, in diversity of species of trees, this forest does not compare with that of the tropics. The type which contains the most physiognomic prominents is the low altitude forest of North Auckland where 14 species are listed as occurring constantly (MCKELVEY & NICHOLLS, 1957). Even if we double this number to account for less important or smaller species it does not compare with the 179 species of trees, 10 cm or more in diameter, counted in a 3.5 hectare plot in a virgin *terra firme* forest in Brazil. (PIRES, DOBZHANSKY & BLACK, 1953).

The following trees in various combinations play an important part over all or part of the range of the forests:

Agathis australis (Araucariaceae). Kauri (Fig. 14). Perhaps the most famous timber tree in New Zealand, attaining a massive size, with unbranched and untapering bole. It was originally prominent in forest of the North Auckland Peninsula the Barrier Islands and Coromandel. It does not occur south of 38 °.

Beilschmiedia tarairi (Lauraceae). Taraire. One of the commonest trees in the northern forests extending southwards only to 38 °S.

Beilschmiedia tawa (Lauraceac). Tawa. A species of great importance, often dominant in much of the North Island, but very local in northern South Island where it reaches 42 °S.

Metrosideros robusta (Myrtaceae). Northern Rata. This tree is important in North Island, and on the west coast of South Island to Hokitika. It often begins life as an epiphyte, and by fusion of large aerial roots the irregular trunk is formed, within which the original supporting tree finally dies.

Metrosideros umbellata (Myrtaceae). Southern Rata. A smaller tree than the Northern Rata, and an important member of montane forest in

Fig. 14. Agathis australis (Araucariaceae), the Kauri pine, in the Omahuta Reserve, northern North Island. Measurements: 17 m to first limb, and 14 m girth at breast height. (New Zealand Forest Service Photo J. H. G. Johns, A.R.P.S.)

Fig. 15. Regeneration of totara *(Podocarpus totara)* on flats and hills north of Taumarunui near central North Island. In the foreground dairy pasture with scattered *Bellis perennis*. (Photo: E. J. GODLEY.)

western districts of the South Island. It also occurs on Stewart Island, the Auckland Islands, and sparingly in North Island.

Weinmannia racemosa (Cunoniaceae). Kamahi. A very common tree in the New Zealand forests, extending from about 38 °S through all three islands, and important in both lowland and montane forests. North of 38 °S it is replaced by *W. sylvicola*.

Podocarpus totara (Podocarpaceae). Totara. This species is widely distributed in New Zealand and is particularly prominent in central North Island. It now commonly regenerates on open hill sides in North Island (Fig. 15). In Westland and on Stewart Island totara is replaced by the next species.

Podocarpus hallii (Podocarpaceae). A widespread species, closely related to *P. totara* but usually ascending to higher altitudes.

Podocarpus ferrugineus (Podocarpaceae). Miro. A common tree in forest areas throughout North, South, and Stewart Islands with a distribution similar to rimu and thus of very local occurrence in parts of eastern South Island.

Podocarpus spicatus (Podocarpaceae). Matai. This species is widely distrib-

Fig. 16. A stand of Kahikatea, *Podocarpus dacrydioides* in Westland, South Island, with *Coprosma* scrub in the foreground. (Photo: E. J. GODLEY.)

uted in North and South Islands but is rare on Stewart Island.

Podocarpus dacrydioides (Podocarpaceae). Kahikatea or white pine (Fig. 16). This species is widely distributed in the North and South Islands, but is very rare on Stewart Island. Pure stands on wet low lying areas form the Kahikatea swamp forest.

Dacrydium cupressinum (Podocarpaceae), Rimu, or red pine. This is widespread in New Zealand, but of very local occurrence in parts of eastern South Island. In some areas such as hill slope forest in Westland, rimu is clearly out of phase with the present climate and consists of over-mature trees with little regeneration. On the lowland terraces however, it thrives. (HOLLOWAY, 1954b).

Libocedrus bidwillii (Cupressaceae). Kaikawaka. Often prominent in subalpine forests. All these species are absent from the Chatham Islands.

The middle layer of the forest is dense, and quite diverse in species. COCKAYNE (1958) has given the following list of small trees and shrubs which may commonly occur. *Carpodetus serratus* (Escalloniaceae), *Aristotelia serrata*, *Elaeocarpus hookerianus* (Elaeocarpaceae), *Plagianthus betulinus* (Malvaceae), *Melicytus ramiflorus* (Violaceae), *Fuchsia excorticata* (Onagraceae), *Schefflera digitata*, *Pseudopanax crassifolium* (Araliaceae), *Griselinia littoralis* (Cornaceae), *Myrsine australis* (Myrsinaceae), and *Coprosma lucida* (Rubiaceae).

The following species are listed as common on the forest floor: *Microlaena avenacea* (Gramineae), *Carex dissita* (Cyperaceae), *Astelia fragrans* (Liliaceae), *Ranunculus hirtus* (Ranunculaceae) and species of *Uncinia* (Cyperaceae), *Nertera* (Rubiaceae), *Hydrocotyle* (Umbelliferae), and *Cardamine* (Cruciferae).

The following fern genera are well represented: *Asplenium*, *Blechnum*, *Phymatodes*, members of the Hymenophyllaceae, and the tree ferns *Cyathea smithii* and *Dicksonia squarrosa*. The epiphytes *Lycopodium billardieri* and *Tmesipteris tannensis* are found throughout.

OLIVER (1930) listed 50 vascular species in New Zealand forests which are habitually epiphytic, and classifies them as follows: semi-woody pteridophytes, (3), filmy ferns, (20) tufted ferns, (7), creeping ferns, (5), succulent herbs, (2), orchids, (7), shrubs, (4), and tussock plants (2), *(Collospermum hastatum*, *Astelia solandri)*. Most of these are derived from the tropical element in the flora. The four shrubs are *Senecio kirkii* (Compositae), *Griselinia lucida* (Cornaceae), *Pittosporum kirkii* and *P. cornifolium* (Pittosporaceae). The epiphytic vascular flora decreases proceeding south. Of the above shrubs only *G. lucida* crosses Cook Strait, and many species do not penetrate far into the South Island.

The climbing plants of these forests and also of the non-forest coastal and lowland vegetation are found in the following genera:

Scramblers: *Gleichenia* (Gleicheniaceae), *Carmichaelia* (Papilionaceae), *Fuchsia* (Onagraceae), *Helichrysum* (Compositae), *Rubus* (Rosaceae), *Scandia* (Umbelliferae), *Tetragonia* (Aizoaceae).

Root climbers: *Arthropteris* (Davalliaceae), *Blechnum* (Blechnaceae), *Phymatodes* (Polypodiaceae), *Freycinetia* (Pandanaceae), *Metrosideros* (Myrtaceae).

Winding lianes: *Lycopodium* (Lycopodiaceae), *Lygodium* (Schizaeaceae), *Calystegia*, *Ipomoea* (Convolvulaceae), *Muehlenbeckia* (Polygonaceae), *Parsonsia* (Apocynaceae), *Ripogonum* (Smilacaceae), (Fig. 17), *Senecio* (Compositae), *Tecomanthe* (Bignoniaceae).

Tendril climbers: *Sicyos* (Cucurbitaceae), *Passiflora* (Passifloraceae), *Clematis* – sensitive petioles (Ranunculaceae).

A series of useful forest profiles is given by ROBBINS (1962), and phycomycetous mycorrhizas are discussed bv BAYLIS (1967).

Fig. 17. Entanglement formed by the stems of the supplejack (*Ripogonum scandens*, Smilacaceae) in forest at Wairoa Gorge, northern South Island. (Photo: E. J. GODLEY.)

BEECH FORESTS

Beech species are not found on Stewart Island or over large areas of North Island, and are also missing from a large segment of western South Island. (Fig. 18). Here in the south the beech forest stops at the Paringa River, with several outlying areas of silver beech about 13 miles to the north in the upper Karangarua River. In the north the boundary runs along the Taramakau River, the Arnold River and the Grey River to the coast. There are some outlying patches of beech a little south of the Arnold and in the Otira. Various authors agree that this segment corresponds roughly with that covered by the last major advance of the Pleistocene ice, and that the more easily distributed podocarp-dicotylous species were able to recolonise it first.

HOLLOWAY (1948), has emphasised the wide range of conditions under which beech forests are found . . . 'in regions with an annual rainfall as high as 250 inches, or as low as 25 inches' . . . and of the many different types of beech forest which exist. The dominants may be either one or more of the four species. Usually at lower altitudes conifers and hardwoods

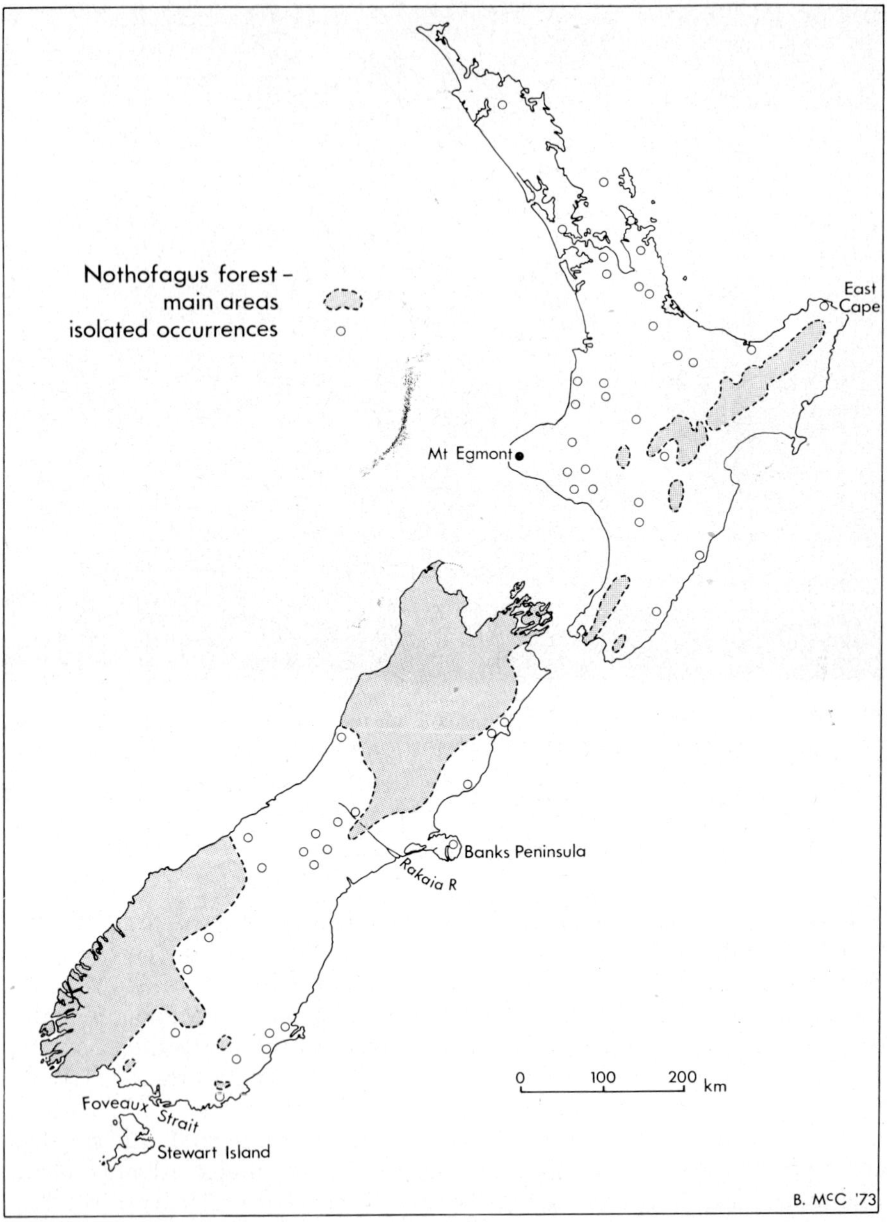

Fig. 18. Distribution of southern beech *(Nothofagus)* in New Zealand. (From P. Wardle, 1964.)

are associated with the beeches and 'beech-podocarp' forest is a common classification. Of the main trees listed in the previous section all, except perhaps *Beilschmiedia tarairi*, can be found associated with *Nothofagus*. At higher altitudes a canopy of pure beech is more common.

Nothofagus truncata (Hard beech) is the only species found north of about latitude 37°. It was common at one time at sea-level on small islands near Auckland, and isolated patches are known in North Auckland. It is important in the southern Tararua Range in the Marlborough Sounds area and in western Nelson, but does not go further south than the Taramakau River. It is usually found on poorer soils than red beech.

Nothofagus fusca (Red beech), is closely related to *N. truncata*. Both species have singly serrate leaves which are larger than the other New Zealand species. Red beech can endure a lower temperature than hard beech, and extends from about 37°S to Lake Te Anau (45½°S), with some stands further south.

Nothofagus menziesii (Silver beech), stands apart from the other species in several characters of flower and fruit, and by the small doubly serrate leaves. It is frequently found at timberline. It ranges from about 37°S to the southern limit of the genus in New Zealand and is the only Nothofagus found in the patches of beech forest in the extreme south east of South Island.

Nothofagus solandri with small entire leaves, consists of two sub-species, *N. solandri* var. *solandri* (black beech) and *N. solandri* var. *cliffortioides* (mountain beech). It is found from near East Cape in North Island to the northern shore of Foveaux Strait, but is local in the southern Ruahine and northern Tararua Ranges of North Island. It is a species with 'a wide range of tolerance for harsh conditions but with a poor competitive ability –' (J. WARDLE, 1970a).

The interior of beech forest is in general more open than that of podocarp-dicotylous forest. In wetter areas the undergrowth is at its greatest profusion, while in the driest forest there is a paucity of vegetation under the canopy. J. WARDLE (1970b) has described associations of *N. solandri* ranging from pure beech forest with a mean number of vascular plant species (plus moss) of only 5.2, to a complex association with a mean number of 30.0. Shrubs and small trees common in many areas of beech are: *Griselinia littoralis* (Cornaceae), *Pseudowintera colorata* (Winteraceae), *Coprosma foetidissima*, *C. pseudocuneata* (Rubiaceae), *Pseudopanax colensoi*, *P. simplex* (Araliaceae), *Phyllocladus alpinus* (Podocarpaceae), *Myrsine divaricata* (Myrsinaceae), and small-leaved *Pittosporum* species. Given sufficient moisture the forest floor may be thickly carpeted with bryophytes, in which will be such species as *Microlaena avenacea* (Gramineae), *Luzuriaga parviflora* (Liliaceae), species of *Nertera* (Rubiaceae), and ferns such as *Blechnum discolor*, *B. procerum*, *Hymenophyllum multifidum* and *Polystichum vestitum*. Epipytes and lianes are not important except in beech-podocarp forest.

Fig. 19. Part of the Kaingaroa Forest, central North Island, showing clear felling of *Pinus radiata* in the Waihura cutting section. (New Zealand Forest Service Photo by J. H. G. Johns, A.R.P.S.)

Exotic conifer forests

In 1972 exotic forests planted for timber covered 583,000 ha, of which 306,832 ha was State Forest. By far the most important species is *Pinus radiata*, a native of California with rapid diameter and height growth. Douglas fir *(Pseudotsuga menziesii)* a tree for favourable sites is next in importance, followed by Corsican Pine *(P. nigra)* used on poor sites where *P. radiata* does not thrive. During the 1920's and 1930's much land considered unsuitable for agriculture was planted in pines and forms the nucleus of the present day exotic forest. A major planting was on the Kaingaroa plains, central North Island. (Fig. 19). This was an area of rolling pumice soils supporting tussock and scrub where stock ill-thrift ruled out agriculture until, in 1937, the 'bush-sickness was traced to a cobalt deficient soil. Other plantings were at Karioi, Central North Island, above the then accepted limit for agriculture; on the Moutere gravel soils of Nelson; on the shallow shingle soils on the north banks of the Hurunui and Waimakiriri rivers, north Canterbury; and on the severely podzolised soils left after the felling of Kauri in North Auckland

and terrace rimu in Westland, (POOLE, 1969). The latter plantings were unsuccessful.

Pine forests are now a characteristic feature of New Zealand vegetation and in some places support quite a diverse undergrowth of indigenous plants (MCQUEEN, 1961).

Scrub and Timberline

Perhaps the most important lowland scrub genus in the country is *Leptospermum* (Myrtaceae) with either *L. scoparium* (manuka) or *L. ericoides* (kanuka), singly or mixed, forming stands from 3–4 m to 9–10 m high. The ecology of the two species was compared in south-eastern South Island by Burrell (1965) who distinguished seral stands where rainfall exceeded 75 cm, climax stands where rainfall is about 60 cm (with bushes quite widely spaced) and sub-climax stands with intermediate rainfall. *L. scoparium* tolerated boggy ground, whereas *L. ericoides* required good drainage, is the longer-lived species and, in time, predominates in mixed stands. Both species are important 'nurses' for forest trees, and in northern New Zealand for example protect the early life of the Kauri. Here they are often joined by the spiny introduced *Hakea sericea* (syn. *H. acicularis*) (Proteaceae) and these two species cover large areas of poor gumland, so called because of the Kauri gum once dug there.

Other lowland scrub genera can be illustrated from the dry hills of eastern South Island. Common here are the spiny *Discaria toumatou* (Rhamnaceae), *Coprosma propinqua* and *C. virescens* (Rubiaceae), *Muehlenbeckia complexa* (Polygonaceae) as well as the local endemics *Clematis afoliata* (Ranunculaceae) and *Sophora prostrata* (Papilionaceae) (Fig. 20). Broom *(Cytisus scoparius)* and gorse *(Ulex europaeus)* are common introduced scrub species (cf. BURROWS, 1969a).

At high altitudes *Nothofagus* forest often stretches uninterrupted to the alpine vegetation, but where the forest is not beech there is usually a belt of subalpine scrub between the forest below and alpine vegetation above. This scrub can form an almost impenetrable barrier of prostrate trunks and stiff interlocking branches. A main type, classified by COCKAYNE (1958) as 'shrub-composite' is dominated by species of *Senecio* and *Olearia*. In more detail, P. WARDLE (1960) described, in the Hokitika catchment (wet western South Island) low forest dominated either by *Hoheria glabrata* (Malvaceae), or *Olearia*, and scrub dominated either by *Olearia colensoi*, *Dracophyllum uniflorum* (Epacridaceae) or *Dacrydium biforme* (Podocarpaceae).

In parts of the Tararua Range the upper limit of beech is lower than expected, with a belt of scrub above it. ZOTOV (1938) suggested that this depression is due to frequent fogs which reduce light intensity so that *Nothofagus menziesii* cannot compete with scrub species such as *Olearia colensoi*. P. WARDLE (1964) modified this proposal slightly to fit

Fig. 20. View across the shingle river-bed of the Hurunui River (eastern South Island) which is lined by introduced trees: weeping willow *(Salix babylonica)* on right; *Pinus radiata* centre; and crack willow *(S. fragilis)* on left. On the dry hills above are the rounded dark shrubs of *Sophora prostrata*, with patches of bracken in the gullys and scrub lower down, as well as scattered short tussocks of *Festuca* and *Poa*. (Photo: E. J. GODLEY.)

other areas, by suggesting that *Nothofagus* invades more slowly under high rainfall or frequent fogs than under drier sunnier conditions.

New Zealand is no exception to the general rule that the upper limit of continuous woody vegetation is lower in the southern than the northern hemisphere. In North Island from East Cape to the Ruahine Ranges it is about 1350 m dropping to about 1200 m in the Tararua Ranges (41 °S). In western South Island at 41 °S timberline is at about 1200–1300 m declining to about 900–980 m at 45 °S, while in central and eastern South Island timberline can be 150–300 m higher than at a similar latitude in the west (P. WARDLE, 1965). On Stewart Island (47 °S) timberline is about 740 m and on Campbell Island (52 °S) it is 150 m. ZOTOV (1938) has suggested a correlation between timberline and a mean temperature of 10.5 °C in the mid-summer month (January), and P. WARDLE (1971) has discussed a hypothesis that alpine timberline represents 'the highest altitude at which shoots of woody plants can grow and ripen under the air temperatures which prevail at the height of tree canopies'.

The Grasslands

General

The indigenous grasslands are dominated by species with the tussock (bunch grass) habit. These associations have their closest relation to the *Duriherbosa* formation class of Rübel (1938) in that the vegetative organs are more strengthened by mechanical tissue than by turgescence. But an important difference is that the leaves of the New Zealand tussock species do not die back completely in winter. Tussock associations are (or were) found from near sea-level to the alpine zone, and there are two main types. Tall-tussock grassland is dominated by one or another of the larger species of snow-grass *(Chionochloa)* ranging in height from about 180 to 50 cm. This grassland is found more at higher altitudes, either above forest in higher rainfall areas or above short-tussock grassland in drier areas and is important in North, South, Auckland and Campbell Islands, but not Stewart Island (where *Chionochloa* is found) or Chatham Islands (where *Chionochloa* is not found). Short-tussock grassland is mainly found in the lower and drier areas of South Island, and is usually dominated by species of *Festuca* and *Poa*, up to about a half metre in height and yellow-brown in general appearance. However as with the two main forest types the dominant species of these grasslands are not mutually exclusive and *Festuca* can be an important associate of some species of *Chionochloa*, (Fig. 21).

Short-tussock grassland

A century ago this covered a considerable area of eastern South Island from sea-level to between 800 and 900 m. Since then the margins have been extended in some areas where beech forest has been cleared. Within the area as found by the first settlers Zotov (1939) proposed two sub-regions differentiated on climatic grounds. The first, called 'true steppe' was considered to have an upper limit of about 600 m and to lack a forest history. Grasslands outside this area, or within it but between 600 and 1200 m were described as 'induced steppe' and considered capable of returning to forest. However fossil evidence has since shown that woody vegetation occurred throughout the 'true steppe' area (Molloy *et al.*, 1963). The decline of the forests through natural and Polynesian fires, the rise to predominance of tall-tussock grassland, and the change to short-tussock grassland through grazing and fire have been discussed by Connor (1964), Connor & Macrae (1969) and Molloy (1969b).

A detailed study was made by Barker (1953) of an area of tussock grassland on the Hunters Hills, South Canterbury, which is fairly typical of much of the middle South Island. The dominants in the short-tussock grassland were *Festuca novae-zelandiae*, (hard tussock, yellow tussock), *Poa*

Fig. 21. Tall tussocks of *Chionochloa rubra* in the Havelock Valley, eastern South Island. The nearest tussock is about 2 m tall. In the foreground are short tussocks of *Festuca novae-zelandiae.* (New Zealand Forest Service. Photo by J. H. G. JOHNS, A.R.P.S.)

caespitosa, (silver tussock), and *Poa colensoi* (blue tussock). Of these, *Poa caespitosa* was a differential species, in that, unlike the other two, it was rarely found in the tall tussock grassland above. It chooses the more mesophytic habitats, and '. . . is plentiful over most of the lower and easier slopes, assuming dominance on damp shady slopes, in depressions and concave slopes and on river terraces where there are the usual low tussock-accompanying species.' It was '. . . comparatively shallow rooted, its roots seldom penetrating deeper than one metre with the majority concentrated in the upper 40 cm.' *(ibid.)*.

Festuca novae-zelandiae was the most abundant tussock and '. . . is the sole dominant on steep, strongly isolated exposed slopes, and a higher elevations of the community. *F. novae-zelandiae* is characteristically deep-rooted, (down to 2 metres) with a well developed system of laterals in the top-soil.' *Poa colensoi* '... is often intermingled with the other tussock grasses over the lower slopes of shady and intermediate aspect. It assumes local dominance where soil depth diminishes as on ridges and rocky outcrops ...' *(ibid.)*.

Short-tussock grassland is floristically diverse. Barker listed 37 ground stratum species including 13 adventives. In fescue grassland throughout Canterbury (east central South Island) CONNOR & MACRAE (1969) listed 28 species (10 adventives) characteristic of short-tussock grasslands, 56 species (12 adventives) classified as widespread in tussock grasslands, and 6 species characteristic of various kinds of tall-tussock. As well as this, 52 species occurred only once in the samples and 22 species occurred twice. Examples of characteristic short-tussock species are the small composites *Raoulia monroi*, *Lagenophora pumila*, and *Vittadinia australis*, and adventives such as *Crepis capillaris* and *Trifolium arvense*. Widespread grassland species include the spiny shrub, *Discaria toumatou*, the grasses *Agropyron scabrum* and *Poa caespitosa* as well as adventives such as *Rumex acetosella*, *Hypochoeris radicata* and *Anthoxanthum odoratum*.

Hillside pastures of danthonia (particularly species of Section Notodanthonia) were once extensive.

The original short (and tall) tussock grassland has been modified in greater or lesser degree during a century of stocking. The extreme was seen in the driest areas such as Central Otago, and the upper Awatere Valley in southern and northern South Island respectively. Here stocking and burning opened up the association and made an ideal home for rabbits, whose depredations accelerated the vegetation changes. Deserts of the scab-weed, *Raoulia australis*, a composite mat-plant were induced. Attempts to reverse the trend have involved extermination of rabbits, control of burning, and reduction of grazing by sheep.

TALL-TUSSOCK GRASSLAND

Tall-tussock grassland is found above timberline to a greater or lesser extent along the axial range of North Island, with *Chionochloa pallens* occurring throughout and *C. flavescens* appearing on the Tararua Ranges in the south. In South Island *C. pallens* and a widespread un-named species are usually found on the alpine slopes above the forest of the more westerly and wetter mountains (Fig. 22), although both species have some eastward extensions. On the foothills and mountains of eastern South Island, grassland dominated by *C. rigida* covers large areas extending down into Otago. In Canterbury, where *C. rigida* occurs south of the Rakaia river and on Banks Peninsula, the grassland usually occurs from about 600 m upwards until it meets *C. macra* grassland or continues up into the alpine zone. North of the Rakaia *C. macra* (Fig. 24) is extensive from lower to higher altitudes (CONNOR & MACRAE, 1969). *Chionochloa rubra*, which

Fig. 22. Tall tussock grassland of *Chionochloa pallens* above Lewis Pass, northern South Island, with *Celmisia coriacea* in foreground. Interspersed are low shrubs of *Hebe*, and mats of small-leaved *celmisias*. Nothofagus forest is in the distance, with the Spenser Mountains beyond. (Photo: G. C. KELLY.)

tolerates poorly drained conditions, is important in central North Island on areas devastated by volcanic eruptions several centuries ago, and is also found on Mount Egmont, ZOTOV (1963). *Chionochloa rubra* grassland is widespread, but scattered, in South Island but was once of considerable importance on poorly drained land in Southland. On Secretary Island, Fiordland, *Chionochloa acicularis* is important in snow-tussock herb-field above timberline (Fig. 23) (WARDLE, MARK & BAYLIS (1970)). The remaining tall snow-grass, *C. antarctica* of the Auckland and Campbell Islands, will be discussed later. There are other smaller species of *Chionochloa* which are locally important such as the alpine carpet grass,

Fig. 23. Chionochloa acicularis grassland with clumps of *Dracophyllum fiordense* (taller) and *D. menziesii* (shorter) at tree limit (about 1000 m) on Secretary Island, Fiordland National Park, south-west South Island (From WARDLE, MARK & BAYLIS, 1970). (Photo: G. T. S. BAYLIS.)

C. australis of northern South Island, and *C. crassiuscula* and *C. oreophila* of the wetter western alpine zone of South Island. On Stewart Island, *C. crassiuscula* with *C. pungens* forms the meagre high altitude grassland. More detailed aspects of the biology of the snow-grass species have been discussed by MARK (1969).

On the Hunters Hills, BARKER recorded the boundary between short and tall tussock as between 700 and 750 m on east-facing slopes and lower on south-facing slopes. Here there was also a change from the Hurunui soil type of the short tussock, to the Kaikoura soil type of the tall tussock. The physiognomic dominants were the *Chionochloa* tussocks and beneath them the cotton plant, *Celmisia spectabilis* (Compositae). The latter forms tufts of stout, linear-oblong leaves with dense tomentum beneath, and has well protected growing points enabling it to survive tussock fires. In South Canterbury it is nearing its southern limit, but from here north it is an important plant of the tall tussock, and in many areas has increased considerably. Also common were short tussocks of *F. novae-zelandiae* and *Poa colensoi*. CONNOR (1961) emphasised the importance of *Celmisia spectabilis*, and also *Festuca matthewsii*, for classifying tall-tussock grasslands in Canterbury, and later described *Chionochloa rigida/F. novae-zelandiae/Celmisia spectabilis* grassland; *C. rigida/F. novae-*

Fig. 24. Vegetable sheep *(Raoulia eximia)* and tussocks of *Chionochloa macra*, on a deteriorating greywacke shingle slope on the Torlesse Range, eastern South Island. Remnants of the original surface can be seen on Foggy Peak in the distance (altitude 1750 m). (Photo: E. J. Godley.)

zelandiae grassland; and *C. rigida/F. matthewsii* grassland.

Like the short-tussock grassland the tall tussock is floristically diverse but with adventive species usually less common. Barker listed 42 ground stratum species as important for the Hunters Hills but with only 5 adventives. Connor & Macrae (1969) describing many samples from *Chionochloa rigida* grassland in Canterbury list 31 species characteristic of this grassland, 8 species characteristic of tall-tussock grasslands in general, 57 species (13 adventives) widespread in tussock grasslands, 9 species (3 adventives) characteristic of short-tussock grassland, and one species characteristic of *Chionochloa rubra* grassland. Examples of the first group are *Aciphylla aurea, Cyathodes colensoi, Pentachondra pumila, Celmisia lyallii;* and of the last *Bulbinella angustifolia*.

In northern South Island Wraight (1963) has described culturally induced associations involving deterioration of *Chionochloa* grassland and rise to prominence of *Notodanthonia setifolia* – yet another example of the vulnerability of the tall tussocks to the combined onslaughts of fire and grazing.

NON-INDIGENOUS GRASSLANDS

The indigenous grasslands of New Zealand were used for grazing, and still are, but where feasible this 'permanent grass of low productivity' (LEVY, 1970) was replaced by introduced grasses, as was the forest over much of the country. This new grassland is now characteristic of much of New Zealand and the basis of agricultural production. The basic species used to produce a variety of local strains are the rye-grasses (*Lolium perenne* and *L. italicum*) with cocksfoot *(Dactylis glomerata)*, timothy *(Phleum pratense)* and paspalum *(P. dilatatum)* also valuable, the latter particularly in the warmer north. No less important in the sward are the bred strains of white clover *(Trifolium repens)* and red clover *(T. pratense)*. The story of the 'grasslands revolution' has been told by LEVY *(op. cit.)* and SMALLFIELD (1970).

Swamps and bogs

Burning and drainage have drastically modified much of the wet-lands of New Zealand. Three striking dominants in different kinds of swamps are the New Zealand flax, *Phormium tenax* (Agavaceae); the raupo, *Typha orientalis* (Typhaceae); and the nigger-head, *Carex secta*, a sedge standing in the water on a fibrous trunk. Extensive 'semi-swamp' forests were once dominated by the large tree *Podocarpus dacrydioides*, while the cabbage-tree *(Cordyline australis)* increases considerably when the forests are felled and drained.

Of the various kinds of bogs found in New Zealand (CRANWELL, 1953; HARRIS, 1953) three are of exceptional interest. The restiad peat bogs dominated by the monotypic endemic *Sporadanthus* are found only in two widely separated localities in northern New Zealand and on distant Chatham Island. CAMPBELL (1964) described examples with an upper tier of *Sporadanthus traversii* and a lower sheltered tier of *Calorophus minor (Hypolaena lateriflora)*, both Restionaceae. The latter species with its water-retentive roots was considered the equivalent of *Sphagnum* in the standard type of raised bog in the northern hemisphere. On extremely leached lowland soils on the high rainfall west coast of South Island there are fern-sedge wastelands known locally as 'pakihis'. Important plants are the low-growing *Gleichenia circinnata*, *Baumea teretifolia (Cladium teretifolium)* and *Tetraria capillaris (Cladium capillaceum)* (RIGG, 1962). It is thought that pakihi soils developed under podocarp forest and that a change in soil or climate caused the change to bog. In the wet mountains, and in the lowlands of the south are cushion bogs so called because of the predominance of cushion plants of such genera as *Oreobolus* (Cyperaceae), *Phyllachne* (Stylidiaceae), *Gaimardia*, *Centrolepis* (Centrolepidaceae) and *Donatia* (Donatiaceae). Mats of small-leaved *Astelia* spp. often occur in these bogs, which are of considerable biogeographical interest, occurring in many southern lands and having similar generic composition. Sphagnum species may be present in both lowland and highland bogs throughout the main islands, and on Chatham and Campbell Islands but are extremely rare on the Auckland Islands (GODLEY, 1969a). In New Zealand the genus is not an important peat-former.

Alpine Vegetation

New Zealand is well isolated from other lands, but even more isolated from anything approaching it in height or extent is its mountain system. To find something comparable we must look to South America. Some 500 species do not descend to the lowlands in New Zealand. The main diversity is in South Island, but alpine species are found along the North Island mountain chain and on isolated peaks further north to 37°S. We have already mentioned the subalpine scrub, the tall tussock-grassland of the lower alpine zone, and the alpine cushion bogs. Here we will first note some of the important alpine genera (cf. PHILIPSON & HEARN, 1962).

One of the most characteristic New Zealand genera is *Celmisia* (Compositae) which has evolved some 60 species in New Zealand and four in Australia (GIVEN, 1969), and is predominantly a genus of higher altitudes. In *C. coriacea*, the largest species, the leaves are about 60 cm long, oblong-lanceolate with a silvery pellicle, and aggregated into large tufted rosettes (Fig. 22). *C. spectabilis* is also constructed on this plan but has smaller leaves with tomentum beneath. *C. lyallii* and *C. armstrongii* are also tufted but have long stiff, narrow sharply pointed leaves. Species such as *C. walkeri* and *C. novae-zelandiae* form diffuse clumps owing to the prostrate stems and have small, spaced leaves. In *C. laricifolia* there are tufts of tiny needle-like leaves and in *C. sessiliflora* all parts are compacted into a dense cushion. *C. gracilenta* can be a common plant of lower altitudes, particularly in tussock grassland, and some species even grow on the coast or in sheltered valleys nearby, e.g. *C. major* (Auckland west coast), *C. monroi* var. *conspicua* (Marlborough), *C. mackaui*, (Banks Peninsula), *C. lindsayi*, (South Otago coast), and *C. vernicosa* and *C. campbellensis*, (Subantarctic Islands). Other composite genera of the mountains are *Helichrysum*, *Leucogenes* (the New Zealand edelweiss), and *Haastia* and *Raoulia* which have been previously noted.

Scrophulariaceae is also an important family in alpine New Zealand with shrubs and sub-shrubs in *Hebe* and *Parahebe* (including the whipcord hebes mentioned earlier) and cushions in *Pygmea*. In the herb-fields and by stream-sides *Ourisia* and *Euphrasia* contribute some of the most attractive flowers.

The gentians are also prominent, as well as *Ranunculus* which has an important centre of development in the New Zealand mountains (FISHER, 1965) culminating in *R. lyallii* with deep green peltate leaves up to

25 cm in diameter and large inflorescences of white-petalled flowers.

Of great interest are the 24 obligate scree species, found only on the inhospitable scree slopes of the drier South Island mountains (FISHER, 1952) and derived from 17 genera in 12 families. This specialisation for such a habitat is something not so far paralleled in North America or Europe.

The vegetation of the central South Island mountains is discussed by BURROWS (1969b). The greywacke here is easily fragmented and the ridges sharp, in contrast to the more stable rounded schistose mountains in parts of Central Otago further to the south. Here the high alpine vegetation was classified by MARK & BLISS (1970) into four major habitat-physiognomic types: fell-field (particularly non-schistose mountains), herb-field, cushion and snow-bank.

The origin of the New Zealand alpine flora is a problem of considerable interest. Mountain building in the Pliocene was preceded in the Miocene by a low-lying subtropical archipelago. From whence then came the plants which colonised the mountains, and what has led to their diversity in so relatively short a time? COCKAYNE (1967) summarised his views on the mid-Tertiary warm period as follows: 'Under such conditions many subantarctic species would go to the wall, and the few survivors, except *Fuchsia* and some others, would be confined to the summits of the hills, faces of rock and such other stations as offered tolerable conditions – similar stations, indeed, to those occupied even yet at low levels by certain alpine plants.' And of the retreat of the glaciers he wrote: '. . . much bare ground would gradually become available for recolonisation. Seeds from species of the lowlands, and from plants of those rocks and peaks which projected above the ice, would be carried to this bare ground. . . . Some of these plants would be ancient enough, though many archaic types must have perished; others would have been evolved under the new conditions, hybridism probably playing a notable and, quite likely, a rapid part. Thus has possibly come about that extreme diversity of form shown by *Hebe*, *Celmisia* and many other genera, and it may be that much of the high-mountain flora is of quite recent origin.'

P. WARDLE (1968) has suggested that if during the period of Tertiary warmth there were infertile soils on peneplaned uplands drenched by persistent mist and rain some cool climate plants could undoubtedly exist there, and also (1963b) that ancestors of alpine plants could have escaped the warm period by existing in more temperate conditions at higher latitudes on a southward extension of New Zealand. RAVEN (1973), on the other hand, proposes that many of the ancestors of our alpine plants entered *via* Australia following the late Pliocene and subsequent elevation of mountains in Malaysia, New Guinea, Australia and New Zealand.

Adventive plants

Some 1700 species not native to New Zealand have now been recorded in the wild, but only part of these can be regarded as fully naturalised. ALLAN (1937b) analysed the 500 species which he considered to be in this category and found that 80% were from the Old World, 11% from the Americas, (North, 26 species; tropics, 10; South, 8;), 6% from Australia and 3% from South Africa. At least 56% of these species were considered to have entered as impurities in seed, 24% escaped from cultivation, 15% may have entered adhering to animals, and 5% in ballast or by other means. A similar diversity of origins was shown for the adventive flora of Canterbury in eastern South Island, by HEALY (1969).

This new vegetation has entered areas disturbed by man or animals, and in some places the species are now integral parts of an association as in the short tussock grassland. However, the alien species have not been successful in competing with the natives unless there has been some preliminary disturbance of the habitat, or unless the habitat was originally open, as in river beds or fresh waters. The invasion of the latter by adventives has been catalogued by MASON (1970a, 1970b).

There are over 80 adventives which are so aggressive that they are listed as noxious weeds. Some of the most notorious are blackberry *(Rubus fruticosus)*; californian thistle *(Cirsium arvense)*; ragwort *(Senecio jacobaea)*; and sweet brier *(Rosa rubiginosa)*. Also listed is the native *Cassinia leptophylla*, a composite bush particularly aggressive on open hillsides in southern North Island and north-eastern South Island.

Some adventive species which are physiognomically prominent have already been mentioned – the widespread gorse, *Ulex europaeus*, the broom, *Cytisus scoparius*, and *Hakea sericea* (syn. *H. acicularis*). Other examples are *Lupinus arboreus* on dunes and river beds and *Ammophila arenaria* much of it planted on dunes. In eastern South Island there is a serious infestation of *Nassella trichotoma* an unpalatable South American tussock grass, and special legislation has been necessary to combat this pest of pastoral land.

The Outlying Islands

The Kermadec Islands (29°15′–31°24′S) have some 117 species of ferns and flowering plants 12% of which are endemic and 76% found in New Zealand. There are no endemic genera and no gymnosperms. On Raoul Island *Metrosideros kermadecensis* dominates the lower altitude dry forest, in which also grow *Cyathea milnei*, *Myrsine kermadecensis*, *Coprosma petiolata* and the mainland species *Corynocarpus laevigatus* and *Myoporum laetum*. In the wet forest above are found *Ascarina lanceolata*, *Cyathea kermadecensis*, *Rhopalostylis cheesemanii* and the mainland species *Melicytus ramiflorus* (OLIVER, 1910). The effect of the 1964 eruption and of goats

Fig. 25. Upper limit of *Dracophyllum* scrub on Campbell Island, at about 150 m, merging into *Poa litorosa* grassland, Col Peak in distance. (Photo: E. J. GODLEY.)

on the vegetation as well as the spread of introduced plants such as the aroid *Alocasia macrorrhiza* are discussed by SYKES (1965).

The Chatham Islands support some 300 species of ferns and flowering plants of which 14% are endemic and 85% found in New Zealand. The one endemic genus is *Myosotidium* (Boraginaceae). There are no gymnosperms or other large mainland trees, and much of the original forest has been destroyed. In coastal forest *Corynocarpus laevigata* and *Rhopalostylis sapida* were common but did not extend to the 'tableland' forest where the endemic *Dracophyllum arboreum* was dominant. WRIGHT (1959) suggested that this species 'became the dominant plant of habitats that, on the mainland, would have been occupied by podocarp-broadleaved forest, manuka *(Leptospermum)* scrubland or tussock grassland.'

In the subantarctic islands there are about 193 species, of which 31% are endemic and 63% found in New Zealand proper. The one endemic genus is *Pleurophyllum*. The Snares (48°S) has low forest of *Olearia lyallii* and *Senecio stewartiae* and patches of *Poa litorosa* and *P. foliosa* grassland (FINERAN, 1964). On the Auckland Islands (50°32′S) a narrow belt of *Metrosideros umbellata* coastal forest in sheltered eastern inlets is replaced by a tangled scrub of *Myrsine divaricata*, *Cassinia vauvilliersii*, *Coprosma* spp. *Pseudopanax simplex*, *Dracophyllum longifolium*, and stunted *M. umbellata* to about 300 m. Windswept lanes supporting cushion and mat plants pass through this scrub. Above this is *Chionochloa antarctica* grassland (COCKAYNE, 1909; MOAR, 1958; GODLEY, 1965). The interaction between the introduced animals confined to Enderby Island and the vegetation there is described by TAYLOR (1971). On Campbell Island (52°22′S) there is a belt of *Dracophyllum* scrub to about 150 m in sheltered situations, above which is *Poa litorosa* grassland. *Chionochloa antarctica* grassland is found on the floors of certain valleys and on cushion bogs and is more extensive than indicated by OLIVER & SORENSEN (1951). Antipodes Island (49°41′S) is predominantly a grassland island. *Poa litorosa* is the dominant tussock and of the 62 species of vascular plants only the three coprosmas are woody, one of them a mat plant. There are scattered swamps and bogs with species of *Blechnum* and *Carex* prominent (GODLEY, 1969b). On Macquarie Island (54°30′S) there are no woody plants and only 31 species of flowering plants, 3 ferns and 1 lycopod. *Poa foliosa* grassland is common.

Acknowledgements

I am very grateful to Miss D. H. SMITH for her help in assembling material for this article, and to Dr. I. A. E. ATKINSON, Mr. H. E. CONNOR, Mr. A. P. DRUCE and Dr. P. WARDLE for critical comments. Mr. K. R. WEST and BEVERLEY McCULLOCH kindly drew the figures.

For Figures 11 and 12 I am indebted to the Director-General, New Zealand Forest Service and to the Ministry of Works.

REFERENCES

ALLAN, H. H. 1937a. A consideration of the biological spectra of New Zealand. *Journal of Ecology* 25: 116–52.

ALLAN, H. H. 1937b. The origin and distribution of the naturalised plants of New Zealand. *Proceedings of the Linnean Society of London*. Session 150: 25–46.

Allan, H. H. 1940. 'A handbook of the naturalized flora of New Zealand'. Government Printer, Wellington. 344 pp.
Allan, H. H. 1961. 'Flora of New Zealand'. Vol. I. Government Printer, Wellington. 1085 pp.
Atkinson, I. A. E. 1964. The flora, vegetation and soils of Middle and Green Islands, Mercury Islands Group. *New Zealand Journal of Botany* 2: 385–402.
Banks, J. 1963. 'The Endeavour Journal of Joseph Banks 1768–1771'. 2 vols. 2nd ed. Beaglehole, J. C. ed. Angus & Robertson. 476 pp. & 406 pp.
Barker, A. P. 1953. An ecological study of tussock grassland. New Zealand Department of Scientific and Industrial Research Bulletin 107.
Barton, I. L. 1972. On the vegetation of the Hunua Ranges, Auckland. *New Zealand Journal of Botany* 10: 8–26.
Baylis, G. T. S. 1958. A botanical survey of the small islands of the Three Kings Group. *Records of the Auckland Institute and Museum* 5: 1–12.
Baylis, G. T. S. 1967. Experiments on the ecological significance of phycomycetous mycorrhizas. *New Phytologist* 66: 231–43.
Brownlie, G. 1962. Geographical relationships of the New Zealand fern flora. *Pacific Science*, 16, 363–65.
Burrell, J. 1965. Ecology of *Leptospermum* in Otago. *New Zealand Journal of Botany* 3: 3–16.
Burrows, C. J. 1969a. Lowland and upland scrub. In Knox, G. A. ed., 'The Natural History of Canterbury'. A. H. & A. W. Reed, Wellington. 620 pp.
Burrows, C. J. 1969b. Alpine Grasslands. In Knox, G. A. ed., 'The Natural History of Canterbury'. A. H. & A. W. Reed, Wellington. 620 pp.
Bussell, W. T. 1968. The growth of some New Zealand trees. *New Zealand Journal of Botany* 6: 63–85.
Campbell, E. O. 1964. The Restiad peat bogs at Motumaoho and Moanatuatua. *Transactions of the Royal Society of New Zealand.* Botany 2: 219–27.
Clapham, A. R., Tutin, T. G. & Warburg, E. F. 1962. 'Flora of the British Isles'. Cambridge University Press. 2nd ed. 1269 pp.
Clarke, C. M. H. 1968. Flowering periods of alpine plants at Cupola Basin, Nelson, New Zealand. *New Zealand Journal of Botany* 6: 205–20.
Cockayne, L. 1909. The ecological botany of the subantarctic islands of New Zealand. In 'The Subantarctic Islands of New Zealand'. I: 182–235. Government Printer, Wellington.
Cockayne, L. 1912. Observations concerning evolution, derived from ecological studies in New Zealand. *Transactions and Proceedings of the New Zealand Institute* 44: 1–50.
Cockayne, L. 1926. Monograph of the New Zealand Beech Forests. part 1. The ecology of the forests and the taxonomy of the beeches. New Zealand State Forest Service. Bulletin No. 4.
Cockayne, L. 1958. 'The Vegetation of New Zealand'. 3rd (reprint) ed. H. R. Engelmann (J. Cramer), London, New York. 456 pp.
Cockayne, L. 1967. 'New Zealand Plants and their Story'. 4th ed. Godley, E. J. ed. Government Printer, Wellington. 269 pp.
Connor, H. E. 1957. Breeding systems in some New Zealand grasses. *New Zealand Journal of Science and Technology*, Ser. A. 38: 742–51.
Connor, H. E. 1960. Breeding systems in New Zealand grasses III. Festuceae Aveneae and Agrostideae. *New Zealand Journal of Agricultural Research*, 3: 728–33.
Connor, H. E. 1961. A tall-tussock grassland community in New Zealand. *New Zealand Journal of Science* 4: 825–35.
Connor, H. E. 1964. Tussock grassland communities in the Mackenzie Country, South Canterbury, New Zealand. *New Zealand Journal of Botany* 2: 325–51.
Connor, H. E. 1966. Breeding systems in New Zealand grasses. 7. Periodic flowering of snow tussock, *Chionochloa rigida*. *New Zealand Journal of Botany* 4: 392–7.
Connor, H. E. & Macrae, A. H. 1969. Montane and subalpine tussock grasslands in

Canterbury. In KNOX, G. A. ed., 'The Natural History of Canterbury'. A. H. & A. W. Reed, Wellington. 620 pp.

CRANWELL, L. M. 1953. An outline of New Zealand peat deposits. Proceedings 7th Pacific Science Congress, 1949. 5: 186–208.

CRANWELL, L. M. & MOORE, L. B. 1935. Botanical notes on the Hen & Chicken Islands. *Records of the Auckland Institute and Museum* 1: 302–18.

CUMBERLAND, K. B. 1962. Climatic change or cultural interference? – New Zealand in Moahunter times. pp. 88–142. In 'Land and Livelihood'. New Zealand Geographical Society, Caxton Press, Christchurch.

DAWSON, J. W. 1964. Unisexuality in the New Zealand Umbelliferae. *Tuatara* 12: 67–68.

DAWSON, J. W. 1967a. The New Zealand species of *Gingidium* (Umbelliferae). *New Zealand Journal of Botany* 5: 84–116.

DAWSON, J. W. 1967b. New Zealand Umbelliferae *Lignocarpa* gen. nov. and *Scandia* gen. nov. *New Zealand Journal of Botany* 5: 400–17.

DAWSON, J. W. 1968. New Zealand Umbelliferae. A leaf comparison of *Aciphylla* and *Anisotome*. *New Zealand Journal of Botany* 6: 450–8.

ESLER, A. E. 1967. The vegetation of Kapiti Island. *New Zealand Journal of Botany* 5: 353–93.

FINERAN, B. A. 1964. The flora of the Snares Islands, New Zealand. *Transactions of the Royal Society of New Zealand. Botany* 3: 237–70.

FINERAN, B. A. 1966. The vegetation and flora of Bird Island, Foveaux Strait. *New Zealand Journal of Botany* 4: 133–46.

FISHER, F. J. F. 1952. Observations on the vegetation of screes in Canterbury, New Zealand. *Journal of Ecology* 40: 156–67.

FISHER, F. J. F. 1965. The Alpine Ranunculi of New Zealand. New Zealand Department of Scientific and Industrial Research Bulletin 165.

GARNIER, B. J. 1958. 'The Climate of New Zealand'. Edward Arnold, London, 191 pp.

GIVEN, D. R. 1969. A synopsis of infrageneric categories in *Celmisia* (Astereae-Compositae). *New Zealand Journal of Botany* 7: 400–18.

GODLEY, E. J. 1961. Fauna and Flora. In SINCLAIR, K. ed., 'Distance looks our way'. Paul's Book Arcade for the University of Auckland.

GODLEY, E. J. 1965. Notes on the vegetation of the Auckland Islands. *Proceedings of the New Zealand Ecological Society* 12: 57–63.

GODLEY, E. J. 1966. Breeding systems in New Zealand plants. 4. Selfsterility in *Pentachondra pumila*. *New Zealand Journal of Botany* 4: 249–54.

GODLEY, E. J. 1968. A plant list from the Cordillera de San Pedro, Chiloe. *Revista Universitaria (Universidad Catolica de Chile)*. Año 53, 65–77.

GODLEY, E. J. 1969a. Additions and corrections to the flora of the Auckland and Campbell Islands. *New Zealand Journal of Botany* 7: 336–48.

GODLEY, E. J. 1969b. Subantarctic botany. pp. 41–2. In 'Triennial Report 1966–68 Botany Division DSIR'. Government Printer, Wellington. 64 pp.

GREEN, P. S. 1972. Passiflora in Australasia and the Pacific. *Kew Bulletin*, 26: 539–58.

GUTHRIE-SMITH, H. 1936. 'Sorrows and Joys of a New Zealand Naturalist.' A. H. & A. W. Reed. 252 pp.

HARRIS, W. F. 1953. Palynology as a key to the history of New Zealand vegetation. Proceedings 7th Pacific Science Congress, 1949 5: 164–78.

HEALY, A. J. 1969. The adventive flora of Canterbury pp. 261–333. In KNOX, G. A. ed., 'The Natural History of Canterbury'. A. H. & A. W. Reed, Wellington. 620 pp.

HEALY, A. J. 1970. 'Identification of Weeds and Clovers.' Editorial Services Limited, Wellington. 191 pp.

HEINE, E. M. 1938. Observations on the pollination of New Zeland flowering plants. *Transactions and Proceedings of the New Zealand Institute* 67: 133–48.

HOLLOWAY, J. T. 1946. Forest associations of the Longwood Range, Southland. *New Zealand Journal of Forestry* 5: 199–209.

Holloway, J. T. 1948. Ecological investigations in the *Nothofagus* forests in New Zealand. *New Zealand Journal of Forestry* 5: 1–10.

Holloway, J. T. 1954a. Forests and climate in the South Island of New Zealand. *Transactions of the Royal Society of New Zealand* 82: 329–410.

Holloway, J. T. 1954b. An Ecological classification of the forest types of the Westland podocarp region. *New Zealand Journal of Forestry* 7: 23–33.

Jepson, W. L. 1951. 'A manual of the flowering plants of California'. University of California Press, Berkeley and Los Angeles. 1238 pp.

Levy, E. B. 1970. 'Grasslands of New Zealand'. 3rd ed. Government Printer, Wellington. 374 pp.

McKelvey, P. J. & Nicholls, J. L. 1957. A provisional classification of North Island forests. *New Zealand Journal of Forestry* 7: No. 4: 84–101.

McQueen, D. R. 1961. Indigenous-induced vegetation and *Pinus radiata* on volcanic ash soils. *Proceedings of the New Zealand Ecological Society*. 8: 1–14.

Mark, A. F. 1965. Flowering, seeding, and seedling establishment of narrow-leaved snow tussock, *Chionochloa rigida. New Zealand Journal of Botany* 3: 180–193.

Mark, A. F. 1970. Floral initiation and development in New Zealand alpine plants. *New Zealand Journal of Botany* 8: 67–75.

Mark, A. F. & Bliss, L. C. 1970. The high-alpine vegetation of Central Otago, New Zealand. *New Zealand Journal of Botany* 8: 381–451.

Mason, R. 1970a. Macrophytic water plants. Proceedings of New Zealand Water Conference. part 1: 9.1–9.11.

Mason, R. 1970b. Aquatic weeds. pp. 159–191. In Healy, A. J. 'Identification of Weeds and Clovers'. Editorial Services Limited, Wellington. 191 pp.

Melville, R. 1969. Leaf venation patterns and the origin of the Angiosperms. *Nature* 224: 121–25.

Millener, L. H. 1960. Our plant world. New Zealand Junior Encyclopedia 1: 310–36.

Moar, N. T. 1958. Contributions to the quaternary history of the New Zealand flora. 1. Auckland Island peat studies. *New Zealand Journal of Science* 1: 449–65.

Moar, N. T. 1971. Contributions to the Quaternary History of the New Zealand Flora. 6. Aranuian pollen diagrams from Canterbury, Nelson, and North Westland, South Island. *New Zealand Journal of Botany* 9: 80–145.

Molloy, B. P. J. 1969a. Evidence for post-glacial climatic changes in New Zealand. *Journal of Hydrology (N.Z.)* 8: 56–67.

Molloy, B. P. J. 1969b. Recent history of the vegetation. pp. 340–60. In Knox, G. A. ed., 'The Natural History of Canterbury'. A. H. & A. W. Reed, Wellington. 620 pp.

Molloy, B. P. J. *et al.* 1963. Distribution of subfossil forest remains, eastern South Island, New Zealand. *New Zealand Journal of Botany* 1: 68–77.

Moore, L. B. & Adams, N. M. 1963. 'Plants of the New Zealand Coast'. Pauls Book Arcade, Auckland and Hamilton. 113 pp.

Moore, L. B. & Edgar, E. 1970. 'Flora of New Zealand'. Vol. 2. Government Printer, Wellington. 354 pp.

Nicholls, J. L. 1956. The historical ecology of the indigenous forest of the Taranaki upland. *New Zealand Journal of Forestry* 7: 17–34.

Ohwi, J. 1965. 'Flora of Japan'. Smithsonian Institution, Washington, D.C. 1067 pp.

Oliver, W. R. B. 1910. The vegetation of the Kermadec Islands. *Transactions and Proceedings of the New Zealand Institute* 42: 118–75.

Oliver, W. R. B. 1930. New Zealand epiphytes. *Journal of Ecology* 18: 1–50.

Oliver, R. L. & Sorensen, J. H. 1951. Botanical investigations on Campbell Island. Cape Expedition Series Bulletin No. 7.

Philipson, W. R. 1965. The New Zealand genera of the Araliaceae. *New Zealand Journal of Botany*. 3: 333–41.

Philipson, W. R. & Hearn, D. 1962. Rock garden plants of the southern alps. The Caxton Press, Christchurch. 167 pp.

PIRES, J. M., DOBZHANSKY, TH. & BLACK, G. A. 1953. An estimate of the number of species in an Amazonian forest community. *Botanical Gazette* 114 No. 4: 467–77.
POOLE, A. L. 1969. 'Forestry in New Zealand. The shaping of policy'. Hodder & Stoughton, Auckland. 112 pp.
POOLE, A. L. & ADAMS, N. M. 1963. 'Trees and shrubs of New Zealand'. Government Printer, Wellington. 250 pp.
RATKOWSKY, D. & A. 1974. New plant discoveries in Tasmania. *Australian Plants* 7: 384–6.
RAUNKIAER, C. 1934. 'The life form of plants and statistical plant geography'. Clarendon Press, Oxford. pp. 632.
RAVEN, P. H. 1973. Evolution of subalpine and alpine plant groups in New Zealand. *New Zealand Journal of Botany* 11, 177–200.
RIGG, H. H. 1962. The Pakihi bogs of Westport, New Zealand. *Transactions of the Royal Society of New Zealand. Botany* 1: 91–108.
ROBBINS, R. G. 1961. The montane vegetation of New Guinea. *Tuatara*, 8, 121–34.
ROBBINS, R. G. 1962. The podocarp-broadleaf forest of New Zealand. *Transactions of the Royal Society of New Zealand. Botany* 1: 34–75.
ROYEN, P. VAN & EHRENDORFER, F. 1970. New combinations in *Parahebe* (Scrophulariaceae). *Taxon* 19: 483.
RÜBEL, E. 1936. Plant communities of the world. pp. 263–90. In GOODSPEED, T. H. ed., 'Essays in geobotany'. University of California Press. 319 pp.
RUSSEL, R. S. 1936. The mechanism of leaf-fall in certain New Zealand trees. *Transactions and Proceedings of the New Zealand Institute* 65: 407–21.
SCHIMPER, A. F. W. 1903. 'Plant geography upon a physiological basis'. Oxford. 839 pp.
SCHWEINFURTH, U. 1966. Neuseeland. Bonner Geographische Abhandlungen Heft 36.
SHORE, B. F. 1969. Dioecism in New Zealand Escalloniaceae. *New Zealand Journal of Botany* 7: 113–24.
SLEUMER, H. 1970. The identity of *Plectomirtha* Oliv. with *Pennantia* J. R. & G. Forster (Icacinaceae). *Blumea* 18: 217–8.
SMALLFIELD, P. W. 1970. 'The Grasslands Revolution in New Zealand'. Hodder & Stoughton, Auckland. 151 pp.
SYKES, W. R. 1965. Notes on the 1964 eruption and the vegetation of Raoul Island. *Proceedings of the New Zealand Ecological Society* 12: 20–2.
SYKES, W. R. & GODLEY, E. J. 1968. Transoceanic dispersal in *Sophora* and other genera. *Nature* 218: 495–6.
TAYLOR, B. W. 1955. The flora, vegetation and soils of Macquarie Island. Australian National Antarctic Research Expeditions Reports. Series B Vol. 11.
TAYLOR, R. H. 1971. Influence of man on vegetation and wildlife of Enderby and Rose Islands, Auckland Islands. *New Zealand Journal of Botany* 9: 225–68.
TROLL, C. 1948. Der asymetrische Aufbau der Vegetations-zonen und Vegetationstufen auf der Nord- und Südhalbkugel. Geobotische Forschungsinstitut Rüebel. Bericht. 1947. 46–83.
WARDLE, J. 1970a. The ecology of *Nothofagus solandri* 1. The distribution and relationship with other major forest and scrub species. *New Zealand Journal of Botany* 8: 494–531.
WARDLE, J. 1970b. The ecology of *Nothofagus solandri* 2. The Associations. *New Zealand Journal of Botany* 8: 532–70.
WARDLE, P. 1960. The subalpine scrub of the Hokitika Catchment, Westland. *Transactions of the Royal Society of New Zealand* 88: 47–61.
WARDLE, P. 1963a. Growth habits of New Zealand subalpine shrubs and trees. *New Zealand Journal of Botany* 1: 18–47.
WARDLE, P. 1963b. Evolution and distribution of the New Zealand flora, as affected by quaternary climates. *New Zealand Journal of Botany* 1: 3–17.
WARDLE, P. 1964. Facets of the distribution of forest vegetation in New Zealand. *New Zealand Journal of Botany* 2: 352–66.

WARDLE, P. 1965. A comparison of alpine timberlines in New Zealand and North America. *New Zealand Journal of Botany* 3: 113–35.
WARDLE, P. 1968. Evidence for an indigenous pre-quaternary element in the mountain flora of New Zealand. *New Zealand Journal of Botany* 6: 120–25.
WARDLE, P. 1971. An explanation for alpine timberline. *New Zealand Journal of Botany* 9: 371–402.
WARDLE, P., MARK, A. F. & BAYLIS, G. T. S. 1970. Vegetation studies on Secretary Island, Fiordland. Part 9: Additions to Parts 1, 2, 4 and 6. *New Zealand Journal of Botany* 8: 3–21.
WARMING, E. & VAHL, M. 1909. 'Oecology of Plants'. Oxford, Clarendon Press. 422 pp.
WEBB, L. J. 1959. Classification of Australian rain forests. *Journal of Ecology* 47: 556.
WENT, F. W. 1971. Parallel Evolution. *Taxon* 20: 197–226.
WRAIGHT, M. J. 1963. The alpine and upper montane grasslands of the Wairau River Catchment, Marlborough. *New Zealand Journal of Botany* 1: 351–76.
WRIGHT, A. C. S. 1959. Soils of Chatham Island (Rekohu). *New Zealand Department of Scientific and Industrial Research Soil Bureau Bulletin* 19: pp. 61.
WYLIE, R. B. 1954. Leaf organisation of some woody dicotyledons from New Zealand. *American Journal of Botany* 41: 186–91.
ZOTOV, V. D. 1938. Some correlations between vegetation and climate in New Zealand. *New Zealand Journal of Science and Technology* 19: 474–87.
ZOTOV, V. D. 1939. Survey of the tussock-grasslands of the South Island, New Zealand. *New Zealand Journal of Science and Technology* 20A: 212A–244A.

V. THE AMPHIBIANS, REPTILES, BIRDS AND MAMMALS

by

P. C. BULL & A. H. WHITAKER

Introduction

About 100 species of native, terrestrial vertebrates live in New Zealand or its outlying islands which extend from the Kermadecs (Lat. 30°S) to Macquarie Island (Lat. 54½°S). There are three species of frogs, about 30 reptiles (all but one being lizards), some 65 land birds and two mammals (both bats); a further 80 species of breeding birds are littoral or marine forms, and three species of seals breed in the region. Taxonomic revisions and new information concerning a few species on the verge of extinction may require some minor adjustment to the above figures; also, for some birds, the distinction between littoral and terrestrial species is somewhat arbitrary. Finally, another 70 species of terrestrial vertebrates, mainly birds and mammals, have been brought to New Zealand from other countries by man and are now well established in the wild.

The native fauna has been strongly influenced by the geological history of the region. The separation of New Zealand from other lands by wide areas of ocean since at least Cretaceous times has resulted in a somewhat depauperate fauna of terrestrial vertebrates; for instance, there are no snakes and no native land mammals except bats. On the other hand, the fauna includes some interesting animals of very ancient lineage such as the only surviving rhynchocephalian reptile *(Sphenodon punctatus)*, an endemic genus of primitive frogs *(Leiopelma)* and two endemic orders of ratite birds (Apterygiformes and the now extinct Dinornithiformes). Some of these groups such as the moas (Dinornithiformes) probably underwent adaptive radiation during the Tertiary when land in the New Zealand region seems to have been in the form of a changing archipelago which would favour the formation of new species (FLEMING, 1962a). Other non-marine vertebrates reached New Zealand during the Tertiary and later, probably on rafts of driftwood (lizards) or by flying (birds and bats). Many species susceptible to cold may have become extinct during the Pleistocene because there were no northern lands to which they could retreat during glacial periods; evidence for such extinctions is, however, mainly indirect (FLEMING, 1962b) because terrestrial vertebrates are rare as fossils in New Zealand.

The four classes of terrestrial vertebrates are discussed separately below, the amphibians and reptiles by one author (AHW) and the birds and mammals by the other (PCB). The classes vary greatly in the number of New Zealand species they contain and in the thoroughness with which they have been studied. The following sections are therefore of unequal length, but each presents the available data in the same order; viz. representation of the class in New Zealand, endemics of special interest, origin, evolutionary trends, and distribution and ecology.

The Amphibians

New Zealand's native amphibians comprise only three species of small frogs in the endemic genus *Leiopelma* (F. Leiopelmatidae). In addition there have been several introductions of frogs and toads to New Zealand in the past hundred years, but only two, *Litoria aurea* and *L. ewingi* (F. Hylidae) from Australia, have established successfully (McCann, 1961).

Leiopelma and *Ascaphus*, the latter from western North America, are the only genera in the family Leiopelmatidae (Fawcett & Smith, 1971). This, in turn, is the only family in the sub-order Amphicoela and includes the most primitive members of the order Salientia, distinguished from all other frogs by their amphicoelous vertebrae, nine (instead of eight) presacral vertebrae, and the possession of the tail-wagging muscles *Musculus pyriformis* and *M. caudalipuboischiotibialis*. Apart from these characters the primitive position of *Leiopelma* is further indicated by the cranial morphology, the skeleton, the circulatory system and the development. In addition, *Leiopelma* lacks tympanic membranes, eustachian tubes and vocal sacs (Stephenson & Stephenson, 1957). All three species of *Leiopelma* are small (up to 47 mm body length) brown or brown and green frogs.

Ancestral *Leiopelma* stock were probably in the New Zealand region as early as the Jurassic, and hence frogs form a part of the Archaic element of the New Zealand fauna (Fleming, 1962a). According to Raven & Axelrod (1972), they probably entered New Zealand overland when it was joined to Antarctica. The relationships of *Leiopelma* and *Ascaphus* are uncertain, but the distribution of the Leiopelmatidae, if not an artifact of taxonomy, is probably a relict one.

The relationships of the three species of *Leiopelma* are virtually unknown. *L. hamiltoni* may be closely related to *L. archeyi* as both have well-developed parotoid glands and lack webbing between the toes, and *L. archeyi* has the skeletal characteristics of a juvenile *L. hamiltoni*. These similarities suggest that *L. archeyi* may be a neotenic form of *L. hamiltoni*. The most obvious morphological difference between the two species is in the maximum size; 41 mm for *L. archeyi* and 47 mm for *L. hamiltoni*. Neoteny has probably played a part in the evolution of *L. hochstetteri* (Stephenson, 1960).

In keeping with its occupation of areas without standing water

Plate 1. Upper: Habitat of *Leiopelma hamiltoni* on Maud I., Cook Strait; (Photo: F. O'Leary, National Museum); Lower left: *Leiopelma archeyi* from Tokotea Ridge, Coromandel, with egg cluster. The cryptic colouration is typical of all species of *Leiopelma;* (Photo: Dr. B. D. Bell, Ecology Division, D.S.I.R.); Middle Right: Developing embryos of *Leiopelma archeyi*; (Photo: Dr. B. D. Bell, Ecology Division, D.S.I.R.); Lower Right: Hatching eggs of *Leiopelma archeyi;* (Photo: Dr. B. D. Bell, (Ecology Division, D.S.I.R.).

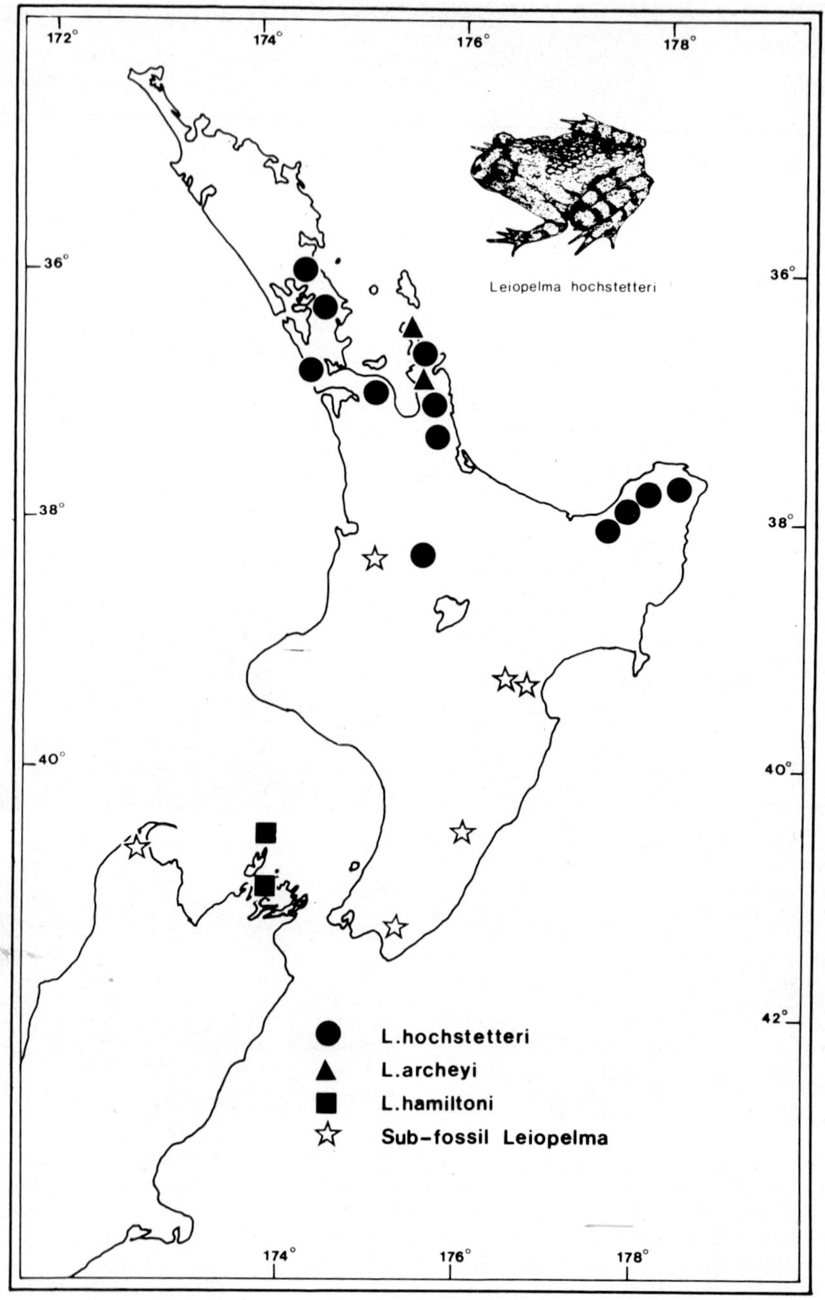

Fig. 1. Present and subfossil distribution of *Leiopelma*.

Leiopelma exhibits a specialised method of reproduction in which there is no free-swimming tadpole stage, development being entirely intra-capsular (Plate 1). Amplexus is in the primitive pelvic position as in *Ascaphus* (Dr. B. D. BELL, pers. comm.). The eggs are unpigmented, heavily-yolked and surrounded by a gelatinous capsule about 5 mm in diameter. They are laid in spring in clusters of up to eleven in dark, damp and sheltered sites (such as under stones or logs, in dense decaying vegetation or in water-logged soils), and hatch in early summer. At hatching the young frog still has a tail which it retains for 3–4 weeks and which it uses to break free from the egg capsule (STEPHENSON, 1961).

The present distribution of *Leiopelma* is localised (Fig. 1), but in the recent past, probably within the last 1000 years, the genus was much more widespread. Subfossil skeletons similar to those of living species, but about twice the size, have been found at Paterau (J. TOWNSEND & P. JOHNS, pers. comm.) in Nelson province; at Martinborough and Coonoor (S. MARKHAM, pers. comm.) in Wellington province; at Patoka and Hukanui (S. MARKHAM, pers. comm.) in Hawke's Bay province; and at Waitomo (C. TEMPLER, pers. comm.) in Auckland province. These specimens have all been taken from limestone caves and sink-holes, and have been associated with a recent forest fauna. The Martinborough specimens were also associated with bones of the Polynesian rat or kiore *(Rattus exulans)*, believed to have been in New Zealand for about 1000 years, an indication of their very recent origin.

L. hochstetteri is the most common species, being found quite widely in Auckland province, from Whangarei south to the Waikato and East Cape. It lives in native forests, usually near streams or where surface moisture is high. Specimens have been collected from near sea-level up to 800 m. *L. hochstetteri* is nocturnal, usually passing the day in the shelter of a log or stone. On the Coromandel Peninsula *L. hochstetteri* occurs sympatrically with *L. archeyi* from which it can readily be distinguished by its webbed toes and the absence of parotoid glands. *L. hochstetteri* also differs from *L. archeyi* in several skeletal characters (STEPHENSON & STEPHENSON, 1957).

L. archeyi is known only from the ranges of the Coromandel Peninsula where it has been found from about 250 m up to 1000 m. Though occasionally living near streams, *L. archeyi* appears to prefer the mist-enshrouded forests and subalpine grasslands of the high ridges. In some areas on Mt Moehau *L. archeyi* reaches a density equivalent to one per 10 m^2 (STEPHENSON & STEPHENSON, 1957).

L. hamiltoni is the rarest of the three species and perhaps one of the rarest frogs in the world. For many years the only known locality was a small (0.25 hectare) heap of stones near the summit of Stephens Island, Cook Strait, though it is now known to occur in a 15 hectare remnant of coastal forest on nearby Maud Island (Plate 1). Like the other two species, *L. hamiltoni* is nocturnal, spending the day in a damp and shaded site, usually under logs or stones. At night it frequently climbs as much as five metres up into the vegetation in search of food. *L. hamiltoni* is perhaps the most terrestrial species; surface water is present in its habitat only after rain (CROOK *et al.*, 1971).

The Reptiles

COMPOSITION OF THE REPTILE FAUNA

Apart from the well-known tuatara (*Sphenodon punctatus:* F. Sphenodontidae), the indigenous terrestrial reptile fauna of New Zealand consists of approximately thirty endemic species of small lizards in the families Gekkonidae and Scincidae. Four species of marine turtles (*Dermochelys coriacea:* F. Dermochelyidae; and *Chelonia mydas*, *Eretomochelys imbricata*, *Caretta caretta:* F. Cheloniidae) and two species of marine snakes (*Laticauda colubrina*, *Pelamis platurus:* F. Hydrophiidae) come ashore occasionally, usually on the northern coasts (MCCANN, 1966) and the salt-water crocodile (*Crocodylus porosus:* F. Crocodylidae) has been reported within 480 km of the New Zealand coast (ROBB, 1973).

Three endemic genera of geckos are currently recognised, *Hoplodactylus* with four species, *Heteropholis* with six and the monotypic *Naultinus* (WERMUTH, 1965). The two indigenous genera of skinks *Leiolopisma* (17 spp.) and *Sphenomorphus* (1 sp.) (MCCANN, 1955, ROBB, 1970), also occur widely beyond New Zealand. The taxonomy of the New Zealand lizards is far from settled and considerable rearrangements will be necessary.

Only one exotic reptile is known to be established in New Zealand; the lizard *Leiolopisma delicata* (F. Scincidae) from Australia is locally abundant in the Auckland region.

ENDEMIC REPTILES OF SPECIAL INTEREST

One of the most interesting reptiles in the world is the tuatara. This large (up to 70 cm) lizard-like animal is the sole survivor of the once diverse lepidosaurian order Rhynchocephalia, that was most abundant and widespread during the Mesozoic. The tuatara is discussed at length in Chapter VIII.

The New Zealand geckos are also of particular interest being the only members of that large and diverse family that give birth to live young rather than laying eggs. Though strictly ovoviviparous, as the foetus develops from a fully yolked egg, placental attachment has been demonstrated (Boyd, 1942). The New Zealand species are the only geckos that lack specialised egg teeth.

The most likely explanation for this method of reproduction is that it is a recent adaptation to enable the geckos to survive the cold climates of the Pleistocene. The primitive method of gecko reproduction is believed to be by laying pliable-shelled eggs (Werner, 1972) and as all other species in the sub-family Diplodactylinae do so it can be assumed that the geckos which colonised New Zealand were oviparous. In regions with cold climates reptiles that give birth to live young have a selective advantage over oviparous species because they can protect the embryo from climatic extremes and seek out the warmth essential for development (Bellairs, 1969). The evolution from oviparity to ovoviviparity in the New Zealand geckos could easily have been achieved by retention of the eggs within the oviduct for the whole period of development, with a subsequent loss of shell membranes. Such evolution probably began with the decreasing temperatures of the early Pleistocene, a time when temperatures were too low for oviparous species to survive.

Several features, mainly osteological, suggest the New Zealand species are amongst the most primitive geckos (Kluge, 1967a).

Origins of the New Zealand lizards

Apart from the tuatara, which is known from Holocene deposits, there are no fossils of terrestrial reptiles from New Zealand. The few subfossil lizard remains that have been collected all appear to be of living species and give no indication that the lizard fauna may have been more diverse than it is at present.

The origin and evolution of geckos has recently been studied by Kluge (1967a and b). The family Gekkonidae is believed to have originated in Asia in the Upper Jurassic. The subfamily Diplodactylinae, which includes the New Zealand species, probably separated from ancestral gekkonid stock as early as the Cretaceous and dispersed to the Australian region via the Indo-Australian Archipelago during the Upper Cretaceous arriving in Australia by the Palaeocene. During the Eocene the Diplodactylinae differentiated into two 'genus groups', the Carphodactylini and Diplodactylini, the former retaining the more primitive features. This initial evolution probably took place in the north-eastern Australia/south-eastern New Guinea region with the Carphodactylini spreading through New Caledonia to New Zealand. The Carphodactylini include all the New Zealand and New Caledonian genera plus four that occur in Australia. The Diplodactylini are known only from the Australian continent. The New Zealand geckos are most closely related to the New Caledonian genera, and appear to be directly derived from them.

The New Zealand skinks are part of the subfamily Lygosominae, a relatively recent and rapidly expanding derivative of the most primitive subfamily of skinks, the Scincinae. Like the diplodactyline geckos, the Lygosominae originated in the South-east Asia/Australian region from which they have since spread to tropical and subtropical areas throughout the world. The Australian region remains the stronghold of the Lygosominae both in the number of species and the variety of adaptations (GREER, 1970).

Both skink genera presently recognised from New Zealand *(Leiolopisma, Sphenomorphus)* are widely distributed in Australasia and the western Pacific and almost certainly evolved in this region. Dispersal to New Zealand was either direct from Australia or, like the geckos, via New Caledonia. Except for *Leiolopisma suteri*, which is apparently related to *Leiolopisma lichenigera* of Lord Howe Island (D. R. TOWNS, pers. comm.), the New Zealand skinks show no close relationships with overseas species. Nevertheless, cytogenetical work has revealed similarities between the karyotypes of some New Zealand skinks and Australian species in the same genera (O'MALLEY, 1971).

The long isolation of New Zealand since the Mesozoic means that dispersal of lizards to New Zealand must have been trans-oceanic and that the families involved have been capable of crossing about 1500 km of sea. The Gekkonidae and Scincidae are the only lizards with cosmopolitan distributions and are renowned for their vagility. They occur on many remote islands and are the only lizard families that are widely distributed in the Pacific. The absence of other terrestrial reptiles, especially snakes, can be attributed to the isolation of New Zealand since the Mesozoic.

Dispersal of geckos and skinks to New Zealand was presumably by rafting. Present surface currents in the Tasman Sea (e.g. East Australian Current), aided by the prevailing westerly winds could transport rafting lizards from New Caledonia or Australia to New Zealand within weeks. Chances of such colonisation are greatly increased by the long coast line of New Zealand (34–47 °S) which is orientated across these sea currents and winds. Most summer hurricanes on the other hand travel in a south-easterly direction, i.e. directly from New Caledonia to New Zealand. Past currents and winds are likely to have been just as suitable for dispersal of lizards from Australia to New Zealand.

The high level of endemism of the geckos compared with the skinks suggests they have been isolated in New Zealand for much longer and that transoceanic dispersal from the north for geckos is neither as easy nor as frequent as for skinks. Though many species in both families have fairly broad habitat tolerances which help them to colonise new areas, the differences in ability to disperse could be related in part to the opportunities and abilities for rafting. Several species of skinks occupy littoral habitats (in New Zealand *Leiolopisma zelandica*, *L. smithi*, *L. suteri* for example) and often forage on the intertidal zone. Not only do these species risk being washed out to sea in flotsam but their higher salt tolerances give them better chances of survival at sea than the geckos, which prefer supra-littoral habitats.

Compared with the tuatara and frogs *(Leiopelma)* which have probably been in the New Zealand region since the Jurassic, the lizards are relative newcomers. The geckos form part of the Malayo-Pacific element of the New Zealand fauna and probably entered New Zealand in the Miocene when the climate was warmer and the land more extensive than now. Because of their generic ties with Australia the skinks are considered more recent arrivals than the geckos. Present distribution patterns suggest *Leiolopisma* was in New Zealand by the Pleistocene and that *Sphenomorphus* arrived at the same time or slightly later.

Evolutionary trends within the New Zealand region

The New Zealand geckos and skinks are relatively homogeneous and within New Zealand speciation in both families involves in most cases only minor morphological differences and variations in colour patterns. The New Zealand geckos are all typical gekkoniform lizards ranging in size from *Hoplodactylus pacificus* (average total length of adults 14 cm, weight 8 g) to *Hoplodactylus duvauceli* (28 cm, 80 g). The skinks are typical lacertiform lizards varying from *Leiolopisma smithi* (12 cm, 5 g) to *Leiolopisma grande* (28 cm, 50 g).

As all three of the New Zealand gecko genera are endemic and morphologically much closer to each other than to the New Caledonian stock only one invasion is postulated. The geckos that originally colonised New Zealand may well have been similar to the present day *H. pacificus*, the most widespread and least specialised species. Though subsequent radiation has given rise to three genera the species fall easily and naturally into two phylogenetic groups. The first group comprises the four species of *Hoplodactylus* which are largely nocturnal, have the proximal portion of the toe slightly expanded to a lamellar pad, are either terrestrial or arboreal, and are coloured brown or grey (Plate 2). The second group are diurnal, have a tapering series of lamellae from sole to claw, are arboreal, and are predominantly bright green in colour. This group includes all the species of *Naultinus* and *Heteropholis* (Plate 2). This very close relationship of *Naultinus* and *Heteropholis* was confirmed by the recent demonstration that *Naultinus* × *Heteropholis* hybrids of various species combinations are fertile (M. J. Meads, pers. comm.). Unlike the *Hoplodactylus* species, many of which occur sympatrically, the species of *Naultinus* and *Heteropholis* are allopatric.

Most authorities believe skinks reached New Zealand on at least two, probably three occasions to account for *Sphenomorphus* and both oviparous and ovoviviparous *Leiolopisma* species. However, after a comprehensive

Plate 2. Upper: The large nocturnal gecko, *Hoplodactylus duvauceli;* Upper middle: The diurnal tree gecko, *Heteropholis stellatus;* Lower middle: The nocturnal skink, *Leiolopisma alani;* Lower: The diurnal skink, *Leiolopisma infrapunctatum;* (All Photos: D. L. Homer, Geological Survey, D.S.I.R.).

revision of the higher taxonomy of skinks (Greer, 1970), in which seven of the New Zealand species including both genera were examined, Greer (pers. comm.) believes that the New Zealand *Sphenomorphus* species are more closely related to the New Zealand *Leiolopisma* species than to other members of the genus. If so, the New Zealand skinks have either evolved from a single invasion or are the result of speciation from two invasions, at different times, of the same parent stock; *L. suteri* which not only differs markedly from the other New Zealand skinks, but also resembles *L. lichenigera* of Lord Howe Island, probably arrived separately. If this is correct then the New Zealand species of *Sphenomorphus* will possibly require new generic status.

The New Zealand skinks (excluding *L. suteri*) can be separated into two phylogenetic groups on both morphological and ecological characters, and this division has received recent support (G. Hardy, pers. comm.) from research on chromosomes and hemipenis morphology. This separation involves grouping five of the present *Leiolopisma* species *(L. aeneum, L. alani, L. homalonotum, L. oliveri, L. ornatum)* with the *Sphenomorphus* species; the second group comprises the remaining species of *Leiolopisma* (Plate 2). Skinks of the *Leiolopisma* group are fast-moving, diurnal inhabitants of open, generally non-forested habitats. The body is oval in section, the limbs and digits are relatively long and slender, the snout is pointed, and the lower eyelid has a transparent palpebral disc. In contrast skinks of the *Sphenomorphus* group are slow-moving, crepuscular or nocturnal inhabitants of forested or heavily vegetated habitats. The body is quadrangular in section, the limbs and toes are short, the snout blunt and the central portion of the lower eyelid is scaly or at least covered by one or two large, opaque scales.

Though the New Zealand lizard genera have evolved to fill different niches (arboreal/terrestrial, nocturnal/diurnal, forested/non-forested), the species show few of the specialised morphological features that have arisen elsewhere. The arboreal geckos *Naultinus* and *Heteropholis*, unlike all the other diplodactyline genera with prehensile tails, lack subcaudal lamellae. None of the New Zealand geckos has developed complex scansorial lamellae but have instead the primitive, narrow, clawed toe with a single series of transverse lamellae. Species such as *Hoplodactylus granulatus*, which pass the day clinging to trunks or limbs of trees, have not evolved the lateral 'fringes' of skin that occur in some species (e.g. *Phyllurus*, *Uroplatus*) to assist camouflage. The *Sphenomorphus* skinks, though spending much of their time burrowing through leaf-litter or matted vegetation, do not show any reduction in the number of digits or limbs.

Lizards living in cold climates near the northern and southern limits of distribution are usually live-bearing and the ovoviviparity of most New Zealand lizards has been cited as an adaptation to such environments (Darlington, 1957). Though present climates especially in the North Island are warm enough for oviparous lizards (evidence *L. suteri* and *L. delicata*), conditions during the Pleistocene glaciations were probably too cold for their survival. The reproduction of geckos has already been discussed. Though the genera *Leiolopisma* and *Sphenomorphus* contain both oviparous and ovoviviparous members in other parts of their range, all but one of the New Zealand species are live-bearing. The exception is

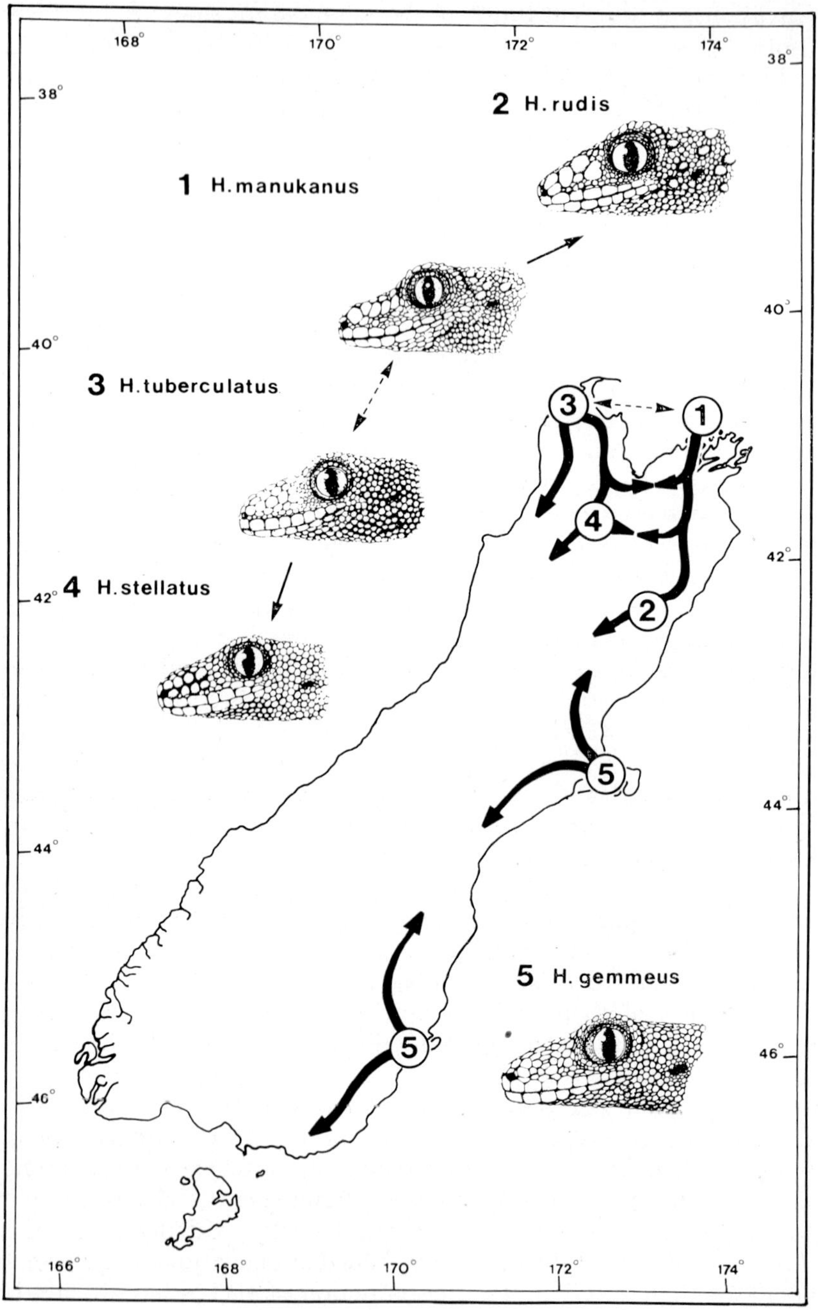

Fig. 2. Post-Pleistocene radiation of the genus *Heteropholis* and the relationship of the different species.

L. suteri which is distributed north of 37 °S and is probably a post-Pleistocene arrival.

Many New Zealand lizard species are very variable, particularly those with wide latitudinal ranges. During the last Pleistocene glaciation suitable lizard habitats were restricted to the north of the North Island and narrow coastal strips perhaps as far south as 46 °S. Following this glaciation there was probably a southward and upward spread of the lizard fauna which has given rise to some quite distinct clines (e.g. *Naultinus elegans*, *Leiolopisma zelandica*) or speciation (e.g. *Heteropholis* spp. as detailed below).

The genus *Heteropholis* has radiated in the South Island only (Fig. 2), its North Island equivalent being *Naultinus*. There appear to be three lines of speciation stemming from *H. tuberculatus*, *H. manukanus* and *H. gemmeus* respectively. *H. tuberculatus* and *H. manukanus* at the northernmost extremities of their ranges (northwest Nelson and Marlborough Sounds) not only resemble each other very closely but are also very similar to *N. elegans*. From these two areas, which were probably refuges during the glaciation, there has been a southward radiation; *H. tuberculatus* to *H. stellatus* and *H. manukanus* to *H. rudis*. *H. gemmeus*, which also shows close affinities with *N. elegans*, probably remained in the vicinity of Banks Peninsula, and possibly Otago Peninsula, during the glaciation and later spread throughout Canterbury and eastern Otago. Further evidence for a refuge for lizards in the region of the Otago Peninsula is provided by *L. grande*, a very distinct skink known only from Otago and South Canterbury.

H. manukanus is known from the Marlborough Sounds and elsewhere in northern Marlborough province. To the west of its range, on Whangamoa Hill near Nelson, populations of *H. manukanus* abut those of *H. stellatus*. Over a distance of 10–15 km populations of typical *H. manukanus*, showing distinctive dorso-lateral rows of raised scales, are replaced by populations of *H. stellatus*. A few lizards that appear to be natural hybrids have been collected in this area. There is no obvious geographical barrier or vegetation change at this point. South of the Marlborough Sounds *H. manukanus* grades slowly into the *H. rudis* populations of eastern Nelson province and the Kaikoura coast. *H. rudis* and *H. stellatus* occur within 18 km of each other in the upper Wairau River, and *H. rudis* and *H. gemmeus* are about 30 km apart on the Waiau River near Hanmer. *H. tuberculatus* has been recorded from a few scattered localities in north-west Nelson and on the West Coast. Populations of *H. tuberculatus* merge with those of *H. stellatus* in Golden Bay, the latter species being distributed over most of Nelson west of the Alps. *H. gemmeus* has a very wide distribution throughout the eastern regions of Canterbury and Otago provinces.

Some speciation has occurred where lizard populations have been isolated on remote islands. For example *Leiolopisma fallai* of the Three Kings Islands resembles *L. smithi*, and *Leiolopisma dendyi* of the Chatham

Islands is probably derived from *L. zelandica*. No species is peculiar to inshore islands.

Distribution and ecology of New Zealand lizards

Both geckos and skinks are widely distributed throughout the three main islands of New Zealand. The four species on Stewart Island *(Hoplodactylus granulatus, H. pacificus, Leiolopisma zelandica, L. lineoocellatum)* at 47°16′S are very near the southernmost limits of their families. As expected there is a decrease in the number of species with increasing latitude, north of Auckland are ten species (4 geckos, 6 skinks), in Marlborough eight (4 geckos, 4 skinks), in Otago six (3 geckos, 3 skinks) and on Stewart Island four (2 geckos, 2 skinks). The only indigenous oviparous species is distributed north of 37°S. Latitudinal limits for the genera are as follows: *Hoplodactylus* 34°08′–47°16′S, *Naultinus* 34°23′–41°36′S, *Heteropholis* 40°30′–46°41′S, *Leiolopisma* 34°08′-47°16′S and *Sphenomorphus* 34°08′–41°36′S.

Two geckos *(H. granulatus, H. pacificus)* and two skinks *(L. zelandica, L. lineoocellatum)* have been found at altitudes up to 1700 m in the South Island, but the number of species decreases with increasing altitude. There are no alpine lizard species; all New Zealand lizards occur from sea-level, or within 200 m of it, to an upper limit varying between genera as follows: *Sphenomorphus* less than 400 m, *Naultinus* 900 m, *Heteropholis* 1500 m, *Hoplodactylus* and *Leiolopisma* 1700 m.

Geckos and skinks have been recorded from most of the small inshore islands around New Zealand though few of the outlying islands have been colonised. Both families have reached the Three Kings Islands, 58 km north-west of North Cape, on several occasions and at least two of the species there *(Leiolopisma smithi, L. suteri)* are virtually indistinguishable from mainland forms. Skinks (but seemingly not geckos) have dispersed to the Chatham Islands 800 km east of the main islands accounting for the endemic species. Lizards have never been found on the Kermadec Islands 800 km north-east of New Zealand nor on any of the subantarctic islands.

Two distinct faunal breaks show in the distribution patterns of New Zealand lizards. Cook Strait has formed a barrier to dispersal for about 15,000 years. This 22 km wide seaway limits three of the genera: *Sphenomorphus* and *Naultinus* are only found north of the Strait whilst *Heteropholis* only occurs south of it. Both *Hoplodactylus* and *Leiolopisma* span this barrier and each of these genera has three species which occur in both the North and South Islands. The ecology of *Sphenomorphus* suggests that if it was in New Zealand during the last glaciation the only suitable habitat for it would have been north of Auckland; subsequent southward dispersal would have been halted at Cook Strait. *Naultinus* and *Heteropholis*

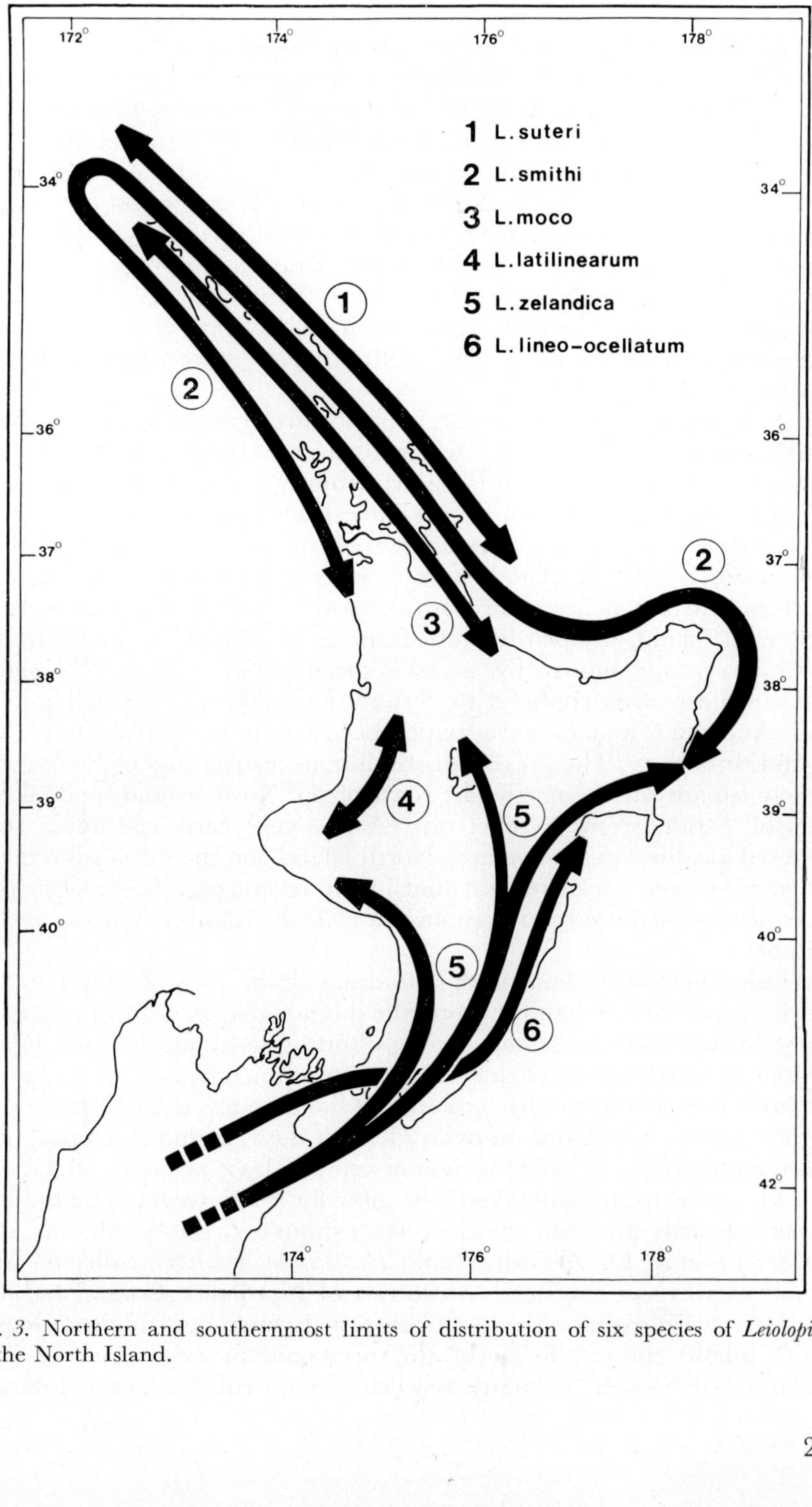

Fig. 3. Northern and southernmost limits of distribution of six species of *Leiolopisma* in the North Island.

have evolved from the same ancestral stock on different islands. The second faunal break occurs in the central North Island and affects only the genus *Leiolopisma* (Fig. 3). McCann (1955) was the first to recognise this distribution pattern though he tied it rather too closely to the 38 °S latitude. No single line can be drawn to fit the distribution of all species but the ranges of four lie wholly north of 38 °42′S and the ranges of six lie wholly south of 37 °30′S. No species of *Leiolopisma* occurs both north and south of these latitudes. Of five widespread species, two have their northern limit and three their southern limit in this region. Though present climates probably maintain this faunal break, it may have originated during the early Pleistocene when interglacial sea-levels formed an extensive seaway (Fleming, 1962a) between Taranaki and Hawke's Bay.

Specific patterns of distribution fall into three categories: widespread, restricted and relict. The most widespread and abundant lizard in New Zealand is *H. pacificus* which is known from many localities throughout the entire range over which geckos have been reported. Alpine regions above 1700 m are the only extensive areas of the main islands in which this species is lacking. Another very widespread species is *L. zelandica* which is distributed from the central North Island to the islands south of Stewart Island, in open habitats from sea-level to 1700 m. Restricted distributions are shown by several species, for example *Leiolopisma latilinearum* which is confined to eastern Taranaki and a small part of south Auckland, and *L. grande* which occurs only in central Otago and South Canterbury. The present discontinuous distribution of *H. duvauceli*, namely islands off the north-east coast of the North Island, and islands in Cook Strait, seems a relict one. A few very early specimens from scattered localities in the northern North Island and the sub-fossil remains of a similar-sized gecko suggest that in the recent past *H. duvauceli* may have been distributed throughout the North Island, at least at low altitudes.

Within their altitudinal and latitudinal limits New Zealand lizards live in every major habitat from the high-water mark (some species forage in the intertidal zone), through both grassed and forested inland habitats to subalpine herbfields. The areas occupied include such climatic extremes as 320 mm rain per annum in Otago on less than 100 rain days to 7500 mm in Fiordland on over 250 rain days; summer temperatures that commonly reach 30 °C to winter ones at least as cold —6 °C; and frost free areas to those blanketed in snow for many weeks. The habitats of some lizards are very specific, for example *L. suteri* inhabits only boulder beaches. On the other hand *H. pacificus* has been collected from all habitats except the very wet forests of Fiordland. Coastal habitats support the lizard populations of greatest diversity and highest density whilst inland forests (formerly the predominant vegetation of New Zealand) are poorest. As many as seven species with a combined density of about one per 3 m^2 (locally up to 4 per m^2) have been reported from some coastal localities.

All New Zealand lizards are primarily insectivorous and eat a wide variety of invertebrates. The diurnal geckos *Naultinus* and *Heteropholis* feed predominantly on flies, beetles and other insects; the nocturnal species *(Hoplodactylus)* on caterpillars, moths and spiders. Gut contents of *H. pacificus* include significant numbers of very small food items such as collembola and soil mites. The skinks not only eat live animals but will take carrion. Skinks living in the nesting colonies of seabirds (petrels, gannets) have been observed feeding on regurgitated stomach oils and partly digested fish. *L. suteri* is at least partially aquatic and enters rock pools or even the sea in search of small crustacea. Both geckos and skinks also eat soft fruit and berries, and geckos have been seen taking nectar.

Breeding data for New Zealand lizards are poor, though it seems all the species except *L. suteri* are ovoviviparous (BARWICK, 1959, BOYD, 1942). The geckos almost invariably have two young. The gestation period is between five and six months in all species for which data are available; mating takes place in spring or summer, and the young are born in autumn or winter. There is some evidence for sperm storage in *H. pacificus*, with mating in autumn and fertilisation in spring. With skinks the number of young born varies from 2–8 in *Sphenomorphus* and 4–10 in *Leiolopisma*. Mating takes place in spring and the young are born approximately four months later in summer or early autumn. For *L. suteri* the pattern is much the same with egg-laying in early summer and hatching in late summer. The normal clutch size is 2–3, and the eggs are laid in damp sand just above the high-tide line. Communal nesting sometimes occurs.

New Zealand lizards have few natural predators. There are no indigenous mammals to eat them and though fourteen species of birds have been reported with lizard prey only the kingfisher *(Halcyon sancta)* takes many. The tuatara, which is known to take both geckos and skinks, has a very restricted distribution. Although these indigenous predators probably have a negligible effect on lizard numbers the impact of several introduced mammals, mainly rats, cats and mustelids, now widely established in New Zealand, is more important. Three rodents have been recorded eating New Zealand lizards: *Rattus norvegicus* (BETTESWORTH & ANDERSON, 1972), *R. exulans* (BETTESWORTH, 1972; D. R. TOWNS, pers. comm.), and *Mus musculus*. Islands without rats support many more lizards, both in species and numbers of individuals, than do adjacent islands occupied by *Rattus exulans* (WHITAKER, 1973). Nocturnal ground-dwelling lizards, especially those that forage in the open (e.g. *L. oliveri* and *H. duvauceli*) are particularly vulnerable. In the presence of rats nocturnal lizards are often found only in places affording some protection, such as stone heaps, cliff-faces, or extremely dense, matted vegetation. The occurrence in the Wairarapa of bones of *Rattus exulans*, associated with those of lizards now extinct in the district, suggests that rats may have been important in causing the relict distributions of lizards such as *H. duvauceli*.

A number of parasites have been reported from New Zealand geckos and skinks including a haemosporidian, haemogregarines, a gut protozoan (*Trichomonas*), a nematode, a trematode and several mites (HARDY, 1972), and recently a tick has been collected from *H. duvauceli*. New Zealand lizards are also known to carry salmonella bacteria and *L. zelandica* and *H. pacificus* are believed to be vectors in human salmonellosis (DE HAMEL & MCINNES, 1970).

The Birds

COMPOSITION OF THE AVIFAUNA

New Zealand has relatively few species of native land birds but they include representatives of two endemic orders of ratites (one extinct) and three endemic families of passerines. Of 65 native species of land and freshwater birds now living in New Zealand, 37 (57%) are endemic; the percentage of endemics is much higher (some 70%) if subfossil and recently extinct species are included. In contrast, only about 30% of the numerous species of sea and shore birds are endemic, though many others have distinct subspecies in the New Zealand region.

A recent checklist (KINSKY, 1970) records 328 living and subfossil birds from New Zealand and its outlying islands (Kermadecs, Chathams, Bounty, Antipodes, Auckland, Campbell and Macquarie). Of 96 listed species that do not breed in the New Zealand region, nearly half are migrant shore birds from the Northern Hemisphere (many occurring irregularly and in small numbers) and the rest are either wide-ranging oceanic species or wind-blown strays mainly from Australia.

The breeding species include many seabirds, especially albatrosses, petrels and shearwaters (Procellariiformes) of which no less than 41 species breed in the region. Other breeding sea and shore birds comprise: penguins (8 spp.), a tropic bird, a gannet, a booby, cormorants (5 marine spp.), a heron, oystercatchers (3), plovers (4), stilts (2), a skua, gulls (3) and terns (9 spp.). Four of the terns, as well as the tropic bird and booby, breed only at the subtropical Kermadec Islands. Ten shore birds (the gulls, three plovers, the stilts, an oystercatcher and a tern)

---

*. Data mainly from KINSKY (1970) and RIDPATH & MOREAU (1966).

†. Four additional species are confined to outlying islands: a warbler and flycatcher ('robin') to Chatham Islands, a parakeet to Antipodes and a snipe to Chathams and subantarctic islands.

‡. One species common in 19th century now extinct (quail) or possibly so (thrush).

§. Known from subfossil bones in New Zealand.

Table 1. The Non-Marine Avifaunas of New Zealand and Tasmania Compared*.

		Number of Extant Species			
		Native		Introduced by Man	
		N.Z.	Tasmania	N.Z.	Tasmania
Dromiceiidae	emus		1		
Apterygidae	kiwis	3			
Podicipedidae	grebes	2	2		
Phalacrocoracidae	shags	3	1		
Ardeidae	herons	3	2		
Threskornithidae	spoonbills	1			
Anatidae	ducks, geese	7	7	4	
Accipitridae	eagles, harriers	1	6		
Falconidae	falcons	1	2		
Phasianidae	pheasants, quail	‡	2	7	
Turnicidae	painted quail		1		
Rallidae	rails	8	7		
Charadriidae	plovers	2	4		
Columbidae	pigeons, doves	1	2	2	2
Psittacidae	parrots	6	8	3	
Cuculidae	cuckoos	2	4		
Strigidae	owls	1	3	1	
Podargidae	frogmouths		1		
Aegothelidae	owlet-nightjars	§	1		
Alcedinidae	kingfishers	1	2	1	1
Xenicidae	N.Z. wrens	3			
Menuridae	lyrebirds				1
Alaudidae	larks			1	1
Hirundinidae	swallows	1	2		
Motacillidae	pipits	1	1		
Artamidae	wood-swallows		1		
Campephagidae	cuckoo-shrikes		1		
Prunellidae	accentors			1	
Muscicapidae	flycatchers	3	9		
Sylviidae	warblers	5	11		
Turdidae	thrushes		2	2	1
Timaliidae	quail-thrushes		1		
Dicaeidae	pardalotes		3		
Zosteropidae	white-eyes	1	1		
Meliphagidae	honeyeaters	3	10		
Emberizidae	buntings			2	
Fringillidae	finches			4	2
Estrildidae	weaver finches		1		
Ploceidae	weavers			1	1
Sturnidae	starlings			2	1
Callaeatidae	wattle-birds	2			
Cracticidae	bell-magpies		4	1	
Turnagridae	N.Z. thrushes	‡			
Corvidae	crows	§	1	1	
Totals		61†	104	33	10

nest inland, some occasionally others invariably, but most of these species, especially the waders, winter on the coast.

The native land and freshwater birds consist of kiwis (3 spp.), grebes (2), cormorants (3), herons (2), a bittern, a spoonbill, ducks (7), raptors (2), rails (8), plovers (2), a snipe, a fruit pigeon, parrots (3), parakeets (4), cuckoos (2), an owl, a kingfisher, New Zealand wrens (3), a swallow, a pipit, warblers and flycatchers (10), a white-eye, honey-eaters (3) and wattle-birds (2). Six other species (a quail, a rail, an owl, a wren, a wattle-bird and the New Zealand 'thrush') present within the last hundred years, are now either extinct or possibly so; and the Auckland Island merganser *(Mergus australis)*, a partly marine species, is also extinct. Finally, the non-marine breeding birds now include 33 species brought to New Zealand by man from Europe (18 spp.), Australia (7), Asia (5) and North America (3). The introduced species comprise four waterfowl (duck, goose and swans), seven gallinaceous birds (partridge, quail, pheasant), three parrots, a dove, a pigeon, an owl, a kingfisher and 15 passerines; two other species (Cape Barren goose and red-vented bulbul), formerly present in small numbers, seem now extinct (KINSKY, 1970).

The 65 living species of native land and freshwater birds listed above seem a modest number for a country the size of New Zealand (268,674 km^2, including outlying islands). Tasmania, in contrast, with an area only about a quarter that of New Zealand has 104 native species of breeding land birds (RIDPATH & MOREAU, 1966). Compared with New Zealand, Tasmania is particularly rich in raptors, flycatchers, warblers and honeyeaters; these groups collectively have 38 species in Tasmania but only 13 in New Zealand (Table 1). Three of the families listed in Table 1 as occurring only in Tasmania, formerly had representatives in New Zealand also; a quail became extinct during the early days of European settlement, and a crow and an owlet-nightjar are known from subfossil bones. On the other hand, the colonisation of New Zealand by white-eyes *(Zosterops lateralis)* and swallows *(Hirundo tahitica)* during the past 120 years, added two families previously unrepresented in New Zealand. Tasmania's proximity to the Australian mainland, to which it was intermittently connected during the Pleistocene, probably allowed the entry of species that could not make the long oceanic journey to New Zealand, some 1600 km to the east. Further, at least 20 Tasmanian species regularly migrate to the Australian mainland in winter (RIDPATH & MOREAU, 1966), but in New Zealand the two cuckoos are the only land birds that migrate across the wide seas separating New Zealand from tropical islands to the north.

ENDEMIC BIRDS OF SPECIAL INTEREST

New Zealand's two endemic orders of ratite birds are the Apterygiformes or kiwis (see Chapter VII) and the Dinornithiformes or moas (Fig. 4 and

Fig. 4. Artist's impression of one of the larger moas (*Dinornis* sp.).

Plate 3). The moas were a group of large to very large flightless birds known from subfossil remains on the three main islands of New Zealand (ARCHEY, 1941, OLIVER, 1949); the authenticity of a reported moa bone from Australia is discounted (SCARLETT, 1969). Apart from numerous bones from caves, swamps and dunes, evidence of moas is provided by gizzard stones, gut contents, eggs (a few nearly intact), tracheal rings, fossil foot-prints, skin and feathers; unusually dry caves have occasionally yielded bones with muscles and skin still attached, and a few feathers contain pigment. Moas, more closely related to kiwis than to other ratites, had raft-shaped sterna with no keels, and wing bones were completely lacking; the feathers had after-shafts but no barbicels. KINSKY's (1970) checklist admits 24 species of moas in seven genera and two families. The Dinornithidae, or greater moas (Fig. 4), with eight species in a single genus, were characterised by a long metatarsus, a wide flattened skull and a broad triangular beak; the largest species *Dinornis maximus* stood some three metres high. The Anomalopterygidae, or lesser moas, with 16 species in six genera, tended to be shorter and stouter birds with more rounded skulls and narrower beaks (Plate 3). Most genera had one or more species in each island and usually the North Island forms were smaller than their South Island equivalents; a few species occurred in both islands. The classification of moas is complicated by the birds' great variation in size (some of it probably a sex difference) and by the poor preservation of some of the material. While a few moa bones date

Plate 3. Upper left: Skeleton of a moa, *Emeus crassus*, which probably stood nearly 1.5 m tall; (Photo: MUIR and MOODIE, National Museum); Upper right: The rifleman, *Acanthisitta chloris*, New Zealand's smallest bird; (Photo: K. V. BIGWOOD, National Publicity Studios); Lower left: The Snares Island (black) race of the tit *Petroica macrocephala;* (Photo: R. ANDERSON, N.Z. Wildlife Service); Lower right: The flightless weka, *Gallirallus australis*, which perhaps evolved from an early invasion of *Rallus philippensis*-like stock; (Photo: J. E. C. FLUX, Ecology Division, D.S.I.R.).

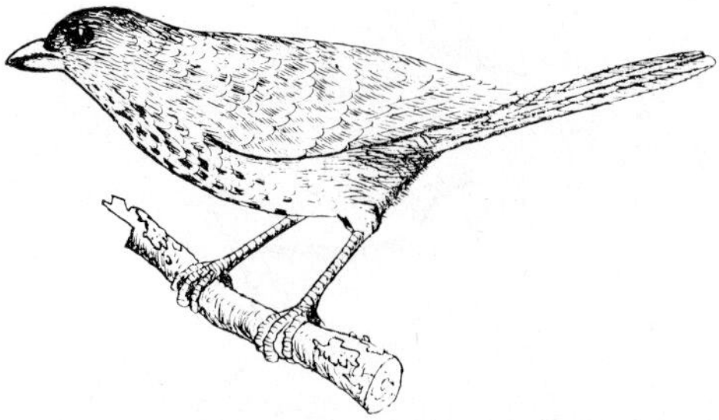

Fig. 5. New Zealand thrush *Turnagra capensis.*

Fig. 6. New Zealand wattle-birds. Left, kokako *(Callaeas cinerea)*; centre, huia *(Heteralocha acutirostris)* head of female above, male below; right, saddleback *(Philesturnus carunculatus)*.

from the Pliocene and one from the Upper Miocene, most are only a few thousand years old and some much less (FLEMING, 1962c). A few of the smaller moas may have persisted in the South Island until about the early eighteenth century and conceivably even a little later in remote areas.

The three endemic families of passerines, the wattle-birds (Callaeatidae), 'thrushes' (Turnagridae) and 'wrens' (Xenicidae), probably represent at least three original colonisations during the Tertiary (FLEMING, 1962b). The relationships of these three families are uncertain; MAYR & AMADON (1951) place the wattle-birds with a group of Australian and Papuan corvid-like families and the New Zealand thrush may have similar

Fig. 7. Rock wren *(Xenicus gilviventris)*.

affinities (FALLA *et al.*, 1966); the New Zealand wrens show similarities to both the Old World pittas and the South American antpipits. The family Turnagridae has only one species *Turnagra capensis* (Fig. 5) and its two subspecies, one in each of the main islands, though formerly widespread in forest areas, are now extremely rare and perhaps extinct. The wattle-birds (Fig. 6), named from the small, fleshy, coloured wattles that hang on either side of the gape, consist of three monotypic genera. *Callaeas* and *Philesturnus* formerly occurred on both the main islands, with a separate subspecies on each, but *Heteralocha* (the huia) was confined to the North Island. Though formerly quite numerous in forest areas, all the wattle-birds greatly declined soon after European settlement and *Heteralocha*, always less numerous than the others, seems to have been extinct since about 1907 (FALLA *et al.*, 1966). The two subspecies of *Philesturnus* were reduced to relict populations on a few small islands; the probability of early extinction was, however, reduced when conservationists succeeded in spreading the birds to several other islands free of European rats and cats. The North Island subspecies of *Callaeas* remains quite numerous in a few localities but is threatened by further destruction of its forest habitat; the South Island subspecies is now extremely rare. *Philesturnus* has a mixed diet of insects and fruits, but *Callaeas* tends to feed more on leaves and shoots. *Heteralocha* apparently fed mostly on insects and was especially notable for the extreme sexual dimorphism of the bill (Fig. 6); the birds were said to feed in pairs, the female probing holes with its long curved bill and the male chipping away rotten wood with its short stout bill.

The New Zealand 'wrens' comprise two genera of small greenish birds with very short tails: *Acanthisitta* (Plate 3) with one species and *Xenicus* (Fig. 7) with three. The two subspecies of *A. chloris*, one in each main island, are New Zealand's smallest birds and, unlike all other members of the endemic families, are still common, especially in beech *(Nothofagus)* forest. Of the three species of *Xenicus*, the rock wren *(X. gilviventris)* remains locally common in alpine areas of the South Island but the other species are rare or extinct. The Stephen Island wren *(X. lyalli)* is known only from 11 specimens brought in by the lighthouse keeper's cat in 1894.

Plate 4. Upper left: The kea *(Nestor notabilis)* on alpine parrot, in snow; (Photo: J. E. C. FLUX, Ecology Division, D.S.I.R.); Upper right: Red-crowned parakeet, *Cyanoramphus novaezelandiae*; (Photo: K. V. BIGWOOD, National Publicity Studios); Lower: The kakapo *(Strigops habroptilus)*, a flightless, nocturnal parrot; (Photo: P. MORRISON, N.Z. Wildlife Service).

Plate 5. Upper left: The blue duck, *Hymenolaimus malacorhynchos*, an inhabitant of fast-flowing mountain streams; (Photo: J. H. G. JOHNS, N.Z. Forest Service); Upper right: The wrybill, *Anarhynchus frontalis*, which has its bill curved to the right; (Photo: H. GUTHRIE-SMITH, National Museum); Lower: The short-tailed bat, *Mystacina tuberculata;* (Photo: J. L. KENDRICK, N.Z. Wildlife Service).

Two endemic genera of parrots *Nestor* and *Strigops* are of special interest. *Nestor* includes one of New Zealand's few alpine birds, *N. notabilis*, the kea (Plate 4), which, like its forest relative the kaka *(N. occidentalis)*, normally feeds on insects, berries and leaves. Some individuals, however, have acquired the habit of settling on sheep immobilised by snow or weakness and pecking through the skin of the back to reach the fat below. Keas, which are confined to the South Island high country, have remained quite numerous despite bounties formerly paid by sheep farmers for their destruction. Another large parrot *Strigops habroptilus*, the kakapo (Plate 4), now extremely rare, is of interest because it is nocturnal and almost flightless; it has some similarities to the Australian night parrot *Geopsittacus*. New Zealand parakeets (Plate 4) consist of four species of *Cyanoramphus*, one confined to the mainland, one to Antipodes Island, and two occurring on both the mainland and outlying islands.

The three endemic genera of waders include a small snipe *(Coenocorypha)* which persists on subantarctic islands, the shore plover *(Thinornis)* now confined to South-East Island in the Chathams, and the wrybill *(Anarhynchus)* a small plover unique in having its rather long bill curved to the right (Plate 5). Among other noteworthy birds, with no close relatives beyond New Zealand, are *Hymenolaimus malacorhynchos* the blue duck (Plate 5) which feeds mainly on insect larvae in mountain streams, and the giant rail-like birds *Diaphorapteryx* and *Aptornis* known only from subfossil remains. The several genera of living rails include *Gallirallus* the flightless weka (Plate 3) and *Notornis* or takahe (Fig. 9) which is also flightless. *N. mantelli*, long known from only four specimens collected between 1848 and 1898, was believed extinct until, in 1948, a small and diminishing population was discovered in a few mountain valleys in the south-west of the South Island.

Origins of the avifauna

Three main geographic elements can be recognised in the New Zealand avifauna, namely: Australian, Holarctic and Austral. A fourth element, the Endemic or Archaic, contains forms with a long evolutionary history in New Zealand and no close relatives elsewhere; it comprises the moas and other endemic groups already discussed.

The Australian element includes, among other groups, warblers, flycatchers, honeyeaters and many water birds (ducks, rails, herons, etc.); it is by far the strongest element among New Zealand's non-marine birds (Falla, 1953) and continues to increase. During the past 120 years, no less than eight Australian birds have established themselves as breeding species in New Zealand (a heron, a spoonbill, a duck, a coot, two plovers, a swallow and a white-eye). Eleven other species cannot be distinguished from Australian subspecies and are probably recent colonists (Fleming, 1962b); still more Australian species are represented in New Zealand by distinct subspecies, and several of New Zealand's endemic species belong to Australian genera. This colonisation of New Zealand by birds from Australia seems to have been going on for a very long time for, as pointed out by FLEMING (1962b), even some endemic genera of New Zealand birds have clear relationships to Australian genera (*Anthornis* to *Meliphaga*, *Notiomystis* to *Meliornis*, and *Cnemiornis* to *Cereopsis*).

Some New Zealand birds of Australian origin became extinct in pre-European times. For instance, bones of a swan *(Cygnus sumnerensis)*, closely related to the Australian *C. atratus* but larger, are abundant in subfossil deposits, especially at the Chatham Islands. Its extinction is the more remarkable since *C. atratus* introduced from Australia by European man, has become very abundant. Other extinct birds include a large flightless goose *(Cnemiornis)* related to the Australian *Cereopsis*, a crow *(Palaeocorax)*, ducks (*Pachyanas*, *Euryanas*, *Malacorhynchus* and *Biziura*, the last two still living in Australia), a coot *(Nesophalaris)*, and an eagle *(Harpagornis)*. The eagle, related to *Haliaeetus*, had strong legs and comparatively short wings and was perhaps a predator or scavenger on moas and other large flightless birds.

A few New Zealand birds have their nearest relatives in the Northern Hemisphere (FALLA, 1953); for example the South Island pied oyster-catcher *(Haematopus ostralegus finschi)*, the black-billed gull (*Larus bulleri*, one of the *L. ridibundus* assemblage), the scaup *(Aythya novaeseelandiae)*, the extinct Auckland Island merganser *(Mergus australis)* and, perhaps, the snipe *(Coenocorypha aucklandica)*. It is possible, of course, that some of the species reached New Zealand by way of Australia where they have since become extinct, but there is no direct evidence supporting this. On the other hand, two recent records of the northern shoveler *(Anas clypeata)* in New Zealand show that non-marine birds from the Northern Hemisphere do occasionally reach New Zealand. Further, among sea and shore birds, the arctic skua *(Stercorarius parasiticus)* and many migrant waders regularly spend the northern winter in New Zealand, some such as the eastern bar-tailed godwit *(Limosa lapponica baueri)* in very large numbers.

The Southern or Austral element in the New Zealand avifauna consists exclusively of seabirds, especially penguins, albatrosses and petrels. Eight species of penguins (in five genera) breed in New Zealand or its sub-antarctic islands, and a subfamily of extinct penguins (11 species in 7 genera) is known from fossils of Eocene to Pliocene age. Many of the petrels and albatrosses have a circumpolar distribution and several, for example the fulmar prion *(Pachyptila crassirostris)*, have distinct subspecies breeding on islands in the New Zealand region (some New Zealand petrels, *Pterodroma hypoleuca* for instance, seem to have a tropical rather than a southern origin). Other birds included in the Austral element are the southern black-backed gull *(Larus dominicanus)* which occurs in New Zealand, South America and South Africa and perhaps the marine shags *(Leucocarbo)* which have several distinct races in New Zealand and its off-lying islands and relatives in South America.

While the geographic origin of many New Zealand birds seems reasonably clear the scarcity of fossils means there is little direct evidence as to when the various groups first colonised New Zealand. The bones recovered from swamps and caves, or from the middens of Polynesian man, are all geologically recent and many are only a few hundred years old. Apart from penguins (Upper Eocene to Pliocene), New Zealand's bird fossils consist only of a few bones of moas from the Upper Miocene onwards, an albatross from the Lower Oligocene, a pelecaniform bird *(Pseudontornis)* from the Pliocene and a goose *(Cnemiornis)* from the Lower Pleistocene.

FLEMING (1962b) has used the differences that distinguish New Zealand birds from their overseas relatives as a rough indication of the time that has lapsed since their colonisation, 'unreliable in particular cases, but useful in default of other evidence'. Strong subspecies have developed in New Zealand since Cook Strait last became a barrier to dispersal some 15,000 years ago, and since forests came to the formerly glaciated Auckland Islands, perhaps less than 10,000 years ago. On this basis, the 17 New Zealand subspecies of Australian species have probably developed since the late Pleistocene. If strong subspecies can form within 15,000 years, the 13 New Zealand species in overseas genera need have arrived in New Zealand no earlier than the early Pleistocene, say a million years ago. FLEMING suggests the endemic genera (representing 24 colonisations) perhaps arrived during the 25 million years of the late Tertiary, the endemic families (at least 3 colonisations) during the early Tertiary, and the two endemic orders (moas and kiwis) during the Upper Cretaceous, some 70 million years ago. As pointed out by FLEMING, such 'educated guessing' has many pitfalls; the data can change with changes in the systematic ranking given to the endemic group, and differences in the speed of evolution will confuse the issue.

Some groups of New Zealand birds include forms which seem to represent two or more invasions by birds of the same ancestral stock, the descendants of the first invasion having evolved into distinct species before the arrival of the second. The Australian flycatcher genus *Petroica*, for instance, is represented in New Zealand by two subgenera (Fig. 8). The New Zealand tits are grouped with Australian and Pacific Island species in *Petroica s. str.*, and the New Zealand robins, larger birds with longer tarsi, in the subgenus *Miro*. The robins seem to be derived from an early (perhaps Pliocene) invasion of *Petroica* stock and the tits from a much later one, perhaps early Pleistocene (FLEMING, 1950). Among other pairs perhaps representing successive invasions are the black and pied stilts, the former *(Himantopus novaezealandiae)* a New Zealand endemic and the latter *(H. h. leucocephalus)* probably a recent arrival from Australia. At the generic level, the takahe *(Notornis mantelli)* probably originated

Fig. 8. Two New Zealand flycatchers: a robin *Petroica (Miro) australis* at left and a male pied tit *(Petroica macrocephala toitoi)* at right.

Fig. 9. Pukeko *(Porphyrio p. melanotus)* at left and takahe *(Notornis mantelli)* at right.

from an early invasion of stock which, much later, produced the pukeko *(Porphyrio p. melanotus)*, another recent arrival in New Zealand judging by the lack of subspeciation (Fig. 9). Similarly, *Gallirallus* the New Zealand woodhen or weka (Plate 3) seems to be derived from the genus *Rallus* later represented in New Zealand by *R. philippensis* and other species.

EVOLUTIONARY TRENDS WITHIN THE NEW ZEALAND REGION

There is surprisingly little evidence in New Zealand of the adaptive radiation that is such a striking feature of other isolated archipelagos such as the Galapagos and Hawaiian Islands. Some adaptive radiation is apparent particularly in the moas and, to a lesser extent, in the wattle-birds, wrens and parrots. Possibly other specialised forms, particularly those adapted to forests, became extinct during the cold periods of the Pleistocene (FLEMING, 1962c).

As might be expected in an archipelago, several New Zealand birds have developed island races, but most of these are distinguished from each other by relatively minor differences in size or colour. On the other hand, a few evolutionary trends, not related to particular islands, are more marked, and often characteristic of several different groups of New Zealand birds, for instance trends towards large size, flightlessness and melanism (OLIVER, 1955). Island races and these other evolutionary trends are discussed separately below.

a. Island Races

Cook Strait, separating the two main islands, is only 22 km wide, yet many forest birds have developed distinct races on either side of the Strait, presumably since the sea last broke through some 15,000 years ago (FLEMING, 1962b). Such birds include a kiwi *(Apteryx australis)*, a rail *(Gallirallus australis)*, a parrot *(Nestor meridionalis)*, the possibly extinct owl *(Sceloglaux albifacies)*, two wrens (*Acanthisitta chloris* and *Xenicus longipes)*, three flycatchers (*Rhipidura fuliginosa*, *Petroica macrocephala* and *P. australis*), two wattle-birds (*Philesturnus carunculatus* and *Callaeas cinerea)*, the possibly extinct New Zealand 'thrush' *(Turnagra capensis)* and, outside the forest, the fernbird *(Bowdleria punctata)*.

Foveaux Strait, some 26 km of shallow seaway separating Stewart Island from the South Island, has several small islands which perhaps act as stepping stones and the Strait seems much less of a barrier to the dispersal of birds than does Cook Strait. Nearly all living South Island birds occur, or formerly occurred, also on Stewart Island. The few species that have developed distinct races on Stewart Island include a kiwi *(Apteryx australis)*, a rail *(Gallirallus australis)*, a snipe *(Coenocorypha aucklandica)*, a wren *(Xenicus longipes)*, the fernbird *(Bowdleria punctata)* and a 'robin' *(Petroica australis)*.

Other birds, presumably with greater powers of dispersal, have not developed obviously distinct races on the main islands, but some have done so on more remote islands such as the Chathams some 800 km east of New Zealand. Such species are the New Zealand pigeon *(Hemiphaga novaeseelandiae)*, two parakeets (*Cyanoramphus novaezelandiae* and *C. auriceps*) and two honeyeaters (*Anthornis melanura* and *Prosthemadera novaeseelandiae*); the Chatham Island warbler (*Gerygone albofrontata*) is specifically distinct from the mainland *G. igata*.

Finally, a few widely distributed New Zealand birds have not yet evolved distinctive island races. Apart from several water birds (shags, herons, ducks, rails, etc.) such species include a harrier (*Circus approximans*), a falcon (*Falco novaeseelandiae*), an owl (*Ninox novaeseelandiae*) and a kingfisher (*Halcyon sancta*). Many of these birds are strong fliers and some may have colonised New Zealand comparatively recently. The eight Australian birds that have established themselves in New Zealand during the past 120 years have all been recorded from both the North and South Islands and five breed in both islands.

Although there is some evidence that birds on small islands have smaller clutches than do their mainland relatives (CODY, 1966), the number of New Zealand clutches analysed is too small to be conclusive. This trend is shown by the duck *Anas castanea;* the Australian race lays an average clutch of 9 eggs, the New Zealand one (given specific status as *Anas aucklandica* in KINSKY (1970)) a clutch of 6 and the Auckland Island race only 3–4 eggs (LACK, 1970).

b. Changes in Size

The more ancient elements of the New Zealand avifauna include several birds of unusually large size. The largest of all were the moas which were among the world's largest birds. In several other groups of smaller birds, the New Zealand representatives were, nevertheless, much larger than their nearest overseas relatives. For instance, the endemic goose *Cnemiornis*, standing about one metre tall, was larger than its Australian relative *Cereopsis* and, among the parrots, *Strigops* is much larger than *Geopsittacus*. When birds of the same stock invaded New Zealand, or its outlying islands, on two or more widely separated occasions, the descendants of the first invasion are usually larger in size than those of the second. For instance, robins (*Miro*) are larger than tits (*Petroica s. str.*) (Fig. 8), the parakeet *Cyanoramphus unicolor* is larger than *C. novaezelandiae* or *C. auriceps* and, among rails, *Gallirallus* is larger than *Rallus* and *Notornis* than *Porphyrio* (Fig. 9). This suggests that evolution towards larger size has occurred within the New Zealand region.

On the other hand, a few New Zealand birds are smaller than their presumed ancestral stocks. The Auckland Island merganser, for instance was smaller than its subfossil relative on the New Zealand mainland and much smaller than its Northern Hemisphere relatives. Likewise, the modern Chatham Island race of the snipe *Coenocorypha aucklandica* is smaller than its extinct relative *C. chathamica*.

In some New Zealand birds, particularly the extinct flightless species, the South Island form tended to be larger than the North Island one, thus conforming to Bergmann's Rule. This applied to several species of moas, to the extinct rail *Aptornis* and perhaps to the extinct goose *Cnemiornis*. Some of the bones, for example those of *Cnemiornis*, were from deposits of different age and, with species evolving from one size to another, the size of the bones may reflect when rather than where the birds were living.

c. Flightlessness

Large size and flightlessness tend to go together, and New Zealand is particularly rich in birds that are flightless or nearly so. Apart from the 24 species of moas, which had lost all outward trace of wings, the list includes the kiwis, several genera of rails (*Gallirallus*, *Notornis*, *Aptornis*, and *Diaphorapteryx*), a parrot *(Strigops)*, a goose *(Cnemiornis)* and the Auckland Island race of the duck *Anas aucklandica*. While the lack of mammalian predators in New Zealand (until Polynesian man brought rats and dogs) may have facilitated the evolution of flightless birds, it is noteworthy that the flightless rail *Tribonyx mortieri* flourishes in Tasmania despite the presence of several carnivorous marsupials (RIDPATH & MOREAU, 1966).

Of the several groups of flightless birds, only the moas and kiwis seem ancient enough to have reached New Zealand during the Cretaceous, the last period during which New Zealand was joined, or nearly joined, to other lands (FLEMING, 1962a, RAVEN & AXELROD, 1972). The other groups of flightless birds must therefore be derived from ancestors that flew to New Zealand.

The seeming absence of bones of moas or kiwis at the Chatham Islands is consistent with the lack of geological evidence for land connections between these islands and the New Zealand mainland. It is surprising therefore, that bones of a giant, flightless rail *(Diaphorapteryx hawkinsi)* and of a flightless parrot *(Strigops habroptilus)* have been found at the Chatham Islands. Positively identified remains of *Diaphorapteryx hawkinsi* have not yet been found on the New Zealand mainland so the species may be derived from some flying ancestor which long ago colonised the Chatham Islands where it subsequently became flightless. The parrot *(Strigops habroptilus)*, known only from a few fragments but seemingly the same species as occurs on the mainland, may have been brought to the Chathams by early Polynesian voyagers, perhaps as food.

d. Melanism

Melanism has developed in several different groups of New Zealand birds, and especially in the flycatchers. The South Island race of the fantail *(Rhipidura fuliginosa)* is dimorphic, and in some districts up to 20% of the birds are black instead of pied as in other races of the species throughout its wide range (but black fantails occasionally appear in the North Island). Likewise, the tits *(Petroica)* and their early derivatives the robins *(Miro)* are usually dark above and whitish (or yellowish) below, but the Snares Island tit (Plate 3) and the Chatham Island robin are black. The shag *Phalacrocorax melanoleucos* is dimorphic; a few individuals are white below (like the parent stock in Australia) but in most New Zealand birds the white is confined to the throat and the bird is predominantly black. New Zealand also has a black stilt *(Himantopus novaezealandiae)* and a black oystercatcher *(Haematopus unicolor)* and both seem to be derived from pied forms. Finally, the western weka *(Gallirallus australis australis)* and the Stewart Island shag *(Leucocarbo chalconotus)* each has two colour phases one of which, though not black, is nevertheless very much darker than in any other race.

Distribution and ecology of New Zealand birds

New Zealand's six endemic families of birds are all confined to the three main islands and smaller inshore islands, and so are the two endemic genera (*Finschia* and *Mohoua*) of Australian warblers (Subfamily Malurinae). The moas and kiwis were, of course, flightless, and the other groups were probably so specialised for life in the mainland forests that they were unable to survive in, or even reach, the Auckland and Chatham Islands when these became forested after the cold periods of the Pleistocene.

A few species are even more restricted and occur on only one of the three main islands: *Mohoua albicilla*, *Notiomystis cincta* and *Heteralocha acutirostris* are known only from the North Island and *Apteryx haastii*, *Nestor notabilis*, *Xenicus gilviventris* and *Mohoua ochrocephala* only from the South Island; *Finschia novaeseelandiae*, occurring in both the South Island and Stewart Island, is absent from the North Island. Although *Apteryx oweni*, *Podiceps cristatus*, *Strigops habroptilus* and *Notornis mantelli* are now confined to the South Island, they once lived in the North Island as well, and doubtless several of the other species listed above were formerly more widespread.

The Auckland and Chatham Islands lack many birds characteristic of the New Zealand mainland, but they have sometimes provided a last refuge for species elsewhere extinct. The merganser *(Mergus australis)*, for instance, was confined to the Auckland Islands until it became extinct about 1905. Recent discoveries of subfossil remains of a slightly larger merganser on the New Zealand mainland (Kear & Scarlett, 1970) suggest that the Auckland Island population was the last outpost of a species that formerly had a much wider distribution. The shore plover *(Thinornis novaeseelandiae)* has shown a comparable contraction of range within historic times and is now confined to South-East Island in the

Chathams. Finally, the New Zealand snipe *(Coenocorypha aucklandica)* which occurs as distinct races on Snares, Chathams, Antipodes and Auckland Islands and on small islands near Stewart Island, was once present on Little Barrier Island near Auckland (OLIVER, 1955) and in Taranaki where subfossil remains have been found (MEDWAY, 1971).

A few distinctive New Zealand species have spread to other parts of the southwest Pacific. The parakeet genus *Cyanoramphus*, for instance, with four species and nine subspecies in New Zealand and its outlying islands (Kermadecs, Chathams and the subantarctic islands including Macquarie in the far south) has spread to New Caledonia, Tahiti, Norfolk Island and Lord Howe Island. Each of these islands has (or had) its own race of *C. novaezelandiae* or else a distinct species as in Tahiti *(C. zealandicus)*, and Norfolk Island was further enriched by having representatives of a New Zealand parrot *(Nestor)* and pigeon *(Hemiphaga)*.

Only two New Zealand breeding land birds migrate far outside the region. The two cuckoos winter on tropical Pacific islands ranging from the Bismarck Archipelago and the Solomon Islands in the west (mainly *Chalcites lucidus*) to Fiji and the Society Islands in the east *(Eudynamis taitensis)*. Among seabirds several of the petrels and shearwaters that breed in New Zealand range widely over the southern oceans and some, such as *Puffinus griseus*, regularly migrate to the north Pacific. The young of a few New Zealand sea and shore birds regularly cross the Tasman Sea to Australia. This habit is particularly marked in young gannets *(Sula bassana serrator)* which appear in Australian waters soon after leaving their nesting colonies in New Zealand (gannets also breed in Australia though in much smaller numbers than in New Zealand); a few young gannets from New Zealand even reach Western Australia some 6700 km from where they were reared (WODZICKI, 1967). Two other New Zealand species, a tern *(Sterna striata)* and a dotterel *(Charadrius bicinctus)* appear in Australia each winter, but the birds, apparently mostly juveniles, are not very numerous; neither species nests in Australia.

The abundance and diversity of New Zealand's seabirds probably result from two main facts: first, the land areas provide the only available nesting places for the numerous birds that feed over the vast areas of plankton-rich ocean and, second, the region extends from the Kermadec Islands (Lat. 30°S), where tropical seabirds occur, to Macquarie Island (Lat. 54½°S) where subantarctic and even antarctic species are found. The land birds, on the other hand, are rather lacking in diversity and an important consequence of this is that the few species present are able to occupy relatively broad niches. For instance, until recently, the fantail *(Rhipidura fuliginosa)* was the only species feeding regularly on flying insects and it does so in a wide range of habitats from one end of the country to the other, and from sea level to subalpine scrub. In some habitats fantails no longer have a monopoly of taking flying insects; chaffinches *(Fringilla coelebs)* brought from Europe by man, frequently hawk for insects over water, and welcome swallows *(Hirundo tahitica)*, recent colonists from Australia, regularly do so.

Several groups of New Zealand birds have fed on the leaves and shoots of plants. Apart from the extinct moas (24 species), leaf-eating birds included the extinct swan *(Cygnus sumnerensis)* and goose *(Cnemiornis calcitrans)*, the takahe *(Notornis mantelli)* and kakapo *(Strigops habroptilus)* and, in the forest canopy, the kokako *(Callaeas cinerea)* and pigeon *(Hemiphaga novaeseelandiae)*. In view of their size, numbers and vegetable diet the moas and to a lesser degree the other leaf-eating species, must have browsed quite heavily on the native vegetation. To some extent they were the ecological equivalents of the browsing mammals of other countries.

Reflecting the dominant vegetation of pre-Polynesian times, the majority of New Zealand land birds are forest species. The quite extensive tussock grasslands, especially in the South Island, are now occupied mainly by pipits *(Anthus novaeseelandiae)*, harriers *(Circus approximans)* and introduced skylarks *(Alauda arvensis)*, but formerly some of these areas supported wekas *(Gallirallus australis)* and large numbers of the now extinct quail *(Coturnix novaezealandiae)*; seeding grasses also provided food for parakeets from nearby forests. In spring and early summer, the South Island tussock grasslands and their associated wide shingly river beds are important nesting areas for oystercatchers *(Haemotopus ostralegus)*, stilts (*Himantopus leucocephalus* and *H. novaezealandiae*), banded dotterel *(Charadrius bicinctus)*, wrybill *(Anarhynchus frontalis)*, black-fronted terns *(Chlidonias hybrida)* and black-billed gulls *(Larus bulleri)*.

The only species largely confined to alpine areas at or above the tree line are the rock wren *(Xenicus gilviventris)* and the kea *(Nestor notabilis)* though the latter quite frequently descends to lower altitudes, especially in winter. The brown creeper *(Finschia novaeseelandiae)*, another South Island species, occurs from sea level to subalpine scrub, but it seems particularly at home in the latter habitat. Fleming (pers. comm.) has suggested the creeper may be an alpine equivalent of the related yellow-head *(Mohoua ochrocephala)* of lowland forests just as the kea and rock wren are the alpine equivalents of the kaka *(Nestor meridionalis)* and bush wren *(Xenicus longipes)*, respectively. The few remaining kakapo *(Strigops habroptilus)* and takahe *(Notornis mantelli)*, formerly widespread species, now occur only in a few mountainous localities in the South Island.

Most of the introduced birds are common only on farmland or other habitats strongly modified by man, but a few are more widespread. Blackbirds *(Turdus merula)* and chaffinches *(Fringilla coelebs)* and, to a lesser extent song thrushes *(Turdus philomelos)* penetrate deeply into native forests, and hedgesparrows *(Prunella modularis)* and redpolls *(Acanthis flammea)* are quite common in native alpine vegetation as well as at lower altitudes. Many introduced species have spread to outlying islands in the subantarctic and elsewhere (Williams, 1953).

Many New Zealand birds have a fairly long breeding season extending from late winter (August) until well into summer (January or later), and passerines often rear two or more broods during this time; at Gisborne (Latitude 38°40'S) a fantail reared five broods, each of three young, in a season extending from August to March (Blackburn, 1965). Some shags (*Phalacrocorax carbo*, for instance) seem to lay at almost any season but usually there is a peak in spring and, at least in *P. varius*, often another in autumn (March–April). Most petrels (Procellariidae) have rather rigid laying dates but these may differ between related species; for instance among the smaller shearwaters (*Puffinus* species), *P. assimilis* lays in May-June (winter), *P. gavia* in September–October (spring) and *P. huttoni* in November–December (early summer). Apart from *Puffinus assimilis*, winter breeders include *Procellaria cinerea* (May), *P. westlandica* (May, but the closely related *P. parkinsoni* lays in December) and *Pterodroma macroptera* (May–June); the majority of petrels, however, lay in spring.

The clutches of some of New Zealand's endemic birds seem small. The few records available for the New Zealand 'thrush' *(Turnagra capensis)* and wattle-birds (Callaeatidae) suggest that these species laid clutches of only two or three eggs and kiwis (*Apteryx* spp.) normally lay but a single egg (occasionally two in *A. australis* and *A. haasti*). The takahe *(Notornis mantelli)* lays only one or two eggs, an extremely small clutch for a rail.

The Mammals

Composition of the Mammal Fauna

Until the arrival of Polynesian man, New Zealand had only two species of land mammals and both were small insectivorous bats. Reports of an otter-like animal in the South Island during the early days of European settlement (Watson, 1960) lack confirmation and probably arose from myth and misidentification. The absence of native land mammals, other than bats, is a striking feature of the New Zealand fauna and reflects the country's long isolation from other lands. The introduction, spread and status of the many species of mammals brought to New Zealand from other lands by man were documented by Thomson (1922), Wodzicki (1950, 1961) and Wodzicki & Flux (1967), and the ecology of these and of native mammals, including marine ones, was recently reviewed by Gibb & Flux (1973).

Polynesian man brought only a now extinct race of dog *(Canis familiaris)*, and the Polynesian rat *(Rattus exulans)*, once widespread but now mainly on small islands perhaps because of inability to compete with European rats especially *Rattus rattus*. European man, on the other hand, brought many mammals to New Zealand. Rats (*Rattus norvegicus* and, probably later, *R. rattus*) and mice *(Mus musculus)* came as stowaways in the ships of the early explorers, sealers and whalers, who also brought pigs *(Sus scrofa)* and goats *(Capra hircus)* for food. The first European settlers introduced more domestic animals: sheep *(Ovis aries)*, cattle *(Bos taurus)*, horses *(Equus caballus)*, dogs *(Canis familiaris)* and cats *(Felis catus)*; and all of these later became feral. Rabbits *(Oryctolagus cuniculus)*, and later hares *(Lepus europaeus)*, were introduced for food and sport, and rabbits increased to such numbers that stoats *(Mustela erminea)*, weasels *(M. nivalis)* and ferrets *(M. putorius)* were released to control them. Later still, deer (*Cervus elaphus*, *C. canadensis*, *C. unicolor*, *C. nippon*, *C. timorensis*, *Dama dama*, *Alces americana* and *Odocoileus virginianus*), chamois *(Rupicapra rupicapra)* and thar *(Hemitragus jemlahicus)* were introduced for sport, and Australian opossums *(Trichosurus vulpecula)* for fur. Hedgehogs *(Erinaceus europaeus)* and wallabies (*Macropus rufogriseus*, *M. eugenii*, *M. parma*, *M. dorsalis*, *Wallabia bicolor* and *Petrogale penicillata*) also became established in the wild.

The seas that for so long protected New Zealand from invasion by land mammals are the home of many whales, dolphins and seals, and these marine mammals are discussed briefly below since the seals, at least, come ashore to breed. The whales and dolphins, reviewed by GASKIN (1968), comprise 27 species of which most are wide-ranging or cosmopolitan. Although seven species of seals have been recorded from the New Zealand region, only four are of regular occurrence, mainly in the subantarctic. Only the fur seal *(Arctocephalus forsteri)* is at all common on mainland coasts; it breeds in the south and west of the South Island and in the subantarctic. The sea lion *(Phocarctos hookeri)* and the elephant seal *(Mirounga leonina)* breed on the subantarctic islands where the leopard seal *(Hydrurga leptonyx)* also occurs as a regular visitor from the south.

ENDEMIC MAMMALS OF SPECIAL INTEREST

One of the bats, *Mystacina tuberculata* (Plate 5), is the sole member of the Mystacinidae, an endemic family of uncertain relationships; it shows some similarity with the Molassidae, a widespread tropical group of free-tailed bats. The other bat, *Chalinolobus tuberculatus*, is a member of the cosmopolitan family Vespertilionidae and is very closely related to other species of *Chalinolobus* in eastern Australia and New Caledonia (DWYER, 1960).

Mystacina is a most remarkable animal. While quite a strong flier in the usual manner of bats, it is also an agile climber and can scramble quickly over flat surfaces. The wings are folded in a peculiar way so that the delicate portions are protected from injury while the limbs are left relatively free for walking and climbing, activities that are assisted by the large adhesive soles of the feet and by the talons at the bases of the claws (DWYER, 1962). Another remarkable feature of *Mystacina* is that its range in size is greater than that reported for any other species of bat (DWYER,

1970). The significance of this variation is uncertain. An earlier separation into a smaller northern subspecies and a larger southern one was abandoned when skeletons of adult bats from caves in the north were found to include both large and small forms. The ages of the skeletons have, however, not yet been established, so possibly the two forms lived at different times and, perhaps, in different climates.

Distribution and ecology of New Zealand mammals

Both kinds of bats were formerly widespread in the North and South Islands, but only *Mystacina* has been recorded from Stewart Island. Bats have disappeared from many districts following the clearing of forests; they have not adapted to farmland and, even in native forests seem scarcer than formerly, perhaps because of predation by the introduced ship rat *(Rattus rattus)*. Although *Mystacina* pursues insects on the wing, it may take some of its food by climbing along branches and twigs. It can eat quite large insects and, at Stewart Island, has been reported as scraping flesh from the bodies of muttonbirds *(Puffinus griseus)* hung up to dry; G. Kuschel (pers. comm.) reports finding seeds of *Freycinetia* and *Collospermum* in *Mystacina* droppings. *Chalinolobus*, on the other hand, seems to feed mainly on small insects taken on the wing, and it hibernates when cold weather reduces the availability of such prey. *Mystacina*, apparentlyless dependent on flying insects, has been recorded in all months except March and September and may not hibernate as regularly as does *Chalinolobus*.

The beaked whale *Tasmacetus shepherdi* and the dolphin *Cephalorhynchus hectori* have been recorded only from New Zealand and may be endemic. Of the seals, *Phocarctos hookeri* is confined to the New Zealand region, *Arctocephalus forsteri* occurs also in Australia (King, 1969) and *Mirounga leonina* is widely distributed in subantarctic latitudes of the Atlantic, Indian and Pacific Oceans (Turbott, 1952).

Recent Changes in the Vertebrate Fauna

New Zealand's fauna of terrestrial vertebrates has changed greatly during the past thousand years or so. Many species have declined in numbers and range, and several have become extinct; on the other hand, new species have arrived from other countries and some of these are now abundant throughout the land. These processes of extinction and colonisation are examined separately below, the former in some detail the latter more briefly.

Forty-five species of birds (24 of them moas) became extinct before about 1800 A.D. and are known only from subfossil remains (Kinsky, 1970). In addition, at least five species and six subspecies of native birds have become extinct since 1800 A.D. (Williams, 1962), and a few others such as the laughing owl, kakapo and New Zealand thrush, are extremely rare or perhaps extinct. The native frogs, some lizards, the tuatara and the two bats are much reduced in numbers and range, and an undescribed frog (larger than living species of *Leiopelma*) is known only from subfossil bones in caves. Bats, marine mammals and many birds have declined in numbers within the last two hundred years, but the reductions in the ranges of native frogs, some lizards, the tuatara and the large flightless birds began much earlier.

Although the underlying causes of these extinctions and changes in distribution have been much discussed, particularly in relation to birds, they are still not well understood. Some authorities look to non-human factors, such as climatic change, to explain the extinctions, but others blame the impact of man. Proponents of climatic change have usually stressed the adverse effects on moas of flooding and the encroachment of forests during a pluvial period following the last Pleistocene glaciation (ARCHEY, 1941, OLIVER, 1949); but FALLA (1955), presumably with a later period in mind, suggested that 'significant periods of drought' may have contributed to the decline of moas and other large birds. FLEMING (1962c) however, claims that the magnitude of climatic changes during the Holocene was very small compared to that survived by moas during the Pleistocene and, in support, he points to marine invertebrates which suffered many extinctions during the Pleistocene but few since. While OLIVER (1949) thought that a period of high rainfall was possibly the most important factor contributing to the extinction of moas, he considered that orthogenetic trends – particularly towards very large size – were perhaps also involved.

On the other hand, FLEMING (1962c, 1969) thought that the simplest explanation for the extinction of many flightless birds, and the reductions in the ranges of other terrestrial animals such as the tuatara, was to attribute them all to the great ecological changes brought about by the coming of the early Polynesian settlers with fire, rats *(Rattus exulans)* and dogs. This view is supported by the fact that of the 45 species of birds that became extinct before 1800 A.D., the bones of 34 species, including 21 species of moas, have been found in association with the middens or camp sites of Polynesian man (KINSKY, 1970). This direct predation by man on moas and other large birds was perhaps supplemented by the effects of rats (FLEMING, 1969). *Rattus exulans* has an adverse effect on tuatara and lizard populations (CROOK, 1973, WHITAKER, 1973). Perhaps more important than predation by either man or rats were changes in the habitat caused by fires started by man. Soil profiles provide ample evidence of such fires which no doubt also explain the reductions in range suffered by several terrestrial molluscs about this time (FLEMING, 1962c).

Still greater environmental changes followed the arrival of Europeans who not only converted much of the lowland forest to farmland but also imported a great assortment of foreign animals some of which became widely distributed and abundant. Thus the native fauna had to contend not only with loss of lowland forest habitat but also with foreign species that competed with them, preyed upon them, brought new diseases or greatly changed the character of the remaining forests. The relative importance of these factors probably varied from one species to another but, in many instances, loss of habitats and predation are clearly significant. Large areas of indigenous forest, particularly in the North Island, have been replaced by exotic grasses and trees, and the fauna of such areas has likewise become predominantly exotic. Extensive native forests remain in mountainous areas of both islands but, as pointed out by FALLA (1955), some endemic birds may require a combination of high and low altitude forest to provide a long seasonal succession of blossoms and ripening fruits; the loss of so much lowland forest may explain, at least in part, the seeming scarcity of birds in many mountain forests.

The evidence for predation as a limiting factor on the native fauna is convincing. The influence of the early Polynesian settlers and of the rats and dogs that accompanied them has already been mentioned. Europeans killed birds for food (pigeons and ducks) or as collectors' specimens (huias) until this was prevented by law. Sportsmen still hunt several species of native waterfowl under seasonal licence, and certain Maori families have the right to collect muttonbirds (the young of the sooty shearwater, *Puffinus griseus*, in the Stewart Island area, and of the grey-faced petrel *Pterodroma macroptera* on small islands off the north-east coast of the North Island). Perhaps the worst example of predation by European man was on fur seals which, because of the value of their skins, were hunted almost to extinction; under strict protection, fur seals are now rapidly increasing in numbers. Many native species have suffered predation by the animals introduced by man, rather than by man himself. Remains of lizards and birds are common in the dens or in the scats of mustelids and cats, and a lighthouse keeper's cat apparently exterminated the Stephen Island wren. Species such as parakeets, now rare in many mainland areas, nevertheless remain abundant on islands free of predatory mammals. The importance of predation by European rats (especially *Rattus rattus*) and cats in depleting native faunas was demonstrated only too clearly when these animals became established on some small islands off Stewart Island (ATKINSON & BELL, 1973).

Claims that introduced diseases or competition from exotic birds have been important in the decline of native species are poorly documented. A poxvirus, or pox-like lesions, have been found in pipits, white-eyes and robins but, as yet, there is no evidence that the infections are associated with any marked decline in the numbers of these species (AUSTIN *et al.*, 1973). On the other hand, the rapid disappearance of bellbirds from North Auckland last century, while the species survived on nearby islands and in mainland districts further south, suggests disease (TURBOTT, 1961), as does the sudden disappearance of wekas (sometimes coinciding with reports of sick birds) from extensive areas while the species remained numerous, and even increased, in other mainland areas seemingly equally subject to predation and modification of habitat.

Suggestions that native species are adversely affected by competition from exotic ones are largely speculative. Introduced frogs occupy different habitats from native ones, the only introduced lizard has a very localised distribution, and there are no introduced bats; such competition as may exist is therefore more likely to be found among birds. Introduced mallards and black swans may reduce the food available to native ducks, and starlings *(Sturnus vulgaris)* frequently take nectar from native flax *(Phormium tenax)* and pohutukawa *(Metrosideros excelsa)* which locally are important sources of food for native honeyeaters. Comparatively few introduced birds penetrate deeply into native forests; only blackbirds and, to a lesser extent, chaffinches and hedgesparrows are sufficiently numerous to be credible competitors with native insect-eating birds in this habitat. Among water birds, however, the recent establishment and dramatic increase of the white-faced heron *(Ardea novaehollandiae)* seems to have been followed by some reduction in the range of habitats occupied by the longer established reef heron *(Egretta sacra)*. Conceivably the endemic takahe and black stilt also suffered reductions in range when their respective relatives, the pukeko and pied stilt, became established in pre-European times.

Perhaps the most serious competitor faced by some species of native vertebrates in certain localities is the German wasp *(Vespula germanica)*. Accidentally introduced in 1945, this predatory insect is now widespread in New Zealand and, locally, becomes enormously abundant in some seasons. At such times, wasps must remove a high proportion of the insect and nectar supplies on which several species of birds and lizards depend for food.

The number of extinctions in Polynesian times seems high (even allowing that the period covered about a thousand years), but this probably reflects the vulnerability of the fauna rather than the severity of predation or the degree of environmental change. In the complete absence of mammalian predators some New Zealand species may have evolved habits that made them particularly vulnerable when first exposed to mammalian predators. This suggestion is in accord with the fact that extinctions of birds were more frequent among endemic species (especially those in endemic genera) than among those, presumably more recent arrivals, with close relatives in Australia (McDowall, 1969).

Fewer extinctions could be expected in European times if the most vulnerable species had already been exterminated by the Polynesians. A further factor is that, at least since the end of the nineteenth century, conservationists have been increasingly successful in achieving legal protection of rare species and in establishing sanctuaries for native fauna. Where it not for offshore islands (some free of mammals and only moderately influenced by man) the number of extinctions would be considerably higher than it is.

The extinctions and reductions in range, discussed above, have been offset to some extent by the arrival and spread of new species from other lands. Apart from animals brought by man, himself a relatively new colonist, the newly established vertebrates are exclusively birds, although earlier Holocene colonists may have included the long-tailed bat and, possibly the lizard *Leiolopisma suteri*. Within the past 120 years no less than eight species of Australian birds have reached New Zealand unassisted and some, such as the white-eye and white-faced heron, are now abundant and widespread. Other birds, such as the pukeko and white heron are indistinguishable from their Australian counterparts and may be regarded as recent colonists (FLEMING, 1962b). The activities of man, while leading to the extinction of some endemic species, may nevertheless have created new and favourable conditions for the establishment of new arrivals from Australia. The terrestrial vertebrates transported to New Zealand by European man, and successfully established there, include two species of frogs, a lizard, 33 birds and 33 mammals; a further mammal, the rat *Rattus exulans*, arrived much earlier with the Polynesians. The most dramatic change in the vertebrate fauna of New Zealand during the past thousand years is undoubtedly the decline and extinction of many of the large ground feeding birds and their replacement by a diverse assortment of mammals.

While many native animals have suffered as a result of human activities, a few have greatly benefited. Certain skinks of the genus *Leiolopisma*, for instance, prefer an open habitat and there is now much more of this as a result of man's clearing of forest and scrublands. Likewise the harrier now has much more open country to hunt over, and rabbits and other introduced mammals provide new and abundant food. All three species of gulls, benefiting from abundant food at rubbish dumps and the outfalls of sewers and meatworks, have greatly increased in numbers.

Although the clearing of forests has meant loss of habitat for many native animals, some forest birds have been able to exist and even flourish in the new habitats. Grey warblers and fantails, for instance, are now common in farmland plantations, hedgerows, orchards and gardens. Other species, while still mainly living in native forests, nevertheless visit parks and gardens when some favoured food, often provided by exotic trees, is available; tuis and bellbirds come for nectar and pigeons for fruits, blossoms and young shoots. Finally plantations of exotic pines provide a new and acceptable habitat for several native birds (JACKSON, 1971) and the populations of a few avian species may be as high or even higher in pine forests as in native beech *(Nothofagus)* forests (GIBB, 1961). Thus for these species, the extensive pine forests planted in areas formerly in scrub or tussock grassland compensate to some extent for the felling of indigenous lowland forest. Unfortunately modern trends in forest management often involve clear felling of pine plantations before they have reached their optimum condition as a habitat of native birds.

Acknowledgements

We wish to thank the following people for their constructive criticism of parts of the manuscript; Drs. B. D. Bell, D. G. Dawson, C. A. Fleming, J. E. C. Flux, J. A. Gibb and M. R. Rudge and Messrs I. G. Crook and R. H. Taylor. The illustrations of the frog and lizards were kindly drawn by Dr. A. E. Bell and those of the birds by Mr. R. E. Brockie. The permission of the New Zealand Herpetological Society to reprint the drawing of *Leiopelma* is acknowledged.

Addenda

A revision of *Leiolopisma* and its relatives (Greer, A. E. 1974. *Aust. J. Zool., Suppl. Ser.*, No. 31, 1–67) places all N.Z. skinks in the genus *Leiolopisma*. By restricting *Leiolopisma* to 33 species distributed from Mauritius through southern Australia to N.Z., New Caledonia and the New Hebrides, a southerly relict distribution is implied.

REFERENCES

Archey, G. 1941. The Moa – A study of the Dinornithiformes. *Bull. Auck. Inst. Mus.* 1: 1–145.

Atkinson, I. A. E. & Bell, B. D. 1973. 'Offshore and Outlying Islands'. pp. 372–392 in 'The Natural History of New Zealand. An Ecological Survey'. Ed. G. R. Williams. A. H. and A. W. Reed, Wellington.

Austin, F. J., Bull, P. C. & Chaudry, M. A. 1973. A Poxvirus isolated from Silvereyes (*Zosterops lateralis*) from Lower Hutt, New Zealand. *J. Wildlife Diseases* 9: 111–114.

Barwick, R. E. 1959. The Life History of the Common New Zealand Skink, *Leiolopisma zelandica* (Gray, 1843). *Trans. Roy. Soc. N.Z.* 86: 331–380.

Bellairs, A. 1969. 'The Life of Reptiles'. 590 pp. Weidenfeld and Nicholson, London.

Bettesworth, D. J. 1972. *Rattus exulans* on Red Mercury Island. *Tane* 18: 117–118.

Bettesworth, D. J. & Anderson, G. R. V. 1972. Diet of *Rattus norvegicus* on Whale Island, Bay of Plenty, New Zealand. *Tane* 18: 189–195.

Blackburn, A. 1965. Breeding of the North Island Fantail. *Notornis* 12: 127–137.

Boyd, M. M. M. 1942. The Oviduct, Foetal Membranes, and Placentation in *Hoplodactylus maculatus* Gray. *Proc. Zool. Soc. Lond. Ser. A.* 112: 65–104.

Cody, M. L. 1966. A General Theory of Clutch Size. *Evolution* 20: 174–184.

Crook, I. G. 1973. The Tuatara, *Sphenodon punctatus* Gray, on Islands with and without Populations of the Polynesian Rat, *Rattus exulans* (Peale). *Proc. N.Z. Ecol. Soc.* 20: 115–120.

Crook, I. G., Atkinson, I. A. E. & Bell, B. D. 1971. Habitats of the Frog *Leiopelma*

hamiltoni McCulloch on Stephens and Maud Islands, Cook Strait. *N.Z. Jl. Sci.* 14: 1082–1093.
DARLINGTON, P. J. 1957. 'Zoogeography: The Geographical Distribution of Animals'. 675 pp. John Wiley and Sons, New York.
DWYER, P. D. 1960. New Zealand Bats. *Tuatara* 8: 61–71.
DWYER, P. D. 1962. Studies on the two New Zealand Bats. *Zool. Publ. Victoria Univ. Wellington* 28: 1–28.
DWYER, P. D. 1970. Size variation in the New Zealand Short-tailed Bat. *Trans. Roy. Soc. N.Z., Biol. Sci.*, 12: 239–243.
FALLA, R. A. 1953. The Australian Element in the Avifauna of New Zealand. *Emu* 53: 36–46.
FALLA, R. A. 1955. New Zealand Bird Life Past and Present: Cawthron Lecture No. 291-14, Cawthron Institute of Scientific Research, Nelson.
FALLA, R. A., SIBSON, R. B. & TURBOTT, E. G. 1966. 'A Field Guide to the Birds of New Zealand'. 254 pp. Collins, London.
FAWCETT, J. D., & SMITH, H. M. 1971. Liopelmatina Mivart, 1869 (Amphibia: Salientia): Proposed Emendation under the Plenary Powers to Leiopelmatidae, Z.N. (S.) 1936. *Bull. Zool. Nomencl.* 28: 50–52.
FLEMING, C. A. 1950. New Zealand Flycatchers of the Genus *Petroica* Swainson (Aves), Part II. *Trans. Roy. Soc. N.Z.* 78: 127–160.
FLEMING, C. A. 1962a. New Zealand Biogeography – a Paleontologist's Approach. *Tuatara* 10: 53–108.
FLEMING, C. A. 1962b. History of the New Zealand Land Bird Fauna. *Notornis* 9: 270–274.
FLEMING, C. A. 1962c. The Extinction of Moas and other Animals during the Holocene Period. *Notornis* 10: 113–117.
FLEMING, C. A. 1969. Rats and Moa Extinction. *Notornis* 16: 210–211.
GASKIN, D. E. 1968. The New Zealand Cetacea. *N.Z. Marine Dept. Fisheries Res. Bull.* 1 (New Series): 1–92.
GIBB, J. A. 1961. Ecology of the Birds of Kaingaroa Forest. *Proc. N.Z. Ecol. Soc.* 8: 29–38.
GIBB, J. A. & FLUX, J. E. C. 1973. 'Mammals'. Pp. 334–371 in 'The Natural History of New Zealand. An Ecological Survey'. Ed. G. R. WILLIAMS. A. H. and A. W. Reed, Wellington.
GREER, A. E. 1970. A Subfamilial Classification of Scincid Lizards. *Bull. Mus. Comp. Zool.* 139: 151–183.
DE HAMEL, F. A. & MCINNES, H. M. 1971. Lizards as Vectors in Human Salmonellosis. *J. Hyg. Camb.* 69: 247–253.
HARDY, G. S. 1972. A Review of the Parasites of New Zealand Reptiles. *Tuatara* 19: 165–167.
JACKSON, R. W. 1971. Birds in Exotic Forests in New Zealand. *N.Z. Jl. Forestry* 16: 61–68.
KEAR, J. & SCARLETT, R. J. 1970. The Auckland Islands Merganser. *Wildfowl* 21: 78–86.
KING, J. E. 1969. The Identity of the Fur Seals of Australia. *Aust. Jl. Zool.* 17: 841–853.
KINSKY, F. C. 1970. 'Annotated Checklist of the Birds of New Zealand'. 96 pp. Ed. F. C. KINSKY, Convener Checklist Committee, Ornithological Society of New Zealand. A. H. and A. W. Reed, Wellington.
KLUGE, A. G. 1967a. Higher Taxonomic Categories of Gekkonid Lizards and their Evolution. *Bull. Amer. Mus. Nat. Hist.* 135: 1–60.
KLUGE, A. G. 1967b. Systematics, Phylogeny, and Zoogeography of the Lizard Genus *Diplodactylus* Gray (Gekkonidae). *Aust. Jl. Zool.* 15: 1007–1108.
LACK, D. 1970. The Endemic Ducks of Remote Islands. *Wildfowl* 21: 5–10.
MAYR, E. & AMADON, D. 1951. A Classification of Recent Birds. *Amer. Mus. Novitates* 1496: 1–42.

McCann, C. 1955. The Lizards of New Zealand, Gekkonidae and Scincidae. *Dom. Mus. Bull.* 17: 1–127.
McCann, C. 1961. The Introduced Frogs of New Zealand. *Tuatara* 8: 107–120.
McCann, C. 1966. The Marine Turtles and Snakes occurring in New Zealand. *Rec. Dom. Mus.* 5: 201–215.
McDowall, R. M. 1969. Extinction and Endemism in New Zealand Land Birds. *Tuatara* 17: 1–12.
Medway, D. 1971. Subfossil Avian Remains from the Awakino-Mahoenui Area. *Notornis* 18: 218–219.
Oliver, W. R. B. 1949. The Moas of New Zealand and Australia. *Dom. Mus. Bull.* 15: 1–204.
Oliver, W. R. B. 1955. 'New Zealand Birds'. 661 pp. A. H. & A. W. Reed, Wellington.
O'Malley, F. M. 1971. The Karyotypes of Three Species of New Zealand Lizards (Scincidae): A Preliminary Study. Unpubl. B.Sc. Thesis, Victoria Univ. Wellington.
Raven, P. H. & Axelrod, D. I. 1972. Plate Tectonics and Australasian Paleobiogeography. *Science* 176: 1379–1386.
Ridpath, M. G. & Moreau, R. E. 1966. The Birds of Tasmania: Ecology and Evolution. *Ibis* 108: 348–393.
Robb, J. 1970. A New Skink of the Genus *Leiolopisma* from New Zealand. *Proc. Kon. Ned. Akad. Wetensch. Ser. C. Biol. Med. Sci.* 73: 228–229.
Robb, J. 1973. 'Reptiles and Amphibia'. pp. 285–303 in 'The Natural History of New Zealand. An Ecological Survey'. Ed. G. R. Williams, A. H. & A. W. Reed, Wellington.
Scarlett, R. 1969. On the Alleged Queensland Moa, *Dinornis queenslandiae* De Vis. *Mem. Queensland Mus.* 15: 207–212.
Stephenson, E. M. 1960. The Skeletal Characters of *Leiopelma hamiltoni* McCulloch, with Particular Reference to the Effects of Heterochrony on the Genus. *Trans. Roy. Soc. N.Z.* 88: 473–488.
Stephenson, E. M. 1961. New Zealand Native Frogs. *Tuatara* 8: 99–106.
Stephenson, E. M. & Stephenson, N. G. 1957. Field Observations on the New Zealand Frog, *Leiopelma* Fitzinger. *Trans. Roy. Soc. N.Z.* 84: 867–882.
Thomson, G. M. 1922. 'The Naturalisation of Animals and Plants in New Zealand'. 607 pp. Cambridge University Press, Cambridge.
Turbott, E. G. 1952. 'Seals of the Southern Ocean'. pp. 195–215 in 'The Antarctic Today'. Ed. F. A. Simpson. A. H. & A. W. Reed, Wellington.
Turbott, E. G. 1961. The Interaction of Native and Introduced Birds in New Zealand. *Proc. N.Z. Ecol. Soc.* 8: 62–66.
Watson, J. S. 1960. The New Zealand 'Otter'. *Rec. Cant. Mus.* 7: 175–183.
Wermuth, H. 1965. Liste der rezenten Amphibien und Reptilien: Gekkonidae, Pygopodidae, Xantusiidae. *Das Tierreich* 80: 1–246.
Werner, Y. L. 1972. Observations on eggs of Eublepharid Lizards, with comments on the Evolution of the Gekkonoidea. *Zool. Meded. Leiden.* 47: 211–224.
Whitaker, A. H. 1973. Lizard Populations on Islands With and Without Polynesian Rats, *Rattus exulans* (Peale). *Proc. N.Z. Ecol. Soc.* 20: 121–130.
Williams, G. R. 1953. The Dispersal from New Zealand and Australia of some Introduced European Passerines. *Ibis* 95: 676–692.
Williams, G. R. 1962. Extinction and the Land and Freshwater-inhabiting Birds of New Zealand. *Notornis* 10: 15–32.
Wodzicki, K. A. 1950. Introduced Mammals of New Zealand. *N.Z. Dept. Sci. Ind. Res. Bull.* 98: 1–255.
Wodzicki, K. 1961. Ecology and Management of Introduced Ungulates in New Zealand. *La Terre et la Vie* 15: 130–157.
Wodzicki, K. 1967. The Gannets at Cape Kidnappers. 2. Dispersal and Movements. *Trans. Roy. Soc. N.Z., Zool.*, 9: 17–31.
Wodzicki, K. & Flux, J. E. C. 1967. Guide to Introduced Wallabies in New Zealand. *Tuatara* 15: 47–59.

VI. THE FRESHWATER FISHES

by

R. M. McDOWALL & A. H. WHITAKER

Introduction

New Zealand has a small freshwater fish fauna, by any standard of comparison. A precise enumeration is not yet possible as some taxonomic problems remain to be resolved, but there are about 30 native species. Eight families are represented, none of which is endemic. There are 10 genera of which only two are endemic, while all except four of 27 species formally recognised in recent studies are endemic. Thus while endemism is high at the species level, it is very low at higher taxonomic levels, suggesting that the fauna is probably not of great age.

In a discussion of the ecology and zoogeography of these fishes it is important to note that the families represented in New Zealand are neither primary nor secondary freshwater fishes, but all are peripheral fishes, so-called because of the ease with which they move between fresh and salt water (Darlington, 1957).

In the pages that follow we have considered the distribution, relationships and ecology of each family, and these family discussions are then synthesized in a biogeography of the fauna.

Family geotriidae

The family Geotriidae is one of two southern temperate families of lampreys. (The other family Mordaciidae occurs in Australia and South America.) These two families, and the family Petromyzonidae in the northern temperate, make up the agnathan order Petromyzoniformes. All three families consist primarily of species that are anadromous and parasitic, although there are a few species that are non-anadromous and/or non-parasitic (Hubbs & Potter, 1971). The family Geotriidae contains one species *Geotria australis* Gray (Fig. 1), found in Australia, Tasmania, New Zealand, and southern South America (Patagonian Argentina and Chile). In New Zealand *G. australis* is found mostly in coastal streams and rivers. It is anadromous, migrating upstream from the sea to spawn, and its inland distribution probably is limited by its inability to penetrate the swiftly flowing and often turbulent rivers characteristic of New Zealand's topography. For this reason, *G. australis* is found mainly in the lowlands. However, *G. australis* adults are found in some small, swiftly-flowing rocky streams.

Little is known of its ecology. Although once common, adult lampreys

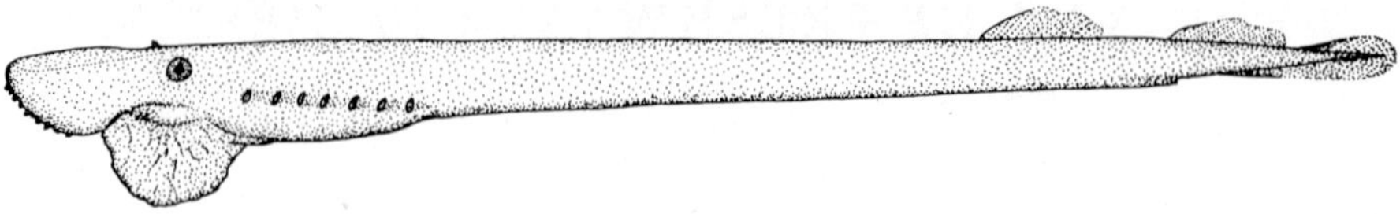

Fig. 1. Geotria australis Gray; pouched adult male.

are not often seen nowadays, possibly because they migrate mostly during the night (Maskell, 1929). The upstream pre-spawning migration of adult lampreys seems to occur in the spring, but at this time they are not ready to spawn. It appears that they spend the summer in fresh water, during which time the gonads develop at the expense of body tissues and fat reserves and they probably do not feed once they leave the sea. Spawning has not been observed and nothing is known of the spawning site, habits or behaviour. The larvae are typical lamprey ammocoetes, which inhabit muddy creeks and drains, or silty backwaters of larger rivers and streams, and, although they are not often found, they may occur in dense concentrations. The ammocoetes live in burrows in the mud and are believed to feed by filtering micro-organisms from the water. When about 100 mm long after an undetermined growth period, the ammocoetes metamorphose into the macrophthalmic stage, which is a miniature of the adult. This metamorphosis involves the development of eyes, changes in the fins and oral sucker, and a change in colour from drab brown to bright silver with blue markings. These changes occur during the autumn and winter, and the macrophthalmic migrates down to the sea, where growth to the adult occurs. There are very few reports of lampreys from the sea, and apart from the fact that it is a parasite on larger marine fishes, nothing is known about its marine life.

Family retropinnidae

The family Retropinnidae is a small family found only in south-eastern Australia, Tasmania, and New Zealand. There are two genera: *Retropinna*, which is shared with the Australian region, and *Stokellia*, which is endemic to New Zealand. The taxonomy of *Retropinna* is confused and unstable in both Australia and New Zealand; there are probably about three species in Australia, of which some are partly marine. In New Zealand there is one anadromous species *Retropinna retropinna* (Richardson) (fig. 2). (Another commonly listed anadromous species – *R. osmeroides* Hector – is a growth stage of *R. retropinna* and thus a junior synonym, McDowall, 1972b.) *R. retropinna* has become landlocked in many lakes, and four lake species have been described, viz. *R. lacustris* Stokell, *R. obtusirostris* Stokell, *R. chathamensis* Stokell and *R. abbreviata* McDowall. The taxonomic status of these lake species is in some doubt; Woods (1968) treated them all as lake races of *R. retropinna*.

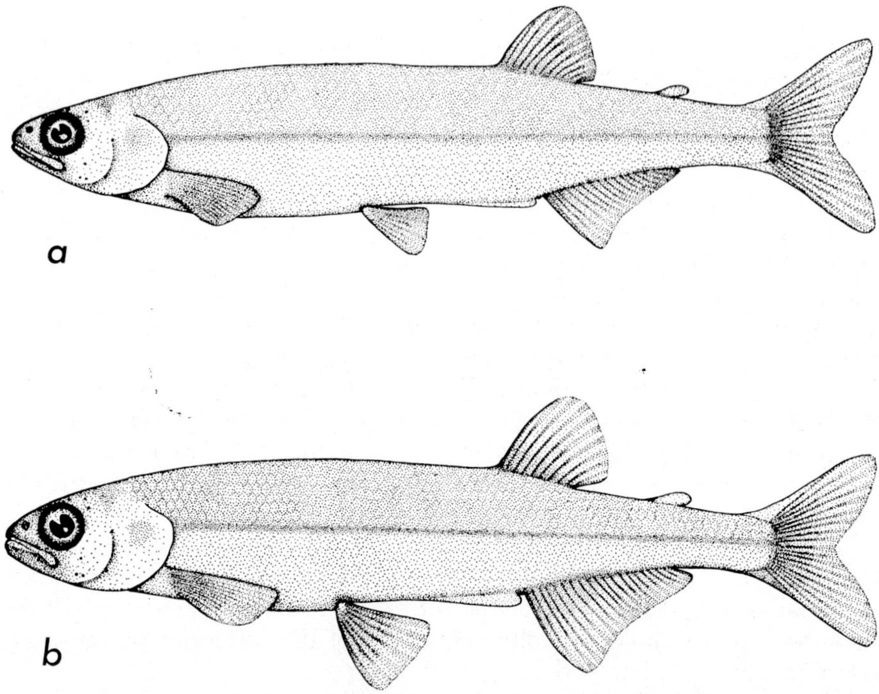

Fig. 2. Retropinna retropinna (Richardson); a. female; b. male (note larger fins and deeper abdominal keel of male).

R. retropinna is found in coastal seas and in rivers, streams and lakes it can reach from the sea. Its inability to ascend swiftly flowing, turbulent rivers has resulted in little inland penetration. Nevertheless, *R. retropinna* has been found great distances inland in the more placid Waikato and Wanganui Rivers. *R. retropinna* has not been recorded from inshore islands, but occurs in rivers and lakes of the Chatham Islands.

The ecology of *R. retropinna* is largely unstudied, but what is known of the life history pattern shows that little of the life cycle is spent in fresh water. Large shoals of adults migrate from the sea into estuaries and lowland rivers during spring and summer. At migration, the spawning shoals comprise mainly ripe two years olds, with a few immature two year olds and a very few ripe three year olds. Ripe fishes of both sexes develop small nodules or nuptial tubercles on the body, and the sexes can be distinguished by the presence of these tubercles on the fins of males only; in addition the fins of males are noticeably larger than those of females (fig. 2). Spawning shoals are usually found in slow-flowing lowland waters, backwaters, lagoons and lakes. Spawning is undescribed, but probably occurs either in tidal estuaries or a little further upstream in fresh water. Like many salmoniform fishes, adult *Retropinna* probably die after spawning. Several thousand small eggs about 0.5 mm in diameter are laid, probably over sandy substrates in gently flowing or still waters. Larval or postlarval *R. retropinna* are not known from fresh water but juveniles are sometimes caught in the sea, near the shore, in bays or harbours. So it seems probable that freshly hatched larvae are carried out to sea, where they grow to maturity.

Lacustrine populations of *Retropinna* are found in many lakes that vary from warm, eutrophic, lowland coastal dune lakes to cold, oligotrophic alpine and sub-alpine glacial lakes. The fish spend their entire lives in relatively still waters of lakes, rarely entering lake tributaries. The eggs are demersal and are laid over silty beds at river or stream mouths or on the lake bed in sandy bays; the larvae are planktonic in the open waters of lakes. Lake *Retropinna* are usually dwarfed, reaching about half the size of diadromous fishes; JOLLY (1967) considered that they take two years to reach maturity. Lake *Retropinna* occur naturally in many lakes and were introduced into others to provide a forage fish for introduced salmonid game fishes. These retropinnid introductions were successful in several lakes in the Rotorua area and in Lake Taupo, but otherwise they have generally failed. Although there is some evidence that the smelt are valuable as a food resource for the introduced salmonids (SMITH, 1959), their role in the ecology of the lakes, and of the salmonid populations, remains poorly understood.

The genus *Stokellia* is monotypic, comprising only *S. anisodon* (Stokell), which is endemic to New Zealand. *S. anisodon* inhabits coastal seas, and lower reaches of rivers in Canterbury, Otago and Southland and is the only diadromous fish species in the fauna that is not found all around New Zealand. Reasons for this restricted range are not understood.

The life history of *S. anisodon* was studied by MCMILLAN (1961) and proved to be similar to that of *R. retropinna*. The sexes are dimorphic in the same characters as those reported for *R. retropinna*, and in spring and summer shoals of ripe adults move from the sea through estuaries to spawn in fresh water. The eggs are small and are laid in gently flowing water over silty bottoms. They hatch within three weeks, depending on temperatures, and the newly-hatched larvae, about 5 mm long, are carried out to sea. After spawning the adults die. MCMILLAN (1961) commented on the importance of this species to the ecology of river estuaries, pointing out that the migrating and spawning fishes form an important food resource for gulls and terns that sometimes nest nearby, and for predatory estuarine and freshwater fishes.

FAMILY PROTOTROCTIDAE

The family Prototroctidae contains only the genus *Prototroctes*, with one species in south-eastern Australia and Tasmania, and another in New Zealand. The Australian species is now rare, and the New Zealand species *P. oxyrhynchus* Günther has not been reported authoritatively since the 1920's and is considered to be extinct.

Little is known about *P. oxyrhynchus*, and published information is meagre and inconsistent. *P. oxyrhynchus* was formerly present around the whole of New Zealand in lowland streams and rivers. It was a shoaling fish of rivers and streams, mostly with rocky or gravel rather than muddy beds. What little is known of the life history suggests that the fish migrated upstream from the sea to spawn, and that they occurred in fresh water during summer, autumn, and early winter. Where they spawned, the nature of spawning, eggs, development, etc. remain unknown, but it seems that the larvae went to sea, as fish smaller than about 125 mm long were never reported from fresh water.

Family Galaxiidae

The family Galaxiidae contains 13 New Zealand species (McDowall, 1970) – about half the native freshwater fish fauna. In addition to New Zealand, the family occurs in Australia and Tasmania (15–20 species), New Caledonia (one species), Chile, Argentina, Tierra del Fuego and the Falkland Islands (four species) and South Africa (one species). The record of *Galaxias* from India (Day, 1888) is best discounted (McDowall, 1973c).

Galaxiids in New Zealand are included in two very distinct genera: *Galaxias*, with ten species, and *Neochanna*, an endemic genus with three species. The ten species of *Galaxias* fall easily and naturally into a series of species groups which are discussed below. These groups vary in their compactness, but each group brings together species that are more closely related to each other than to species in other groups (McDowall, 1970).

Three very similar, large, stout bodied species – *G. fasciatus* Gray (Fig. 3a), *G. argenteus* (Gmelin) and *G. postvectis* Clarke – form a morphologically compact group in which the species are separable most easily by colouration. They are all diadromous, and all are found in lowland habitats around New Zealand (*G. fasciatus* and *G. argenteus* occur on the Chatham Islands), having almost identical ranges and often being found together in a stream or creek. They all spawn in autumn, have numerous (several thousand) small eggs, and as far as we know, the larvae go to sea soon after hatching where they spend the winter. In spring, they migrate into fresh water in huge mixed-species shoals, as highly characteristic slender, transparent juveniles which form a part of a small seasonal 'white bait' fishery. The migrations relate closely to flooding in the rivers, and movements of the juveniles of each species can be related to characteristics of the water into which they migrate. Factors like pH, water colour and turbidity, oxygen and temperature appear to be important (McDowall, unpublished data). Adults of *G. argenteus* are

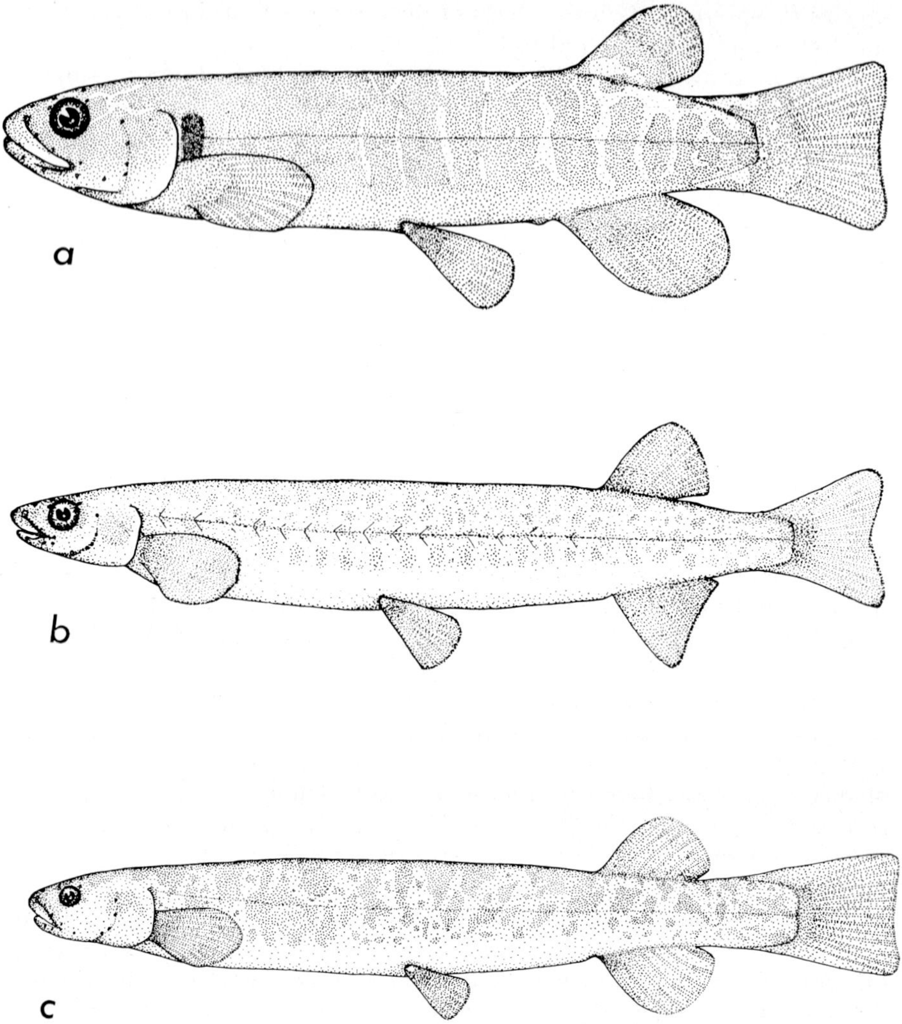

Fig. 3. Galaxias species; a. *Galaxias fasciatus* Gray; one of the stout bodied species; b. *Galaxias maculatus* (Jenyns); a slender, shoaling whitebait species; c. *Galaxias divergens* Stokell; one of the slender alpine species.

found in swampy streams, flax swamps and some lakes, those of *G. fasciatus* in podocarp forest bogs and pools of small forest streams, while *G. postvectis* seems also to be characteristic of these small streams.

A second species group contains two small, slender shoaling species (*G. maculatus* (Jenyns) (fig. 3b) and *G. gracilis* McDowall) which are distinctive in having forked caudal fins (in other species the fin is truncated to rounded). *G. maculatus* is one of the more distinctive New Zealand galaxiids, and except for *G. gracilis* (which is a land-locked derivative) probably stands far apart from other New Zealand galaxiids. *G. maculatus* is found in southeastern Australia, Tasmania, New Zealand and the Chatham Islands, Chile, Argentina and Tierra del Fuego, and the Falkland Islands (*G. maculatus* and *Geotria australis* are probably amongst the most widely dispersed freshwater fishes in the world). Stocks of *Galaxias maculatus* in all these widely separated areas exhibit no signs of evolving separate subspecies or even races (McDowall, 1972a).

G. maculatus has a unique life history. Like those of the stout bodied species discussed above, larvae of *G. maculatus* are found in the sea. The juveniles migrate into fresh water with those of *G. fasciatus* etc., and growth to maturity occurs in fresh water in open, lowland streams and rivers, usually beyond tidal influence and often many miles upstream. Breeding occurs in autumn when shoals of ripe fish migrate down stream into tidal estuaries at about the full and new moons. How the fish are able to respond to the lunar cycle is unknown, but the migrating fish reach the estuaries during the spring tides, and shoals of ripe fish move up onto grassy estuarine flats covered by water only at the high spring tides. The eggs are deposited amongst the grasses and are eventually washed down amongst the bases of the grass clumps where humidity remains high and temperatures low and stable. The eggs are stranded out of the water during the subsequent two weeks and usually hatch when the next spring tide covers the grasses. Although actual spawning is related directly to lunar-tidal stimuli, the spawning migration is one of very few well authenticated examples of response of an animal living outside the marine environment to lunar stimuli (McDowall, 1969a).

G. gracilis is a polytopic species derived from *G. maculatus* by land-locking in a series of small lakes on or near the northern head of the Kaipara Harbour, Northland. Landlocking has resulted in elimination of the marine juvenile from the life cycle and a simplified lake-limited life cycle has evolved. *G. gracilis* is a dwarf species reaching only about 55 mm length (cf. 85–150 mm in *G. maculatus*) and differs from *G. maculatus* in a striking reduction in vertebral number (47–53, mean about 49 in *G. gracilis;* 57–66, mean about 60 in diadromous *G. maculatus;* McDowall, 1972a). Little is known of its biology.

A third species group comprises a further diadromous species, *G. brevipinnis* Günther (fig. 4) and what appears to be an entirely freshwater derivative, *G. vulgaris* Stokell. *G. brevipinnis* is found in Tasmania, New Zealand, the Chatham Islands, and the Auckland and Campbell Islands to the south of New Zealand. It is found around the whole of New Zealand. Although its young migrate upstream from the sea, adult *G. brevipinnis* are found great distances inland and the species shows remarkable capacity for climbing waterfalls and cascading torrents. *G. brevipinnis* shares many characteristics of the stout-bodied species, having a similar life history pattern, with the same type of marine-freshwater, migratory

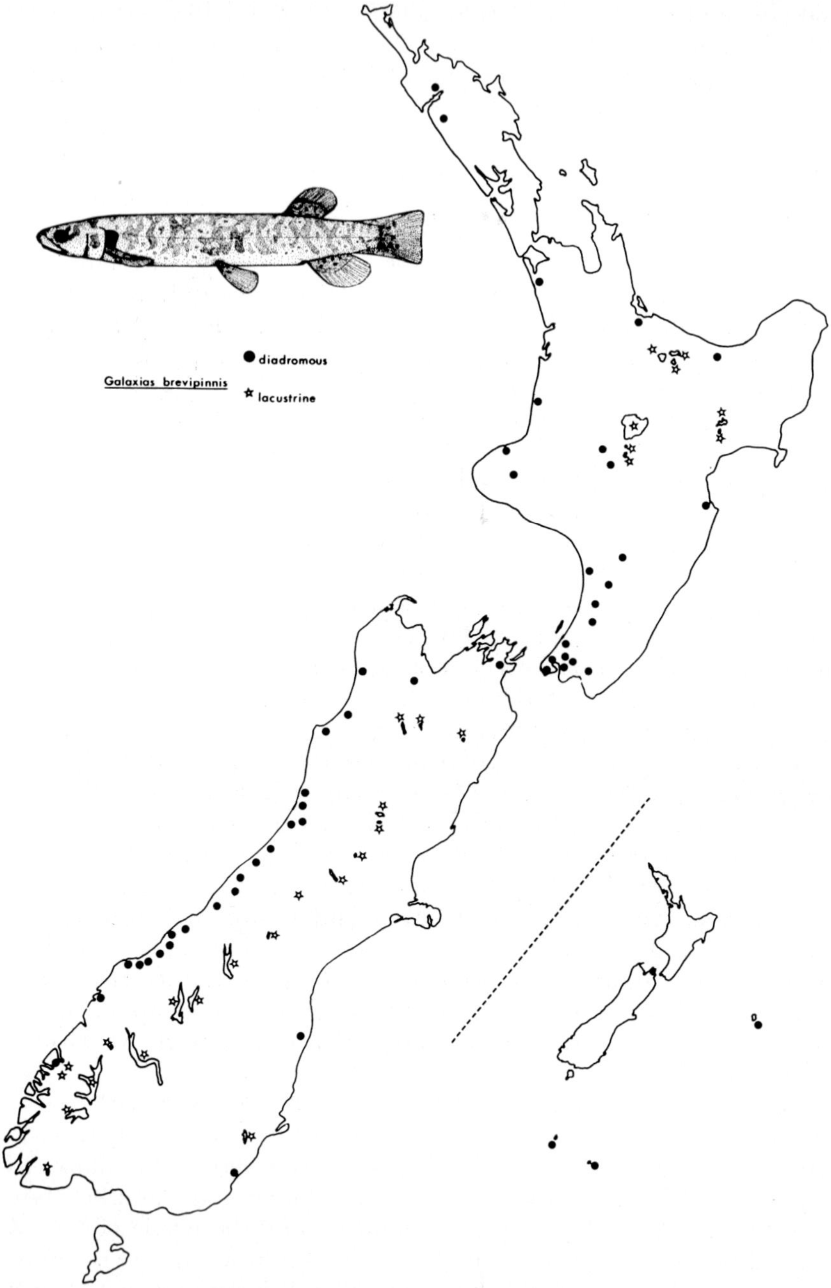

Fig. 4. Distribution of a diadromous galaxiid, *Galaxias brevipinnis* Günther (note broad distribution in lowland, coastal areas, and on Chatham, Auckland and Campbell Islands).

juvenile stage; but it is much more slender in build. It is broadly sympatric with the stout-bodied species, but occupies swiftly flowing, bouldery streams whereas the stout-bodied species are more often pool-dwellers. Like the stout-bodied species, *G. brevipinnis* is confined largely to streams that have a good forest canopy.

In addition to diadromous populations, *G. brevipinnis* has many secondarily lacustrine populations both in lakes of the volcanic plateau of the central North Island and in glacial, alpine and subalpine lakes of the Southern Alps of the South Island (fig. 4). Lacustrine populations have adults that live in lake tributaries, and larval and juvenile stages that live in the lakes. There is a distinct migration of juveniles from the lakes to the tributaries, comparable with migrations in diadromous populations from the sea to rivers. Separate taxonomic status has been ascribed to some of these lake populations, but this seems unwarranted (McDowall, 1970).

G. vulgaris is very similar in form and appearance to *G. brevipinnis* and it differs from it primarily in its life history pattern, in that the whole life is spent in fresh water. It is found in gravelly or bouldery inland streams and rivers, usually in moderately swiftly flowing waters. The rather fewer, larger eggs are laid in swift rapids near the customary adult habitat, and when the larvae hatch they move into quietly flowing marginal waters where they may be found in small shoals (Benzie, 1968). As they grow, they become benthic and secretive in habit. *G. vulgaris* is known almost exclusively from eastern flanks of the Southern Alps, from North Canterbury to Southland, but one population is known from the western side of the alps in the upper Buller River system.

A fourth species group comprises three small, slender, alpine and subalpine species (*G. divergens* Stokell, *G. paucispondylus* Stokell, *G. prognathus* Stokell), all endemic to New Zealand. Characters like a very long, slender caudal peduncle, and a small head with broad, shallow gape, bring these species together in a very compact group. They are confined to moderately to swiftly flowing, cold, fresh waters, and they do not have the obvious juvenile stage characteristic of diadromous species. The group is widespread in the southern half of the North Island, the north-west of the South Island, and extending over into the eastern Southern Alps. The three species are sympatric in one restricted area of their ranges, all three having been found in the upper tributaries of the Buller River near Maruia.

G. divergens (fig. 3c) is known from one locality in the Urewera Mountains, Bay of Plenty, also Hawkes Bay, Wairarapa, Wellington, and throughout the north-west of the South Island from Marlborough to about the Taramakau River in Westland. *G. paucispondylus* is found in tributaries of the Maruia River in the upper Buller River system, and in several rivers along the eastern flanks of the Southern Alps, from Canterbury to Southland. *G. prognathus* is so far known only from tributaries of

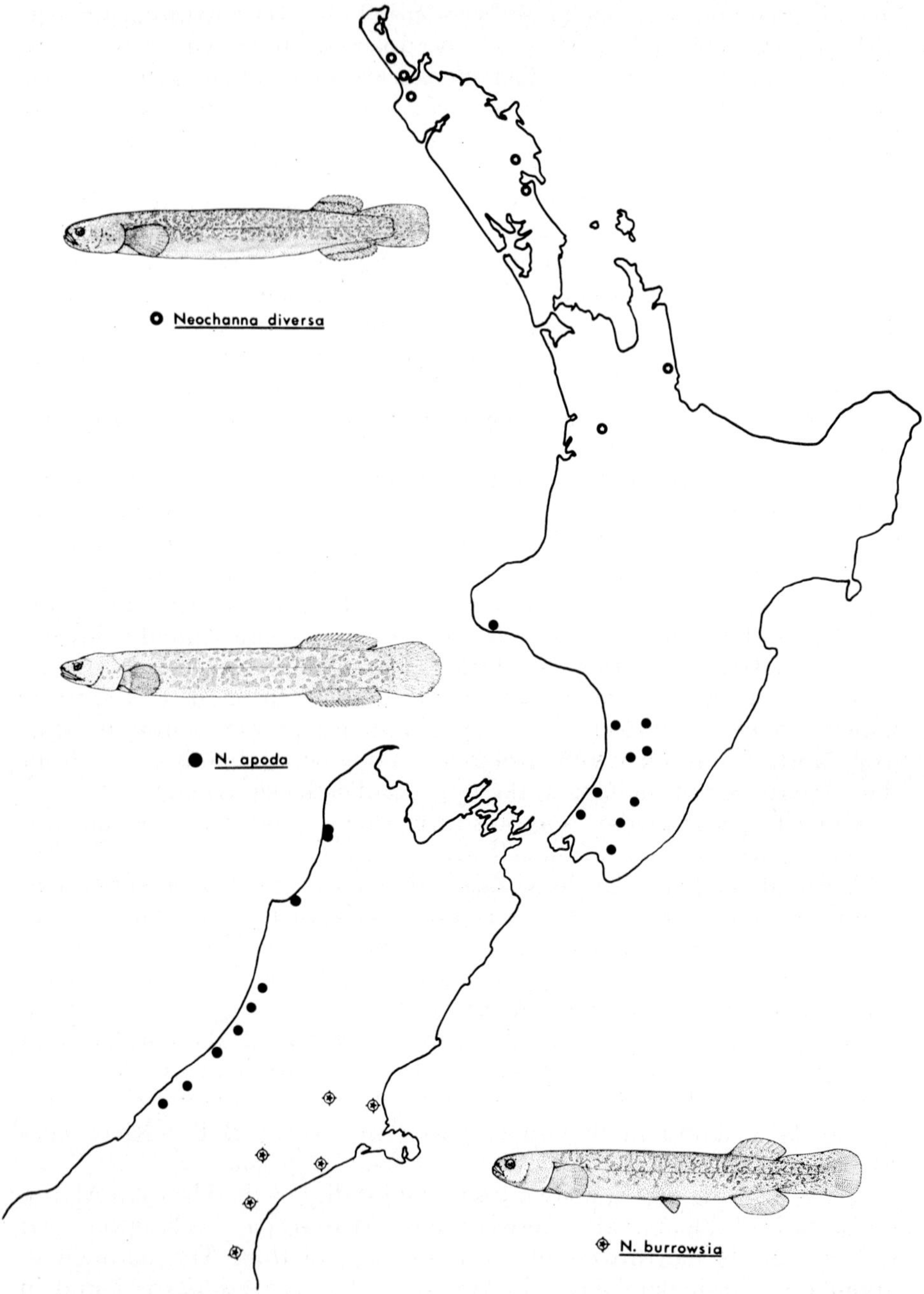

Fig. 5. Distribution of three related non-diadromous galaxiids, *Neochanna diversa* Stokell, *N. apoda* Günther, and *N. burrowsia* (Phillipps) (note restricted ranges of each species, and allopatry of all three).

the Maruia River and alpine tributaries of the Rakaia River in central Canterbury. These species spawn in spring and summer, have very few, large eggs (a few hundred up to 2 mm in diameter) and probably spend the entire life history in or near typical adult habitats. Reproduction of *G. prognathus* and *G. paucispondylus* is unstudied. HOPKINS (1971) studied the life history of *G. divergens* and although he found newly hatched larvae in large numbers near typical adult habitats, only two unhatched eggs were discovered, and the spawning site remains unknown.

Three New Zealand galaxiids belong in the highly distinctive endemic genus *Neochanna*. It is characterised by elongate, almost eel-like form, with long, low, fleshy, dorsal and anal fins that are almost confluent with rounded caudal fin, small eyes, and elongate tubular nostrils that project forwards over the upper lip. Two species, *N. apoda* Günther and *N. diversa* Stokell lack the pelvic fins and girdle, and in *N. burrowsia* (Phillipps) they are much reduced (fig. 5). *N. diversa* is found from Northland, to Waikato, *N. apoda* is widespread in the southern half of the North Island from Taranaki southwards, and also along the west coast of the South Island to about as far south as Whataroa, and *N. burrowsia* occurs in Canterbury, east of the Southern Alps (fig. 5). *Neochanna* species, collectively known as mudfishes, are found in open and forest-covered swamps, bogs, and swampy creeks and drains. They are characteristic of water bodies that dry up during periods of drought, and when this happens, the fish may be found aestivating in pockets in the mud, under logs or leaf litter, or in natural moist depressions or holes in the ground. They seem able to endure prolonged periods without water. Little is known of their biology, but it is likely that all three species spawn during much of the year. No spawning migrations are suspected, and the moderately large eggs are probably laid in or near typical adult habitats, where the young may also be found. ELDON (1971) discussed the possibility that the eggs are laid out of water.

FAMILY ANGUILLIDAE

The cosmopolitan freshwater eel family Anguillidae has two New Zealand representatives. *Anguilla dieffenbachi* Gray is endemic to New Zealand, and is found throughout New Zealand from river estuaries great distances inland. Records of this species from the Auckland Islands are probably erroneous, although it is known from the Chatham Islands. *A. dieffenbachi* is the eel of rivers and streams in New Zealand, being much less common in lagoons and lakes. It is remarkable for the great size it reaches – there are records of eels weighing up to 50 kg, which may be mythical – but it certainly reaches more than 20 kg and a length of nearly 2 m. *Anguilla australis* Richardson is found in Australia, and some islands to the north of New Zealand – Lord Howe, Norfolk, New Caledonia, Fiji, and probably others. In New Zealand it is more lowland in range than *A.*

dieffenbachi, but is found around the whole of New Zealand and on the Chatham Islands. It is commonest in lagoons, coastal lakes, swamps, and sluggish streams and rivers near the sea. Two subspecies of *A. australis* have been recognised (SCHMIDT, 1928; GRIFFIN, 1936), but the taxonomic and biological significance of these subspecies is not clear.

Life histories of the two species of eel are essentially similar. Breeding grounds of western Pacific eels are not yet known, although it is thought that New Zealand eels may spawn in the tropical or sub-tropical Pacific somewhere northeast of Australia (CASTLE, 1963). The leptocephali reach New Zealand coasts during late winter and early spring, when they metamorphose into the glass eel, 60–70 mm long. Glass eels migrate from the sea into river estuaries and settle on the bottom, becoming pigmented, and some time later they move upstream into the rivers penetrating even the smallest tributaries. They may take many years to reach maturity – CAIRNS (1941) reported *A. dieffenbachi* up to 19 years old. During spring and summer before their spawning migrations, adult eels undergo a second metamorphosis; they lose the bulbousness of lips and head, the eyes become enlarged and the colour pattern changes. In autumn and early winter, the adults move downstream into the sea to begin their long migration to the spawning grounds.

FAMILY PLEURONECTIDAE

New Zealand has a freshwater flounder, *Rhombosolea retiaria* Hutton, which is endemic. Although it is often omitted from lists and discussions of freshwater fishes it is nevertheless as much a freshwater species as many others normally included. *R. retiaria* is commonly found in the sea, but may also occur many miles upstream in quite swiftly flowing, gravelly rivers, and it is clearly able to spend long periods in freshwater. It is found around the whole of New Zealand.

The biology of *R. retiaria* is wholly unstudied. It seems that either the eggs are laid in fresh water and are carried out to sea, or the adults migrate downstream into the sea to spawn. No larval or juvenile stages are known from fresh water.

FAMILY ELEOTRIDAE

The family Eleotridae is a widespread Indo-Pacific marine and freshwater family which reaches its southernmost limits in New Zealand. The taxonomy of New Zealand representatives has recently been reviewed (MCDOWALL, 1975). One genus *Gobiomorphus* Gill is currently used although formerly two species were included in *Philypnodon* Bleeker. All species are endemic to New Zealand, although species of *Gobiomorphus* are found in south-eastern Australia.

Six species of *Gobiomorphus* are recognised, viz. *G. huttoni* (Ogilby),

G. gobioides (Valenciennes), *G. basalis* (Gray), *G. cotidianus* McDowall, *G. hubbsi* (Stokell) and *G. breviceps* (Stokell). This taxonomy (after McDowall, 1975) differs from previous arrangements (e.g. Stokell, 1941, 1962) in several features. The name *G. basalis* is applied to a species not recognised in any recent studies and *G. cotidianus* is used for the so-called common bully, previously referred to as *G. basalis* (Gray), by Stokell (1941) and others. *G. alpinus* Stokell is regarded as a junior synonym of *G. basalis*.

G. gobioides is found only in the lower reaches of rivers or in estuaries; it is rarely found in large numbers but seems to favour bouldery streams that drop quickly into the sea. Breeding is unknown.

G. cotidianus is common in quieter reaches of rivers and streams, either in the open, or beneath marginal cover, and on open beds of lakes and lagoons. Its biology is little studied but its reproduction resembles that of *G. huttoni*, discussed below. Diadromous populations of *G. cotidianus* have marine larvae, while larvae of lacustrine populations may be found in the lake plankton.

The biology of *G. huttoni* is moderately well understood (McDowall, 1964, 1965a, b). Larval life is marine, and juveniles, about 20–30 mm long, migrate from the sea into fresh water during spring and summer. They penetrate considerable distances inland, although falls and large rapids provide effective barriers to migration. In fresh water, *G. huttoni* is found in streams under rocks towards the lower ends of pools where the water is gathering pace before entering rapids. It is a secretive species, spending much time beneath cover. Spawning occurs in winter and early spring. The brightly coloured males assume a melanic nuptial colouration and establish defended territories beneath chosen rocks. Courting behaviour has not been observed, but the sexes pair for spawning and the female deposits her small, ovoid eggs in a uniform mat beneath a rock chosen by the male. After spawning, the female is driven from the territory, although the male may reproduce with several females in the same site. The male continues to defend the territory and guard the primitive nest until the eggs hatch. The newly hatched larvae are carried out to sea, returning some months later to complete the life cycle.

G. hubbsi (Stokell) is found all around New Zealand and inhabits very swiftly flowing waters in bouldery rapids, often in large rivers; it is diadromous, and has a life cycle essentially similar to that of *G. huttoni*.

G. breviceps (Stokell) differs from most other New Zealand eleotrids in being restricted to freshwaters. It is found in rivers of the southern North Island, the northwest of the South Island and the whole of the South Island east of the Southern Alps. Mostly, *G. breviceps* is a fluviatile species, usually found in pools where the water is gently flowing, but it is also abundant in some upland lakes where it inhabits weed beds or the open lake floor. *G. breviceps* lays fewer, much larger eggs than the other eleotrids. Its reproduction is essentially similar to that of *G. huttoni*,

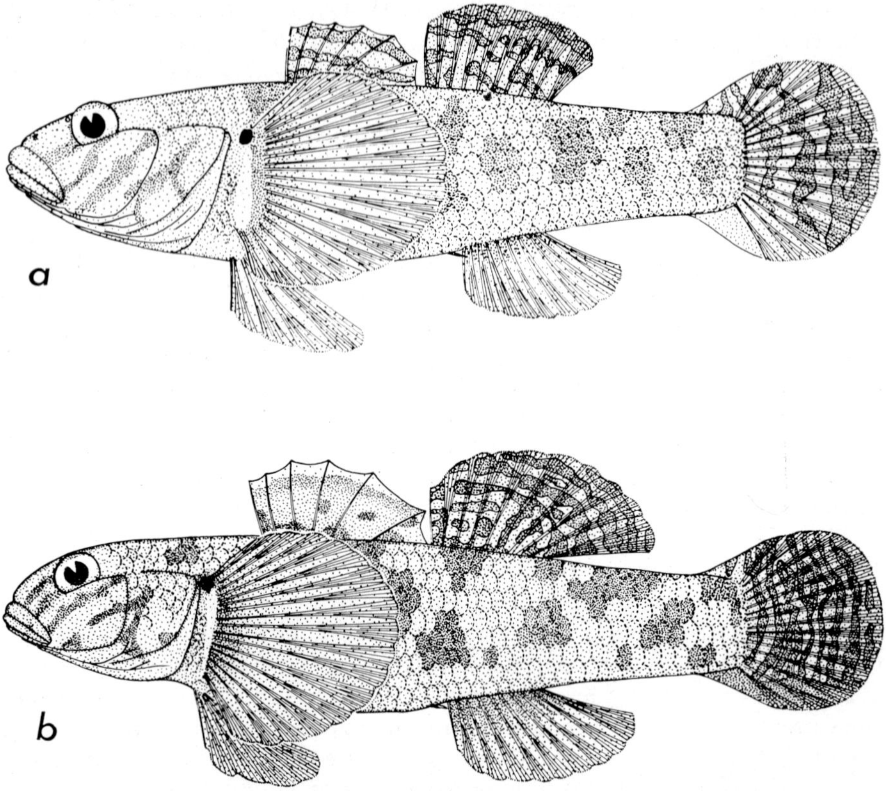

Fig. 6. Gobiomorphus huttoni (Ogilby); a. female; b. male (note more bulbous snout and larger fins of male).

with a melanic male guarding a territory, and eggs being deposited by the female on the under side of a flat rock. When the larvae emerge, they live in quiet marginal waters of their natal stream where they grow to maturity.

G. basalis (Gray) is, like *G. breviceps*, a large-egged species, but is found primarily in the North Island, where it is widely distributed. *G. basalis* and *G. breviceps* become sympatric in the southern North Island. It is found primarily in gently flowing reaches of rocky-boulder streams. Little is known about it.

Family mugiloididae

The family Mugiloididae is a widespread Indo-Pacific family of marine fishes which has one representative in freshwaters – *Cheimarrichthys fosteri* Haast (fig. 7), a New Zealand endemic. *C. fosteri* is a diadromous

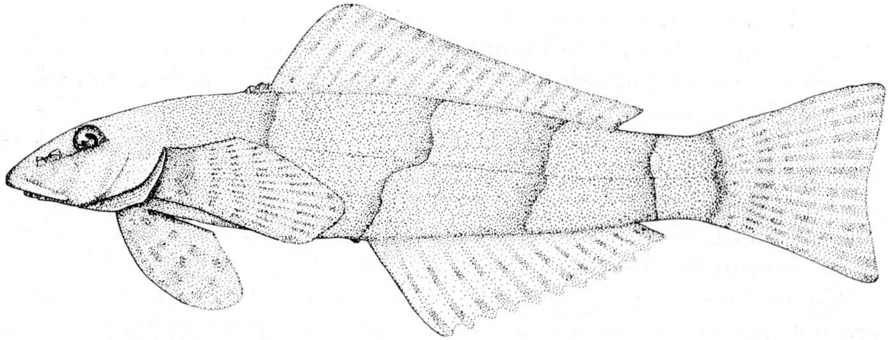

Fig. 7. Cheimarrichthys fosteri (Haast).

species that is very widespread in New Zealand occurring in most coastal rivers, especially larger, more unstable ones; in some rivers it penetrates long distances inland. Its habitat is the most swiftly flowing, turbulent water of river rapids, where it shelters beneath and amongst rocks on the river bed. Mature to ripe adults are found in typical adult habitats during spring and summer, and as they are found in typical adult habitats, sometimes a great distance from the sea, it seems unlikely that there is a substantial spawning migration, but rather that spawning occurs near where adults are customarily found. The eggs are very small – about 0.5 mm in diameter – and number many thousands. Neither the eggs nor larvae of *C. fosteri* are known from fresh water, but juveniles about 25 mm long are occasionally captured in the sea near the shore and similar fish are sometimes caught in river estuaries, apparently making their way upstream. At this stage they resemble the adults in form and colouration. It seems, therefore, that the eggs or larvae are carried downstream into the sea after spawning.

Food of the freshwater fishes

The food of few of the native freshwater fishes has been studied in detail, but there is a consistent pattern. All the species studied in detail (*Anguilla australis* and *A. dieffenbachi*, Cairns, 1942; Hopkins, 1965, 1970; *Galaxias maculatus*, McDowall, 1968; *Galaxias divergens*, Hopkins, 1971; *Gobiomorphus breviceps*, Hopkins, 1965, 1970; *Gobiomorphus huttoni*, McDowall, 1965b), have proved to be primarily invertebrate carnivores, feeding mainly on chironomid larvae when small, and extending the range of foods to larger elements of the invertebrate bottom fauna as they grow; Ephemeroptera, Trichoptera, some Mollusca and Crustacea become important in the diet, with the addition of terrestrial insects in some species. All the studies reached the same conclusion; that the fish tend to eat anything in their habitat with which they can cope. As a result the diet varies widely with habitat. Chironomids, ephemeropterans and trichopterans tend to be important in rocky stream habitats while copepods and amphipods attain importance where there is abundant aquatic vegetation.

Few of the native fishes can be regarded as piscivores. The work of Cairns (1942) showed that eels prey on fish only when they reach large size; most of the other species are too small to be fish predators, although the river flounder, *R. retiaria*, may prey heavily on the migratory juveniles of the various diadromous galaxiid species.

The feeding habits of the now extinct grayling, *Prototroctes oxyrhynchus*, are not known, although it has been stated that they used to graze on algal growth on the rocks in rivers. If this was true, *P. oxyrhynchus* would have been the only herbivorous species in the fauna. As New Zealand's rivers tend to be rocky, and swiftly flowing, and were to a great extent enclosed in forest, aquatic vegetation is not a dominant part of the river biota of New Zealand. It is therefore not surprising that herbivorous fishes are largely absent from the fauna. In contrast, New Zealand rivers support a very rich invertebrate fauna and it is to be expected that these would provide the principal diet of the freshwater fishes.

A Zoogeographic Synthesis

Ecological zoogeography

New Zealand's freshwater fishes exhibit two distinct life history patterns – a diadromous pattern in which a part of the life is spent at sea, and a non-migratory exclusively freshwater pattern. Within New Zealand, distribution is closely related to life history patterns. Many species spend a part of their lives in the sea, and existence of the marine stage affects the life history strategy and the distribution pattern. Species whose eggs or larvae occur at sea tend to have numerous, small eggs, e.g. *Retropinna retropinna*, the five diadromous *Galaxias* species, *Gobiomorphus huttoni*, *Cheimarrichthys fosteri*, and almost certainly *Rhombosolea retiaria* and the two species of *Anguilla*. Clearly, eggs or larvae of species going to sea tend to be dispersed widely by ocean currents. To ensure that sufficient progeny return to fresh water to grow to maturity there is a selective advantage in having many eggs. Small larvae hatch from small eggs but this is no great disadvantage when the newly hatched larvae have the very rich food resource of the marine plankton on which to feed.

Species that are confined to fresh water tend to have fewer, larger eggs, e.g. *Galaxias divergens*, *Galaxias prognathus*, *Gobiomorphus breviceps*. The larvae are not dispersed by ocean currents, making large numbers of eggs unnecessary to offset this dispersal. But the larvae tend to be carried downstream by the stream flow; as larger eggs produce larger and presumably stronger-swimming larvae, there is an advantage for the larvae to be larger when they hatch. In addition, flowing fresh waters do not carry rich food supplies for small, newly-hatched larval fishes. So larger larvae, carrying bigger yolk reserves, are at an advantage in fresh waters.

With the exception of *Stokellia anisodon*, all those species which are diadromous are found around the whole of New Zealand e.g. *Galaxias brevipinnis* (fig. 4). Those species found outside New Zealand, those occurring on outlying islands of New Zealand (Auckland, Campbell and Chatham Islands), and those found on a few inshore islands (Great and Little Barrier, Marotiri, Cavalli, Kapiti, and others), are also species that spend a part of their life in the sea. It is obvious that existence of this marine stage facilitates broad dispersal and wide range of these species. However, the fact that they must penetrate New Zealand's turbulent river systems from the sea, means that many of them are prevented from penetrating far upstream, and tend to be found only in lowland and coastal streams.

In contrast, species that do not go to sea tend to be much more restricted in their distribution, e.g. *Neochanna* species (fig. 5), are endemic to New Zealand, and all are confined to the two main islands. Clearly restriction of species to fresh water limits their ability to disperse, and so they have more limited ranges than diadromous species. However, they also tend to be more inland to alpine in distribution.

Historical zoogeography

The Southern-temperate Element: The principal, and oldest element in the freshwater fish fauna of New Zealand comprises groups of fishes that are characteristic of cool temperate lands of the Southern Hemisphere. The freshwater fish fauna that can be regarded as typically southern temperate is very small, comprising only two families of lampreys (Geotriidae and Modaciidae) and the sub-order Galaxioidei (families Galaxiidae, Prototroctidae, Retropinnidae, and Aplochitonidae). The family Galaxiidae is by far the largest and most widespread and is closely related to the family Aplochitonidae (Australia – the monotypic genus *Lovettia*, and South America and the Falkland Islands – two species of *Aplochiton* Jenyns). Standing rather apart from these two families, but nevertheless showing decided affinities with them, are the closely related families Retropinnidae and Prototroctidae (both confined to Australia and New Zealand). The families Geotriidae and Mordaciidae form a small and probably ancient radiation in the south corresponding to a larger, northern radiation of lampreys in the family Petromyzonidae. The sub-order Galaxioidei forms a larger but probably equally ancient radiation in the south that is obviously related to and has an origin common with the major northern hemisphere radiation of salmonoid-osmeroid fishes. The lampreys are clearly an ancient, relict group, and the galaxioid-salmonoid-osmeroid complex is also ancient, forming the base of much of the teleost lineage (GREENWOOD *et al.*, 1966). Thus it seems that these groups may have had early origins in the united continental masses associated with Gondwanaland in the Mesozoic and that northern and southern radiations in each taxon-pair evolved, as the northern hemisphere continents moved towards their present positions around the north pole, and southern hemisphere land masses radiated from the south pole.

It would be easy to assume that the present broad range of lampreys and galaxioid fishes is also a result of the former existence of Gondwanaland. But there is much that militates against such a simple explanation. *Geotria australis* is present in Australia, Tasmania, New Zealand and South America without any reported subspeciation or phyletic evolution in any of these highly isolated areas. The implication of a Gondwanaland distribution for *G. australis* is that it has remained phenotypically stable in each area since fragmentation of Gondwanaland. Such an hypothesis is unacceptable, even more so when it is remembered that *G. australis* has a marine life, almost certainly of several years. Clearly, transoceanic dispersal is the most appropriate explanation for the present range of *G. australis*, even though ancient origins of the Geotriidae may have been Gondwanian.

For the sub-order Galaxioidei a similar explanation seems broadly applicable. Within the family Galaxiidae are six species that certainly have a marine stage lasting about six months and it is these species that have broad ranges, often spanning wide ocean gaps. The occurrence of *Galaxias brevipinnis* in Tasmania, New Zealand, Chatham, Auckland and Campbell Islands and of *G. maculatus* in Australia, Tasmania, New Zealand, Chatham Islands, Chile, Argentina, Tierra del Fuego and Falkland Islands is due to recent and perhaps continuing transoceanic dispersal. From osteological and other studies (McDOWALL, 1969b) it seems that diadromous species, which we suggest can disperse across oceans, are also the most primitive and generalised galaxiids. A rather high proportion of galaxiids, e.g. 8 of 13 New Zealand species and probably 2 of 4 South American species, do not enter the sea at any stage. These species seem to be more specialised and therefore are probably freshwater derivatives of the diadromous species.

In general it would seem that transoceanic dispersal is a sufficient explanation for the zoogeography of galaxiid fishes, but there are some problems. *Neochanna* shows great similarity to the Tasmanian mudfishes (*Saxilaga* Scott) but osteological studies are required to show whether they are closely related, or whether their similarities are convergent. If *Neochanna* and *Saxilaga* prove to be closely related we have then either to hypothesize that an entirely freshwater species has dispersed across the Tasman Sea, or that *Saxilaga* and *Neochanna* have a very ancient common origin in Gondwanaland. The same sort of problem exists with the Chilean *Brachygalaxias bullocki* (Regan) and South African *Galaxias zebratus* (Castelnau) (McDowall, 1973a).

Looking at the galaxiid species groups, we find that *G. argenteus*, *G. fasciatus* and *G. postvectis* are very similar to and closely related to the diadromous Australian species *G. truttaceus* (Valenciennes); the derivation of the three New Zealand species from *G. truttaceus* seems likely. *G. brevipinnis* is common to Tasmania, New Zealand and the Auckland and Campbell Islands. The New Zealand and Auckland and Campbell Island stocks could have been derived independently from Tasmania as ocean currents favour dispersal of fishes from Tasmania to the Auckland and Campbell Islands more than from New Zealand to these islands. *G. maculatus*, present in Australia, New Zealand and South America, has clearly dispersed across the sea, and the existing directions of ocean currents favour a dispersal from Australia, eastwards to New Zealand, and thence to South America.

The origins and relationships of the small, slender alpine species (*G. divergens*, *G. paucispondylus* and *G. prognathus*) are not understood, but these three species are obviously very closely related to one another. Their present range suggests that their ancestral form had spread over the southern North Island and much of the South Island, perhaps before uplift of the Southern Alps in the late Tertiary (Fleming, 1975). *G. divergens* occurs in the Wellington-Marlborough-Nelson-West Coast area, spanning Cook Strait. *G. prognathus* is known only in the Maruia System in the headwaters of the Buller River and in the Rakaia System on the eastern side of the Southern Alps. *G. paucispondylus* is also found in the Maruia System and in several widespread localities on the eastern side of the alps. It seems that a single species, probably similar to *G. paucispondylus*, occurred widely throughout the range of the species group, and that its range was fragmented by uplift of the Southern Alps. Phyletic evolution produced *G. divergens* on one side and *G. paucispondylus* on the other. *G. prognathus* is the most specialised of these species and this suggests that it may have been in competition with a similar species, like *G. divergens* or *G. paucispondylus*, this competition producing initially character displacement, and finally speciation. It seems likely that after reproductive barriers between *G. paucispondylus* and *G. divergens* had developed, one of these species reinvaded the range of the other, and the resulting intense competition produced *G. prognathus*. Since this time there has obviously been further recrossing of the Alps, and it is notable that the only locality where both *G. paucispondylus* and *G. vulgaris* are known west of the divide is in the vicinity of the Lewis Pass where a swampy area actually spans the divide and stream capture is likely.

The three neochannoid species are allopatric. It seems that their common ancestor was formerly widespread over much of New Zealand. Populations that we now recognise as *N. burrowsia* were probably isolated in Canterbury by uplift of the Southern Alps and perhaps restricted to small pockets in Canterbury by subsequent Pleistocene glaciations. The more northern isolate continued to evolve towards the more specialised condition seen in *N. apoda* and *N. diversa* – primarily a loss of the pelvic fins and girdle. During the Pliocene, a marine transgression covered most of the southern half of the North Island (FLEMING, 1975) and this would have isolated populations in the north of the North Island – leading by phyletic evolution to *N. diversa* – and in the Nelson-Buller-West Coast area of the South Island – leading to *N. apoda*. Land connections between the North and South Islands following retreat of the marine transgression and lasting well into the Pleistocene (FLEMING, 1975) would have allowed *N. apoda* to reach the North Island, if it was not already there.

The family Prototroctidae has one Australian species and had one New Zealand species, both of which probably spend (or spent) a part of their lives in the sea, and it seems natural to explain distribution of the genus by dispersal across the Tasman Sea. The same is true for the family Retropinnidae. Lake-limited isolates of *Retropinna* occur in both Australia and New Zealand, but these seem to be derived from euryhaline diadromous species, and the Australia-New Zealand distribution of *Retropinna* can satisfactorily be explained by dispersal through the sea. *Stokellia* is obviously a retropinnid, but how it evolved, and why it is limited to Canterbury, Otago and Southland streams is unknown.

The Indo-Pacific Element: The second major element in the fauna comprises fishes derived from the great tropical and sub-tropical Indo-West Pacific fish fauna. This element in New Zealand is derived in two ways – either from freshwater species elsewhere or from local marine species.

The New Zealand eels and eleotrids are related to freshwater species elsewhere. New Zealand has two eel species, one shared with Australia, the other endemic but similar and probably closely related to Australian species. The presence of freshwater eels in New Zealand can be attributed to extension of the range of Australian species further south into the western temperate Pacific, and, with the prolonged marine larval life of anguillids, can be explained satisfactorily by dispersal through the sea.

The eleotrid genus presently recognised in New Zealand – *Gobiomorphus* – is shared with Australia and as it contains species in New Zealand that have marine living juveniles, they could easily have reached New Zealand through the sea. In New Zealand all the recognised species except two have marine larvae, and all these are New Zealand-wide in range. The two exceptions are *G. basalis* which is widespread primarily in the North Island, and *G. breviceps* which is widespread primarily in the South Island and the southern North Island. *G. breviceps* has been the most successful freshwater-limited species in dispersing around New Zealand.

Two species in the fauna have no relatives in freshwaters elsewhere. *Rhombosolea retiaria* is derived from marine species of *Rhombosolea* found in New Zealand coastal seas. *Cheimarrichthys fosteri* has formerly been included in a monotypic family, but some workers (PHILLIPPS, 1929; ALLEN, 1957) suggested that it is closely related to the New Zealand mugiloidid *Parapercis colias*. Recent osteological studies (McDOWALL, 1973b) have shown that this is so. *C. fosteri* appears to be a freshwater derivative of *Parapercis* or a *Parapercis*-like ancestor, and belongs, with *Parapercis*, in the family Mugiloididae.

Conclusion: The zoogeography of New Zealand's freshwater fishes conforms to a clear pattern. Even though elements of the fauna may have had ancient evolutionary origins in Gondwanaland, derivation of the fauna seems clearly to be by transoceanic dispersal followed by phyletic evolution and speciation in New Zealand.

Comparison of the New Zealand and Australian faunas shows that New Zealand freshwater fishes have strong affinities with those of south-eastern Australia and Tasmania. Existence of the East-Australian Current which sweeps down the east coast of Australia, past Tasmania, and then turns across the south Tasman Sea to impinge on the west coast of New Zealand provides a mechanism whereby euryhaline Australian fresh-water fishes could be carried to New Zealand, and we suggest that this is the principal means of derivation of the New Zealand freshwater fish fauna.

REFERENCES

ALLEN, K. R. 1957. Freshwater fish. In Science in New Zealand, pp. 3–12. Reed, Wellington.

BENZIE, V. 1968. The life history of *Galaxias vulgaris* Stokell, with a comparison with *G. maculatus attenuatus*. *N.Z. Jl. Mar. Freshwat. Res.* 2: 628–653.

CAIRNS, D. 1941. Life history of the two species of New Zealand fresh water eel. Part I. Taxonomy, age and growth, migration and distribution. *N.Z. Jl. Sci. Technol.* 23 (2): 53–72.

CAIRNS, D. 1942. Life history of the two species of fresh water eel in New Zealand. II. Food and inter-relationships with trout. *N.Z. Jl. Sci. Technol.* 23 (4): 132–148.

CASTLE, P. H. J. 1963. Anguillid leptocephali in the southwest Pacific. *Zoology Publs. Vict. Univ. Wellington* 33: 1–14.

DARLINGTON, P. J. 1957. Zoogeography, the geographical distribution of animals. Wiley, New York. 675 pp.

DAY, F. 1888. Supplement to the fishes of India, being a natural history of the fishes known to inhabit the seas and freshwaters of India, Burma, and Ceylon. Williams and Norgate, London.

ELDON, G. A. 1971. Suspected terrestrial deposition of eggs by *Neochanna apoda* Günther, (Pisces: Salmoniformes: Galaxiidae). *Jl. Fish. Biol.* 3 (2): 247–249.

Fleming, C. A. 1975. The geological history of New Zealand and its biota. In Kuschel, W. (Ed.), Biogeography and ecology in New Zealand, pp. 1–86, Junk, The Hague.

Greenwood, P. H., Rosen, D. E., Weitzman, S. H. & Myers, G. S. 1966. Phyletic studies of teleostean fishes with a provisional classification of living forms. *Bull. Amer. Mus. Nat. Hist.* 131 (4): 339–456.

Griffin, L. T. 1936. Revision of the eels of New Zealand. *Trans. Proc. R. Soc. N.Z.* 66 (1): 12–66.

Hopkins, C. L. 1965. Feeding relationships in a mixed population of freshwater fish. *N.Z. Jl. Sci.* 8 (2): 149–157.

Hopkins, C. L. 1970. Some aspects of the bionomics of fish in a brown trout nursery stream. *Fish. Bull. N.Z. Mar. Dep.* 4: 1–38.

Hopkins, C. L. 1971. Life history of *Galaxias divergens* (Salmonoidea: Galaxiidae). *N.Z. Jl. Mar. Freshwat. Res.* 5 (1): 41–57.

Hubbs, C. L. & Potter, I. C. 1971. Distribution, phylogeny and taxonomy. In Hardisty, M. W. & Potter, I. C. (Eds.). The biology of lampreys 1: 1–65. Academic Press, London. 2 vols.

Jolly, V. H. 1967. Observations on the smelt *Retropinna lacustris* Stokell. *N.Z. Jl. Sci.* 10 (1): 330–355.

McDowall, R. M. 1962. A redescription of the freshwater red-finned bully *Gobiomorphus huttoni* (Ogilby). *Trans. R. Soc. N.Z. Zool.* 3 (2): 3–15.

McDowall, R. M. 1964. Studies on the biology of the red-finned bully *Gobiomorphus huttoni* (Ogilby). Part I. Habitat and species inter-relationships. *Trans. R. Soc. N.Z. Zool.* 4 (12): 175–182.

McDowall, R. M. 1965a. Studies on the biology of the red-finned bully *Gobiomorphus huttoni* (Ogilby). Part II. Breeding and life history. *Trans. R. Soc. N.Z. Zool.* 5 (14): 177–196.

McDowall, R. M. 1965b. Studies on the biology of the red-finned bully *Gobiomorphus huttoni* (Ogilby). Part III. Food studies. *Trans. R. Soc. N.Z. Zool.* 5 (17): 233–254.

McDowall, R. M. 1968. *Galaxias maculatus* – the New Zealand whitebait. *Fish. Bull. N.Z. Mar. Dep.* 2: 1–84.

McDowall, R. M. 1969a. Lunar rhythms in aquatic animals – a general review. *Tuatara* 17 (3): 133–144.

McDowall, R. M. 1969b. Relationships of galaxioid fishes with a further discussion of salmoniform classification. *Copeia* 1969 (4): 796–824.

McDowall, R. M. 1970. The galaxiid fishes of New Zealand. *Bull. Mus. Comp. Zool. Harv.* 139 (7): 341–431.

McDowall, R. M. 1972a. The species problem in freshwater fishes and the taxonomy of diadromous and lacustrine populations of *Galaxias maculatus* (Jenyns). *Jl. R. Soc. N.Z.* 2 (3): 325–367.

McDowall, R. M. 1972b. The taxonomy of estuarine and brackish-lake *Retropinna* from New Zealand (Galaxioidei: Retropinnidae). *Jl. R. Soc. N.Z.* 2 (4): 501–531.

McDowall, R. M. 1973a. The status of the South African galaxiid (Pisces: Galaxiidae); *Ann. Cape Prov. Mus.* 9 (5): 91–101.

McDowall, R. M. 1973b. Relationships and taxonomy of the New Zealand torrent fish *Cheimarrichthys fosteri* Haast (Pisces: Mugiloididae). *Jl. R. Soc. N.Z.* 3 (2): 199–217.

McDowall, R. M. 1973c. *Galaxias indicus* Day, 1888 – a *nomen dubium*. *Jl. R. Soc. N.Z.* 3 (2): 191–192.

McDowall, R. M. 1975. A revision of the New Zealand species of *Gobiomorphus* (Pisces: Eleotridae). Rec. Natn. Mus. Wellington.

McMillan, H. M. 1961. An addition to the knowledge of the fish *Retropinna anisodon* Stokell. *Trans. R. Soc. N.Z. Zool.* 1 (10): 139–144.

Maskell, F. G. 1929. On the New Zealand lamprey, *Geotria australis* Gray. Part I. Biology and life history. *Trans. Proc. N.Z. Inst.* 60 (1): 167–201.

Phillipps, W. J. 1929. Notes on the freshwater papanoko (*Cheimarrichthys fosteri* Haast). *N.Z. Jl. Sci. Technol.* 11 (3): 166–168.

Schmidt, J. 1928. The freshwater eels of Australia with some remarks on the short finned species of *Anguilla*. *Rec. Aust. Mus.* 16 (4): 179–210.

Smith, D. C. W. 1969. The biology of the rainbow trout *(Salmo gairdneri)* in the lakes of the Rotorua district, North Island. *N.Z. Jl. Sci.* 2 (3): 275–312.

Stokell, G. 1941. A revision of the genus *Gobiomorphus*. *Trans. R. Soc. N.Z.* 70 (4): 265–276.

Stokell, G. 1962. A new species of *Gobiomorphus*. *Trans. R. Soc. N.Z. Zool.* 2 (3): 31–34.

Woods, C. S. 1968. Variation and taxonomic changes in the family Retropinnidae (Salmonoidea). *N.Z. Jl. Mar. Freshwat. Res.* 2 (3): 398–425.

VII. THE KIWI

by

BRIAN REID & G. R. WILLIAMS

Introduction

Kiwis comprise an order (Apterygiformes) of ratite birds peculiar to New Zealand. They are by far the smallest and most aberrant of this rather heterogeneous flightless group; and when the first specimen was described in 1813 sceptics, feeling they were victims of a hoax, demanded to see so strange a creature.

Diverse studies have, for a long time, tried to establish the ancestry of kiwis. Some have asserted that the ratites have evolved from a common ancestor (i.e. that they are monophyletic), others that there has been more than one ancestor (i.e. that they are polyphyletic) and that resemblances may be attributed to convergent evolution. Others again have claimed that kiwis are more closely related to some carinate (flying) birds than to the other ratites. Some have maintained that all ratites are descended from an ancestor (or ancestors) that could fly, others that their ancestors were flightless; and whereas some believe that ratites are 'primitive', others claim they exhibit 'arrested' or neotenous characteristics.

Whatever the true evolutionary history of kiwis may be, the end result is a bizarre flightless, nocturnal creature – tailless and bewhiskered and almost devoid of wings; and New Zealanders (also colloquially called 'Kiwis'), whilst knowing little about the bird and seldom seeing or even hearing it, nevertheless have a great affection for, and take pride in associating themselves closely with, this unique avian enigma, now their unofficial national emblem. (Plate 6).

Discovery

According to OLIVER (1955) the description which was greeted with such incredulity was that by SHAW and NODDER of a South Island brown or common kiwi obtained about 1813 in Dusky Sound on the south-west coast of the South Island. In 1827 DUMONT D'URVILLE's expedition encountered the North Island race of the same species at Tolaga Bay in the East Cape district (LESSON, 1839) – and this race was officially described 23 years later (BARTLETT, 1850). A specimen of the Stewart Island race of the brown kiwi was first exhibited in 1890 and subsequently described by ROTHSCHILD (1893). The little spotted kiwi was first described from a specimen taken at the northern end of the South Island (GOULD, 1847) and the great spotted kiwi was first described from two specimens killed in Westland during 1871 (POTTS, 1872), although there were many stories about large spotted grey kiwis circulating prior to that time. An interesting account covering the early history of the scientific investigations of kiwis was given by NEWTON (1893).

Description

There are three species which show a range in size similar to that of domestic poultry. The largest are as big as Rhode Island Red or Australorp hens and the smallest, which are approximately one-third the weight of the largest, are about the size of bantams.

Distinctive features include a long and slightly curved bill with nostrils near the tip, a cone-shaped body (because of reduced pectoral development) that tapers markedly to a strong neck and comparatively small head, powerful muscular legs which make up one third of the total weight, small eyes, large ear apertures and many long tactile bristles about the face and base of the bill. The wings, which end in a claw, are very small (40–50 mm) and there is no external tail.

The plumage is loose and hair-like and does not change in form throughout life. Its neotenous characteristic of having weak barbs and lacking aftershafts gives kiwis a permanently shaggy appearance. Depending upon condition of the bird and time of year, plumage represents between 4.7 and 6.8 per cent of the body weight.

Little spotted kiwis are pale grey with irregular bands of brownish-black; great spotted kiwis are similar but darker and often have a chestnut tinge on the back. Brown kiwis are brown with black striations, and the plumage of the North Island race is stiffer and more hair-like than that of the two southern subspecies. Albinism, which is unlikely to be disadvantageous to a nocturnal species, appears to be common, and BULLER (OLIVER, 1955) recorded that five of 300 North Island brown kiwis taken as specimens from the Kaimanawa Range were albinos.

In all species females are larger and about 20 per cent heavier than males. In little spotted kiwis females weigh about 1220 g and males 1000 g. Males of the North Island brown kiwi weigh from 1,400 to 2,600 (mean 1930) g and females weigh 1,800–3,300 (mean 2360) g. These variations reflect both the wide range in size and the fat-storage capacity of adults; – birds in average condition carry about 300 g of subcutaneous fat. The South Island and Stewart Island races of the brown kiwi are generally larger than the North Island form, and females in good condition (but not excessively fat) may weigh more than 3,500 g. A significant series of weights is not available for the great spotted kiwi but this species is thought to be about the same size as the South Island brown kiwi: Two presumed females in captivity each weighed 2,700 g and a presumed male 2,200 g (RODERICK, pers. comm.).

In all species the length of the bill is very variable; but usually the female's is 25–30 per cent longer than the male's. North Island brown kiwi adult females have a mean bill length of 126 mm (109–145), males 97.5 mm (87–116).

Though primarily birds of the indigenous rain forests, kiwis may also be found in scrub, native grassland and even in exotic forests and pastoral country; and they occur from sea level to alpine meadows. More will be said of their habitat requirements in a later section.

Calls

Although each species and subspecies has contact calls of characteristic quality, all are essentially similar: Males utter a shrill prolonged whistle, slightly ascending and descending (from which the bird's name is derived); females have a lower, hoarser cry. These calls are heard at dusk, during the night and in the early morning. When alarmed or aggressive kiwis growl, hiss and loudly snap their mandibles. Chicks also snap their mandibles – apparently as a contact call – and, according to ROBSON (1958), do not make the adult call until about 12 months old, but squeak instead.

The other common noise, which is hardly a vocalisation, is a loud snuffling made by expelling air forcibly through the nostrils while feeding. This is clearly audible at many metres distance as the bird wanders about in the forest at night, probing continuously with its bill.

Classification and the Geographical Distribution of Races

CLASSIFICATION

Three species are recognised. One occurs on North, South and Stewart Islands, two are found only in the South Island and one of these once occurred in the North Island but apparently has been extinct there for some time. There are no records of kiwis ever having occurred naturally on any of the offshore islands and no recent extinct species are known.

The following scheme, a modification of that proposed by MATHEWS (1931), is that given in the Annotated Checklist of the Birds of New Zealand (1970):

Family: APTERYGIDAE
Species:

1. *Apteryx australis* Shaw and Nodder 1813, the brown – or common – kiwi.
   Subspecies: *A. australis mantelli* Bartlett 1850 (North Island, north of 40°S). *A. a. australis* Shaw and Nodder 1813 (South Island, mainly southern Westland) Plate 1. *A. a. lawryi* Rothschild 1893 (Stewart Island) Plate 2.
2. *Apteryx oweni* Gould 1847, the little spotted kiwi. (North Island, presumed extinct, South Island, west coast mainly.) Plate 3.
3. *Apteryx haasti* Potts 1872, the great spotted kiwi. (South Island, north-west mainly).

Plate 1. South Island brown kiwi *(Apteryx a. australis)*. Mount Bruce Native Bird Reserve. N.Z. Wildlife Service.

Separation of the species is based on differences in size and coloration and is supported by the clear differences in calls. VERHEYEN (1960) considered the osteological differences between the brown kiwi and the little spotted enough to warrant generic distinction, and he proposed the genus *Stictapteryx* for the latter; however, this proposal has not as yet prompted other workers to express recent opinions on relationships within the family.

GEOGRAPHICAL DISTRIBUTION

Not only does the brown kiwi occur on all three main islands, it has been introduced and established on some offshore ones as well. The species is still widespread and common in suitable habitat over most of the North Island and Stewart Island and the west coast of the South Island, but here, apparently, it is common only in the south. It is known from subfossil remains on the eastern side of the Southern Alps (see Plate 1 and Fig. 1).

The little spotted kiwi once occurred widespread in both the North and South Islands. It has not been reported from the North for almost a century, and all other records from this island are subfossil. In the South

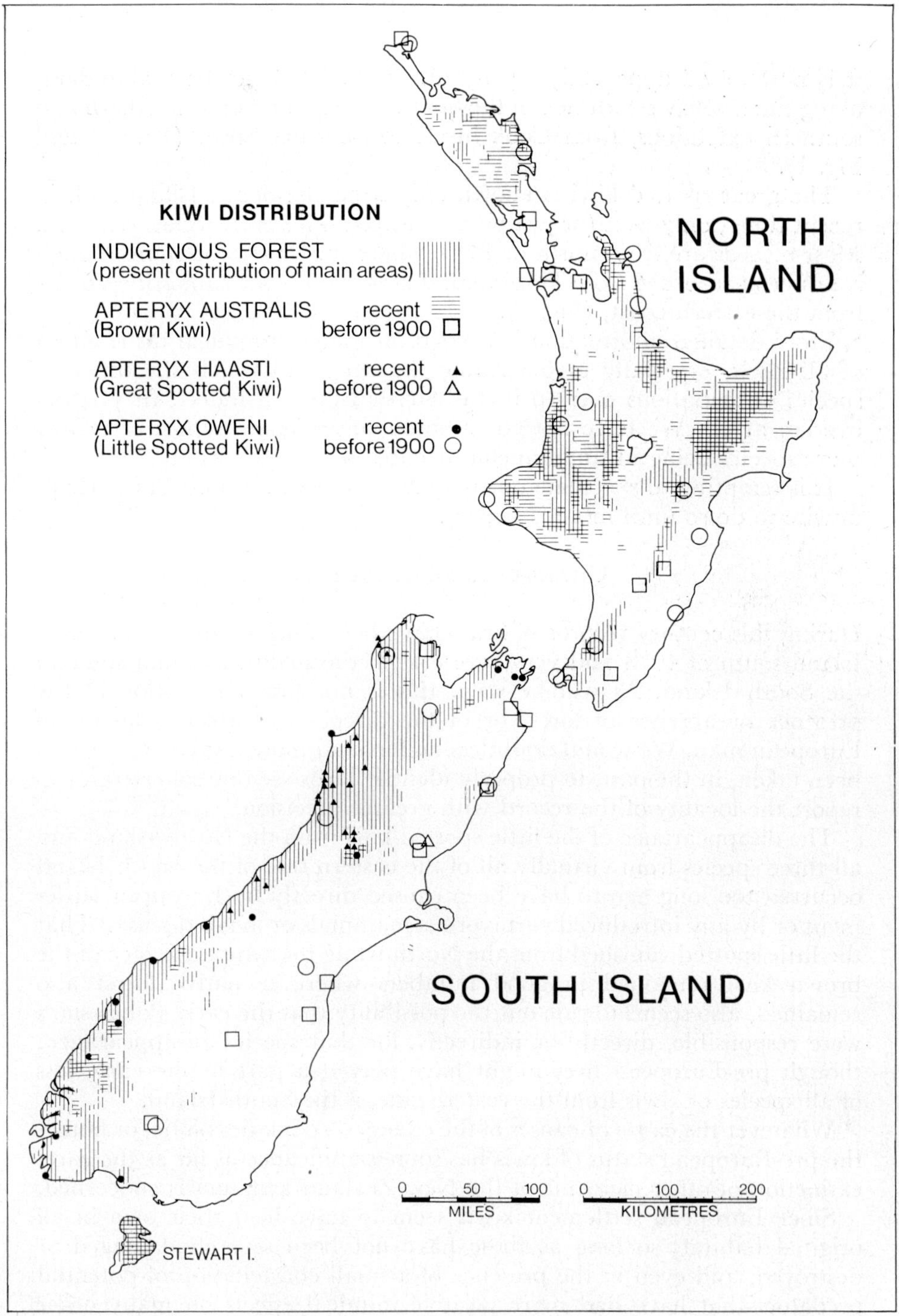

Fig. 1. The geographical distribution of the three species of kiwi. Subfossil distributions beyond the present ranges are shown for each species.

it is now found apparently evenly distributed, but nowhere abundant, along the whole western side of the Southern Alps and their northern and southern extensions, from Cook Strait to Foveaux Strait (Plate 3 and Fig. 1).

The great spotted kiwi is known only from the South Island and, in recent times, only west (or just on the eastern side) of the Southern Alps. Most records are from north of 43 °S, where it is fairly common, though it has been recorded from Fiordland. There is only one subfossil record – from the eastern coast, (Fig. 1).

More detailed information is needed on the geographical distribution of all kiwis, especially to determine the extent to which the different species' distributions overlap in the South Island, if indeed they really overlap at all. We have not come across any reliable report of two or more species being in simultaneous occupation of one locality.

It is tempting to speculate on the pattern of distribution; but perhaps unwise to do so until more is known.

Changes in Distribution

During this century records of kiwis have been uncommon in the North Island south of 40 °S and very rare east of the main mountain spine of the South Island. As will be seen, this is not just a reflection of the presence or absence of forest or of the direct or indirect influence of European man. We would doubtless understand more if greater care had been taken, in the past, to properly identify birds seen or collected, or to report the locality of the record with greater precision.

The disappearance of the little spotted kiwi from the North Island and all three species from virtually all of the eastern side of the South Island occurred too long ago to have been caused directly by European settlement or by any introduced carnivorous mammals or avian diseases. That the little spotted vanished from the North while the sympatric race of the brown kiwi remained in good numbers wherever native forest also remained, also seems to rule out the possibility that the early Polynesians were responsible, directly or indirectly, for that species' disappearance; though pre-European fires might have played a part in the early loss of all species of kiwis from the eastern side of the South Island.

Whatever the cause or causes of the changes – climatic, biotic, or both – the pre-European status of kiwis has some significance as far as the early extinction of other elements of the New Zealand avifauna is concerned.

Since European settlement kiwis seem to have held their own in all original habitats so long as these have not been severely damaged or destroyed, and even in the presence of a small constellation of potential predators that have had more positive inimical effects on many other native birds (Atkinson, 1973, Gibb & Flux, 1973, Williams, 1973, and see later under Predators).

Much more recently, there have been indications that the brown kiwi, at least, is spreading into rough marginal farmland and parts of the forests of introduced pines. This upsurge of adaptability may be more apparent than real or may be, to some extent, a measure of our ignorance of the habitat requirements of the birds – or both. Among the critical factors of their environment may be such things as soil texture, humus content, high humidity and dense overhead cover rather than native vegetation itself. Certain values and combinations of these may be essential for suitable ground for burrowing, soil soft enough for probing and the development of a soil fauna suitable for food. The common assumption that kiwis have gradually become adapted to changed conditions is certainly a likely one; but at least equally as likely is the one that conditions in novel habitats have gradually become more favourable for kiwis. Yet another possibility is that some indigenous forests may in places, be becoming less suitable (through milling operations, destruction of food-bearing vegetation, modifications caused by introduced browsing and grazing mammals and so on), thus forcing resident birds to range further afield. Nothing reliable is known of the differences in habitat requirements that separate the three species.

Evolution and Affinities

Any discussion on the relationships of the kiwi must refer to the other ratites – the moas of New Zealand (Dinornithiformes), emus and cassowaries of Australia and New Guinea (Casuariiformes), ostriches of Africa (Struthioniformes) and rheas of South America (Rheiformes) – as well as to the tinamous of South America (Tinamiformes), and the cosmopolitan rails (Rallidae).

Geological record

The geological record for New Zealand, Australian and South American species is poorly documented. Several species of rails are known from Upper Cretaceous deposits (63–72 million years ago) in the northern hemisphere (Brodkorb, 1971); but in New Zealand the earliest members of this order date from the Quaternary – i.e. within the last two million years (Scarlett, 1970). Although the earliest known kiwi *(Pseudapteryx gracilis)* also dates from the Quaternary (Storer, 1960), footprints attributed to this family are present in mudstone of Upper Miocene age (about 10 million years ago) from near Murchison in the South Island (Mildenhall, 1974). The tinamous enter the record in the late Pliocene of Patagonia and have a known history of only two to four million years, although their archaic morphology suggests a much earlier origin.

Larger fossils tend to have a more detailed record as they are more likely to be salvaged from quarries or during mining and engineering

Plate 2. Stewart Island brown kiwi *(Apteryx australis lawryi)*. Note the difference in feather texture between this species and the little spotted kiwi (Plate 3). B. J. HARCOURT.

excavations. The earliest known moa *(Anomalopteryx antiquus)* and the earliest Casuariiformes date from the upper Miocene (8–10 million years ago), whereas both the Rheiformes and the Struthioniformes are known from the Eocene (40–60 million years ago). It is widely accepted that most, if not all, orders of birds had become established by the end of the Eocene (BRODKORB, 1971) and although we can only guess when the kiwis and moas diverged from each other, FLEMING (1962) believes the ancestor or ancestors of these two endemic orders colonised New Zealand during the Upper Cretaceous.

RELATIONSHIP TO OTHER RATITES

The kiwis and the moas have each been depicted by some investigators as separate and isolated offshoots in the evolutionary scheme between the rhea and the ostrich on one branch and the cassowary and the emu on another (WELTY, 1962). PYCRAFT (1900) and ARCHEY (1941), however, believed the kiwis and moas were more closely related to each other and that both are perhaps closer to the South American rheas than they are to the ratites of Australia, New Guinea and Africa.

Investigations on spreading of the sea floor and continental drift

(Christoffel & Falconer, 1972; Christoffel, pers. comm.) show that both New Zealand and Australia were at higher latitudes and contiguous with Antarctica during the Lower Cretaceous. Although New Zealand had become separated from Australia by about 80 million years ago, some believe that it remained connected to Australia by a land bridge (the Lord Howe Rise) until much more recently. The connections between the other austral land masses were more ancient. Southern Africa and Australia were linked earlier in the Triassic and it is thought that South America was joined to Antarctica during the same period; and it is generally accepted that Africa and South America drifted apart about 140 million years ago in the Jurassic.

From the relative age of these various land connections we may infer that the New Zealand and Australian ratites should have the closest kinship and that the level of affinity between these Australasian forms and the South American and African ratites would be about the same, and that the link between ostriches and the rhea-tinamou group would be the most remote – if all connecting areas in this 'Gondwana' complex were (and had remained) equally free from geographic or climatic barriers.

Immunochemical studies of ratite egg-white proteins by Osuga & Feeney (1967) showed that some biochemical properties link the kiwi to the cassowary and emu while other properties link it with the ostrich and rhea. Although these workers believe the relationship between the large cursorial ratites and the kiwi is fairly remote, they state that 'the kiwi is without question a ratite' and consider that their studies endorse the belief that the ratites are monophyletic.

Sibley (1967 and pers. comm.) is less sure about the kiwi's position. While his biochemical studies show the ostrich is truly related to the emu and cassowaries and also suggest the rheas may have their closest links with the ostrich, they indicate no obvious affinities of the kiwi. There does seem, however, to be one protein that links *Apteryx* with the other ratites, but if there is a phylogenetic relationship it would undoubtedly be very ancient.

Falla (1964) noted that in several respects kiwis do not conform with the general ratite pattern. Although the pectoral and forelimb anatomy of the kiwi indicates its descent from a flying progenitor, its skeleton is very poorly pneumatised when compared with the other living ratites and this permits the speculation that, of this group, kiwis were probably the first to diverge from their ancestral stock and become flightless.

Relationship to carinates

a. Rails

Several authorities believe the kiwi is related to some extant carinate species. VERHEYEN (1960), following LOWE (1933) thought it was allied to the penguins but this view has few supporters and recent immunochemical comparisons of the egg-white proteins of the kiwi and Adelie penguin show no such close relationship.

In 1915 HARRISON reported that the feather lice from three species of kiwi were 'robust' members of the genus *Rallicola* (=*Aptericola*). Large-sized hosts tend to harbour large mallophagans and the seemingly disproportionately large size – and the generic position – of these lice led him to believe kiwis were an offshoot of a larger-bodied (moa-like) stock which, in turn, was more closely related to the *Rallidae* (rails) than to any other living birds. However, a comparison of the measurements of *Rallicola takahe* (HOLLOWAY, 1955) with species of the same genus from kiwis shows the former to be about twice as big as the latter, although its host species, the flightless rail, *Notornis mantelli*, is approximately the same size as the brown kiwi. Hence HARRISON's theory that *Rallicola* represents the type of Ischnoceran mallophaga found upon extinct moas is now less convincing.

The relationship between kiwis and rails, which had previously been postulated by FÜRBRINGER (1888), still finds wide support – particularly among entomologists – because the genus *Rallicola* is the universal feather louse (infesting all genera) of rails; and, in general, phyletically related mallophagan parasites commonly occur on similarly related host species. CLAY (1949) has since shown that mallophaga of the *Rallicola* complex (i.e. *Aptericola*, *Wilsoniella*, *Parricola*, *Furnaricola*) occur on hosts belonging to several orders (Apterygiformes, Galliformes, Gruiformes, Charadriiformes, Cuculiformes and Passeriformes). More recently CLAY (1961) found a second genus, *Apterygon*, infesting kiwis. The affinities of this genus are uncertain but it may belong to a rather generalised *Menacanthus* group which is found on a wide range of hosts, including the Galliformes, Tinamiformes, Columbiformes and Passeriformes (CLAY, pers. comm.). The genus *Apterygon* has many distinctive characters and like *Aptericola*, is represented by at least three species. These features suggest that the ancestors of both genera were parasitic on the ancestral kiwi and diverged with the divergence of the host species. CLAY (pers. comm) believes that *Apterygon* is probably derived from a primitive Menoponid ancestor and that *Aptericola* stems from an ancestral *Rallicola* stock which was either parasitic on a rallid stock ancestral to the kiwi or secondarily acquired by an ancestral kiwi.

Although related birds tend to have related mallophagan parasites, the widespread occurrence of *Rallicola* and the somewhat uncertain affinities of *Apterygon* preclude their use as a totally dependable guide to the phylogeny of the host species – especially as the distribution of the former may be partly due to secondary infestations (CLAY, 1949, 1950, 1953).

Notwithstanding these provisos some anatomical features are also

Plate 3. Little spotted kiwi *(Apteryx oweni)*. Mount Bruce Native Bird Reserve. N.Z. Wildlife Service.

shared by kiwis and rails. OLSON (1973) recently observed that in young rails the ilium and ischium are not fused posteriorly. This condition of the pelvis which does not persist in the adult stage of any known rail, even the neotenic flightless species, is very similar to that found in kiwis and moas. Furthermore, the span of time needed to evolve flightlessness in this ancient family, in OLSON's view, can probably be measured in generations rather than in millennia because the process does not involve the development of new structures but merely the retention of juvenile skeletal characters and proportions. OLSON believes New Zealand was invaded twice by *Rallus philippensis* and that the first invasion gave rise to the large flightless weka *(Gallirallus australis)* which, excluding its well developed tail, superficially more closely resembles the brown kiwi than any other bird. As *Gallirallus* and *Apteryx* share the same habitat and have similar feeding habits the likeness may be due more to convergence than to kinship, for SIBLEY's researches show no clear evidence of relationship between rails and kiwis.

b. Tinamous

Tinamous have certain features in common with ratites: a palaeognathous

(= dromaeognathus of Huxley) palate in which the pterygoids come into firm contact with the prevomers, heavy legs, a penis and the absence of a pygostyle; and Garrod's classification (1874) placed tinamous, kiwis and the large ratites together in the 'Cohort Struthiones'. Although McDowell (1948) and Hofer (1949) independently showed that, in its definitive form, the palaeognathous palate consists of four different types, further investigation of embryological material is required since their findings do not entirely preclude possible ancestral relationships between any of the palaeognathine species; and several workers believe there is some kind of affiliation between the volant tinamous and the ratites. Parker (1862) noted a close resemblance between the rhamphothecae (the horny covering of the bills) of tinamous, ostriches and kiwis. This structure has since been studied in more detail by Parkes & Clark (1966) who have shown the remarkably consistent pattern displayed in all the ratites and in the tinamous is not matched in the other orders of birds. While admitting this character on its own proves nothing, Parkes & Clark believe it is additional evidence for a phylogenetic, rather than a convergent, relationship and they consider that while the ratite-tinamou assemblage is polyphyletic at the generic level, it is probably monophyletic at the family level.

The rheas have attained the ratite grade most recently. Their wings are the least reduced and their similarities to the tinamous led McDowell (1948) to combine the Rheidae and Tinamidae into a single order. The kiwi, on the other hand, has the most reduced wing and while this, along with other features, indicates an early deviation from carinate stock it still has many features more in common with tinamous than with the cursorial ratites. In both families the hallux is elevated, the egg is large relative to body size, females are larger than males and both are omnivores. Tinamous, although able to fly, are adapted to a life on the ground, conceal themselves in burrows, indulge in some nocturnal activity and the male is responsible for incubation and defending the young (Sick, 1964). Tinamous tend to be smaller than kiwis but there is some overlap in body size.

Osuga & Feeney (1967) conclude from their immunochemical studies that although the tinamous are quite closely related to the ratite group, this relationship is less close than that of the ratites to each other; and Tyler & Simkiss (1959), from investigations on the structure and chemical composition of eggshells, believe tinamous probably have a closer kinship to kiwis than to the other ratites. Although Sibley (1967 and pers. comm.) obtained very similar starch gel and electrophoretic patterns of egg-white proteins from tinamous and kiwis, his subsequent studies using the more sophisticated 'isoelectric focusing' technique indicate that a close apterygid-tinamid relationship is more apparent than real and that, at most, these two families may be only distantly related.

The kiwi remains a mystery. Though some of the features it shares in common with the cursorial ratites can be attributed to convergence, other features are more indicative of a phylogenetic relationship. Similarly, the kiwi shares certain features with both rails and tinamous but there seems little evidence of a rallid-tinamid relationship. One family of kiwi feather-lice supports a rallid relationship, but the other family of lice suggests a possible affinity with the tinamous and Galliformes. Although some galliform species and tinamous (in common with kiwis) have a penis, and SIBLEY's investigations indicate a relationship between these two groups, the Galliformes lack the rhamphothecal structure characteristic of ratites and tinamous – and it is difficult to conceive of a phylogenetic bill character which would appear in tinamous and kiwis, but not in similar-feeding Galliformes if all three are related.

Although JEHL (1971), from a study of the colour patterns of downy young, found no evidence for a close relationship between the tinamous and any of the ratites (including the kiwi), McFARLANE (1971), in his study of avian spermatozoa, found that the spermatozoa of the Galliformes and of the tinamous have an axonemal sheath – a character not present in any other birds including the one ratite (ostrich) so far studied. This possible relationship contradicts the findings of GYSELS (1970), whose studies on eye lens proteins, led him to conclude that the tinamous are more closely related to the ratites, and particularly the rhea, than they are to the Galliformes.

The kiwi shows affinities with rails, tinamous and the large ratites, and although we can only conjecture as to which is the most convincing relationship the present state of knowledge leads us to concur with the recommendations of SIBLEY & FRELIN (1972), which is: that the kiwis be treated as by WETMORE (1960), namely, as the Order Apterygiformes, placed next to the Tinamiformes; and that the Tinamiformes be placed near the Galliformes.

Ratites and their Relationships

The fossil record does not give any clues on the phylogenetic relationship of the ratites to each other; nor does it provide any positive links between any of these orders and any carinate orders, and for more than a century, workers have argued as to whether these flightless birds have had a common ancestry, or have evolved convergently from unrelated stock.

Superimposed on this argument was the debate on whether the ratites were 'degenerate' offspring of flying predecessors (as believed by OWEN, 1841 and PARKER, 1891), or were primitive birds descended directly from a non-flying reptilian ancestor. The latter view which was endorsed by LOWE (1928, 1933, 1942), HOLMGREN (1955) and BLOTZHEIM (1958) has since been shown to be untenable (see DE BEER, 1956 and 1964 and papers cited therein) and today few would dispute that the ratites descended from carinate stock.

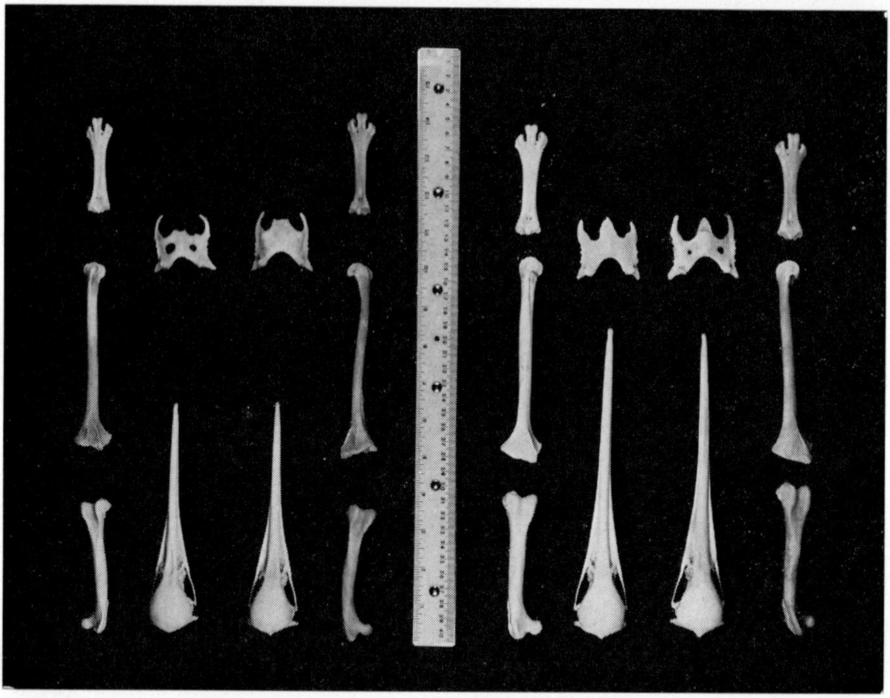

Plate 4. Skulls, sterna, and leg bones of two adult males and two adult females of the North Island brown kiwi. The true size difference between the sexes is shown by the relative lengths of the leg bones – the bill length of the female is disproportionately long. The sternum, which lacks a keel, is very reduced and irregularly shaped. N.Z. Wildlife Service.

The debate about monophyletic versus polyphyletic origin of the ratites, however, has continued with outspoken adherents to either position ever since 1831 when LESSON first included the kiwi in this group. LOWE's belief that the ratites represent a monophyletic, natural group was shared by CHANDLER (1916). More recently BOCK (1963) from cranial evidence and MEISE (1963) from behavioural characters, both concluded that the large ratites, kiwis and tinamous should be regarded as a monophyletic group. Detailed anatomical studies, however, had led FÜRBRINGER (1888) to conclude the ratites were not a uniform group and McDOWELL (1948), HOFER (1949), MAYR & AMADON (1951) and STRESEMAN (1959) also support a polyphyletic origin.

Several workers (HARRISON, 1915; OLIVER, 1945; EICHLER, 1948; CLAY, 1950 and 1957; TYLER & SIMKISS, 1959; SIBLEY, 1960 and OSUGA & FEENEY, 1967), although acknowledging that many similarities in the ratites are the result of convergent evolution, still consider that some of these flightless birds may have shared a common ancestor because of their closer similarities to each other.

Anatomy

Skeleton

The skeleton of other ratites is distinctly pneumatized (Bellairs & Jenkins, 1960); but with the exception of the skull the skeleton of the kiwi lacks air sacs. The dried skeleton comprises between 6.2 and 8.0 per cent of the total live weight (Plate 4). Whereas the wing bones represent less than one per cent, the leg bones make up more than 45 per cent by weight of skeletal material. There are 46 vertebrae – 15 cervical, 8 thoracic, 14 fused synsacral (of which three are thought to be true sacral vertebrae) and nine caudal. None of the latter are fused to form a pygostyle.

The skull is remarkable in having no inter-orbital septum as such – the enlarged nasal capsules intervene between the orbits, as in most mammals.

The vertebral parts of the ribs are very wide and, as in penguins, the uncinate processes remain unfused. Both the coracoid and scapular are reduced and fused to form a single bone. There is no clavicle. In kiwis and cassowaries the process of digital reduction in the forelimb has gone further than in other living birds and all digits except one (probably the third) are rudimentary or absent. In the foot one of the tarsal elements (the astragalus) remains unfused and separate. In contrast with other ratites which have only three toes (or in ostriches, two) the kiwi has four.

Soft parts

A pronounced reduction in the mass of the musculature of the pectoral area is associated with the meagre wing development, and some muscles are absent (Beddard, 1898, Berger, 1960).

The heart, which represents on average 0.66 per cent of the body weight, has a simple sinus venosus similar to that in reptiles (Adams, 1937). Glenny (1963) noted that in the patterns of its main cervical and thoracic arteries, the kiwi does not show particularly close affinities to the other ratites.

In common with them, however, kiwis have an archecentric gut (Mitchell, 1901) which is highly modified and includes a supra-duodenal loop. The proventriculus is small, the gizzard moderately muscular and there are two well-developed caeca of equal length. The total length of the alimentary canal (including the caeca, each of 19 cm) varies between 215 and 245 cm. Verheyen (1960) reports that the uropygial gland, which is absent in other ratites, is large but lacks a nipple.

The brain, which weighs between 9.4 and 10.9 g, was first described by Owen (1871). It has been studied subsequently by Hunter (1923),

Craigie (1929, 1930, 1935) Durward (1932) and Krabbe (1957, 1959). In brief, the structure of the forebrain is typically avian, except that the sessile olfactory bulbs – each of which measures 12 mm in diameter (Bang & Cobb, 1968) – are exceptionally and notably developed and have a structure similar to that in mammals. The tactile system is also extensively developed, especially the dorsal sensory centres of the fifth cranial nerve. On the other hand, the cerebellum is small, its deep nuclei are simple and the optic lobes and related parts are very reduced.

Physiology

Sight

With the exception of the southernmost (Stewart Island) race of the brown kiwi, members of which frequently emerge from their burrows to forage at early morning, evening or on overcast days, kiwis are essentially nocturnal. They avoid bright light; and simple observations of both captive and wild birds suggest that they are at even more of a disadvantage in daylight than are owls, and perhaps have vision akin to that of moles. Their small, degenerate eyes – like those of all other birds (and contrary to the findings of Owen, 1841) – contain a pecten. This organ, which characteristically is extremely vascular and extensively pleated with numerous folds or vanes, is reduced to a smooth narrow cone (Sillman, 1973) and closely resembles the conus papillaris of some reptiles (Walls, 1942). Barlow & Ostwald (1972) have pointed out that the pecten, generally, is in a good position to shade the area centralis from the sun's glare; but its unusual alignment and very reduced size would preclude this function in the kiwi. Wenzel's (1968) experiments have clearly shown that vision is not needed for efficient feeding.

Hearing

The presence of large ear openings and the fact that the kiwi cocks its head with its ear pointing in the direction of distant or soft noises suggest hearing is probably important and fairly well-developed, but no critical work has yet been done on this sense.

Smell

Most birds are thought to have little or no sense of smell, but kiwis have long been considered an exception. Not only are the olfactory bulbs well developed, but the olfactory chamber is particularly large and the functional epithelial surface area is further increased by the presence of about five transverse folds on the concha (Portmann, 1961). Furthermore, kiwis are the only birds with external nares at the tip of the bill.

There are several reports of kiwis in captivity, after sniffing their standard daily ration, refusing to eat if it was prepared by other than their usual zookeeper; and of generally placid birds either attacking or running from people in their enclosure if their greeting sniffs gave no clues of the intruders' identity. Numerous observations indicate that kiwis search for food by using the bill (e.g. BULLER, 1888, BENHAM, 1906, STRONG, 1911); but WENZEL's (1968) experiments on captive birds demonstrated conclusively that they are capable of locating foods by olfactory clues alone.

Body temperature

FARNER *et al.* (1956) measured body temperatures and found that those at night (38.2–39.9 °C) were higher than those during the day (36.4–37.2 °C). The weighted mean difference of 1.76 °C was statistically highly significant; and these workers considered that captive North Island brown kiwis, at least, have lower body temperatures than those of other birds of about the same body size.

Blood and blood formation

POOLE (pers. comm.) from a single observation, reported an erythrocyte count of $1.9 \times 10^6/mm^3$. This compares closely with the 1.89×10^6 found in ostrich blood but is much less than the mean of about 3×10^6 found in carinate birds (STURKIE, 1954). Also, POOLE reported the corpuscular volume to be about 38 per cent of total blood volume. This implies that kiwi erythrocytes must approach in size those of the ostrich which has the largest so far reported for birds (WELTY, 1962).

In a study of red blood cell formation in eight species of birds ENGELBERT & YOUNG (1970) found that the usual clone cell method of erythropoeisis was well developed in kiwis. Though one of the two kiwis involved had a higher percentage of new cells than any of the other birds sampled, it would be unwise to attach any significance to this single observation.

Longevity

The average life span for kiwis in the wild is not known since no banding studies have yet been undertaken, but they are probably long-lived. According to RUTGERS & NORRIS (1970) a little spotted kiwi lived for 20 years at London Zoo; and an adult female brown kiwi sent to Sydney Zoo in 1956 (i.e. 18 years ago) is still active and in good health in at least her 20th year.

Displays

No formal displays have been reported, and while it can be argued that kiwis have little to display with – no tail, no visible wings and no obvious colour patches – the use of the long bill and flexible neck in conjunction with various trunk movements would permit many complex movements. If displays are not part of the species' social repertoire it is probably because their nocturnal habits and poor vision annul the value of these rituals.

Nothing is known of the size of the territory (if any), duration of pair bonds or population ecology of any species of kiwi. For general and anecdoted accounts of behaviour of wild and captive birds, BULLER's of 1888 and 1905 and POTTS' of 1872 are still the most interesting and informative.

Breeding Biology

It is not known what external stimulus triggers breeding. Neither light nor temperature seem to be the primary initiators for the peak egg-laying period extends from July to February (OLIVER, 1955, FALLA *et al.*, 1966) and laying occurs in all months of the year (KINSKY, 1971).

The male, in common with all other ratites, has a well-developed muscular penis (OWEN, 1849, CAITHNESS, 1971) and the female has a clitoris. In nearly all birds only the left oviduct and ovary develop to maturity; but KINSKY (1971) has shown a functional right ovary is always present in kiwis and, if more than one egg is laid in a season, ovulation occurs alternately in each with the eggs being laid through the left oviduct. Some adult females even retain a vestigial right oviduct.

The age of sexual maturity is not known. One North Island brown kiwi captured as a chick first laid at the age of four years and nine months (ROBSON, 1958) and a male at Sydney Zoo successfully fertilized eggs when it was about 30–32 months old (CLAYTON, 1972). At the other extreme, an adult female of unknown age when taken to Sydney produced nine chicks over a five-year period, and the last of these was hatched during her fifteenth year at that zoo.

REISCHEK (1930) reported that, in the wild, males fight vigorously when competing for a female – but observations on captive birds suggest the female, who will occasionally vigorously kick her smaller mate if she is not feeling receptive to his advances, is usually the more assertive.

After a partnership is formed there is considerable burrowing, or otherwise preparing a nest site, by both birds for about two months before the egg is laid. The nest may be in a hollow log, a rock fissure, or may be excavated underground. The floor and sides of the chosen cavity are usually lined with fern fronds and other vegetation. ROBSON (1958), who bred North Island brown kiwis in captivity, noted that the male did more of the initial nest building and was thereafter almost exclusively responsible for its subsequent maintenance.

Males approaching breeding condition in captivity may indulge in a series of vigorous antics – leaping and hopping in great strides around their enclosure and then lying on their backs and kicking their feet in the air; or else rolling, either 'head over heels' or sideways down banks and inclines.

Copulations start about 18–22 days before the first egg of the season is laid and, at the peak of activity, occur three or more times a day. The male initiates mating by either tapping or stroking the female on her back near the base of the neck with his bill. She responds by crouching low with her neck and head extended forward and resting on the ground. The male then mounts and, lacking the stabilising role of wing movements, often grasps the hen's back feathers in his bill to maintain balance. There are no post-copulatory displays and the entire procedure lasts about five minutes.

Captive females may lay up to five eggs in a continuous series, but in the wild clutches contain one or two or, very rarely, three eggs. The interval between two consecutive eggs ranges from 11 to 57 days, the mean being 33 days (Robson, 1958, Kinsky, 1971). For several days prior to laying the female's food intake increases greatly but she then fasts during the one or two days immediately preceding laying. During this period the presence of the egg causes her to walk with legs wide apart.

The routine during early incubation varies. The first egg of the season may sometimes be left unattended for 2 to 6 days after laying or the female may share in the incubation for the first week or two. Thereafter, the male, who carries ample fat reserves to supplement reduced and irregular feeding during the nesting period, alone sits on the eggs which are warmed by an extensive incubation patch that extends from the anterior margin of the sternum to the cloaca and includes the inner sides of the thighs. The incubation period ranges from 74 to 84 days and this range probably reflects the amount of cooling which results from absenteeism by the sitting bird which may either fast for periods of up to seven days, or leave the nest once or twice on most nights to seek food. On leaving, the male frequently covers the eggs and burrow entrance with litter and, on returning, often brings in extra vegetation to maintain the nest lining. The male of a captive pair which sat on a five-egg series remained so employed for seven months and lost about 900 g in body weight in the process (Robson, 1958).

Females have been seen to enter the nest burrow with live food in their bills and this suggests they may occasionally feed the incubating male. Sometimes the female will briefly cover the eggs while the male is out feeding, and she has also been observed to insulate the temporarily-vacated nest entrance with a covering of debris.

It seems from marked eggs that, although they are not touched during the early part of incubation, they are turned or rolled over daily by the male during the latter half; and when the clutch is greater than one, a viable egg may become completely buried and lost under the nest

Plate 5. North Island brown kiwi *Apteryx a. mantelli*: Female and egg. National Museum.

materials. Addled eggs and dead chicks are buried in the burrow. FRANCIS (1971) observed that the incubating male, while out feeding, may bathe to dampen its plumage 'so as to maintain the correct humidity in the burrow'.

The eggs of all species have smooth, thin, unmarked white or greenish-white shells. They are, especially in the little spotted kiwi, proportionately larger relative to the size of the adult female than the eggs of any other bird – with the doubtful exception of those of some of the smallest storm petrels. (Plate 5, Fig. 2). Eggs of the little spotted kiwi range from 275–370 g with a mean of 310 g (108.5 mm × 71.0 mm) and are 25 per cent of the female's body weight. The North Island brown kiwi's egg weighs between 330 g and 520 g (mean 440 g, 125.5 mm × 78.5 mm) and is from 18.0 to 22.0 percent of the female's body weight (REID, 1971a, 1972b).

It is also probable that kiwis' eggs are the most nutritive laid by birds and this is very likely correlated with the advanced state in which the chick hatches. The contents of a fresh egg were found to be 61 per cent yolk and 39 per cent albumen (REID, 1971b). For comparison, the eggs of precocial species (which hatch active, open-eyed, downy young) contain from 35–45 per cent yolk; those laid by altricial species (which hatch

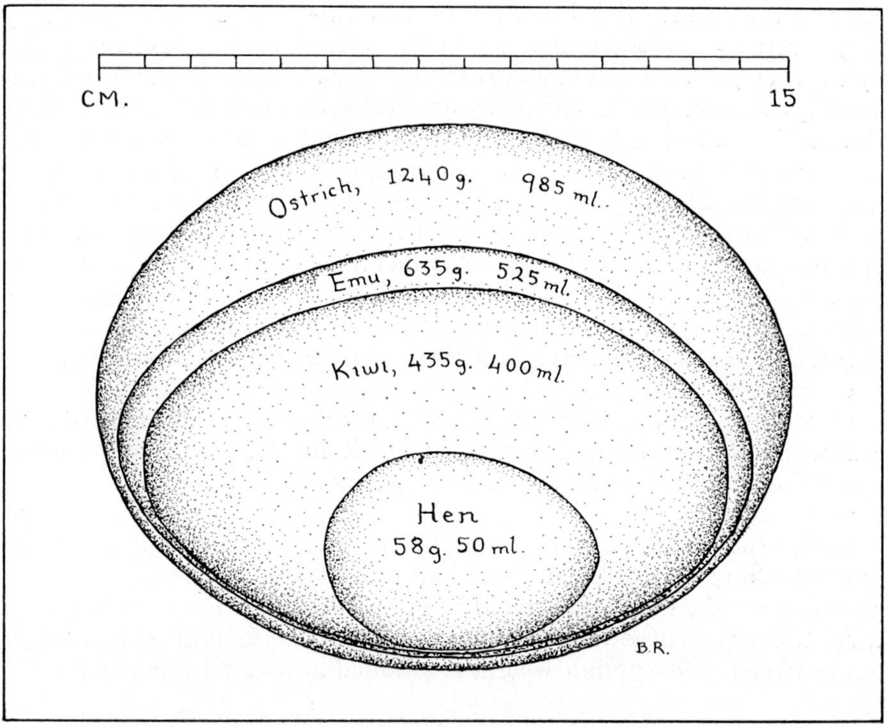

Fig. 2. Eggs of various ratites and of the domestic fowl – weights are for fresh eggs and include the shell, but volumes are internal. The shell of the emu egg is about four times, and the shell of the ostrich egg about eight times the weight of the kiwi egg shell. Total weights, therefore, exaggerate the size differences between the kiwi egg, and eggs of the larger ratites, as the weight of the ostrich egg is nearly three times the weight of the kiwi egg, but its internal volume is only about two-and-a-half times as great. The 58 g hen egg and the 435 g kiwi egg are laid by birds weighing about 2.4 kg. N.Z. Wildlife Service.

helpless, blind, naked young) contain from 13–28 per cent yolk (ROMANOFF & ROMANOFF, 1949).

At hatching, the male blocks the entrance to the nest and, after the chick has emerged, eats some of the shell. The remainder is stamped down, broken and buried in the nest lining. The newly-hatched brown kiwi chick, which weighs about two-thirds of the weight of the fresh egg, is unable to stand. Its abdomen, grossly distended with yolk, rests on the ground and its legs are splayed out sideways. It takes its first two or three ungainly steps when about 60 hours old – shuffling with its toes and metatarsi resting on the ground. On the fourth day it can stand, on its fifth day it can walk unsteadily on its toes alone and two days later it walks quickly and confidently.

Chicks probably do not feed during their first 72–84 hours; but thereafter and until they first leave the burrow two or three days later, they may receive some food which is carried in by the parent. Accompanied by the male, the chick first leaves the burrow to forage lightly when it is 5–8 days old and first eats with vigour when it is approximately 9–12 days old. At the end of its first 10–14 days its weight may be about one-third less than at hatching, but it is well equipped to withstand nearly two weeks of partial fasting. In one chick, which died a few hours after hatching and at death weighed 319 g, the food reserves (112 g of yolk in a fully enclosed sac and at least 43 g of body fat) amounted to about half of the total weight at hatching (REID, 1972a). Another chick that was killed outside its burrow (and was therefore probably at least five or six days old) weighed 281 g, including a still very substantial food reserve of 54 g of enclosed yolk.

In contrast to those of the brown kiwi, chicks of the little spotted kiwi are capable of standing and moving freely within 24–30 hours of hatching; but notwithstanding this early mobility, they also remain in the burrow for several days.

Unlike their parents, chicks up to the age of 30–35 days may leave the burrow at any time of the day or night to feed; but when they reach the age of about five weeks they avoid bright light. In captivity, the hatching weight is regained at the age of three to four weeks and is doubled at ten to twelve weeks; adult weight is attained at about 12 months.

Food Habits

All species of kiwis have the same general food habits but little is known about them in detail. BULLER's (1888) comments are typical of those of most other authors: earthworms are the principal item of diet and insects, land snails and berries from a variety of forest trees make up the rest.

The following list, which combines the findings from more recent investigations of the stomach contents of 38 North Island brown kiwis, indicates that this subspecies, at least, is truly omnivorous (GURR, 1952; BULL, 1959; WATT, 1971; BULFIN, pers. comm.). Although several remnants were not identified (and in two instances examination was restricted to insects and seeds) the variety of foods recorded is considerable: Invertebrates included earthworms, woodlice, millipedes and centipedes, slugs and snails, at least four species of spiders and about 32 species of insects. The latter consisted of crickets and wetas, ants, cicada nymphs, caterpillars and pupae of moths and adults and larvae of various beetles – including some which produce repellent secretions and smells unpleasant to man (WATT, 1971). Plant material included seeds and berries from one or more species belonging to the following indigenous or introduced genera – *Albizzia*, *Carex*, *Cirsium*, *Coprosma*, *Cordyline*, *Cyathodes*, *Dacrydium*, *Elaeocarpus*, *Gahnia*, *Geniostoma*, *Hebe*, *Leptospermum*, *Leycesteria*, *Melianthus*,

Mida, *Nestegis*, *Podocarpus*, *Pseudopanax*, *Rhopalostylis*, *Rubus*, *Solanum*, *Stellaria*, *Uncinia*, *Urtica* and *Vitis* – and leaves of *Coprosma* and *Cyathodes*.

A cursory inspection of another two stomachs showed that the bulk of the contents consisted of fibrous, grass-like, plant material. Observations on the performance of kiwis kept in captivity indicate that, although earthworms may form a very important part of their diet, some vegetable matter (berries and leaves) is essential for their long-term well-being (Reid, 1970); and most zoos keeping kiwis now supplement the meat and worm ration with raisins and nuts and berries of such plants as *Cotoneaster* and *Pyracantha*. Nothing is known in detail of what constitutes a balanced diet; but the importance of plant material to the feral kiwi may be inferred from the composition of its skin fat (Shorland & Gass, 1961): Those unsaturated fatty acids which are more characteristic of seeds and green leaves than earthworms are present in such quantity (16% of the total) that it is obvious that there is a significant vegetable component in the diet.

Studies on kiwis in captivity have shown individual birds to have very different food preferences. Some, rejecting natural foods such as earthworms or huhu grubs (Coleoptera), readily accept alien foods such as meat, breakfast cereals, spaghetti or dried fruits. Others ignore these in favour of natural items.

When feeding on worms kiwis probe their bills into the animals' tunnels and, by a back-and-forth levering action, enlarge them into a conical hole that may be more than 12 cm deep and up to 10 cm wide at ground level. This type of feeding is accompanied by frequent snuffling as the kiwis exhale forcefully to perhaps clear soil from their nasal passages.

Between 1.5 and 20.5 g of grit have been found in all gizzards inspected. The number of stones present has varied from four to 225 and they ranged in diameter from about 2 mm to more than 10 mm. Their presence emphasises the importance to kiwis of foods other than soft-bodied earthworms.

Parasites and Diseases

Parasites

a. Ectoparasites: Two genera of mallophagan feather lice – *Rallicola* (= *Aptericola* of Harrison, 1915) and *Apterygon* – have been recorded on kiwis (Pilgrim, pers. comm., Clay, 1972; Tandan, 1972). Both genera contain three known species and although there is some uncertainty about the identification of some host material, in those examples where no confusion exists the parasites have been found to be host – specific, with each species of louse confined to a particular species or subspecies of *Apteryx:*

Host	Mallophagan parasite	
	Rallicola	*Apterygon*
A. a. mantelli	*R. gadowi* (Harrison)	*A. mirum* (Clay)
A. a. australis	*R. gadowi* (Harrison)	?
A. a. lawryi	*R. gadowi* (Harrison)	*A. dumosum* (Tandan)
A. oweni	*R. pilgrimi* (Clay)	?
A. haasti	*R. gracilentus* (Clay)	*A. hintoni* (Clay)

Of the host species, there can be little doubt that *Apteryx oweni* and *A. haasti* are more closely related than either is to *A. australis* and CLAY (1972) believes that *R. pilgrimi* and *R. gracilentus* are more nearly related than either is to *R. gadowi*.

MASKELL (1896) described two species of ticks – *Ixodes apteridis* (from North Island brown kiwis) and *I. aptericola* (from South Island brown kiwis); but recent investigations (G. RAMSAY, pers. comm.) show the two are, in fact, one species and endorse the opinion of DUMBLETON (1953) that the kiwi tick *(I. apteridis)* is the same as *I. anatis* which infests some native ducks.

Mites of the genus *Megninia* have been identified (GARDNER, pers. comm.) from captive North Island kiwis and two new genera, *Kiwialges* and *Kiwilichus* have recently been described by GAUD & ATYEO (1970) – the former from all three species of *Apteryx*, the latter from *A. australis*.

b. Endoparasites: CHATIN (1885) identified the nematode *Apterycis* from the intestine of brown kiwis. In 1900 two tapeworms (*Drepanidotaenia minuta* and *Diapterygis*), belonging to the family Hymenolepididae, were described from the gut of North Island kiwis by BENHAM. Subsequent investigations have shown that other cestodes of the family Davaineidae (*Raillietina* sp. and *Davainea* sp.) also infest the North Island race which, in addition, harbours acanthocephalans (*Echinorhynchus* sp.) and at least four families of nematodes (*Tricharidae*, *Heterakidae*, *Acuariidae* and *Spiruridae*); but so far no trematodes have been reported (D. ELLITO, pers. comm.).

DISEASES

Although post-mortems have shown free-living kiwis are sometimes highly infected with parasites, there have been very few investigations on the effect of these on the host. Cestodes (*Raillietina* sp.) were considered to be the cause of focal catarrhal enteritis in one wild bird; and a three-month old captive kiwi died following invasion of its brain tissue by nematodes *(Spiruridea)*. The brain lesions in this chick were similar to those seen in the brain, liver and lungs of other kiwis – and it is now thought that these lesions were also due to parasitism (D. RUTHERFORD & C. SMITH, pers. comm.).

The distal position of the external nares and the long olfactory passage may make kiwis vulnerable to certain respiratory, or even pulmonary,

Plate 6. A New Zealand 20 cent coin showing the kiwi motif on its reverse side. Department of Scientific and Industrial Research.

diseases. SMITH *et al.*, (in prep.) have recorded pneumoconiosis in several that died in captivity; but it is likely that this condition would also develop in wild birds, as during dry periods, particularly, dust and dirt would be inhaled while feeding.

Predators

Prior to the advent of man to New Zealand some 1000 years ago kiwis were probably largely free from predation. The indigenous fauna lacks terrestrial mammals and effective avian predators as far as kiwis are concerned.

The Polynesians introduced the Polynesian rat *(Rattus exulans)* and the domestic dog; and the Europeans such potential predators as the ship and Norway rats (*R. rattus* and *R. norvegicus*), three species of mustelids and the domestic cat. Although all three mammals (and others) soon became widespread and often abundant, there is no evidence at present to indicate that they, separately or collectively, have seriously reduced the numbers and range of any species of kiwi, which persist in good numbers in many parts of New Zealand where various combinations of these mammals are known to occur.

Direct human predation does not appear to have been significant either, except occasionally and locally. The Maoris probably used the kuri or 'native' dog to locate occupied nests and it is said that they also attracted the birds by making light rustling (wormlike) noises in the undergrowth (NATUSCH, 1974); but, in general, kiwis were not taken in quantity for food; neither have they been exploited in this way by Europeans. However, they were taken in large numbers during the latter part of the 19th century for museums and zoos and for their feathers (the Maoris also used their feathers for making cloaks). One hunter is said to have taken over 2000 in once season, (POTTS, 1872) and there must have been many similar instances. Nowadays, kiwis are fully protected by law; nevertheless large numbers are killed accidentally each year by being caught in traps set for the introduced Australian possum, or destroyed in scrub- or forest-clearing schemes. In spite of all such exactions, no species or subspecies is in danger of extinction; and this reasonably satisfactory state of affairs is likely to continue so long as large areas of relatively unmodified indigenous forest and scrub remain.

Acknowledgements

We gratefully acknowledge the assistance of all those who supplied local data on distribution which Mr. M. HARRISON transformed into the distribution map.

We also thank the following – Mrs. F. R. ALLISON, Mrs. M. J. BULFIN, D. ELLIOTT, R. P. HANSEN, T. R. HARTREE, F. C. KINSKY, R. G. ORDISH, Prof. R. L. C. PILGRIM, W. S. H. POOLE, Dr. G. W. RAMSAY, Mrs. L. L. ROCKELL, C. D. RODERICK, B. ROWE, Dr. F. B. SHORLAND, Prof. C. G. SIBLEY, C. F. SMITH, E. G. TURBOTT and Prof. P. VELA – for giving advice, providing information or commenting on parts of the text.

REFERENCES

ADAMS, W. E. 1937. A contribution to the anatomy of the avian heart as seen in the kiwi and the yellow-crested penguin. *Proc. Zool. Soc., London.* 107B: 417–441.

Annotated Checklist of the Birds of New Zealand 1970. Ornithological Society of N.Z., A. H. & A. W. Reed, Wellington.

ARCHEY, G. 1941. The moa, a study of the Dinornithiformes. *Auckland Inst. and Mus. Bull.* 1: 1–119.

ATKINSON, I. A. E. 1973. Spread of the ship rat (*Rattus r. rattus* L.) in New Zealand. *J. Roy. Soc. N.Z.* 3: 457–472.

Bang, B. G. & Cobb, S. 1968. The size of the olfactory bulb in 108 species of birds. *Auk* 85: 55–61.

Barlow, H. B. & Ostwald, T. J. 1972. Pecten of the pigeon's eye as an inter-ocular eye shade. *Nature (New Biology)* 236: 88–90.

Bartlett, A. D. 1850. On the Genus *Apteryx*. *Proc. Zool. Soc. London*. 1850: 274–276.

Beddard, F. E. 1898. 'The Structure and Classification of Birds' Longmans, Green & Co., London.

Bellairs, A. D. & Jenkins, C. R. 1960. in Marshall, A. J. (Ed.), 'Biology and Comparative Physiology of Birds.' Vol. 1, Academic Press, New York.

Benham, W. B. 1900. The structure of the rostellum in two new species of tapeworm from *Apteryx*. *Quart. Micros. J.* 43: 83.

Benham, W. B. 1906. The olfactory sense in *Apteryx*. *Nature*, 74: 222–223.

Berger, A. J. 1960. in Marshall, A. J. (Ed.), 'Biology and Comparative Physiology of Birds'. Vol. 1. Academic Press, New York.

Blotzheim, U, Glutz von. 1958. Zur Morphologie und Ontogenese von Schultergürtel, Sternum und Becken von *Struthio*, *Rhea* und *Dromiceius*. Ein Beitrag zur Phylogenese der Ratiten. *Rev. Suisse Zool.* 65: 609–722.

Bock, W. J. 1963. The cranial evidence for ratite affinities. Proc., 13th Internat. Orn. Congr. Ithaca 1962: 39–54.

Brodkorb, P. 1971. in Farner, D. S. and King, J. R. (Eds.) 'Avian Biology' Vol. 1. Academic Press, New York.

Bull, P. C. 1959. Stomach contents of a North Island kiwi. *Notornis* 8: 143–145.

Buller, W. L. 1888. 'A History of the Birds of New Zealand.' 2nd Ed., Vol. 2. London: 318–319.

Buller, W. L. 1905. Supplement to 'A History of the Birds of New Zealand'. London.

Caithness, T. A. 1971. Sexing kiwis. *Internat. Zoo Yearbook* 11: 206–208. Zoological Society, London.

Chandler, A. C. 1916. A study of the structure of feathers with reference to their taxonomic significance. *Univ. Calif. Zool. Publ.* 13: 243–446.

Chatin, M. J. 1885. Parasites de l'Apterix. Comptes Rendus de la Soc. de Biol., Paris. 1: 770–771.

Christoffel, D. A. & Falconer, R. K. H. 1972. Marine magnetic movements in the south-west Pacific Ocean and the identification of new tectonic features. Antarctic Research Series, American Geophysical Union.

Clay, Theresa. 1949. Some problems in the evolution of a group of ectoparasites. *Evolution* 3: 279–299.

Clay, Theresa. 1950. A preliminary survey of the distribution of the Mallophaga in the class Aves. *J. Bombay Nat. Hist. Soc.* 49: 430–443.

Clay, Theresa. 1953. Revisions of the genera of Mallophaga – I. The *Rallicola*-complex. *Proc. Zool. Soc., London* 123: 563–587.

Clay, Theresa. 1957. The Mallophaga of birds. 1st Symposium on host specificity among parasites of vertebrates, Neuchatel: 120–155.

Clay, Theresa. 1961. A new genus and species of Menoponidae (Mallophaga: Insecta) from *Apteryx*. *Ann. Mag. Nat. Hist.* 3: 571–576.

Clay, Theresa. 1972. The species of *Rallicola* parasitic on Kiwis. *N.Z.J. Sci.* 15: 70–76.

Clayton, L. J. 1972. Breeding and behaviour of the kiwi at Sydney Zoo. *Internat. Zoo Yearbook* 12: 134–136. Zoological Society of London.

Craigie, E. H. 1929. The vascularity of the cerebral cortex in a specimen of *Apteryx*. Anatomical Rec., Philadelphia 43: 209–214.

Craigie, E. H. 1930. Studies on the brain of the kiwi. *Comparat. Neurol. Philadelphia* 49: 223–357.

Craigie, E. H. 1935. The cerebral hemispheres of the kiwi and of the emu. *J. Anat. London* 69: 380–392.

de Beer, G. 1956. The evolution of ratites. *Bull. Brit. Mus. (Nat. Hist.), Zool. Series* 4: 57–76.

de Beer, G. 1964. Phylogeny of the ratites. in Thomson, A. L. (Ed.), 'New Dictionary of Birds', Nelson London.
Dumbleton, L. J. 1953. The ticks of the New Zealand sub-region. *Bull. Cape Expedition, D.S.I.R.* 14. Govt. Printer, Wellington.
Durward, A. 1932. Observations on the cell masses in the cerebral hemisphere of the kiwi. *J. Anat. London* 66: 437–477.
Eichler, W. 1948. Some rules in ectoparasitism. *Amer. Mag. Nat. Hist. (Ser. 12)* 1: 588–598.
Engelbert, V. E. & Young, A. D. 1970. Erythropoiesis in peripheral blood of seven species of New Zealand and one species of Canadian birds. *Canad. J. Zool.* 48: 227–230.
Falla, R. A. 1964. Kiwi. in Thomson, A. L. (Ed.), 'New Dictionary of Birds'. Nelson. London.
Falla, R. A.; Sibson, R. B. & Turbott, E. G. 1966. 'A Field Guide to the Birds of New Zealand'. Collins, London.
Farner, D. S., Chivers, N. & Riney, T. 1956. The body temperatures of the North Island kiwi. *Emu* 56: 199–206.
Feeney, R. E. & Allison, R. G. 1969. 'Evolutionary Biochemistry of Proteins'. Wiley-Interscience, John Wiley, New York.
Fleming, C. A. 1962. The extinction of moas and other animals during the Holocene period. *Notornis* 10: 113–117.
Francis, K. E. 1971. 'The New Zealand Kiwi'. Whitcombe and Tombs, Wellington.
Fürbringer, M. 1888. 'Untersuchungen zur Morphologie und Systematik der Vögel'. Van Holkema, Amsterdam.
Garrod, A. H. 1874. On certain muscles of birds and their value in classification. Part II. *Proc. Zool. Soc. London* 111–123.
Gaud, J. & Atyeo, W. T. 1970. Acariens sarcoptiformes plumicoles parasites des Apterygiformes. *Acarologia* 12: 402–414.
Gibb, J. A. & Flux, J. E. C. 1973. Mammals, in Williams G. R. (Ed.) 'The Natural History of New Zealand: An Ecological Survey' 334–371. A. H. & A. W. Reed, Wellington.
Glenny, F. H. 1963. Patterns of the main cervical and thoracic arteries in six genera of flightless birds. *Amer. Zoologist.* 3: 488.
Gould, J. 1847. *Proc. Zool. Soc., London.* 1847: 93.
Gurr, L. 1952. Some food of the North Island kiwi. *Notornis* 4: 209–210.
Gysels, M. 1970. Some ideas about the phylogenetic relationships of the Tinamiformes, based on protein characters. *Acta Zool. Pathol. Antwerp.* 50: 3–13.
Harrison, L. 1915. Mallophaga from *Apteryx;* with a note on the genus *Rallicola. Parasitology*, 8: 88–100.
Hofer, H. 1949. Die Gaumenlucken der Vögel. *Acta. Zool. Internat. Tidskr. Zool.*, 30: 209–248.
Holloway, B. A. 1955. A new species of *Rallicola* from *Notornis mantelli. Rec. Dominion Mus. Wellington, Zool.* 2: 113–119.
Holmgren, N. 1955. Studies on the phylogeny of birds. *Acta Zoologica.* 36: 243–328.
Hunter, J. I. 1923. The forebrain of *Apteryx australis. Proc. K. Nederlandse Akad. Wetensch.* 26: 807–824.
Jehl, J. R. 1971. The colour patterns of downy young ratites and tinamous. *Trans. Soc. Nat. Hist. San Diego.* 16: 291–302.
Kinsky, F. C. 1971. The consistent presence of paired ovaries in the kiwi. *J. für Ornithol.* 112: 334–357.
Krabbe, K. H. 1957. On the unique formation of the brain in an embryo of the kiwi. *Encephale*, 52: 612–622.
Krabbe, K. H. 1959. Studies on the brain development of the kiwi. 'Morphogenesis of the Vertebrate Brain' VIII. Ejnar Munksgaard, Copenhagen.
Lesson, R. P. 1831. 'Traité d'Ornithologie'. Vol. 1. Paris.

LESSON, R. P. 1839. 'Voyage autour du monde . . . sur la corvette la Coquille.' N. J. Gregoir, Brussels.
LOVERN, J. A. 1940. The lipids of the earthworm. *Biochem. J.* 34: 709–711.
LOWE, P. R. 1928. Studies and observations on the phylogeny of the ostrich and its allies. I. *Proc. Zool. Soc. London.* 98B: 185–247.
LOWE, P. R. 1933. On the primitive characters of the penguins, and their bearing on the phylogeny of birds. *Proc. Zool. Soc. London* 103B: 483–539.
LOWE, P. R. 1942. Some additional factors bearing on the phylogeny of the Struthiones. *Proc. Zool. Soc. London.* 112B: 1–20.
McDOWELL, S. 1948. The bony palate of birds – the Palaeognathae. *Auk* 65: 520–549.
McFARLANE, R. W. 1971. The ultrastructure and phylogenetic significance of avian spermatozoa. Thesis, University of Florida, Graduate School.
MASKELL, W. M. 1896. On some tick parasites of the kiwi. *Trans. N.Z. Inst.* 29: 290–293.
MATHEWS, G. M. 1931. 'List of the birds of Australasia'. Taylor and Francis, London.
MAYR, E. & AMADON, D. 1951. A classification of recent birds. *Amer. Mus. Novitates,* 1496: 1–42.
MEISE, W. 1963. Verhalten der Straussartigen Vögel und Monophylie der Ratitae. *Proc. 13th Internat. Ornithol. Congr.*: 115–125.
MILDENHALL, D. 1974. The record of the rocks: in 'New Zealand's Nature Heritage', Vol. 1(2): 43–46, Paul Hamlyn, Wellington.
MITCHELL, P. C. 1901. On the intestinal tract of birds. *Trans. Linnean Soc. London, (Zoology)* 8: 173–275.
NEWTON, A. 1893. 'A Dictionary of Birds'. Adam and Charles Black, London.
OLIVER, W. R. B. 1945. Avian evolution in New Zealand and Australia. *Emu* 45: 55–77, 119–152.
OLIVER, W. R. B. 1955. 'New Zealand Birds'. A. H. & A. W. Reed, Wellington.
OLSON, S. L. 1973. Evolution of the rails of the South Atlantic Islands. *Smithsonian Contrib. Zool.* 152: 1–53.
OSUGA, D. T. & FEENEY, R. E. 1967. Biochemistry of the egg-white proteins of the ratite group. *Arch. Biochem. Biophys.* 124: 560–574.
OWEN, R. 1841. On the anatomy of the southern apteryx *(A. australis)*, I. *Trans. Zool. Soc. London.* 2: 257–296.
OWEN, R. 1849. On the anatomy of the *Apteryx australis*, II. *Trans. Zool. Soc. London.* 3: 277–300.
OWEN, R. 1871. On Dinornis (Part 16). *Trans. Zool. Soc. London.* 7: 381–394.
PARKER, T. J. 1891. Observations of the anatomy and development of Apteryx. *Phil. Trans. Roy. Soc. London* 182: 25–134.
PARKER, W. K. 1862. On the osteology of gallinaceous birds and tinamous. *Trans. Zool. Soc. London,* 5: 149–241.
PARKES, K. C. & CLARK, G. A. 1966. An additional character linking ratites and tinamous, and an interpretation of their monophyly. *Condor,* 68: 459–471.
PORTMANN, A. 1961. in MARSHALL, A. J. (Ed.), 'Biology and Comparative Physiology of Birds'. Vol. 2. Academic Press, New York.
POTTS, T. H. 1872. On the birds of New Zealand. *Trans. N.Z. Inst.* 5: 186–196.
PYCRAFT, W. P. 1900. On the morphology and phylogeny of the Palaeognathae and Neognathae. *Trans. Zool. Soc. London.* 15: 149–290.
REID, BRIAN, 1970. Feeding kiwis in captivity. Wildlife 1970 – A Review. Dept. Internal Affairs, Wellington.
REID, BRIAN, 1971a. The weight of the kiwi and its egg. *Notornis* 18: 245–249.
REID, BRIAN, 1971b. Composition of a kiwi egg. *Notornis* 18: 250–252.
REID, BRIAN, 1972a. North Island brown kiwi *(Apteryx a. mantelli)* Measurements and weights of a young chick. *Notornis* 19: 261–266.
REID, BRIAN, 1972b. Kiwi eggs laid at Wellington Zoo. *Notornis* 19: 276–277.
REISCHEK, A. 1930. 'Yesterdays in Maoriland' Whitcombe and Tombs, Wellington.
ROBSON, F. D. 1958. Kiwis in captivity. Bull. Hawkes Bay Mus. Napier.

ROMANOFF, A. L. & ROMANOFF, A. J. 1949. 'The Avian Egg'. John Wiley, New York.
ROTHSCHILD, L. W. 1893. Note on *Apteryx haasti*. *Ann. Mag. Nat. Hist. (Ser 6)*. 11: 43–45, 99, 300.
RUTGERS, A. & NORRIS, K. A. 1970. 'Encyclopaedia of Aviculture'. Vol. 1. Blandford, London.
SCARLETT, R. J. 1970. A small woodhen from New Zealand. *Notornis*, 17: 68–74.
SHAW, G. & NODDER, F. P. 1813. 'The Naturalist's Miscellany'. Vol. 24. Pl. 1057.
SHORLAND, F. B. & GASS, J. P. 1961. Fatty acid composition of the depot fats of the kiwi. *J. Sci. Food Agric.*, 3: 174–177.
SIBLEY, C. G. 1960. The electrophoretic patterns of avian egg-white proteins as taxonomic characters. *Ibis*, 102: 215–284.
SIBLEY, C. G. 1967. Proteins: history books of evolution. *Discovery* 3(1): 5–20.
SIBLEY, C. G. & FRELIN, C. 1972. The egg-white protein evidence for ratite affinities. *Ibis* 114: 377–387.
SICK, H. 1964. Tinamou in THOMSON, A. L. (Ed.), 'A New Dictionary of Birds'. Nelson, London.
SILLMAN, A. J. 1973. In FARNER, D.S . & KING, J. R. (Eds.) 'Avian Biology'. Vol. 3. Academic Press. New York.
STORER, R. W. 1960. in MARSHALL, A. J. (Ed.), 'Biology and Comparative Physiology of Birds'. Vol. 1, Academic Press, New York.
STRESEMAN, E. 1959. The status of avian systematics and its unsolved problems. *Auk.* 76: 269–280.
STRONG, R. M. 1911. On the olfactory organs and the sense of smell in birds. *J. Morphol.* 22: 619–661.
STURKIE, P. D. 1954. 'Avian Physiology'. Comstock, New York.
TANDAN, B. K. 1972. The species of *Apterygon* parasitic on kiwis. *N.Z. J. Sci.* 15: 52–69.
TYLER, C. & SIMKISS, K. 1959. A study of the egg-shells of ratite birds. *Proc. Zool. Soc. London* 133: 201–242.
VERHEYEN, R. 1960. Les Kiwis (Apterygiformes) dans les systèmes de classification. *Bull. Soc. Roy. Zool. d'Anvers* 15.
WALLS, G. 1942. 'The Vertebrate Eye and its Adaptive Radiation', Cranbrooke, Bloomfield Hills, Michigan.
WATT, J. C. 1971. The North Island kiwi: a predator of pasture insects. *N.Z. Entomol.* 5(1): 25–27.
WELTY, J. C. 1962. 'The Life of Birds'. Saunders, Philadelphia.
WENZEL, B. M. 1968. Olfactory prowess in the kiwi. *Nature* 220: 1133–1134.
WETMORE, A. 1960. A classification for the birds of the world. *Smithsonian. Misc. Coll.* 139(11): 1–37.
WILLIAMS, G. R. 1973. Birds in WILLIAMS, G. R. (Ed.) 'The Natural History of New Zealand. An Ecological Survey': 304–333. A. H. & A. W. Reed, Wellington.

VIII. THE TUATARA

by

I. G. CROOK

The interest which the tuatara *(Sphenodon punctatus)* (Figs 1–4) has aroused amongst scientists stems from its affinities with an otherwise extinct order of reptiles, the Rhynchocephalia. Rhynchocephalians occur in fossil beds originating from the lower Triassic (*ca* 200 million years B.P.) through the Jurassic and into the lower Cretaceous (to about 135 million years B.P.). They seem to have been at their most abundant in the latter part of the Triassic and during the Jurassic because fossils from these periods are both varied and common, particularly in South America and Africa. However the fossil record peters out during the Cretaceous period. The rich Cretaceous beds of Europe and South America have yielded only a few small members of the order, and consequently paleontologists consider it to have become extinct sometime during this period, or at least by the time it ended 60 million years ago.

However there was one exception. The living tuatara undeniably points to the survival of the order, even though it is outwardly hard to distinguish from many lizards. Indeed JOHN GRAY, who first described the tuatara, from a skull sent to him from New Zealand, declared that it was a lizard (GRAY, 1831). He called it *Sphaenodon* (now *Sphenodon*) because of the large wedge shaped teeth in the front of the upper jaw. These overhang the lower jaw and give the skull a beak-like appearance.

Then, about 1840, Dr. E. DIEFFENBACH, a naturalist who travelled widely in New Zealand in the mid-nineteenth century, brought GRAY some whole specimens, but GRAY did not connect these with the skull he had described ten years previously. He named this 'new' animal *Hatteria punctata* (GRAY, 1842) and described it as an Agamid lizard.

To cap the confusion, Sir RICHARD OWEN compared features of the skulls of the tuatara and some fossil reptiles from South Africa (OWEN, 1845), and in the process gave the tuatara a third name, *Rhynchocephalus*. This comes from the Greek *rhynchos* 'snout' and *kephale* 'head', recognising the same feature that prompted GRAY's first name.

It was another 22 years before Dr. ALBERT GÜNTHER, GRAY's successor as zoologist at the British Museum, recognised that GRAY's and OWEN's animals were one and the same and announced the tuatara was not a lizard! He proposed it should be placed in a separate order and erected a new one for the purpose – Rhynchocephalia (GÜNTHER, 1867).

Fig. 1. Adult male tuatara, length from snout to vent ca. 260 mm, overall length (if tail unbroken) would be a little more than double this.

Fig. 2. Adult male tuatara at the entrance of its burrow.

Fig. 3. Adult male tuatara eating a gecko *(Hoplodactylus pacificus)*.

Fig. 4. Hatchling tuataras.

GÜNTHER based his assessment on features of the skeleton of the tuatara, one of the most important of which is the division of the temporal fossae, large openings in each side of the skull behind the orbit, into two parts by a rigid bony arch. Another is the presence of uncinate or hook-like processes on the ribs similar to those found in the crocodiles and birds but which are absent in the lizards.

Several features of the skeleton which are characteristic of the order are also primitive ones, for example, the bony protuberances from mandible and palate which serve as teeth, the form of the palate itself, the parietal foramen on the dorsal midline of the skull at the junction of the frontal and parietal bones and the amphicoelous nature of the vertebrae. In the Rhynchocephalia the bodies of the vertebrae have concave articulating surfaces at each end instead of the more advanced form in which the convex surface of one articulates with the concave posterior surface of the one in front of it.

However other features are more advanced; and these include a reduction in the number of bones of the skull, the acquisition of single- instead of double-headed ribs and the presence of a large opening in the ventral part of the pelvis (ROBB, 1973).

GÜNTHER's conclusion about the tuatara was reinforced by comparison with a fossil from the upper Jurassic found in Germany and England which is known as *Homoeosaurus*. Fossils of its skeleton and impressions of its skin show it to be almost identical to the tuatara. Little wonder the tuatara has been nominated 'living fossil'; VON WETTSTEIN (1931) summarised the significance of the tuatara thus (translated from the original German):

'... The Rhynchocephalia have preserved so many primitive and ancient characteristics respectively taken over from the Permo-Triassic Thecodontia* and have themselves altered in the main characteristics so little since the Triassic that no objections can be raised to designating the still living *Sphenodon punctatus* as the little-changed type of a Triassic reptile or even as a relic of the Triassic. If we still consider the very close relationship of the Rhynchocephalia to the Thecodontia, rich in forms and possessing many phyla, which presumably gave rise via the Theromorpha to the Mammalia and via the Dinosaura to the Aves, the great phylogenetic importance of this order and the great interest devoted to its last living representatives is thereby elucidated.'

Scientists were not slow to follow up GÜNTHER's declaration and to take advantage of the unprecedented opportunity the tuatara offered – to carry detailed studies back, along the evolutionary line, to a group

* A group known generally as 'stem reptiles' from which all reptiles are presumed to be descended.

otherwise extinct for millions of years. A group whose immediate predecessors also gave rise to the birds and mammals. To fully appreciate the impact this must have had on the scientific community of the time it must be realised that it came only a few years after CHARLES DARWIN published On the Origin of Species and propounded the basis of present day ideas of evolution.

'Primitive' and Adaptive Features

Much of the early work on the tuatara covered things like anatomy and development in considerable detail and it has been surveyed in a comprehensive review by VON WETTSTEIN (1931). Since then research has concentrated more on physiology and finer points of structure (ROBB, 1973), but the underlying motives have remained much the same as those of earlier studies: to find out whether, and then how, a particular structure, tissue, organ system or physiological process varies from that of more 'modern' animals. In so doing the researcher hopes to elucidate the very processes of evolution itself.

This has sometimes had positive results. Primitive or less specialised features of the tuatara, and others that may be so, have come to light. For example, SYED MUZAMMIL ALI (1941) found the tuatara to have reached a less-advanced stage in the evolution of autotomy, (the ability to shed part of its tail) than some of the fragile-tailed lizards. Both morphologically and as an effective defence against predators, autotomy is relatively poorly developed in the tuatara.

MARCHALONIS *et al.* (1969) found the lymphoid system of the tuatara to be particularly primitive. Organs such as lymph nodes and tonsils and lymphoid tissue associated with the alimentary canal do not occur in the tuatara, nor do even rudimentary lymphoid nodules such as the jugular bodies of toads. The only organ with discernible lymphoid characteristics is the spleen although phagocytic activity was detected in the liver.

Of the other physiological systems the tuatara's methods of reproduction can also be classed as primitive.

Eggs which are about 30 mm long and have a leathery shell, are laid in early summer, usually November, in clutches of 8–12 or 14 in holes in the ground. The female excavates the hole over a period of two or three nights and it may take the form of a shallow depression or short blind tunnel. When laid the eggs are covered and abandoned. They do not hatch for at least twelve months, a period including one of cold weather in which development stops. Consequently, the overall length of the incubation period is longer than that of any other reptile.

During an animal's life cycle there is usually a time when it is relatively vulnerable to attack or chance variations in its physical environment – for instance, the period between laying and hatching of the eggs of reptiles. Abandonment or even reduction of this period would be a successful adaptation to an environment which includes predators capable of digging up the eggs or which is subject to extremes of temperatures and degrees of dessication likely to kill the embryos. Such an adaptation is found in the viviparous lizards which are so well represented in New Zealand (WHITAKER & BULL, this volume), but the very long period of external development of the tuatara egg and the time added to it by the hibernation of the embryo is just the reverse and can be regarded as a primitive feature.

However, the significance of many other features, some of which are unique, such as the lack of an intromittent organ in the male, is less easy to interpret in an evolutionary sense. MILLIGAN (1924) found the tuatara to have a very low rate of respiratory metabolism (as measured by rate of carbon dioxide excretion), and WILSON & LEE (1970) showed the oxygen consumption of the inactive tuatara at a temperature of 30 °C to be 55 percent of the predicted value for lizards of the same weight. However, under the same conditions the heart rate of the tuatara was a little more than twice the predicted value for a lizard. The tuatara is also active at fairly low temperatures compared with many lizards. My observations suggest that maximum activity occurs at ambient temperatures of between 17 and 20 °C, which agrees closely with an estimate of 16–18 °C made on captive animals by WOJTUSIAK (1973). Activity declines steadily as temperature falls, becoming negligible below 7 °C. BOGERT (1953) suggested that tolerance of low temperatures may be a primitive feature, but WILSON & LEE applied the term 'cold adaptation' to the tuatara, which has involved a shift in the temperature relations of heart rate and oxygen consumption. Considering the tuatara's nocturnal habits and the climate of the region where it occurs 'cold adaptation' also has an ecological significance.

Other features of the species are almost certainly specialised. The retina of the eye of the tuatara includes a plate of cells which reflect light. Light entering the eye through the lens is reflected from this plate (the *tapetum lucidum*) back to the posterior surface of the lens where it is re-reflected to the retina. It is a device for intensifying the amount of light impinging on the retina under conditions of poor object illumination, such as at dusk or at night. Many nocturnal animals have a *tapetum* and even in the tuatara it must be regarded as having adaptive significance.

The tuatara's skull may also be thought of as adaptive, even though in this instance the adaptation involves features which link the tuatara with the Rhynchocephalians and differentiate it from the lizards. Because of the rigid bars of bone separating each temporal fossa into two parts and the solid union between the quadrate bone and the rest of the skull, the skull of the tuatara is a particularly strong and rigid structure. The skulls of lizards and snakes are much less rigid, and in many the two parts of the skull (the anterior or maxillary and posterior or occipital part) can move in relation to one another. In some the quadrate bone can also move and such skulls are termed 'kinetic'; they allow the mouth to be opened to remarkable angles. KÜHN-SCHNYDER (1954) pointed out that ancestral reptiles have kinetic skulls and that articulated connections between the cranial segments are well developed in the embryo tuatara, but, with their associated muscles, they regress during development. On this basis KÜHN-SCHNYDER suggested that the rigid Rhynchocephalian skull may be a modification of the primitive form and an adaptation for a particular manner of feeding. He went on to cite mussels as an example of the type of food the tuatara crushes and eats. I cannot confirm this observation but tuataras frequently do eat the small petrels (burrow-nesting sea-birds) with which they co-exist. The often kill them by crushing the skull between their jaws, and the decapitated bodies of petrels are a common sight on tuatara islands during the birds' breeding season.

Finally there are features of the tuatara which can certainly be classed as primitive but which are present in a more or less reduced or vestigial form. The prime example of this is the 'third' or pineal eye. Some literature has placed a great deal of emphasis on its presence in the tuatara but it is comparably well-developed in a number of lizards (DAWBIN, 1962) and functions to control exposure to sunlight in some of them (STEBBINS & EAKIN, 1958).

Its function in the tuatara, if it has one, has not been determined. STEBBINS (1958) was unable to demonstrate any reaction to either light or heat, which produced response from lizards, and numerous attempts to elicit reponses to light of various wavelengths in very young tuataras have been inconclusive (DAWBIN, 1962).

DENDY (1910) made a detailed study of the pineal eye of the tuatara. The organ itself is about 0.5 mm in diameter and lies directly beneath the parietal foramen. It possesses traces of a retina and lens and the surface of the lens is covered by a plug of translucent tissue which fills the parietal foramen. This in turn is covered with skin and scales which are opaque except in the very young animal. The transparent skin and scales directly over the parietal foramen in the young tuatara was likened to a 'cornea' by STEBBINS (1958) and was the reason for earlier attempts to elicit reactions from very young animals. The nerve connections of the organ have degenerated, at least in the adult, and the lens and chamber contain masses of pigmented tissue which is further evidence of degeneration, but Dendy also described secretory activity in some of these structures.

Although his experiments were inconclusive STEBBINS (1958) considered the state of development of the organ to be consistent with it having some sort of function. Although it has apparently degenerated in some respects, as DENDY described, STEBBINS considered that it may have a role as a regulator of autonomic functions (such as blood pressure) in relation to changes in solar radiation. These it could fulfil as an endocrine gland, and the secretions described by DENDY could be hormonal.

For the moment these possibilities are purely speculative, and whatever function the pineal eye may have had originally seems to have been lost or at least modified. This helps to demonstrate the difficulties of assessing whether or not a particular feature is a primitive one that has been retained in its original form or whether it is in the process of being lost or modified to fulfil some new function. Considering this further, the apparent conservativeness of the tuatara, at least in skeletal morphology, cannot always be equated with lack of specialisation or primitiveness in either anatomic or physiological systems, and the degeneration of the pineal eye and existence of specific anatomical and physiological adaptations to its present environment even suggest that conservativeness in the framework may not have extended to its contents.

Distribution and Status

The tuatara or its predecessors probably arrived in the New Zealand region during the Jurassic, coming overland from Antarctica about 90 million years B.P. (RAVEN & AXELROD, 1972). This was before the country became isolated from all other land masses, and the tuatara is considered part of the archaic element of the New Zealand fauna (FLEMING, 1962). However there is virtually no fossil record of the tuatara's history in New Zealand, the only fossils known are recently discovered ones from the Miocene (Dr. THOMAS RICH, pers. comm.) and some from the last interglacial period of the Pleistocene (Mr. R. J. SCARLETT pers. comm.). On the other hand sub-fossil remains (those post-dating the Pleistocene) are abundant and show that the tuatara once occurred on both of the main islands (Fig. 5).

Sub-fossils have been found in coastal sand dunes, peat bogs, caves and Maori middens. These records have been tabulated (CROOK & MORAN, in prep.), but equating presence in middens with details of a species' distribution is difficult. There is always a possibility that remains may have been moved, sometimes long distances, from the original habitats. For example, some middens many miles from the nearest sea-coast contain the remains of marine animals, thus establishing that Maoris sometimes collected their food in one place and ate it in another. This included transfer from offshore islands to mainland, and some middens with tuatara remains are on the coasts directly opposite small islands where the tuatara is still found.

The same objections cannot be levelled at remains in peat bogs or many of those scattered far and wide in sand dunes or piled with the bones of other animals which have fallen down pot-holes in limestone country. These show the tuatara once occurred in places from close to North Cape at the northern extremity of the North Island (34°26′S)

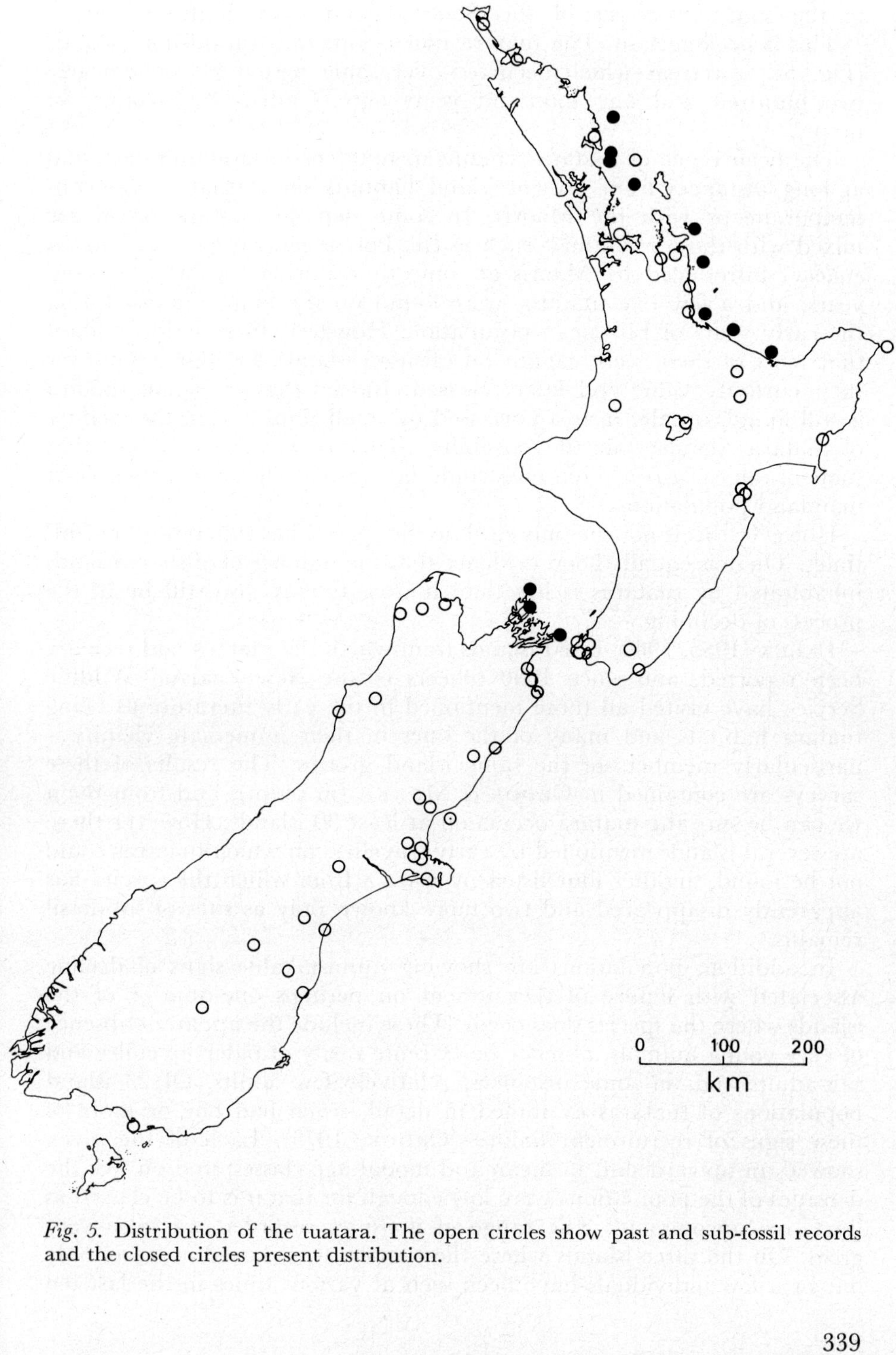

Fig. 5. Distribution of the tuatara. The open circles show past and sub-fossil records and the closed circles present distribution.

to the southern coasts of the South Island near Bluff (46°36′S).

This is no longer so. The tuatara now occurs only on offshore islands (Fig. 5), a retreat which occurred over some period between nearly two hundred and one thousand years ago (Crook & Moran, in prep.).

The occurrence of tuatara remains in middens in large numbers and at long distances from present island habitats shows that it was contemporaneous with the Maoris. In some deposits tuatara bones are mixed with those of animals such as the Polynesian rat or kiore *(Rattus exulans)* introduced by Maoris at some time during the last thousand years, and a few live tuataras were found on the main islands during the early years of European occupation. However, there is little doubt that most of these were caught on offshore islands, brought ashore for their curiosity value and later released. Indeed the occasional tuatara is still found, usually near a port used by small ships and in the vicinity of tuatara islands; but the possibility remains, however remote, that some of the early-recovered ones could have been relics of nearly extinct mainland populations.

However this is not the only decline the species has suffered in 'recent' times. There is equally good evidence that the number of offshore islands inhabitated by tuataras is less than it was. It may also still be in the process of declining.

Falla (1935, 1960) listed islands from which the tuatara had recently been reported, and since 1950 officers of the New Zealand Wildlife Service have visited all those mentioned in the early literature as being tuatara habitats and many of the ones in their immediate vicinity – particularly members of the same island groups. The results of these surveys are contained in Crook & Moran (in prep.), and from them we can be sure the tuatara occurs on at least 30 islands. However there are several islands mentioned by early travellers on which tuataras could not be found, another four listed by Falla from which the species has apparently disappeared and two more known only as sites of sub-fossil remains.

In addition, populations are showing unmistakable signs of decline associated with failure of recruitment on perhaps one-quarter of the islands where the species does occur. These include the apparent absence of very young animals, absence or extreme rarity of older juveniles and sub-adults and, in some instances, relatively few adults. Of 23 island populations of tuataras examined in detail, seven had one or more of these signs of recruitment failure (Crook, 1973). Each of the seven showed an upward shift in mean and modal age classes, and on five the densities of the populations were low enough for tuataras to be classed as 'rare' or 'uncommon'. The range of densities recorded was also very great. On the three islands where the tuatara was classed as 'rare' only one or a few individuals have been seen at various times in the last ten years, but on islands where the species is common counts of 20/hour of search have been achieved.

All except one of the islands where the tuatara is declining also have introduced populations of the Polynesian rat, and there is strong circumstantial evidence that the tuatara cannot persist in its presence: 1. Seven other islands inhabited by the rat in the series examined are in groups which otherwise support tuataras, but they lack any tuataras themselves. 2. The majority of islands reported in the early literature where tuataras apparently no longer occur have also had rats introduced to them. 3. Not one island has been found where the Polynesian rat has established which also supports a clearly self-maintaining population of tuataras.

The conclusion that rats are inimical to tuataras would also provide an explanation for the disappearance from the main islands, occurring as it did after introduction of the Polynesian rat, but introduction of rats is not the sole cause of decline, and possibly extinction, of tuatara populations or of radical variations in population structure.

As indications of this there is firstly the one island lacking rats which now supports only a very small population of aged tuataras. Secondly there are several islands without rats which apparently could support tuataras, and may have in the past, but do not do so now. Amongst these are rat-free islands in groups which otherwise support tuataras, those mentioned in the early literature as having tuataras which do not have them now and one island with sub-fossil remains of tuataras which now has populations of neither tuataras nor rats. Finally, several populations of tuataras on rat-free islands consist almost totally of young animals. On these islands tuataras were classed as being 'common' (Crook, 1973), capture rates of between five and 12/hour of search being recorded, but animals over 250 mm in length from snout to vent were not found and the mean and modal age classes were lower than for other rat-free islands. (Use of snout-vent length as an index of age is based on the observations of Dawbin (1962) that the tuatara has a very slow growth rate and that measurable increments occur during at least the first 50 years of life and probably for a good deal longer. In addition, measurements of young animals on Stephens Island in Cook Strait (40°40'S, 176°00'E) made in successive years since 1971 to complement data collected by Dawbin show them to grow at rates of about 10 mm per year in snout to vent length declining to 5 mm and less as length reaches and exceeds 200 mm. Therefore snout-vent length is a useful indicator of age, within limitations imposed by the difficulty of obtaining accurate measurements, and tuataras probably reach a length of *ca.* 180 mm in 20 years. After this sexual differences in size and rate of growth start to become apparent, females being smaller; but even males may take a further 20 years or longer to achieve a length of 250 mm. Provisional growth curves prepared by Dawbin *(in litt.)* from his more substantial data confirm these broad approximations).

Although some of the instances of 'disappearance' of tuataras could be the result of mistakes in the early literature rather than extinctions of previously existing populations, the obvious lack of recruitment and the aging adult population on the one island closely follows what appears to be happening on the islands with rats. However at first sight the occurrence of normal-sized populations of only young animals appears to be a totally different picture, perhaps one of catastrophic and massive decline in numbers followed by a rapid return to normal density through higher than normal rates of recruitment. CROOK & MORAN (in prep.) discuss this and the possible causes of such a catastrophe, including the burning of island vegetation by Maori mutton-birders and direct predation by Maoris and possibly by European collectors.

On some islands collection of specimens was fairly intensive. Karewa Island, for example, (37°32′S, 176°09′E, the type locality of the species and only five hectares in area) was a source of many of the specimens used by the early anatomists and was subject to frequent visits in the latter part of the nineteenth century (e.g. BULLER, 1870, 1876; GOLD-SMITH, 1884; REISCHEK, 1885 and THOMAS, 1891). It is also one of those with a large population of only young animals. However, there does not appear to have been a catastrophic decline on Karewa. The early visitors generally reported good numbers of tuataras but SLADDEN (pers. comm. to SCHMIDT, 1953) who knew the island well and published a description of it (SLADDEN, 1924) could find only large animals and expressed fears for the population. This resembles the first signs of decline seen on the rat-inhabited islands, but unlike those on islands with rats the population has recovered its ability to reproduce, and the rate of recruitment has been high enough to result in a major shift in age distribution in the intervening years.

CROOK & MORAN (in prep.) speculate on the possible causes and significance of what has happened on Karewa Island. Although more investigations are needed one thing is certainly clear: existing populations of tuataras, of which between 20 and 25 are apparently maintaining their numbers successfully, cannot be regarded as secure from influences potentially capable of causing their decline even if invasions of islands by rats are prevented.

The possibility of serious decline of populations which are maintaining themselves successfully at the moment is heightened by the size of many of the islands on which they occur and hence the numbers of animals involved. Of 22 'normal' tuatara populations only two are on islands over 100 ha in size, four on islands between 10 and 100 ha and the remaining 16 on islands less than 10 ha. Research on Stephens Island has shown that tuataras reach remarkable population densities, as many as 30 individuals occurring on an area of 625 m^2. This indicates that tuataras could reach densities approaching 500 animals/ha (a biomass of *ca.* 250 kg/ha) although average densities may be considerably lower than

Fig. 6. Stephens Island. The settlement is associated with its use as a lighthouse station.

this. Thus a typical island habitat of 10 ha is not likely to support more than 5,000 tuataras.

Habitat and Ecology

The islands on which tuataras are found are in two main groups, one off the north-east coast of the North Island between North and East Capes and the second in Cook Strait (Fig. 5), but members of both groups have similar features.

1. They are nearly all isolated and difficult of access being wholly or partly cliff-bound (Figs 6 and 7) and subject to rapid changes in weather and frequent storms characteristic of the maritime climate of the region.
2. Consequently they are frequently exposed to strong winds and inundation with salt spray which together exert a strong influence on their vegetation. The less-modified ones are covered mainly by scrub and coastal forest dominated by salt- and wind-tolerant species on exposed coasts such as pohutukawa *(Metrosideros excelsa)* in the northern group (Fig. 8) and ngaio *(Myoporum laetum)* (Fig. 9) and taupata *(Coprosma repens)* which are more prominent on the southern ones (Atkinson & Bell, 1973).

Fig. 7. Landing on Stephens Island, a lesson in the inaccessibility of tuatara habitats.

Fig. 8. Forest vegetation on Cuvier Island (Hauraki Gulf). Dominated by pohutukawa, note the dense understorey and ground cover which is characteristic of vegetation on islands where petrels are rare or uncommon.

Fig. 9. Principally ngaio forest on Stephens Island (Cook Strait). Note the virtual absence of ground cover vegetation which is characteristic of islands with dense populations of petrels.

3. However, despite isolation and difficult access human influence is readily apparent on many. Until relatively recently Maoris burned the vegetation of these islands, and many bear the signs of this in areas of flax *(Phormium)* and manuka *(Leptospermum scoparium)* and kanuka *(L. ericoides)* scrubland and regenerating forests dominated by pioneer species. Where recovery is almost complete evidence of past fires remains in the frequent occurrence of charcoal in soil profiles.

Other islands have been used by Europeans, usually as lighthouse stations, and part of the forest has been destroyed to make way for grazing land on some of these. Stephens Island in Cook Strait is the prime example (Figs. 6 and 7) and it demonstrates that the tuatara can survive in an environment considerably modified by farming (Fig. 10 and 11).

4. However, the major common factor on all tuatara-inhabited islands is the presence of breeding populations of petrels and shearwaters of the families Hydrobatidae, Procellariidae and Pelecanoididae. These vary in size from Buller's shearwater *(Puffinus bulleri)* (450 mm) to the tiny white-faced storm petrel *(Pelagodroma marina)* (200 mm). Most islands support several different petrel species, for example, Main Trio Island in the Trio group in Cook Strait (40°50′S, 176°00′E) has sooty shear-

Fig. 10. The grazed part of Stephens Island.

water *(Puffinis griseus)*, flesh-footed shearwater *(P. carneipes)*, fluttering shearwater *(P. gavia)*, fairy prion *(Pachyptila turtur)* and diving petrel *(Pelecanoides urinatrix)* (CAMPBELL, 1967). However, islands where the tuatara is maintaining its numbers almost invariably have very large numbers of either or both the fairy prion (Fig. 12) and diving petrel, the two smallest, widely-distributed petrels in the region. The only smaller species in the tuatara's range (the white-faced storm petrel) is present in any numbers on only one island where the tuatara also occurs.

Petrel burrows honeycomb much of the soil surface on the tuatara islands and can reach densities approaching four per square metre. Petrels therefore constitute a major factor in the environment, exerting an influence over soils, vegetation and animals. CAMPBELL (1967), for example, found that dense populations of petrels inhibited development of vegetation strata beneath the canopies of island forest and scrub communities, and WARD (1961) showed that soils burrowed by petrels are very friable, have a low pH and high levels of nitrogen and phosphorus.

Petrels very effectively cultivate island soils and incorporate litter and introduce large quantities of mineral and organic material. This may encourage development of a high biomass of ground-dwelling invertebrate animals (DAWBIN, 1962) and, through food chains, of small lizards. Data confirming this suggestion remains to be collected, but the

Fig. 11. The ungrazed part of Stephens Island showing also small plantations of taupata which have been planted to encourage regeneration of the original forest vegetation.

overall effect of petrels is abundantly evident on first impressions alone once islands with and without large petrel populations are compared (Figs. 8 and 9).

The strict coincidence of tuataras and petrels and the possible relationships between them have frequently been causes of comment in the literature. Large populations of petrels provide the burrows in which many tuataras live, though tuataras can and frequently do construct their own. Under these circumstances the enhanced friability of soils burrowed by petrels would materially assist the tuatara. Tuataras also eat the adults, chicks and eggs of the smaller species of petrels. The remainder of their food consists of invertebrates and small reptiles (Fig. 3), which conditions created by the petrels may encourage, and finally the open ground characteristic of a petrel island (Fig. 9) may provide ideal conditions for the foraging tuatara (Dawbin, 1962).

From this the question arises as to whether the tuatara may be dependent upon co-existence with a population of diving petrels or fairy prions for its survival. The restriction of self-maintaining populations of tuataras to islands where these birds occur is certainly too complete to

Fig. 12. A fairy prion in the hand.

be purely coincidental, but the survival of both is in some way linked with the absence of rats (Crook & Moran, in prep.). Islands to which rats have been introduced have very few of the smaller petrels and frequently they are absent altogether; and there is little doubt that predation of their eggs and chicks is the cause of their disappearance.

Extinction of tuataras on islands with rats could equally be a matter of direct predation (or competition for food). On the other hand it may follow the extermination of the small petrels on which the tuatara is dependent.

This problem has led me to examine more closely the relationships between tuataras and the small petrels. If their presence on the same islands was purely coincidental, and the relationship between them unnecessary for the tuatara's survival, their distributions on larger islands where they both occur might be expected to be independent of one another.

A series of seven plots, each consisting of two to four contiguous 625 m^2 quadrats (25 × 25 m), have been set up on various parts of Stephens Island. The series covers areas in which densities of burrows range from 0 to 3.7/m^2. Tuatara numbers on these plots were estimated from the accumulated total of different individuals caught during 16 night searches. This has shown that over lower densities of petrel bur-

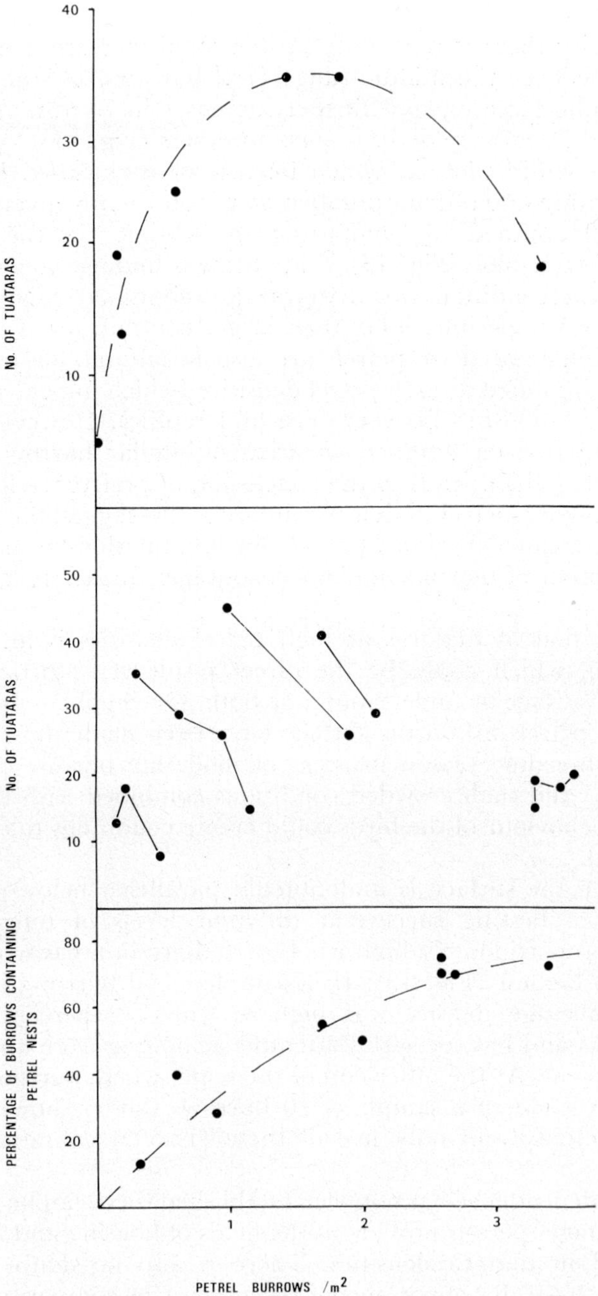

Fig. 13. A. Relationship between tuatara numbers and burrow density on Stephens Island (averages on plots consisting of 2–4 contiguous 625 m^2 quadrats). B. Variation in tuatara numbers with burrow density on contiguous quadrats of the same plots. C. Variation in the proportion of burrows containing breeding fairy prions with overall burrow density.

rowing activity there is a roughly proportional increase in tuatara and petrel numbers. In the middle range (1–2 burrows/m^2) the increase in tuatara numbers ceases, and further increases in burrow densities are accompanied by a decrease in tuatara numbers (Fig. 13).

However, within plots in which burrow density varied between 0.2 and 2/m^2 tuatara and burrow numbers were consistently inversely related. This was determined by comparing the regimes in the contiguous quadrats of each plot (Fig. 13), and the two findings together suggest that optimum conditions for tuataras are found at moderate petrel densities or at the peripheries of their larger concentrations or both.

Areas uninhabitated by petrels are also unsuitable for tuataras, but tuataras are inhibited at high petrel densities (which indicates some form of negative relationship between bird and reptile). However, the wide differences in tuatara numbers in areas of similar burrow density on different plots (which leads to the conclusion of preference for the edges of areas of higher petrel activity) reinforces the suggestion that petrels in some way create conditions suitable for inhabitation by tuataras. This is far from proof of dependence but encourages further investigation of the hypothesis.

The inhibition of tuataras at high petrel densities is an interesting complication, which could be the direct result of disturbance by the birds on the surface or underground or both. On Stephens Island counts of up to 1.6 petrels/m^2 on the surface have been made during the early part of the breeding season in areas of moderate burrow density (1–2 burrows/m^2), and such crowded conditions combined with the courting and sexual behaviour of the birds could create conditions tuataras would seek to avoid.

Activity on the surface is undoubtedly paralleled below-ground and comparison of nesting success at different levels of burrow density suggests disproportionately high levels of activity in areas where burrow densities are highest (Fig. 13). In a sample of 50 burrows in one area where the average density was high (3.4/m^2) 72 percent contained fledged chicks and less successful attempts at nesting were apparent in a further 6 percent. At the other end of the scale, where burrows averaged 0.23/m^2, only eight of a sample of 40 burrows (20%) showed any sign of nesting activity and only five of these (12.5%) contained fledging chicks.

Although dependence has not been established there can be no question of the advantages petrels provide in the fields of housing and food, which are reflected in tuatara densities. There is also no doubt that these relationships have developed in the recent past in comparison with the history of the tuatara, or its immediate predecessors, because it is not until the Eocene (54 million years B.P.) that birds of the Order Procellariiformes first appear in the fossil record.

Acknowledgments

Figures 1, 2, 4, 6, 7, 9, 10, 11 and 12 are by Mr J. H. Johns A.R.P.S. and Figure 3 is by Mr A. H. Whitaker. I also wish to thank all those who contributed to the surveys and original research reported here and especially Mr L. R. Moran of the N.Z. Wildlife Service.

REFERENCES

Atkinson, I. A. E. & Bell, B. D. 1973. Offshore and outlying islands. In Williams, G. R. (Ed.), The natural history of New Zealand, 372–392. A. H. &. A. W. Reed, Wellington.

Bogert, C. M. 1953. The tuatara: why is it a lone survivor? *Scientific Monthly* 76: 163–170.

Buller, W. 1870. A list of the lizards inhabiting New Zealand with descriptions. *Trans. and Proc. of the N.Z. Inst.* 3: 9–10.

Buller, W. 1876. Notes on the tuatara lizard *(Sphenodon punctatum)*, with a description of a supposed new species. *Trans. and Proc. of the N.Z. Inst.* 9: 317–325.

Campbell, D. J. 1967. The Trio Islands, Marlborough Sounds; an ecological study of a bird-modified island. Unpublished dissertation for the degree of Master of Science, Victoria University of Wellington, Wellington, New Zealand.

Crook, I. G. 1973. The tuatara, *Sphenodon punctatus* Gray, on islands with and without populations of the Polynesian rat, *Rattus exulans* (Peale). *Proc. of the N.Z. Eco. Soc.* 20: 115–120.

Dawbin, W. H. 1962. The tuatara in its natural habitat. *Endeavour* 21: 16–24.

Dendy, A. 1910. On the structure, development and morphological interpretation of the pineal organs and adjacent parts of the brain in the tuatara *(Sphenodon punctatus)*. *Phil. Trans. of the R. Soc., London* 210B: 227–331.

Falla, R. A. 1935. The tuatara *(Sphenodon punctatus)*. *Bulletin of the Auckland Zoological Society* 2: 3–5.

Falla, R. A. 1960. The fauna of New Zealand, New Zealand reptiles. In McClintock, A. H. (Ed.), A descriptive atlas of New Zealand. R. E. Owen, Government Printer, Wellington.

Fleming, C. A. 1962. New Zealand biogeography. A paleontologist's approach. *Tuatara* 10: 53–108.

Gold-Smith, E. C. 1884. Description of Mayor Island. *Trans. and Proc. of the N.Z. Inst.* 17: 417–427.

Gray, J. E. 1831. Note on a peculiar structure in the head of an Agama. *Zoological Miscellany* 1: 13–14.

Gray, J. E. 1842. Description of two hitherto unrecorded species of reptiles from New Zealand (*Hatteria punctata* and *Naultinus elegans*) *Zoological Miscellany* 4: 72.

Günther, A. 1867. Contribution to the anatomy of *Hatteria* (*Rhynchocephalus* Owen). *Phil. Trans. of the R. Soc. of London* 157: 595–629.

Kühn-Schnyder, E. 1954. The origin of lizards. *Endeavour* 13: 213–219.

Marchalonis, J. J., Ealey, E. H. M. & Diener, E. 1969. Immune response of the tuatara, *Sphenodon punctatum*. *Austr. J. of Exp. Biol. and Med. Sci.* 47: 367–380.

Milligan, R. R. D. 1924. The respiration and metabolism of the tuatara. *Austr. Ass. for the Adv. of Sci. Rep.* 16: 404–406.

Owen, R. O. 1845. Report on the reptilian fossils of South Africa. *Trans. of the Geo. Soc. of London* 7: 59–84.
Raven, P. H. & Axelrod, D. I. 1972. Plate tectonics and Australasian Paleobiogeography. *Science* 176: 1379–1386.
Reischek, A. 1885. Observations on *Sphenodon punctatum*, fringe-back lizard (tuatara). *Trans. and Proc. of the N.Z. Inst.* 18: 108–110.
Robb, J. 1973. Reptiles and amphibia. In Williams, G. R. (Ed.). The natural history of New Zealand 285–303. A. H. & A. W. Reed, Wellington.
Schmidt, K. P. 1953. A visit to Karewa Island, home of the tuatara. *Fieldiana Zoology* 34: 153–164.
Sladden, B. 1924. Karewa: an island sanctuary. *N.Z. Jl. of Sci. and Technol.* 7: 182–187.
Stebbins, R. C. 1958. An experimental study of the 'third eye' of the tuatara. *Copeia* 3: 183–190.
Stebbins, R. C. & Eakin. 1958. The role of the third eye in reptilian behaviour. American Museum Nova 1870: 1–40.
Syed Muzammil Ali. 1941. Studies on the comparative anatomy of the tail in Sauria and Rhynchocephalia I. *Sphenodon punctatus* Gray. *Proc. of the Indian Acad. of Sci.* 13: 171–192.
Thomas, A. P. W. 1891. Preliminary note on the development of the tuatara *(Sphenodon punctatum)*. *N.Z. Jl of Sci.* 1: 27–30.
Ward, W. T. 1961. Soils of Stephens Island. *N.Z. Jl of Sci.* 4: 493–505.
Wettstein, O. Von. 1931. Rhynchocephalia. Kukenthal and Krumbach's Handbuch der Zoologie 7: 1–128, 215–235.
Wilson, K. J. & Lee, A. K. 1970. Changes in oxygen consumption and heart-rate with activity and body temperature in the tuatara *Sphenodon punctatum*. *Comp. Biochem. and Phys.* 33: 311–322.
Wojtusiak, R. J. 1973. Some ethological and biological observations on the tuatara in laboratory conditions. *Tuatara* 20: 97–109.

IX. THE MARINE BENTHIC ECOLOGY AND BIOGEOGRAPHY

by

G. A. KNOX

Introduction

In 1960 I reviewed the littoral ecology and biogeography of the southern oceans (KNOX, 1960) and three years later the biogeography and intertidal ecology of the Australasia coasts (KNOX, 1963). In a decade since the latter of these reviews considerable advances have been made in our knowledge of the composition, distribution and systematics of the New Zealand benthic marine flora and fauna.

In 1960 the New Zealand Oceanographic Institute began a systematic survey of the New Zealand shelf to a depth of 200 m (DAWSON, 1970). About 400 benthic stations were sampled in lines spaced latitudinally at about 40-mile intervals round the New Zealand coast (MCKNIGHT, 1969, Fig. 2). The distribution, ecology, sediment preferences, and zoogeographical relationships of selected groups and species are being analysed. The first detailed analysis on the echinoids has recently been published (MCKNIGHT, 1969) while DAWSON 1965, 1970) has given some of the results in his analysis of the biogeography of the New Zealand subantarctic and the faunal relationships (based on echinoderm distribution) between the New Zealand plateau and the New Zealand sector of Antarctica.

A similar survey of the slope faunas beyond the edge of the continental shelf has virtually been completed (DAWSON, 1970). In this survey special attention has been paid to the Chatham Rise because of its critical position in relation to the Subtropical Convergence; the earlier Chatham Islands 1954 Expedition (KNOX, 1957; FELL, 1960) had indicated the richness of the fauna on this feature.

Systematic sampling of some 300 benthic stations to the South of New Zealand have been carried out as part of the New Zealand Antarctic Research Programme (DAWSON, 1965, 1970).

In addition to the above programme of benthic sampling University biologists have been engaged in some limited detailed sampling of selected areas. Of note is the survey of the faunas of the slope and canyon systems off the Otago coast by the Portobello Marine Biological Station (BATHAM, pers. comm.).

Our knowledge of the distribution and zonation patterns of the flora and fauna of the rocky shores of the New Zealand region has also been much extended in the last decade. My own work has been concentrated in the southern New Zealand region and has included the northern half of the Auckland Islands (KNOX, 1968a), the Snares Islands and the southern and western coasts of Stewart Island. The shores of the Antipodes Islands have been surveyed by MANNERING, a member of the 1969 Antipodes Islands Expedition (WARHAM & JOHNS, in press). MORTON & MILLER (1968) have published their detailed account of the ecology of New Zealand seashores with special reference to the coasts of the northern half of the North Island.

All this activity has resulted in numerous systematic publications and a number of reviews of the distribution and relationships of selected taxonomic groups. The Scleractinium corals have been reviewed by SQUIRES & KEYES (1967), the algal-living littoral Gammarid amphipods by BARNARD (1972), the Brachyuran crabs by BENNETT (1964), GRIFFEN (1969) and DELL (1968) and the echinoderms by PAWSON (1961, 1965a, 1965b and 1968), McKNIGHT (1969) and DAWSON (1970).

For the purposes of this review the New Zealand region is considered as extending from the Kermadec Islands in the north to Campbell Island in the south and including the Challenger Plateau, the Chatham Rise, the Campbell Plateau and the Bounty Platform (see Fig. 1).

Bathymetry

According to BRODIE (1964) the morphology of the area of ocean floor in the vicinity of New Zealand is complex. The seabed exhibits a considerable relief seaward of the continental slope, and ridges, rises, and plateaus rising above the deep ocean floor extend many hundreds of kilometres into the Pacific and Southern Oceans and the Tasman Sea.

THE NEW ZEALAND CONTINENTAL SHELF

The continental shelf is nowhere very extensive. Its outer edge, where the depth increases very rapidly, lies in an average depth of about 130 m; this limit is variable, the shelf edge being much deeper (about 180 m) to the south of Stewart Island, and also westward off Cape Egmont and Wanganui (McKNIGHT, 1969). It is commonly less than 80 km in width and its outer edge generally plunges steeply to the upper continental slope.

Several banks lie close to but separated from the main shelf; these occur around the Three Kings Islands in the north and off the south-west corner of the South Island (Puysegur Bank). Other isolated shallow areas, Mernoo and Veryan Banks, lie on the Chatham Rise and another Pukaki Bank, is on the southern border of the Campbell Plateau. Smaller shelf areas surround the outlying islands, Chatham, Auckland and Campbell Islands.

McKnight (1969) has described the characteristics of the shelf sediments. Generally, in nearshore shallow water on the open coast, coarse-grained sediments of gravel or coarse sand grades predominate, the sediment becoming finer offshore (McDougall & Brodie, 1967). Much of the shelf is covered by finer-grained sediments such as fine sand, muddy-sand, or sandy mud (Pantin, 1966). Mud is uncommon except off river mouths and in coastal embayments and protected harbours with little wave action. Thus most of the shelf area is covered by fine-grained mud, muddy sands and sandy muds. Coarse sediments cover appreciable local areas such as much of the shelf from off Otago Peninsula southwards to the southern shelf edge on Puysegur Bank, westward off Wanganui and the Three Kings Islands (McKnight, 1969) and in Foveaux Strait (Cullen, 1967).

The archibenthal and abyssal regions

The bathymetry of the sea floor round New Zealand beyond the continental shelf is the result of three structural trends (Brodie, 1964). In the northern region to the west the trend is north-west–south-east including features such as the Norfolk Ridge, the Norfolk Trough, the Lord Howe Rise and the Challenger Plateau. To the south of these features lies the deep Tasman Basin. On the east in this northern region the dominant trend is north-north-east–south-south-west. The dominant feature here is the deep Kermadec Trench which continues along the east coast of the North Island as the Hikurangi Trench. This terminates in a series of canyons and troughs in Cook Strait and off the Kaikoura Peninsula.

To the east of Banks Peninsula the dominant trend is east–west. The main feature is the Chatham Rise, a broad elongate submarine ridge extending some 960 km to the east of the Chatham Islands. The crest of the rise is generally shoaler than 500 m and the northern flank is markedly steeper than the southern. To the south of this feature lies the Bounty Trough.

South of the Bounty Trough and connected to the west with the continental shelf surrounding Stewart Island and the Snares Islands is an extensive submarine flat, the Campbell Plateau with an average depth of about 700 m; Campbell Island and the Auckland Islands and the shallow Pukaki Bank rise from this Plateau. The Bounty and Antipodes Islands to the east form a platform of their own separated from the plateau by depths of about 2,000 m and from each other by more than 4,000 m. Macquarie Island is linked to the New Zealand land mass by a long intermittent submarine ridge (Brodie & Dawson, 1965).

From the biogeographical viewpoint the seafloor beyond the edge of the continental shelf is divided into the archibenthal zone, defined by Brodie (1964) as 'the area of the seafloor immediately surrounding

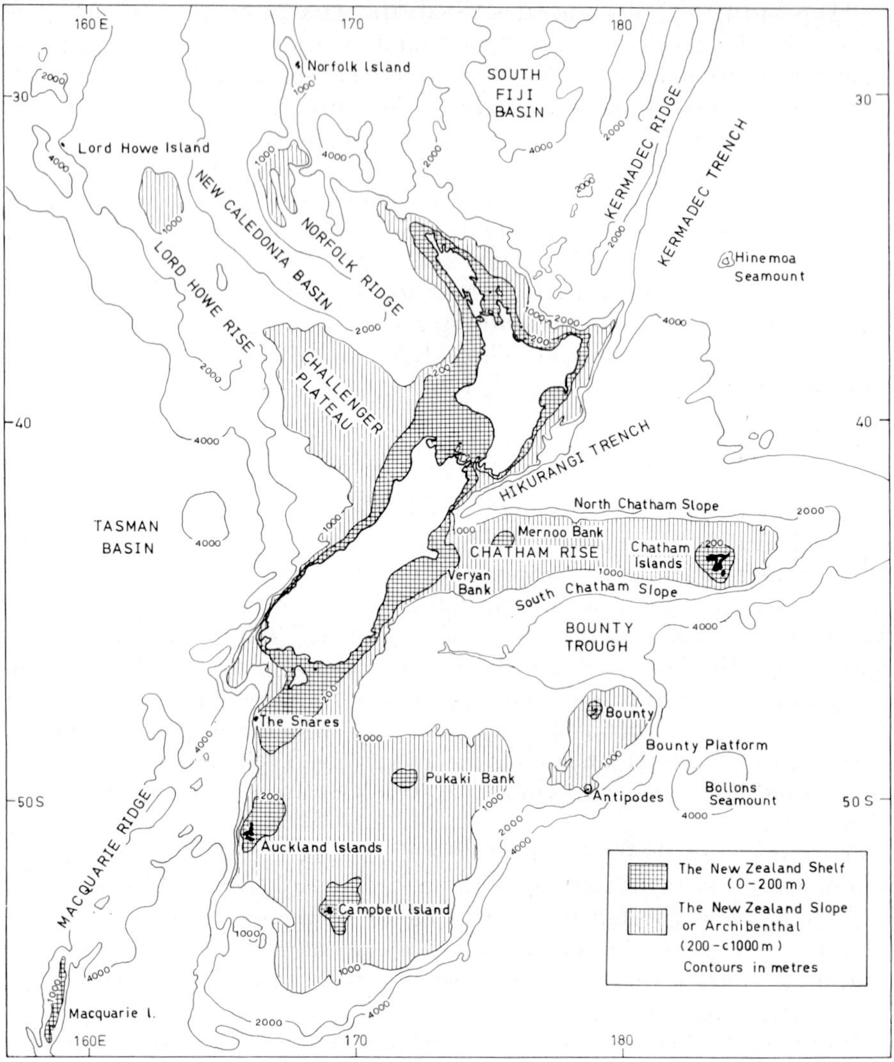

Fig. 1. The bathymetry of the New Zealand region including the Sub-antarctic.

New Zealand extending to the mid-depths of the marginal continental slopes', and the abyssal zone. The extent of the archibenthal zone extending from about 200 to 1000 m is shown in Fig. 1. It includes the Challenger Plateau and the Lord Howe Rise to the north-west of Cook Strait, the Chatham Rise to the east of the South Island and the Campbell Plateau to the south.

The sediments of the New Zealand seafloor beyond the continental shelf have not yet been described in detail. BRODIE (1964) has discussed sedimentation in the New Zealand region in general terms and NORRIS (1964) has given a detailed account of the sediments of the Chatham Rise. The sediments of the Rise include shell gravels in the Mernoo and Veryan Banks and concentrated greensands on the Reserve Bank. Foraminiferal oozes cover most of the rest of the area shallower than 500 m and silty and clayey sediments, including abundant faecal pellets blanket the deeper slopes and adjacent basins. Glauconite comprises a considerable proportion of some samples and phosphatised Miocene foraminiferal oozes in nodular form are widely distributed on the surface of the Rise.

According to BRODIE foraminiferal oozes dominate the sediments of the continental slope and basins off New Zealand. On the highs of the ridges clastic sediments such as greensands, fine black sands, rock and shell are recorded.

Oceanography

In recent years the main features of the hydrological environment round New Zealand have been defined. GARNER (1961, 1962) has summarized the results of hydrological observations in the New Zealand region in 1955 and from the period 1877–1955 to give a general picture of the oceanographic 'climate' on the shelf and further offshore. The hydrology of the water masses of the southern New Zealand region and the Southern Ocean to the south have been described by BURLING (1961) and HOUTMAN (1967). Other regional studies have included the seasonal range of temperatures on the shelf (GARNER, 1969), the southern Hikurangi Trench region (GARNER, 1967a), the southern Kermadec Trench region (RIDGWAY, 1970), the region north-west of New Zealand and the north-east Tasman Sea (GARNER, 1970), and the south-east Tasman Sea (GARNER, 1967b).

CURRENTS AND CONVERGENCES (Fig. 2)

The New Zealand region is influenced by both subtropical and subantarctic water masses. Within these three principal surface water movements give rise to the New Zealand coastal circulation. To the north of the North Island lies the warm subtropical Trade Wind Drift moving in a predominantly south-west direction; to the south of the South Island is the cold West Wind Drift moving towards the north-east. To the west is the Tasman Current of modified subtropical water, broadly south- and east-flowing in the Tasman Sea.

The southern part of the New Zealand mainland lies across the boundary between the subtropical and subantarctic waters, the Subtropical Convergence. A 'convergence' may be defined as an area of the sea surface towards which the surrounding surface water moves and at which it sinks. It is characterized by rather steep gradients in both temperature

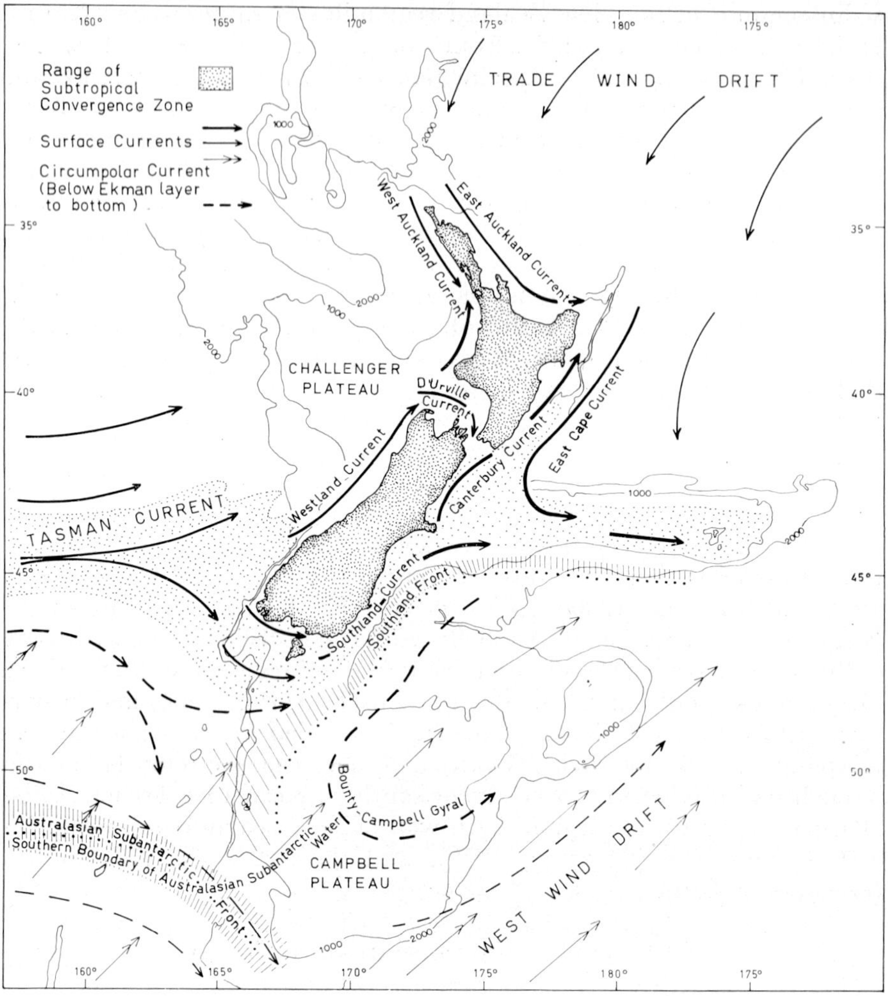

Fig. 2. Distribution of surface currents and convergences in the New Zealand region.

and salinity. The Subtropical Convergence has been variously defined by different workers (Garner, 1959), and recently Burling (1961) has recognized and delimited a broad zone within which the convergence lies, the exact position varying throughout the year. This convergence region lies off South Westland and Fiordland on the west coast, and extends past Stewart Island. Off the east coast it forms a band running east-west across the Chatham Rise, tending rather sharply northwards, east of the Chatham Islands (Garner, 1962).

Details of the coastal current system are now reasonably well known from hydrological observations and drift-card releases (Garner, 1959, 1961, 1962; Brodie, 1960; Burling, 1961). This coastal current pattern can be derived from the direct effect of the regional influences discussed above, combined with the effects of the form of the New Zealand landmass and nearshore bottom topography, and the varying effects of extremes of local winds.

The following water movements can be detected: 1. The warm East Auckland Current from the North Cape down the Auckland east coast to the Bay of Plenty. 2. The warm West Auckland Current from north and west of Cape Reinga down the Auckland west coast a variable distance towards Cape Egmont. 3. The warm East Cape Current derived from the Trade Wind Drift passing south along the coast from East Cape and south of Gisborne on the outer side of the cold Canterbury Current. Off the Cook Strait region this current swings towards the east. 4. The Southland Current containing Subtropical Convergence water and an admixture of Australasian Subantarctic Water (GARNER, 1961, 1962) moves through Foveaux Strait and round Stewart Island to move north-east along the Otago coast. Near its northern limit this current turns east to join the Subtropical Water transported southward in the East Cape Current. According to BURLING (1961) this Southland Current reaches to the bottom in a depth of several hundreds of metres and between it and the West Wind Drift Circumpolar Subantarctic Water is developed the Southland Front. 5. Off the Canterbury coast the Canterbury Current of cold subantarctic water flows from Banks Peninsula north past Cook Strait and along the east coast of the North Island as far as Gisborne. Shoreward transfers of upwelled Subantarctic Water often contribute cold, poorly saline water to this current. 6. The Westland Current of Subtropical Convergence Water flows along the west coast from the Fiordland region to meet the southward flowing West Auckland Current at the East Tasman Convergence. 7. The D'Urville Current flows from off the western entrance of Cook Strait east into the strait and thence south towards the narrows of Cook Strait.

The deep water circulation is imperfectly known. However the predominant movement is in a north-easterly direction. BURLING (1961) has identified an anticlockwise flow of subsurface water in a gyral extending over the Bounty Trough and the north-eastern area of the Campbell Plateau. He has named this feature the Bounty-Campbell Gyral.

SURFACE WATER MASSES

KNOX (1960) proposed a terminology for the coastal water masses of the temperate and Antarctic regions of the Southern Hemisphere. The water masses recognized in the New Zealand area are:

a. Subantarctic Cold Temperate Water

This lies between the Antarctic and Subtropical Convergences and is characterized by an even meridional temperature gradient, with surface temperatures increasing from south to north from 3.0° to 11.5 °C in the winter and from 5.5° to 14.5 °C in the summer, and with a salinity of about 34.5‰ or higher. Macquarie Island lies on or near the boundary between this water mass and the colder antarctic surface water.

Two distinct types of Subantarctic Water to a depth of about 600 m have been recognized in the southern New Zealand region (BURLING, 1961; HOUTMAN, 1967). They are Circumpolar Subantarctic Water of salinity below 34.5‰ and warmer more saline Australasian Subantarctic Water, with salinities sometimes >34.5‰. The latter lies to the east of New Zealand south of the Subtropical Convergence Zone, to the south of Stewart Island and over the greater part of the Campbell Plateau. To the east of this water mass lies the Circumpolar Subantarctic Water. The boundary between these two masses forms a laterally continuous steep gradient of properties extending to depths of 1000 m and was called by BURLING (1961) the Australasian Subantarctic Front. HOUTMAN (1967) has renamed this feature the Subantarctic Divergence.

The Subantarctic Islands lie within this zone and at times the eastern coasts of Stewart Island, the South Island and the southern part of the North Island are influenced by water of this type.

b. Cold Temperate Mixed Waters

GARNER (1959) points out that the terms subtropical and subantarctic in the hydrological sense require qualification when near coastal waters are considered. He states that: 'Modifications of nearshore properties may occur. Shallow or semi-enclosed waters may undergo large changes in temperature under local climatic effects, turbulent flows in tidal waters may induce intense vertical mixing of colder subsurface and near-surface layers and the run-off of fresh water from the adjacent land will modify the temperature and salinity pattern in some areas. The formation of 'slope' water over the continental shelf through mixing of oceanic and coastal shelf waters may be another source of complex surface gradients and areas of cold water may be found by upwelling.'

Upwelling of locally colder water is known to occur near the Three Kings Islands, off East Cape, off Kaikoura Peninsula, off the Canterbury coast, off Cape Farewell and generally off the west coast of the North Island (MCKNIGHT, 1969).

In these Cold Temperate Mixed Waters winter mean temperature varies from 7.0° to 14.5 °C and summer mean temperatures from 11.0° to 18.0 °C, while salinities, although variable, usually range from 34.0 and 34.8‰.

Along the east coast of New Zealand these waters occur on the cold side of the Subtropical Convergence in inshore waters between Banks Peninsula and Castle Point. Off the west coast while the water is poorly mixed water from the Convergence region, inshore waters are affected both by upwelling and riverine dilution. Foveaux Strait, Stewart Island and the eastern coast of the South Island south of Banks Peninsula form a transition zone between the Subantarctic Cold Temperate Water and the Cold Temperate Mixed Waters.

c. Transitional Warm Temperate Waters

The northern part of the North Island, north of East Cape on the east coast and Herekino on the west lie to the north of the Subtropical Convergence and could be considered to lie within the warm temperate ('subtropical') zone, but faunistically and floristically they retain many features of the cold temperate areas to the south. In these waters temperatures generally do not rise much above 20.0 °C. Salinities are normally greater than 35.0‰.

d. Warm Temperate (Subtropical) Water

These waters lie north of the Subtropical Convergence. Salinities are generally high (34.9 to 35.5‰) and temperatures range up to 25.0 °C. The Kermadec Islands lie on the boundary between this Warm Temperate Water and Tropical Water to the north.

SUBSURFACE WATER MASSES

Mixed water sinks at the Antarctic Convergence to form Antarctic Intermediate Water which then moves north and eastwards. It can be defined by the intermediate salinity minimum generally lying between 34.2 and 34.4‰. The depth of this salinity minimum varies from 800 m to 1000 m north of the Subtropical Convergence and approaches the surface near the Antarctic Convergence. Temperatures within this water mass, depending on latitude and depth, vary from 2.0° to 7.0 °C.

GARNER (1961) found that in general bottom water at the edge of the North Island shelf has a temperature of 12.0° to 15.0 °C and a salinity of 34.9 to 35.4‰. In the deeper water offshore from the shelf the 10.0 °C isotherm is found in general at a depth of 400 to 500 m, 5.0 °C in 1000 to 1200 m and at a depth of 2600 m the temperature is 2.0 °C.

Most of the archibenthal region throughout the area under consideration is influenced by this Antarctic Intermediate Water.

Biogeographical Nomenclature

As HEDGPETH (1953) points out biogeography, the branch of biology devoted to the summarizing and analysing of the distribution of organisms, has two divergent phases or aspects. Classical or traditional biogeography is concerned with the distribution patterns of species, genera, or larger systematic categories or of total faunas. It is concerned with the internal distribution patterns within a region as well as the relationships of the group or fauna with other regions. In addition it often includes speculation, sometimes aided by fossil records, as to the origins and dispersal of the group or fauna under consideration. In such discussions single records and occurrences are often given as much weight as records from localities where the species occurs in abundance. It is well known for littoral animals that isolated non-breeding populations may occur beyond the normal distribution range for the species (KNOX, 1954). Unfortunately in this connection many taxonomic accounts do not give adequate information as to the abundance of the species records and for many organisms, especially for molluscs and echinoderms, they do not distinguish between live and dead specimens.

We need more critical distribution studies along the lines of that of McKNIGHT (1969) for the New Zealand echinoid fauna where records of live and dead or fragmented specimens are recorded separately. McKNIGHT points out that while records of dead material fall almost completely within the geographic range of living specimens there are cases where dead material is recorded some distance from the living. In some cases this may be due to the transport of material offshore to deeper water. Where a species is restricted to a particular range of sediment type the occurrence of dead material in a different sediment type can confidentally be ascribed to transport from elsewhere.

Because different groups of organisms differ in their responses to the same range of environmental conditions it is not surprising that the distributional patterns established for one group are not reflected in another. Nevertheless as we shall see below, there is a considerable measure of agreement in the overall broad patterns of distribution.

The second approach to biogeography is through the ecological relationships of the organisms typified by the work of ZENKEVICH (1947, 1949, 1963) and HEDGPETH (1953, 1971). Where sufficient information is available comparisons can be made between the communities of organisms occupying similar environments in different regions as well as comparisons within a region of the communities found on or in the same substrate at different latitudes. Littoral communities in the New Zealand region are now sufficiently well-known to enable such comparisons to be made (KNOX, 1960, 1963, 1968b; MORTON & MILLER, 1968). In the following account there is a mixture of both approaches dependent upon the amount of information that is available.

FLEMING (1962a, 1962b) has put forward a scheme of nomenclature which is followed below in discussions on the external relationships and origins of the New Zealand marine benthic biota. These elements are: Endemic; Malayo-Pacific; Australian; Neoaustral; Paleoaustral; Holarctic; Cosmopolitan.

The terms Endemic, Australian and Cosmopolitan are self-explanatory. The term Malayo-Pacific covers the New Zealand biotic element of apparently tropical derivation. It includes the northern ex-tropical element of other writers variously referred to as Malayan, Melanesian, Malaysian, Palaeotropic and Indo-Pacific. The term Austral used in its primary sense of southern refers to the element that previously has been referred to as Antarctic, Fuegian, South American or Subantarctic; it includes plants and animals distributed by the West Wind Drift and its contribution to the biota has been greatest before the land masses had drifted apart and later when climatic zones moved northward (in the Pliocene and Pleistocene).

The older Paleoaustral element consists of representatives of genera or endemic genera of northern families which have not recolonized the littoral and shelf of formerly glaciated islands in post-glacial times and which appear to be poorly adapted to trans-oceanic dispersal. In contrast the Neoaustral element includes species that have either not differentiated into geographic races since they became widely distributed in a circumpolar subantarctic zone, or have merely subspeciated to form easily reorganizable representative forms (FLEMING, 1963). Their post-glacial dispersal has been across the sea or an eastward direction down the West Wind Drift as FELL (1962) has shown for the echinoderm genera from the northern land areas.

The Holarctic element is a relatively minor one consisting of species with northern temperate affinities.

Distribution Patterns and Relationships of Selected Groups

Within the limits of this review it is not possible to encompass all the major groups of marine organisms. Those which are discussed have been selected because they have been the subject of recent taxonomic reviews and where there is adequate distributional data upon which sound biogeographical subdivisions can be based.

MARINE ALGAE

For the New Zealand region the Chlorophyta (CHAPMAN, 1956) and the Phaeophyta (LINDAUER, CHAPMAN & AIKEN, 1961) have been the subject of recent reviews. Earlier check-list of the New Zealand marine algae were published by LAING (1927a, 1929, 1939), while LINDAUER (1947) has published an annotated checklist of the brown seaweeds. Comprehensive lists for the Dunedin district and the Hauraki Gulf have been published by NAYLOR (1954) and DELLOW (1955) respectively. However the red seaweeds are as yet imperfectly known and data on this group must be used with caution (MOORE, 1961). MOORE (1949,

1953, 1961) has assembled valuable data on the distribution of New Zealand seaweeds.

Internal distribution

Moore (1949, 1961) has discussed the internal distribution of the New Zealand marine algae based on a detailed study of the ranges of some 200 species of apparently restricted distribution. The following main distribution types were recorded.

A. Not found south of East Cape, or off Albatross Point on the west coast (39 species), for example: (a) East coast only, *Carpophyllum elongatum, Halopteris hordacea, Perithalia capillaris;* (b) East coast and north of Kaipara Harbour, *Xiphophora chondrophylla* var. *minus;* (c) both coasts, *Gigartina alveata, Pachymenia himatophora* (the same distribution as mangroves). *Vidalia colensoi* extends to Albatross Point on the west and Mahia Peninsula on the east. *Carpophyllum plumosum*, chiefly, if not entirely, east coast reaches Castle Point.

B. Northern but reaching to or just beyond Cook Strait (15 species), for example: *Pterocladia lucida, Apophloea sinclairii, Caulerpa sedoides, C. hyperoides, C. articulata, Melanthalia abscissa.*

C. Southern but extending as far as Albatross Point on the west and East Cape (9 species), for example: *Halopteris congesta, H. funicularis, Chaetomorpha darwinii.*

D. Southern but extending as far north as Castle Point (5 species), for example: *Caulerpa brownii, Adenocystis utricularis, Macrocystis pyrifera.*

E. Southern but extending as far as Wellington (27 species), for example: *Marginariella urvilleana, M. boryana, Desmarestia firma, Xiphophora chondrophylla* var. *maxima.*

F. Not extending north of Timaru and Milford Sound (10 species), for example: *Apophloea lyalii, Pachymenia lusoria, Delesseria crassinervia, Myriogramme crispata, Brogniatella australis, Cladophora pellucida.*

On the basis of algal distributions Moore recognizes the following algal provinces: 1. A Kermadec Province; 2. An Auckland Province, the North Island north of East Cape and Albatross Point; 3. An Intermediate Province, the rest of the North Island, excluding the Cook Strait region; 4. A Central Province, the Cook Strait region and the South Island north of Timaru and Milford Sound; 5. A Forsterian Province, the rest of the South Island and Stewart Island; 6. A Chatham Province, the Chatham Islands; and 7. A Rossian Province, the subantarctic islands.

Of the 200 species listed by Moore at least ten from the Kermadec Province, including several belonging to semitropical genera, do not reach New Zealand proper, forty or more are found only in the Auckland Province, some twenty northern species reach about north of Cook Strait unlike nearly thirty southern species that do not extend beyond the northern boundary of the Central Province. At least a dozen species are not known north of the Forsterian Province and the otherwise dominant lower eulittoral brown alga, *Carpophyllum maschalocarpum* is lacking from the same area.

External relationships

LAING (1927b) has analysed the external relationships of the New Zealand marine algae, but his account is in need of revision in the light of recent taxonomic work. The relationships appear to be chiefly with southern Australia and through Australia to some slight extent with the Indian and South Atlantic Oceans.

Comparisons can best be made with the temperate coasts of South Australia stretching from about the southwest of Western Australia to about the Victoria-New South Wales border and including Tasmania. For this region WOMERSLEY (1961) lists 1,010 species as compared with 649 for the New Zealand region (see Table 1). Two important points revealed by this table are first, the greater richness of the southern Australia flora, and secondly, its higher degree of endemism, 71 per cent as contrasted to 42 per cent for the New Zealand flora.

Table 2 gives the total number of genera and the number of endemic genera found in southern Australia and New Zealand as well as the number of genera endemic to temperate Australasia.

In the Phaeophyta Australia has twelve endemic genera whereas New Zealand has only three. The latter include *Marginariella* Richard, a distinctive and handsome genus with two distinct species and *Landsburgia* Harvey also with two species, *L. quercifolia* growing on the coasts north of Banks Peninsula and *L. myricaefolia* confined to the Chatham Islands. In addition the following five distinctive genera, *Hormosira* Endlicher, *Notheia* Bailey and Harvey, *Perithalia* J. Agardh, *Ptilopogon* Reinhe and *Xiphophora* Montagne are found only along southern Australia, in New Zealand and the associated islands. Where there are common genera the number of species in Australia is in nearly all cases much greater than in New Zealand, for example, *Cystophora*, J. Adargh, 23 species compared with 6, and *Sargassum* C. Adargh, 30 species compared with 6. The characteristic cold temperate southern genus *Durvillea*, however, is represented in Australia by the single endemic species, *D. potatorum*, whereas in New Zealand there are three species, the endemic *D. willana*, and the widely distributed *D. antarctica* and *D. caespestipes*.

The Rhodophyta show a similar picture with 72 endemic genera in Australia and only 8 in New Zealand. Of the non-endemic New Zealand genera, 103 out of 115 occur in Australia. Again the number of species in a genus in Australia greatly exceeds that of the New Zealand species. Notable exceptions, however, are the genera *Polysiphoria* Greville with 23 species in New Zealand and 7 in Australia and *Gigartina* Stackhouse with 29 species in New Zealand and 14 in Australia.

Table 1. Specific composition of the algal flora of temperate Australia and New Zealand.

Group	Total number of species		Number and percentage of endemic species		Number and percentage of species endemic to Australasia		
	Aust.	N.Z.	Aust.	N.Z.	No.	% Aust.	% N.Z.
Chlorophyta	94	148	43 (46%)	67 (38%)	4	4.2%	1.7%
Phaeophyta	191	131	134 (70%)	40 (30%)	11	6 %	8.4%
Rhodophyta	725	403	538 (75%)	174 (43%)	59	8 %	14.6%
Total	1,010	649	715 (71%)	281 (42%)	74	7.3%	11 %

Table 2. Generic composition of the algal flora of temperate Australia and New Zealand.

Group	Total number of genera		Number and percentage of endemic genera		Number and percentage of genera endemic to Australasia		
	Aust.	N.Z.	Aust.	N.Z.	No.	% Aust.	% N.Z.
Chlorophyta	27	29	3 (11%)	2 (7 %)	0	0 %	0 %
Phaeophyta	63	66	12 (19%)	3 (4.5%)	5	8 %	7.2%
Rhodophyta	239	131	72 (30%)	8 (6 %)	8	3.8%	6.1%
Total	329	226	87 (26%)	13 (5.7%)	13	4 %	6 %

From the above it is clear that the Australian element in the predominant are in the New Zealand region. The other important element is an Austral one with an Antarctic-subantarctic group of species such as *Chaetomorpha darwinii*, *Scytothamnus australis*, *Adenocystis utricularis*, *Lessonia variegata*, *Durvillea antarctica*, *Desmarestia willii*, *Delisea pulchra*, *Ballia callitricha* and *Griffithsia antarctica*. This Austral element is better represented in New Zealand than in Australia, 35 species as compared with 21 species. Due to the lack of a fossil record the number of species which can be considered as Neoaustral as opposed to Paleoaustral cannot be determined with any confidence.

Scleractinia corals

Squires & Keyes (1967) have recently reviewed the distribution and relationships of the New Zealand Scleractinian corals. The scleractinian fauna comprises 25 species in 16 genera. Few species are stenobathic but rather are found through a considerable depth range with little or no discernible break at the edge of the shelf.

Internal distribution

SQUIRES & KEYES (1967) distinguish five internal distribution patterns. One represented by the endemic *Flabellum rubrum* consists of species widely distributed round the New Zealand shelf. A second group of species, including *Oculina virgosa*, *Kionotrochus (K.) suteri*, *Flabellum aoteara* and *F. gracile*, are largely restricted to the east coast of the North Island north of East Cape. A third miscellaneous group is largely northern in its distribution. A fourth group of four colonial species occur round the North Island or the South Island east coast and on the Chatham Rise. Their affinities are mainly to the north. A fifth group including *Caryophyllia profunda and Desmophyllum cristogalli* are widely distributed in the New Zealand region.

External relationships

New Zealand corals are all ahermatypic, or non-reef building, the temperatures in winter, even in the extreme north, falling below the critical 18 °C for hermatypic, or reef-building corals. SQUIRES & KEYES (1967) state: 'The ahermatypic corals of New Zealand are related to those of the South Pacific and Antarctica, and show little relationship to those of Australian waters. Less than half (12) of the known species (25) are endemic, while (7) are autochthonous, and traceable to forms of the New Zealand Neogene.' They consider the fauna to be not an old one, but a new one, despite the high level of endemism.

Oculina virgosa and *Sphenotrochus ralphae* appear to be remnants of a great invasion of New Zealand by northern corals during the Otaian (Lower Miocene) when sub-tropical water moved south. While most of these invading species disappeared during the late Tertiary in cooling seas these two species adapted and survived. On the other hand, *Culicia rubeola*, *Caryophylla profunda*, *Desmophyllum cristagalli* and *Flabellum rubrum* are all late Tertiary immigrants which arrived when the diversity of the New Zealand coral fauna was undergoing a sharp reduction. These are Neoaustral species.

Of the 25 New Zealand coral species only two, *Paracynthus conceptus* and *Letepsammia* have Australian affinities; four species, *Caryophyllia profunda*, *Flabellum apertum*, *F. lowekeyesi*, *F. knoxi*, are of southern derivation (Neoaustral), or as now conceived are members of the Southern Ocean fauna. Other species which occur to the north of New Zealand (Malayo-Pacific) include *Dendrophyllia japonica*, *Notocyathus orientalis*, *Caryophylla maculata* and *Flabellum aotearoa*.

According to SQUIRES & KEYES (1967) the New Zealand coral fauna shows a pattern of an older fauna that has had superimposed on it a late Tertiary cold-water element that is now being modified by immigration from the north.

Gammarid amphipods

Barnard (1972) has recently reviewed the taxonomy and distribution of the New Zealand gammarideans living in algae in littoral habitats. He recorded a total of 113 species in 61 genera of which 80 species and 9 genera are endemic. This degree of endemism is low at the generic level but high at the species level.

Internal distribution

Barnard's conclusions can be summarized as follows. Little evidence of biogeographic divisions within New Zealand is apparent in the Gammaridea. The northern part of New Zealand, in contrast to the situation in some other invertebrate groups, appears to lack a warm-temperate amphipodean fauna distinct from that of the remainder of New Zealand. It is, however, characterised by a low frequency of individuals in the intertidal and marked reductions in individuals of a few species abounding in southern waters. The southern cold-temperate waters show this in reverse. There is a similar diversity of species in the north and south. Only four of the common intertidal species of the two regions have distributions limited to warm waters and seven to cold waters.

It should be pointed out that the material examined by Barnard came almost entirely from the east coast and that the west coast gammaridean fauna is very incompletely known.

External relationships

Barnard concludes that the entire New Zealand gammaridean fauna is dominated by species with warm temperate Australian affinities. In general these species are eurythermic and there is no evidence of a stenothermic warm-temperate element in the transitional warm-temperate waters on the eastern coast of the North Island north of East Cape.

The fauna is as well balanced phyletically as that of any other warm-temperate region, except perhaps Australia, in spite of New Zealand's long isolation. This isolation is however reflected in the paucity of intertidal members of the families Amphithoidae, Leucothoidae, Phliantidae (Australian), the low genetic diversity of the Podoceridae, and the low numbers of aorids (Isaeidae). The families Colomastigidae, Cyproidenae, Eusiridae and Liljeborgiidae on the other hand are exceptionally diverse.

Fourteen species are cosmopolitan and 15 species have unknown affinities. Of the rest 27 have Austral affinities, 26 have Australian affinities and 23 can be considered to have Indo-Malayan affinities. There are 8 unique New Zealand endemic taxa which are probably derived from warm-water immigrants.

If the cosmopolitan species are disregarded 27 percent of the remaining 99 have cold-water affinities and 73 percent warm-water ones. Of the 27 cold-water Austral species, 10 have a circum-subantarctic distribution and 17 have Antarctic or Subantarctic affinities.

It is clear that the West Wind Drift has contributed a significant but small cold-temperate component throughout the New Zealand region. Since the main relationships are with Australia, BARNARD considers that New Zealand is an outlyer of Australia marked by some impoverishment suggesting that viable immigrants have necessarily been eurythermans.

BRACHYURA

As a result of the work of BENNETT (1964) and GRIFFEN (1966) the New Zealand crab fauna is now reasonably well known. DELL (1968) has reviewed the composition and distribution of the Brachyuran fauna as a whole. The total number of species in the fauna (excluding the Hymenosomidae) is 51 species in 39 genera. Thus compared with Chile (77 species) and South Australia (93 species) the fauna is relatively restricted. Thirty-two (82 percent) of the genera are represented by only one species compared with 74 percent for South Australia and 71 percent for Chile.

The spider-crabs (Majidae) are well represented with 11 genera but the only genera with any considerable number of species are *Leptomithrax* with six species and *Notomithrax* with three species.

Internal distribution

DELL (1968) has analysed the pattern of distribution of 38 intertidal and shelf species along the east coast (Fig. 3). The general pattern which emerges is as follows. 1. A large group of species, including the genera *Notomithrax*, *Cancer*, *Nectocarcinus*, *Ovalipes*, *Heterozius*, *Pilumnus*, *Pinnotheres*, *Hemigrapsus*, *Cyclograpsus*, *Helice* and *Hemiplax*, ranges over most of the North and South Islands, the majority extending as far south as Stewart Island. Two genera *Cancer* and *Chlorinoides* extend to the Auckland Islands and one, *Nectocarcinus*, to Campbell Island. There is more diminution in numbers towards the north.

Apart from three species the rest are endemic and DELL concludes that they may be presumed to be species which have been in the New Zealand area for some considerable time, or in some cases even to have developed here. 2. Another major group is predominantly northern in

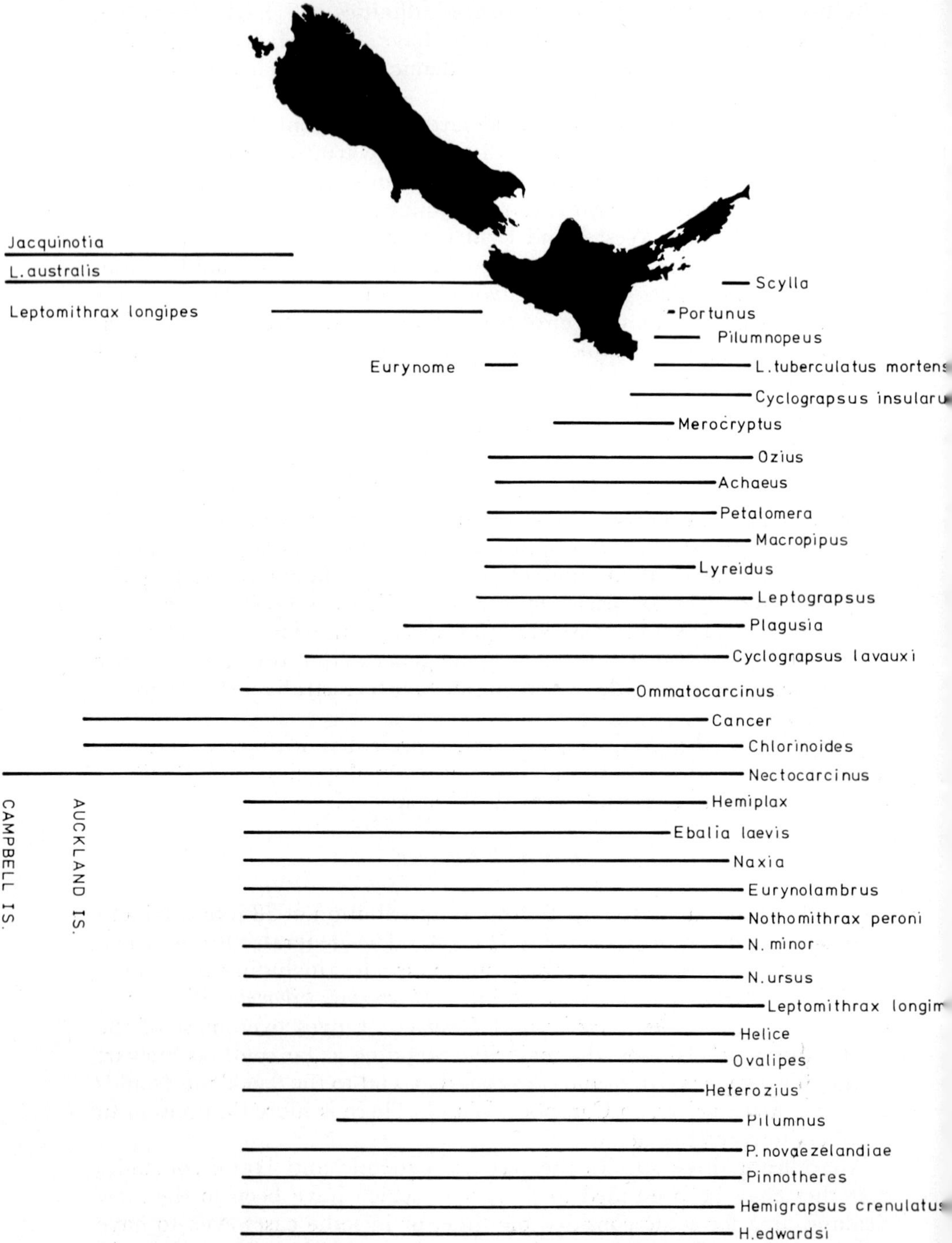

Fig. 3. Known distribution of 38 species of intertidal and shelf crabs (After DELL, 1968).

distribution with the species extending to a varying distance south. A group which does not extend much further south than Cook Strait includes *Merocryptus*, *Achaeus*, *Ozius*, *Petalomera*, *Macropipus*, *Lyreidus* and *Leptograpsus* all of which also occur in Australia and can perhaps be considered to be recent eurythermal warm-water immigrants. 3. A relatively small group has a southern distribution pattern within the New Zealand area. It includes the endemic genus *Jacquinotia* and *Leptomithrax australis*. 4. There is another very small group comprising *Leptomithrax longipes*, *Eurynome* and perhaps *Ommatocarcinus* which have restricted distribution ranges which do not coincide but which are grouped towards the centre of the mainland.

Of the spider crabs five species are commonly found intertidally and with one exception are universally distributed round the mainland. Three species are restricted to the archibenthal and also seem to be widely distributed. The 10 species found on the continental shelf all extend varying degrees into the archibenthal.

There are anomalies of distribution patterns in the three peripheral areas, the far north, the subantarctic islands and the Chatham Islands.

In the north from the Bay of Islands northward there is a steady falling off in both the presence and abundance of intertidal crab species.

The southern islands of New Zealand, especially Campbell Island, approach the limit for Brachyura in the South Pacific and South Indian Oceans. Some 21 species have been recorded from Stewart Island and Foveaux Strait. Seven species are known from the Auckland Islands (including *Halicarcinus planatus* which is not known from the mainland) while six species are recorded from Campbell Island. The occurrences of *Chlorinoides*, *Leptomithrax*, *Cancer* and *Nectocarcinus* are extensions of range of typically mainland faunas. *Halicarcinus planatus* is is widely distributed round the Southern Ocean and is the only southern high latitude species of crab. *Jacquinotia* is an endemic genus extending north to the southern mainland. It seems likely that this monospecific genus evolved in southern New Zealand.

Sixteen crab species are known from the Chatham Islands. The main characteristics of this crab fauna (Dell, 1968) are: 1. The general paucity of species. 2. The absence of graspoid crabs (except for the oceanic *Planes*). 3. The lack of intertidal crabs except near low water. No crab species are endemic to the Chathams.

External relationships

The fauna can conveniently be grouped into the categories outlined above. 1. Endemic element. Of the 51 species no less than 27 (53 percent) are believed to be endemic. Among the spider crabs 12 species (66 percent) are not found outside New Zealand waters. 2. Malayo-Pacific element. Nine of the species found in New Zealand are also known from various localities from Japan to the Indian Ocean to the south-west Pacific; six of these are also known from Australia. 3. Australian element. Eight species of New Zealand crabs are known elsewhere only from Australia and the New Zealand representatives have certainly been distributed across the Tasman. 4. Austral element. Only two species, *Hemigrapsus crenulatus* and *Plagusia chabrus*, can be considered as Austral in origin. The former is known only elsewhere from Chile and the latter is widely distributed round the southern extremities of the land masses of the Southern Ocean. This Austral element is low compared with other animal groups but may be explained in part by the lack of crabs in the southern part of the Southern Hemisphere and the marked fall off in crab species along land masses which extend into comparatively high latitudes. 5. Cosmopolitan element. Only three species of limited occurrence can in any way be considered to be cosmopolitan. At the specific level the external relationships can be summarized as follows:

	Number of species	Percentage
Endemic	27	53
Malayo-Pacific	10	20
Australian	9	17
Austral	2	4
Cosmopolitan	3	6

DELL (1968) has also analysed the relationships of the species occurring in various bathymetric ranges. The percentages of the total fauna in each bathymetric group (depths in metres) is given below.

	Intertidal	0–90	0–180	0–180+	over 180
Endemic	47	57	33	73	50
Malayo-Pacific	27	15	67	9	0
Australian	13	21	0	18	33
Austral	13	0	0	0	0
Cosmopolitan	0	7	0	0	17

It can be seen that the endemic element is least represented among the species with wide benthic ranges on the shelf, amongst which the Malayo-Pacific element is not strongly represented. The Australian and Cosmopolitan components are most marked amongst the archibenthal crabs. The Austral element is confined to the intertidal zone.

On the basis of the above consideration of the composition, internal distribution and external relationships of the New Zealand crab fauna DELL makes the following generalisations.

First the fauna is 'unbalanced'. With few exceptions each genus is represented by only one or two species.

Comparison with other areas shows a close similarity in faunal composition with oceanic island faunas. There is a strong endemic element, but there is also evidence that species are still being acquired (such as

Scylla serrata, *Portunus pelagicus* and *Cyclograpsus insularum*). There are strong Australian and Malayo-Pacific elements but it seems probable that the main migration route has been from Australia across the Tasman Sea.

The internal distribution of crabs follows the pattern seen in other animal groups with a northern and southern group of species, each one extending southwards or northwards in an apparently independent fashion. In the crabs the northern element is overwhelmingly stronger. In the extremities (the Chatham Islands, the subantarctic islands and to a lesser extent the far north) the fauna becomes impoverished.

ECHINODERMATA

As DAWSON (1965) points out echinoderms are especially suitable for ecological and biogeographical studies. 'They are relatively sedentary and usually have limited bathymetrical ranges, often showing definite sediment preferences and correlations with salinity and temperature. The phenomena of abbreviated larval stages and direct development are also important in the distribution of certain species.' Due to the work of FELL (1953a, 1953b, 1958, 1960, 1962, 1963), PAWSON (1961, 1963, 1965a, 1965b, 1965c, 1965d, 1968, 1970), DAWSON (1965, 1970) and MCKNIGHT (1969) their taxonomy and detailed distribution is probably better known than for any other phylum.

At the present time approximately 237 species of echinoderms are known from the New Zealand region (DAWSON, 1970). The known asteroid fauna is 65 species of which 34 occur on the shelf. The ophiuroid fauna now comprises at least 87 species of which 45 are found in shallow water and about 80 occur in deeper water. The echinoids are represented by 49 species, 28 of which occur on the shelf (BAKER, 1968; MCKNIGHT, 1969), while the holothurian fauna comprises 36 species (PAWSON, 1970).

Internal distribution

1. Intertidal and shelf fauna. PAWSON (1964) has analysed the distribution of echinoderms along the east coasts of New Zealand and DAWSON (1965) and MCKNIGHT (1969) have given additional information. Within this fauna the four elements listed below can be found. The percentages within each of the four echinoderm groups are given.

	Northern	Central	Southern	Widespread
Asteroidea	21%	21%	26%	17%
Ophiuroidea	48%	12%	22%	—
Echinoidea	55%	—	28%	17%
Holothuroidea	15%	19%	41%	25%

From the above it can be seen that the proportions of the northern and southern elements differ for the four groups. The greatest variation is in the northern element which is least pronounced in the Asteroidea.

The distribution of some selected groups, notably the shelf echinoids and the holothurians will be discussed, followed by a consideration of the fauna of the subantarctic islands and the Campbell Plateau and finally a general review of the internal patterns of echinoderm distribution.

Shelf echinoids

McKnight (1969) records 28 species as occurring on the New Zealand shelf and intertidal zone. Of these 12 (43 percent) are rare species known from fewer than five records, four species (14 percent) are moderately rare known from fewer than 20 but more than five records, and nine (33 percent) are moderately common and three (10 percent), *Pseudechinus albocinctus*, *Evechinus chloroticus* and *Echinocardium cordatum* are very common.

No species appears to be confined to the intertidal zone although 12 species occur intertidally. Five species are restricted to the intertidal zone and the adjacent shallow water down to 50 m. Nine species are restricted to the continental shelf and a further 13 occur on but are not restricted to the shelf.

The northern element of 16 species (55 percent of the shelf fauna) is characterized by a large proportion of the rarer species (11 species). The southern element of eight species (28 percent), on the other hand, contains only two species with restricted distributions.

On the east coast northern restricted elements occur in the area between the Three Kings Islands and East Cape, while the other northern species occur more or less north of the Cook Strait area; southern elements occur more or less in and south of the Cook Strait area, with the restricted elements occurring on the southern shelf. On the west coast, northern elements extend as far south as Cape Foulwind while most of the southern elements are confined to southern Fiordland.

McKnight found no substantial differences in the distribution patterns of species with pelagic larvae compared with those having direct development, indicating that the mode of development does not appear to be important in distribution.

Of the older members of the fauna, *Apatopygus* is southern in distribution and *Fellaster* is more northern, but other species such as *Goniocidaris* spp., *Pseudechinus* spp., *Peronella hinemoae* and *Evechinus chloroticus*, occur both in the north and south. Recent or Pleistocene immigrants to the fauna, such as *Holopneustes*, *Heliocidaris*, *Clypeaster*, *Laganum*, *Brissus* and *Brisopsis* are all included in the northern element. Fell (1953) has suggested that these species have come from Australia via the Tasman Current. If they did we would, as pointed out by McKnight (1969), expect them to be found on the west coast especially in the Fiordland area where this current first impinges on the New Zealand coast. Their absence from this area suggests that a more northern migration route should be sought.

Holothuroidea

PAWSON (1965, 1970) has recently reviewed the taxonomy and distribution of the New Zealand sea cucumbers. Forty-nine species in 36 genera are recognized with a ratio of species per genus of approximately 1.5; this approximates the ratio for Antarctica and Southern Chile, but is slightly less than that for Australia. Recent work (PAWSON & FELL, 1965; PAWSON, 1966) indicates that perhaps U-shaped holothurians reflect a primitive stage in the evolution of the group. It is of interest therefore that no less than 10 of those commonly encountered in New Zealand are U-shaped and this high incidence of primitive forms is a notable feature of the New Zealand fauna.

Another feature is the high percentage of dendrochirotids which are characterized by direct development with no pelagic larval stage. These holothurians are especially well adapted to life in cold waters, comprising 70 percent of the Antarctic fauna (PAWSON, 1969). In the warmer waters of the New Zealand shelf 45 percent of the species are dendrochirotids.

Twelve species occur intertidally. Of the 27 species that occur on the shelf 13 (approximately 50 percent) appear to be restricted to the shelf while the remainder extend into the archibenthal. Of the entire echinoderm fauna of New Zealand approximately 58 percent are stenobathic shelf species. In the archibenthal zone 31 species occur, of which 17 (approximately 55 percent) are restricted, with the remaining 14 extending onto the shelf.

According to PAWSON (1970) the holothurians do not show as convincingly as for other groups the patterns of distribution into widespread, northern and southern groups. No species are typically northern and the southern group of species differs from that in the central region mainly by the absence of some species found there.

The echinoderm fauna of the offlying islands. 1. Kermadec Islands; Only two species of the Kermadec fauna are shared with the New Zealand fauna. However one of the restricted species, *Patieriella oliveri*, is very closely related to *P. regularis* of New Zealand and the two almost certainly share a common ancestry (PAWSON, 1961). In addition the Kermadecs starfish, *Astrostole rodolphi*, is closely related to the *A. scabra* of the New Zealand mainland.

Of the two species shared with New Zealand the ophiuroid *Ophidiaster kermadecensis* is found only elsewhere in northern New Zealand and the other, the ubiquitous urchin *Evechinus chloroticus*, is not found outside the New Zealand region and occurs as a fossil at least the Pliocene.

2. The Subantarctic Islands; DAWSON (1965, 1970), FELL (1953a) and PAWSON (1961, 1965) have all discussed the echinoderm fauna of the New Zealand subantarctic islands and Macquarie Island. The following

table adapted from DAWSON (1965) summarizes the echinoderm distribution in this region.

Number of species and biogeographic distribution of the major groups of echinoderms (excluding Ophiuroids and Crinoids) from the New Zealand subantarctic islands (shelf and shore) and the Campbell Plateau (archibenthal).

	No. of species	Common with N.Z. shelf and/or slope	Endemic in Subantarctic	Extra limited distribution
AUCKLAND IS.				
Asteroids	11	7	2	2
Echinoids	4	4	—	—
Holothurians	8	6	2	—
CAMPBELL IS.				
Asteroids	5	4	1	—
Echinoids	2	1	1	—
Holothurians	4	3	1	—
ANTIPODES IS.				
Asteroids	2	2	—	—
Echinoids	4	3	1	—
Holothurians	2	2	—	—
BOUNTY IS.				
Asteroids	1	1	—	—
Echinoids	5	4	1	—
Holothurians	2	1	1	—
SNARES IS.				
Asteroids	10	10	—	—
Echinoids	9	9	—	—
Holothurians	2	1	1	—
MACQUARIE IS.				
Asteroids	14	6	6	2
Echinoids	1	1	—	—
Holothurians	6	4	1	1
CAMPBELL PLATEAU (Archibenthal)				
Asteroids	17	13	?4	?
Echinoids	5	5	—	—
Total species	70	48	19	5

Sixty-eight percent of the echinoderms from the New Zealand Subantarctic are shared with the New Zealand Shelf or Slope. Endemic species comprise 27 percent and only seven percent of the subantarctic species are also known from the Magellanic or other extra-limited areas.

DAWSON (1965) considers that epiplanktonic drift, as postulated by FELL (1962), occurring down the wind-driven surface layers of the West Wind Drift, has contributed no species to the New Zealand Subantarctic.

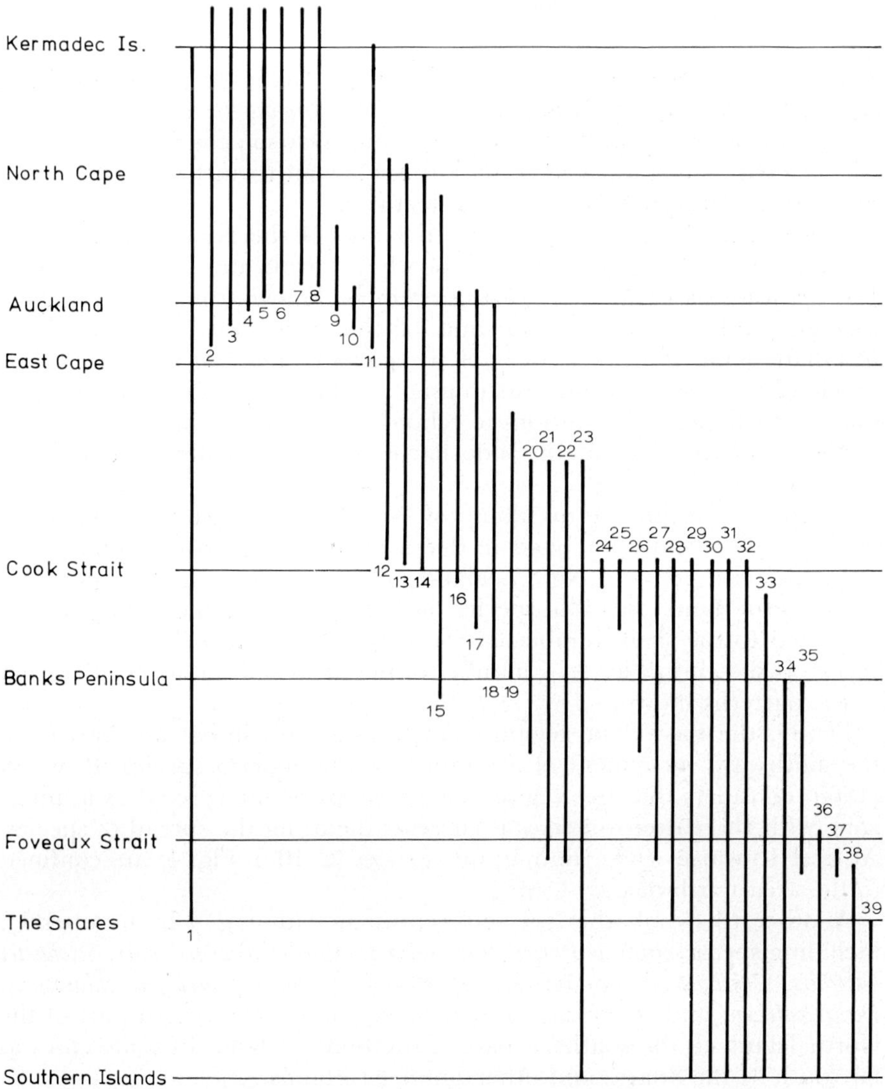

Fig. 4. Horizontal distribution of stenobathic shelf echinoderms along the east coast of New Zealand. Key: 1, an assemblage of 13 species which ranges the entire east coast; 2, *Astropecten polyacanthus;* 3, *Ophiocentrus pilosa;* 4, *Asterodiscus truncatus;* 5, *Holopneustes inflatus;* 6, *Heliocidaris tuberculata;* 7, *Centrostephanus rodgersii;* 8, *Laganum depressum tonganense;* 9, *Brissus gigas;* 10, *Luidia varia;* 11, *Ophidiaster kermadecensis;* 12, *Pectinura cylindrica;* 13, *Ophiozonoida picta;* 14, *Ophioceres huttoni;* 15, *Stegnaster inflatus;* 16, *Amphiura rosea;* 17, *Ophiopteris antipodum;* 18, *Heterothyone ocnoides;* 19, *Astrostole scabra;* 20, *Stolus huttoni;* 21, *Pentagonaster pulchellus;* 22, *Asterodon miliaris;* 23, *Pseudechinus novaezealandiae;* 24, *Hemilepis abernethyi;* 25, *Eurygonias hylacanthus;* 26, *Neothyonidium dearmatum;* 27, *Pectinura maculata;* 28, *Asterodon dilatatus;* 29, *Apatopygus recens;* 30, *Monamphiura magellanica;* 31, *Calvasterias suteri;* 32, *Trochodota dunedinensis;* 33, *Amphioplus basilicus;* 34, *Paranepanthia aucklandensis;* 35, *Psolidiella nigra;* 36, *Trachythyone amokurae;* 37, *Ophionephthys stewartensis;* 38, *Pseudopsolus macquariensis;* 39, *Pseudocnus leoninoides*. Those species known from but a single record are not included. (After PAWSON, 1965).

The widespread Australian, Indo-West Pacific and cosmopolitan genera *Trochodota*, *Chirodota*, *Spatangus*, *Paramaretia*, *Briaster*, and *Psolus* contribute at least eight species to the fauna. Dawson considers that their bathymetrical tolerances have probably allowed them to migrate along submarine ridges and other bottom features.

A surprising feature of the starfish fauna of the New Zealand Subantarctic is the presence of archibenthal warmwater species such as *Benthopecten pentacanthus* and *Zoroaster spinulosa* known from northern New Zealand and the Indo-Pacific and of the deep-water starfish of the Chatham Rise. Recent records of the genera *Lithosoma* and *Ceramaster*, previously unknown from Australasia are further evidence suggesting that the principal contributions to echinoderm fauna of the New Zealand Subantarctic are Australian, Indo-Pacific and cosmopolitan elements.

Distribution of echinoderms along the New Zealand east coast; PAWSON (1965b) has recently analysed in detail the distribution of echinoderms along the east coast of New Zealand. The east coast shelf supports an echinoderm fauna of 145 species in 92 genera. Stenobathic species restricted to the shelf number 83 (56 genera). The distribution of these species is shown in Fig. 4. A number of points arise from a consideration of the patterns shown.

Firstly, the East Cape region constitutes a most important barrier to the southward movement of a number of echinoderm species. PAWSON (1965) considers that this region may perhaps be interpreted as a 'filter' zone with the pattern of coastal currents dictating the spread of species. At least 10 warm-water 'immigrant' species (2–10 in Fig. 4) are confined to the area north of East Cape.

Another 13 stenobathic species (represented in Fig. 4 by number 1), including species such as *Patieriella regularis*, *Allostichaster polyplax*, *Stichaster australis*, *Coscinasterias calamaria*, *Ophionereis fasciata*, *Evechinus chloroticus*, *Ocnus calcaria* and *Trochodota dendyi*, range from the northern part of the North Island to the southern part of the South Island, in some cases as far south as the Snares and other southern islands.

South of East Cape, the shelf fauna has a different aspect. Its most notable feature includes the presence of some southern species, virtual absence of stenobathic northern immigrants and an overlap of endemic species which have either a northern or a southern distribution within New Zealand but whose individual ranges do not embrace the entire coast. Cook Strait is an important region of overlap of northern and southern species giving the fauna a 'mixed' aspect.

Foveaux Strait has many species in common with Cook Strait but differs in possessing the following features: 1. Presence of some species which are not known further north. 2. Peaking of populations in several species including *Apatopygus recens*, *Asterodon dilatatus*, *Pseudochinus novaezealandiae*, *Goniocidaris umbraculum*, *Calvasterias suteri* and *Pentagonaster pulchellus*. 3. Impoverishment in some Families, such as the Ludiidae and Astropectinidae. 4. Lack of a pronounced overlap of northern and southern species, a characteristic of Cook Strait.

In contrast to the shelf fauna the archibenthal or slope fauna does not show the north-south distribution patterns. This archibenthal fauna comprises 100 species. Of these 57 percent are endemic compared with 85 percent on the shelf, indicating the ability of deeper water fauna to spread more widely.

Thirty-three species are common to the shelf and slope and of these seven (20 percent) are known from elsewhere. Thirty species are common to the shelf, slope and abyss; of these 14 (46 percent) are endemic but widely distributed and 16 (54 percent) are known from elsewhere. Twenty-four species are common to the slope and abyss, and of these 18 (75 percent) are known from elsewhere. Eleven species are restricted to the slope, and of these two (18 percent) are known from elsewhere.

There is no restricted archibenthal fauna as such. In the first three categories listed above the percentage of immigrant species found elsewhere rises from 20 percent through 54 percent to 75 percent, indicating how the fauna loses its 'New Zealand character' with increase in depth.

External relationships

In his analysis of the stenobathic echinoderm shelf fauna PAWSON (1965) concludes that its generic content indicates strong affinities with the Indo-West Pacific region (Malayo-Pacific element). At the specific level 83 percent of the species are endemic, 15 percent have warmer-water distribution, especially in Australia (Australian element) and two (three percent) are circumpolar (Austral element). This shelf fauna has no true cosmopolitan elements; most of the cosmopolitan species in the echinoderm fauna are eurybathic forms. The two circumpolar elements are *Ocnus calcareus* and *Monamphium magellanica*, both of which are commonly found in the holdfasts of seaweeds such as *Macrocystis* and hence could easily have been transported in surface drift as postulated by FELL (1962).

The Malayo-Pacific element is an important one. These are species known from the Australian-Indo-Pacific region, and warmer northern waters. Two of these species, *Allostichaster polyplax* and *Coscinasterius calamaria* are widely distributed along the coasts of New Zealand. Nine other species (2–11 in Fig. 4) have, as pointed out above, a restricted northern distribution.

Of the 100 archibenthal species 57 (57 percent) are endemic, 27 (27 percent) are Australian or Indo-Malayan elements, 12 (12 percent) are cosmopolitan, and only four (4 percent) are Austral elements.

In his analysis of the echinoid shelf fauna, MCKNIGHT (1969) concludes that the major external relationship is with Australia but that there are two components to this relationship. Firstly there are species

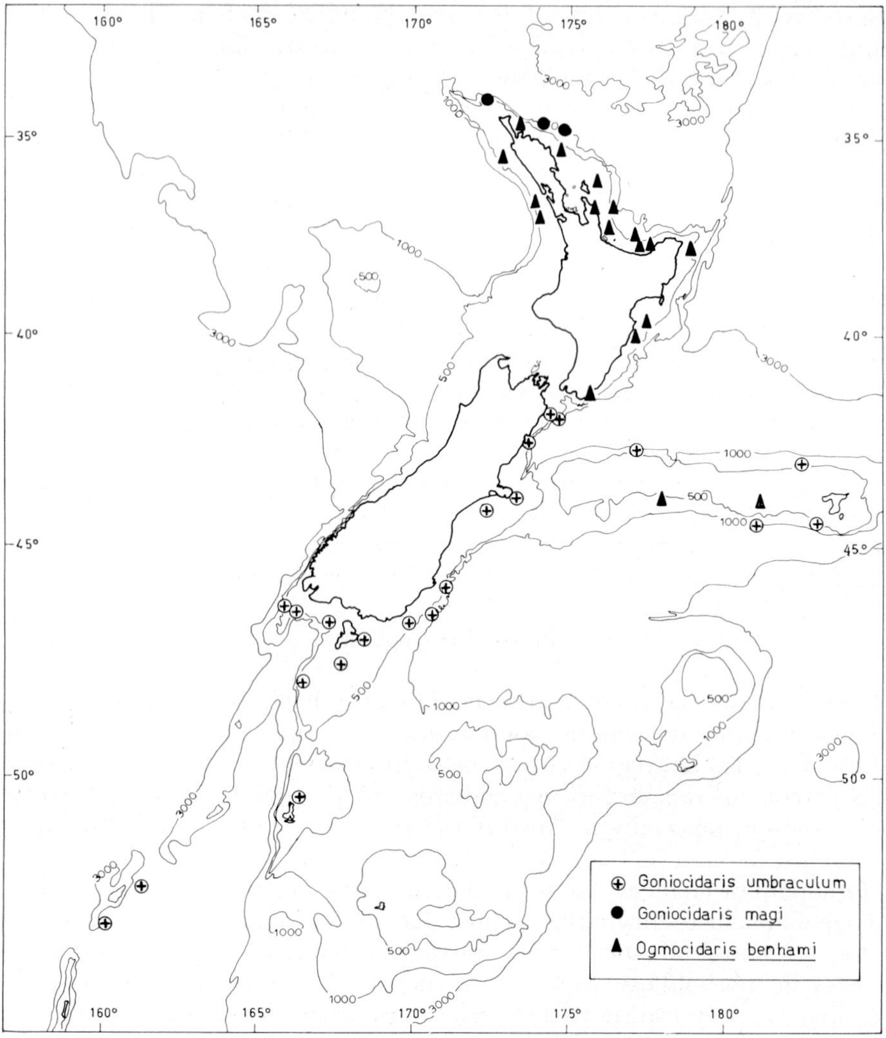

Fig. 5. Distribution patterns of the New Zealand shelf echinoids: (a) *Goniocidaris magi*, a subtropical species, (b) *Ogmocidaris benhami*, a northern warmwater species, (c) *Goniocidaris umbraculum*, a southern cold-water species. (Based on data from DAWSON, 1965 and MCKNIGHT, 1969).

such as *Araeosoma thetidis*, *Amblypneustes pachistus*, *Paramaretia multituberculata* and *Pseudechinus huttoni*, which are shared only by the two areas. Secondly, there are species and genera common to Australia, New Zealand and other areas. Malayo-Pacific relationships are especially strong, including species such as *Clypeaster virescens*, *Brissus gigas*, *Salenocidaris hastigera*, *Brissopsis oldhami* and *Gymnopatagus magnus*.

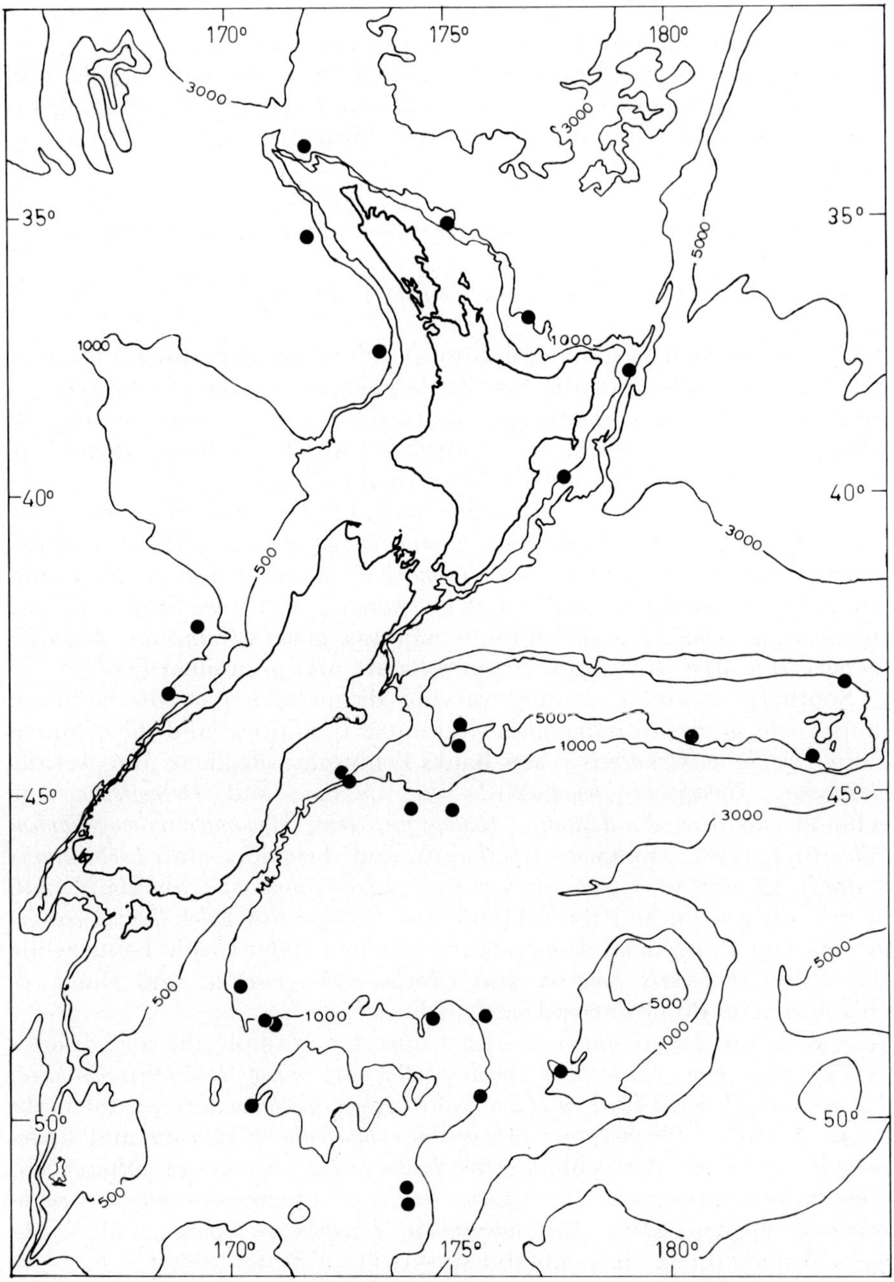

Fig. 6. The distribution of the echinoid *Phormosoma bursarium*, a widespread slope (archibenthal) species. (Based on data from DAWSON, 1965 and MCKNIGHT, 1969).

Figs 5 and 6 show some typical echinoid distribution patterns. Fig. 5 shows the distribution of three shelf cidarids, *Goniocidaris magi*, a subtropical species, *Ogmocidaris benhami*, a northern warm-water species and *Goniocidaris umbraculum*, a southern cold-water species. Fig. 6 contrasts the widespread distribution of a typical archibenthal echinoid, *Phormosoma bursarium.*

Internal distribution patterns – general conclusions

a. Littoral and shelf flora and fauna

The distribution of the dominant littoral (intertidal) and upper sublittoral plants and animals along the New Zealand east coasts have been analysed by Knox (1960). Ten main types of distribution are recognized (Fig. 7). These can be grouped into five categories which are here extended to include the internal shelf distributions discussed above.

1. Species not extending in quantity north of Dunedin: for example the algae *Apophloea lyallii*, *Pachymenia lusoria*, *Brogniatella australis* and *Myriogramme crispata*, the spider crab, *Jacquinotia edwardsii*, the echinoderms, *Trachythyone amoukurae* and *Ocnus brevidentis*, and the limpet *Cellana strigilis redimiculum.* The subantarctic mollusca genera *Gaimardia*, *Kidderia*, *Costokidderia*, *Margarella* and *Kerguelenella* are also prominent here.

2. Southern species extending varying distances north: the southern limpet *Cellana strigilis redimiculum* (in limited quantities) and the echinoid *Paranepanthia aucklandensis* reach Banks Peninsula; the algae *Marginariella urvilleana*, *Xiphophora chondrophylla* var. *maxima* and *Desmarestia*, the echinoderms *Asterodon dilatatus*, *Apatopygus recens*, *Monamphium magellanica*, *Calvasterias suteri*, *Trochodota dunedinensis*, and the spider crab *Leptomithrax australis* extend to Cook Strait; the algae *Adenocystis utricularis*, and *Macrocystis pyrifera* and the echinoderms *Pentogonaster pulchellus*, *Asterodon miliaris* and *Pseudechinus novaezealandiae* reach to about Castle Point, while the algae *Halopteris congesta* and *Chaetomorpha darwinii* and the crab *Ommatocarcinus* do not extend beyond East Cape.

3. Species not found south of East Cape: for example the algae *Carpophyllum elongatum*, *Xiphophora chondrophylla* var. *minor*, *Halopteris hordacea*, *Perithalia capillaris*, *Gigartina alveata* and *Pachymenia himantophora;* the crabs *Scylla*, *Portunus*, *Pilumnopeus*, *Leptomithrax tuberculatus mortensi* and *Cyclograpsus insularum;* the echinoderms *Iuidia varia*, *Astropecten polyacanthus*, *Ophidiaster kermadecensis*, *Asterodiscus truncatus*, *Ophiocentrus pilosa*, *Centrostephanus rodgersii*, *Heliocidaris tuberculata*, *Holopneustes inflatus* and *Brissus gigas*. Among the molluscs notable species are *Xenophora neozelanica*, *Tonna haurakiensis*, *Pteronotus eos*, *Gomphina maorum*, *Alcithoe depressa*, *Venericardia reinga* and *Haliotis virginea crispata*, and several species of the Cymatiidae.

4. Northern species extending varying distances south: for example, amongst the algae, *Vidalia colensoi* extends as far as Mahia Peninsula,

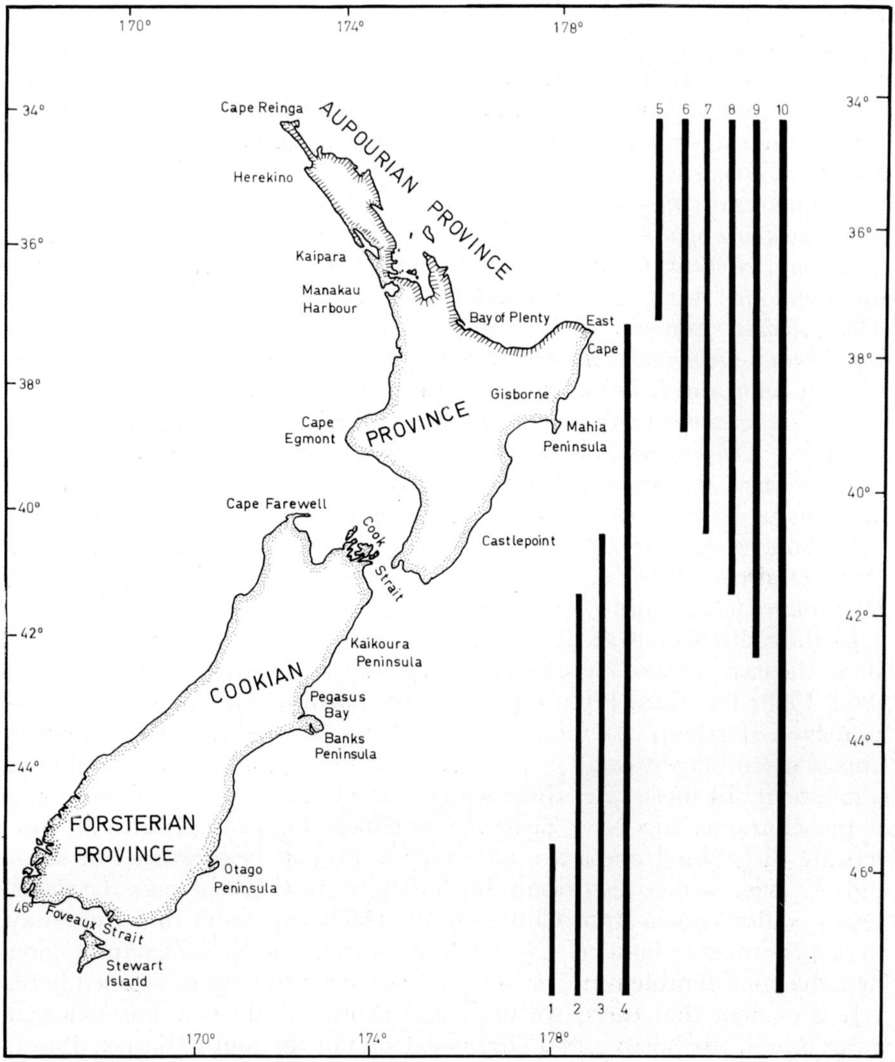

Fig. 7. Distribution patterns of the dominant intertidal plants and animals along the New Zealand east coast. (After KNOX, 1963).

Carpophyllum plumosum reaches Castle Point, *Apophloea sinclairii*, *Caulerpa sedoides* and *C. hypenoides* reach to or just beyond Cook Strait and *Pterocladia lucida* and *Melanthalia abscissa* extend as far as the Kaikoura Peninsula. Among the crabs *Merocryptus* extends to Castle Point, *Ozius truncatus*, *Achaeus fissifrons*, *Petalomera wilsoni*, *Macropipus corrugatus*, *Lyreidus tridentatus* and *Leptograpsus variegatus* reach to about Cook Strait and *Plagusia capensis* extends as far as Kaikoura Peninsula. In the echinoderms

Pectinura cylindrica, *Ophiozonoida picta*, *Ophioceres huttoni*, and *Amphiura rosea* reach the Cook Strait region, *Stegnaster inflatus*, *Heterothyone ocnoides*, and *Astrostole scabra* extend as far as Banks Peninsula. The molluscs *Novastoa lamellosa* and *Nerita melanotragus* extend to Cook Strait and the limpet *Cellana denticulata* and the barnacle *Chamaesipho columna* to the Kaikoura coast.

5. A universal component, including the algae *Codium adhaerens*, *Glossophora kunthii*, *Splachnidium rugosum*, *Lessonia variegata*, *Ecklonia radiata*, *Cystophora retroflexa*, *C. torulosa*, *Hormosira banksii* and *Bostrychia arbuscula;* the anemones *Actinia tenebrosa* and *Oulactis mucosa;* the barnacles *Chaemosipho columna*, *Elminius plicatus* and *E. modestus;* the molluscs *Modiolus neozelanicus*, *Melagraphia aethiops*, *Lunella smargada*, *Cellana radians*, *C. ornata*, *Littorina cincta* and *L. unifasciata;* the crabs *Ebalia laevis*, *Naxia*, the endemic genus *Eurynolambrus*, *Notomithrax peronii*, *N. minor*, *N. ursus*, *Leptomithrax longmanus*, *Chlorinoides*, *Cancer*, *Nectocarcinus*, *Ovalipes*, *Heterozius*, *Pilumnus novaezelandiae*, *P. lumpinus*, *Pinnotheres*, *Hemigrapsus crenutatus*, *H. edwardsii*, *Cyclograpsus lavauxii*, *Helice crassa* and *Hemiplax hirtipes;* the echinoderms *Astropecten polycanthus*, *Ophidiaster kermadecensis*, *Asterodiscus truncatus*, *Ludia varia*, *Ophiocentrus pilosa*, *Centrostephanus rodgersii*, *Heliocidaris tuberculata*, *Hoploneustes inflatus* and *Brissus gigas*.

In their discussions of the distribution patterns of plants and animals along the east coast of New Zealand most authors (Moore, 1961; Dell, 1962, 1968; Pawson, 1965) recognize three distinct faunal assemblages – namely a northern assemblage, a southern assemblage and a middle (mixed assemblage). In all groups there is a strong universally distributed component. In most cases these widely distributed species include many of the characteristic New Zealand endemics; for example the 20 universally distributed crabs are all endemic except for *Nothomithrax minor* and *N. ursus* which also occur in Australia and *Hemigrapsus crenulatus*, which is also known from Chile. Dell (1968) concludes that they may all be presumed to be species which have been in the New Zealand region for some considerable time, or in some cases even to have developed here.

It is evident that there are two main centres of distribution, a warm water one in the north and a cold water one in the south (Knox, 1963). From each of these centres characterised by endemic and non-endemic species confined to them, numbers of species extend varying distances north and south forming an extensive transition zone which is related to the fluctuations in hydrological conditions of the mixed cold-temperate waters along the central east coast. The proportions of northern, southern and widespread species vary from group to group but nevertheless the pattern is a consistent one except perhaps for the gammarid amphipods (Barnard, 1972). However further collecting in the extreme north and south may alter the picture for the amphipods.

The west coast distributions are not so well documented although there are indications of a somewhat similar pattern. The major components of the flora and fauna on an internal basis then are: a warm-temperate component (categories 3 and 4 above), a cold temperate component (categories 1 and 2 above, and a universal component (category 5 above).

Within the cold temperate component two subdivisions can perhaps be recognized, the subantarctic and the New Zealand mainland cold temperate. Many northern species have their southern limit in the Foveaux Strait-Stewart Island region, others extend to the Snares Islands and still others to the subantarctic islands to the south. In the crabs the subantarctic element is small compared to the New Zealand mainland cold-temperate element but in the echinoderms the subantarctic element is a strong one. However, it is possible that some of the subantarctic species are eurythermal cold-water species that have spread south from the New Zealand mainland.

In most groups the northern warm water element is a strong one and there is also evidence that species are still being acquired (DELL, 1968). Many of these northern species are known from elsewhere especially Australia.

b. Biogeographical subdivisions and faunal provinces

KNOX (1963) has traced the history of biogeographical concepts in the New Zealand region. Currently the following subdivisions are recognized for the littoral and shallow shelf faunas: a Kermadec Province (tropical–sub-tropical), an Aupourian Province (transitional warm temperate), a Cookian (cold temperate mixed waters), a Forsterian (transitional between the subantarctic cold temperate and the cold temperate mixed waters) an Antipodean Province (subantarctic cold temperate) and a Moriorian Province (cold temperate mixed waters). The approximate boundaries of these biogeographical provinces are shown in Fig. 7.

DELL (1962) has reviewed the status of New Zealand marine provinces in the light of additional information on the distribution of the New Zealand marine mollusca. He concluded that no clear boundaries could be defined for any of the mainland littoral provinces, and that shelf distribution patterns did not by any means coincide with those for the littoral faunas. He suggests that a more dynamic approach is required that recognizes the viewpoint advanced by KNOX in 1960 and which has been emphasized in this review; this envisages the fauna as being composed of the three elements discussed above.

In spite of this criticism the biogeographic province concept is nevertheless a useful one in any discussion of the biogeography and ecology of the New Zealand region. Previous discussions have revolved around the criteria upon which the Provinces are erected and some authors have considered that a high degree of endemism is a necessary criterion. On this basis the Aupourian which has a high degree of endemism (approximately 40 percent for the molluscs), would easily qualify; the Cookian province, however, which is a region of overlap between northern and southern forms and with few species having restricted ranges within its borders and even fewer endemic species, would not qualify on this basis.

When the distribution patterns of the flora and fauna as a whole are considered and combined with a consideration of the patterns of community structure the usefulness of the Province concept is more firmly established. This is especially so for the littoral and upper sub-littoral down to the limits of algal growth. The dominant algae and animals show vertical zonation patterns that are characteristic of the different biogeographic regions such that a typical Aupourian, Cookian or Forsterian pattern can be recognized.

Tropical–sub-tropical coasts. The Kermadec Islands; DELL (1968) has discussed the relationship of the Kermadec Island mollusca, CHAPMAN (1961) that of the algae and PAWSON (1961) that of the echinoderms. The echinoderm relationships have already been discussed (p. 375). Chapman records 12 species of Myxophyceae, 20 Chlorophyceae, 10 Phaeophyceae and 35 Rhodophyceae. As to be expected, the algae show a strong tropical influence and a number of species (10 of the 200 species upon which MOORE (1949, 1961) based her analysis of algal distributions) not otherwise found in the New Zealand region, including several belonging to semi-tropical genera. Typical New Zealand genera such as *Xiphophora*, *Carpophyllum*, *Cystophora* and *Hormosira* are not represented.

DELL (1958) has discussed the relationship of the Kermadec Island mollusca. The endemic element comprises 34 percent (85 species), the essentially New Zealand element is small – only eight species – and in addition 16 of the species occur in both New Zealand and Australia. DELL concludes that it is not probable that these species have been derived from Australia rather than from New Zealand. A further 50 species occur also in Australia. A number of the genera occurring on the Kermadecs are otherwise endemic to New Zealand, for example, *Eudoxochiton*, *Onithochiton*, *Maorichiton*, *Austronotoa*, *Haurakiopsis* and *Pinnoctopus*. The ancestors of the species belonging to these genera may have colonized the Kermadecs during the period of maximum northward movement of the Subtropical Convergence.

It thus appears that the extent of the New Zealand influence varies from group to group. DELL concludes that: 'it is difficult to assign the group (the Kermadecs) to any zoogeographic region and a similar difficulty occurs with the neighbouring Lord Howe and Norfolk Islands. The marine fauna of these three marginal areas represent an interplay of factors such as geological age, distance from neighbouring land, the direction of water surface movements, and larval mortality and duration.'

Transitional warm temperate coasts. The Aupourian Province; Through the work of BERQUIST (1961), CARNAHAN (1952), CHAPMAN (1950), CRANWELL & MOORE (1938), BEVERIDGE & CHAPMAN (1950), DELLOW (1950, 1955), MOORE (1944), and MORTON & MILLER (1968), the general features of the littoral zonation of the Aupourian region are reasonably well known.

It is generally considered that the southern boundary of the Aupourian Province on the east coast lies in the vicinity of East Cape, a number of typical Aupourian endemics reaching their limits at various points along the coast to the north and other Aupourian dominants decreasing in abundance to the south. Details of some of these distributions have been discussed above in the section on 'Internal distribution patterns'. POWELL (1940) has listed 649 species of mollusca for the Aupourian Province of which 261 (40 percent) are endemic. This is the largest percentage of endemics for any New Zealand biogeographical province, but it should be noted that his list includes deep-water species. The limits of the Province on the west coast are not so well defined, since the current system (Fig. 2) is more uniform on the west with northward moving mixed water extending well up the Auckland Peninsula with an absence of any well marked geographical boundary. The transition zone between the Aupourian and Cookian Provinces on the west is somewhat extended, although distributional data on the extent of the transition is lacking. On the evidence for echinoderm distribution PAWSON (1961) would fix the southern limit on the west as far south as Cape Egmont. The Aupourian Province then is a region with a strong and characteristically northern (warm temperate) endemic element in its major eulittoral biota (DELLOW, 1955) and with a fairly prominent Australian element. Zonation patterns are also highly characteristic especially in the lower eulittoral and upper sublittoral which are dominated by species of *Carpophyllum* on the east coast and *Pachymenia himantophora* and *Gigartina marginifera* on the west.

Features of the Aupourian Province discussed above include: 1. The high percentage of endemic species. 2. The restricted distribution patterns of many of the species. 3. The large proportion of northern warm water species. 4. The relative absence of cold-temperate species with subantarctic affinities.

Cold temperate mixed waters; The central coasts. The Cookian Province; The only accounts of the littoral zonation of this Province are those of OLIVER (1923) and KNOX (1953, 1963, 1968) together with some notes in MOORE (1944) and WOMERSLEY (1960). As pointed out above the Cookian Province is a region of overlap between northern warm-water species and southern cold-water species, changes in zonal dominants occurring as some species drop out and others take their place.

Zonation patterns in the Cookian region differ from those in the Aupourian in the following respects: (1) The absence of a band of *Apophloea* in the upper eulittoral. (2) The gastropod *Nerita melanotragus* which is a prominent feature of Aupourian shores is of limited occurrence south of East Cape and does not extend beyond Cook Strait. (3) The mixed algal mat of the lower eulittoral of the Aupourian is not a conspicuous feature of the Cookian zonation. (4) *Cellana denticulata*, a dominant limpet from Kaikoura north, is of limited local occurrence in the Aupourian Province and *Cellana flava* is restricted to the Cookian east coast from Pegasus Bay to East Cape. (5) The band forming Aupourian red algal species *Pachymenia himantophora*, *Gigartina alveata* and *G. marginifera* are not found in the Cookian Province. (6) Considerable differences, especially on the east coast, are to be seen in the lower eulittoral and upper sublittoral. The large brown algae occupy a lower level on the shore in the Aupourian Province. Species of *Cystophora* play a less important role and those of *Carpophyllum* a more important one in the Aupourian than in the Cookian, *C. elongatum* being confined to the Aupourian Province and *C. plumosum* extending south as far as Castle Point.

The Cookian Province then is a region of wide overlap between the three main biogeographic elements, northern, southern and widespread. In addition there are few Cookian endemics and the number of restricted species is also small.

The Southern Coasts. The Forsterian Province; this region is transitional between the sub-antarctic cold temperate and the cold temperate mixed waters combining features of the zonation patterns of central New Zealand with those of the sub-antarctic. BATHAM (1956, 1958, 1965) has described two areas on the north-eastern coast of this region and a sheltered shore in Dusky Sound on the west.

The Forsterian Province zonation patterns show some considerable differences from those in other parts of New Zealand, of which the major are: (1) The increased development of the lichen banding of an essentially sub-antarctic type. (2) The presence of the sub-antarctic *Hildenbrandia* in the littoral fringe. (3) The greater development of a *Bostrychia* band in the upper eulittoral. (4) While it resembles the Aupourian Province in the presence of an *Apophloea* band a different species occurs and it differs from the Cookian Province where such a band does not occur. (5) A similar situation is found in the development of a *Pachymenia* band, a feature absent from the Cookian Province, but present in the Aupourian where a different species occurs. (6) The southern limpet *Cellana strigilis redimiculum* is confined to the Forsterian Province and the Cookian limpets *C. flava* and *C. denticulata* are absent. (7) In the lower eulittoral the Aupourian and Cookian *Carpophyllum maschalocarpum* is absent. (8) In the upper sublittoral species of *Marginariella* are of greater importance than they are in other parts of New Zealand. (9) The almost complete algal covering found throughout the littoral in some localities is not paralleled elsewhere in New Zealand. (10) Among the mussels, *Perna canaliculus* is absent from the Forsterian Province and the subantarctic

Aulacomya maoriana becomes a dominant zone forming species. The latter species extends throughout New Zealand but in the Cookian and Aupourian Provinces occurs as a subordinate species mixed with *Mytilus edulis aoteanus* and *Perna canaliculus*.

The Forsterian Province then is a region with a strong endemic element, especially among the algae, and a pronounced subantarctic element either of wide ranging circum-sub-antarctic species or of New Zealand subantarctic endemics. Among the algae the endemic species include *Hymenena semicostata*, *Brogniatella australis* and *Apophloea lyallii* and the sub antarctic species *Pachymenia lusoria*, *Delesseria crassinerva*, *Myriogramme crispata* and *Codium dimorphum*. The following circumpolar sub-antarctic species, which also extend varying degrees into the Cookian Province, are well developed in the Forsterian; *Adenocystis utricularis*, *Ballia scoparia*, *Chaetomorpha darwinii*, *Chaetangium fastigiatum*, *Desmarestia firma*, *Halopteris funicularis*, *Macrocystis pyrifera*, *Phycodrys quercifolia*, *Schizoseris davisii*, and *Scytothamnus fasciculatus* (MOORE, 1961).

The Chatham Islands. The Moriorian Province; The Chatham Islands lie some 560 miles due east of Banks Peninsula. Biologically they are a very interesting group since they lie approximately on the Sub-tropical Convergence and hence are subjected to the influence of both cold sub-antarctic and warm sub-tropical waters to varying degrees. The only published information on the littoral ecology is that of KNOX (1954, 1960, 1963). This is here supplemented by further unpublished observations.

A striking feature of the intertidal ecology of the islands to one familiar with New Zealand shores is the odd mixture of species characteristic of the Aupourian and Forsterian Provinces. For example, algae such as *Apophloea lyallii* and *Pachymenia lusoria*, found only in the southern part of the South Island and Stewart Island, grow alongside other species such as *Carpophyllum plumosum* which are limited to the northern parts of the North Island (KNOX, 1954). In a like manner the northern mollusc *Novastoa lamellosa* is found alongside the southern species *Cellana strigilis*. Another outstanding feature is the complete absence of many of the commoner species of the New Zealand shores. Notable examples are the barnacles *Chamaesipho columna*, *Elminius modestus* and *Pollicipes spinosus* and the molluscs *Perna canaliculus*, *Modiolus neozelanicus*, *Mactra discors*, *Dosinia anus*, *D. lambata*, *D. subrosea*, *Myadora striata*, *Nerita melanotragus*, *Scutus breviculus*, *Patelloida* sp., *Notoacmea* sp., *Lunella smargada*, *Struthiolaria* sp., *Alcithoe* sp., *Benhamina obliquata* and *Amaurochiton glaucus*. Other species such as *Saxostrea glomerata*, and *Mytilus edulis aoteanus*, which are dominant band-forming species on New Zealand shores are of extremely rare occurrence. Others recorded by FINLAY (1926), such as *Trochus huttoni*, *Micrelenchus tenebrosus* and *Xymene plebejus*, do not appear to have been collected since. It is probably that these are forms which only occasionally reach the Chathams from the mainland (DELL, 1960). There is also some evidence of fluctuations in the relative abundance of some of the species, for example, *Cantharidus opalus cannoni* was recorded by Powell (1933) as being common throughout the Chathams, yet by 1954 it was found to be extremely rare (Knox, 1957; Dell, 1962). In addition to the barnacles and molluscs mentioned above Grapsoid crabs are absent, as well as the Anomuran *Petrolisthes elongatus*, one of the commonest and most ubiquitous animals on New Zealand shores. The absence of some of these species has led to some interesting changes in the vertical zonation of the others when compared with their zonation on New Zealand shores. One of the most interesting of these absences is that of the commonest New Zealand barnacle *Chamaesipho columna*, which occurs throughout New Zealand, the Kermadec Islands and temperate Australia and Tasmania. The dominant Chatham Islands barnacle is *Chamaesipho brunnea*, which on northern New Zealand shores generally occupies a comparatively narrow zone from about M.H.W.N. to M.H.W.S. above the optimum development of *C. columna*. On the Chathams, in the absence of the latter, *C. brunnea* extends throughout the upper and mid-eulittoral zones down to the upper limit of the *Durvillea* zone at about M.L.W.N.

The zonation patterns are most distinctive and show resemblances to those found in the northern and southern parts of the New Zealand mainland. They resemble those of northern New Zealand in the bands of *Chamaesipho brunnea*, *Novastoa lamellosa* and *Carpophyllum plumosum* and those of southern New Zealand in the bands of *Apophloea lyallii*, *Pachymenia lusoria* and *Xiphophora chondrophylla* var. *maxima*. On the other hand, there are resemblances to sub-antarctic shores in the absence of many mainland dominants and in the presence of *Cellana strigilis* as the dominant limpet and *Aulacomya maoriana* as the dominant museel. Moore (1949, 1961) has given some details concerning the Chatham Island algae. Of some 190 species recorded from the group 10 species are not otherwise recorded from the New Zealand region, the most striking of which is the endemic *Landsbergia myricaefolia*. The others are a mixture of northern and southern species with the latter the predominant element. Dell (1960) records 47 endemic species or sub-species of molluscs from Chatham Island waters under 50 fathoms. His analysis of 320 species of molluscs has revealed that no less than 204 occur in the Cook Strait area, 6 are pelagic and 49 endemic. Only 13 are confined elsewhere to the north while 38 occur in the south with 5 in the Sub-antarctic Islands. A truer picture of the relationships would be obtained if, as Dell has suggested, the Cook Strait forms were divided into those with northern and southern affinities. Pawson (1961) and Fell (1960) have discussed the relationships of the Chatham Island echinoderms. Of the 24 shelf species 8 (33 percent) are recorded from the Chathams, the Chatham Rise and the New Zealand mainland shelf, especially in the Cook Strait region. Two species (8 percent) *Henricia lukinsii* and *Calvasterias suteri* occur in the Subantarctic Islands and clearly show a West Wind Drift pattern.

One species (4 percent) is circumpolar. The remaining 13 species (55 percent) are recorded for the New Zealand coast and are particularly well known in the Cook Strait region. Since no species can be regarded as typical of the Maorian Province, PAWSON would amalgamate it with the Cookian, but in view of the highly characteristic zonation patterns, the mixture of northern and southern species and 'the peculiar characteristics of the fauna and flora due to the elements missing as much as to those present' this province is as distinct as any in the New Zealand region (DELL, 1960).

The present littoral flora and fauna of the Chathams has obviously been derived from the New Zealand mainland, a conclusion supported by geological evidence. FINLAY (1929) states that the whole of the recent molluscan fauna of the Chathams seems to be independent of the Tertiary faunas, the most characteristic shells inhabiting the shores being absent from Pliocene beds. He concluded that 'the Recent molluscan fauna of the Chatham Islands is not a remnant or an evolution of the Tertiary fauna found there, but a repopulation from the mainland, in Post-Pliocene times, but yet long enough ago for characteristic regional species and sub-species to have evolved'. Ocean currents are favourable for the transport of larvae and epiplanktonic adults from New Zealand to the Chathams (KNOX, 1960).

The Sub-antarctic Islands. The Antipodean Province; This province includes the following island groups, with distances from the New Zealand mainland: Auckland Islands 300 km; Antipodes Islands 780 km; Campbell Island 510 km; Bounty Islands 780 km. The Snares Islands some 105 km south of Stewart Island are generally included in the Forsterian Province. According to POWELL (1961) the molluscan fauna is clearly Forsterian rather than Antipodean and FELL (1953a) on the basis of echinoderm evidence has also included the Snares in the Forsterian Province. Zonation patterns resemble those of the Subantarctic with some Forsterian features such as the presence of a band of *Apophloea* on the less exposed shores. Barnacles are practically absent being represented by a few scattered clumps of *Elminius plicatus*. The only limpets are the sub-antarctic *Cellana strigilis flemingi* and *Notoacmea pileopsis sternus*. Typical Forsterian algae such as *Cystophora torulosa*, *C. scalaris* and *Hormosira banksii* are absent. The Snares then appear to be an outlyer of the Forsterian Province with a pronounced subantarctic affinity. Full assessment of their status awaits detailed analysis of the flora and fauna. The richness of the flora and fauna of the other subantarctic islands depends on size and distance from the New Zealand mainland; the island groups in order of decreasing floral and faunal diversity are: Auckland Islands, Campbell Island, Antipodes Islands and Bounty Islands.

The littoral zonation patterns show resemblances to those described from Macquarie Island (KENNY & HAYSON, 1962). They differ from the New Zealand mainland in the following respects. (1) The absence of littorinids except for *Littorina cincta* on the Auckland Islands. (2) The absence of barnacles except for the reduced barnacle cover on the Auckland Islands. (3) In the lower littoral the typical New Zealand species of the genus *Cystophora* are missing. (4) Many of the common band forming animals of the New Zealand shores such as *Cellana radians*, *C. ornata*, *Sypharochiton pelliserpentis*, *Melagraphia aethiops*, *Lepsiella scobina*, *L. albomarginata*, *Patelloida corticata*, *Siphonaria zelandica*, *Perna canaliculus* and *Pomatoceros cariniferus* are absent. Some such as *Zediloma digna*, *Zediloma arida*, *Littorina cincta*, *Lepsithais lacunosus*, *Benhamina obliquata* are found only on the near and larger Auckland Islands.

According to MOORE (1961) the Antipodean Province has a fairly long list of endemic species and at least seven very common mainland species have not been recorded – *Carpophyllum maschalocarpum*, *Ecklonia radiata*, *Euzoniella incisa*, *Glossophora kunthii*, *Hormosira banksii*, *Splachnidium rugosum* and *Zonaria angustata*. An analysis of the echinoderm fauna of the southern islands (FELL, 1953a; PAWSON, 1961, 1968) demonstrates that the bulk of the fauna is of New Zealand derivation. PAWSON (1968) records 15 echinoderms (echinoids and holothurians) from this Province and of these 10 (66 percent) are common to the Antipodean Province and the New Zealand mainland. DAWSON (1965) concludes that 18 percent of the known echinoderms from the New Zealand Sub-antarctic are shared with the New Zealand shelf or slope; endemic species comprise 27 percent and seven percent are known from the Magellanic or other extra-limital areas. FELL (1953a) concluded that the greater part of the echinoderm fauna has been derived from an assemblage of species which has been the common heritage of all parts of the New Zealand submarine plateau. A small percentage of the fauna, on the other hand, has possibly been derived from some originally southern species which have achieved a circumpolar distribution on account of their epiplanktonic habit, influenced by the West Wind Drift. DELL (1962) on molluscan evidence concludes that: 'There seems little doubt that there is a well defined faunal break between the southern islands of New Zealand (CAMPBELL, the Aucklands; Bounties and Antipodes) and the New Zealand mainland including the Snares. At the same time many of the really characteristic Antipodean genera and species cross this faunal break and extend to a varying degree to the north.' He cites the following examples. The very characteristic sub-antarctic genus *Gaimardia* has one species on Macquarie, two on the southern islands and one on the southern part of the South Island. *Kidderia* which also appears sub-antarctic in origin, has three species on Macquarie, two on the southern islands and five in New Zealand. The genus *Margarella* which is widely represented in the Antarctic and Sub-antarctic has one species on Macquarie, two on the southern islands, one on the Chathams and three on Stewart Island and the southern part of the South Island. Limpets of the *Cellana strigilis* group have one sub-species on the Aucklands and Campbell, another on the Antipodes, one on the Bounties, one on the Snares, and one on the Chathams while one is the dominant limpet of the Forsterian Province. *Macquariella* has one species on Macquarie, and one at the Antipodes, while one has been recorded from the Aucklands, the Chathams and Stewart Island. *Kerguelenella* has one species on Macquarie, one on Campbell, the Aucklands and the Antipodes, and another at the Snares and Stewart Island.

The New Zealand Subantarctic Islands then exhibit many of the features of the sub-antarctic as a whole, particularly in the zonation patterns. However, they differ from other subantarctic islands such as Macquarie and Kerguelen in that there is a strong New Zealand endemic element in the flora and fauna. Recent work on the echinoderms from the Campbell Plateau (DAWSON, 1965, 1968; PAWSON, 1968) lends support to the view that Australian, Indo-Pacific and cosmopolitan elements have contributed to the fauna and that the subantarctic and Antarctic elements are of lesser importance.

The position of Macquarie Island. In the past (DELL, 1952; POWELL, 1951, 1961; PAWSON, 1961; KNOX, 1960, 1963) the affinities of the flora and fauna of Macquarie Island with the New Zealand region, and the Antipodean Province in particular, have been considered to be weak and it has generally been regarded part of the Kerguelenian Province with Kerguelen, the Crozets and Prince Edward and Marion Islands. PAWSON (1961) stated that: '... whereas there is obviously a very close echinoderm faunal relationship between New Zealand and the Antipodean Province islands, the echinoderm faunas of the other islands mentioned (including Macquarie) are of a very different character, comprising mainly circumpolar species, together with some Antarctic groups.' In 1968, however, PAWSON in discussing the echinozoan fauna of these islands concluded that close relationships between the Macquarie Island fauna and that of New Zealand are evident. This change in viewpoint is due to the increased information then available largely as a result of the investigations carried out by the New Zealand Oceanographic Institute (DAWSON, 1965, 1970).

Macquarie Island was for a long time thought to be separated from the New Zealand Plateau by water of 4,000 m or deeper (cf. ADIE, 1963, p. 460, Fig. 2), which led zoogeographers to look away from New Zealand for faunal affinities (DAWSON, 1970). However, a detailed bathymetric and geological survey of Macquarie Island and the associated seafloor (BRODIE & DAWSON, 1965) has shown that: 1. A submarine ridge extends north of Macquarie Island to the south-west corner of the New Zealand shelf. 2. Much of this ridge is shallower than 2,000 m and has individual high areas which form shallow banks at 100 m or less. 3. The maximum distance between the shallower parts of the ridge is about 55 kilometres.

Thus stenobathic animals with restricted larval life could have spread between New Zealand and Macquarie Island (DAWSON, 1970).

DELL (1964) in an analysis of the then recorded 42 species of mollusca from Macquarie Island concluded that they represented a faunal mixture derived from various sources, and he suggested that Macquarie's inclusion within the Kerguelenian Province loses sight of the very distinctive character of the Macquarie Island fauna. DELL's analysis indicated the following relationships: Kerguelenian, 12 percent; circum-subantarctic, 12 percent; Heard Island, 5 percent; New Zealand 5 percent; of no particular affinity 2 percent.

KENNY & HAYSON (1962) in their account of the shore fauna gave the following analysis for 58 species: eight endemic, 14 New Zealand, 12 Kerguelen, 18 wide-ranging subantarctic and six Antarctic. They concluded that the relationships were stronger with Antarctica than with New Zealand.

PAWSON (1968) in his analysis of the echinozoan fauna of the Subantarctic Islands and Macquarie Island re-examined the relationships in the light of the new information on the Macquarie Ridge. The sole echinoid recorded from Macquarie Island, *Pseudechinus novaezealandiae*, is widely distributed on the Campbell Plateau and along the east coast from Cook Strait south (Fig. 4). Although this species has a pelagic larva distribution from New Zealand to Macquarie would be against prevailing water movements and it probably reached Macquarie by migration across the ridges. Among the holothurians *Trochodota dunedinesis*, *Ocnus brevidentis* and *Pseudopsolus macquariensis* are also known from New Zealand the former two species being widely distributed in southern New Zealand waters. The asteroids on the other hand are more varied in their relationships (DAWSON, 1965) with the circum-Subantarctic genera *Anasterias*, *Hymenaster*, *Similasterias* and *Cycethra* and the Antarctic species *Porania antarctica*. On the other hand three of the seven species of *Henricia* occurring in the New Zealand region, *H. aucklandiae*, *H. lukinsi* and *H. ralphae*, are found at Macquarie Island.

c. *The archibenthal fauna*

As pointed out above in the accounts of distribution patterns for selected groups of organisms the archibenthal fauna is characterized by: 1. the wide distribution of many of the species; 2. the comparative paucity of restricted species; 3. the absence of a clearcut division into groups with predominantly northern or southern affinities. DELL (1963), however, notes that some northern archibenthal mollusca are not yet known south of East Cape.

In his account of the archibenthal mollusca of New Zealand DELL

(1956) notes that beyond the shelf edge there is a gradual change from a typically shelf fauna to a typical archibenthal fauna, though some shelf forms continue well into the archibenthal. A similar situation pertains in other groups especially the echinoderms.

DELL also found that a small group of essentially subantarctic species, which occurred in moderate depths (0–200 m) in the south of the Forsterian Province and in the Antipodean Province, were present in deeper water further north. He also found that the overwhelming majority of the genera, 241 out of 280 (86 percent) were also represented in the shelf fauna. Another 13 genera were previously known only from the New Zealand Tertiary, two genera were known elsewhere only from the Subantarctic and nine genera were endemic to the New Zealand archibenthal.

The Past History of the Fauna

The past history of the New Zealand fauna based to a large extent on the fossil record of the mollusca is now well known primarily due to a series of papers by FLEMING (1944, 1949, 1952, 1960, 1962b) culminating in that masterly synthesis entitled 'New Zealand Biogeography. A Paleontologist's Approach' (FLEMING, 1962b). KNOX (1963) has also discussed the past history of the New Zealand marine fauna with special reference to problems of speciation in intertidal animals. As FLEMING (1952) points out the persistence of the concept of biogeographic regions implies a considerable degree of agreement between the distribution areas of different groups of organisms; faunal boundaries based on one group are often valid for another. Thus in studies of historical biogeography we may infer that faunal changes in one group were paralleled in others, that the factors controlling invasions, changes in distribution and extinctions, determined by study of fossils in one phylum, may apply to another phylum which has left little or no fossil record. Few organisms have left as complete a record of their past distribution as the marine shelled mollusca and the New Zealand molluscan fossil record is a particularly complete one.

There is evidence that temperatures increased during the early Tertiary to a maximum in the mid-Tertiary followed by a decline that was particularly marked at the end of this period. FLEMING has outlined the evidence for the marine molluscan faunas, FELL (1953b) has treated the echinoderms, SQUIRES (1958) the corals, while HORNIBROOK (1951, 1953) has dealt with the marine micro-fossils. FLEMING (1959, 1962b) in reviewing the relationships through time of the Australian-New Zealand area pictures New Zealand and Australia in the lower Cretaceous as a connected land mass lying outside the equatorial region characterized by coral reefs; later in the Upper Cretaceous New Zealand was isolated, continuing thereafter as a constantly changing archipelago. The nucleus of the New Zealand molluscan fauna dates from the late Cretaceous and

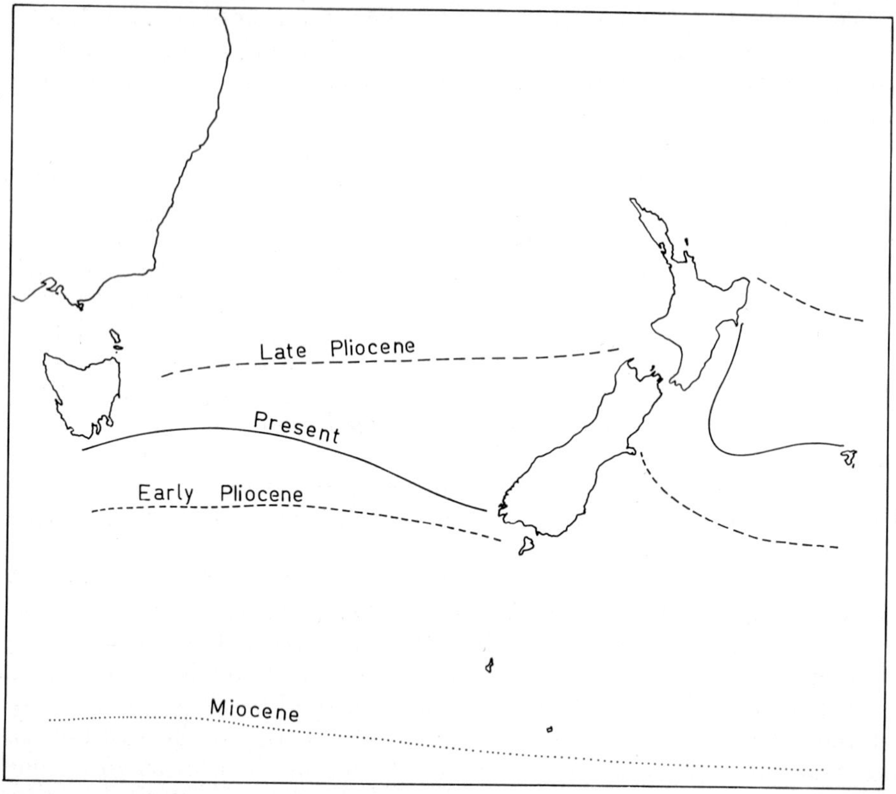

Fig. 8. Reconstruction of the approximate positions of the warm-water cold-water boundary (Subtropical Convergence) from the Miocene to the present. (After KNOX, 1968).

by the beginning of the Tertiary it included many endemic groups indicative of isolation for some considerable time (FINLAY, 1925, 1926; MARWICK, 1953). Throughout the Tertiary there have been substantial additions and subtractions at various times; of warm-water Indo-Pacific types with a maximum influx in the Mid-Miocene, of southern cold water species in the Mid-Pliocene, and of some Australian species, particularly in the late Pliocene. Striking differences between the Miocene molluscs of Australia and New Zealand are accentuated by the more tropical nature of the former (FLEMING, 1957). It appears that most sedentary coastal groups (for example, venerids, volutes) have evolved independently on either side of the Tasman Sea during most of the Tertiary, and only groups with long-lived larvae (for example, Cymatiidae, Conidae) show close relationships due to trans-Tasman crossings. The characteristic New Zealand molluscs (*Neilo*, *Struthiolaria*, many volutes, venerids, turritellids, Naticids, buccinulids) have fossil ancestry extending back to Mid-Tertiary, Eocene, and even late Cretaceous time and Australian lineages developed there without reaching New Zealand. Brachiopods (of limited dispersal ability) also had an independent Tertiary listing on either side of the Tasman (ALLAN, 1935, 1947, 1960).

Analysis of middle and late Tertiary faunal trends leads to the following conclusions (FLEMING, 1944, 1962b): Firstly, in the Lower-Miocene, the New Zealand area lay wholly in the Sub-tropical Zone of surface waters and there is no indication of faunal zoning due to hydrological differences (Fig. 8). About 175 mollusc genera appeared for the first time in this period, most of Indo-Pacific affinity, but some distinctly Australian and a few of Austral (southern) distribution. This marked faunal influx is also shown by corals, Foraminifera and Ostracoda. Secondly, during the middle and late Miocene, successive faunas record a progressive cooling of the seas around New Zealand by the extinction of sub-tropical genera that had marked the Mid-Tertiary thermal maximum, such as the corals *Platyhelia*, *Dendrophyllia* and *Stephanocyathus ixine* (still living in the Indo-Pacific) and the commonest Tertiary crab (*Tumidocarcinus*, endemic to Australasia). Immigrant forms from the Indo-Pacific, Australia and the south continued to arrive, some having a short-lived existence and others, such as the molluscs *Zeacumantus*, *Atrina* and *Mactra* and the spider crab *Leptomithrax*, remained to become characteristic features of the recent fauna. Thirdly, in the early Pliocene subantarctic waters approached New Zealand; a tongue of cold water had reached North Canterbury on the east coast but did not affect areas to the west and north. The most dramatic faunal event was the progressive extinction of Miocene warm-water groups. Fourthly, by late early Pleistocene (Nukumaruan) subantarctic water extended through North Canterbury, Marlborough, Wellington and southern Hawkes Bay to approximately 40 °S. Marine invertebrate groups now confined to subantarctic waters came north to the southern North Island; the molluscs *Chlamys delicatula*, *Tawera subsulcata*, *Zeacolpus symmetricus* and the crab *Jacquinotia*. Fifthly, continued advance of southern waters was inhibited by the increasing development in the Mid-Pleistocene of the Tasman Current carrying sub-tropical water south from their position in the Nukumaruan and preventing New Zealand seas from reflecting the further cooling which occurred in air temperatures and culminated in the Pleistocene glaciation. The immigrant mollusca were mainly Australian forms.

It is clear that the major factor involved in the faunal changes outlined above has been the migrations which have occurred in the portion of the warm-water cold-water boundary (FLEMING, 1944; KNOX, 1963). At present this boundary, the Subtropical Convergence, straddles New Zealand (Fig. 2). Analysis of faunal trends from the Miocene (FLEMING, 1944) leads to the following conclusions:

(1) In the Miocene the New Zealand area lay wholly in the Subtropical

Zone of surface waters and there is no indication of faunal zoning due to hydrological differences.

(2) In the early Pliocene subantarctic (cold-temperate) waters approached New Zealand; a tongue of cold water reached North Canterbury on the east coast but did not affect the more northern coasts and the west coasts.

(3) By the Lower Pleistocene subantarctic water had extended up the east coast to approximately 40 °S and to Cook Strait in the west.

(4) Continued advance of these cold waters was inhibited in the Mid-Pleistocene by the development of the Tasman Current and the East Cape current.

The fluctuations which have occurred in the mean position of the Subtropical Convergence are shown in Fig. 8. FLEMING (1944, 1949) has outlined molluscan faunal changes which have accompanied these fluctuations (KNOX, 1963).

External Relationships – General Conclusions

Although there are differences in detail between the various taxonomic groups as far as their external relationships are concerned there is nevertheless a general similarity in the patterns which have been described above. These general features can be summarized as follows:

1. The faunas tend to show a pattern of an older fauna (mainly of warm-water origin) that has had superimposed on it a late Tertiary cold-water element that in some recent times was modified by immigration from the north.
2. Immigration of warm-water species of Australian and Malayo-Pacific origin is a process that is active at present.
3. As would be expected in an island archipelago that has been isolated for some considerable time from other land masses by large expanses of ocean, the degree of endemism is high, generally exceeding 50 percent at the specific level in most groups. This endemism is however not as marked at the generic level.
4. All groups show a predominance of species of Malayo-Pacific origin.
5. There is a significant Australian element but species belonging to this group are mostly ones of Indo-Malayan affinity.
6. The Paleoaustral element is small.
7. There is a significant but small cold-temperate (Neoaustral), or subantarctic component, probably distributed via the West Wind Drift.

REFERENCES

ADIE, R. J. 1933. Geological evidence on possible Antarctic land connections. Pp. 455–63 in 'Pacific Basin Biogeography. A Symposium. Tenth Pacific Science Congress, Honolulu, Hawaii, 1961.' GRESSITT, J. L. (ed.). Bishop Museum, Honolulu.

ALLAN, R. S. 1935. The fauna of the Reefton Beds (Devonian, New Zealand). *N.Z. Geol. Sur. Pal. Bull.* No. 14, 72 pp.

ALLAN, R. S. 1947. A revision of the Brachiopoda of the Lower Devonian strata of Reefton, New Zealand. *J. Pal.* 21: 436–452.

ALLAN, R. S. 1960. The succession of Tertiary brachiopod faunas in New Zealand. *Rec. Cant. Mus.* 7: 233–268.

BAKER, A. N. 1968. The echinoid fauna of north-eastern New Zealand. *Trans. Roy. Soc. N.Z., Biol. Sci.* 11(1): 1–7.

BARNARD, L. J. 1972. The marine fauna of New Zealand: Algae living littoral Gammaridea (Crustacea, Amphipoda). *Bull. N.Z. Dep. scient. ind. Res.* 210. 216 pp.

BATHAM, E. J. 1956. Ecology of southern New Zealand sheltered rocky shore. *Trans. Roy. Soc. N.Z.* 84(2): 447–465.

BATHAM, E. J. 1958. Ecology of southern New Zealand exposed shore at Little Papanui, Otago Peninsula. *Trans. Roy. Soc. N.Z.* 85(4): 647–658.

BATHAM, E. 1965. Rocky shore ecology of a southern New Zealand fiord. *Trans. Roy. Soc. N.Z., Zool.* 6: 215–227.

BENNETT, E. W. 1964. The marine fauna of New Zealand: Crustacea Brachyura. *N.Z. Dep. scient. ind. Res. Bull.* 153. 120 pp.

BERQUIST, P. L. 1961. Notes on the marine algal ecology of some exposed rocky shores of Northland, New Zealand. *Botanica Marina* 1: 86–94.

BRODIE, J. W. 1964. Bathymetry of the New Zealand region. *Bull. N.Z. Dep. scient. ind. Res.* 161: 55 pp.

BRODIE, J. W. & DAWSON, E. W. 1965. Morphology of the north Macquarie Ridge. *Nature, Lond.* 207(4999): 844–855.

BURLING, R. W. 1961. Hydrology of circumpolar waters south of New Zealand. *Bull. N.Z. Dep. scient. ind. Res.* 143: 66 pp.

CARNAHAN, J. A. 1952. Intertidal zonation at Rangitoto Island, New Zealand. *Pacific Sci.* 6: 35–46.

CHAPMAN, V. J. 1956. Revision of the marine algae of New Zealand, Part 1. Myxophyceae and Chlorophyceae. *J. Linn. Soc. (Bot.)* 55: 330–350.

CULLEN, D. J. 1967. The submarine geology of Foveaux Strait. *Bull. N.Z. Dep. scient. ind. Res.* 184: 68 pp.

DAWSON, E. W. 1959. An interpretation of the molluscan genus *Amphidesma* in New Zealand. *J. Malacol. Soc. Aust.* 3: 39–58.

DAWSON, E. W. 1965. Oceanography and marine zoology of the New Zealand Subantarctic. *Proc. N.Z. ecol. Soc.* 12: 44–57.

DAWSON, E. W. 1970. Faunal relationships between the New Zealand Plateau and the New Zealand sector of Antarctica based on echinoderm distribution. *N.Z. J. mar. freshw. Res.* 4(2): 126–140.

DELL, R. K. 1952. Marine biology. Pp. 129–150 in 'The Antarctica Today'. SIMPSON, F. A. (ed.). A. H. & A. W. Reed, Wellington.

DELL, R. K. 1956. The archibenthal Mollusca of New Zealand. *Dominion Mus. Bull.* 18: 1–235.

DELL, R. K. 1958. The marine mollusca of the Kermadec Islands in relation to the molluscan faunas in the south-west Pacific. Proc. 8th Pacific Sci. Congr. 3: 499–502.

DELL, R. K. 1960. Crabs (Decapoda, Brachyura) of the Chatham Islands 1954 Expedition. *N.Z. Dep. scient. ind. Res. Bull.* 139: 1–7.

DELL, R. K. 1962. New Zealand marine provinces – do they exist? *Tuatara* 10(1): 43–52.

DELL, R. K. 1964. Marine mollusca from Macquarie and Heard Islands. *Rec. Dom. Mus.* 4: 267–301.
DELL, R. K. 1968. Composition and distribution of the New Zealand Brachyuran fauna. *Trans. Roy. Soc. N.Z., Zool.* 10(25): 225–240.
DELL, R. K., GRIFFEN, D. J. G. & YALDWYN, J. C. 1970. A new swimming crab from the New Zealand Subantarctic and a review of the Genus *Nectocarcinus* A. Milne-Edwards. *Trans. Roy. Soc. N.Z., Biol. Sci* 12(7): 49–68.
DELLOW, V. 1950. Intertidal ecology at Narrow Neck Reef, New Zealand. (Studies in intertidal zonation 3.) *Pacific Sci.* 4(4): 255–274.
DELLOW, V. 1955. Marine algal ecology of the Hauraki Gulf. *Trans. Roy. Soc. N.Z.* 83: 1–91.
FELL, H. B. 1949. The constitution and relations of the New Zealand echinoderm fauna. *Trans. Roy. Soc. N.Z.* 77(5): 208–211.
FELL, H. B. 1953a. Echinoderms from the subantarctic islands of New Zealand: Asteroidea, Ophiuroidea and Echinoidea. *Rec. Dom. Mus., Wellington* 2(2): 73–111.
FELL, H. B. 1953b. The origins and migrations of Australasian echinoderm fauna since the Mesozoic. *Trans. Roy. Soc. N.Z.* 81(1): 245–255.
FELL, H. B. 1958. Deep-sea echinoderms of New Zealand. *Zool. Publ. Vict. Univ. Wellington* 24: 1–40.
FELL, H. B. 1960. Biological results of the Chatham Islands 1954 Expedition. Part 2. Archibenthal and littoral echinoderms. *Bull. N.Z. Dep. scient. ind. Res.* 139. 76 pp.
FELL, H. B. 1962. West-wind-drift dispersal of echinoderms in the Southern Hemisphere. *Nature* 193(4817): 759–761.
FELL, H. B. 1963. The spatangid echinoids of New Zealand. *Zool. Publ. Vict. Univ. Wellington* 32: 1–8.
FINLAY, H. J. 1925. Some modern concepts applied to the study of Cainozoic mollusca from New Zealand. Verbeek. Mem. Birthday Vol.: 161–172.
FINLAY, H. J. 1926. A further commentary on New Zealand molluscan systematics. *Trans. N.Z. Inst.* 57: 320–485.
FLEMING, C. A. 1944. Molluscan evidence of Pliocene climatic change in New Zealand. *Trans. Roy. Soc. N.Z.* 74: 207–230.
FLEMING, C. A. 1949. The geological history of New Zealand. *Tuatara* 2(2): 72–89.
FLEMING, C. A. 1952. The Post-Miocene evolution of marine faunas in the South-west Pacific and its bearing on the problem of marine faunal provinces. Proc. 7th Pacific Sci Congr. 3: 309–318.
FLEMING, C. A. 1959. Notes on New Zealand Recent and Tertiary mussels (Mytilidae). *Trans. Roy. Soc. N.Z.* 87: 165–178.
FLEMING, C. A. 1960. Trans-Tasman relationships in natural history. Pp. 228–246 in 'Science in New Zealand', CALLAHAN, F. R. (ed.). Aust. and N.Z. Ass. Advanc. Sci., Wellington.
FLEMING, C. A. 1962a. The nomenclature of biogeographic elements in the New Zealand biota. *Trans. Roy. Soc. N.Z. Gen.* 1(2): 13–22.
FLEMING, C. A. 1962b. New Zealand biogeography: A paleontologists approach. *Tuatara* 10(2): 53–108.
GARNER, D. M. 1959. The subtropical convergence in New Zealand surface waters. *N.Z. J. Geol. Geophys.* 2: 315–337.
GARNER, D. M. 1961. Hydrology of New Zealand coastal waters: 1955. *Bull. N.Z. Dep. scient. ind. Res.* 138: 85 pp.
GARNER, D. M. 1962. Analysis of hydrological observations in the New Zealand region, 1874–1955. *Bull. N.Z. Dep. scient. ind. Res.* 144: 45 pp.
GARNER, D. M. 1967a. Hydrology of the Southern Hikurangi Trench Region. *Bull. N.Z. Dep. scient. ind. Res.* 177.
GARNER, D. M. 1967b. Hydrology of the south-east Tasman Sea. *Bull. N.Z. Dep. scient. ind. Res.*, 181: 40 pp.

GARNER, D. M. 1969. Seasonal range of sea temperature on the New Zealand Shelf. *N.Z. Jl mar. freshw. Res.* 3: 209–219.
GARNER, D. M. 1970. Hydrological studies in the New Zealand region, 1966 and 1967. Oceanic hydrology north-west of New Zealand. Hydrology of the north-east Tasman Sea. *Bull. N.Z. Dep. scient. ind. Res.* 202: 50 pp.
GRIFFEN, D. J. G. 1969. The marine fauna of New Zealand: spider crabs, Family Majidae (Crustacea, Brachyura). *Bull. N.Z. Dep. scient. ind. Res.* 172: 111 pp.
HEDGPETH, J. W. 1953. An introduction to the zoogeography of the northwestern Gulf of Mexico with reference to the invertebrate fauna. *Publ. Inst. mar. Sci.* 3(1): 111–224.
HEDGPETH, J. W. 1971. Perspectives of benthic ecology in Antarctica. Pp. 93–136 in 'Research on the Antarctica'. Amer. Assoc. Adv. Sci., Washington, D.C.
HORNIBROOK, N. DE B. 1951. Tertiary and Recent marine Ostracoda of New Zealand. *N.Z. Geol. Sur. Pal. Bull.* No. 18: 82 pp.
HORNIBROOK, N. DE B. 1953. Faunal immigrations to New Zealand: I. Immigration of Foraminifera to New Zealand in the Upper Cretaceous and Tertiary. *N.Z. J. Sci. Tech., B* 34: 436–444.
HOUTMAN, TH. J. 1967. Water masses and fronts in the Southern Ocean, south of New Zealand. *N.Z. Dep. scient. ind. Res. Bull.* 174.
KENNY, R. & HAYSON, N. 1962. Ecology of rocky shore organisms at Macquarie Island. *Pacific Sci.* 10: 245–263.
KNOX, G. A. 1954. The intertidal flora and fauna of the Chatham Islands. *Nature* 174: 877.
KNOX, G. A. 1957. General account of the Chatham Islands 1954 Expedition. *Bull. N.Z. Dep. scient. ind. Res.* 122. 37 pp.
KNOX, G. A. 1960. Littoral ecology and biogeography of the southern oceans. *Proc. Roy. Soc. B* 152(949): 577–624.
KNOX, G. A. 1963. The biogeography and intertidal ecology of the Australasian coasts. *Oceanogr. Mar. Biol. Ann. Rev.* 1: 341–404.
KNOX, G. A. 1963. Problems of speciation in intertidal animals with special reference to New Zealand shores. Pp. 7–29 in 'Speciation in the Sea', Systematics Assoc. (U.K.), Publ. No. 5.
KNOX, G. A. 1968a. Tides and intertidal zones. Pp. 131–146 in Symposium on Antarctic Oceanography, Santiago, Chile, 13–16 September, 1966. Scott Polar Research Institute, Cambridge, England.
KNOX, G. A. 1968b. The plants and animals of the rocky shores. Pp. 536–552 in 'Natural History of Canterbury', KNOX, G. A. (ed.). A. H. & A. W. Reed, Wellington.
KNOX, G. A. 1970. Biological oceanography of the South Pacific. Pp. 155–182 in 'Scientific Exploration of the South Pacific'. W. S. WOOSTER (ed.). Nat. Acad. Sci., Washington, D.C.
LAING, R. M. 1927. A reference list of New Zealand marine algae. *Trans. Proc. N.Z. Inst.* 57: 126–185.
LAING, R. M. 1929. A reference list of New Zealand marine algae. Supplement 1. *Trans. Proc. N.Z. Inst.* 60: 575–583.
LAING, R. M. 1939. New Zealand seaweeds – Reference List No. II. *Trans. Roy. Soc. N.Z.* 69: 134–164.
LINDAUER, V. J. 1947. An annotated list of the brown seaweeds, Phaeophyceae, of New Zealand. *Trans. Roy. Soc. N.Z.* 76: 542–566.
LINDAUER, V. W., CHAPMAN, V. J. & AIKEN, M. 1961. The marine algae of New Zealand. Part II. Phaeophyceae. *Nova Hedwegia* 3(2/3): 129–350.
LUCKENS, PENELOPE A. 1972. Distribution of *Divaricella huttoniana* (Mollusca: Bivalvia: Lucinidae) in the New Zealand Region. *N.Z. J. mar. freshw. Res.* 6(3): 259–276.
MARWICK, J. 1953. Faunal migrations in New Zealand seas during the Triassic and Jurassic. *N.Z. J. Sci. Tech., B* 34: 317–321.

McDougall, J. C. & Brodie, J. W. 1967. Sediments of the western shelf of the North Island, New Zealand. *Bull. N.Z. Dep. scient. ind. Res.* 179: 54 pp.
McKnight, D. G. 1969. Infaunal benthic communities of the New Zealand continental shelf. *N.Z. J. mar. freshw. Res.* 3: 409–444.
McKnight, D. G. 1969. An outline distribution of the New Zealand shelf fauna: Benthos survey, station list and distribution of the Echinoidea. *Bull. N.Z. Dep. scient. ind. Res.* 195: 91 pp.
Moore, L. B. 1944. New Zealand seaweed for agar manufacture. *N.Z. J. Sci. Tech., B* 25: 183–209.
Moore, L. B. 1949. The marine algal provinces of New Zealand. *Trans. Roy. Soc. N.Z.* 77: 187–189.
Moore, L. B. 1953. Some distribution problems illustrated from brown algae of the genus *Halopteris*. Proc. 7th Pacific Sci. Congr., 5: 13–18.
Moore, L. B. 1961. Distribution patterns in New Zealand seaweeds. *Tuatara* 9: 18–23.
Morton, J. & Miller, M. 1968. 'The New Zealand Seashore'. Collins, London-Auckland. 638 pp.
Naylor, M. 1954a. The check list of the marine algae of the Dunedin district. *Trans. Roy. Soc. N.Z.* 82: 645–663.
Naylor, M. 1954b. Occurrence of *Durvillea caepestipes* on Chatham Island. *Nature* 173: 1099–1100.
Neall, V. E. 1970. Notes on the ecology and paleoecology of *Neothyris*, an endemic New Zealand brachiopod. *N.Z. J. mar. freshw. Res.* 4: 117–225.
Norris, R. M. 1964. Sediments of the Chatham Rise. *Bull. N.Z. Dep. scient. ind. Res.* 159: 39 pp.
Oliver, W. R. B. 1923. Marine littoral plant and animal communities in New Zealand. *Trans. N.Z. Inst.* 54: 495–545.
Pantin, H. M. 1966. Sedimentation in Hawke Bay. *Bull. N.Z. Dep. scient. ind. Res.* 171: 70 pp.
Pawson, D. L. 1961. Distribution patterns of New Zealand echinoderms. *Tuatara* 9(1): 9–18.
Pawson, D. L. 1963. The Holothurian fauna of Cook Strait, New Zealand. *Zool. Publ. Vict. Univ. Wellington* 36: 1–38.
Pawson, D. L. 1965a. The bathyal holothurians of the New Zealand region. *Zool. Publ. Vict. Univ. Wellington* 39: 1–33.
Pawson, D. L. 1965b. The distribution of echinoderms along the east coast of New Zealand. *Trans. Roy. Soc. N.Z., Zool.* 6(24): 245–252.
Pawson, D. L. 1965c. Some echinozoans from north of New Zealand. *Trans. Roy. Soc. N.Z., Zool.* 5(15): 197–224.
Pawson, D. L. 1965d. New records of echinoderms from the Snares Islands to the south of New Zealand. *Trans. Roy. Soc. N.Z., Zool.* 6(25): 253–260.
Pawson, D. L. 1966. Phylogeny and evolution of the holothuroids. pp. 641–646 in Treatise on Invertebrate Paleontology (U) Echinodermata 3(2). Moore, R. C. (ed.). University of Kansas Press, Lawrence.
Pawson, D. L. 1968. The Echinozoan fauna of the New Zealand Subantarctic Islands, Macquarie Island and the Chatham Rise. *Bull. N.Z. Dep. scient. ind. Res.* 187: 52 pp.
Pawson, D. L. 1969. Holothuroidea. In American Geographical Society Map Folio Series, 36–38, Folio 11.
Pawson, D. L. 1970. The marine fauna of New Zealand: Holothuroidea. *Bull. N.Z. Dep. scient. ind. Res.* 52: 69 pp.
Pawson, D. L. & Fell, H. B. 1965. A revised classification of the dendrochirate holothurians. *Brevoria* 214: 1–7.
Powell, A. W. B. 1933. The marine Mollusca of the Chatham Islands. *Rec. Auck. Inst. Mus.* 1: 181–208.
Powell, A. W. B. 1940. The marine mollusca of the Aupourian Province, New Zealand. *Trans. Roy. Soc. N.Z.* 70: 205–248.

Powell, A. W. B. 1951. Antarctic and subantarctic mollusca. *Discovery Reports* 26: 49–196.
Powell, A. W. B. 1961. New Zealand biotic provinces. *Tuatara* 9: 1–8.
Ralph, P. M. 1961. The New Zealand thecate hydroids. Part V. The distribution of New Zealand thecate hydroids. *Trans. Roy. Soc. N.Z., Zool.* 1(7): 103–111.
Ridgway, V. M. 1970. Hydrology of the Southern Kermadec Trench region. *Bull. N.Z. Dep. scient. ind. Res.* 205: 1–28.
Squires, D. F. 1958. The Cretaceous and Tertiary corals of New Zealand. *N.Z. Geol. Sur. Pal. Bull.* 29: 107 pp.
Squires, D. F. & Keyes, I. W. 1967. The marine fauna of New Zealand: Scleractinian corals. *Bull. N.Z. Dep. scient. ind. Res.* 185: 46 pp.
Stringer, B. L. 1969. The species of New Zealand Onchiidae (Mollusca, Gastropoda) and their distribution. *N.Z. J. mar. freshw. Res.* 3(1): 29–45.
Warham, J. & Johns, P. M. 1975. The University of Canterbury Antipodes Island Expedition 1969. *J. Roy. Soc. N.Z.*
Womersley, H. B. 1960. The marine algae of Australia. *Bot. Rev.* 25: 545–614.
Yaldwyn, J. C. 1960. Crustacea Decapoda Natantia from the Chatham Rise: a deep water bottom fauna from New Zealand. *N.Z. Dep. scient. ind. Res. Bull.* 139: 13–53.
Yaldwyn, J. C. 1965. Antarctic and Subantarctic Decapod Crustacea. Monographiae Biologicae, 15: 324–332.
Zenkevich, L. A. 1947. О задачах, об'екте и методе морской биогесграфии. Зоол. Жур., 26(3): 201–218, 16 figs. The task, the object and the methods of marine biogeography, Summary. *Zool. Zh. (Zoologicheskii Zhurnal)* 26(3): 219–220.
Zenkevich, L. A. 1949. La structure biologique de l'océan. C.R. XIIIe Congrès Internat. Zool., pp. 522–529.
Zenkevich, L. A. 1963. 'Biology of the Seas of the U.S.S.R.' (Translated by S. Botcharskaya.) Allen and Unwin, London.

X. THE LIMNOLOGY

by

V. M. STOUT

The limnology of New Zealand reflects in many ways the age, size and geographical position of the country. New Zealand is geologically young, and in comparatively recent times the topography has been influenced by extensive earth movements, volcanic activity and glaciation.

There are many lakes, often oligotrophic and in a young eroding landscape. The rivers are frequently swift flowing, with gravel beds and few placid stretches so that extensive beds of rooted vegetation are uncommon. The rocks surrounding the lakes and rivers are often slow leaching so that the ionic content of the water is low.

New Zealand is a small country, and long and narrow, with the long axis at right angles to the prevailing wind. The lakes and rivers are never far from the sea. They are frequently subjected to strong winds, and may receive contributions of salts from the sea. The oceanic climatic regime influences the temperature relations in the lakes. The country lies in the southern temperate zone between the latitudes of approximately 35° and 45° South, and the latitudinal range appears to be sufficient to be reflected in temperature differences in different lakes. New Zealand is in the path of the strong west wind drift, which blows from the direction of the nearest land mass, Australia. However, New Zealand is separated from Australia by nearly 2000 km and its geographical isolation and insular nature has influenced the composition of the freshwater flora and fauna. The country is in some cases a refuge which retains groups rare or absent in other countries, but it also lacks representatives of other groups and the total number of freshwater species is small.

The lakes are thus often at least partly volcanic, glacial or coastal in origin; they are usually warm monomictic, not thermally stratified or with a deep thermocline; and frequently have a low content of nutrient ions. The biota show some general similarities to similar lakes elsewhere but their composition also reflects the geographical position of New Zealand.

Early investigations

Before the arrival of european man, the maoris explored the lakes and rivers extensively and developed many legends connected with them (Beattie, 1945). They also fished for eels, the small native galaxiid fish and the freshwater crayfish. The earliest scientific work was on the bathymetry of the lakes, firstly of the country's largest lake, Taupo (Fig. 1)

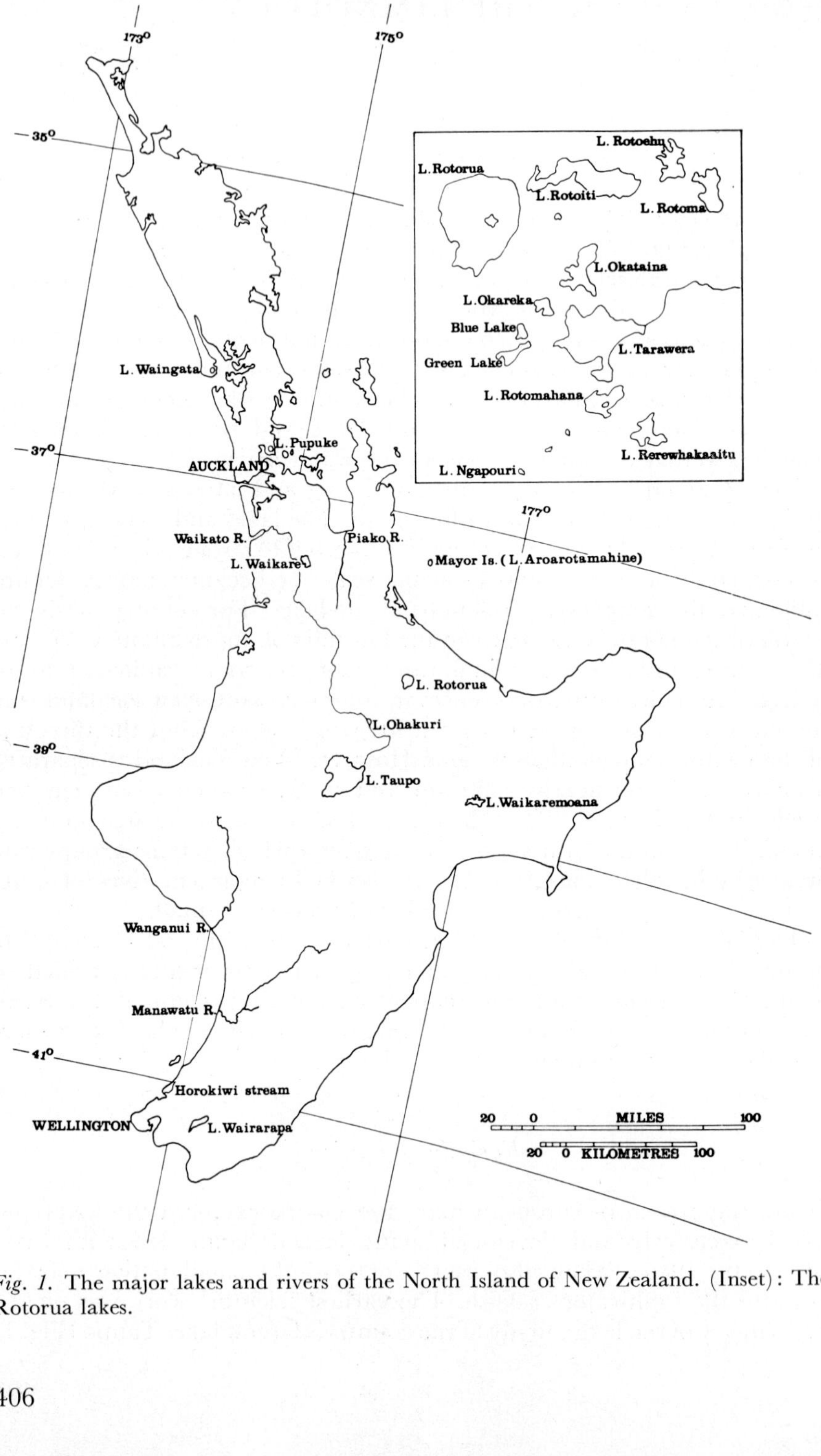

Fig. 1. The major lakes and rivers of the North Island of New Zealand. (Inset): The Rotorua lakes.

(Cussen, 1887) and then of seven North Island and two South Island lakes by an expedition from Cambridge University (Lucas, 1904). The expedition also collected benthic fauna from the lakes, including several samples of oligochaetes from depths down to 360 m (Benham, 1903, 1904). Further bathymetric work was done in the course of geological surveys, and one chemical investigation was undertaken (Phillipps & Grigg, 1922), but during the first half of the present century most of the freshwater work was on the systematics of the fauna. The only lake studies were of Taupo (Armstrong, 1935) and three small South Island lakes, Sarah (Flint, 1938) and Lyndon and Pearson (Percival, 1951) (Figure 6, Inset). More extensive work on lakes began in the 1950's with the studies especially of Jolly, Bayly and Fish. The early work on rivers was principally concerned with the introduced trout and the availability of food organisms for them, but included some investigations of the biology of the invertebrates.

Lakes

Most of the lakes have a multiple origin; but many in the North Island were formed predominantly by volcanic activity (Table 1), and many in the South Island by glacial action (Table 2).

Lakes of volcanic origin are found in calderas and explosion craters, often modified by later lava flows, and usually lie in a circular basin with a relatively flat bottom. Lake Taupo (Fig. 2) has an area of 616 km^2 and is surrounded by soils derived from pumice which are either still covered by indigenous forest, or have been converted to pasture land. A little work has been done on the lake by Armstrong (1935) and Jolly (1965, 1968). The fifteen Rotorua lakes (Fig. 3) lie in the thermal region northeast of Taupo and vary in size and depth. They are surrounded by indigenous forest and agriculture land, and like Lake Taupo, are increasingly used for recreation. They are the most intensively studied lakes in New Zealand; the main published accounts are by Jolly (1968) and others and Fish (1963–1970). There are several other small lakes of volcanic origin, often close to the sea. The two which have been most studied are Lake Pupuke in Auckland (Barker, 1966, 1970 and Green, 1967) and Lake Aroarotamahine, on Mayor Island, off the east coast of the North Island (Bayly, 1962). Both are saucer-shaped crater lakes.

The glacially formed lakes include most of the larger South Island lakes, such as Lake Wakatipu (Fig. 4). They are usually typical deep, steep-sided fjord-type lakes with flat floors and dammed by moraines. The troughs are sometimes over-deepened to lie below sea level and form a cryptodepression. In Lake Manapouri, New Zealand's deepest lake (444 m), the bottom is 265 m below sea level (Irwin, 1971). The lakes are situated in the mountains, frequently in sedimentary or metamorphic rock, and surrounded by indigenous forest and, in lower rainfall areas, by

Fig. 2. Lake Taupo, a large volcanic lake in the North Island. (Whites Aviation Ltd.)

Fig. 3. Lake Okareka and Lake Tarawera, near Rotorua. (Whites Aviation Ltd.)

Fig. 4. Lake Wakatipu, a large glacially-formed lake in the South Island. (Whites Aviation Ltd.)

Fig. 5. Lake Rotoroa, a glacially-formed lake in the north of the South Island. (Whites Aviation Ltd.)

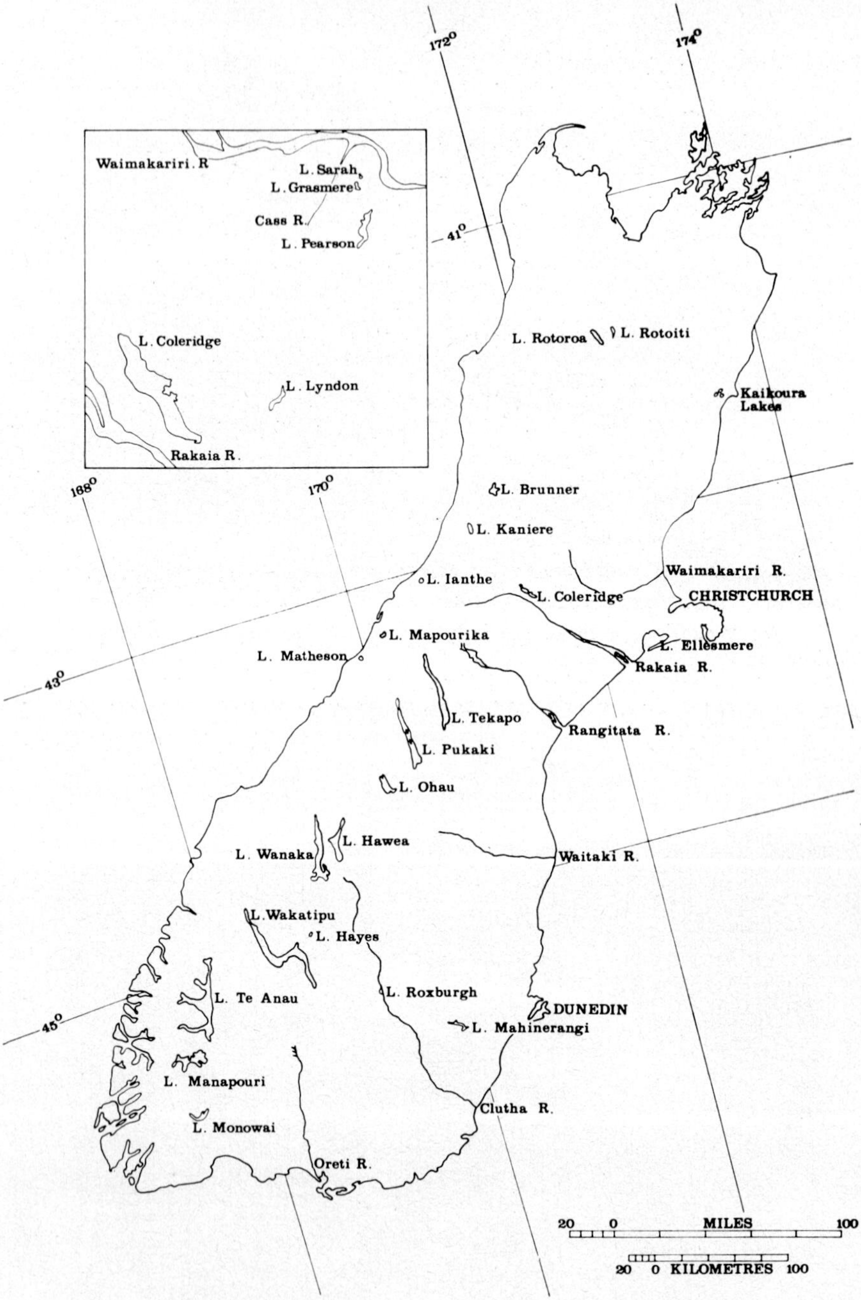

Fig. 6. The major lakes and rivers of the South Island of New Zealand. (Inset): Some of the inland Canterbury lakes and rivers.

Fig. 7. Lake Coleridge and some of the smaller Canterbury lakes, with the Rakaia River entering from the right. (V. C. BROWNE)

tussock grassland which is usually not intensively farmed. The sediments of Lake Wakatipu are mostly mud with a low organic matter content (BRODIE & IRWIN, 1970) and those of Lake Manapouri mostly mud, except for sandy areas out from the river mouths (IRWIN, 1971). The two large lakes in the Nelson area near the north of the South Island, lakes Rotoroa (Fig. 5) and Rotoiti, have been included in this group although they differ slightly in some features from the more southern lakes. The large South Island lakes have been studied principally by JOLLY (1968) and recently by STOUT (in preparation). The smaller lakes in the mountain valleys on the east of the Southern Alps are mostly in glacial troughs or depressions in glacial debris, which may be blocked by alluvial fans, as in the case of Lake Pearson (Fig. 6, Inset) (GAGE, 1959). They are mostly shallow and surrounded to a large extent by tussock grassland. The Canterbury lakes (for example, Fig. 7) have been studied by FLINT (1938), PERCIVAL (1951) and STOUT (1969a and c, 1972). Some of the lakes further south are situated in improved pasture land and one of these, Lake Hayes (Table 2), which at one time received the effluent from a cheese factory, was studied from 1947 to 1954 by JOLLY

Fig. 8. Lake Mapourika, a glacially-formed lake in Westland with a view of the Franz Josef glacier. (Whites Aviation Ltd.)

(1952, 1968). Investigations have only recently begun on some of the numerous small high altitude lakes. The larger lakes on the western side of the Southern Alps, Brunner and Kanieri, appear to be similar to the other large South Island lakes, but the smaller Westland lakes, such as lakes Ianthe, Mapourika (Fig. 8) and Matheson have several distinctive features. They are often situated in depressions in glacial debris and surrounded by native forest, and they lie in a region of high rainfall. Some recent investigations have been made on these lakes (STOUT, 1973 and in preparation).

A number of lakes in both islands have been formed by the blocking of river valleys, principally by landslides or by fluviatile action. The deepest North Island lake, Waikaremoana (Table 1), originated by a landslide damming a river system. The markedly dendritic lake has a large drainage area, mostly covered with indigenous forest. A little work has been done on the lake, principally by JOLLY (1968). Two other large North Island lakes, Waikare and Whangape, were formed on an aggrading flood plain and are very shallow, 3.6 m and 2.7 m respectively. Small peaty lakes also in the Waikato region may have had similar origins. Three small lakes in agricultural land near the coast in the Kaikoura region on the east coast of the South Island (Table 2) have

Fig. 9. Lake Benmore, a hydro-electric impoundment on the Waitaki River in the South Island. (Whites Aviation Ltd.)

been formed by rivers blocking the mouths of tributary streams, and were studied for several years (STOUT, in preparation).

Coastal lakes have been formed in sand dunes or by a bar or spit cutting off part of the sea or deflecting a river. Sand dune lakes are common on the west coast of the North Island, and include basin lakes, such as Lake Waingata, and blocked valleys. They are small and shallow, usually with a sandy bottom but sometimes with soft organic muds in deeper water. Two series of these lakes (Table 1), one mostly north of Auckland and the other north of Wellington were visited several times between 1949 and 1952 (CUNNINGHAM *et al.*, 1953) and later by FISH (1966a). Lakes formed by bars or spits include Lake Wairarapa at the southern tip of the North Island and Lake Ellesmere, near Christchurch in the South Island. Both are large, shallow brackish lakes with fluctuating water levels and drainage from rich pasture land.

Artificial lakes consist chiefly of hydro-electric impoundments and water supply reservoirs. There are a number of hydro-electric lakes on the Waikato River, draining Lake Taupo, and on the Waitaki and Clutha Rivers, draining some of the large glacial lakes. The Waikato lakes are surrounded by rich agricultural pasture land and there has been extensive growth of aquatic macrophytes, particularly introduced species (CHAP-

MAN, 1970). The South Island hydro-electric lakes lie chiefly in tussock grasslands, and at the present time they mostly do not support extensive beds of macrophytes. The Waitaki lakes (Table 2), between about four and forty years old, are currently being investigated, and Lake Roxburgh on the Clutha River was studied by WINTER (1964). Two water supply reservoirs near Wellington have been studied (STEVENSON, 1952).

Other types of lakes, such as those originating principally from tectonic movements or the solution of limestone, are not well represented although Lake Hayes (Table 2) was formed partly by faulting.

TEMPERATURE RELATIONSHIPS

The temperature of New Zealand lakes is strongly influenced by the oceanic climate and young topography of the country. The lakes are often exposed to strong winds, especially in coastal regions and in the rugged mountains of the South Island. In oceanic climates, the temperature of the epilimnion tends to be relatively low, which reduces the stability of the thermocline and the lakes may show an indefinite stratification (HUTCHINSON, 1957). The size, shape and amount of exposure to wind influence the thermal type. New Zealand lakes appear principally to belong to one of three intergrading types. They may be warm monomictic lakes, in which the temperature does not fall below 4 °C in winter, and have a relatively shallow or a deep thermocline; or they may be polymictic (homothermous). The warm monomictic lakes with a shallow thermocline are usually protected from strong winds and basin shaped, whereas those with a deep thermocline may be exposed to the wind but are deep lakes. These form the two extremes of a continuous series, and some intermediate lakes are found. The polymictic or holothermous lakes are mostly shallow and wind-exposed.

The warm monomictic lakes include many of the North Island lakes, which may develop thermal stratification for several months over the summer (Table 1). Those with a shallow thermocline include several of the Rotorua lakes such as Okaro (IRWIN, 1968, FISH, 1970). Most of the Westland lakes in the South Island are protected from frequent strong winds and also have well developed shallow thermoclines (Table 2) (STOUT, 1973). Deep thermoclines have been found in a large number of New Zealand lakes, especially in the South Island. In the North Island they include some of the Rotorua lakes, such as Okataina (IRWIN, 1968, FISH, 1970), whose morphology renders them prone to mixing by the wind, and Lakes Pupuke and Aroarotamahine which are close to the sea and exposed to strong winds. In Lake Taupo and the large glacially-formed lakes of the South Island there is a flat bottom and thus a greater mean depth. According to HUTCHINSON *(loc. cit.)*, when the mean depth is greater there is a higher temperature of the hypolimnion, a more gentle thermocline and a cooler epilimnion, conditions which are

Table 1. North Island lakes: General features.

Lake	Origin	Maximum depth (m)	Temperature stratification	Thermocline depth (m)	Maximum surface temperature (°C)	Minimum bottom temperature (°C)	Secchi disc transparency (m)	Minimum oxygen in bottom water (% saturation)
Pupuke	Volcanic	55	Present	14–28	24.2	11.3	1.0–5.2	24
Rotorua lakes	Usually volcanic	112	Usually 5–6 months	8–20	22.0	7.0	1.7–14	0
Taupo	Volcanic	163	About 5 months	25–30	19.1	10.6	14 (average) 18 (maximum)	78
Waikaremoana	Landslide	258	Present	—	17.1 (March)	8.5	7 – 9	85
Sand dune lakes	Sand barriers	29 (usually less than 10)	Occasional and temporary	3– 5	24.0	10.8	Usually 0.5– 3 9 (maximum)	0

Table 2. South Island lakes: General features.

Lake	Origin	Maximum depth (m)	Temperature stratification	Thermocline depth (m)	Maximum surface temperature (°C)	Minimum bottom temperature (°C)	Secchi disc transparency (m)	Minimum oxygen in bottom water (% saturation)
Large glacial lakes	Glacial	444	Usually present	15–150	20.4	7.4	9 –12	70
Waitaki hydro-electric lakes	Artificial impoundments	93	Rarely present	20– 30	21.7	4.0	1.2– 3.5	90
Smaller Canterbury lakes	Predominantly glacial	30	Rarely present	Variable	21.0	3.0	0.5– 8.0	40 (rare)
Hayes	Glacial and faulting	33	Present	14– 18	18.4	8.2	6.0	13
Westland lakes	Glacial	107	Present	4– 12	22.0	8.0	1.5– 7.5	2.9
Small coastal lakes	Blocked streams	3	Temperature gradient	—	24.4	5.0	0.4– 2.0	50

shown by these lakes (Tables 1 and 2). The thermocline in the large southern lakes is often broad and without a pronounced slope. In Lake Manapouri, for example, the thermocline in March is from 10° to 14°C and extends between approximately 20 m and 60 m (IRWIN, 1971). IRWIN recorded a maximum surface temperature of 16.25°C and a minimum temperature of 7.77°C. Thermal stratification in the lake lasts for about six months. In at least some of these lakes the thermocline deepens during the summer. In Lake Wakatipu the thermocline is at about 54 to 57 m in February and 96 to 100 m in June, and may extend down to 150 m (IRWIN & JOLLY, 1970). The isotherms slope downwards towards the southern, or downwind, end. Some of the lakes develop multiple thermoclines. Deep thermoclines are also found in some of the smaller but less wind exposed South Island lakes, such as Lake Hayes (Table 2).

Polymictic or homothermous lakes, in which thermal stratification is normally absent, include most of the coastal lakes of the North Island, two of the Rotorua lakes (Rotorua and Rotoehu, Fig. 1, Inset) and most of the smaller Canterbury lakes. In some of the coastal and Canterbury lakes, a temporary thermal stratification may be established for a few days and then broken down by the wind.

Some of the smaller South Island lakes may be dimictic, with two circulation periods during the year, although no such lake has been fully investigated. However some small Canterbury lakes, such as Lake Lyndon (Fig. 6, Inset) and Lake Ida, often develop an ice cover and inverse thermal stratification in winter (STOUT, 1969a and c). In summer temperature stratification appears not to be established in these lakes, although the water may reach a temperature of 18°C.

In the hydro-electric lakes, there is frequently a regular temperature gradient between the surface and bottom water in summer but no sharp thermocline. This condition has been reported for Lake Maraetai on the Waikato River (REID, 1966) and in the Waitaki lakes, although the latter occasionally develop a weak, temporary thermocline. A pronounced regular temperature gradient is also found in summer in the shallow coastal lakes near Kaikoura (Table 2).

No meromictic lakes have been recorded in New Zealand. In Lake Manapouri, for example, IRWIN (1971) considers that although the water temperature at depths below 200 m changes little (7.77°–8.0°C) throughout the year there is periodic turnover of the lake.

Temperatures are generally lower in the more southern lakes, and in those at higher altitudes. In the deeper lakes, such as Taupo and the large glacially-formed lakes, the seasonal temperature range is usually less than in shallower lakes, such as the Rotorua, Canterbury or coastal Kaikoura lakes.

Heat budgets have been calculated for only a few New Zealand lakes, and most of the values obtained have been low, probably related to the oceanic climate. For the two small volcanic lakes, Aroarotamahine and Pupuke, the values were respectively 6000 cal/cm² (BAYLY, 1962) and 16000 cal/cm² (BARKER, 1970), both low possibly because the lakes are small and coastal. FISH (1970) calculated heat budgets of 5300, 11000, 11800 and possibly 57600 cal/cm² for four of the Rotorua lakes. In the South Island a small Canterbury lake, Grasmere (Fig. 6, Inset), situated at an altitude of 600 m, has a heat budget which has varied in different years between 8000 and 10000 cal/cm² (STOUT, in preparation). For Lake Manapouri the heat budget has been calculated as approximately 40000 cal/cm², the larger value being presumably related to the large size, deep thermocline and relatively low altitude (181 m) of the lake.

Surface seiches occur in Lake Wakatipu (BOTTOMLEY, 1956, COX, 1965) where they have a fundamental period of 52 minutes, and a first harmonic of 27 minutes, and may persist for several days, probably occurring at most times of the year. There are also smaller transverse seiches. Surface seiches have also been recorded in Lake Taupo (JOLLY, 1957), and internal seiches in two of the Rotorua lakes, Okataina (FISH, 1970) and Rotoiti (GREEN *et al.*, 1968). The observed period in Rotoiti is 19.6 hours.

LIGHT PENETRATION

In New Zealand light penetration has mostly been determined as Secchi disc visibility. High values (usually 10 to 14 m) have been recorded in most of the large glacially-formed lakes, and in Lake Taupo where the highest reading (18 m) (JOLLY, 1968) was recorded. In some of the Rotorua lakes such as Okataina, which are relatively deep and with small inflows, high values are also found.

Low values are mainly caused by phytoplankton or silt. JOLLY (1968) recorded her lowest reading (0.8 m) in a small lake, Rotongaio, near Lake Taupo during an algal bloom. In Lake Pupuke the values are lowest (1.0 m) during the summer growth of phytoplankton (BARKER, 1970), and low values due to algae have been recorded in some of the small Canterbury lakes. In these lakes and in the sand dune lakes and the small lakes near Kaikoura, low Secchi disc readings may be caused by both phytoplankton and inorganic matter. Very low values may be due to the presence of glacial silt, as in Lake Pukaki (0.2 m) and Lake Tekapo, and in the Waitaki hydro-electric impoundments which receive water from these two lakes. Another hydro-electric lake, Roxburgh, also has visibility reduced, at times to 0.3 m, because of glacial silt (WINTER, 1964). Seasonal variations are pronounced in some lakes, depending on the phytoplankton, as in Lake Pupuke, or on the amounts of algae or silt (depending on seasonal rains or snow melt), as in several Canterbury lakes (STOUT, 1972, in preparation). MITCHELL (1971) found a good inverse correlation in Tomahawk Lagoon between variations in Secchi disc readings and the amount of phytoplankton productivity.

Light penetration is reduced by dissolved organic matter leached from the surrounding terrestrial vegetation and colouring the water brown in many of the Westland lakes (STOUT, 1973, in preparation) and the small peaty Waikato lakes.

CHEMICAL CHARACTERISTICS

Most New Zealand lakes are well oxygenated throughout the year (Tables 1 and 2). In Taupo, Waikaremoana and the large glacially-formed lakes the oxygen content of the bottom water remains high even when the lakes stratify. The volume of the hypolimnion is large, thermal stratification weak and the lakes are unproductive. Lakes in which the oxygen content of the hypolimnion is markedly reduced are usually shallow and have a shallow thermocline. They include some of the Rotorua lakes, such as Okaro and Ngapouri (JOLLY, 1968, FISH, 1970); some of the Westland lakes, such as Matheson (STOUT, 1973, in preparation), and Lake Rotongaio, near Taupo (JOLLY, 1968). When the sand dune lakes thermally stratify the bottom water may become deoxygenated, and reduction of oxygen in the bottom water sometimes occurs when Canterbury lakes stratify. In Lake Hayes and Lake Pupuke the thermocline is at a moderate depth, and the oxygen content moderately reduced. FISH (1969a, 1970) calculated oxygen deficits for three Rotorua lakes. The mean seasonal hypolimnetic oxygen deficit was 10.0 g O_2/m^3 in Okaro, 9.7 g O_2/m^3 in Ngapouri and 2.3 g O_2/m^3 in the unproductive Okataina, which has a deep thermocline.

The chemical content of the lake water (Tables 3–6) is generally low, especially in most of the glacially-formed and seepage lakes. The glacially-formed lakes are similar to glacial lakes in other parts of the world, such as the English Lake District, and have a low ionic concentration and a relatively high silicate content. In volcanic lakes which are fed by thermal waters the ionic concentration may be high, especially chloride, sulphate, silicate and phosphate (JOLLY, 1968, FISH, 1970). A very high value of silicate (112 g/m^3) in Lake Rotomahana is thought to be due to leaching from volcanic ash. Proximity to the sea, and the related strong winds carrying salts, influences the chemical content of a number of lakes, particularly the sand dune lakes, the small volcanic lakes Pupuke and Aroarotamahine, and the lakes near Kaikoura. All of these have high chloride, sodium and magnesium values, sometimes high silicate, and usually high alkalinity. The sodium concentration may be relatively high even in some more inland lakes (STOUT, 1969c). Many of the Westland lakes (such as Matheson), and probably also the little known peaty Waikato lakes, are acid, dystrophic, with a high content of dissolved organic matter presumably related, in Westland, to the high rainfall and to leaching from the abundant terrestrial vegetation.

Lakes in which thermal stratification is established during the summer

Table 3. North Island lakes: Major anions (g/m^3) and pH.

Lake	pH	Alkalinity (as $CaCO_3$)	Cl	SO_4	Maximum NO_3—N	Maximum NH_4—N in bottom water	Maximum soluble P
Pupuke	7.8–9.5	76– 80.8	43– 45	19.7–28.8	0.31	0.95	—
Rotorua lakes	5.8–9.0	4–166	30	9 –32	Trace	64.0	1.7
Taupo	7.0–7.6	40	—	—	Trace	0.02	0.02
Waikaremoana	—	42	—	—	—	—	—
Sand dune lakes	6.2–7.1 (Northland series) 7.9–8.4 (Wellington series)	—	29–126.4	—	0.24	—	0.01

Table 4. South Island lakes: Major anions (g/m³) and pH.

Lake	pH	Alkalinity (as $CaCO_3$)	Cl	SO_4	Maximum NO_3—N	Maximum NH_4—N in bottom water	Maximum total organic N	Maximum soluble P	Maximum total P
Large glacial lakes	6.5–7.1	8.5–41	1– 4	1– 3	0.15	0.06	0.54	0.20	0.03
Waitaki hydro-electric lakes	6.5–7.5	11 –25	< 1– 1	5–10	0.04	0.05	1.15	0.02	0.15
Smaller Canterbury lakes	6.4–8.9	15 –54	<1– 8	3– 9	0.1	0.08	1.2	0.87	0.21
Hayes	7.2–8.0	—	<1– 3	—	0.0	0.149	—	1.30	—
Westland lakes	3.3–6.8	0 –27	<1– 5	—	0.1	0.02	0.87	0.30	0.18
Small coastal lakes	6.4–9.3	7 –60	13–63	3– 4	2.0	1.57	2.8	0.04	0.295

Table 5. North Island lakes: Major cations (g/m³), ionic concentration and conductivity.

Lake	Na^+	K^+	Ca^{++}	Mg^{++}	SiO_2	Ionic concentration (meq/l)	Conductivity at 25 °C (mS/m)
Pupuke	33–36	1.85–2.38	6.88– 9.76	13.96–15.52	—	3.30	—
Rotorua lakes	—	—	3 –12	0.5 – 2.7	0.05–112	0.28–14.32	15.8–19.2
Taupo	—	—	—	—	20 – 29	1.21	—
Waikaremoana	—	—	—	—	—	0.89	—
Sand dune lakes	28–52	3.0	4.6 – 8.4	5.8 – 8.2	0.2 – 20.0	—	—

Table 6. South Island lakes: Major cations (g/m³), ionic concentration and conductivity.

Lake	Na^+	K^+	Ca^{++}	Mg^{++}	SiO_2	Maximum Fe	Ionic concentration (meq/l)	Conductivity at 25 °C (mS/m)
Large glacial lakes	1.5– 3.2	0.2 –1.2	4.1– 7.3	0.2 –1.0	2– 6	2.8	0.4–1.0	2.5– 6.5
Waitaki hydroelectric lakes	1.8– 2.6	0.48–1.96	2.8– 9.6	0.2 –1.9	3–10	1.7	0.3–0.7	4.3– 7.8
Smaller Canterbury lakes	1.8– 8.1	0.05–1.2	2.2–16	0.2 –3.8	0– 8	2.0	0.2–1.5	3.4–11.4
Hayes	—	—	—	—	2– 3	0.25	1.3	—
Westland lakes	1.9– 4.4	0.16–2.5	0.8– 8.3	0.24–1.6	1– 6	4.8	0.0–0.8	2.1– 6.9
Small coastal lakes	5.5–20.0	0.35–2.7	2.8–12.9	0.2 –3.6	0–20	8.0	0.4–3.0	8.5–18.2

may show an increase in certain ions, particularly nitrogen (as NH_4-N), phosphorus and occasionally silicate, in the hypolimnion. Increased values have been recorded in the Rotorua lakes Okaro and Ngapouri (Fish, 1970), Westland lakes such as Matheson (Stout, 1973, in preparation), Lake Pupuke (Barker, 1970) and Lake Hayes (Jolly, 1968). Lake Pupuke has a pronounced pH and alkalinity stratification during the summer.

Marked differences between neighbouring lakes and seasonal variations have been recorded particularly in the two groups of lakes for which most chemical information is available, the Rotorua lakes (Jolly, 1968, Fish, 1970) and the small Canterbury lakes (Stout, 1969c). Smaller lakes tend to have a higher ionic concentration than larger lakes. Horizontal differences in phosphate were found in a synoptic survey of the large shallow Lake Rotorua (Fig. 1, Inset), but not in the adjacent elongate and deeper Lake Rotoiti (Fish & Chapman, 1969).

High phytoplankton responses to the addition of trace elements were found in several South Island lakes, in comparison with similar work on lakes in North America (Goldman, 1964, 1972). The four large glacial lakes responded to cobalt, and sometimes to molybdenum and iron; while the three small Canterbury lakes responded to most additions, particularly manganese, iron, cobalt and molybdenum. A small Otago lake, Moke, had a marked response to the addition of molybdenum. Low values of these trace elements are also recorded from a number of New Zealand soils, and cobalt and molybdenum in particular are often added to phosphatic fertilizers applied to the land. Fertilizer application, frequently from the air, is now common practice including applications to rugged terrain. A number of amino acids, peptides and free sugars, forming the major fraction of the dissolved organic matter in the surface waters of eight Rotorua lakes, were identified by Briggs (1962a and b).

A nutrient budget calculated for Lake Rotorua (Fish, 1969b) indicated that run-off from agricultural land is probably the major source of nutrients, but that some of the thermal inflows provide the main source of ammonium salts. It has been suggested (Fish, 1969a) that the low content of dissolved salts, especially calcium, in the lake water may influence the effect of increased nutrients, since some plant nutrients would normally be held at low concentrations by the buffering and absorbing properties of calcium salts. If calcium and other dissolved salts are low, as they are in many New Zealand lakes, there may be less buffering effect and the nutrients would become more rapidly available. High nitrogen and phosphorus values in Lake Hayes and the small lakes near Kaikoura are also probably due to agricultural run-off. Inundation of former pasture, together with run-off from the surrounding land, have probably been at least partly responsible for increases particularly in phosphorus found in the hydro-electric lakes of the Waikato River when new impoundments have been formed (Reid, 1966). Probably the main source of nitrogen and phosphorus in many of the lakes is from agriculture. New Zealand is principally a farming country and a large number of lakes have at least part of their catchment used for agriculture, particularly pasture for stock.

Plankton

The amount of plankton present is influenced by the morphology and the temperature, light and chemical conditions. In the numerous New Zealand lakes in which the ionic concentration is low, the plankton is usually sparse. Plankton is particularly low in numbers in some South Island lakes, such as Pukaki and Tekapo, in which the water is turbid due to large amounts of glacial silt. In the large clear lakes the plankton is also sparse, but distributed over a wide depth range. In small homothermous lakes, such as some of the Canterbury lakes, the plankton may be almost equally abundant at all depths, but where thermal stratification is well developed, the zooplankton at least may show a sharply defined lower limit at the thermocline (Bayly, 1962). The smaller lakes appear to be often richer in the amount of plankton, and with more variation in species composition between lakes, than the larger lakes. Occasionally the zooplankton may be apparently relatively much more abundant than the phytoplankton (Stout, 1969c). Average organic dry weights, taken from values obtained over at least a year, can be compared for a number of South Island lakes (Stout, 1969c, 1970 and in preparation). In a large glacial lake, Manapouri, the average organic dry weight was 0.58 g/m^3; in a Westland lake, Mapourika, 1.33 g/m^3; in a small Canterbury lake, Grasmere, 3.28 g/m^3, and in two of the Kaikoura lakes, 13.02 g/m^3 and 12.75 g/m^3 respectively. Values are always low in the large glacial lakes, but more variable in the small Canterbury lakes (0.04–3.3 g/m^3, Stout, 1969c).

Seasonal variations appear to be more marked in the South Island than the North Island lakes. The small Canterbury lakes, such as Sarah and Grasmere, may have pronounced spring and autumn peaks, and usually a small summer increase in numbers (Flint, 1938, Stout, 1972, in preparation). In Lake Hayes, Jolly (1952) recorded a large spring burst of phytoplankton and summer and autumn peaks of zooplankton. Cassie (1969) found distinct autumn and also spring peaks in the phytoplankton in the North Island lakes Rotorua and Rotoiti, and seasonal variations have been reported in Lake Pupuke (Green, 1967, Barker, 1970). There are differences in the extent and times of the peaks in different lakes and in different years.

Phytoplankton

The planktonic algae usually belong to widespread species found also

Table 7. North Island lakes: Plankton

Lake	Phytoplankton – general	Phytoplankton – dominants	Phytoplankton – less abundant	Zooplankton – dominants	Zooplankton – less abundant
Pupuke	'Blooms' occur	*Closterium* *Staurastrum*	*Ceratium hirundinella*	*Calamoecia lucasi*	*Asplanchna brightwellii* *Bosmina meridionalis* *Keratella cochlearis*
Rotorua lakes	Diatoms usually dominant; green algae sometimes common	*Melosira granulata* *M. distans* *Asterionella formosa*	*Staurastrum* *Staurodesmus* *Cosmarium* *Closterium*	*Bosmina meridionalis* *Ceriodaphnia dubia* *Calamoecia lucasi*	*Asplanchna priodonta* *Keratella cochlearis*
Taupo	Sparse	*Melosira* *Staurastrum*	—	*Boeckella propinqua* *Bosmina meridionalis* *Ceriodaphnia dubia*	*Daphnia carinata* *Synchaeta pectinata*
Waikaremoana	Green algae	*Staurastrum* *Volvox* *Ceratium*	—	*Daphnia carinata* *Ceriodaphnia dubia* *Boeckella hamata*	—
Sand dune lakes	Usually mostly diatoms, peridinians, green algae	*Peridinium* (Northland series) Wellington series – variable	—	copepods *Keratella cochlearis*	*Brachionus caudatus* *Filinia longiseta*

Table 8. South Island lakes: Plankton.

Lake	Chlorophyll a content	Phytoplankton – general	Phytoplankton – dominants	Zooplankton – dominants	Zooplankton – less abundant
Large glacial lakes	low	Diatoms, desmids and small green algae	*Cyclotella* *Tabellaria* *Diatoma elongatum* *Synedra* *Staurastrum*	*Bosmina meridionalis* *Ceriodaphnia dubia* *Boeckella dilatata*	*Daphnia carinata* *Synchaeta pectinata* *Polyarthra vulgaris*
Waitaki hydro-electric lakes	low	Diatoms	*Synedra* *Melosira distans* *Diatoma elongatum* *Asterionella formosa*	*Boeckella dilatata* *Bosmina meridionalis*	*Asplanchna priodonta* *Synchaeta pectinata*
Smaller Canterbury lakes	low-medium	Diatoms and green algae	*Melosira granulata* *Asterionella formosa* *Diatoma elongatum*	*Ceriodaphnia dubia* *Bosmina meridionalis* *Boeckella hamata*	*Synchaeta pectinata* *Keratella cochlearis* *Filinia longiseta*
Hayes	—	Diatoms and green algae	*Melosira* *Closteriopsis*	*Ceriodaphnia dubia*	*Bosmina meridionalis* *Daphnia carinata* *Boeckella dilatata*
Westland lakes	low-medium	Diatoms, desmids, small flagellates	*Cyclotella* *Staurastrum* *Vacuolaria* *Cryptomonas*	*Boeckella hamata* *Ceriodaphnia dubia*	*Bosmina meridionalis* *Macrocyclops albidus* *Asplanchna priodonta* *Keratella cochlearis*
Small coastal lakes	high	Blue-green and green algae	*Closterium aciculare* *Melosira granulata* *Trachelomonas* *Anabaena*	*Boeckella hamata* *Keratella quadrata* *Brachionus* sp.	*Ceriodaphnia dubia* *Filinia longiseta* *Bosmina meridionalis*

in the northern hemisphere. The phytoplankton is often dominated by diatoms or green algae, with desmids occasionally common (Tables 7 and 8).

Algal blooms occur in several lakes. Blooms of blue-green algae (*Anabaena circinalis*, *A. flos aquae*, *Nodularia spumigena*) have been recorded in some sand dune lakes such as Waiparera (FLINT, 1970); the Rotorua lakes Rotorua and Rotoiti (REID, 1964, CASSIE, 1969); Lake Tutira, south of Waikaremoana; the Kaikoura lakes and another South Island coastal lake, Forsyth (FLINT, 1970); a small Canterbury lake (STOUT, 1969c, FLINT, 1970), and a small Otago coastal lake (FLINT, *loc. cit.*) and in some of the Waikato hydro-electric lakes when they were first impounded (REID, 1966, CASSIE, 1969). Blooms of the desmid *Closterium aciculare* were found in Lake Pupuke in late May to June (GREEN, 1967), in one of the Kaikoura lakes in July (FLINT, 1970) and in a sand dune lake in August (CUNNINGHAM *et al.*, 1953). This species is also common in Lakes Rotorua and Rotoiti (CASSIE, 1969). Rotorua, Rotoiti and two of the Waikato hydro-electric lakes (Maraetai and Ohakuri) have dense growths of the diatoms *Melosira granulata* and *Asterionella formosa* (REID, 1966, CASSIE, 1969, FLINT, 1970). Dinoflagellate blooms have been reported in a number of small North Island lakes, including some of the northern sand dune lakes (CUNNINGHAM *et al.*, 1953) and Lake Pupuke (GREEN, 1967).

Diatoms are dominant and green algae are common in Lakes Rotorua and Rotoiti (CASSIE, 1969) and in many of the small Canterbury lakes, such as Sarah and Grasmere (FLINT, 1938, 1970; STOUT, 1969c). The most common species are *Melosira granulata* and *Asterionella formosa*, and *Diatoma elongatum* is also abundant in several of the Canterbury lakes. *Melosira granulata* may also be prominent in the Kaikoura lakes and JOLLY (1952) reported a species of *Melosira* in Lake Hayes. *Asterionella formosa* is widespread in South Island lakes. *Crucigenia truncata* is common in some small Canterbury and Otago lakes, including Lake Pearson, and other green algae abundant in some of the Canterbury lakes include *Pediastrum boryanum*, *P. tetras*, *Gleotila pelagica*, *Mougeotia* (FLINT, 1970) and *Botryococcus braunii*. Diatoms, including *Synedra* sp., are dominant in the Waitaki hydro-electric lakes.

In the large glacial lakes diatoms are dominant, especially *Cyclotella stelligera*, *Tabellaria* and *Diatoma elongatum*, but there may also be green algae, particularly those with small cells and desmids such as *Staurastrum limnaticum*. Lake Taupo also has mostly diatoms and *Staurastrum*, and species of *Staurastrum* are common in a number of New Zealand lakes (THOMASSON, 1960). Some of the Westland lakes, such as Mapourika, have diatoms and desmids whereas in others, such as Matheson, the plankton is dominated by small flagellates. One small Canterbury lake, Vagabonds Inn, has a large number of desmid genera.

Chlorophyll values, obtained over at least a year, are available for a

Table 9. Chlorophyll a content of some South Island lakes.

5–140 mg/m³ (Eutrophic lakes)	small coastal Kaikoura lakes. Tomahawk lagoon (small coastal Otago lake)
1–15 mg/m³ (Mesotrophic lakes)	Grasmere, Sarah, Tripp (small Canterbury lakes)
0.3–2.5 mg/m³ (Oligotrophic lakes)	Lyndon, Hawdon (small Canterbury lakes) Mapourika (Westland lake) Manapouri, Te Anau, Nelson lakes (large glacial lakes)

number of South Island lakes (STOUT, 1969c, in preparation; MITCHELL, 1971) and some of these lakes have been classified according to their chlorophyll a content following the arrangement of VOLLENWEIDER (1968) (Table 9). Values over 100 mg/m^3 of chlorophyll a have frequently been recorded in the Kaikoura lakes. The chlorophyll a content is higher in the smaller, shallower lakes and seasonal fluctuations appear to be most marked in these lakes, including Canterbury lakes such as Grasmere (STOUT, 1972) and Tomahawk lagoon in Otago (MITCHELL *loc. cit.*).

ZOOPLANKTON

The zooplankton of New Zealand lakes contains relatively few species, particularly of crustaceans. There are only three common Cladocera: *Bosmina meridionalis*, *Ceriodaphnia dubia* and *Daphnia carinata;* although species more common in shallower water, such as *Simocephalus vetulus* and *Chydorus sphaericus*, are occasionally found in the plankton. All species appear to be distributed throughout New Zealand. No predaceous Cladocera have been recorded. Three genera of calanoid copepods are found. The most common is the southern hemisphere genus *Boeckella* containing seven species showing different distribution patterns. *B. propinqua* is found at low altitudes in the North Island and northern part of the South Island; *B. hamata* is found in the eastern parts of both islands, often in smaller lakes; *B. delicata* has only been recorded in Lake Omapere in the north of the North Island and Lake Brunner on the west coast of the South Island; *B. dilatata* is found in the middle and south of the South Island, possibly prefering low temperatures; *B. triarticulata* is found in ponds and small lakes in the South Island; *B. symmetrica* has been recorded only from an Auckland pond and *B. minuta* from a Wellington reservoir (BAYLY, 1964, 1967a; VIDAL, 1968). *Calamoecia* has only one species in New Zealand, *C. lucasi*, found in the north and east of the North Island (BAYLY, 1961) and sometimes common. *Gladioferens spinosus* and *G.*

pectinatus are recorded only from coastal bodies of water (BAYLY, 1963a). There are few pelagic adult cyclopoid copepods, although the younger stages of species whose adults live in the littoral zone may be more common. *Macrocyclops albidus*, *Mesocyclops leuckarti*, *Eucyclops serrulatus* and *Tropocyclops prasinus* have all been found as adults in the open water, but seldom abundant.

The rotifers in the plankton have been less studied but appear mainly to belong to species widespread in other parts of the world (see RUSSELL, 1960) and distributed throughout both islands. Predaceous species are among the few predators in New Zealand plankton. A carnivorous water mite, *Piona uncata exigua*, is also found in the plankton of a number of lakes, particularly small South Island lakes (STOUT, 1972). A freshwater medusa, *Craspedacusta sowerbyi* has been reported occasionally from a few North Island lakes, including Taupo, Maraetai and several Rotorua lakes (FISH, 1971). It appears to be favoured by high water temperatures. A heliozoan belonging to the genus *Actinosphaerium* has been found, sometimes in large numbers, in several of the large glacially-formed lakes such as Manapouri (STOUT, 1973, in preparation). The young stages of a small fish, the smelt *Retropinna lacustris*, is occasionally present in the plankton but there are few other fish feeding on the zooplankton.

The zooplankton is often dominated by two species of Cladocera and one species of calanoid copepod (Tables 7 and 8). The Cladocera are usually *Bosmina meridionalis* and *Ceriodaphnia dubia;* although in some lakes (such as Waikaremoana) *Daphnia carinata* appears to replace *B. meridionalis*, and in others (such as HAYES) (JOLLY, 1952) to be present in equivalent numbers. *C. dubia* is apparently absent from Lake Pupuke. The calanoid copepod in North Island lakes is usually *Calamoecia lucasi*, as in the Rotorua and some of the northern sand dune lakes and Lake Pupuke; or *Boeckella propinqua*, as in Lake Taupo and Lake Aroarotamahine. However, *B. hamata* is present in some of the east coast lakes, such as Waikaremoana. *B. hamata* and *B. dilatata* are the dominant calanoids in South Island lakes; *B. hamata* mostly in the smaller lakes, including many in Canterbury and Westland, and *B. dilatata* in the large glacially-formed lakes, the Waitaki hydro-electric lakes and Lake Hayes. In the small coastal lakes, the sand dune lakes in the North Island and the lakes near Kaikoura in the South Island, the Cladocera may be less common (with *Bosmina meridionalis* often the main species) and rotifers, together with a calanoid copepod, more abundant.

Several species of rotifer may be common, apparently especially in the smaller lakes. *Keratella cochlearis* is prominent in a number of lakes in different parts of the country – Pupuke, sand dune lakes, Rotorua lakes, small Canterbury lakes, and Westland lakes – whereas *K. quadrata* occurs in the Kaikoura lakes. *Synchaeta pectinata* is possibly most common in the less productive lakes, being relatively abundant in Lake Taupo, the large glacially-formed lakes, the Waitaki hydro-electric lakes and some small Canterbury lakes. *Polyarthra vulgaris* is also common in some large glacially-formed lakes. Two species of *Asplanchna* are sometimes abundant, *A. brightwelli* (for example in Lake Pupuke) and *A. priodonta* (for example in Rotorua, Westland and Waitaki hydro-electric lakes). In the sand dune lakes and coastal lakes near Kaikoura, species of *Brachionus* and *Filinia longiseta* may be prominent, and the latter species is also common in some small Canterbury lakes (STOUT, 1972).

There are pronounced differences in the amount of zooplankton present in different lakes, as is indicated by the organic dry weight values for the total plankton in some South Island lakes (p. 423). Zooplankton is sparse in the large glacial lakes, and to a lesser extent in Lake Taupo, some of the Rotorua lakes and probably a number of the sand dune lakes (CUNNINGHAM *et al.*, 1953). Greater quantities are present in the Westland, small Canterbury and probably several Rotorua lakes, and Lake Hayes, and zooplankton is abundant in the small Kaikoura coastal lakes and in small volcanic lakes like Pupuke (GREEN, 1967). However these are indications only and there may be considerable variations between adjacent lakes besides variations in different seasons and years.

Different times of seasonal abundance are shown by the various components of the zooplankton. The Cladocera are often dominant during the late spring and summer, as in Lake Hayes (JOLLY, 1952). In several Canterbury lakes (Grasmere, Pearson, Lyndon) *Ceriodaphnia dubia* may be more abundant in November and December and *Bosmina meridionalis* from late November to March (PERCIVAL, 1951, STOUT, 1972). In Lake Aroarotamahine *Daphnia carinata* was most common in spring (September) and *B. meridionalis* in April and May, but sampling was possible only in five different months (BAYLY, 1962). Calanoid copepods may be common throughout the year, or until at least the middle of the winter. In Hayes, *Boeckella dilatata* was abundant in autumn (April), in Canterbury lakes *B. hamata* is often common in autumn and early winter and in Pupuke *Calamoecia lucasi* was abundant through the early winter. *B. propinqua* was present in large numbers in all the samples taken from Aroarotamahine. Rotifers are dominant in winter in many South Island lakes. The water mite *Piona uncata exigua* disappears from the plankton during the winter months in at least some Canterbury lakes.

Diurnal vertical migrations of the zooplankton have been recorded in several New Zealand lakes. Increases in numbers in the surface waters at dusk and at dawn were found in the copepods and *Daphnia carinata* in Lake Hayes, the extent of migration varying at different times of the year and under different weather conditions. Similar dusk and dawn rises were found in Lake Pupuke in several species, including *Bosmina meridionalis*, in July. A single later evening rise was recorded in Lake Taupo in *B. meridionalis* and *Ceriodaphnia dubia* at several times of the year (JOLLY, 1965) and in the large glacially-formed Nelson lake Rotoiti in *B. meridionalis* and mature cyclopoid copepods in January (STOUT, in preparation);

and maximum numbers of *C. dubia* and *Boeckella hamata* at the surface at 0300 hours in late December in the small Canterbury lake Lyndon (PERCIVAL, 1951). A reversed diurnal migration was found in *Daphnia carinata* and male *B. propinqua* in Lake Aroarotamahine, the animals rising to the surface during the day, although *Bosmina meridionalis* did not migrate in this lake (BAYLY, 1963b). A reversed migration has been reported from a lake in Africa with similar alkalinity and general chemical composition.

LITTORAL AND BENTHIC REGIONS

The littoral and benthic plants and animals present depend partly on the type of shore and substrate, and partly on the degree of invasion by introduced species of aquatic macrophytes. In many of the lakes, especially in the mountain regions, at least part of the shore is wind-exposed and rocky, so that rooted vegetation is sparse. In more sheltered areas the native species of macrophytes have frequently been replaced by exotic species, especially in the North Island.

Vegetation

Rocky and stony shores may have seasonal growths of filamentous algae, sometimes together with charophytes as in the Rotorua lake, Rotoiti (CHAPMAN *et al.*, 1971a). For example the filamentous green algae *Bulbochaete* and *Ulothrix* may be abundant in the littoral region of some Canterbury and Westland lakes (FLINT, 1970).

Rooted vegetation is sparse in many of the large glacially-formed lakes, which have steep rocky sides, and in the small Kaikoura lakes, in which turbidity of the water restricts submerged plant growth. Native macrophytes are still, or have been until recently, predominant in a number of lakes, such as Waikaremoana, many of the sand dune lakes, some of the Canterbury lakes (MASON, 1969), the Waitaki hydro-electric lakes and Lake Roxburgh and, in sparse quantities, in the Westland lakes. The principal submerged species are *Myriophyllum propinquum*, *M. elatinoides*, *Potamogeton cheesemanii* and *P. ochreatus*, often with the introduced *Ranunculus fluitans*. Several species, particularly *Isoetes alpinus* in the South Island, may be common in shallow water and the bulrush *Typha muelleri* at the shore. *Phragmites* is not found in New Zealand. Charophytes are often abundant in deeper water.

Elodea canadensis was introduced into New Zealand about 1868 and occurs fairly extensively in both islands (CHAPMAN, 1970). It forms dense beds in a number of lakes, including the Rotorua lakes, the Waikato hydro-electric lakes, several Canterbury lakes (Grasmere, Lyndon, Pearson) and Otago lakes (Waipori), and recently in the large glacial Nelson lakes. A more recent introduction, *Lagarosiphon major*, is found mostly in the North Island, but also at the northern tip of the South Island. In the Rotorua lakes it occurs with *E. canadensis* as extensive weed beds which interfere with angling, boating and other activities (FISH, 1966b, FISH & WILL, 1966). In the submerged vegetation of Lake Rotoiti there are a number of communities related to the type of substrate and the depth of water (CHAPMAN *et al.*, 1971a and b). In shallow water to a depth of 2 m and at least partly protected from exposure to wind there is a squat growth of several species, mostly indigenous and often forming mounds. In depths of 2 m to 6 m *L. major* predominates, together with *E. canadensis* especially at the upper and lower limits of the beds. *L. major* is more common on silty sand and *E. canadensis* and *Nitella hookeri* on pure silt. *N. hookeri* often extends into deeper water. CHAPMAN *et al.* consider that light penetration probably limits the depth of *L. major* to about 6 m, and of *E. canadensis* to about 7.5 m. They suggest that *E. canadensis* and probably the mixed indigenous community occupied larger areas before the invasion of *L. major*, and that the original vegetation was probably the mixed indigenous species in shallow water and *N. hookeri* and species of *Potamogeton* in deeper water. *Elodea* then replaced the *Potamogeton*, and later *Lagarosiphon* has partly replaced the *Elodea*. The chemical composition of both *E. canadensis* and *L. major* vary in different seasons and in different lakes, apparently related to the nutrient content of the water (FISH & WILL, 1966). Attempts have been made to eradicate the *Lagarosiphon* by applications of diquat. In a bay of Lake Rotoiti, destruction of the *Lagarosiphon* was followed by a phytoplankton bloom and then by extensive growth of *Nitella* (FISH, 1966b). Two other introduced macrophytes, *Egeria densa* and *Ceratophyllum demersum*, have recently invaded several lakes in the North Island, particularly the Waikato hydro-electric lakes (CHAPMAN, 1970). Extensive beds of the four introduced species cause considerable problems in the latter lakes.

Fauna

The animals on exposed stony shores, such as are found in some of the Canterbury lakes (STOUT, in preparation), are primarily insects usually found in running waters, particularly larvae of Plecoptera, leptophlebiid Ephemeroptera, sericostomatid and other Trichoptera, and Coleoptera. There are also a number of animals more typical of lakes, especially the gastropods *Potamopyrgus antipodarum* and *Gyraulus corinna*, hydroptilid Trichoptera (*Oxyethira* sp.), the planarian *Cura pinguis*, *Plumatella repens* and sponges. This second group increases in number and variety with decreasing exposure to wind.

Among the emergent vegetation the fauna is sparse and the main animals are Collembola, *Microvelia macgregori* (Heteroptera) and spiders. The water strider, *Hydrometra ribesci* appears to be present only at the edges of northern lakes.

The fauna of the aquatic macrophytes has been investigated in some of the Canterbury lakes (STOUT, 1969a, in preparation), the sand dune lakes (CUNNINGHAM *et al.*, 1953, FISH, 1966a), Lake Taupo (ARMSTRONG, 1935) and Lake Roxburgh (WINTER, 1964). Prominent in and among the vegetation in Canterbury lakes are the gastropods *Potamopyrgus antipodarum*, *Gyraulus corinna* and *Physastra variabilis;* insects, particularly Heteroptera *(Sigara arguta)*, various Trichoptera (*Oxyethira*, *Paroxyethira*, and leptocerids), Zygoptera (*Xanthocnemis zealandica* and *Lestes colensonis*), Anisoptera (*Procordulia grayi* and *Hemicordulia australiae*), a lepidopteran *(Nymphula nitens)*, dytiscids (*Liodessus plicatus* and *Rhantus pulverosus*), and orthocladine Chironomidae; and crustaceans, such as Cladocera (*Simocephalus vetulus*, *Chydorus sphaericus* and *Alona guttata*), cyclopoid copepoda (*Eucyclops serrulatus* and *Tropocyclops prasinus*) and ostracods *(Cypridopsis vidua)*. Other animals which may be common are *Chlorohydra viridissima*, *Cura pinguis*, nematodes, tubificids, naiads, leeches (*Glossiphonia* spp., *Placobdella maorica*) and water mites *(Piona uncata exigua)*. The sand dune lakes also contain abundant molluscs (especially *Potamopyrgus* and bivalves) and chironomid larvae, but the freshwater crayfish *Paranephrops planifrons* and freshwater crab *Hymenosoma lacustre* are also sometimes plentiful, although the latter appears to have disappeared from at least one lake in recent years (FISH, 1966a). In the hydro-electric Lake Roxburgh the fluctuating water level influences the distribution of the fauna and numbers are highest just below the lower limit of fluctuations probably partly due to a change in the type of sediment. *P. antipodarum*, *G. corinna*, *Sphaerium*, oligochaetes and chironomid larvae are common, the numbers varying with the sediment type. Other animals which may be common among weed beds in New Zealand lakes are the gastropods *Lymnaea tomentosa* (native) and *L. stagnalis* (introduced as fish food), the heteroptera *Anisops assimilis* and *A. wakefieldi*, the hydracarines *Hydrachna maramauensis* and *Eylais waikawae* and the amphipod *Paracalliope fluviatilis*. There is a noticeable absence of assellid isopods. The fauna of the weed beds is thus a varied one, with a mixture of native and widespread species.

On and in the mud, both in shallow water and extending into deeper water are *P. antipodarum* and *G. corinna*, sphaeriid bivalves and the large freshwater mussel *Hyridella menziesi*, chironomid larvae (FORSYTH, 1971), and oligochaetes. Of the latter, a lumbricid *Lumbriculus variegatus* is widely distributed and nine species of Tubificidae plus three species of Naididae have been reported from New Zealand lakes (BRINKHURST, 1971), including several rare and interesting species from considerable depths in the large glacial lakes Manapouri and Wakatipu (BENHAM, 1903, 1904). The numbers of *Chironomus zealandicus* decrease with depth in Lake Pupuke, probably due to decreasing oxygen or increasing sulphide in the water, or both (BARKER, 1966). Some of the fauna extends to depths of about 91 m on the soft brown sediment of Lake Taupo.

The large brackish Lake Ellesmere, near Christchurch, has an unusual fauna which includes two species of mysids, *Tenagomysis chiltoni* and *T. novaezealandiae* (Stout, 1969a).

The vertebrate fauna of New Zealand lakes consists mostly of fish; there are no reptiles and virtually no mammals. The native fish are mainly secondarily freshwater, often highly euryhaline, and many particularly of the lowland fish retain a marine stage in their life history (McDowall, 1964, 1968, 1970a). The approximately thirty species (mostly endemic) chiefly belong to the Family Galaxiidae, but also common are the Family Eleotridae and two species of *Anguilla* (one endemic) (for example Burnet *et al.*, 1969). There are several species of introduced fish, the most successful being Salmonidae, particularly two species of trout (*Salmo trutta* and *S. gairdnerii*) and, in a restricted area, the quinnat salmon (*Oncorhynchus tschawytscha*). The fauna of a lake will often include one or both species of trout and of *Anguilla*, a galaxiid (such as *Galaxias lynx*) and an eleotrid (such as *Gobiomorphus breviceps*). The native fish usually feed principally on invertebrates; apparently none are completely herbivorous or feed entirely or even dominantly on other fish (except perhaps the eels). Because of the shortage of fish-eating predators the native fish tend to lack defence mechanisms (McDowall, 1968). The introduced fish are the basis of important sport fisheries. Rapid growth, of for example *S. gairdnerii*, is found in many lakes and has been investigated in several of the Rotorua lakes (Smith, 1959) where the main food of the larger fish is smelt, together with eleotrids and freshwater crayfish and of the smaller fish, terrestrial and aquatic insects. There is some variation in growth between lakes and fish production appears to be correlated with the trophic status of the lake (especially temperature relations and oxygen content); production decreasing as the lakes become more eutrophic (Fish, 1968). There is an inverse relationship between numbers and weight, and in some other lakes, such as Lyndon, slower growth and smaller size has been attributed to high population numbers (especially of juveniles) and a harsh environment (Percival & Burnet, 1963).

Productivity and trophic status

High primary productivity values have been recorded in Tomahawk Lagoon, a shallow coastal lake near Dunedin in which blooms of blue-green algae (*Anabaena flos aquae*) and *Gymnodinium* occur (Mitchell, 1971). Mean rates of phytoplankton productivity for daylight hours (estimated by ^{14}C and oxygen light and dark bottle methods) were 157 mg C/m^3/hour (P max) and 741 mg C/m^2/day. Periods of high phytoplankton productivity during which rooted vegetation was sparse alternated with times of abundant growth of *Myriophyllum elatinoides* and *Ranunculus fluitans* when the phytoplankton productivity was much lower,

2.6 mg C/m³/hour (P max) and 25 mg C/m²/day during daylight hours. By contrast, in an Otago hydro-electric lake, Mahinerangi, the phytoplankton productivity (estimated by ^{14}C) was comparable to some of the least productive lakes in the temperate zone. Mean rates in daylight hours of 2.02 mg C/m³/hour (P max) and 76 C/m²/day were estimated. The low values were considered to be due to the physical conditions and changes in water level. In Lake Waipori, a very shallow lake on the same river system as Mahinerangi, the productivity was slightly higher per unit volume (3.32 mg C/m³/hour (P max)) but lower for the water column (26 mg C/m²/day). High biomasses of rooted vegetation are found particularly in some of the Rotorua lakes. *Lagarosiphon* dry weight values of 10 kg/m² were recorded in Lake Rotoiti, with 5.5 kg/m² in shallow water and 0.5 kg/m² in deep water in Lake Rotorua (Fish, 1963).

The zooplankton may indicate the trophic status of a lake not only by the relative organic dry weight (see p. 429) but also by the dominant groups present. From preliminary observations in some South Island lakes it appears that in oligotrophic lakes *Boeckella* is often dominant, with *Bosmina* less abundant; in mesotrophic lakes in which diatoms and green algae are sometimes abundant, *Bosmina* and *Ceriodaphnia* are usually dominant, with fewer copepods; in mesotrophic lakes particularly with green algae, *Ceriodaphnia* is most often dominant with *Bosmina* and *Boeckella* less conspicuous; and in eutrophic lakes *Bosmina*, *Ceriodaphnia* and copepods may all be abundant at different times of the year. Rotifers appear to show a general increase in number of species from oligotrophic to eutrophic lakes, but the same species may occur in a wide range of conditions.

The trophic status of a lake may depend partly on the situation and morphology, and partly on the extent of human influence on the lake and its catchment. Some New Zealand lakes are tentatively grouped according to their trophic condition in Table 10.

Many New Zealand lakes appear to be basically oligotrophic, probably at least partly related to the youth, geological history and nature of the rock. Small and shallow lakes are usually more productive, as are lakes situated close to centres of population or more intensive agriculture. Coastal lakes tend to be more productive than those further inland. Some lakes, such as the large glacial Nelson lakes, are apparently oligotrophic according to the chemical characteristics of the water and the plankton but support thick beds of rooted vegetation. Several lakes appear to be changing; in particular, recent unpublished research on Lake Hayes indicates increased eutrophication (Mitchell & Burns, pers. comm.).

A paleolimnological study of the moa swamp at Pyramid Valley, near Christchurch, indicates that the former lake was probably small and shallow as the commonest cladoceran *(Pleuroxus hastirostris)* is a species now found in shallow water among rooted vegetation (Deevey, 1955). The ostracods *Limnicythere duffi* and *L. mowbrayensis* alternate in abundance

Table 10. Trophic status of some New Zealand lakes

Eutrophic:
Pupuke
some Rotorua lakes (e.g. Okaro)
some sand dune lakes (especially Wellington series)
small coastal Kaikoura lakes
Forsyth
Tomahawk Lagoon
Mesotrophic:
some Rotorua lakes (e.g. Rotoehu)
some sand dune lakes (especially Northland series)
some small Canterbury lakes (e.g. Grasmere)
Hayes (1947–54) (?)
Oligotrophic:
some Rotorua lakes (e.g. Okataina)
Taupo
Waikaremoana
some sand dune lakes (?)
large glacial South Island lakes
Waitaki hydro-electric lakes
some small Canterbury lakes (e.g. Lyndon?)
some Westland lakes (e.g. Mapourika?)
Dystrophic:
Waikato peat lakes
Ngahewa (Rotorua) (Fish, 1968)
some Westland lakes (e.g. Matheson)

suggesting fluctuating conditions, and *Potamopyrgus* and other molluscs occur as well as the midge *Dryadotanytarsus duffi*, a member of a genus otherwise recorded only from Denmark, Greenland, Alaska and South Africa.

Ponds

The most numerous ponds in New Zealand are the small farm ponds excavated in pasture land throughout the country and mostly used for the watering of stock (and occasionally to encourage waterfowl). They are shallow and may become dry over the summer months. Three of these ponds, all at approximately the same height above sea level but at different latitudes and with different degrees of permanence are compared in the first three columns of Table 11. The most southern pond, on Saddle Hill, is permanent but has little rooted aquatic vegetation; the other two ponds both have abundant vegetation, but may become dry. Probably because of the presence of stock and their small size, all the ponds are often turbid and usually have a high content of nutrients,

Table 11. Ponds

Pond	Saddle Hill, Dunedin	Marley's Hill, Banks Peninsula	Ardmore, Auckland	Kettle-hole, near Coleridge
Description	Seepage farm pond	Seepage farm pond	Farm pond	Seepage kettle-hole pond
Origin	Landslide in 1929	Excavated about 1920	—	Glacial depression
Permanence	Permanent	Occasionally dries	Temporary	Dries once every 10–20 years
Maximum depth (m)	2.2– 3.5	1.0	1.1	1.6– 2.4
Water temperature (°C)	2.8–20.0 (ice)	3.0–30.0 (ice)	9.0–29.0	0.0–31.0 (ice)
Secchi disc readings (m)	1.0– 2.5	— (often turbid)	—	—
Oxygen in bottom water	Always well oxygenated	Low when water level low	Low when water level low	Always well oxygenated
pH	6.3– 7.0	6.0– 7.2	5.6– 6.4	6.6– 9.7
Nutrient content	Albuminoid nitrogen high, phosphate low	Nitrogen high	Nitrogen high soon after pond formed	Nitrogen usually high, phosphate sometimes high
Special chemical features	High chloride	High chloride	High chloride	—

Table 11. continued

Pond	Saddle Hill, Dunedin	Marley's Hill, Banks Peninsula	Ardmore, Auckland	Kettle-hole, near Coleridge
Phytoplankton				
– dominants	Green algae and a blue-green alga	Flagellates	*Eudorina*	Green algae
– less abundant	*Gloeocystis*, diatoms	Green algae		Blue-green algal bloom
Zooplankton				
– dominants	*Boeckella triarticulata*	*Daphnia carinata*	Ostracods	*Boeckella triarticulata*
– less abundant	*Chydorus sphaericus*, rotifers	*Boeckella triarticulata*, *Chydorus sphaericus*, rotifers	*Acanthocyclops vernalis*, *Simocephalus exspinosus*	*Chydorus sphaericus*, *Simocephalus vetulus*, rotifers
Rooted vegetation	*Glycera fluitans*, *Callitriche verna*	Abundant, *Callitriche verna*	Abundant, *Polygonum persicaria*	Abundant, *Myriophyllum elatinoides*
Fauna of rooted vegetation	*Potamopyrgus* (abundant), ostracods, leeches, oligochaetes, flatworms, *Chlorohydra*	Diverse, including insects (especially *Anisops*, *Sigara*, beetles), *Physastra*, frogs	Ostracods, insects (especially *Anisops*, *Sigara*, beetles), *Physa*, frogs	Abundant, ostracods, insects (especially damselflies, *Sigara*, beetles), *Lymnaea*
Bottom fauna	Oligochaetes, *Potamopyrgus*, *Sphaerium*, ostracods, chironomid larvae, beetle larvae	Oligochaetes, chironomid larvae, ostracods, *Sphaerium*	Oligochaetes, chironomid larvae	Oligochaetes, chironomid larvae, nematodes, *Sphaerium*
Main references	Byars, 1960	Stout, 1964, 1969a, unpub.	Barclay, 1966	Stout, 1969a

particularly nitrogen. The chloride content is also high in these three ponds, all of which are not far from the sea. The phytoplankton differs in the ponds, but is usually composed of green algae. Occasionally there are blooms of blue-green algae, and toxic algae have been reported in a number of farm ponds (Flint, 1966). The amount of rooted vegetation present strongly influences the ponds, particularly the diversity of the weed fauna. In Saddle Hill pond the vegetation is relatively sparse and there are few aquatic insects; in the other two ponds with abundant vegetation the fauna is similar in the two ponds, more diverse and in particular includes a variety of insects. The effect of latitude is seen in the temperatures of the ponds: the most northerly pond has the highest minimum temperature, and the southern pond has the lowest maximum temperature. The relative permanence of a pond may influence in particular the oxygen conditions and the zooplankton. The oxygen content of the water is often low when the depth decreases or a pond first fills with water. The zooplankton of the two more permanent ponds is dominated by *Boeckella triarticulata*, *Daphnia carinata*, *Chydorus sphaericus* and rotifers, but in the Ardmore pond the three former are noticeably absent and ostracods predominate. Barclay (1966) compared the Ardmore pond with other more permanent Auckland ponds and considered that the main differences were in the amount and type of rooted vegetation and in the greater abundance of Crustacea in temporary ponds. Other ponds on Marley's Hill with varying degrees of permanence have differences in the nutrient content of the water and an increasing number of species (together with some changes in kinds of species) in the more permanent ponds (Stout, 1964).

In parts of the South Island kettle-hole ponds are common, formed in depressions in mounds of glacial debris. They may be permanent or temporary, and have markedly fluctuating water levels, a wide temperature range and considerable seasonal differences in the chemical content of the water. A kettle-hole pond near Lake Coleridge (Table 11) is similar in size to the Saddle Hill pond, but at a higher altitude (approximately 600 m above sea level). In summer it has a dense growth of rooted vegetation and occasional algal blooms (with a blue-green algal bloom in one year), and the water becomes markedly alkaline. There is a rich weed fauna similar to the Marley's Hill and Ardmore ponds. *Boeckella triarticulata* is common, as in the other two more permanent ponds.

Ponds in old river beds, dammed small streams and artificial ponds in shingle pits are found in various parts of the country, and may have a rich flora and fauna depending on the age and size of the pond.

Springs and Subterranean waters

There are a number of large cold water springs in New Zealand. Waikoropupu Springs, in the north of the South Island, are among the largest known cold water springs in the world, with possibly the largest discharge in the southern hemisphere. The water has a constant temperature of 11.7 °C and is extremely clear (WELLS pers. comm.). There is some rooted vegetation. Western Springs, in Auckland, form a small lake which supports a rich growth of macrophytes; former beds of *Elodea canadensis* have now been almost replaced by the recently introduced *Egeria densa* (JOHNSTONE, 1972). There is a high nutrient inflow especially of nitrate and silicate (with values in the lake up to 4.4 g/m^3 and 20 g/m^3 respectively), originating partly from the springs and partly from waterfowl. There are many smaller springs and a number of lakes are to some extent spring-fed, including Rotorua, Grasmere and Hayes.

New Zealand has the only major thermal area in the southern hemisphere and there are numerous thermal springs especially in the centre of the North Island and along the alpine fault in the Southern Alps. The upper temperature limit for blue-green algae (principally *Mastigocladus laminosus*) in a number of North Island thermal waters is 60–65 °C, and for bacteria 99–101 °C in neutral and alkaline waters, but lower if the water is acid (BROCK & BROCK, 1971). These temperature limits for blue-green algae are similar to those found in Iceland where the same species occurs, but lower than those found in Yellowstone Park which has a different species. Several species of *Ephydrella* are found, as in thermal streams.

The wide plains of the shingle rivers, especially in Canterbury, Nelson and Hawkes Bay, contain extensive subterranean water sometimes rising to the surface in artesian wells. The waters have an interesting specialised fauna, usually small (1–2 mm), and often delicate, unpigmented, apparently without eyes and with special sensory organs. The group best represented are the Crustacea. Four species of isopods (*Cruregens fontanus* and three members of the Family Phreatoicidae) and three species of amphipods have been recorded from wells (CHILTON, 1894, 1909, NICHOLLS, 1944), and more recently Stygocaridacea and both families of Bathynellacea have been found in interstitial waters in Canterbury and Nelson (SCHMINKE & NOODT, 1968). A planarian, an annelid *(Phreodrilus subterraneus)* and *Potamopyrgus subterraneus* are also recorded from Canterbury wells (CHILTON, 1894). Studies currently being undertaken have found over 60 subterranean aquatic species, many of great zoogeographical interest, and including two species of turbellarians, at least six oligochaetes, thirty crustaceans (particularly amphipods), three dytiscid beetles, about twelve water mites, and ten molluscs (KUSCHEL pers. comm., CLIMO, this volume). A similar fauna has been found in limestone cave streams, with sometimes even the same species. Insects and mites which normally live mainly in the gravel beds of streams have been discovered underground, together with the strictly phreatic fauna, up to 150 m away from the river bed (KUSCHEL, pers. comm.).

Because of the youth of the country, New Zealand rivers often flow on relatively straight courses towards the sea; they frequently have uniformly stony conditions with a swift current and are liable to flooding. Vegetation, even moss, is usually sparse. There may be a greater density of invertebrates than in, for example, English rivers (PERCIVAL, 1932) but a smaller variety of genera, and in particular fewer chironomid larvae and water mites. The most abundant invertebrates in the predominantly stony river beds are probably the members of the Family Leptophlebiidae and *Coloburiscus humeralis* (Ephemeroptera); Trichoptera such as *Pycnocentria*, *Hydrobiosis* and *Hydropsyche;* the prosobranch *Potamopyrgus antipodarum* and beetles (possibly belonging to the genus *Hydora*) (PERCIVAL, undated). Chironomid larvae are less abundant, and more variable in occurrence are the mayfly *Nesameletus ornatus* (particularly found in swift currents), the neuropteran *Archichauliodes diversus* (often in silty conditions), simuliids of the genus *Austrosimulium* (in swift waters), other Trichoptera such as *Helicopsyche* and *Oxyethira* and Plecoptera such as *Zelandobius*. Rarer forms include turbellarians, tubificids and naiads, the freshwater crayfish *Paranephrops*, shrimp *Paratya*, amphipod *Paracalliope*, large primitive plecopteran *Stenoperla*, Anisoptera, Blepharoceridae and the corixid *Sigara*. Mayflies belonging to the Family Heptageniidae are absent, but New Zealand Leptophlebiids closely resemble Heptageniids in the flattened body and brushing mouth parts. *Archichauliodes* appears to replace *Sialis*. On the plains, where there is a more moderate flow and the river bed is more stable with some vegetation, shrimps, amphipods, ostracods, gastropods, and larvae of Trichoptera, Ephemeroptera, Chironomidae and Anisoptera are the most common; with small Sphaerid bivalves and numerous tubificids in the soft sediments (PERCIVAL, undated). The few large deep rivers, such as parts of the Waikato and the Clutha, have not been investigated.

In the North Island many of the streams have a fast-flowing section in hilly country and a slower flow across the plains. In the Wellington area mayflies and stoneflies are abundant in swift waters, with different species of mayflies according to the current speed (PHILLIPS, 1929). The fauna is more abundant when the stream is surrounded by terrestrial vegetation. The streams often contain populations of galaxiids, eleotrids and eels (MCDOWALL, 1964, 1965a and b). The Horokiwi stream, just north of Wellington (Fig. 1) has a bed of shingle, stones and boulders and is subject to floods (ALLEN, 1951). The catchment is mostly grassland and the upper part of the stream is in hilly country while the lower part crosses an alluvial valley. The water temperature is 5° to 21 °C and pH 6.9. The water has a relatively high silicate, nitrate and phosphate content, and 30 g $CaCO_3$ per m^3. Sericostomatidae (Trichoptera) and Leptoperlidae (Plecoptera) are abundant in the upper part of the stream, chironomid larvae and *Austrosimulium* in the middle reaches and *Potamopyrgus*, beetles and oligochaetes near the mouth. Several small native fish and *Anguilla dieffenbachi* are present, and the trout production is high.

The Hinaki and Hinau streams in the Wairarapa, with a current speed of 1 to 8 cusecs (except during floods), have beds of coarse gravel and large stones and flow through pasture. Ephemeroptera (especially leptophlebiids belonging to the genus *Deleatidium*), and Trichoptera (*Olinga feredayi* and *Pycnocentrodes* spp.) are the most abundant invertebrates, and crayfish *(Paranephrops planifrons)* are common (HOPKINS, 1970, 1971). The invertebrate fauna is more numerous and has a higher biomass (10.71–19.75 g/m²) than in the Horokiwi stream. Juvenile *Salmo trutta*, *Gobiomorphus breviceps*, *Anguilla dieffenbachi* and *A. australis schmidti* are present in the main waters of each stream and feed principally on Ephemeroptera, Trichoptera and Diptera with little interspecific competition. The fish production is 73.54 g/m² in the Hinaki and 24.17 g/m² in the Hinau, but only 14.28 g/m² in a branch of the Hinau containing *Galaxias divergens* and a few eels.

In the slower flowing North Island streams abundance of the fauna may depend on the amount of aquatic vegetation (PHILLIPS, 1929). A stream near Auckland flowing partly over limestone had a greater abundance of the trichopteran *Pycnocentria* compared with neighbouring streams (MCLEAN, 1966).

In the South Island, mountain torrents often flow through native beech *(Nothofagus)* and are particularly subject to flooding. They are usually spring fed with a moderately stable temperature (HELSON, 1934) and little algal growth. The fauna consists principally of the larvae of Plecoptera (*Stenoperla prasina*, *Zelandoperla* sp. and *Austroperla cyrene*), Ephemeroptera (leptophlebiids and *Nesameletus*), Trichoptera (such as *Olinga feredayi*), Blepharoceridae, *Austrosimulium* spp. and occasionally chironomid larvae and tipulids.

The streams in the foothills of the Southern Alps and on Banks Peninsula are also frequently subject to sudden floods and the bed is variable with waterfalls, rapids and deep pools (HELSON, 1934, HAMILTON, 1940, WISELY, 1962, STOUT, 1969b). There is often little rooted vegetation but diatoms and filamentous green algae may be apparent. There is a diverse fauna with the larvae of several Ephemeroptera (*Coloburiscus humeralis*, *Atalophlebia versicolor*, *Deleatidium* sp., *Ichthyobotus bicolor*, *Nesameletus ornatus* and *Ameletopsis perscitus*), Plecoptera (*Stenoperla prasina* and three others), Trichoptera (*Olinga feredayi*, *Hydrobiosis umbripennis*, *Hydropsyche colonica*, *Helicopsyche* sp. and *Pycnocentria erecta*), Blepharoceridae *(Neocurupira chiltoni)*, Chironomidae, Tipulidae, *Austrosimulium tillyardi*, Coleoptera, Zygoptera *(Xanthocnemis zelandica)* and several kinds of fish. *Archichauliodes diversus* (HAMILTON, 1940) and the shrimp *Paratya curvirostris* are found particularly in the lower reaches. *Potamopyrgus antipodarum* and *Sphaerium novaezelandiae* are also present.

Fig. 10. The Waitaki River, a large shingle river in the South Island. (Whites Aviation Ltd.)

Shingle rivers are prominent in the South Island, especially in the east (Fig. 10). In the large rivers, such as the Waimakariri, Rakaia, Rangitata and Waitaki (Fig. 6), the bed is very unstable and animals are few due to the movement of the bed, swift current and frequent flooding. There is little growth of moss. The dominant invertebrates are usually *Deleatidium* spp. and a few Trichoptera. Conditions may change markedly in a short distance. When the current is slightly slower, there is increased algal growth and deposition of some silt, and chironomid larvae, tubificids and sometimes tipulids, occur (HIRSCH, 1958). The Cass River (Fig. 6, Inset), at an altitude of approximately 600 m, is smaller but has little vegetation and the dominant invertebrates are the larvae of *Austrosimulium tillyardi*, *Coloburiscus humeralis*, *Oniscigaster distans* (Ephemeroptera), and the caddisflies *Pseudoeconesus mimus*, *Triplectides obsoleta* and *Hydrobiosis umbripennis* (HILGENDORF, 1918).

On the Canterbury plains the upper part of the Selwyn River (which flows into Lake Ellesmere) may have a rich growth of diatoms and *Ulothrix*, and abundant *Pycnocentria* and leptophlebiids, with smaller numbers of chironomid larvae, beetles, tubificids and other caddisflies (PERCIVAL, 1932). In the lower reaches of this river Percival recorded that the fauna was 30 to 55% leptophlebiids and 26–51% Trichoptera (chiefly *Pycnocentria*). *Paracalliope fluviatilis* occurred in pools and areas of slow current and *Coloburiscus humeralis* in the rapids.

Two small branches join the Waimakariri River near its mouth. Parts of the North Branch have a bed of coarse shingle with a growth of diatoms during spring and *Pycnocentria*, *Olinga*, Ephemeroptera, Coleoptera, Plecoptera and *Potamopyrgus antipodarum* (PARROTT, 1929). Other areas have a stable bed with some rooted vegetation and the most abundant invertebrates are *Olinga feredayi* and *P. antipodarum*, with moderate numbers of leptophlebiids, *Helicopsyche* and *Pycnocentrodes*, Hydropsychidae, *Hydora*, Chironomidae, *Paracalliope fluviatilis* and Tubificidae (HIRSCH, 1958). Much of the South Branch has a stable bed with a similar fauna; but there are also regions of slower current with abundant Mollusca (particularly *P. antipodarum* but also *Physastra variabilis* and *Gyraulus corinna*), together with Zygoptera, chironomid and caddisfly larvae. A study of the interrelationships between *Salmo trutta* and the two species of *Anguilla* in this branch showed an increase in abundance but decrease in individual weight of the trout when the number of eels was greatly reduced (BURNET, 1968, 1969).

Further south, the Kakanui River in Otago (MCLAY, 1968) and the Oreti River in Southland (PERCIVAL, 1932) both have coarse gravel beds and flow through farmland, while the Silverstream branch of the Taieri River has a high proportion of silt (SCOTT, 1966).

The thermal streams in the Taupo and Rotorua areas may reach temperatures of at least 45 °C and develop an algal felt with a fauna principally composed of insects, especially Ephydridae and Hydrophilidae, but including eight phyla of macroscopic invertebrates (WINTERBOURN & BROWN, 1967, WINTERBOURN, 1968, 1969a and b).

Possibly the first major human influence on the streams and rivers was the replacement of the original forest cover by pasture land (STOUT, 1973). This led to reduced stability with increased flooding which reduced the fauna (ALLEN, 1960). In small streams the increased amount of light encouraged algal growth with an increase in the animals feeding or living on the algae. More intensive agriculture has apparently often led to an increase of nutrients and a change in the kinds and amount of algae. In the Horokiwi stream the thin algal film was replaced by a diatomaceous mat, and the mayflies and sericostomatid caddisflies were mostly replaced by beetles, *Potamopyrgus antipodarum* and ostracods. Pollution by organic wastes especially from farm discharges, meatworks and dairy factories is widespread in New Zealand and there are pronounced seasonal differences in the amount of discharge. The effects of different degrees of pollution on the fauna has been compared in stable, stony streams (HIRSCH, 1958). The application of DDT to the catchment of a stream caused the almost complete elimination of the amphipods and a reduction in the number of insects, but no change in the annelids or molluscs (HOPKINS *et al.*, 1966). The amount of invertebrate drift at first greatly increased, then fell to almost nothing. Twelve months after the application, there was a return to normal densities but a change in species composition, particularly an increase in *Austrosimulium* and a replacement of *Deleatidium* by grypopteryid stoneflies as the dominant group.

The introduction of trout has probably affected the native fauna. An early study of streams in the Rotorua and Taupo area suggested that 80% of the original plecopteran fauna and 90% of the Trichoptera had been destroyed by trout and the balance of the insect fauna in this region upset, but that South Island rivers were, at that time, much less affected (TILLYARD, 1920). It is difficult to assess the effect of the trout because of parallel changes in the condition of the rivers (ALLEN, 1961) and both have probably been responsible for the extinction of the New Zealand grayling, *Prototroctes oxyrhynchus* (Family Aplochitonidae) and the decline in numbers of some other native fish (McDOWALL, 1968).

Relationships with other Countries

New Zealand freshwaters have some features which they share in common with similar habitats in many other parts of the world; others more distinctly related to those of other southern hemisphere countries, particularly Australia and South America; and some particularly characteristic of New Zealand. Thus, for example, the lakes formed due to glaciation or volcanic activity have relationships with lakes of similar origins elsewhere, such as the English glacially-formed lakes. Many of the plants and animals belong to widespread groups found also in the northern hemisphere, particularly the smaller organisms which may be more readily transported. Many of the algae, protozoa and rotifers apparently belong to widely distributed species, as does *Chlorohydra viridissima*. There are a number of almost cosmopolitan species among the freshwater oligochaetes, especially in the Tubificidae and Naididae, in which most of the genera and several of the species occur widely in other parts of the world; and the only New Zealand representative of the Lumbriculidae is the introduced *Lumbriculus variegatus* (BRINKHURST & JAMIESON, 1971). Several genera of molluscs *(Lymnaea, Gyraulus)* and many genera of the smaller Crustacea are cosmopolitan. In the Cladocera, the genera are often widespread *(Daphnia, Bosmina)* but not the species, and New Zealand lacks any representatives of the northern hemisphere Polyphemidae or *Leptodora*. Of the copepods, the Cyclopoida mainly belong to cosmopolitan genera, and include some cosmopolitan species, such as *Macrocyclops albidus* and *Mesocyclops leuckarti;* and the harpacticoids are probably often cosmopolitan, at least at the generic level, the cosmopolitan *Phyllognathus* and *Elaphoidella* having been reported. The ostracods include a few cosmopolitan species, such as *Cypridopsis vidua*. Among the insects, in the Chironomidae for example many species have

Palaearctic affinities (FREEMAN, 1959). Hydrachnellid genera are often cosmopolitan and New Zealand has at least one species *(Piona uncata)* which is widespread in other parts of the world.

SOUTHERN HEMISPHERE COUNTRIES

A number of groups of freshwater organisms found in New Zealand are particularly characteristic of the southern temperate zone. The flatworm genus *Temnocephala* occurs principally in the southern hemisphere; while the oligochaete family Phreodrilidae occurs elsewhere only in Ceylon (BRINKHURST & JAMIESON, 1971). The freshwater bivalve family Hydriidae (including *Hyridella* and *Cucumerunio*) is essentially southern hemisphere, and the prosobranch family Hydrobiidae (including *Potamopyrgus*) is probably of southern origin (MCMICHAEL, 1967). Among the Crustacea, *Daphnia carinata*, and the genera *Boeckella* (Calanoida), *Attheyella (Delachauxiella)* (BREHM, 1956), *A. (Chappuisiella)* and *Antarctobiotus* (Harpacticoida), *Newnhamia* and *Gomphocythere* (Ostracoda) and *Chiltonia* (Amphipoda) are southern hemisphere. So is the decapod family Parastacidae (BISHOP, 1967) although the New Zealand genus *Paranephrops* is endemic. Of the aquatic insects, the New Zealand members of the Subfamily Podonominae (Chironomidae) are related to forms in Australia and South America, and the plecopteran Family Eusteniidae is southern. The southern hemisphere fish family Galaxiidae has one species, *Galaxias maculatus*, in Australia, New Zealand and South America (MCDOWALL, 1970a). The evidence suggests that several groups may have had a southern origin, with a possible connection by means of Antarctica, or else be relict forms.

Australia

Because of the low rainfall (especially in the interior), many Australian lakes are saline with a high total ionic content and alkalinity and a predominance of sodium and chloride ions (WILLIAMS, 1967, WILLIAMS & WAN, 1972). Phosphorus and nitrogen are also high, except in the few glacially-formed lakes. In New Zealand, a few small Otago lakes are saline and show several similarities to the Australian lakes (BAYLY, 1967b). Freshwater lakes are found in the cooler and more humid regions of Australia, such as Tasmania and parts of Victoria and New South Wales (WILLIAMS, 1967). Those in Tasmania have a low ionic concentration, calcium content (less than 10 g/m^3) and alkalinity and a pH of 6.0 to 8.0 (WILLIAMS, 1964) and are similar in a number of respects to several South Island lakes. Volcanic lakes in southeastern Australia are comparable with Lake Aroarotamahine, but have a higher sodium content (BAYLY & WILLIAMS, 1964). Other freshwater lakes are often either ephemeral or artificial. Australian lakes are often warm monomictic, as in New Zealand, and there is an absence of dimictic lakes, although (as in New Zealand) some mountain lakes become frozen in winter but remain unstratified in summer (WILLIAMS & WAN, 1972).

Australia has a large proportion of endemics in the fauna and has acted as a refuge (WILLIAMS, 1965, WILLIAMS & WAN, 1972). Also, as in New Zealand, several species occur in ecological niches which they would not normally occupy and some common cosmopolitan groups (such as *Asellus*) are absent. The inland saline lakes have a distinctive fauna (BAYLY, 1967b) characterised particularly by centropagid copepods, with *Gladioferens* in brackish water and *Boeckella* in freshwater, both genera being also present in New Zealand. Faunal relationships are often strongest between southeastern Australia and New Zealand. Because of the west wind drift and the fact that Australia is the nearest country to New Zealand, there appears to have been transoceanic dispersal by both the sea and wind of the members of several groups. The oligochaete family Haplotaxidae is relatively abundant in Australia and New Zealand, although it also occurs in Eurasia (BRINKHURST & JAMIESON, 1971). There are several similarities between the crustacean fauna of the two countries, although New Zealand has many fewer Branchiopoda (no Anostraca and only one record of Conchostraca). There is a strong resemblance between the Cladocera with the same genera and often the same species, such as *Neothrix armata*, *Pseudomoina lemnae*, and *Saycia orbicularis* (Macrothricidae) (BREHM, 1956). The two main genera of freshwater calanoids, *Calamoecia* and *Boeckella* are found in both countries, which have six species in common (*C. lucasi*, *B. triarticulata*, *B. propinqua*, *B. delicata*, *B. symmetrica* and *B. minuta*, BAYLY, 1961, 1964, 1967a; VIDAL, 1968). Both New Zealand species of *Gladioferens* are found in Australia (BAYLY, 1963a) and the cyclopoid *Diacyclops crassicaudoides* together with the ostracod *Newnhamia fenestrata* occur in both countries. Phreatoicoidae are common in Tasmania and New Zealand, and Bathynellacea are found in both. The Hymenostomatid crab *Halicarcinus lacustris* occurs in southeastern Australia and the North Island of New Zealand. Among the insects at least two families of Trichoptera (Philanisidae, Philorheithridae), one of Plecoptera (Austroperlidae) and a genus of Odonata *(Uropetala)* are distinct to New Zealand and Australia and the members of the Subfamily Podonominae are more closely related to each other than they are to the South American forms. Two subfamilies of bivalves are found only in Australasia, and the Hydrobiid (Prosobranch) genus *Potamopyrgus* is particularly well developed in southeastern Australia and New Zealand and probably originated in this region (MCMICHAEL, 1967); *Lymnaea tomentosa* and the freshwater limpet genus *Ferrissia* occur in the same area. There are several Australasian genera of freshwater fish in the families Galaxiidae (which probably evolved in Australasia, MCDOWALL, 1970a), Retropinnidae, Aplochitonidae and Eleotridae (MCDOWALL, 1964).

On the west coast of southern South America, particularly in Chile, are a number of large glacially-formed lakes and these are at latitudes similar to that of New Zealand. Some of the large lakes in southern Chile are predominantly cryptodepressions, with steep sides and depths up to 438 m (LÖFFLER, 1960, THOMASSON, 1963, 1964). The lakes are often in pasture land and, like many particularly South Island New Zealand lakes, have an increased inflow in spring due to melting snow and ice and precipitation, and contain introduced salmonids. They are in a region of strong winds and are usually warm monomictic with the thermocline often at 20 to 40 m or polymictic, although a few dimictic lakes have been found. The winter temperatures of 8° to 10°C are similar to those of some large New Zealand lakes. The water is usually clear (with Secchi disc depths recorded up to 25 m) with a low ionic concentration particularly of phosphorus, and well oxygenated. The conductivity (at 18°C) is mostly 20–50 μ S/m, although values as high as 122 μ S/m have been recorded. As with New Zealand lakes, those in South America tend to show a deviation from the standard composition of RODHE (1949) with the sodium, potassium and sometimes magnesium content high. Silica is usually more than 15 g/m^3 and may be as high as 60 g/m^3. Nitrate is usually low and, also as in New Zealand, so are the trace elements.

The lakes of Tierra del Fuego and Valdivia investigated by THOMASSON (1957) occasionally had high conductivity (to 163 μ S/m), phosphorus and nitrogen values.

The phytoplankton in these temperate zone South American lakes tends to be cosmopolitan, with some general similarities to that of New Zealand lakes particularly in the dominance of certain groups but not in the species relationships (LÖFFLER, 1962, THOMASSON, 1963). There is a similar desmid flora (THOMASSON, 1955) and often a predominance of diatoms and green algae (THOMASSON, 1964). The rooted aquatic vegetation is often similar to that of the northern hemisphere (THOMASSON, 1963).

The zooplankton present is related to the paleogeography and the climatic, or thermal, type of the lake (LÖFFLER, 1958). The rotifers are similar to those of the northern hemisphere, but the Crustacea are usually different (THOMASSON, 1963). The warm monomictic lakes contain *Daphnia*, *Diaphanosoma*, *Bosmina* (Cladocera), *Boeckella* and *Diaptomus*, and the polymictic lakes particularly *Daphnia* (LÖFFLER, 1958). They thus have three genera in common with, and two *(Diaphanosoma, Diaptomus)* different from those in New Zealand. Pelagic cyclopoid copepods are rare in the polymictic lakes, as they are in most New Zealand lakes. The Cladocera *Ceriodaphnia dubia*, *Simocephalus vetulus*, *Alona guttata*, *A. affinis*, and *Chydorus sphaericus*, several species of cyclopoid copepods and the genus *Boeckella*, but no calanoid species, are common to both countries, while South America has several genera not found in New Zealand (LÖFFLER, 1962). The species in common are mostly essentially cosmopolitan in distribution, while each country has its own group of endemic species. The profundal fauna shows a general resemblance to that of the Holarctic, and the deep lakes contain a *Tanytarsus* community (BRUNDIN, 1958, THOMASSON, 1964).

The general fauna of the temperate zone of the Andes region shows relationships with that of Australia and New Zealand (FITTKAU, 1969) and many of the endemic species are most closely related to Australasian groups, especially among the aquatic insects, of which the best known are the Plecoptera and Chironomidae. South America has also had a greater invasion of species from the north. However, in common with New Zealand, there are five species of Cladocera and several cyclopoids (although none are restricted to the two regions), the genus *Boeckella*, two species of *Herpetocypris* (Ostracoda) and the Stygocaridacea (Syncarida). The running water Hydrachnellae are related to New Zealand and other southern hemisphere forms, and the genus *Euwandesia* (Hydryphantidae) occurs in both countries (BESCH, 1964). Cold stenothermal insects are related, such as the Plecoptera (ILLIES, 1969), the Subfamily Podonominae (Chironomidae) which has related groups in the two countries, and the Subfamily Diamesinae in which the New Zealand forms are more closely linked with those of South America than with Australian species (BRUNDIN, 1966). *Galaxias maculatus* also occurs in South America and Australasia.

Southern Africa

Most of the southern part of Africa has a low rainfall with pronounced seasonal differences, and water is scarce with small young rivers which often flood (HUTCHINSON *et al.*, 1932). The few lakes include one glacially formed and several coastal lakes, together with farm dams and reservoirs which show pronounced water level fluctuations and often become dry. There are also a number of pans, which have flat and usually circular basins probably formed by wind erosion. They are found both inland and in the coastal region among sand dunes. The water is turbid, and often rich in nutrients with high chloride, silica, and sulphate content. Many are dystrophic, and some alkaline eutrophic. Because the pans are small they have little true plankton. Most of the species are cosmopolitan or with tropical affinities, but some are endemic including all the 'phyllopods', four species of Cladocera, and many of the copepods and ostracods. In western Cape Province there are some shallow lakes associated with sand dunes (HARRISON, 1962). They do not stratify and have a high chloride content, with green and blue-green algae belonging to widespread species. Coastal lakes in northern Zululand are brackish, subtropical, warm and shallow with a high chloride and sodium content (ALLANSON & VAN WYK, 1969).

The freshwater invertebrate fauna includes an 'old element' with relationships to Australasia and the Gondwanaland fauna (HARRISON, 1965). This 'old element' is now cold stenothermal, mostly montane and living in the southern tip of the continent or in refuges. It is a small proportion of the fauna and includes *Phreatoicus*, four genera of Notonemourinae (Plecoptera), two genera of Leptophlebiidae (Ephemeroptera) and six genera of Hydrachnellae. Most of the freshwater fauna has northern affinities. The Cladocera include *Daphnia carinata*, *Simocephalus vetulus* and several other species belonging to cosmopolitan genera (BREHM, 1958). The amphipod *Chiltonia* also occurs in Australasia.

Endemic groups

The endemic flora and fauna includes some relict forms. There are several endemic species of rooted aquatic vegetation (MASON, 1969), some rotifers (BREHM, 1956, RUSSELL, 1960), oligochaetes (BRINKHURST & JAMIESON, 1971) and an endemic family (Lattidae) and several species of molluscs (PONDER, 1964). *Simocephalus obtusatus* is possibly endemic and there are probably a number of endemic genera of harpacticoids. Endemic forms include *Boeckella hamata* and *B. dilatata* (BAYLY, 1964), a number of species of ostracods and the genus *Paranephrops* (HOPKINS, 1970); four genera of Chironomidae (FREEMAN, 1959); most species of Hydrachnellae and at least three genera of running water forms (*Zelandobates*, *Tryssaturus* and *Uralbia*, HOPKINS, 1966, 1967), and all except four species of native freshwater fish (McDOWALL, 1964, 1970b).

Conclusion

The physical and chemical conditions in New Zealand freshwaters are related to the nature and situation of the country. Many of the lakes are either glacially-formed and oligotrophic, with a deep thermocline or homothermous, a high light penetration, well oxygenated water and a low ionic content; or formed by volcanic activity, oligotrophic to eutrophic, usually warm monomictic, sometimes with a higher concentration of nutrient ions and occasionally with hypolimnetic oxygen depletion. Other lakes include a number of small coastal lakes with high concentrations of certain ions and more often eutrophic, several dystrophic lakes and hydro-electric impoundments. The rivers are frequently fast-flowing with shingle beds.

The biological features are influenced both by the physical and chemical conditions and by the isolation and geographical position of New Zealand. The flora and fauna which is present is dependent on an ability to reach the country and on the composition of the flora and fauna of the neighbouring land masses. In many groups of organisms the number of species is relatively small. For example there are apparently fewer genera and species of freshwater oligochaetes than in equivalent regions in the northern hemisphere (BRINKHURST & JAMIESON, 1971) and fewer Cladocera. The zooplankton has few species, with almost no predators including a noticeable absence of predatory Cladocera or Diptera. The species present may be very variable and occupy a wider ecological niche than their counterparts in other countries. The majority are found in both the North and South Islands; and examples of different species of a genus occurring in different parts of the country (as in *Boeckella*) are not common.

New Zealand appears to have been a refuge for some freshwater organisms including deep water oligochaetes, some aquatic insects, the fauna of underground waters such as phreatoicids and several groups of fish, including galaxiids.

A number of groups are absent or poorly represented in the flora or fauna. These include freshwater Anostraca and Conchostraca, Polyphemidae, *Leptodora*, *Diaptomus*, *Asellus*, insects living on the water surface (such as *Nepa*, *Ranatra* and Gyrinidae) and several groups of stream insects. Sometimes a common northern hemisphere animal is replaced by a member of another group in New Zealand: thus to some extent *Diaptomus* is replaced by *Boeckella*, *Gammarus* by *Paracalliope*, the pulmonate *Lymnaea* by the prosobranch *Potamopyrgus*, Heptageniidae by the Leptophlebiidae, and *Sialis* by *Archichauliodes*.

Groups whose relationships are primarily cosmopolitan include the algae, Protozoa and Rotifera (with a few endemic species). The oligochaetes are cosmopolitan and Australasian (BRINKHURST & JAMIESON, 1971) whereas the Mollusca are cosmopolitan, southern hemisphere and endemic (PONDER, 1964). The Cladocera are cosmopolitan and Australasian, the calanoid Copepoda southern hemisphere, Australasian and endemic, the cyclopoids mostly cosmopolitan and the harpacticoids cosmopolitan with one (and possibly more) southern hemisphere genus. The Ostracoda are often endemic, the Amphipoda not usually cosmopolitan, the Phreatoicidae mostly southern hemisphere, the stygocarid Syncarida related to South American forms and the Decapoda often southern hemisphere (with an endemic genus). The aquatic insects have both cosmopolitan and southern hemisphere affinities, and in the Hydrachnellae the genera are usually cosmopolitan and the species endemic. The fish belong principally to endemic species but to more widely distributed genera. There is thus a relatively pronounced cosmopolitan element together with some stronger relationships with southern hemisphere countries, particularly Australia and South America.

Recent changes have been caused principally by human interference, mainly in the form of pollution, especially from agricultural sources; introductions of exotic macrophytes, two species of snails (*Lymnaea stagnalis* and *Physa fontinalis*) and fish; and the use of lakes and rivers for water supplies (including irrigation), the production of hydro-electricity, and recreation (including fishing, swimming and boating).

Thus the limnology of New Zealand is related to that of similar habitats in other parts of the world, but reflects the influence of the peculiar characteristics of the country.

Acknowledgements

I am grateful to the Chemistry Division of the D.S.I.R. in Christchurch for some chemical analyses and to Dr. E. A. FLINT of Ecology Division, D.S.I.R. for algal identifications. Part of this work was financed by research grants from the New Zealand University Grants Committee and the University of Canterbury. Their help is gratefully acknowledged.

REFERENCES

ALLANSON, B. R. & VAN WYK, J. D. 1969. An introduction to the physics and chemistry of some lakes in northern Zululand. *Trans. Roy. Soc. S. Afr.* 38: 217–240.

ALLEN, K. R. 1951. The Horokiwi Stream: a study of a trout population. *Fish. Bull. N.Z. Mar. Dept.* 10: 238 pp.

ALLEN, K. R. 1960. Effect of land development on stream bottom faunas. *Proc. N.Z. Ecol. Soc.* 7: 20–21.

ALLEN, K. R. 1961. Relations between Salmonidae and the native freshwater fauna in New Zealand. *Proc. N.Z. Ecol. Soc.* 8: 66–70.

ARMSTRONG, J. S. 1935. Notes on the biology of Lake Taupo. *Trans. Roy. Soc. N.Z.* 65: 88–94.

BARCLAY, M. H. 1966. An ecological study of a temporary pond near Auckland, New Zealand. *Aust. J. Mar. Freshw. Res.* 17: 239–258.

BARKER, M. A. 1966. The influence of physico-chemical factors on the distribution of *Chironomus zealandicus* in Lake Pupuke. *Tane* 12: 93–95.

BARKER, M. A. 1970. Physico-chemical features of Lake Pupuke, Auckland. *N.Z. J. Mar. Freshw. Res.* 4: 406–430.

BAYLY, I. A. E. 1961. A revision of the inland water genus *Calamoecia* (Copepoda: Calanoida). *Aust. J. Mar. Freshw. Res.* 12: 54–91.

BAYLY, I. A. E. 1962. Ecological studies on New Zealand lacustrine zooplankton with special reference to *Boeckella propinqua* Sars (Copepoda: Calanoida). *Aust. J. Mar. Freshw. Res.* 13: 143–197.

BAYLY, I. A. E. 1963a. A revision of the coastal water genus *Gladioferens* (Copepoda: Calanoida). *Aust. J. Mar. Freshw. Res.* 14: 194–217.

BAYLY, I. A. E. 1963b. Reversed diurnal vertical migration of planktonic Crustacea in inland waters of low hydrogen ion concentration. *Nature, Lond.* 200: 704–705.

BAYLY, I. A. E. 1964. A revision of the Australasian species of the freshwater genera *Boeckella* and *Hemiboeckella* (Copepoda: Calanoida). *Aust. J. Mar. Freshw. Res.* 15: 180–238.

BAYLY, I. A. E. 1967a. A new species of *Boeckella* (Copepoda: Calanoida) and additional comments on some other species of the genus. *Pap. Proc. Roy. Soc. Tasm.* 101: 97–101.
BAYLY, I. A. E. 1967b. The fauna and chemical composition of some athalassic saline waters in New Zealand. *N.Z. J. Mar. Freshw. Res.* 1: 105–117.
BAYLY, I. A. E. & WILLIAMS, W. D. 1964. Chemical and biological observations on some volcanic lakes in the south-east of South Australia. *Aust. J. Mar. Freshw. Res.* 15: 123–132.
BEATTIE, J. H. 1945. Maori lore of lake, alp and fiord: folklore, fairytales, traditions and place names of the scenic wonders of the South Island. Otago Daily Times and Witness Newspaper Co. Ltd, Dunedin: 150 pp.
BENHAM, W. B. 1903. A note on the Oligochaeta of the New Zealand lakes. *Trans. Proc. N.Z. Inst.* 36: 192–198.
BENHAM, W. B. 1904. Some new species of aquatic Oligochaeta from New Zealand. *Proc. Zool. Soc. Lond.* 2: 202–232.
BESCH, W. 1964. Systematik und Verbreitung der südamerikanischen rheophilen Hydrachnellen. *Beitr. z. Neotrop. Fauna* 3: 77–193.
BISHOP, J. A. 1967. The zoogeography of the Australian freshwater Decapod Crustacea. In: A. H. WEATHERLEY (ed.). Australian inland waters and their fauna. Australian National University Press, Canberra: 107–122.
BOTTOMLEY, G. A. 1956. Seiches on Lake Wakatipu, New Zealand. *Trans. Roy. Soc. N.Z.* 83: 579–587.
BREHM, V. 1956. Süsswasserfauna und Tiergeographie. *Öst. Zool. Z.* 6: 250–269.
BREHM, V. 1958. Crustacea: Phyllopoda und Copepoda Calanoida. *South African Animal Life, Uppsala* 5: 10–39.
BRIGGS, M. H. 1962a. The presence of free sugars, peptides and amino acids in filtered lake waters. *Life Sciences* 8: 377–380.
BRIGGS, M. H. 1962b. Further organic constituents of filtered lake waters. *Life Sciences* 11: 649–651.
BRINKHURST, R. O. 1971. The aquatic Oligochaeta known from Australia, New Zealand, Tasmania, and the adjacent islands. *University of Queensland Papers, Department of Zoology* 3: 99–128.
BRINKHURST, R. O. & JAMIESON, B. G. M. 1971. Aquatic Oligochaeta of the world. Oliver and Boyd, Edinburgh: 860 pp.
BROCK, T. D. & BROCK, M. L. 1971. Microbiological studies of thermal habitats of the central volcanic region, North Island, New Zealand. *N.Z. J. Mar. Freshw. Res.* 5: 233–258.
BRODIE, J. W. & IRWIN, J. 1970. Morphology and sedimentation in Lake Wakatipu, New Zealand. *N.Z. J. Mar. Freshw. Res.* 4: 479–496.
BRUNDIN, L. 1958. The bottom faunistical lake type system and its application to the southern hemisphere. Moreover a theory of glacial erosion as a factor of productivity in lakes and oceans. *Verh. Int. Ver. Limnol.* 13: 288–297.
BRUNDIN, L. 1966. Transantarctic relationships and their significance, as evidenced by chironomid midges with a monograph of the subfamilies Podonominae and Aphroteniinae and the austral Heptagyiae. *Kungl. Svenska Vet. Handl.* 11, 1: 472 pp.
BURNET, A. M. R. 1968. A study of the relationships between brown trout and eels in a New Zealand stream. *Fish. Tech. Rep. N.Z. Mar. Dept.* 26: 27 pp.
BURNET, A. M. R. 1969. A study of the inter-relation between eels and trout, the invertebrate fauna and the feeding habits of the fish. *Fish. Tech. Rep. N.Z. Mar. Dept.* 36: 23 pp.
BURNET, A. M. R., CRANFIELD, H. J. & BENZIE, V. L. 1969. The Freshwater Fishes. In: G. A. KNOX (ed.). The Natural History of Canterbury. Reed, Wellington: 498–508.
BYARS, J. A. 1960. A freshwater pond in New Zealand. *Aust. J. Mar. Freshw. Res.* 11: 222–240.
CASSIE, V. 1969. Seasonal variation in phytoplankton from Lake Rotorua and other inland waters, New Zealand, 1966–67. *N.Z. J. Mar. Freshw. Res.* 3: 98–123.

CHAPMAN, V. J. 1970. A history of the lake-weed infestation of the Rotorua lakes and the lakes of the Waikato hydroelectric system. *N.Z. Dept. Sci. Industr. Res. Inf. Ser.* 78: 52 pp.
CHAPMAN, V. J., BROWN, J. M. A., DROMGOOLE, F. I. & COFFEY, B. T. 1971a. Submerged vegetation of the Rotorua and Waikato lakes. 1. Lake Rotoiti. *N.Z. J. Mar. Freshw. Res.* 5: 259–279.
CHAPMAN, V. J., COFFEY, B. T. & BROWN, J. M. A. 1971b. Submerged vegetation of the Rotorua and Waikato lakes. 2. 'Cyclic change' in Lake Rotoiti. *N.Z. J. Mar. Freshw. Res.* 5: 461–482.
CHILTON, C. 1894. The subterranean Crustacea of New Zealand: with some general remarks on the fauna of caves and wells. *Trans. Linn. Soc. Lond. second series, Zool.* 6: 163–284.
CHILTON, C. 1909. The freshwater Amphipoda of New Zealand. *Trans. N.Z. Inst.* 41: 53–59.
COX, B. G. 1965. Oscillations of Lake Wakatipu. *Trans. Roy. Soc. N.Z., Gen.* 1: 183–190.
CUNNINGHAM, B. T., MOAR, N. T., TORRIE, A. W. & PARR, P. J. 1953. A survey of the western coastal dune lakes of the North Island, New Zealand. *Aust. J. Mar. Freshw. Res.* 4: 343–386.
CUSSEN, L. 1887. Notes on the physiography and geology of the King Country. *Trans. Proc. N.Z. Inst.* 20: 316–332.
DEEVEY, E. S. 1955. Palaeolimnology of the Upper Swamp Deposit, Pyramid Valley. *Rec. Cant. Mus.* 6: 291–344.
FISH, G. R. 1963. Observations on excessive weed growth in two lakes in New Zealand. *N.Z. J. Bot.* 1: 410–418.
FISH, G. R. 1966a. An artificially maintained trout population in a Northland lake. *N.Z. J. Sci.* 9: 200–210.
FISH, G. R. 1966b. Some effects of the destruction of aquatic weeds in Lake Rotoiti, New Zealand. *Weed Res.* 6: 350–358.
FISH, G. R. 1968. An examination of the trout population of five lakes near Rotorua, New Zealand. *N.Z. J. Mar. Freshw. Res.* 2: 333–362.
FISH, G. R. 1969a. The oxygen content of some New Zealand lakes. *Verh. Int. Ver. Limnol.* 17: 392–403.
FISH, G. R. 1969b. Lakes: the value of recent research to measure eutrophication and to indicate possible causes. *J. Hydrol. (N.Z.).* 8: 77–85.
FISH, G. R. 1970. A limnological study of four lakes near Rotorua. *N.Z. J. Mar. Freshw. Res.* 4: 165–194.
FISH, G. R. 1971. *Craspedacusta sowerbyi* Lankester (Coelenterata: Limnomedusae) in New Zealand lakes. *N.Z. J. Mar. Freshw. Res.* 5: 66–69.
FISH, G. R. & CHAPMAN, M. A. 1969. Synoptic surveys of Lakes Rotorua and Rotoiti. *N.Z. J. Mar. Freshw. Res.* 3: 571–584.
FISH, G. R. & WILL, G. M. 1966. Fluctuations in the chemical composition of two lake weeds from New Zealand. *Water Res.* 6: 346–349.
FITTKAU, E. J. 1969. The fauna of South America. In: E. J. FITTKAU, J. ILLIES, H. KLINGE, G. H. SCHWABE, H. SIOLI (eds.). Biogeography and ecology in South America, Junk, The Hague, 2: 624–658.
FLINT, E. A. 1938. A preliminary study of the phytoplankton in Lake Sarah (New Zealand). *J. Ecol.* 26: 353–358.
FLINT, E. A. 1966. Toxic algae in some New Zealand freshwater ponds. *N.Z. Vet. J.* 14: 181–185.
FLINT, E. A. 1970. Phytoplankton in some New Zealand surface waters. *N.Z. Water Conf. 1970, Proc. Pt.* 1: 7.1–7.15.
FORSYTH, D. J. 1971. Some New Zealand Chironomidae (Diptera). *J. Roy. Soc. N.Z.* 1: 113–144.

FREEMAN, P. 1959. A study of New Zealand Chironomidae (Diptera, Nematocera). *Bull. Brit. Mus. Nat. Hist., Entomol.* 7: 395–437.

GAGE, M. 1959. On the origin of some lakes in Canterbury. *N. Z. Geographer* 15: 69–75.

GOLDMAN, C. R. 1964. Primary productivity and micro-nutrient limiting factors in some North American and New Zealand lakes. *Verh. Int. Ver. Limnol.* 15: 365–374.

GOLDMAN, C. R. 1972. The role of minor nutrients in limiting the productivity of aquatic ecosystems. In: G. E. LIKENS (ed.). Nutrients and Eutrophication: The limiting-nutrient controversy. Amer. Soc. Limnol. Oceanogr., Kansas; Special Symposia, 1: 21–33.

GREEN, J. D. 1967. Studies on the zooplankton of Lake Pupuke. *Tane* 13: 77–98.

GREEN, J. D., NORRIE, P. H. & CHAPMAN, M. A. 1968. An internal seiche in Lake Rotoiti. *Tane* 14: 3–11.

HAMILTON, A. 1940. The New Zealand dobson-fly (*Archichauliodes diversus* Walk.): life-history and bionomics. *N.Z. J. Sci. Technol.* 22(1A): 44A–55A.

HARRISON, A. D. 1962. Hydrobiological studies on alkaline and acid still waters in the Western Cape Province. *Trans. Roy. Soc. S. Afr.* 36: 213–235.

HARRISON, A. D. 1965. Geographical distribution of riverine invertebrates in Southern Africa. *Arch. Hydrobiol.* 61: 387–394.

HELSON, G. A. H. 1934. The bionomics and anatomy of *Stenoperla prasina* (Newman). *Trans. Proc. Roy. Soc. N.Z.* 64: 214–248.

HILGENDORF, F. W. 1918. Notes from the Canterbury College Mountain Biological Station, Cass. No. 6. The insect life. *Trans. N.Z. Inst.* 50: 135–144.

HIRSCH, A. 1958. Biological evaluation of organic pollution of New Zealand streams. *N.Z. J. Sci.* 1: 500–553.

HOPKINS, C. L. 1966. Two new species of water mite (Acari, Hydrachnellae) from New Zealand. *Trans. Roy. Soc. N.Z. (Zool.)* 8: 111–117.

HOPKINS, C. L. 1967. New genera and species of water mites (Acari, Hydrachnellae) from New Zealand. *Trans. Roy. Soc. N.Z. (Zool.)* 10: 33–44.

HOPKINS, C. L. 1970. Some aspects of the bionomics of fish in a brown trout nursery stream. *Fish. Res. Bull. N.Z. Mar. Dept.* 4: 38 pp.

HOPKINS, C. L. 1971. Production of fish in two small streams in the North Island of New Zealand. *N.Z. J. Mar. Freshw. Res.* 5: 280–290.

HOPKINS, C. L., BREWERTON, H. V. & MCGRATH, H. J. W. 1966. The effect on a stream fauna of an aerial application of DDT prills to pasture land. *N.Z. J. Sci.* 9: 236–248.

HUTCHINSON, G. E. 1957. A treatise on limnology. Volume 1. Geography, Physics and Chemistry. John Wiley and Sons, New York.

HUTCHINSON, G. E., PICKFORD, G. E. & SCHUURMAN, J. F. M. 1932. A contribution to the hydrobiology of pans and other inland waters of South Africa. *Arch. Hydrobiol.* 24: 1–154.

ILLIES, J. 1969. Biogeography and ecology of neotropical freshwater insects, especially those from running waters. In: E. J. FITTKAU, J. ILLIES, H. KLINGE, G. H. SCHWABE & H. SIOLI (eds.). Biogeography and ecology in South America, Junk, The Hague: 685–708.

IRWIN, J. 1968. Observations of temperatures in some Rotorua district lakes. *N.Z. J. Mar. Freshw. Res.* 2: 591–605.

IRWIN, J. 1971. Exploratory limnological studies of Lake Manapouri, South Island, New Zealand. *N.Z. J. Mar. Freshw. Res.* 5: 164–177.

IRWIN, J. & JOLLY, V. H. 1970. Seasonal and areal temperature variation in Lake Wakatipu. *N.Z. J. Mar. Freshw. Res.* 4: 210–216.

JOHNSTONE, I. M. 1972. Limnology of Western Springs, Auckland, New Zealand. *N.Z. J. Mar. Freshw. Res.* 6: 298–328.

JOLLY, V. H. 1952. A preliminary study of the limnology of Lake Hayes. *Aust. J. Mar. Freshw. Res.* 3: 74–91.

Jolly, V. H. 1957. Thermal stratification in some New Zealand lakes. *Proc. N.Z. Ecol. Soc.* 4: 43–44.
Jolly, V. H. 1965. Diurnal surface concentrations of zooplankton in Lake Taupo, New Zealand. *Hydrobiologia* 25: 466–472.
Jolly, V. H. 1968. The comparative limnology of some New Zealand lakes. I. Physical and chemical. *N.Z. J. Mar. Freshw. Res.* 2: 214–259.
Löffler, H. 1958. Die Klimatypen des holomiktischen Sees und ihre Bedeutung für zoogeographische Fragen. *Öst. Akad. Wiss. Math. Nat. Kl., Abt. 1* 167: 1–33.
Löffler, H. 1960. Limnologische Untersuchungen an chilenischen and peruanischen Binnengewässern. 1. Die physikalisch-chemischen Verhältnisse. *Ark. Geofysik* 3: 155–254.
Löffler, H. 1962. Zur Systematik und Ökologie der chilenischen Süsswasserentomostraken. *Beitr. Z. Neotrop. Fauna* 2: 143–248.
Lucas, K. 1904. A bathymetrical survey of the lakes of New Zealand. *Geogr. J.* 23: 645–660, 744–760.
Mason, R. 1969. The vegetation of freshwaters. In: G. A. Knox (ed.). The Natural History of Canterbury, Reed, Wellington: 452–457.
McDowall, R. M. 1964. The affinities and derivation of the New Zealand freshwater fish fauna. *Tuatara* 12: 59–67.
McDowall, R. M. 1965a. Studies on the biology of the red-finned bully *Gobiomorphus huttoni* (Ogilby). II. Breeding and life history. *Trans. Roy. Soc. N.Z., Zool.* 5: 177–196.
McDowall, R. M. 1965b. Studies on the biology of the red-finned bully *Gobiomorphus huttoni* (Ogilby). III. Food studies. *Trans. Roy. Soc. N.Z., Zool.* 5: 233–254.
McDowall, R. M. 1968. Interactions of the native and alien faunas of New Zealand and the problem of fish introductions. *Trans. Amer. Fish. Soc.* 97: 1–11.
McDowall, R. M. 1970a. The Galaxiid fishes of New Zealand. *Bull. Mus. Comp. Zool.* 139: 341–432.
McDowall, R. M. 1970b. A second species of *Galaxias* common to Tasmania and New Zealand (Pisces: Galaxiidae). *Rec. Dom. Mus.* 7: 13–19.
McLay, C. L. 1968. A study of drift in the Kakanui River, New Zealand. *Aust. J. Mar. Freshw. Res.* 19: 139–149.
McLean, J. A. 1966. A comparative ecological study of three stream faunas in the Auckland area. *Tane* 12: 97–102.
McMichael, D. F. 1967. Australian freshwater Mollusca and their probable evolutionary relationships: a summary of present knowledge. In: A. H. Weatherley (ed.). Australian inland waters and their fauna, Australian National University Press, Canberra: 123–149.
Mitchell, S. F. 1971. Phytoplankton productivity in Tomahawk Lagoon, Lake Waipori, and Lake Mahinerangi. *Fish. Res. Bull. N.Z. Mar. Dept.* 3: 87 pp.
Nicholls, G. E. 1944. The Phreatoioidea. Part II. The Phreatoicidae. *Pap. Proc. Roy. Soc. Tasmania* 1943: 1–156.
Parrott, A. W. 1929. *N. Cant. Acclim. Soc.* 65th Ann. Report.
Percival, E. 1932. On the depreciation of trout-fishing in the Oreti (or New river), Southland, with remarks on conditions in other parts of New Zealand. *Fish. Bull. N.Z. Mar. Dept.* 5: 48 pp.
Percival, E. 1951. Planktonic animals in Lakes Lyndon and Pearson. *N. Cant. Acclim. Soc.*, 87th Ann. Report: 22–24.
Percival, E. undated. Remarks on the freshwater fauna of New Zealand. Unpublished manuscript.
Percival, E. & Burnet, A. M. R. 1963. A study of the Lake Lyndon rainbow trout *(Salmo gairdnerii)*. *N.Z. J. Sci.* 6: 273–303.
Phillipps, W. J. & Grigg, F. J. T. 1922. The geochemistry of the thermal lakes, North Island, New Zealand, in relation to problems bearing on the acclimatized Salmonidae. *N.Z. J. Sci. Technol.* 5: 156–165.

PHILLIPS, J. S. 1929. A report on the food of trout and other conditions affecting their well-being in the Wellington district. *Fish. Bull. N.Z. Mar. Dept.* 2: 31 pp.
PONDER, W. F. 1964. The freshwater Mollusca of New Zealand. *Tane* 10: 12–35.
REID, J. 1964. Study of cause and effects of eutrophication of Lake Rotorua. Health Dept. Rept., Rotorua: 4 pp.
REID, L. W. 1966. Wastewater pollution and general eutrophication of a hydroelectric impoundment. *J. Wat. Pollut. Control Fed.* 38: 165–174.
RODHE, W. 1949. The ionic composition of lake waters. *Verh. Int. Ver. Limnol.* 10: 377–386.
RUSSELL, C. R. 1960. An index to the Rotatoria of New Zealand and outlying islands from 1859 to 1959. *Trans. Roy. Soc. N.Z.* 88: 443–461.
SCHMINKE, H. K. & NOODT, W. 1968. Discovery of Bathynellacea, Stygocaridacea and other interstitial Crustacea in New Zealand. *Naturwiss.* 1968: 184–185.
SCOTT, D. 1966. The substrate cover-fraction concept. In: K. W. CUMMINS, C. A. TRYON & R. T. HARTMAN (eds.). Organism-substrate relationships in streams, Pymatuning Symposia in Ecology, Univ. of Pittsburgh, Spec. Public. 4: 75–78.
SMITH, D. C. W. 1959. The biology of the rainbow trout *(Salmo gairdnerii)* in the lakes of the Rotorua district, North Island. *N.Z. J. Sci.* 2: 275–312.
STEVENSON, G. B. 1952. A study of the Wellington city water supply with special reference to plankton growth in storage reservoirs. *N.Z. J. Sci. Technol.* B. 34: 26–45.
STOUT, V. M. 1964. Studies on temporary ponds in Canterbury, New Zealand. *Verh. Int. Ver. Limnol.* 15: 209–214.
STOUT, V. M. 1969a. Life in lakes and ponds. In: G. A. KNOX (ed.). The Natural History of Canterbury, Reed, Wellington: 458–470.
STOUT, V. M. 1969b. The invertebrate fauna of the rivers and streams. In: G. A. KNOX (ed.). The Natural History of Canterbury, Reed, Wellington: 471–497.
STOUT, V. M. 1969c. Lakes in the mountain region of Canterbury, New Zealand. *Verh. Int. Ver. Limnol.* 17: 404–413.
STOUT, V. M. 1970. The invertebrate animals of New Zealand inland waters and factors influencing their distribution. *N.Z. Water Conf. 1970, Proc. Pt I:* 11.1–11.24.
STOUT, V. M. 1972. Plankton composition in relation to nutrient inflow in a small New Zealand lake. *Verh. Int. Ver. Limnol.* 18: 605–612.
STOUT, V. M. 1973. The freshwater environment. In: G. R. WILLIAMS (ed.). The Natural History of New Zealand, Reed, Wellington: 229–250.
THOMASSON, K. 1955. Studies on South American freshwater plankton. 3. Plankton from Tierra del Fuego and Valdivia. *Acta Horti Gotoburgensis* 19: 193–225.
THOMASSON, K. 1957. Studies on South American freshwater plankton. 1. Notes on the plankton from Tierra del Fuego and Valdivia. *Ann. Soc. Nat. Univ. Tartu, N.S.* 1: 52–64.
THOMASSON, K. 1960. Some planktic Staurastra from New Zealand. *Bot. Notiser* 113: 225–245.
THOMASSON, K. 1963. Araucanian Lakes. Plankton studies in North Patagonia, with notes on terrestrial vegetation. *Acta Phytogeogr. Suec.* 47: 139 pp.
THOMASSON, K. 1964. Plankton and environment of North Patagonian lakes. *Ann. Soc. Nat. Univ. Tartu, N.S.* 4: 9–28.
TILLYARD, R. J. 1920. Report on the neuropteroid insects of the hot springs region, New Zealand, in relation to the problem of trout food. *Proc. Linn. Soc. N.S.W.* 45: 205–213.
VIDAL, I. L. 1968. An addition to the genus *Boeckella* (Copepoda) in New Zealand. *N.Z. J. Mar. Freshw. Res.* 2: 684–687.
VOLLENWEIDER, R. A. 1968. Scientific fundamentals of the eutrophication of lakes and flowing waters, with particular reference to nitrogen and phosphorus as factors in eutrophication. *O.E.C.D., Paris.* Mimeographed report.
WILLIAMS, W. D. 1964. Some chemical features of Tasmanian inland waters. *Aust. J. Mar. Freshw. Res.* 15: 107–122.

WILLIAMS, W. D. 1965. The distinctiveness of Australia's inland aquatic fauna. *Aust. Soc. Limnol. Newsl.* 4: 36–42.

WILLIAMS, W. D. 1967. The chemical characteristics of lentic surface waters in Australia. In: A. H. WEATHERLEY (ed.). Australian inland waters and their fauna, Australian National University Press, Canberra: 18–77.

WILLIAMS, W. D. & WAN, H. F. 1972. Some distinctive features of Australian inland waters. *Water Res.* 6: 829–836.

WINTER, J. W. 1964. A survey of Lake Roxburgh, a recent hydro-electric dam. *Proc. N.Z. Ecol. Soc.* 11: 16–23.

WINTERBOURN, M. J. 1968. The faunas of thermal waters in New Zealand. *Tuatara* 16: 111–122.

WINTERBOURN, M. J. 1969a. Water temperature as a factor limiting the distribution of *Potamopyrgus antipodum* (Gastropoda – Prosobranchia) in the New Zealand thermal region. *N.Z. J. Mar. Freshw. Res.* 3: 453–458.

WINTERBOURN, M. J. 1969b. The distribution of algae and insects in hot spring thermal gradients at Waimangu, New Zealand. *N.Z. J. Mar. Freshw. Res.* 3: 459–465.

WINTERBOURN, M. J. & BROWN, T. J. 1967. Observations on the faunas of two warm streams in the Taupo thermal region. *N.Z. J. Mar. Freshw. Res.* 1: 38–50.

WISELY, B. 1962. Studies on Ephemeroptera. II. *Coloburiscus humeralis* (Walker); ecology and distribution of the nymphs. *Trans. Roy. Soc. N.Z., Zool.* 2: 209–220.

XI. THE LAND SNAIL FAUNA

by

F. M. CLIMO

Introduction

RELATIONSHIP OF NEW ZEALAND TO INDO-PACIFIC LAND SNAIL FAUNAL ZONES

The Indo-Pacific region has been divided into land snail faunal zones by a number of authors (FISCHER, 1880–87, 1950; COOK, 1895; SOLEM, 1959). SOLEM (1959) has presented the most recent account giving an illustrated synthesis of the earlier schemes (figs. 29–32) in the course of a major systematic and zoogeographic account of the land and freshwater mollusca of the New Hebrides. He states: 'There are three main faunal assemblages of land snails in the Indo-Pacific region. These are called the Palaeo-Oriental, Pacific Island, and Southern Relict faunas.' New Zealand falls into the area covered by the Southern Relict fauna.

The Palaeo-Oriental fauna, stretching from Java to the Solomons and including N.E. Australia is characterised by the families Camaenidae, Cyclophoridae, Rathoussiidae, and larger helicarionid snails.

The Pacific Island fauna is endemic, but probably derived from the Palaeo-Oriental fauna (SOLEM, 1959: 295).

Southern Australia, New Zealand and New Caledonia are characterised by extensive radiation in the Paryphantidae and Punctidae. This unique feature is characteristic of the Southern Relict faunal zone. The high specific and generic endemism shown by these families in New Zealand is a basis for recognizing New Zealand as a distinct subzone of the Southern Relict zone (SOLEM, 1959: 283, Fig. 29). The endemic nature of the cyclophoracean prosobranch family Liareidae, only recently recognized (CLIMO, 1971), adds weight to the New Zealand subzone. Superimposed upon these basic elements of the New Zealand fauna are widely dispersed minor elements of the Achatinellidae (2 species), Succineidae (1 species), Assimineidae (2 species) and Hydrocenidae (1 species). In addition to the Paryphantidae and Punctidae other archaic elements of the Southern Relict zone represented in New Zealand are the families Bulimulidae and Athoracophoridae, though these are not necessarily diagnostic of the zone.

In view of the relict nature of the fauna to be discussed below, it is significant that the families Punctidae and Paryphantidae are both primitive members of the stylommatophoran order Sigmurethra.

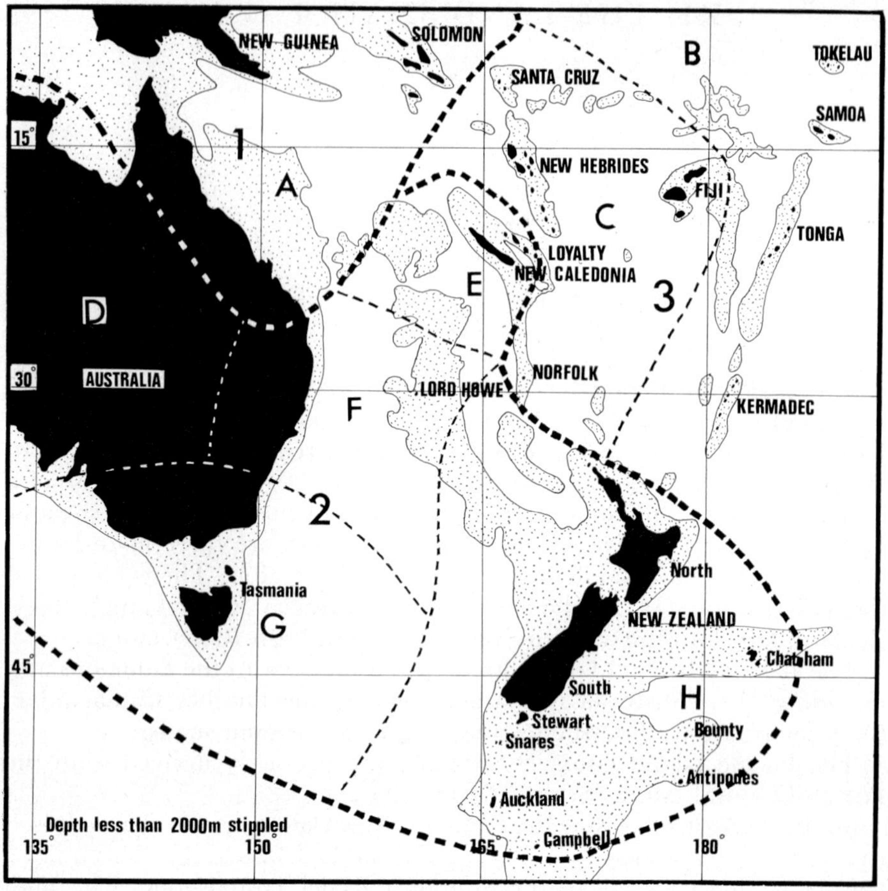

Fig. 1. Land snail regions of the Pacific Ocean (after Solem, 1959: 283, Fig. 29). 1, Palaeo-Oriental; 2, Southern Relict; 3, Pacific Island; A, Papuan; B, Oceania; C, Melanesian; D, Centralian; E, New Caledonia; F, Oxleyan; G, Peronian; H, New Zealand.

The age and origins of the New Zealand land snail fauna

Students of land snail phylogeny and zoogeography are inhibited by a paucity of fossiliferous terrestrial beds older than the Quaternary, and the fact that when these do occur it is only under favourable conditions that the aragonitic shell of land snails survives.

From studies of the fossil land snails that have been collected in different parts of the world (North America; Argentina; Santo Domingo; Jamaica; Marshall and Ellice Islands) it is obvious that many families and genera of land snails were differentiated as early as Cretaceous times and the majority by the early Tertiary (Pilsbry, 1926: 316–19, 1939; Wenz,

1938–44; PARODIZ, 1946, 1949; PILSBRY & OLSSON, 1954; MORRISON, 1955; LADD, 1957, 1958).

Land molluscs (with a few exceptions such as members of the Achatinellidae on the Hawaiian Islands) are extremely conservative and it is the opinion of many writers (e.g. PILSBRY, 1894; MAYR, 1940; HESSE, ALLEE & SCHMIDT, 1951; SOLEM, 1959) that the present families originated in the Mesozoic or late Palaeozoic and that adaptive radiation into particular ecological niches occurred at an even earlier date. It is not too much to postulate, then, that the New Zealand fauna as a whole is at least Cretaceous in origin and that the many generic disjunctions that do occur in the Indo-Pacific are relict patterns (there are a few exceptions) displayed by conservative stock, rather than the results of later oceanic dispersal.

It is widely accepted that New Zealand has been separated from other land masses since at least Cretaceous times. At some stage in the Mesozoic it seems likely that there was semi-continuous land sweeping north-westwards from New Zealand to the New Caledonian and New Hebridean region of the Pacific, and if one accepts continental drift there was continuity of land to Australia and Antarctica until Cretaceous.

The characteristic combination of families present in many parts of the Southern Relict zone (Paryphantidae; Punctidae; Athoracophoridae; Bulimulidae; cyclophoracean families), and particularly in New Zealand, are thought (SOLEM, 1959; 320–21) most easily explained as one 'basic' primitive fauna. This 'basic' fauna has been modified by succeeding waves of land snails in northern Australia, but restricting ourselves to the country considered here, New Zealand has entirely escaped such additional elements. Climatic changes, particularly in Australia, have probably also modified distribution patterns of families markedly within the Southern Relict zone but this factor does not greatly affect New Zealand.

Some writers (e.g. SOLEM, 1959: 322) would consider that the archaic elements of the New Zealand fauna (land snails; lizards; frogs) all suggest that the country has been effectively isolated from land bridge colonization since Jurassic times; others (e.g. OLIVER, 1953) postulate a Cretaceous land bridge. What is undisputed, however, is that the ancestors of the New Zealand land snail fauna, as mentioned above, arrived sometime in the Mesozoic and have evolved in complete isolation since that time.

SOLEM (1959: 255–56) suggested, when discussing statements by earlier authors (e.g. PILSBRY, 1894; VAN DYKE, 1939), that it is the study of land patterns in the Mesozoic which are going to provide the clues for interpreting land snail zoogeography in the Indo-Pacific. In recent years strong evidence has been forth-coming for the recognition of the continent of Gondwanaland (e.g. see symposium in *Antarctic Journal*, 5 (3): 53–80), a late Palaeozoic land mass which split up by plate tectonic movements during the middle Mesozoic. It is postulated that Australasia, Antarctica, Africa, South America and Peninsular India were the basic components of this early continent. Although still a topic of much debate the possibility of the existence of this archaic continent must be considered when discussing relict land snail faunas. I will not try to explain the world distribution of taxa mentioned below, but it is worth bearing Gondwanaland in mind when considering SOLEM's statement that the key to Indo-Pacific land snail zoogeography is in the land patterns of the Mesozoic. Solem was not a supporter of the theory of Continental Drift but much more unrefutable evidence has become available in this field since 1959.

Although most New Zealand land snail genera were probably differentiated well before the Pliocene (and some probably even before the Cretaceous), the distribution of many subgenera and species, particularly in the North Island, correspond with the land patterns of the late Tertiary. Fig. 2 shows the distribution of land in the New Zealand Pliocene, according to FLEMING (1962). As will be shown below, the number of small islands in the North Island region at that time evolved their own species, or even subgenera.

The South Island picture is not so clear because land areas were more continuous in that area of the country in the Pliocene. Unlike the North Island which was divided up into a series of islands (Fig. 2), land was continuous from Cook Strait to the sub-antarctic region through the axis of the South Island. Although one might have expected specific endemism on Banks Peninsula, an island during the Pliocene, this is not evident in the land molluscan fauna. The continual influx of species by rafting from the aggrading edge of the Canterbury Plains probably did not enable genetic divergence on this quite small Pliocene island. The affects of glaciation in the South Island, Stewart Island and the sub-antarctic region during the Pleistocene also resulted in the disjunction of earlier established patterns of distribution, and with few exceptions faunal zones are difficult to define for land mollusca. As will be pointed out, however, there are a number of areas of high specific endemism in the South Island. There is also evidence to support the previous existence of a southern fauna, now represented by a few monotypic genera on the subantarctic islands. The southern New Zealand fauna is a good example of relict distribution patterns with a number of species restricted to alpine pockets and the rest of the restricted species endemic to those areas squeezed into major glacial retreats in the S.W. and N.W. of the South Island.

THE PRESENT STATE OF NEW ZEALAND LAND SNAIL SYSTEMATICS

As is the case with most of the Indo-Pacific region, New Zealand land snail systematics is still in its infancy. Although there are impressive species lists available for most land masses in the area, little work has

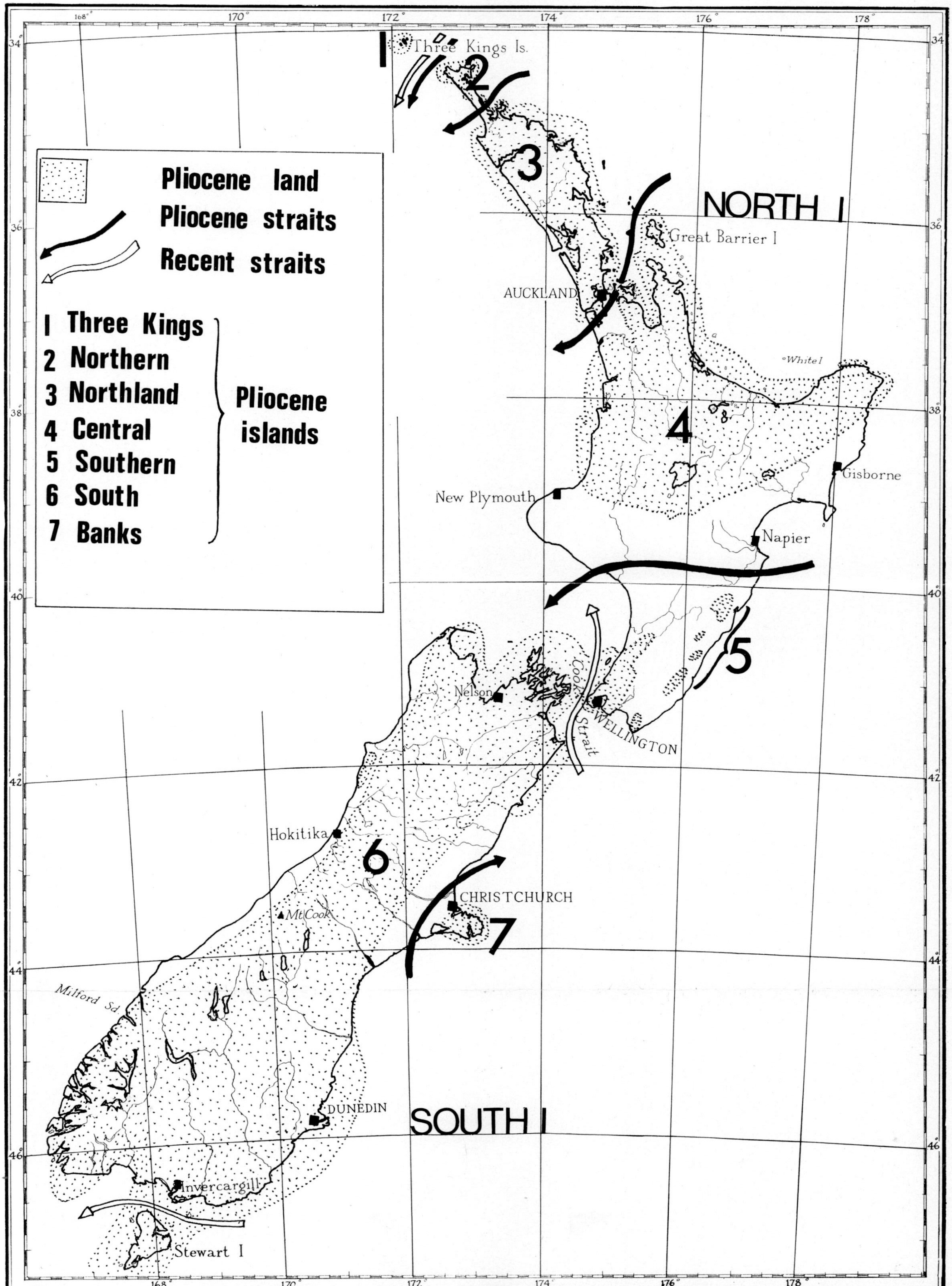

Fig. 2. Distribution of land in New Zealand during the Pliocene (After FLEMING, 1962).

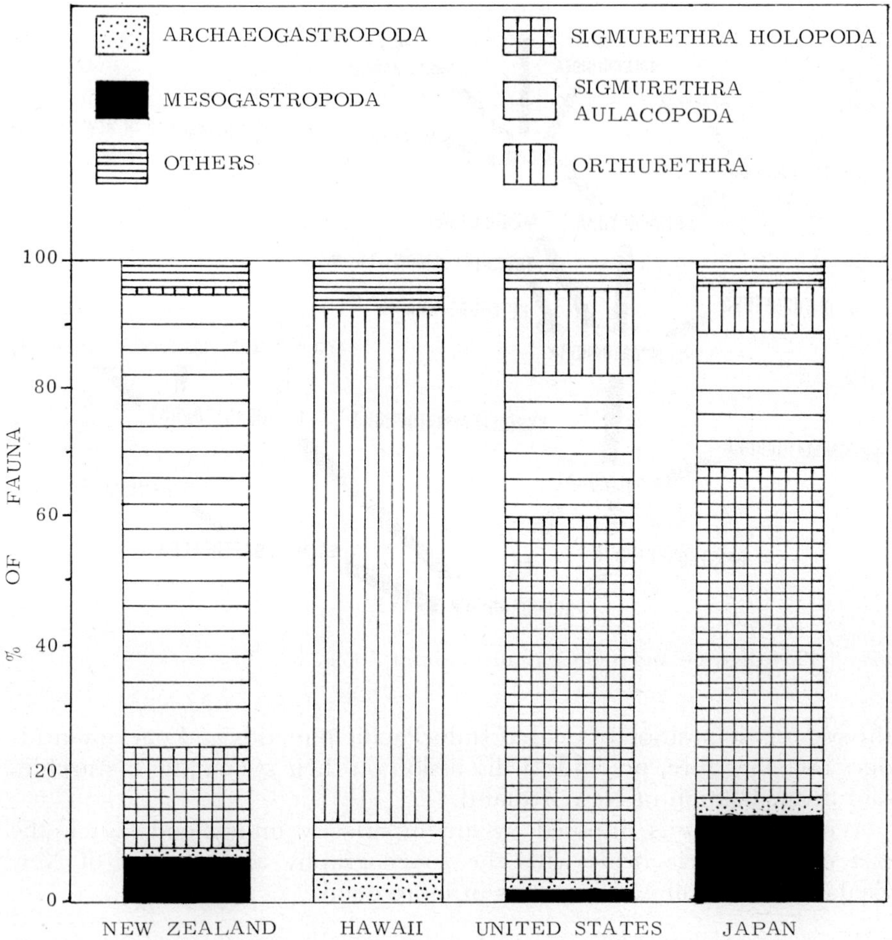

Fig. 3. Proportionate ordinal composition of Japanese, United States, Hawaiian and New Zealand land snail faunas (after SOLEM, 1959: 250, Fig. 4).

been presented which attempts detailed generic comparison between areas.

Until detailed comparative faunal studies are made of land snail elements in the Indo-Pacific (a huge task which must be dealt with in sections by a number of co-operating authors) nothing more than generalized statements can be made about the zoogeographic affinities of one area with any other.

New Zealand malacologists are fortunate that the several available type genera for subfamilies in the Punctidae (*Phenacohelix* Suter; *Charopa* Albers), and the type species for the majority of the genera in the Paryphantidae, are based on New Zealand species. A lack of knowledge of

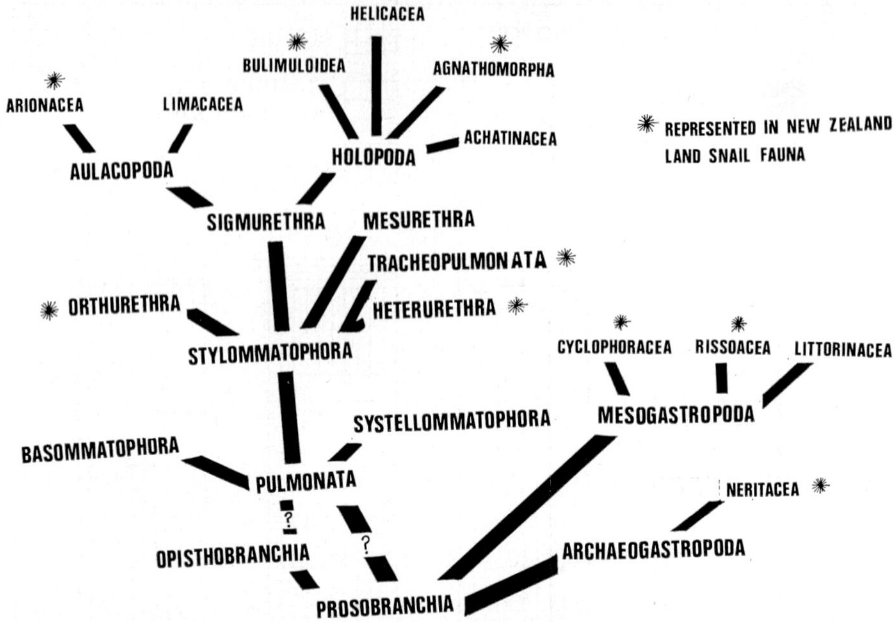

Fig. 4. Phylogeny of land mollusca (after Solem, 1959).

the systematic position of related Indo-Pacific punctids and paryphantids does not, therefore, preclude a discussion on their systematic hierarchies and diversification in New Zealand.

Within the limits imposed by an imperfectly understood fauna, the rest of this article deals with the zoogeography and ecology of New Zealand land molluscan families in more detail.

Composition of the New Zealand Land Snail Fauna

In Fig. 3 the proportionate ordinal composition of the Japanese, United States, Hawaiian, and New Zealand land snail faunas is compared (based on Solem, 1959: 250, Fig. 4).

The most characteristic features of the New Zealand fauna at the ordinal level are the very high percentage composition (70%) of aulacopod sigmurethran pulmonates and the paucity of archaeogastropods and orthurethran pulmonates. The phylogenetic relationships of the land snail taxa mentioned above and graphed in Fig. 3 are shown in Fig. 4 which is based on Solem (1959: 252–53, Figs. 10–11). Brief definitions are given of the Pulmonate orders represented in New Zealand in the major section of this paper which follows.

There are approximately 270 described land snails in New Zealand falling into 9 families. Fig. 5 shows the composition of the New Zealand

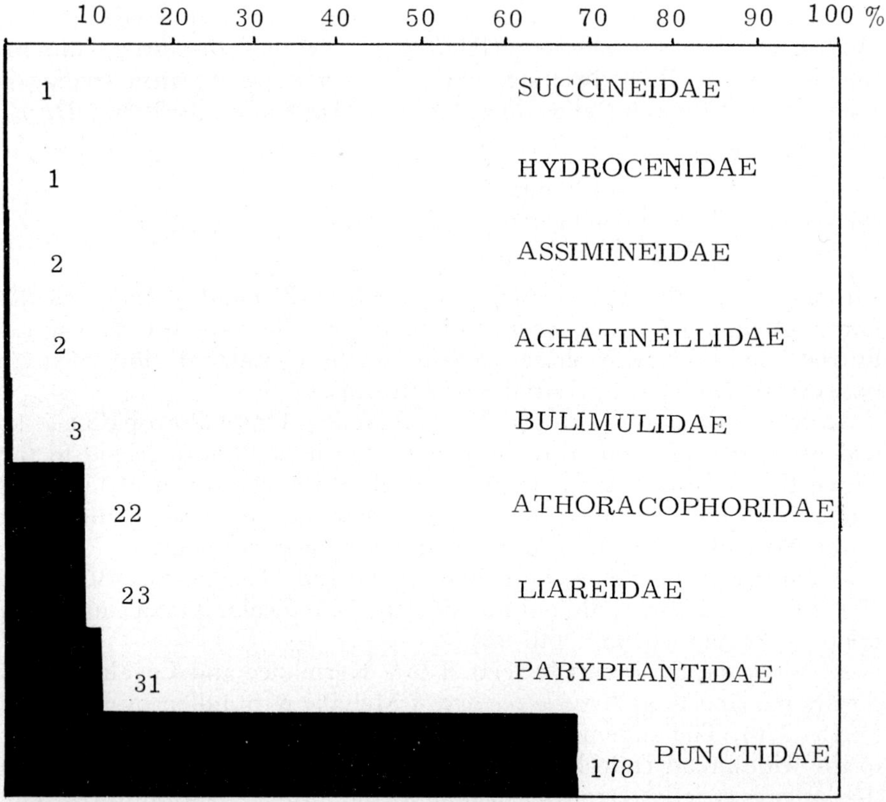

Fig. 5. Percentage composition at the family level of the New Zealand land snail fauna, with numbers of species in each family.

fauna at the family level. Notable by their absence in New Zealand are the superfamilies Helicacea and Limacacea (see Fig. 4) both dominant elements in the European fauna. Although the zonitid snail *Suterella novarae* (Pfeiffer) has been recorded as a New Zealand endemic species for over 100 years, it was collected only once and indications are that the record is nothing more than the introduced European *Oxychilus cellarius* (Mueller).

Detailed Examination of the Land Snail Families

The three Prosobranch families represented in New Zealand are discussed first and then the six Pulmonate families are examined.

Subclass: PROSOBRANCHIA
Order: ARCHAEOGASTROPODA
Superfamily: NERITACEA
Family: HYDROCENIDAE

The only New Zealand representative of the family Hydrocenidae is the species *Omphalorissa purchasi* (Pfeiffer), a widely dispersed, minute (height 2 mm), litter-dwelling snail. The genus *Omphalorissa* Iredale is restricted to Australia and New Zealand (IREDALE, 1933: 57; DELL, 1952: 87–8).

Order: MESOGASTROPODA
Superfamily: RISSOACEA
Family: ASSIMINEIDAE

Although not strictly land snails the two New Zealand members of the family Assimineidae are worthy of mention. They occupy the supralittoral zone of marine shores, which can be considered intermediate between marine and terrestrial environments.

Suterilla neozelanica (Murdoch) (subfamily Omphalotropidinae) is widely distributed from Kaikoura Peninsula in the South Island to the Three Kings Islands at the extreme north of the country, usually living under boulders. *Assiminea vulgaris* (Webster) is only found in the north of the North Island and occupies a similar ecological niche.

When discussing the Assimineidae in the Indo-Pacific, SOLEM (1959: 199–200, 266, Fig. 12) did not mention the New Zealand taxa elaborated upon by POWELL (1933: 146–48).

Assiminea vulgaris is closely related to a Kermadec and Loyalty Island species referred to as *Barleeia chrysomela* Melville & Standen by IREDALE (1913: 370). The endemic New Zealand genus *Suterilla* Thiele is similar to the widespread Indo-Pacific genus *Omphalotropis* Pfeiffer, but whereas the latter is strictly terrestrial in habitat the former is supralittoral. The affinities of the New Zealand genus are not clear but they appear to be with *Omphalotropis*, if one ignores the rather peculiar development of the marginal teeth of the radula.

Superfamily: CYCLOPHORACEA
Family: LIAREIDAE

Cyclophoracean prosobranch land snails are widespread in the Pacific Basin and they have been revised most recently in near entirety by TIELECKE (1940).

In shell features the New Zealand endemic genera *Cytora* Kobelt & Moellendorff and *Liarea* Pfeiffer are very similar to other members of the operculate superfamily, species generally everywhere having circular, flared apertures. MORTON (1952) has shown, however, that anatomically *Cytora* has a distinctive reproductive morphology. CLIMO (1971) showed that *Liarea* had the same type of reproductive morphology as *Cytora*. The penis of these two dioecious genera is provided with an apical intrommitant organ which is pierced by the vas deferens, the most diagnostic feature of the endemic family Liareidae. The relationship between *Cytora* and the Cyclophoridae, the family in which the genus

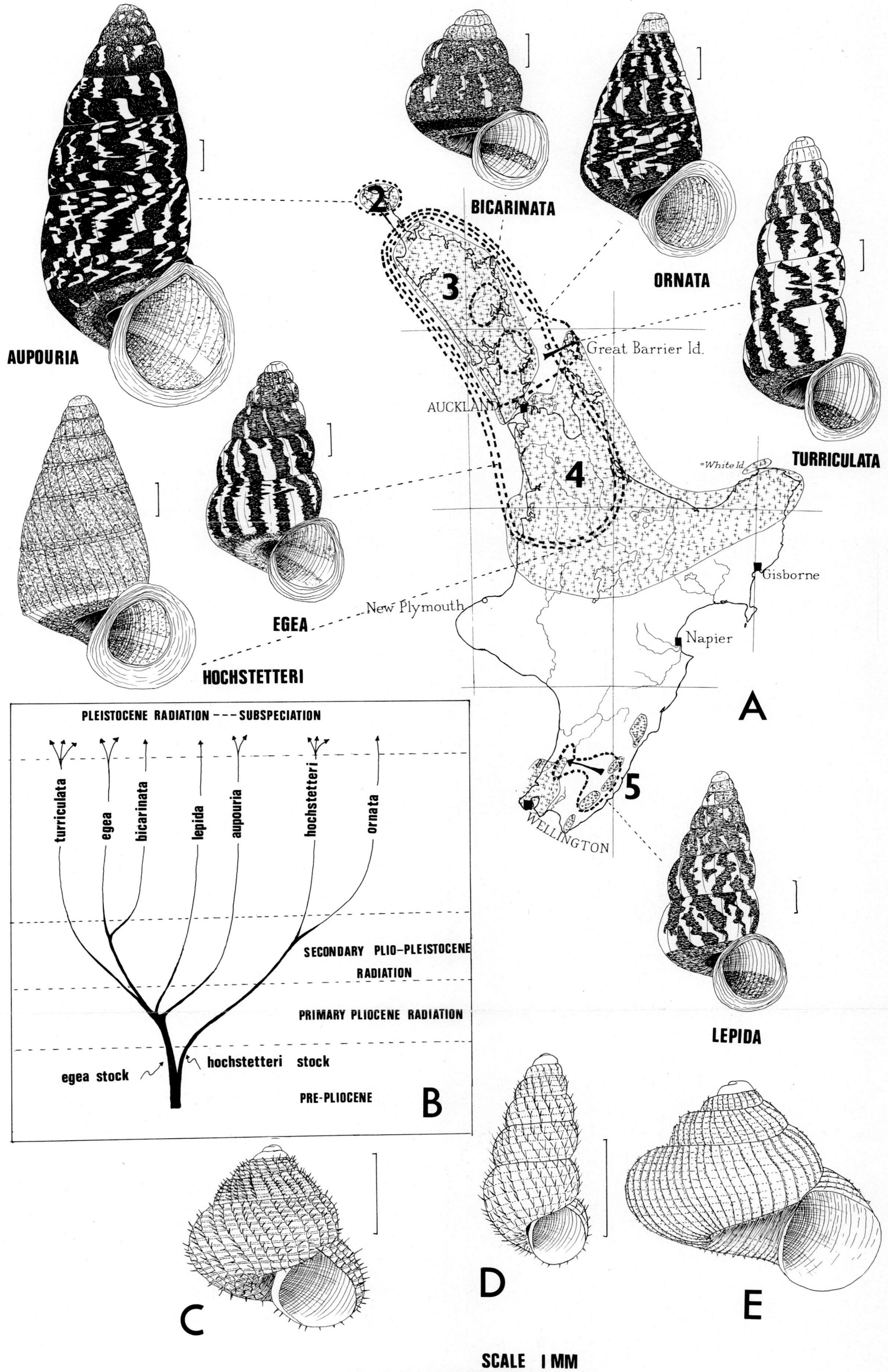

SCALE 1 MM

For explanation of map see Fig. 2

Fig. 6. Family Liareidae. A, distribution and morphological variation in *Liarea* Pfeiffer; B, phylogeny of *Liarea;* C, *Cytora cytora* (Gray); D, *C. hispida* Gardner; E, *C. lignaria* (Pfeiffer).

was previously placed, is a morphologically superficial one. *Cytora* and *Liarea* have more significant taxonomical features in common than either have with exotic taxa. The Cyclophoracea are conchologically rather conservative and taxa in the superfamily can often only be differentiated with certainty by anatomical characters.

Liarea is conchologically very conservative, composed of small (up to 15 mm in height), turretted, varyingly flammulated shells which have a double continuous peristome (Fig. 6). *Cytora* does not have a double continuous peristome and is variable in shape and shell ornamentation (Fig. 6). There are several groups of species within this latter genus which warrant taxonomic separation at the species-group or subgeneric levels, and when this is done the distribution patterns within *Cytora* may be more meaningful.

The presence of an endemic family comprised of two genera and 22 species is indicative of long isolation and the Liareidae can be regarded as a component of the 'basic' New Zealand land snail fauna.

Members of the Liareidae live in leaf-litter and although their feeding habits have not been investigated they are most probably feeders of decaying leaves with their associated bacterial and fungal films.

Perhaps the most specialized member of the family is *Cytora hirsutissima* Powell, from Great King Island, Three Kings Group, to the north of the North Island. The species is restricted to about 10 sq. m. of rock rubble in a very small patch of remnant broadleaf forest. It is found only under large rocks the upper surfaces of which are covered in a particular species of fern (*Arthropteris tenella* Smith). Light intensities are very low in this patch of forest. *Cytora hirsutissima* is totally blind.

The distribution and morphological variation of living specimens of *Liarea* are well known (Powell, 1957) but the southern limits of some species, known from the southern parts of their ranges only from subfossil cave deposits, have yet to be defined. All species are restricted to the North Island, they appear to be warmth-loving and their distribution patterns show a general relationship with Pliocene land patterns (Fig. 6).

It is probable that both *L. egea* and *hochstetteri* were common to both the 'Northland' and 'Central' Pliocene islands, while *lepida* has always been restricted to the 'Southern Island' area. *L. aupouria* evolved in isolation on the 'Northern' island while *bicarinata* and *ornata* did likewise on the 'Northland' island. Some of these Pliocene islands, the northern two in particular, can be more finely subdivided and subspeciation trends have been recorded (Powell, 1957) on smaller areas within these islands (Mountain ranges, small islands now land-locked by sand tombolos etc.).

As mentioned earlier, the internal distribution of *Cytora* is not as clear as for *Liarea*, mainly because the systematics is not completely understood. It is also more widely distributed in New Zealand (Three Kings to Stewart Island), and there are many more species. There is a group of minute, elongate species of *Cytora*, typified by *Cytora hispida* Gardner, restricted to the 'Northern' Pliocene island, a group worthy of subgeneric recognition. A number of other species are restricted in their ranges: *annectens*, *filicosta*, *solitaria* and *hirsutissima* to the Three Kings Islands; *ampla* to the 'Northern' Pliocene island; *depressa* to the N.W. of the South Island.

No further evaluation can be made until the genus is revised in more detail.

Subclass: PULMONATA
Superorder: STYLOMMATOPHORA
Order: TRACHEOPULMONATA

The family to be discussed below is the only member of the above order. Apart from their characteristic lung, tracheopulmonates are characterized by a multi-looped ureter, which opens into the respiratory pore or directly to the exterior (SOLEM, 1959: 33).

Family: ATHORACOPHORIDAE

The Athoracophoridae are a group of predominantly nocturnal pulmonates in which the shell is so reduced that all that remains are a few calcareous granules buried beneath the pallial complex. The most characteristic feature of the family is a dorsal tracheate lung unlike that in any other mollusc. This family of slugs is typically marked by a leaf-vein-like pattern of grooves on the dorsal surface and the common name 'Veined-Slug' is derived from this morphological feature.

Athoracophorid slugs are known from eastern Australia, the Solomon Islands, New Hebrides, New Caledonia and the New Zealand region including the Subantarctic islands (SOLEM, 1959: 44–45, 272, Fig. 18). They are a characteristic element of the Southern Relict zone mentioned earlier, and of the land masses in the geographical range of the family, New Zealand has by far the greatest number of species (23 described). Until quite recently (BURTON, 1963) comparatively little was known of the athoracophorid fauna of New Zealand. Eleven species has been described (GRAY, 1850, 1853; GOULD, 1852; HUTTON, 1878; SIMROTH, 1889; COCKERELL, 1891; SUTER, 1897, 1909; PLATE, 1897; IREDALE, 1915; POWELL, 1937, 1955) prior to BURTON's review, but until this latest paper, which doubled the New Zealand species list, the higher systematics was not well understood. There are still numerous undescribed species in New Zealand.

The Athoracophoridae are not closely related to any other pulmonates and the phylogenetic relationships of the family are obscure. It has become so specialized morphologically, contains so many species, and is so widespread in the New Zealand region that the arrival of the ancestors of the New Zealand species must have coincided with the first immigrants to our archipelago. It seems possible that the New Zealand representatives were derived from the northern part of the Southern Relict zone but have been isolated since pre-Cretaceous times (SOLEM, 1959: 45).

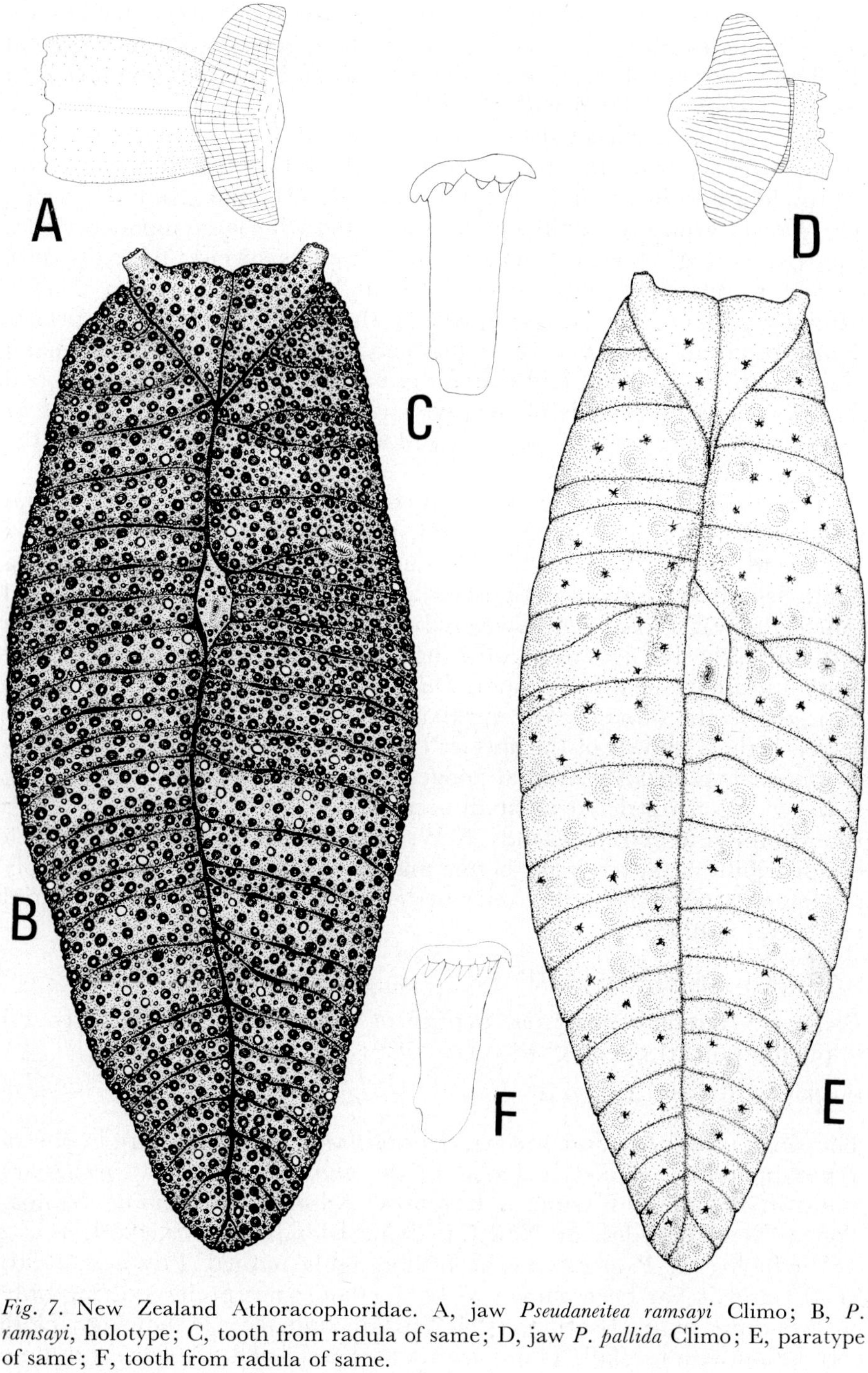

Fig. 7. New Zealand Athoracophoridae. A, jaw *Pseudaneitea ramsayi* Climo; B, *P. ramsayi*, holotype; C, tooth from radula of same; D, jaw *P. pallida* Climo; E, paratype of same; F, tooth from radula of same.

The figured specimens (Fig. 7) of athoracophorid slug are both members of the genus *Pseudaneitea* Cockerell. They have been chosen to represent the family because of their contrasting habitat and external morphological differentiation associated with this.

Athoracophorid slugs can be broadly divided, on a habitat basis, into species which live on the ground surface where they are associated with rotting logs and litter, and a second primarily arboreal group of species. This second group is usually closely associated with large monocotyledonous plants. Fig. 7b represents a ground-dwelling species; typically dark in colour, relatively large (over 4 cm in length) and with a heavily papillate surface. Fig. 7e is typical of the arboreal group of species; semitransparent, with very pale pigment spots, relatively small (maximum length about 3 cm) and with fewer papillae. The illustrated arboreal species spends its periods of inactivity either in the damp leafbases or within the cavity formed between inflorescence stalk and bract of the New Zealand flax, *Phormium tenax*.

Of the number of genera represented in New Zealand, *Athoracophorus* Gould is found only in or near forest, is an arboreal fungivorous feeder but spends its periods of inactivity under logs; *Reflectopallium* Burton, a South Island and subantarctic island genus lives in scrub and grassland (often alpine); *Pseudaneitea* Cockerell is widespread in New Zealand, is very thick skinned, lives in a wide range of habitats and can tolerate lower humidities than the other genera; *Palliopodex* Burton contains one species restricted to the subantarctic Auckland Islands.

Too little is known of the species distributions to formulate any ideas on diversification and internal zoogeographic distribution. Species are, however, often restricted to small areas, for example, The Three Kings Islands, the subantarctic islands, N.W. Nelson in the South Island. When the distribution patterns are better understood they will most probably reinforce patterns shown by other better known New Zealand terrestrial molluscan families.

Order: HETERURETHRA

Diagnosis: 'Ureter along front margin of kidney then following rectum to pneumostomal opening' (Solem, 1959: 33).

Family: SUCCINEIDAE

The single New Zealand species, *Papusuccinea archeyi* (Powell), is one of approximately 7 species included in the genus. The other species are found in Papua New Guinea, Bismarck, Admiralty, Solomon, Samoa, Tonga, New Hebrides and New Caledonia Islands (Solem, 1959: 54).

The biology of *P. archeyi* has been thoroughly studied (Powell, 1950: 61–72) and it has been shown to be a xerophytic species existing only on fore-dune plant associations. It has an annual cycle showing a close correlation with rainfall. The species is now restricted to a few sand dune colonies in the north of the North Island but subfossil remains indicate that its distribution was originally much wider. POWELL (1950: 61) stated: 'The extremely specialized habitat and narrow tolerance of these snails makes the fossil occurrences useful indicators of past zerophytic phases in respect to post-Pleistocene climate.'

Order: ORTHURETHRA

Diagnosis: 'Ureter remote from hind gut, running from tapered anterior end of kidney to anterior border of lung' (SOLEM, 1959: 33).

Family: ACHATINELLIDAE

The Achatinellidae are a Pacific Basin family of orthurethran pulmonates. COOK & KONDO (1960: 21, Fig. 4) postulated three generic centres of evolution: the Hawaiian Islands (8 genera; 52 species), Juan Fernandez (2 genera; 19 species), and the Austral Islands (11 genera; 35 species).

The family is known from only two species in New Zealand, *Tornatellinops novoseelandica* (Pfeiffer) (subfamily Lamellideinae) and *Tornatellides subperforata* (Pfeiffer) (subfamily Tornatellidinae), both of which although endemic are closely related to more northern Pacific species. *T. novoseelandica* is closely related to the widespread central Pacific species *T. variabilis* (Odhner) (COOK & KONDO, 1960: 174, Fig. 78) but the affinities of the other New Zealand representative are more obscure (l.c.: 243–248).

Both species are found only north of the Cook Strait region in New Zealand, are typically minute (height less than 4 mm), tallspired, have lamellate apertures and are to a varying degree arboreal.

Order: SIGMURETHRA

Diagnosis: 'Ureter abruptly reflexed, passing to posterior end of lung cavity, then turning across to the last fold of the gut and following it forward to the mantle edge' (SOLEM, 1959: 33). The Sigmurethra is divided into two suborders on the basis of the relative position and prominence of pedal grooves. In the HOLOPODA the grooves are rather inconspicuous and extend down to or close to the angle of the foot. The foot does not form a major part of the side of the animal. In the AULACOPODA the grooves are conspicuously impressed and situated well above the margin of the foot. The sole of the foot, therefore, is a major part of the side of the animal.

Suborder: HOLOPODA
Family: BULIMULIDAE

The family Bulimulidae is represented in New Zealand by three species of *Placostylus* Beck. The genus is known from the Solomon Islands south of Bougainville, the New Hebrides, the Santa Cruz group, the Fijis exclusive of the Lau Archipelago, New Caledonia, the Loyalty Islands,

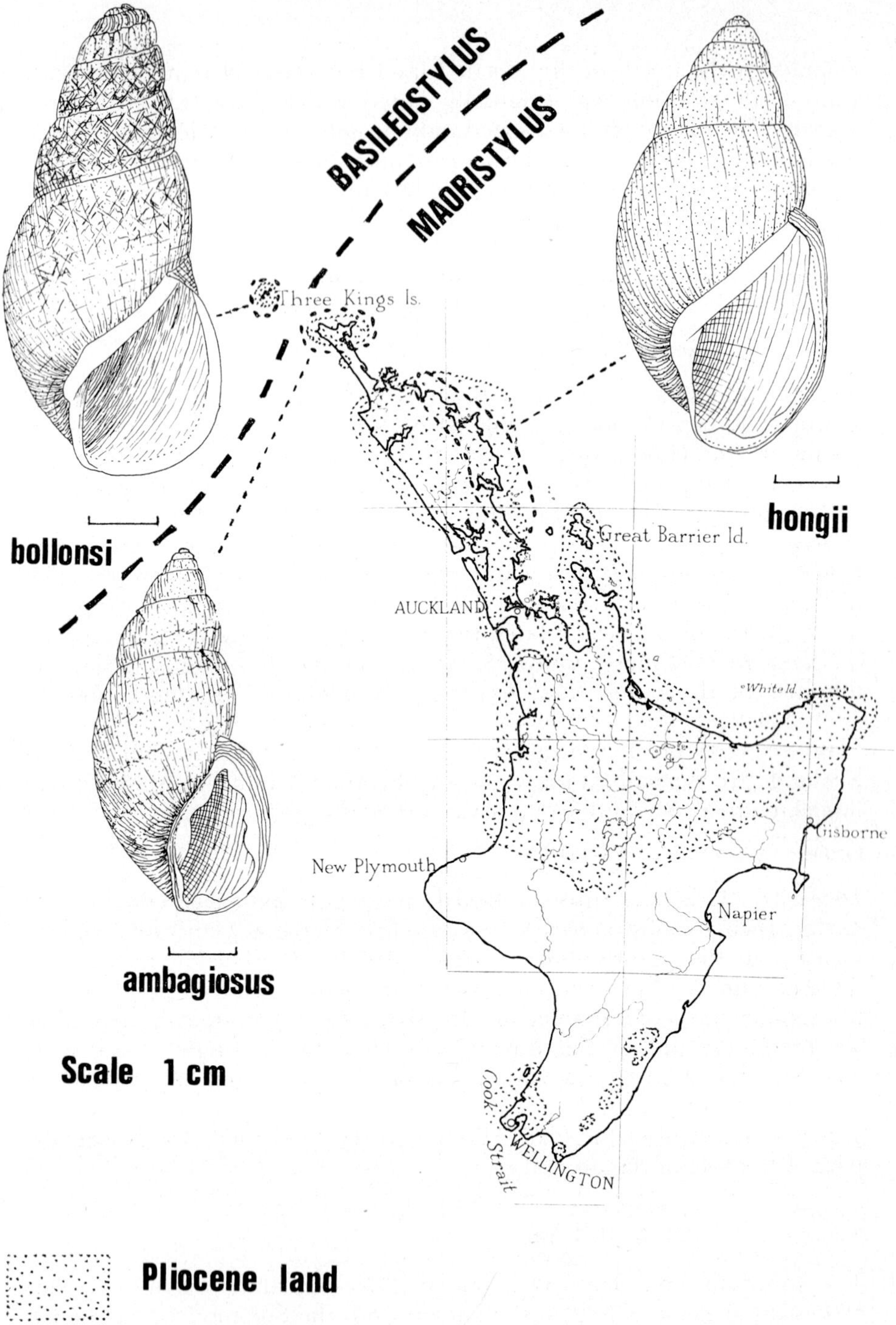

Fig. 8. New Zealand Bulimulidae. Distribution of New Zealand *Placostylus* Beck in relation to Pliocene land.

Lord Howe Island and the northern tip of the North Island in New Zealand (SOLEM, 1959: 126).

The Bulimulidae are primarily a South American family but the subfamily Placostylinae, to which *Placostylus* belongs, is restricted to the Indo-Pacific. All members of the family are warmth-loving and are restricted to tropical or subtropical areas.

Placostylus has a large (over 29 mm), tall-spired shell, large often dentate aperture, lays clutches of shelled eggs and is usually heavy in construction (Fig. 8).

Until a comparative anatomical study is conducted on *Placostylus* the phylogenetic relationships of taxa within the genus will not be known with any certainty. The higher classification of the New Zealand taxa rests at the moment on the work of HAAS (1936), based on embryonic shell features, and discussed by SOLEM, (1959: 127). SOLEM reviewed all phylogenetic evidence for *Placostylus* (pp. 126–31) and came to the conclusion (p. 130, Fig. 5) that the New Hebridean, New Zealand and Lord Howe Island taxa were more closely related than any of these were to New Caledonian, Solomon or Fijian stock. Taxa in the last two areas are considered more advanced and probably derived from Southern Relict stock. *Placostylus* in New Zealand is, then, another example of the 'basic' land snail fauna.

New Zealand species of *Placostylus* are placed in two subgenera. One species, *Placostylus bollonsi* Suter, is placed in subgenus *Basileostylus* Haas. *Basileostylus* is based on the New Zealand *bollonsi* and is restricted to the Three Kings Islands. The other two species, *ambagiosus* Suter and *hongii* Lesson, are classified in subgenus *Maoristylus* Haas as are also the Lord Howe Island species.

The distribution of the New Zealand species is closely correlated with Pliocene land distribution. Subgenus *Basileostylus* evolved in isolation on the Three Kings Islands. *Placostylus (Maoristylus) ambagiosus* evolved in isolation on the 'Northern' Pliocene island and *P. (M). hongii* likewise on the 'Northland' island (Fig. 8).

New Zealand *Placostylus* are always found on seaward slopes and salt-laden air seems to be a prerequisite for their survival. Living in an area of very broken topography as they do in this country such favourable coastal sites are quite scattered. As a consequence, colonies are scattered and subspeciation trends are obvious, particularly in *ambagiosus*. The isolation of colonies on the 'Northern' Pliocene island was enhanced by the previous existence of numerous small islands, now linked by sand tomboles to form one large area which is referred to as the 'Northern' Pliocene island. The role of these islands and ecological requirements in the development of subspeciation patterns has been discussed at length by POWELL (1938, 1947, 1948, 1951). Similar trends have been noted for New Caledonian Placostylus (PAIN, 1955).

All New Zealand species of *Placostylus* are primarily litter-dwelling snails, feeding on freshly-dead fallen leaves. Both *P. ambagiosus* and *P. hongii* live under a variety of plant cover, the former often associated with flax *(Phormium tenax)* and the latter closely associated with the large broadleaf Karaka tree *(Corynocarpus laevigatus)* the leaves of which are apparently a major food item. *P. bollonsi* does not show any close association with particular plant species, but is always found under a broadleaf canopy.

Juveniles of *P. hongii* have developed an arboreal capability but as adults are always found in litter.

Family: PARYPHANTIDAE

The Paryphantidae are restricted to South Africa, the Seychelles, Indonesia, Melanesia, the Caroline Islands, Samoa, Tonga, New Zealand and eastern Australia (SOLEM, 1959: 147, Fig. 18). MOELLENDORFF (1904–04) produced a monographic study on the family and taxa outside New Zealand have been discussed by a number of authors (WATSON, 1934; CONOLLY, 1939; KONDO, 1943; SOLEM, 1959; SMITH, 1969; 1970). In both distribution and anatomy the Paryphantidae appear to be the oldest carnivorous land snail family, having a late Palaeozoic or early Mesozoic origin.

The phylogenetic relationships of taxa within the family must await a comparative anatomical work. My own studies on the New Zealand fauna show that the phylogeny of the New Zealand taxa agrees broadly with the scheme laid out by SOLEM (1959: 148–9) for the Indo-Pacific members of the family (CLIMO, unpublished).

Within New Zealand it is possible to recognize two groups of paryphantids; one containing heavily calcified shells (sometimes with a secondary loss of the calcium carbonate component), a crinkly not very glossy light-brown periostracum and relatively few whorls; and, another group with a tendency to have more tightly coiled shells, unicoloured shells much darker for the small species, the development of a distinctive colour pattern in some subgroups, a very shiny periostracum and a tendency for the shells to contain little lime being composed in the main of pliable conchin. There has been secondary acquisition of a heavily calcified shell in one subgenus of the second group (*Paryphanta* Albers s.str.). The first group of species refers to *Rhytida* Albers, the second to *Paryphanta* and *Wainuia* Powell.

Wainuia is intermediate in shell characters (Fig. 9), radula and reproductive morphology between *Rhytida* s.str. and subgenus *Powelliphanta* O'Connor. *Rhytida*, as POWELL (1930: 18) mentioned, can probably be interpreted as the ancestral form of paryphantid in New Zealand, the other genera being derived from it or from diverging stock with a rhytidid origin. This divergence has possibly already taken place before the Mesozoic colonization of New Zealand by paryphantids.

It is not possible to determine accurately to which of the two groups

Delos Hutton and *Schizoglossa* Hedley belong. The first is widespread in the Indo-Pacific (see below), an archaic paryphantid, and must be near the basal rhytidid stock in origin, while the other is a specialized endemic New Zealand off-shoot from the two main lines of paryphantid development. What remains of the shell of *Schizoglossa* indicates that it is possible closer to the rhytidid line than the paryphantid; the light-brown colouration and heavy shell being the two supporting factors. *Delos*, with its smooth shiny periostracum, variously ornamented by colour markings perhaps indicates closer affinities with the paryphantid line of development.

Reproductive morphology provides few clues for unravelling the New Zealand hierarchy. Reduction in the male organ has occurred in several different paryphantid groups, represented in New Zealand by all members of the genus *Schizoglossa* and some members of the genera *Paryphanta* and *Rhytida*. The reduced penis in *Paryphanta* s. str. is the major criterion for its subgeneric separation from *Powelliphanta* O'Connor. It will be necessary to separate, at the subgeneric level, a group of keeled species of *Rhytida* in New Zealand (Fig. 9). They all have a reduced penis and show a very compact geographic distribution.

The only members of the large-shelled (20–110 mm diameter) *Paryphanta-Wainuia* group of species outside New Zealand are *Victaphanta* Iredale from Victoria, Australia and *Melavitrina* from Tasmania.

Although the medium-sized (12–64 mm in diameter) genus *Rhytida* has been recorded from Tasmania, Australia and New Caledonia, species from those areas show only a broad similarity to the New Zealand type of the genus. They are probably related though not necessarily congeneric.

Until the anatomy of the *Delos*-group of small paryphantids (4–11 mm diameter) is understood outside New Zealand it can only be said that other Indo-Pacific areas (New Hebrides; New Caledonia; Australia; Caroline Islands) have shells with the facies of New Zealand *Delos;* perhaps closely related, perhaps not.

The large shelled-slug (15–40 mm diameter), *Schizoglossa*, is an endemic New Zealand genus whose position in the New Zealand-based phylogeny I have indicated (Fig. 9).

The endemic subgenus of *Delos*, *Delouagapia* Powell, shows affinities with both *Delos* and the exotic genus *Ouagapia* Crosse, the latter recorded from New Caledonia, Solomon Islands, New Guinea, Bismarcks, New Hebrides, Tonga, Samoa and the Fijis by Solem (1959: 149).

All members of the Paryphantidae in New Zealand are dependent upon high humidity and natural plant cover for survival. *Delos* is a normal component of moderately deep forest litter, *Schizoglossa* is most common under logs in rain forest, while *Rhytida*, *Wainuia* and *Paryphanta* are found under larger than average accumulations of litter. With few exceptions all members of the last three genera are forest dwellers but two species of *Paryphanta* are restricted to alpine tussock in the South Island. *Delos (Delouagapia) cordelia* (Hutton) is the only arboreal species.

All New Zealand members of the family lay clutches of shelled eggs in depressions in the soil beneath leaf litter. The number and size of the eggs varies with the species and for a large species like *Paryphanta superba* (Powell) (11 cm in diameter) an individual egg can be 6 mm in greatest dimension.

The main food item of these basically carnivorous snails are earthworms, insects and other snails, however, some species of *Paryphanta* are known to eat fleshy fungi as well. It is noticeable in the field that when ever there is a high concentration of *Rhytida* in an area the numbers of punctid and liareid snails is much reduced. Adult specimens of several species of *Rhytida* are predators of juveniles of their own species. All teeth in the paryphantid radula are long and sabre-like, a probable modification for a predominantly carnivorous mode of life.

All New Zealand paryphantids are nocturnal, but they are often induced onto the litter surface during rain storms or heavy fog when humidity is high and light intensity low.

The distribution of *Paryphanta* in this country is very discontinuous with high concentration in very small areas (colonies). The fate of many colonies of *Paryphanta*, and in some cases species, is virtually sealed; their extinction seems inevitable. The removal or opening-up of lowland forest on fertile alluvial soils, for agricultural purposes, has decimated many colonies. The rat, mice and hedgehog populations associated with human occupation provide a predatory pressure never provided by the flightless weka, which was the only significant endemic predator before the advent of man. Man also introduced pigs, which in their wild state are heavy predators of *Paryphanta.* Probably on a less significant scale, introduced European black-birds and thrushes provide predatory pressure on the smaller species belonging to the genera *Rhytida* and *Wainuia.*

The present distribution of many members of the family in New Zealand appears to be relict one. The ranges of species in this category can be extended by scores of kilometers in some cases by subfossil remains collected in limestone caves or in sand dune deposits.

It appears likely that several paryphantid taxa evolved in isolation on islands within the New Zealand archipelago during the Pliocene, but *Wainuia*, *Paryphanta*, *Delos* and *Rhytida* stock were differentiated well before Pliocene times. One subgenus of *Paryphanta*, one of *Delos*, and two of *Rhytida* (one as yet unnamed) evolved on the islands north of the 'Auckland Pliocene Straits' (Fig. 9); other subgenera speciating or already having speciated south of these straits. As would be expected, the boundaries between taxa are not always clear-cut because of subsequent migrations of species in post-Pliocene times. The Pliocene patterns have been blurred by the retreat and advance of species during the Pleistocene ice-ages, different taxa varying in their tolerance to cold climate conditions and all being adversely affected by lowered precipitation and drier forest conditions.

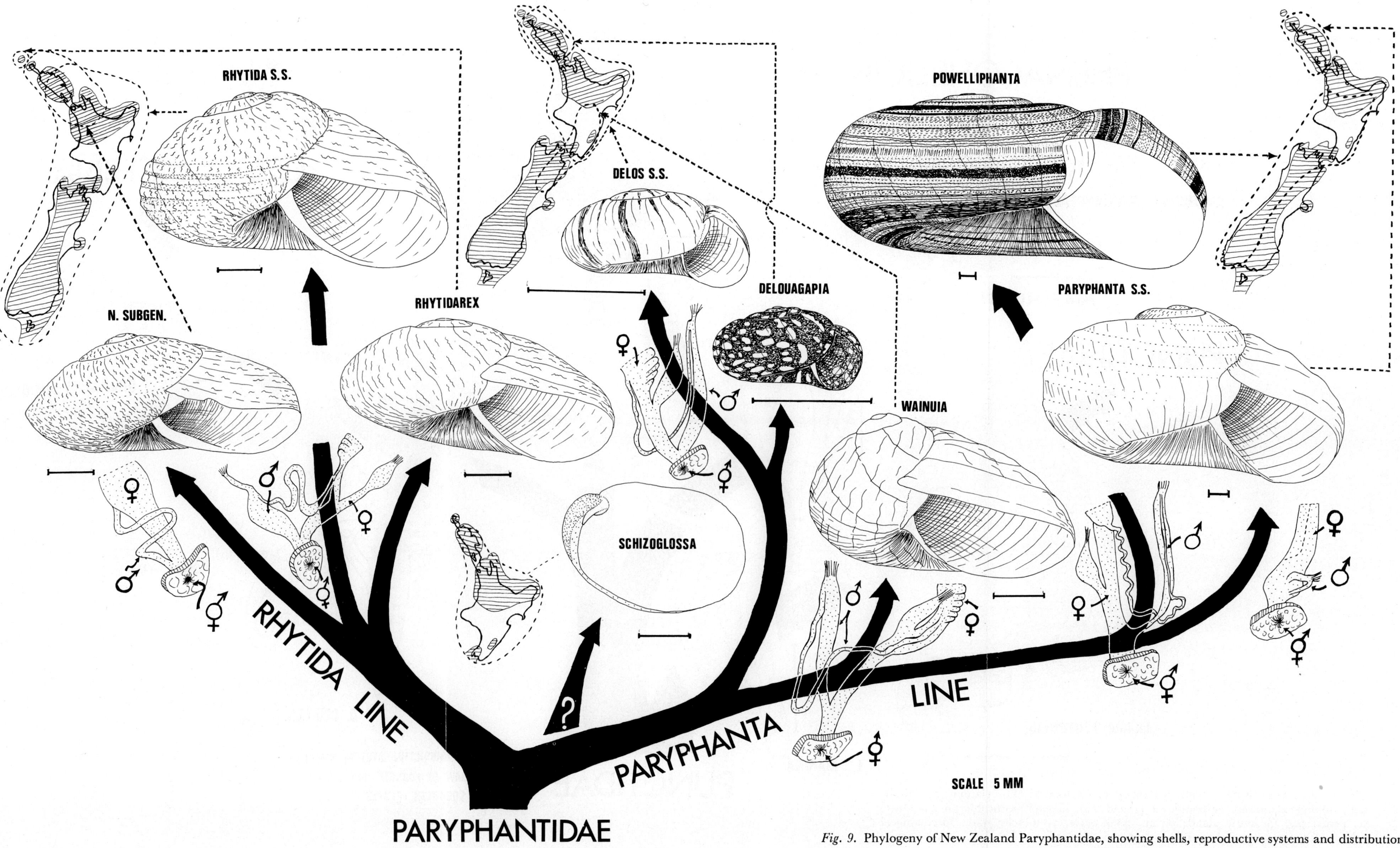

Fig. 9. Phylogeny of New Zealand Paryphantidae, showing shells, reproductive systems and distribution.

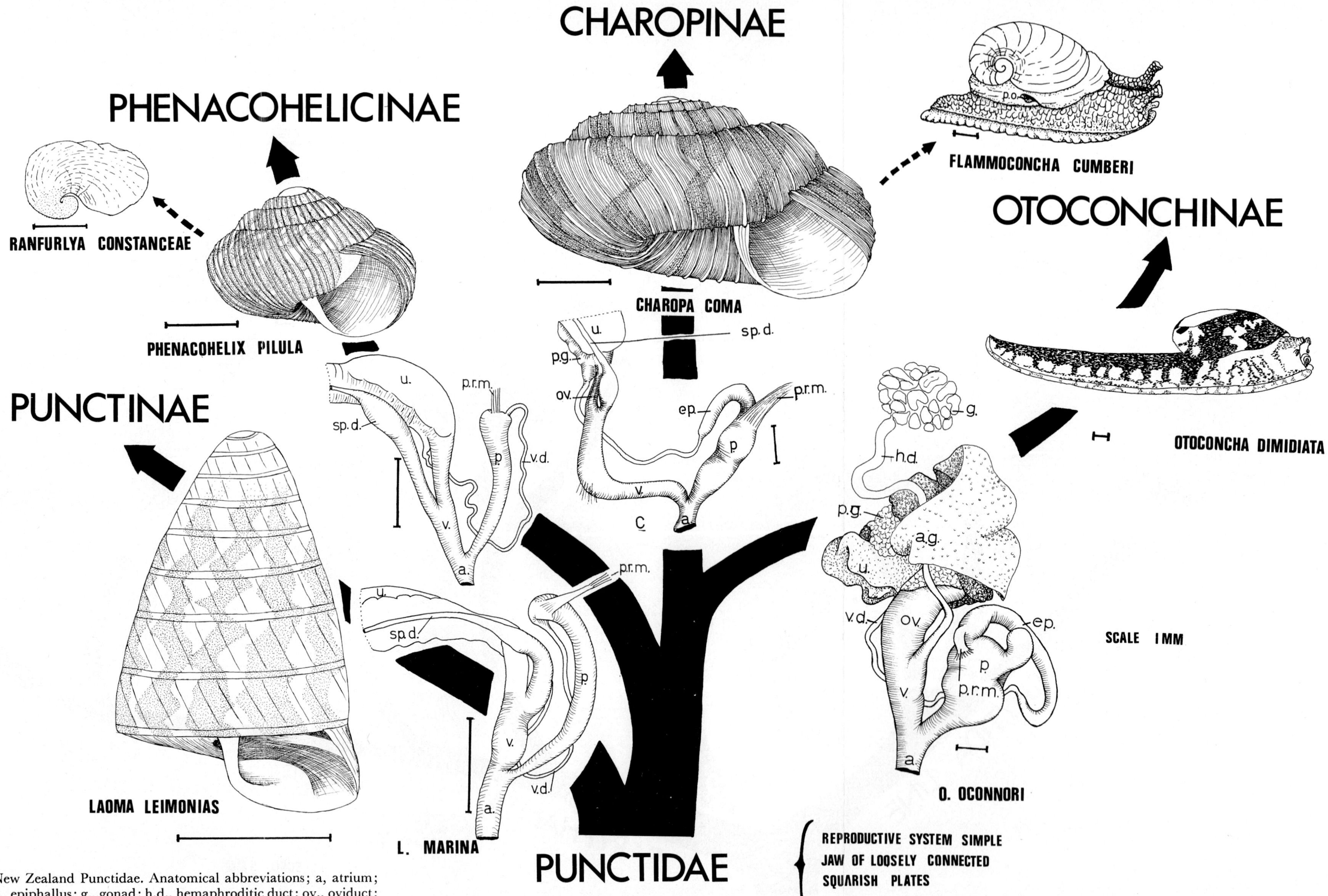

Fig. 10. Phylogeny of New Zealand Punctidae. Anatomical abbreviations; a, atrium; a.g., albumen gland; ep., epiphallus; g., gonad; h.d., hemaphroditic duct; ov., oviduct; p., penis; p.g., prostatic gland; p.r.m., penis retractor muscle; sp.d., spermathecal duct; u., uterus; v., vagina; v.d., vas deferens.

Schizoglossa possibly evolved on one of the scattered Pliocene islands south of the 'Auckland Straits', though it is so widely distributed in the North Island that its origins may predate the Pliocene. *Delos*, widely distributed in the North Island, was too, probably differentiated very early but the speciation patterns obvious now are those of the Pliocene. *Wainuia* probably had a similar evolutionary history to *Delos*, but either developed on, or became restricted to, more southern Pliocene islands, its post-Pliocene distribution not being truncated in the same way by the present Cook Strait as in *Schizoglossa* and *Delos*.

The present distribution of *Paryphanta (Powelliphanta)* (Fig. 9) is a relict pre-Pliocene pattern with the major center of evolution in the north of the South Island. POWELL (1930, 1932, 1936, 1938, 1946, 1949a, 1949b) has written extensively on speciation trends, geographic isolating factors, variation, hybridization and distribution in *Powelliphanta*. I have not the space to delve deeply into this fascinating topic and give below only a brief outline of the species.

Within the genus *Paryphanta*, *Powelliphanta* species and subspecies can be broadly divided into two ecological groups, lowland and highland taxa. In the South Island, *P. spedeni* Powell, *P. rossiana* Powell, *P. superba* Powell and *P. hochstetteri* (Pfeiffer) are exclusively high altitude species; only rarely do colonies exist below 600 m in altitude. The first two species in this group live in alpine tussock and scrub above the bush-line. Of the other South Island species, *P. lignaria* Hutton is predominantly alpine but lowland subspecies (derived from high altitude stock?) do occur, in *P. gilliesi* Smith the reverse is the case and the last species to be described, *P. fiordlandica* Climo, appears to be a lowland species.

There is only one species of *Paryphanta (Powelliphanta)* in the North Island and this is represented by a number of subspecies, some with lowland and some with higher altitudinal ranges.

Within the range of a given species, lowland colonies are generally more continuously distributed than higher altitude colonies. For example, when all subfossil occurrences as well as living colonies are considered the predominantly lowland species, *gilliesi*, in the north of the South Island, has a rather compact distribution. The alpine, west coast South Island species *rossiana*, on the other hand, has a discontinuous distribution stretching down the whole western edge of the Southern Alps. The distribution of alpine species has probably been rendered discontinuous by valley glaciers and local permafrost conditions during the Pleistocene. The south west of the South Island is still biologically poorly investigated and further colonies of *Paryphanta (Powelliphanta)* are likely to be discovered there.

Suborder: AULACOPODA
Family: PUNCTIDAE

In terms of numbers of species this is by far the largest component of the

New Zealand land snail fauna, composed of four subfamilies (Fig. 10), over 20 genera and approximately 180 species. It has, unfortunately, been the least studied of New Zealand families. The minute size (under 4 mm) of most species has probably been a deterrent factor in this respect.

The family is based on the Northern Hemisphere *Punctum* Morse and has a relict global distribution, but nowhere is the species diversification so marked as in New Zealand. It is only in the Southern Relict zone, New Zealand in particular, that the family has been able to radiate in the absence of competition from other faunal elements, for example, the advanced helicids.

The morphological characteristics of the four subfamilies is presented in the following diagram (Fig. 10) in which the type species of genera involved in New Zealand-based subfamily names are figured.

Subfamily: CHAROPINAE

Shell morphology is diverse within this subfamily (Fig. 11) but shells can be artificially grouped as follows: 1. flattened, discoidal to subdiscoidal species, with radial riblets, generally under 5 mm in diameters, typified by the genus *Charopa* Albers; 2. larger taxa such as *Gerontia* Hutton and *Suteria* Pilsbry (diameter up to 10 mm) with a similar shell facies to *Charopa;* 3. species with a tendency towards an auriform shell (Flammo-*concha* Dell), perhaps derived from 2; 4. species with a pyramidal spire, keeled periphery, elaborate peristracal sculpture and under 6 mm in diameter (*Therasiella* Powell).

Of the endemic subgenera of *Charopa*, *Charopa* s.str. is widespread in New Zealand, *Egestula* Iredale is restricted to the northern North Island Pliocene islands mentioned earlier, *Aeschrodomus* Pilsbry is represented by one species with a relict widespread distribution and another restricted to the N.W. of the South Island, and an unnamed subgenus (grouped with *Ptychodon* Ancey by CLIMO, 1969) has one widespread species, one restricted to the N.W. of the South Island and a third restricted to the S.W. of the South Island.

On the basis of present published information on Pacific charopine snails, outside New Zealand, it is not possible to ascertain whether *Ptychodon* as applied to the Pacific fauna (SOLEM, 1959) is congeneric with my New Zealand usage.

Outside New Zealand, *Mocella* Iredale has been recorded from S.W. Australia, and Bismarck, Solomon, Fiji, Samoa, Society, Cook, Tonga, Norfolk and Lord Howe Islands.

Subgenus *Geminoropa* Iredale is found in S.E. Australia and New Zealand.

The above members of the Charopinae are rather conservative taxa with close relatives widely distributed in the Indo-Pacific. The systematics and distribution of the group are too poorly known to speculate on zoogeographic relationships, but that they are a part of the 'basic' Southern Relict fauna is undisputed.

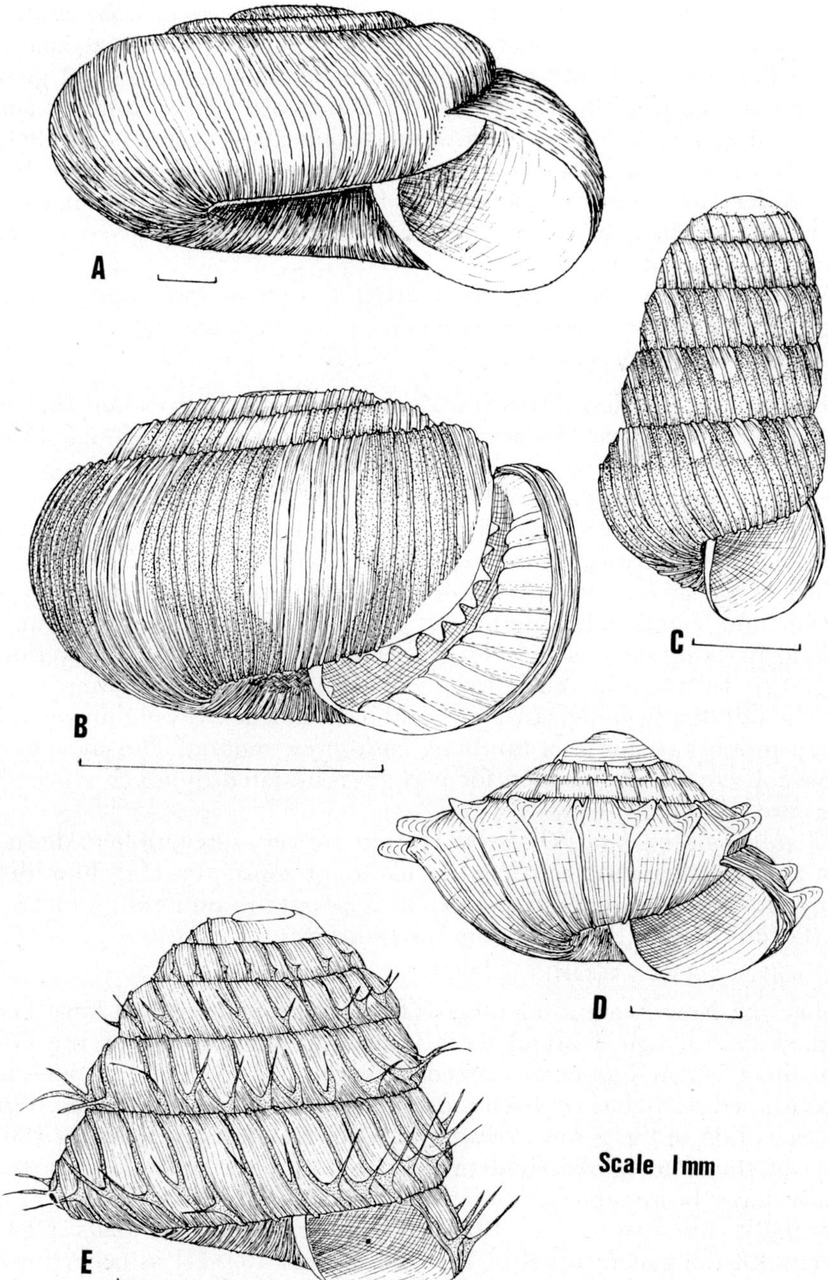

Fig. 11. New Zealand Charopinae. A, *Gerontia pantherina* Hutton; B, *Charopa (Ptychodon) leioda* (Hutton); C, *Phenacharopa novoseelandica (Pfeiffer)*; D, *Therasiella tamora* (Hutton); E, *Charopa (Aeschrodomus) stipulata (Pfeiffer)*.

Of the other endemic charopine taxa, *Therasiella*, *Phenacharopa*, *Flammocharopa* Climo, *Gerontia*, *Suteria* and *Flammoconcha* are all characteristically New Zealand genera. *Flammoconcha* is the most deviant genus in the Charopinae, all species showing a strong tendency towards shell reduction by developing a auriform shell. This development is paralleled in the Otoconchinae Cockerell where all species have an auriform shell, and by *Ranfurlya* Suter in the Phenacohelicinae. Of necessity, taxa with such reduced shells are prone to dessication and are restricted to very humid microhabitats.

Most charopine snails are litter-dwellers though some species have become very flattened and are specialized for life under bark.

Subfamily: OTOCONCHINAE

The systematic position of the New Zealand endemic otoconchine shelled-slugs has been discussed by several authors (BAKER, 1938; DELL, 1952; CLIMO, 1969a, 1971a).

I regard members of the Otoconchinae as specialized derivatives of ancestral 'basic' New Zealand charopine stock. They show a certain external morphological parallel with helicarionid slugs. The shell is reduced, auriform, nearly enclosed by the reflexed mantle in three species and completely internal in the fourth. The visceral hump is anteriorly positioned and a long specialized tail projects posteriorly (Fig. 12). In rest, this tail is wrapped around the visceral hump much like the tail of a sleeping cat. It is capable of violent writhing movements which propel the slug in a tumbling cork-screw motion. This movement is possibly an escape reaction for it is often initiated by touch when the slugs are collected.

Members of the genus *Otoconcha* Hutton are very susceptible to dessication and are therefore, restricted to niches of constantly high humidity. They are most often collected under rotting logs in rain forest, sometimes in association with the onycophorans *Peripatoides* or *Peripatus*.

Subfamily: PHENACOHELICINAE

Unlike the New Zealand members of the Charopinae which have been studied recently on a broad basis (CLIMO, 1969b, 1970), or the Otoconchinae which has been revised recently (CLIMO, 1971a), no such broad-based study has yet been published on the Phenacohelicinae. With the exception of the genus *Phenacohelix* Suter (revised by CUMBER, 1961) none of the other globosely-depressed to turbinate genera in the subfamily have been subjected to critical review. The classification rests, essentially, on IREDALE's (1915) rearrangement of SUTER's (1913) system. Of the genera listed in POWELL (1962: 109–11) as occurring in this subfamily, *Gerontia*, *Therasiella*, and *Suteria* are members of the Charopinae. *Obanella* Dell is a member of the Punctinae and most species in *Flammulina* Martens are in fact members of the conchologically parallel genus *Flammoconcha* in the *Charopinae* (CLIMO, unpublished).

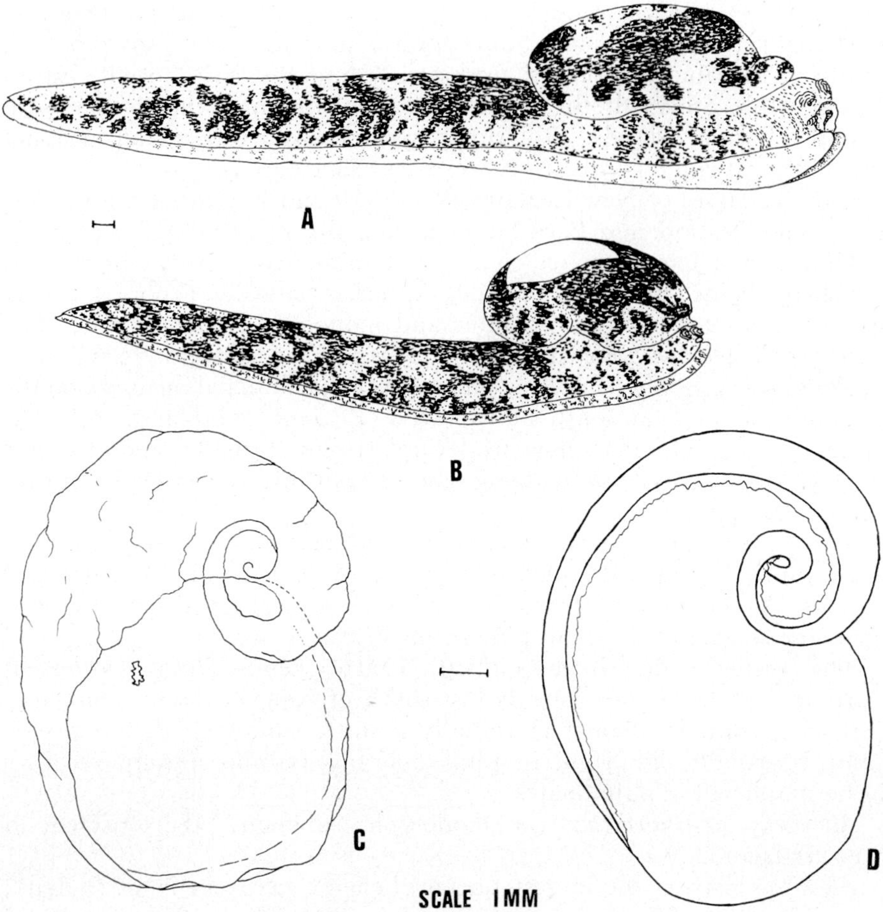

Fig. 12. New Zealand Otoconchinae. A & C, animal and shell of *Otoconcha roscoei* Climo; B & D, animal and shell of *O. oconnori* (Powell).

Of the genera I accept as belonging to the subfamily, *Phelussa*, Iredale, *Thalassohelix* Pilsbry, *Thermia* Hutton, *Serpho* Hutton and *Therasia* Hutton are all probably congeneric, but *Allodiscus* Pilsbry, *Phenacohelix* Suter, *Flammulina* Martens and *Ranfurlya* Suter are all natural taxa (Climo, unpublished). I am engaged in a revision of all New Zealand phenacohelicine taxa and some of this has been published (Climo, 1971b). In this last paper two new genera, *Pseudallodiscus* and *Protoflammulina* and one new subgenus of *Allodiscus*, *Hirsutodiscus*, were described. Until my revision is complete it would be premature to discuss systematics or distribution in any detail and only a few important points are made below.

All New Zealand phenacohelicine taxa, with perhaps the exception of *Flammulina*, are apparently endemic and reflect the general distribution patterns of the Charopinae. Some species are widespread, some have a southern distribution, others are restricted to the N.W. of the South Island while others fit the hypothesis of evolution on northern North Island Pliocene islands. Still other species have widespread disjunct distributions which can be interpreted as relict patterns.

Close relatives of New Zealand taxa are found in Australia and New Caledonia, Norfolk and Lord Howe Island (SOLEM, 1959; 267, Fig. 13).

Unlike the Charopinae which are predominantly litter-dwelling snails, phenacohelicine snails are usually at least partially arboreal, being closely associated with cryptogams and monocotyledonous plants. The two species of '*Serpho*', restricted to the North Island, are entirely arboreal.

There was probably a diverse phenacohelicine fauna at one time on the subantarctic plateau south of the New Zealand mainland. All that remains of this land mass now are several groups of small islands and two of these have monotypic endemic genera, possibly relicts of the earlier diverse fauna.

The genus *Ranfurlya* is a short-tailed phenacohelicine slug with an auriform shell (Fig. 13) which is restricted to the subantarctic Auckland Islands. As mentioned earlier, it is a conchological parallel to the otoconchine slugs and the charopine genus *Flammoconcha*.

The recently described (CLIMO, 1971b) genus *Protoflammulina* is restricted to the Snares Islands just south of New Zealand. This large (up to 10 mm in diameter) radially costate, subdiscoidal, monotypic genus is sexually dimorphic in adult anatomy, possibly a unique feature in hermaphroditic pulmonates.

Allodiscus (Hirsutodiscus), a monotypic subgenus, is restricted to Stewart Island.

Allodiscus s.str. is the largest phenacohelicine taxon in New Zealand, containing over 20 species ranging in size from 3 to 13 mm. Several species are widespread but the majority have limited ranges. In the north of the North Island, two species are restricted to the Three Kings Islands, one to the 'Northern' Pliocene island, one to the 'Northland' Pliocene island, and one to the Poor Knights Islands just east of the northern tip of the North Island. Most of the locally distributed species in the South Island are restricted to very small areas, mainly alpine localities such as the Hooker Valley. All species of *Allodiscus* are globosely-depressed, have spirally striate protoconchs, are radially costate and usually tightly coiled.

Phenacohelix has the shell facies of *Allodiscus* but is recognizable different on anatomical grounds.

Phelussa, *Thalassohelix*, *Serpho*, *Thermia* and *Therasia* are all relatively large (up to 14 mm in diameter), fairly loosely coiled, depressed-turbinate genera, in which the low-relief radial sculpture is primarily periostracal, however, radial riblets are strongly developed in *Phelussa*.

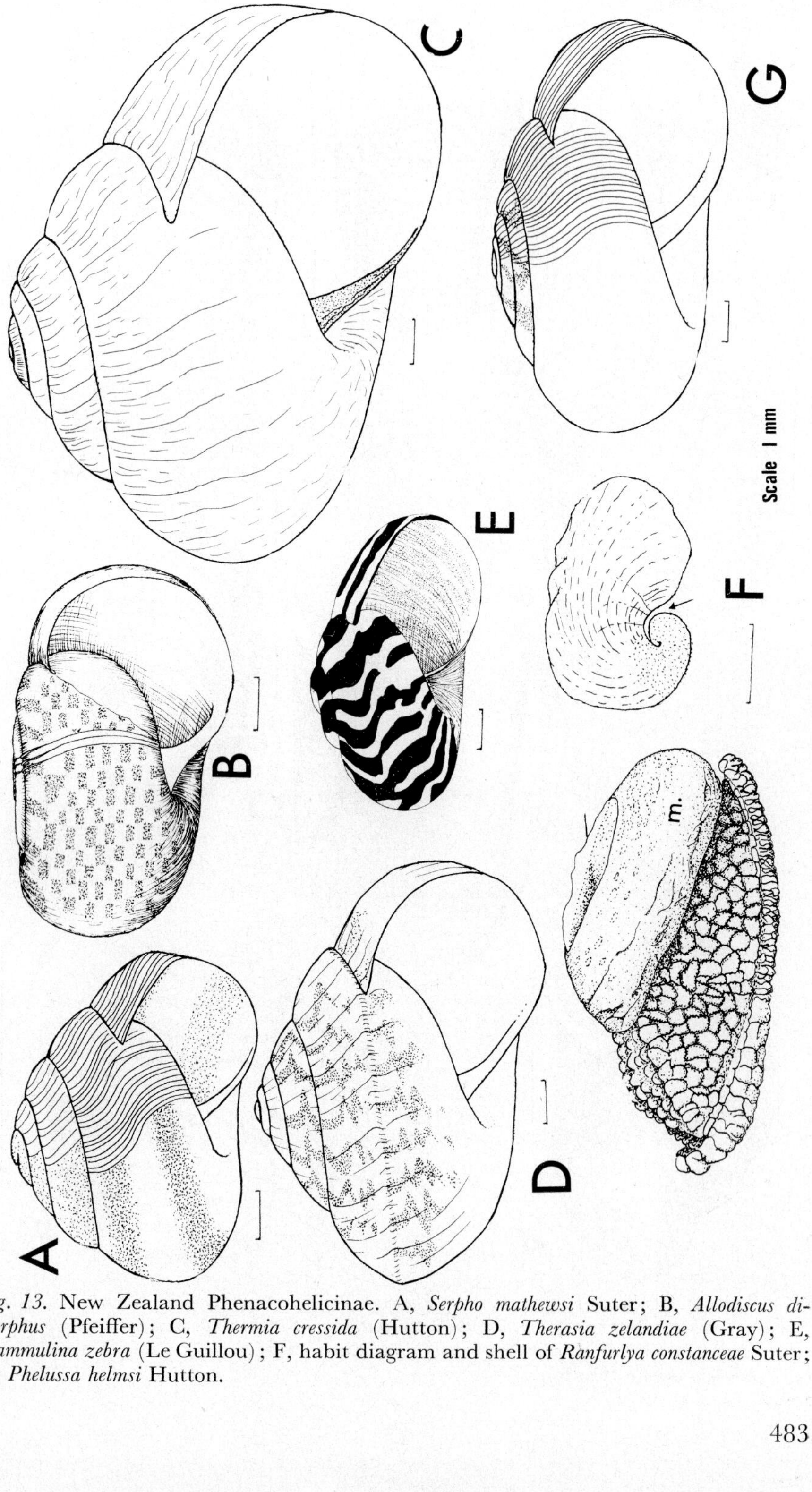

Fig. 13. New Zealand Phenacohelicinae. A, *Serpho mathewsi* Suter; B, *Allodiscus dimorphus* (Pfeiffer); C, *Thermia cressida* (Hutton); D, *Therasia zelandiae* (Gray); E, *Flammulina zebra* (Le Guillou); F, habit diagram and shell of *Ranfurlya constanceae* Suter; G, *Phelussa helmsi* Hutton.

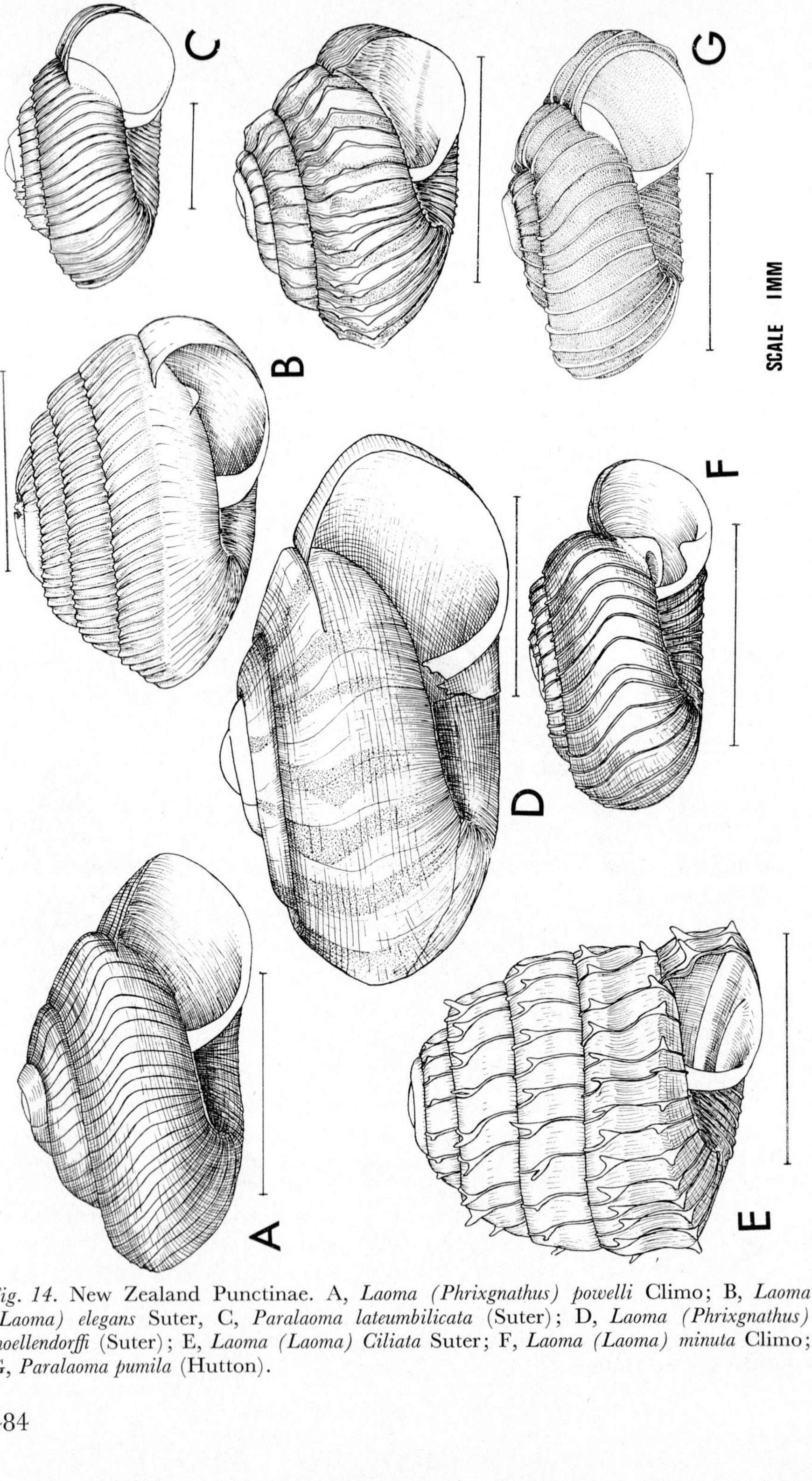

Fig. 14. New Zealand Punctinae. A, *Laoma (Phrixgnathus) powelli* Climo; B, *Laoma (Laoma) elegans* Suter, C, *Paralaoma lateumbilicata* (Suter); D, *Laoma (Phrixgnathus) moellendorffi* (Suter); E, *Laoma (Laoma) Ciliata* Suter; F, *Laoma (Laoma) minuta* Climo; G, *Paralaoma pumila* (Hutton).

Flammulina is very loosely coiled, consisting of small (under 8 mm in diameter), usually smooth shells which are most often collected by beating fern foliage. The type of the genus (*F. zebra* Le Guillou) is strikingly flammulated by reddish-brown markings.

Most phenacohelicine snails feed on litter detritus or living leaf cuticle.

One monotypic genus, *Hydrophrea*, lives aquatically and interstitially in alluvial gravels (see section on the Subterranean Fauna).

Subfamily: PUNCTINAE

The subfamily Punctinae is represented in New Zealand by three genera and approximately 70 species. Many new species have yet to be described.

Members of the subfamily are often called 'Glass Snails'; they are usually very glossy, are minute (less than 2 mm to 6 mm in diameter), vary greatly in shape and sculpture (Fig. 14) and occupy a great range of habitats.

Although the present species list (POWELL, 1962; 112–13) for the subfamily is impressive, the majority of the species will have to be re-evaluated in modern quantitative and anatomical revisions.

The bulk of the New Zealand species are included in the genus *Laoma* Gray which has been subdivided into the three subgenera *Laoma* s.str., *Phrixgnathus* Hutton and *Taguahelix* Powell. The only distinguishing feature, probably an artificial one, between *Laoma* and *Phrixgnathus* is the presence of lamellae in the aperture of the former. Subgenus *Taguahelix*, composed of four species, is restricted to the south of the South Island, Stewart Island and the New Zealand subantarctic islands, and is distinguished by its globosely-depressed shells ornamented by radial costae. It is a southern element in the New Zealand punctine fauna. The first two subgenera are widespread in New Zealand, but nearly all shells with lamellate apertures are restricted to the warmer North Island. The greatest concentration of species of *Laoma* and *Phrixgnathus* is in the north of the North Island but too little is known of their distribution patterns to discuss internal zoogeography.

The genus *Laomarex* Powell is represented by one species restricted to the Three Kings Islands.

Paralaoma Iredale contains the smallest of New Zealand land snails, no species being larger than 2.8 mm in diameter and the smallest measuring 1.8 mm. These litter-dwelling, discoidal to depressed-turbinate (Fig. 14), radially ribbed snails are widespread in New Zealand and are represented by 8 species.

Obanella Dell, composed of three minute (under 2.5 mm in diameter) flattened or depressed-turbinate species has a similar shell facies to *Paralaoma* but the radial sculpture differs in being composed of raised, sparsely distributed, radial lamellae.

Close relatives of the New Zealand punctine genera are found in New Caledonia, the New Hebrides, Australia, the Kermadec and Lord

Howe Islands (*Paralaoma* is based on a Kermadec Island species), but until anatomical work is carried out the significance of conchological similarities between these areas cannot be evaluated.

New Zealand's Geological History and Land Snail Faunal Zones

The distribution patterns of land snails in New Zealand are complex and are the result of several contributing factors, which can be grouped under the following headings: 1. early and middle Tertiary distribution of land; 2. Pleistocene and post-Pleistocene climatic conditions.

During the Cretaceous and Eocene, land was more or less continuous the whole length of New Zealand (FLEMING, 1962: Figs. 4 and 6). Widespread New Zealand snail taxa (e.g. *Therasia; Cytora; Paryphanta (Powelliphanta); Rhytida* s.str.; *Charopa* s.str.; *Charopa (Ptychodon); Laoma (Phrixgnathus); Paralaoma; Omphalorissa*), conservative in nature, probably date from these times, though a number of these could have dispersed and achieved a wide distribution during the Miocene (see below).

During the Oligocene (Fig. 15) New Zealand was divided into a northern land mass roughly corresponding with the present North Island and a southern land mass which included the South Island, Stewart Island and most of the subantarctic plateau. It was probably on these two land masses that the characteristically northern faunal elements (n. subgen. *Rhytida-Rhytidarex* basal stock; *Schizoglossa; Delos; Paryphanta* s.str.; *Charopa (Egestula); Liarea; Laoma (Laoma)* and southern faunal elements *Laoma (Taguahelix); Ranfurlya; Allodiscus (Hirsutodiscus); Protoflammulina; Phelussa; Palliopodex* developed.

During the Miocene there was continuous land stretching along the whole New Zealand archipelago (FLEMING, 1962: Figs. 8 and 9). There must have been some intermixing of northern and southern Oligocene stock during these times but this may have been small due to warmth-adapted snails in the north and cold-adapted in the south.

The affects of the Pliocene distribution of land on the evolution of New Zealand land snail taxa has been discussed in various places above and will not be elaborated upon here. Fig. 15 summarizes all geological and distributional data mentioned in this section of this chapter, and discussions on the Pliocene which took place earlier.

At least four glaciation (low sea-levels) and three interglacial intervals (high sea-levels) occurred in the late Pleistocene (FLEMING, 1962: 87). The shoreline changes that did occur were mainly due to eustatic changes in sea-level. In the last glaciation sea-levels were about 115 m below present levels but in the interglacials, levels were 20 m or more above present levels. The high interglacial sea-levels would have maintained the Pliocene patterns of distribution north of Auckland in the North Island by forming sea-straits between the blocks of land (Pliocene islands) now linked by alluvial lowland.

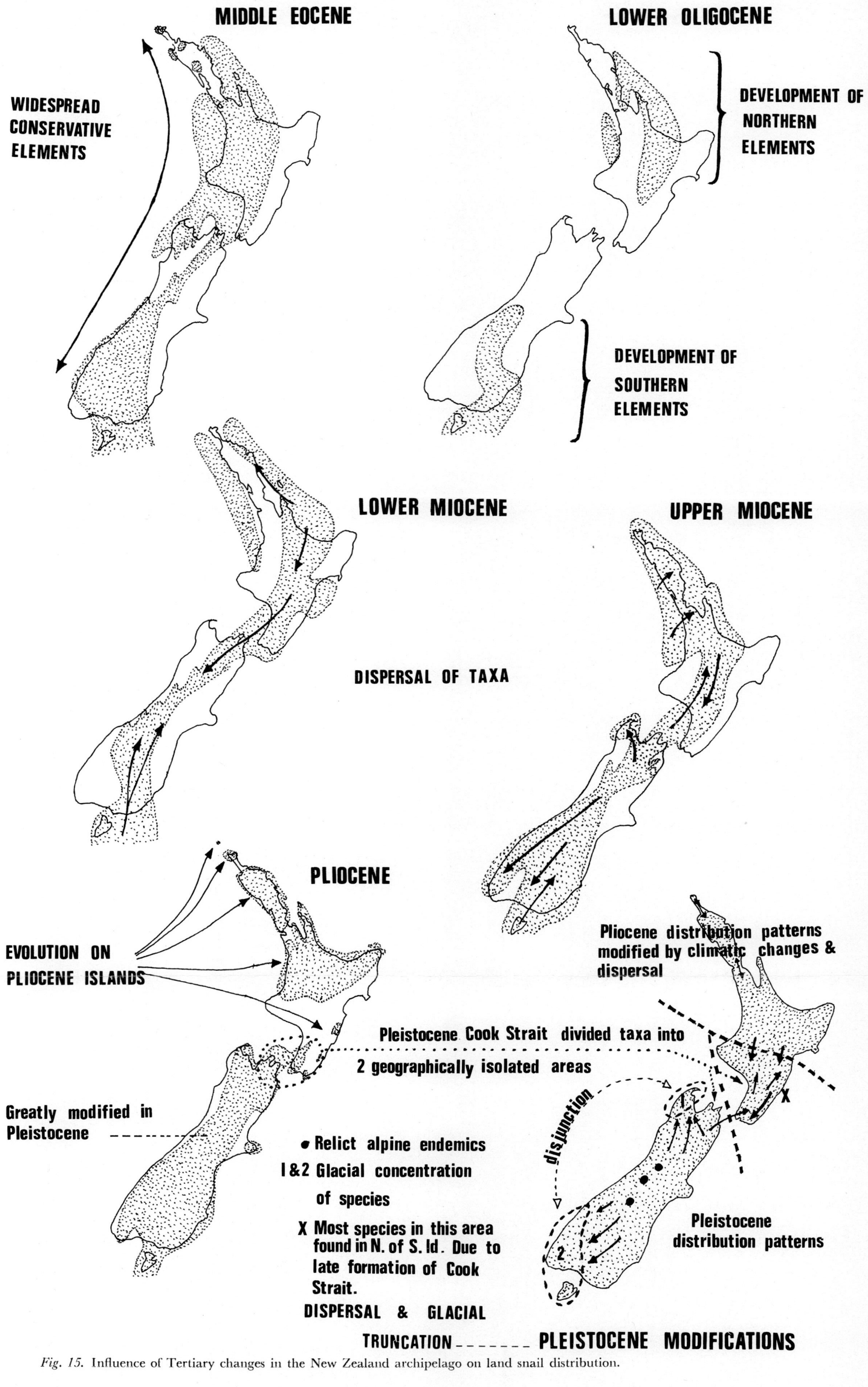

Fig. 15. Influence of Tertiary changes in the New Zealand archipelago on land snail distribution.

During glacial peaks much of the South Island, particularly the narrow central 'neck', was probably a most inhospitable habitat for land snails. During these peaks, apart from local permafrost conditions, there was rapid erosion and periodic deforestation of the mountains. Any snails that were endemic to the central part of the South Island would now only be left in isolated alpine pockets. We find that this is indeed the case.

There are two areas in the South Island that were probably not so adversely affected (from the land snail point of view) during glacial peaks because of the greater area of mountain and valley systems available for pocket retreat and survival: Fiordland-Stewart Island and the extreme N.W. of the island. The mountains were not as high in these areas, the lowland areas were not as completely covered by unstable aggrading alluvium and consequently, there were more niches available for survival.

Many of the endemic species now found in the S.W. or N.W. of the South Island were probably much more widely spread prior to Pleistocene times. These areas of high specific endemism are probably partly due to glacial 'concentration' of species rather than a reflection of earlier Tertiary land distribution. This hypothesis is supported by two species of land snail (*Otoconcha fiordlandica* Dell; *Allodiscus fectoloides* Dell) which are only found in these two areas. Some of the S.W. South Island endemic species, however, are possibly indicative of a southern fauna, though it is difficult to decide which species are glacial relicts and which are derivatives from a southern early Tertiary land mass (and which never extended as far north as central regions of the South Island).

Dell (1957, Fig. 24) provided a map of land snail zones in New Zealand. His scheme is essentially the same as that which I ellaborated upon above. In any scheme of zonation, with respect to land snails, zones are not going to be based on a similar point in New Zealands geological history in the North and South Islands; those in the South are basically Pleistocene, those in the North Pliocene in origin. Superimposed upon this system which is complex enough are the pre-Oligocene conservative widespread taxa, the Oligocene northern and southern elements and the widespread species which dispersed during the Miocene. The final blurring of the picture has been provided by post-glacial migration of species and man's destruction of many natural habitats.

The Subterranean Molluscan Fauna of New Zealand

It has recently been discovered that New Zealand has a quite diverse subterranean molluscan fauna, the snails found in two types of habitat: caves (cavernicolous species); ground-water associated with Pleistocene gravels in river valleys and out-wash plains (phreatic species). Nine of the species (two previously described) are hydrobiids (Hydrobiidae) and one is a stylommatophoran pulmonate (Punctidae; Phenacohelicinae).

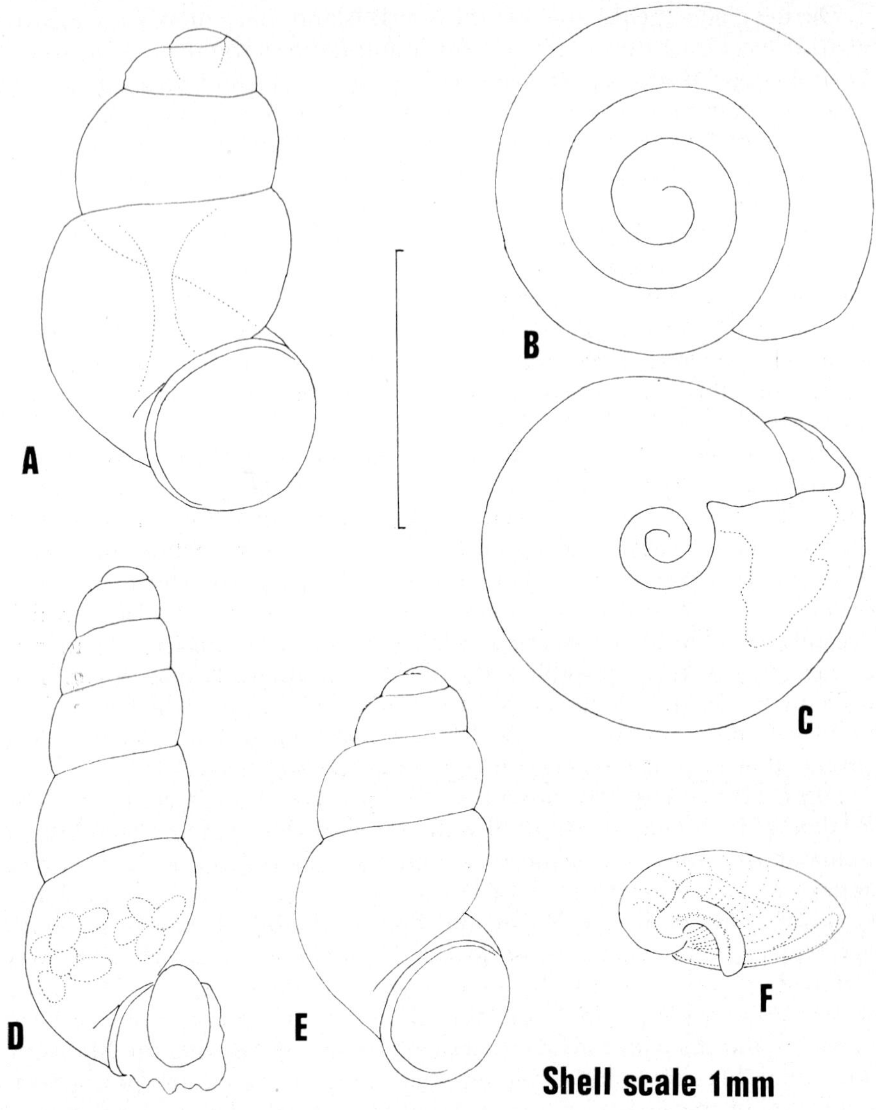

Fig. 16. Subterranean Snails. A, *Hadopyrgus anops* Climo (holotype); B–C, *Hydrophrea academia* (Climo) (holotype); D, *Kuschelita mica* Climo (holotype); E–F, *Catapyrgus spelaeus* Climo (holotype and operculum).

The latter is the first record of an aquatic subterranean stylommatophoran pulmonate. The accommodation of these species has required the erection of four new genera.

A full description of this fauna and its zoogeographic relationships is presented in a paper (Climo, in press) which discusses Australasian

Hydrobiidae, selects both a subfamily for this southern fauna (Hydrobiinae) and three informal tribes for initial orientation purposes. Two of the tribes are strictly Australasian and S.W. Pacific in distribution while the third enigmatic relationship (representing the third tribe) is between a New Zealand and a Dalmation snail.

The following is a summary of the species constituting the above subterranean fauna (the types of new genera are shown in Fig. 16):

	PULMONATA
	STYLOMMATOPHORA
	PUNCTIDAE Morse
	?PHENACOHELICINAE Suter
	Hydrophrea Climo (phreatic)
Type species:	*Laoma (Phrixgnathus) academia* Climo, 1970
	Hydrophrea academia (Climo, 1970)
	PROSOBRANCHIA
	RISSOACEA
	HYDROBIIDAE Fischer
	HYDROBIINAE Fischer
	HEMISTOMIA-TRIBE (Informal)
	Potamopyrgus Stimpson, 1865 (surface-dwelling, phreatic and cavernicolous)
Type species:	*Melania corolla* Gould, 1847
	Potamopyrgus subterraneus Suter, 1905
	Potamopyrgus manneringi Climo
	Hadopyrgus Climo (phreatic and cavernicolous)
Type species:	*Hadopyrgus anops* Climo
	Hadopyrgus anops Climo
	Hadopyrgus brevis Climo
	Kuschelita Climo (phreatic)
Type species:	*Kuschelita mica* Climo
	Kuschelita mica Climo
	Kuschelita inflata Climo
	ISTRIANA-TRIBE (Informal)
	Catapyrgus Climo (cavernicolous)
Type species:	*Catapyrgus spelaeus* Climo
	Catapyrgus spelaeus Climo
	FLUVIOPUPA-TRIBE (Informal)
	Opacuincola Ponder, 1966 (cavernicolous)
Type species:	*Opacuincola caeca* Ponder, 1966
	Opacuincola caeca Ponder, 1966
	Opacuincola troglodytes Climo
	Opacuincola kuscheli Climo

REFERENCES

Baker, H. B. 1938. The Endodont Genus Otochoncha. Proceedings of the Malacological Society of London, 23: 89–91.

Burton, D. W. 1963. A Revision of the New Zealand and Subantarctic Athoracophoridae. *Transactions of the Royal Society of New Zealand (Zoology)*, 3(6): 47–75.

Climo, F. M. 1969a. Classification of New Zealand Arionacea (Mollusca: Pulmonata). I. The Higher Classification. *Records of the Dominion Museum, Wellington*, 6(12): 145–58.

Climo, F. M. 1969b. Classification of New Zealand Arionacea (Mollusca: Pulmonata). II. A Revision of *Charopa* subgenus *Ptychodon* Ancey, 1888. *Records of the Dominion Museum, Wellington*, 6(14): 175–258.

Climo, F. M. 1970. Classification of New Zealand Arionacea (Mollusca: Pulmonata). III. A Revision of the genera *Charopa* Albers, 1860 (excluding subgenus *Ptychodon* Ancey, 1888), *Phenacharopa* Pilsbry, 1893 and *Flammocharopa* n. gen. (Endodontidae: Endodontinae). *Records of the Dominion Museum, Wellington*, 6 (18): 285–366.

Climo, F. M. 1971a. The Systematic Positions of *Cytora* Kobelt and Moellendorff, 1897 and *Liarea* Pfeiffer, 1853 (Mollusca: Mesogastropoda). *Transactions of the Royal Society of New Zealand (Biological Sciences)*, 2(19): 213–16.

Climo, F. M. 1971b. Classification of New Zealand Arionacea (Mollusca: Pulmonata). IV. A Revision of the subfamily Otoconchinae Cockerell (Punctidae Morse). *Records of the Dominion Museum, Wellington*, 7(6): 43–49.

Climo, F. M. 1971c. Classification of New Zealand Arionacea (Mollusca: Pulmonata). V. Descriptions of Some New Phenacohelicinid Taxa (Punctidae: Phenacohelicinae). *Records of the Dominion Museum, Wellington*, 7(11): 95–105.

Climo, F. M. 1974. (in press) Description and affinities of the subterranean molluscan fauna of New Zealand. *New Zealand Journal of Zoology*.

Cockerell, T. D. A. 1891. On the Geographical Distribution of Slugs. Proceedings of the Zoological Society of London, (for 1891): 217.

Connolly, M. 1939. A Monographic survey of South African nonmarine Mollusca. *Annals of the South African Museum*, 33(1): 1–660.

Cook, A. H. 1895. Molluscs. Cambridge Natural History, 3: 1–459.

Cooke, C. M. & Kondo, Y. 1960. Revision of Tornatellinidae and Achatinellidae (Gastropoda, Pulmonata). *Bernice P. Bishop Museum Bulletin*, 221: 3–303.

Craddock, C. *et al.* 1970. *Antarctic Journal of the United States*, 5(3): 53–80.

Cumber, R. A. 1961. A Revision of the Genus *Phenacohelix* Suter, 1892 (Mollusca: Flammulinidae) with Description of a New Species, and Studies on Variation, Distribution and Ecology. *Transactions of the Royal Society of New Zealand*, 1(13): 163–96.

Dell, R. K. 1952a. Otoconcha and its Allies in New Zealand. *Dominion Museum Records in Zoology*, 1(7): 59–69.

Dell, R. K. 1952b. New Species and Genera of New Zealand Snails with a Revision of the Genus Cavellia. *Dominion Museum Records of Zoology*, 1(9): 87–88.

Dell, R. K. 1957. Natural Areas in New Zealand-Land Snails. *Proceedings of the New Zealand Ecological Society*, 4: 13–14.

Fischer, P. H. 1880–87. Manuel de conchyliologique et de paléontologie conchyliologique . . . 1369 pp.

Fleming, C. A. 1962. New Zealand Biogeography. A palaeontologist's approach. *Tuatara*, 10(2): 53–108.

Gould, A. 1852. Mollusca, in United States Exploring Expedition: 2.

Gray, M. E. 1850. Figures of Molluscous Animals, selected from Various Authors, 4: 112.

Gray, J. E. 1853. Descriptions of two new genera (Pfeifferia and Janella) of Land Mollusca. *The Annals and Magazine of Natural History*, (2), 12(72): 412–15.

Haas, F. 1935. Beschreibung neuer Untergattungen und Arten von Mollusken. *Zoologischer anzeiger*, 109(7/8): 188–95.
Hesse, R., Allee, W. C. & Schmidt, K. P. 1951. Ecological animal zoogeography. New York. Wiley. 715 pp.
Hutton, F. W. 1878. Description of some new Slugs. *Transactions and Proceedings of the New Zealand Institute*, 21: 331–332.
Iredale, T. 1913. The land mollusca of the Kermadec Islands. *Proceedings of the Malacological Society of London*, 10: 364–388.
Iredale, T. 1915. A Commentary on Suter's Manual of the New Zealand Mollusca. *Transactions and Proceedings of the New Zealand Institute*, 47: 483.
Iredale, T. 1933. Systematic Notes on Australian Land Shells. *Records of the Australian Museum*, 19: 37–59.
Kondo, Y. 1943. Anatomical studies of three species of *Ouagapia* (Pulmonata, Agnatha, *Paryphantidae*). *Occasional Papers Bernice P. Bishop Museum*, 17(19): 229–48.
Ladd, H. S. 1957. Fossil land shells from deep drill holes on western Pacific atolls. *Deep-Sea Research*, 4: 218–19.
Ladd, H. S. 1958. Fossil land snails from western Pacific atolls. *Journal of Palaeontology*, 32(1): 183–98.
Mayr, E. 1940. Borders and Subdivision of the Polynesian region as based on our knowledge of the distribution of birds. *Proceedings of the 6th Pacific Science Congress*, 4: 191–95.
Moellendorff, O. 1903–04. Rhytididae. *Systematisches Conchylien – Cabinet*, 1, 12B: 1–91.
Morrison, M. P. E. 1955. Notes on American cyclophoroid land snails, with two new names, eight new species, three new genera, and the family Amphicyclotidae, separated on animal characters. *Journal of the Washington Academy of Science*, 45(5): 149–62.
Morton, J. E. 1952. A Preliminary Study of the Land Operculate *Murdochia pallidum* (Cyclophoridae, Mesogastropoda). *Transactions of the Royal Society of New Zealand*, 80(1): 69–79.
Oliver, W. R. B. 1953. Origin of the New Zealand flora. *Proceedings of the 7th Pacific Science Congress*, 5: 131–46.
Pain, T. 1955. Notes on some New Caledonian Placostylus. *Journal de Conchyliologie*, 95(1): 11–19.
Parodiz, J. J. 1946. Bulimulinae fosiles de la Argentina. *Notas del Museo de la Plata*, 11(92): 301–9.
Parodiz, J. J. 1949. Notes. *Physis*, 20(57).
Pilsbry, H. A. 1894. *Manual of Conchology*, (2), 9, XVIII: 366 pp.
Pilsbry, H. A. 1926. *Manual of Conchology*, (2), 27: iv; 177–369.
Pilsbry, H. A. 1939. The Land Mollusca of North America, north of Mexico, 1, part 1: 573 pp.
Pilsbry, H. A. & Olsson, A. A. 1954. Miocene land shell fossils from the Dominican Republic. *Notulae naturae*, 266: 4 pp.
Plate, L. H. 1897. Ueber einen neuen Typus der Lungenathmung, die Niere und ein subcutanes Sinnesorgan bei Nacktschnecken aus der Familie der Janellen. Sitzungsberichte der Gesellschaft naturforschender Freunde zu Berlin. 141–45.
Powell, A. W. B. 1930. The Paryphantidae of New Zealand. *Records of the Auckland Institute and Museum*, 1: 17–56.
Powell, A. W. B. 1932. The Paryphantidae of New Zealand. *Records of Auckland Institute and Museum*. 1: 155–62.
Powell, A. W. B. 1933. The High-Tidal Mollusca of Rangitoto Island, Auckland; with Descriptions of a New Genus and Two New Species. *Transactions of the New Zealand Institute*, 63: 144–52.
Powell, A. W. B. 1936. The Paryphantidae of New Zealand. *Records of the Auckland Institute and Museum*, 2: 29–41.
Powell, A. W. B. 1937. The Shellfish of New Zealand. An illustrated handbook:

91–92. Whitcombe & Tombs, 100 pp. And, 4th edition (1962), Shells of New Zealand: 109–13.

POWELL, A. W. B. 1938. The Paryphantidae of New Zealand, No. 4, and The genus Placostylus in New Zealand. *Records of the Auckland Institute and Museum*, 2: 133–50.

POWELL, A. W. B. 1946. The Paryphantidae of New Zealand, No. 5. Further New Species of Paryphanta, Wainuia and Rhytida. *Records of the Auckland Institute and Museum*, 3(2): 99–136.

POWELL, A. W. B. 1947. Distribution of Placostylus Land Snails in Northernmost New Zealand. *Records of the Auckland Institute and Museum*, 3(3): 173–88.

POWELL, A. W. B. 1948. Land Mollusca of the Three Kings Islands. *Records of the Auckland Institute and Museum*, 3(3): 273–90.

POWELL, A. W. B. 1949a. The Species Problem in New Zealand Land Snails. *Transaction of the Royal Society of New Zealand*, 77(5): 202–8.

POWELL, A. W. B. 1949b. The Paryphantidae of New Zealand, No. 6. Distribution, Hybrids and New Species of Paryphanta, Rhytida, and Schizoglossa. *Records of the Auckland Institute and Museum*, 3: 347–72.

POWELL, A. W. B. 1950. Life History of Austrosuccinea archeyi, an Annual Snail, and its Value as a Post-Glacial Climatic Indicator. *Records of the Auckland Institute and Museum*, 4(1): 61–72.

POWELL, A. W. B. 1951. On Further Colonies of Placostylus Land Snails from Northernmost New Zealand. *Records of the Auckland Institute and Museum*, 4: 134–40.

POWELL, A. W. B. 1955. Mollusca of the Southern Islands of New Zealand. *Cape Expedition Series, Bulletin*, 15: 128–30.

POWELL, A. W. B. 1957. The Molluscan Land Operculate Genus Liarea. *Records of the Auckland Institute and Museum*. 4(5): 271–93.

SIMROTH, H. 1889. Beitrage zur kenntniss der Nacktschnecken. *Nova acta Academiae Caesareae Leopoldino Carolinae germanicae naturae curiosorum*, 54: 69–86.

SMITH, B. J. 1969. A comparative study of two Australian paryphantid snails, with notes on their taxonomic relationships. Proceedings of symposium on Mollusca, India, 1: 164–69.

SMITH, B. J. 1970. Notes on the anatomy of *Victaphanta atramentaria* (Shuttleworth) and *V. compacta* (Cox and Hedley), and the designation of a neotype for *V. atramentaria*. *Journal of the Malacological Society of Australia*, 2(1): 13–21.

SOLEM, A. 1959. Systematics and Zoogeography of the land and freshwater mollusca of the New Hebrides. *Fieldiana: Zoology*, 43: 4–359.

SUTER, H. 1897. A Revision of the New Zealand Athoracophoridae. *Proceedings of the Malacological Society of London:* 217.

SUTER, H. 1909. New Zealand Athoracophoridae. *Proceedings of the Malacological Society of London*, 8(5): 322–28.

SUTER, H. 1913. Manual of the New Zealand Mollusca, with Atlas of Plates, 1915. Govt Printer, Wellington. 1120 pp.

TIELECKE, H. 1940. Anatomie, Phylogenie und Tiergeographie der Cyclophoriden. *Zeitschrift für wissenschaftliche Zoologie* (which is for 1932–44 Abt. B of Archiv für Naturgeschichte), 9(3): 317–71.

VAN DYKE, E. C. 1939. The origin and distribution of the coleopterous insect fauna of North America. *Proceedings of the 6th Pacific Science Congress*, 4: 255–68.

WATSON, H. 1934. *Natalina* and other South African snails. *Proceedings of the Malacological Society of London*, 21(3): 150–96.

WENZ, W. 1938–44. Gastropoda, Prosobranchia, in SCHINDEWOLF, Handbuch der Palaeozoologie, 6: 1–1639.

XII. THE SPIDERS AND HARVESTMEN

by

R. R. FORSTER

Spiders and Harvestmen

Four orders of arachnids are found in New Zealand – mites, false scorpions, spiders and harvestmen. While all are abundantly represented, only the spiders and harvestmen are known well enough to form a basis for discussion.

As with other invertebrate groups, links with European specialists in the nineteenth and early twentieth century led to the early description of some of the more common species but an appreciation of the true diversity and complex speciation of the arachnid fauna in this country has been realised only over the last few decades. The majority of the indigenous spiders and harvestmen are probably now known in collections and it seems that the one hundred and seventy species of harvestmen at present described represent but two thirds while the 700 described species of spiders, a much more diverse and abundant group probably cover one third of the total fauna.

The fact that both of these groups of arachnids are exclusively carnivorous and on the whole show little specificity in the prey they capture, makes them useful subjects for zoogeographic studies. The distribution of forest fauna within New Zealand itself demonstrates a significant disregard for the type of forest cover which suggests that the basic factors influencing distribution patterns have not been necessarily changes in the form of vegetative cover but probably topographical barriers. The harvestmen possess no obvious means for passive distribution and are undoubtedly restricted in their dispersal by their inability to control dessication. The spiders do have a potential for ballooning by silk but it is however characteristic of the great bulk of the New Zealand fauna that this method of dispersal does not seem to be utilized. There is a very strong contrast today between the distribution of those spiders introduced by man, mainly from the northern hemisphere, which are now the common species of pastoral and disturbed land, and the native fauna. The majority of the introduced spiders are found from one end of the country to the other and furthermore some have populated the remote Subantarctic islands, Auckland, Campbell and the Antipodes.

The fact that gossamer is so little used by the native spiders may be related to the insular nature of New Zealand and the restriction of most of the fauna to forest where ballooning has little functional advantage.

Of the two introduced harvestmen, *Nelima doriae* is restricted to North-

Fig. 1. Gradungula sp. (Family Gradungulidae). Peculiar to the rain forest and caves of the South Island of New Zealand, and the southeast coast of Australia. These 4-lunged true spiders do not construct snares but actively hunt their prey.

land while *Phalangium opilio* is found throughout both the North and South Islands. Neither of these harvestmen has managed to colonise native forest although *P. opilio* is found in planted conifer forests.

THE DISTRIBUTION OF HARVESTMEN AND SPIDERS WITHIN NEW ZEALAND

The distribution of harvestmen lends itself more positively to the limitation of faunal areas. Based on the distribution of the suborders Cyphophthalmi and Laniatores, nine faunal areas have been suggested (FORSTER, 1954) characterised by the aggregation of species common to each region. The most distinctive and diverse fauna is found in the Fiordland region at the southern end of the South Island. This fauna shows a close relationship with Stewart Island to the south where the presence of a number of sympatric species related to Fiordland forms suggests a history of repeated invasion and isolation, probably during the Pleistocene. A further extension of the Fiordland fauna is found to the north on both the east and west coasts. The Southern Alps which today form a distinctive barrier limit a clear east-west division of the South Island fauna. The fauna of the Westland region is fairly uniform along its length but becomes more diverse to the north in the Nelson area. Along the east coast a faunal change occurs between Dunedin and Christchurch near the Rangitata River. A number of the species which occur in the northern range of both of the east and west regions are identical with, or closely related to, the fauna found in central and south North Island. The east coast fauna of the South Island shows affinity with the southern North Island area, Wellington and Wairarapa, while that of the west region is related to the Egmont area and the central portion of the North Island. In the North Island there is a further, but less distinct fauna in the East Cape area and a strongly distinctive one in the Northland region north of Auckland. More recently investigation of the Three Kings Islands north of the North Cape, has shown that this small group of islands is also characterised by a strongly endemic fauna and should perhaps be categorised as a separate area in its own right. It has been suggested (FORSTER, 1954) that the basic events which moulded the distribution pattern we find today, took place over the Tertiary during which continuing changes in the disposition of land areas led to the fragmentation of populations with ensuing periods of isolation. That the inclement periods of the Pleistocene must also have played some part in present day distribution, if only in the extinction of some forest forms, seems incontrovertible. It is nevertheless surprising to find that the southern region of Fiordland supports the most varied fauna in the country, strongly suggesting the continuance of forest refugia in this area throughout periods of maximum glaciation.

When the distribution of the fauna as a whole is considered, there is seen a gradual reduction of both genera and species, although not population density, from the south to the north. This trend, being as it is in inverse ratio to the mean temperature, strongly suggests that the present day New Zealand fauna is basically cold adapted. There seems little doubt that there are a number of genera of both harvestmen and spiders today found only in Northland and the Three Kings Islands which could be considered more characteristic of warmer latitudes. Amongst the harvestmen the genus *Triregia* with three species restricted to the region north of Auckland and Three Kings is closely related to the New Caledonian genera *Triconobunus*, and *Diaenobunus* while amongst spiders the genera *Storena* and *Stephanopis* are found. Some species of these latter two genera are today found in Victoria and Tasmania, but the centre of their distribution seems to be in the warmer regions. There is little reason to suggest that these genera represent more recent introductions from Australia when compared with the remaining fauna and their present distribution may reasonably be accounted for by climatic factors which have prevented their spread to the south or alternatively that they represent relics of previously more widespread groups which were exterminated from the more southern regions by climatic factors.

Relationships of the New Zealand fauna

It has been usual when discussing the zoogeographical relationships of the New Zealand fauna to first categorise separate elements which might have colonised the area at different times in the past. With the more widespread acceptance of continental drift, it is perhaps more logical to look at the fauna as a whole and consider whether it could, or at least the bulk of it, be considered to have descended directly from the original Gondwanaland fauna. As far as the spiders and harvestmen are concerned, the contemporary origin for the full fauna in the late Mezozoic is quite feasible and particularly so for the forest fauna. As the fauna of the southern land masses is becoming better known, more and more common features are being brought to light. The close relationship of the New Zealand fauna with a section of the Australian fauna has long been known, but the fact that there is a complete faunal relationship in groups such as the arachnids has been little emphasised. A number of species appear to be common to both countries, suggesting a regular interchange, but some of these species, for example *Ixeuticus martius*, *Lampona cylindrata*, *Hemicloea rogenhoferi* and *Araneus pustulosus* have undoubtedly been recently introduced, probably by man. These spiders are not able to colonise native forest and without the environmental changes brought about by man would probably not have become established. A number of other true forest species which have been considered identical with Australian forms, now however appear to be distinct and endemic to this country. These species belong in the main to genera such as *Cyclosa*, *Celaenia* and *Nanometa* which are composed of species often very similar in appearance and difficult to separate. The relationship between many of these Australian and New Zealand species pairs could, despite their similarity, be quite old. The general picture which does seem to be emerging is that a broad spectrum of the New Zealand fauna is present in Tasmania and to a lesser extent along the eastern coast of Australia where it is found mainly today in rain forest. This broad relationship is more clearly seen from New Zealand, because in Australia there is in addition a portion of the fauna, more obviously so in the spiders than harvestmen which has probably been derived from tropical and sub-tropical regions. The spiders which are now characteristic of the dry sclerophyll forest in Australia are generally not present in New Zealand. The recent discovery of endemic New Zealand species (in the northern part of the North Island) of typical widespread Australian genera such as *Storena* and *Stephanopis* has considerably broadened the link between the two countries as has the discovery of a number of as yet undescribed species from the Australian temperate rain forests which are closely related to New Zealand forest forms. While the relationship with Australia therefore appears to

Fig. 2. Celaenia sp. (Family Araneidae). Found in both Australia and New Zealand these spiders do not construct a web but hang from a pad of silk and grasp male moths which are apparently attracted by scent.

be a very broad one, the relationship with South Africa and South America is more tenuous although still convincing. Amongst the harvestmen the strong representation of Triaenonychids in New Zealand, Australia, South Africa, Madagascar and South America is in itself of

Fig. 3. Amaurobioides picunus Forster (Family Amaurobioididae). Found only along the rocky coastline within the sprayzone this family has been recorded from Tasmania and the southeast coast of Australia, New Zealand and its Subantarctic islands, South Africa and Tierra del Fuego. The spiders live in a silken retreat with an opening at one end constructed on the rock surface.

great interest but there is also a much more significant relationship through the lower taxa. The recent discovery of *Americobunus* in Chile, closely related to *Ankaratrix* (Madagascar), *Traenobunus* (Australia) and *Pristobunus* (New Zealand) parallels similar relationships found in all three sub-orders of harvestmen. One puzzling absence from the harvestmen fauna of New Zealand is however the family Phalangodidae which, while certainly most abundant in the tropics, does extend into the temperate zones of all other southern areas.

Amongst spiders the case for southern relationships has usually been constructed from the distribution of a number of distinctive genera and families but a similar case could probably be established within cosmopolitan genera if the detailed taxonomy of species groups was adequately known. Typical distribution patterns are found in the haplogyne spiders of the families Archaeidae and Mecysmaucheniidae. Archaeids have now been recorded from New Zealand, Australia, South Africa and Madagascar while Mecysmaucheniids are found in New Zealand and South America and will probably be found in Australia. The recent revision of the interesting littoral genus, *Amaurobioides*, (FORSTER, 1970) which constructs a small silken retreat along rocky coasts within the spray zone has shown that this genus has speciated extensively within New Zealand and Australia and is also present in South Africa and South America. The distribution of these spiders before the extent of their speciation was known, had previously been considered an indication of their ability to cross wide stretches of sea but it is probable that this distribution is also an ancient one and indicates the presence in the past of near contiguous shores. The southern distribution of *Myro* has long been used as a typical example of southern distribution, being recorded from New Zealand, Australia, South Africa and South America while it is also found on the Subantarctic islands of Macquarie, Kerguelen and the Crozets.

Ecology

Although little attention has been given to the detailed ecology of harvestmen or spiders in this country, a large body of general information has been gathered together over the last few decades which enables a number of generalised statements to be made.

Surveys of pasture land in recent years have strikingly confirmed the supposition that the fauna of these areas is composed mainly of introduced species originating from the northern hemisphere. The dominant spiders are the small black money spiders (Micryphantidae) which are often present in vast numbers. This family does not occur naturally in New Zealand and probably not in Australia either. Six species have been identified so far, *Erigone prominens*, *Erigone wiltoni*, *Eperigone tridentata*, *Aulacocyba subitanea*, *Araeoncus humilis* and *Diplocephalus cristatus*. These spiders are responsible for the gossamer records in New Zealand. *E. wiltoni* has been described as new from New Zealand and the true home of the species is not known. The harvestmen *Phalangium opilio* is also abundant in these areas as it is in riverbeds and gardens. The larger spiders consist of two or three widespread endemic species of wolf spiders probably derived from the subalpine fauna. In recent years further ground spiders such as the Australian Ctenid *Horioctenoides* and the Clubionid *Supunna* have become established and seem to be rapidly spreading throughout the pastoral country of the North Island and the northern portion of the South Island. Recent surveys of the fauna of planted pinus forest show that these areas also are populated in the main by this introduced fauna and that very few indigenous species have colonised this new habitat.

The fauna of the indigenous lowland tussock land is extremely limited and seems to be composed mainly of species more characteristic of the forest margin, or the forest floor and leaf litter. The most distinctive tussock fauna is found in the inland regions of Otago and to a lesser extent of Canterbury. In these regions a number of spider groups have speciated and a few show special adaptations. The six genera of mygalomorphs recorded from New Zealand are all primarily forest dwellers but five of these genera are well represented in tussockland and modified tussockland, in Central Otago. *Migas* which normally constructs a lidded silken retreat on treetrunks and *Hexathele* which builds a loose silken retreat on the surface of the ground or under stones or bank have been recorded living in burrows excavated in the earth in inland Otago.

The wolf spiders also appear to be more abundant in the Otago region and include the sole burrowing species recorded so far. The paucity of a characteristic open country fauna suggests that other than the inland Otago region, there has been no continuing history of non-forested land. Without exception the present day fauna of lowland open country could have been derived fairly recently from the forest and subalpine regions.

Subalpine and Alpine

The fauna of the subalpine scrub is similar to that of the forest margins but a number of species do appear to be restricted to higher altitudes. The spiders which inhabit subalpine scrub are mainly of the families Araneidae, Thomisidae, and Salticidae. The long legged harvestmen *Megalopsalis* is also often common on foliage. The dominant spiders found in the herb fields are wolf spiders and small Linyphiids. The most abundant fauna is however found on scree slopes or rock faces where many species of wolf and jumping spiders make their home. In contrast to the herb field fauna, some of which is conspicuously patterned, the scree spiders are mainly dark coloured and usually large in size. In addition to the hunting spiders, a considerable number of web building species belonging to the genera *Mamoea* and *Neoramia*, are commonly found. The two species of the family Ctenidae found in New Zealand are both from the subalpine screes in the South Island mountains. While the mountain Lycosidae, Salticidae and Ctenidae show no direct relationship with the forest, the bulk of the other spiders present belong to common forest genera. It does appear that the South Island has the most extensive subalpine fauna and that along the mountain chain this fauna has strongly speciated, but the apparent paucity of a mountain fauna in the North Island may merely reflect the fact that it has not yet been fully studied.

Riverbeds

The shingle riverbeds which are characteristic of this country support an interesting and distinct fauna. It is clear that the majority of this fauna has been derived from high country screes, but in the lower reaches of most river systems consists of quite distinct species. As in the subalpine region the most conspicuous part of this fauna consists of Lycosids and Salticids which live among the stones near the water's edge or on the higher shingle banks. Some of these spiders probably feed on or along the margins of the water but the majority seem to find their prey in amongst the shingle. They are mainly greyish or black in colour and are difficult to see against the grey shingle. Although these spiders have not been extensively studied, it seems that the fauna of each river system is directly related to that of the respective watersheds. The large water spiders (*Dolomedes aquatica* group) are characteristic of these riverbeds but are also found around the shores of clear lakes. These spiders, which are nocturnal, feed mainly on insects trapped on the surface of the water and shelter during the day beneath stones near the water's edge. In contrast to the widespread *Dolomedes minor* group which live mainly away from water and construct nursery webs on low shrubs, the water spiders utilise crevices between the river boulders for the nursery. The egg is attached at the bottom of the crevices and the nursery web consists of a silken sheet spun above the eggsac between the boulders. A number of Theridiids and Linyphiids are also restricted to the riverbeds where they construct small silken snares among the shingle. The Dictynid *Arangina* which is found only in this habitat constructs ladder webs amongst the rocks.

Seashore

A number of dark coloured Salticids and Lycosids are also characteristic of shingly beaches throughout New Zealand, while pale coloured Lycosids are found on sandy beaches. Some of these spiders are directly related to riverbed species and on the West Coast of the South Island where the coastal plain is narrow, single species may extend from the beaches, up the riverbeds to the subalpine screes. The shingle banks above high tide levels also support a characteristic spider fauna which appears to feed mainly on the amphipods and kelp flies which are abundant in these places. The most widespread species is the Desid *Otagoa nova* found throughout New Zealand but in the South Island other spiders such as *Myro marinus*, and *Oramia* are also restricted to this habitat. *Amaurobioides* lives on the rocks and cliff faces above high tide mark but within the spray zone, where they construct silken retreats on the surface of the rock or in crevices. These spiders which have strongly speciated in New Zealand have now been recorded from the rocky coastline of both islands and also occur in the Subantarctic islands. *Desis marina* inhabits the intertidal zone along the rocky shore. As the tide retreats, it actively hunts amphipods and isopods on the exposed rock and as the tide advances, moves back into a crevice to spin a fine silken sheet over the opening to retain the air within the retreat.

The sandy beaches support a very different fauna from the shingle beaches and rocky shore. The two katipo spiders *(Latrodectus katipo, L. atritus)* are restricted to this habitat. Both species construct their snares near the base of grass clumps on the sand or under debris lying on the sand. *L. katipo*, characterised by a bright red stripe on the upper surface of the abdomen is more common and found throughout the country except for the southern portion of the South Island. The uniformly black species, *L. atritus*, is apparently restricted to the northern half of the North Island. While katipos may occasionally be found as far as five km from the sea where sand dunes extend inland, they are usually restricted to a narrow belt within a few hundred metres of the high tide level. A number of other Theridiids, such as *Steatoda*, which is often confused with the katipo, are found in the same habitat as are a number of Linyphiids.

Forest

The forest fauna encompasses the great majority of the spiders and harvestmen indigenous to this country. While the number of species of forest harvestmen will eventually total something over 200 species, the spider fauna will not be far short of 2,000. The harvestmen are represented by only four families, but the family representation of spiders is much wider and includes most of the families characteristic of temperate regions. The predominance of the forest fauna strongly suggests that a substantial forest cover has been a continuous feature throughout the history of the New Zealand land area and that even during the height of the Pleistocene glaciations substantial refugia were maintained. There is no evidence to suggest that altitudinal speciation exists within the forest itself, but it is evident that there is a gradation in both abundance and species diversity related to altitude. The greatest diversity is found from sea level to approximately 600 metres in the South Island and relatively higher in the North Island.

Within the forest three fairly clear divisions may be made between the fauna characteristic of the (1) understory, low shrubs, ferns etc., (2) the forest floor and treetrunks, and (3) the cryptozoic fauna living amongst the leaflitter and moss. No canopy fauna has yet been discovered.

The understory fauna is more diverse near the forest margins and in low altitudes. As well as many web building Theridiids, Linyphiids and Araneids there are numerous hunting spiders, mainly small, which live on the foliage. Characteristic of these are the Dictynoids, *Hapona, Lamina* and *Urquhartia* which have tenent hairs on the tips of the legs which are presumed to improve footing on shiny leaf surfaces. A large number of the understory spiders are structurally modified for the use of insect holes in the branches and twigs as a retreat from which they capture passing prey, (e.g. *Plectophanes*) or construct a snare, (e.g. *Matachia Notomatachia*). A considerable number of the understory spiders seem to rely on attractants

Fig. 4. Hexathele hochstetteri Ausserer. female (Family Dipluridae).
Fig. 5. Porrhothele antipodiana (Walckenaer) female (Family Dipluridae).
Fig. 6. Migas saxitalis Wilton. male (Family Migidae).
Fig. 7. Gradungula sorenseni Forster. female (Family Gradungulidae).
Fig. 8. Phoroncidia sp. male (Family Theridiidae).
Fig. 9. Storena sp. female (Family Zodariidae).
Fig. 10. Plectophanes sp. female (Family Toxopidae).
Fig. 11. Laestrygones sp. female (Family Toxopidae).

for the capture of their prey. Some of these such as *Ulesanus*, *Rhomphaea* and *Episinus* construct single line threads while others such as *Celaenia* merely wait suspended beneath a leaf until male moths are attracted near enough to be captured. Orbweb spiders are abundant near the forest margin but few other than the Metinae and a group of small Tetragnathas penetrate far into the bush.

The forest floor fauna which includes the majority of the forest fauna consists mainly of uniform, rather drab coloured spiders belonging to a wide range of families. These spiders live on the surface of the litter, in and under fallen logs, and also on the trunks of the trees. Most characteristic are genera such as *Mamoea*, *Pakeha*, *Orepukia* and the Mygalomorphs *Porrhothele* and *Hexathele*, which construct silken surface retreats, and the tunnel dwelling *Cantuaria* and *Aparua*. The majority of the forest harvestmen, mostly Triaenonychids also live on the forest floor.

The cryptozoic fauna living in the leaf litter consists almost entirely of small or minute species. The minute Symphytognathidae, only a few millimetres in length, are often quite abundant and seem to take the place of the Micryphantid spiders of the northern hemisphere. While many groups of spiders such as the six-eyed Oonopidae, the Archaeidae and Mecysmaucheniidae are restricted to this habitat, the bulk are related to typical forest floor and understory spiders. Not many of these spiders however appear to construct snares and there is a tendency for cribellate spiders to show a reduction in the cribellum, indicating a reduction in the need for tangling silk. The harvestmen are well represented by a number of small species of Triaenonychids, the small mite-like harvestmen of the suborder Cyphophthalmi and the minute Synthetonychidae which are endemic to New Zealand. In very wet forest and in the upper reaches of most forests where the Bryophyte layer extends up the trunks of the trees, the forest floor and the related treetrunk fauna is reduced while the cryptozoic fauna extends with the moss cover from the forest floor to considerable heights.

REFERENCES

Forster, R. R. 1954. The New Zealand Harvestmen (Suborder Laniatores) Canterbury Museum Bull. No. 2, pp. 1–329.

Forster, R. R. 1956. Terrestrial Environments in New Zealand, *Proc. N. Zeal. Ecological Society*, 3: 23–24.

Forster, R. R. 1961. The New Zealand Fauna and its origins. *Proc. Roy. Soc. N. Zeal.* 89(1): 31–56.

Forster, R. R. 1967. The Spiders of New Zealand, Part 1, *Otago Mus. Bull. No.* 1, pp. 1–124.

Forster, R. R. & Wilton, C. L. 1968. The Spiders of New Zealand Part 2, *Otago Mus. Bull.* No. 2, pp. 1–180.

Forster, R. R. 1970. The Spiders of New Zealand, Part 3. *Otago Mus. Bull.* No. 3, pp. 1–183.

Forster, R. R. & Wilton, C. L. 1973. The Spiders of New Zealand, Part 4, *Otago Mus. Bull.* No. 4, pp. 1–309.

Forster, R. R. & Forster, L. M. 1973. New Zealand Spiders, an introduction. Collins. Auckland and London.

XIII. THE TERRESTRIAL INSECTS

by

J. C. WATT

Introduction

Insects are by far the most specifically numerous group of animals in New Zealand (as elsewhere). They are often enormously abundant as individuals. They include forms vital in processes of litter breakdown and soil formation, in degradation of logs and carcases, and in generally scavenging and cleaning the environment. Many are phytophagous, and some, because they attack commercial crops above or below ground, reduce productivity and are regarded as pests, but others feed on weeds, and are regarded as beneficial. Some insects attack man and his domestic animals. Parasites and predators help regulate populations of both injurious and beneficial organisms. Insects are the dominant food of many birds, lizards, and most freshwater fishes.

Some insects live between high and low tides, and others live above the level of vegetation on high mountains, but most are found between these extremes. Lowland and montane forest and scrubland are the richest insect biotopes in New Zealand. Forest habitats in the northern North Island do not support as rich a fauna as one would expect by comparison with the richness of the fauna in forests in similar latitudes in other southern hemisphere countries, and in cooler forests in New Zealand. Despite lowering of sea-level, there was little extension of land to the northwards beyond its present extremity during glaciations, so many stenothermal warmth-adapted species, being unable to retreat northwards, apparently became extinct during the Pleistocene.

Winter dormancy (diapause) is very unusual in New Zealand insects (Dumbleton, 1967), apparently because, with insularity, the climate during glaciations was considerably less severe than in equivalent latitudes in the northern hemisphere. The present alpine climate is fickle: snow can fall heavily at any time of the year, but conversely there can be relatively warm weather causing thawing in mid-winter. Lowland climates are little more reliable, although milder, in many parts of the country. Consequently, many New Zealand insects must make the best use of favourable periods of weather whenever they occur, so that hibernation or aestivation would be disadvantageous.

Phytophagous insects and freshwater insects are discussed in separate chapters in this book, so they are considered here only in general discussions of the fauna as a whole, or where they demonstrate particular biogeographical or ecological principles. References are restricted almost exclusively to recent papers, largely because of the rapid advances in knowledge of New Zealand insects in the last few years, but partly because older references may be found in the Bibliography of New Zealand Entomology (MILLER, 1956).

The New Zealand Insect Fauna

The known insect fauna of New Zealand is summarised in Table 1, where it is compared with those of Australia, Britain and the world.

The classification, numbers of species for Australia, and the number of families for each order in the world, are taken from 'Insects of Australia' (MACKERRAS, 1970). Numbers of species for New Zealand are derived mainly from WATT (1970). Those for Britain come from KLOET & HINCKS (1945). The numbers of described species of the world have been obtained from several different sources of varying age and accuracy, and in the larger orders must be regarded as only very approximate estimates.

The accuracy of all these figures is dependent on the amount and quality of work which has been done on each group. In the larger orders, each total (except for the British fauna) includes many synonyms which have not yet been established, while many undescribed species are recognised in collections, and there must be others unknown or unrecorded. Thus the figures can provide only a rough comparative guide, and not an accurate count of even the known species.

New Zealand occupies about 0.17% of the total land area of the world, and has about 1.2% of the recorded species of insects. The nearest major land mass, Australia, comprising about 4.91% of the total land area, has about 6.6% of the known world insect fauna. It is scarcely surprising that Australia with its much greater size, latitudinal extent and environmental diversity has a much greater variety of insects than New Zealand, but area for area, New Zealand seems to have a relatively more diverse insect fauna.

There are no island groups anywhere else directly comparable with New Zealand in size, latitude, climate and isolation. The British Isles are roughly comparable in size, but are less diverse, much further from the equator, and separated from the largest continental mass by a strait only 35 km wide, which was formed after the close of the last glaciation by eustatic rise of sea-level. The 20024 species of insects known from the British Isles in 1945 constitute about 2.4% of the world fauna, but the difference between this and the New Zealand total of 9460 can be accounted for entirely in the large orders Diptera, Lepidoptera and Hymenoptera. The British fauna is a large fragment of the European fauna, especially those insects adapted to a cool maritime climate. It has been studied by many more taxonomists for much longer than the faunas of Australia and New Zealand or most tropical areas, so there

Table 1. The Known New Zealand Insect Fauna Summarised and Compared with other Faunas.

Orber	Described species							Recorded families			
	N.Z.	Aust.	Brit.	World	N.Z.% / World	Aust.% / World	Brit.% / World	N.Z.	Aust.	World	N.Z.% / World
Collembola	331	215	261	2000	15.6%	10.8%	13.1%	8	8	8	100%
Protura	2	30	17	170	1.2	17.6	10.0	2	3	3	66
Diplura	8	32	12	660	1.2	4.8	1.8	2	3	4	50
Archaeognatha	1	3	0	250	0.4	1.2	0	1	1	2	50
Thysanura	4	23	11	330	1.2	7.0	3.3	3	3	5	60
Ephemeroptera	26	124	46	1200	2.2	10.3	3.8	4	4	11	36
Odonata	11	248	42	4500	0.2	5.5	0.9	4	16	24	17
Blattodea	26	439	8	3500	1.3	12.5	0.2	3	4	5	60
Isoptera	6*	182	0	2000	0.3*	9.1	0	4*	5	6	66*
Mantodea	1*	118	0	1800	0.06*	6.6	0	1*	2	8	13*
Zoraptera	0	0	0	22	0	0	0	0	0	1	0
Grylloblattodea	0	0	0	12	0	0	0	0	0	1	0
Dermaptera	10	60	9	1200	1.2	5.0	0.8	4	5	7	57
Plecoptera	32	84	32	1250	2.6	6.7	2.6	4	4	8	50
Orthoptera	88	1513	31	20000	0.4	7.6	0.2	7	13	18	39
Phasmatodea	19	132	0	2500	0.8	5.3	0	1	2	2	50
Embioptera	0	65	0	200	0	32.5	0	0	3	8	0
Psocoptera	42	120	68	1700	2.5	7.1	4.0	11	21	28	39
Phthiraptera	192	208	286	2900	6.6	7.1	9.9	9*	10	16	56*
Hemiptera	651	3661	1411	65000	1.0	5.6	2.2	48	87	103	47
Thysanoptera	30*	287	183	4000	0.7*	7.2	4.6	4	4	5	80
Megaloptera	1	16	6	250	0.4	6.4	2.4	1	2	2	50
Neuroptera	11	396	54	4500	0.3	8.8	1.2	5	15	17	29
Coleoptera	4300	19219	3690	320000	1.3	6.0	1.2	88	106	151	58
Strepsiptera	1	93	17	300	0.3	31.0	5.7	1	5	5	20
Mecoptera	1	20	4	400	0.2	5.0	1.0	1	5	7	14
Siphonaptera	20*	68	47	1370	1.5*	5.0	3.4	6*	9	17	35*
Diptera	1870	6256	5199	150000	1.2	4.2	3.5	55	86	114	49
Trichoptera	130	260	188	4000	3.3	6.5	4.7	13	18	24	54
Lepidoptera	1490	11221	2187	112000	1.3	10.0	2.0	33	75	105	32
Hymenoptera	400	8834	6191	110000	0.4	8.0	5.6	30	59	73	41
Totals & Means	9460	53927	20024	818014	1.16%	6.59%	2.45%	353	577	788	44.8%

* Figures comprising at least 50% non-endemic elements (mostly artificial introductions).

are few unknown synonyms and few unrecorded species. Endemism is very low, even at the specific level.

Only three insect orders are entirely absent from New Zealand (Zoraptera, Grylloblattodea and Embioptera). The only mantid (*Orthodera ministralis*) in New Zealand is not endemic, but is one of our commonest insects, and may be indigenous.

In discussing the insect fauna, it is important to distinguish between diversity at higher levels of classification, which reflects the number of original immigrants in the older established groups, and diversity of endemic species, which reflects the amount of adaptive radiation which has taken place in New Zealand. In the aquatic orders Ephemeroptera, Plecoptera and Trichoptera a number of families which are strongly represented overseas are entirely lacking, but in each order the proportion of the world species represented in New Zealand is considerably higher than the mean of about 1.2% for all insects. New Zealand is a diverse, mountainous country, with high precipitation and many streams and rivers, and numerous discrete watersheds. These conditions have apparently favoured adaptive radiation in insects inhabiting flowing waters. By contrast, insects inhabiting still waters (e.g. Odonata, Dytiscidae) are very poorly represented in New Zealand. All our large lakes are very young, being of either glacial or volcanic origin. The size of smaller lakes and ponds makes them relatively ephemeral.

Among terrestrial insects, Collembola and Phthiraptera seem to be exceptionally numerous in species here. In both orders there is an unusually large proportion of non-endemic but indigenous species, i.e. these insects are unusually dispersible (in the case of Mallophaga by virtue of the dispersibility of their avian hosts). Anoplura, all of whose hosts are mammalian, were introduced with their hosts, except for *Antarctophthirus* spp. which parasitise seals.

Orthoptera and Neuroptera are quite poorly represented in New Zealand, but both orders are predominantly tropical. On the other hand, Hymenoptera, which are strongly represented in temperate countries elsewhere, have relatively very few species here. The New Zealand fauna is particularly deficient in Symphyta (3 endemic species) and Aculeata (see VALENTINE in WATT, 1970). Hymenoptera generally are not well represented on oceanic islands, so their ability to cross oceanic barriers would seem to be fairly limited. This does not explain why the few Symphyta and Aculeata that did reach this country have not radiated much more widely. Valentine estimates that he knows of 409 species of endemic undescribed Hymenoptera in New Zealand, but these are nearly all Parasitica.

Many high categories below the ordinal level are absent from, or very poorly represented in, New Zealand. For example, only two of the four recognised suborders of Coleoptera occur here (Adephaga and Polyphaga), two superfamilies of Polyphaga are lacking (Rhipiceroidea and Lyme-

Fig. 1. Chaetosoma scaritides, a member of the endemic family Chaetosomatidae. The larva is a predator on wood-boring larvae. Photo: B. S. Eykel, DSIR.

Fig. 2. Pachyrrhamma fasciferum, a large flightless Rhaphidophorid, on the wall of a tunnel, Khandallah. Photo: B. S. Eykel, DSIR.

xyloidea), and a few major families are absent (Meloidae) or very poorly represented (Buprestidae). Some important subfamilies are lacking (e.g. Cetoniinae and Rutelinae in Scarabaeidae). However, there is an endemic family (Chaetosomatidae), several endemic subfamilies and tribes, and several families are unusually strongly represented (e.g. Helodidae, Colydiidae, Pselaphidae). Similar patterns are characteristic of the other major orders. In Lepidoptera, seven superfamilies are absent, and imbalance is seen at all levels from superfamily to genus. For instance, within the tortricid subfamily Tortricinae, four out of a possible six tribes are present; of these tribes, one (Archipini) contains ten times as many species as the other three combined (Dugdale in Watt, 1970).

The somewhat fragmentary ('disharmonic') nature of the New Zealand insect fauna is reminiscent of faunas of some of the larger oceanic islands. These however lack old endemic elements, and the large flightless ground-inhabiting insects so characteristic of New Zealand (such as Henicidae, Rhaphidophoridae and the larger apterous Carabidae, Tenebrionidae and Curculionidae). The insect fauna suggests that there may not have been a complete link with any other land mass, although dispersal to New Zealand (probably by 'island hopping') must have been very much easier than it is now. However, there probably was a great deal of extinction of warmth-adapted groups during the Pleistocene, and the New Zealand insect fauna may have been considerably less fragmentary during the Tertiary than it is now. Regrettably, no useful pre-Pleistocene insect fossils are known in New Zealand, but studies of shallow-water marine faunas and terrestrial plant fossils has demonstrated substantial extinction of warmth-adapted elements during the late Pliocene and Pleistocene.

Endemism

New Zealand has been long and effectively isolated from other land, and this is reflected in the endemism in the insect fauna. Endemism is very high at the specific level in many groups of New Zealand insects (e.g. Lepidoptera show 94% endemism at the specific level, Orthoptera 95%, Coleoptera 96%, Phasmatodea, Plecoptera, Ephemeroptera and Trichoptera each 100%). This is such a general rule in these orders that one naturally suspects that species occurring here and overseas as well, have been artificially introduced into New Zealand. It seems that this assumption has been made too readily in the past, e.g. the Black Field Cricket *Teleogryllus commodus* was formerly thought to have been introduced artificially from Australia, where it occurs also. However, there is geographical variation within New Zealand both in structural characters of the proventriculus and in the proportions of eggs which enter diapause at 28 °C, while there are differences in proventricular teeth between New Zealand and Australian populations (BIGELOW, 1964), suggesting that the species has been here for very much longer than the 150 years of European settlement.

The almost cosmopolitan distributions (including New Zealand) of the Dytiscid *Rhantus pulverosus* and the Sphingid *Herse convolvuli*, both of which are very strong fliers, appears to be natural. The Monarch Butterfly *Danaus plexippus*, originally a North American insect, did not become established in New Zealand until its food plants (*Asclepias* spp.) were introduced artificially. The Monarch was known to the pre-European Maoris under the name 'kahuku', and for many years after European settlement it was noted as an occasional visitor (MILLER, 1971).

Recently there have been numerous records of Australian Lepidoptera blown over to New Zealand during westerly gales (see references by FOX, HARDING, DUGDALE, RAMSAY; *N.Z. Ent.* 4(2)–5(1), 1968–71). *Lampides boeticus* has definitely established itself here, and three species of Australian moths may have established also, during this period. Many other Australian moths, especially Pyralidae and Noctuidae, occur naturally in New Zealand (DUGDALE, pers. comm.). The Lygaeid bug *Arocatus rusticus* occurs naturally in both countries (EYLES, pers. comm.).

Other insects with readily dispersible adults (e.g. Aphididae, certain small Diptera, Thysanoptera, Cicadellidae, some Staphylinidae – see

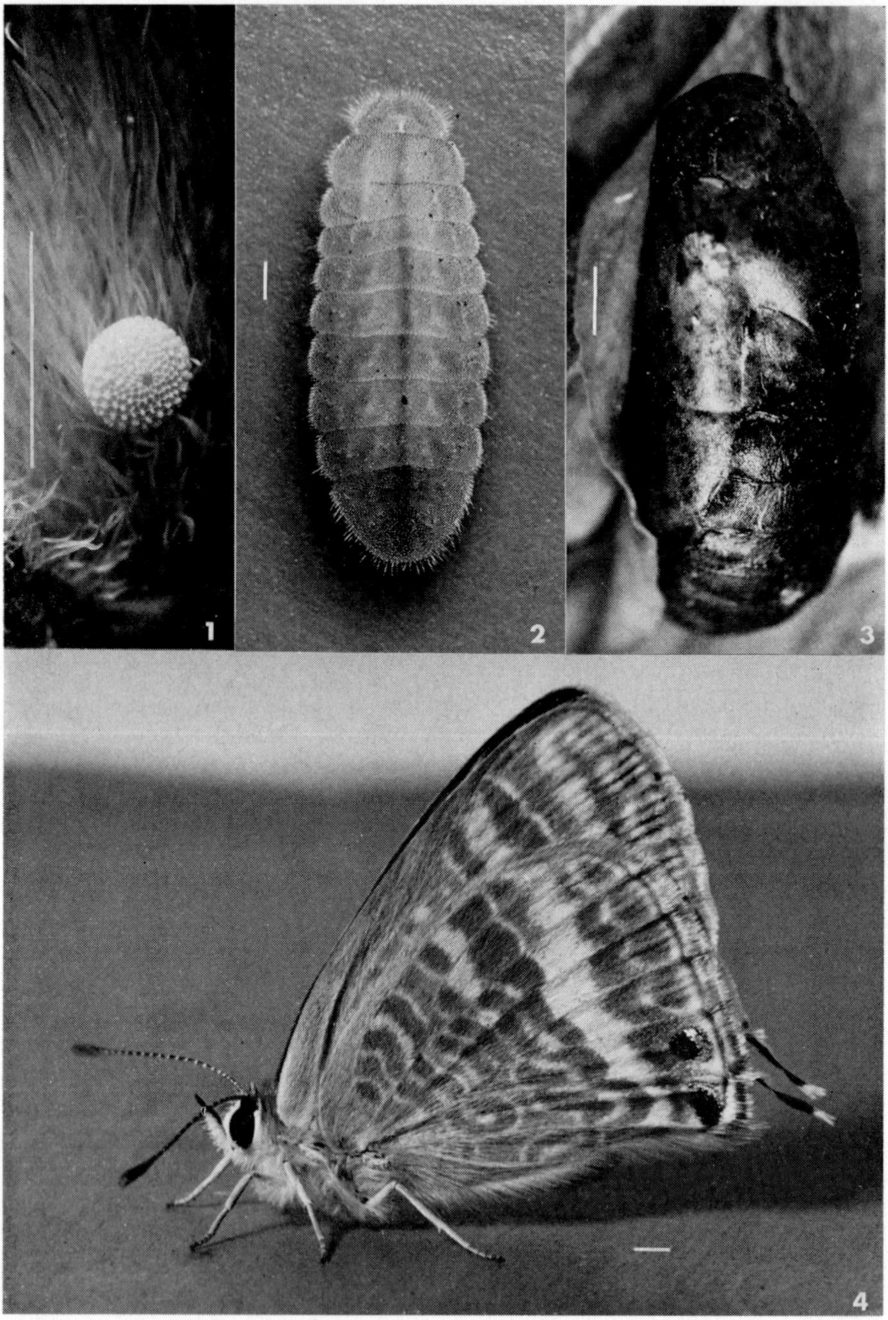

Fig. 3. Lampides boeticus, a Lycaenid butterfly recently self-introduced from Australia. (1) egg, (2) larva, (3) pupa, (4) adult. Photos: B. S. Eykel, DSIR.

Holzapfel *et al.*, 1970, and references therein) which have been trapped at great distances from the nearest land, must frequently arrive in New Zealand carried by air currents from overseas. Presumably many of the non-endemic species of these families have been self-introduced, frequently in pre-human (but post-glacial) times. Specific endemism is especially low in Thysanoptera (19%), Siphonaptera (35%) and Psocoptera (60%).

In Coleoptera the vast majority of non-endemic species have probably been accidentally introduced, mainly in imported goods, packing and dunnage. In recent years detailed records have been kept of overseas wood- and bark-boring insects intercepted at New Zealand ports (Milligan, 1970), and records of general quarantine interceptions are listed by Manson & Ward (1968). A few beneficial insects have been introduced intentionally for biological control of Arthropod pests and weeds (cf. Miller, 1970).

Although the fraction of the New Zealand insect fauna of adventive origin is quite small, non-endemic species are frequently abundant and conspicuous in towns, crops, orchards, pastures and other modified environments. Few native insects have been able to adapt themselves to the drastically changed habitats produced by man, especially by European man, where he has used (and frequently misused) land for agricultural and industrial purposes or habitation. Most endemic insects are confined to natural or slightly modified habitats, and do not seem to be able to adapt themselves successfully to altered environments. There are a few notable exceptions to this rule, e.g. the most serious pests of pastures, including improved pastures consisting entirely of exotic grasses, are *Costelytra zealandica* (Scarabaeidae) and *Wiseana cervinata* (Hepialidae), both of which are endemic at both the specific and the generic level. Most pests of forests, both native and exotic, are endemic insects, and the most important destroyer of old and inadequately treated timber in buildings is the generically endemic *Ambeodontus tristis* (Cerambycidae). Introduced insects which have become established are now part of the New Zealand fauna, however much we may regret the presence of some of them.

The highest level of endemism known in the New Zealand insect fauna is exemplified by endemic families of Lepidoptera (Mnesarchaeidae) and Coleoptera (Chaetosomatidae). In Coleoptera there are several endemic subfamilies and tribes (e.g. Lenacinae of the Rhizophagidae, Euderiinae of Bostrychidae, Agapythinae of Phloeostichidae), and strongly isolated genera (e.g. *Cyclaxyra*, *Priasilpha*) which will probably eventually require their own subfamilies or tribes. Endemism at the generic level is frequent in most orders. Some of the high-level endemics seem to be fairly primitive, relict forms which have survived in isolation in New Zealand, but whose relatives elsewhere have become extinct. Their ancestors probably arrived in New Zealand with *Sphenodon* (the order

Rhynchocephalia does not seem to differ from the order Squamata much, if any, more than the family Chaetosomatidae differs from Cleridae and Trogositidae).

Relationships

The relationships of the diverse groups of the New Zealand insect fauna reflect their place of origin, age, dispersibility, rate of evolution and the time of immigration of their ancestors. There are the few family-group endemics just referred to, with no known members elsewhere, whose geographical affinities are obscure, and which fall into the category variously known as Aboriginal, Endemic or Palaeozelandic.

FLEMING (1963) makes a distinction between primary endemics (due to local evolution) and secondary endemics (due to extinction elsewhere of formerly widespread groups). Neither of our endemic families of insects are monotypic, and they have apparently radiated to some extent in New Zealand. No fossils of either family are known.

Some New Zealand insects have their closest known relatives in Australia. Biogeographically the most interesting of these are families, subfamilies or tribes which are confined to Australia and New Zealand, for example Philorheithridae, Philanisidae (Trichoptera), Phycosecidae, Zolodininae, Ulodinae, Chalcodryidae (Coleoptera). These represent an old Australian element in the New Zealand fauna. Paleoaustral distributions of families or subfamilies covering New Zealand, Australia and southern South America seem to be much more frequent than those restricted to Australia and New Zealand. Examples are: Plecoptera: Eusteniidae, Austroperlidae, Gripopterygidae; Orthoptera: Macropathinae (of Rhaphidophoridae), Deinacridinae (of Henicidae); Hemiptera: Peloridiidae, Isoderminae (of Aradidae); Mecoptera: Nannochoristidae; Neuroptera: Stenosmylinae, Kalosmylinae (of Osmylidae); Diptera: Edwardsininae (of Blepharoceridae), Sciadoceridae, Teratomyzidae; Coleoptera: Migadopinae (of Carabidae), Camiarinae (of Leiodidae), Microsilphinae (of Silphidae), Cavognathidae, Belidae.

These fall into the Paleoaustral category of FLEMING (1963). Both Paleoaustral and old Australian elements comprise mainly primitive groups. Some Paleoaustral groups are thought to have originated in the cooler parts of the southern hemisphere before the break-up of the original southern landmass, and to have remained in their present homes which achieved their present distribution by continental drift. ILLIES (1965) argues strongly in favour of such a history for the Paleoaustral Plecoptera. He provides evidence that the Plecoptera as a whole arose and underwent their early evolution in the temperate southern hemisphere, and subsequently each of the main groups migrated to the northern hemisphere. This view is supported by the presence of the oldest and most primitive plecopteran fossils in Permian deposits of Australia.

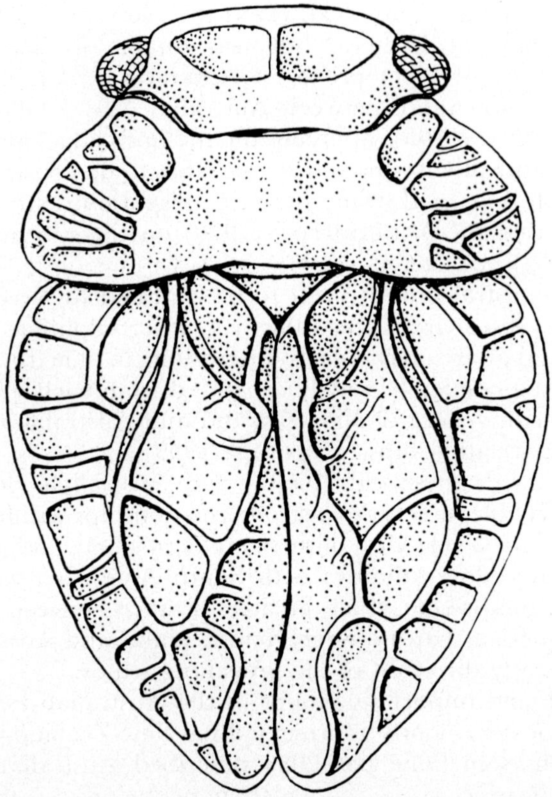

Fig. 4. Xenophyes cascus, a Peloridiid which lives in damp moss (after DRAKE & SALMON, 1948).

Other Paleoaustral groups apparently once had a world-wide distribution, but have since become extinct in the northern continents, probably due to competition from more recently evolved and successful groups. The Sciadoceridae are a notable example of this. The only extant species of this family are *Sciadocera rufomaculata* (occurring in Australia and New Zealand), and *Archiphora patagonica* from the southern tip of South America. Recently a fossil Sciadocerid was found in the Oligocene Baltic Amber of Europe, and even more recently two genera of Sciadoceridae were found in Cretaceous amber from Canada (MCALPINE & MARTIN, 1966).

Some zoogeographers have considered that all Paleoaustral groups originated in the northern hemisphere, and migrated southwards, later becoming extinct on the northern continents (e.g. DARLINGTON, 1965). BRUNDIN (1966) and others have strongly criticised this extreme view. It appears that some, possibly most, Paleoaustral groups originated in the north, but there is abundant evidence that some originated in the south, where they have remained. HENNIG (1960) analysed the relationships of Paleoaustral Diptera, but was unable to demonstrate conclusively that any of them prove an Antarctic evolution centre or an Antarctic migration route between Australia, New Zealand and South America. BRUNDIN (1966), applying the methods advocated by HENNIG to some Paleoaustral Chironomidae, was able to demonstrate an Austral centre of evolution, and transantarctic migration, but his work was strongly criticised by DARLINGTON (1970) on theoretical grounds. See BRUNDIN (1972) however.

A few Paleoaustral groups seem to be confined to New Zealand and Chile, but are absent from Australia (e.g. Microsilphinae and Camiarinae). The Belidae are particularly interesting in that the three genera in Chile are taxonomically widely separated from each other, and their relationships with genera of New Zealand and Australia are closer than their own inter-relationships (KUSCHEL, 1959).

In contrast with these old Austral and Australian elements, many groups of dispersible insects show strong relationships within the southern area at the specific level. Most of these are Australian, and the Neoaustral element showing close affinities with South America is small. This no doubt reflects the present much greater distance between New Zealand and South America than between New Zealand and Australia, and the dominant westerly direction of the prevailing winds.

Many large and important groups in the Australian insect fauna are absent from, or very poorly represented in, New Zealand. For example, the family Chrysomelidae has 2083 described Australian species, but only 143 in New Zealand, of which none belong to the subfamilies Sagrinae, Cassidinae, Hispinae and Criocerinae. The Australian representatives of these subfamilies are predominantly tropical, so it is possible that some were represented in New Zealand during Tertiary times, but became extinct during the Pleistocene.

The New Zealand insect fauna includes a characteristic Malayo-Pacific element not represented in Australia. The Tenebrionid genera *Lorelus* and *Menimus*, for example, are unknown in Australia, but each have several species in New Zealand, and others in New Caledonia, Fiji and other larger islands of the western Pacific, and the Oriental Region. *Menimus* is characteristically found in rain forest in wet rotten wood which has been attacked previously by large 'heart-rot' polypore fungi, a very scarce habitat in Australia. The faunas of the small isolated islands of Lord Howe and Norfolk include a few insect genera which are absent on the Australian mainland, but are present (or have close relatives) in New Zealand and the larger southwestern Pacific islands.

There are a few groups of insects found only in New Zealand and the northern hemisphere, which fall into the Holarctic element of FLEMING (1963). The beetle family Derodontidae shows this type of distribution. The remaining Cosmopolitan element is particularly characteristic of readily dispersible insects such as aphids at the specific level, and of many less dispersible insects at the generic and higher levels of classification. The relationships of the New Zealand insect fauna were also discussed, briefly, by GRESSITT (1961).

It is not possible to make meaningful generalisations on the relative frequency of faunal elements for the New Zealand insect fauna as a whole, because of the enormous biological and historical differences between groups, and even within groups. Stoneflies seem to be predominantly Paleoaustral, but in a large order like Coleoptera, some families, subfamilies, tribes and genera of each of the main faunal elements can be recognised.

Shore Insects

New Zealand shares with Australia the only marine caddis, namely *Philanisus plebeius*. The larva, which inhabits rock pools on rocky coasts, constructs its case with algae, usually the pink encrusting seaweed *Corallina*, and may also incorporate a few small stones. With this case, the larva is exceptionally well disguised and protected from the multitude of intertidal predators. Adults emerge in summer, and may be abundant on suitable rocky coasts. This genus has its own family Philanisidae, whose affinities are somewhat obscure (MOSELY & KIMMINS, 1953). It probably represents an old but primary endemic element in Australia and New Zealand.

Opifex fuscus is a large black mosquito whose biology and mating behaviour are unique. Its larvae breed in saline and hypersaline pools in the splash zone on rocky coasts. Eggs are laid in small cracks and crannies on the moist margin above the water surface, and are very resistant to desiccation. Populations may reach very high densities, the surface of some rock pools being almost covered by swarms of adults, mostly males. The latter spend most of their time on the water surface, hunting and competing for surfacing pupae. If the emerging pupa is a female, genital contact is effected rapidly and the male runs to the edge, dragging the female. *Opifex* is the most primitive member of the tribe Aedini (BELKIN, 1968). New Zealand's only haematophagous Ceratopogonid *(Styloconops myersi)* breeds on the sandy margins of brackish estuaries (DUMBLETON, 1971).

There are a number of marine Chironomid midges ranging from typical Clunioninae in the north to wingless intertidal forms on subantarctic islands. The subantarctic Staphylinid beetle *Baeostethus chiltoni* is found exclusively below high water mark, both as adults and larvae, and extends so far down the shore that some individuals seem to be almost perpetually submerged. Other Staphylinids, Tipulids, a Carabid *Kenodactylus capito* and '*Orchesia*' *rennelli* (Melandryidae) feed in the upper littoral zone, but are rarely submerged for any length of time, and are

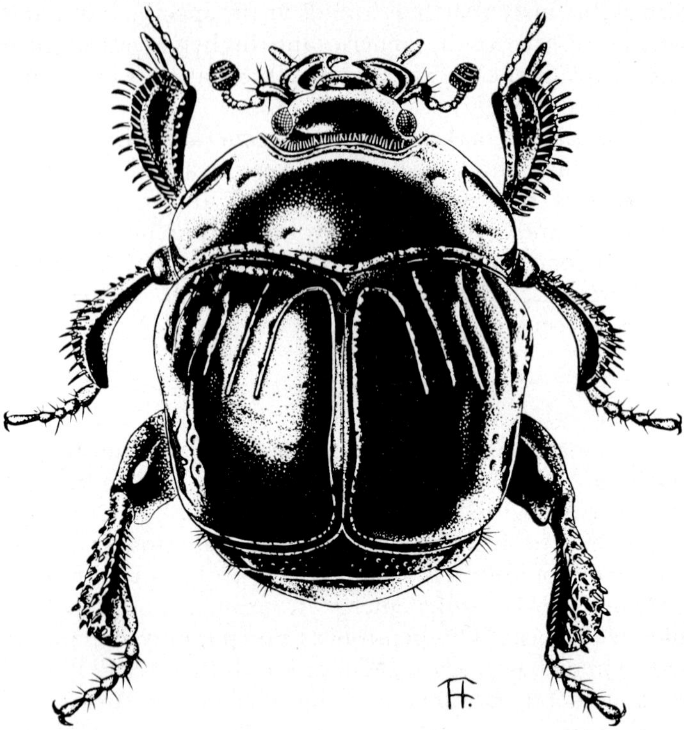

Fig. 5. Reichardtia pedator, a psammophilous Histerid beetle. Del. A. C. HARRIS.

probably best regarded as predominantly supralittoral insects (Kuschel, pers. comm.). *Antarcticodomus* spp. (Salpingidae) and *Lichenobius littoralis* (Anthribidae) are associated with lichens in the spray zone of some subantarctic islands and Stewart Island.

New Zealand has an interesting fauna of sandy beach insects, some of which, especially fossorial Coleoptera, are strongly modified for life in sand (HARRIS, 1970). Genera which have no known close relatives elsewhere, and have apparently evolved their psammophilous adaptations in New Zealand, are *Pericoptus* (Scarabaeidae), *Brullea* (Carabidae), *Reichardtia* (Histeridae), *Actizeta* (Tenebrionidae) and *Cecyropa* (Curculionidae). *Brullea* is apparently derived from the same stock as the endemic *Mecodema* and *Metaglymma*, but none of the other genera have close relatives anywhere. Other sand-dwelling beetle genera are shared with, or have close relatives in, Australia, e.g. *Phycosecis*, *Chaerodes* and *Phycochus*.

Psammophilous burrowing beetles differ considerably from other burrowing beetles, their distinctive features evolving in widely different families, and resulting in strong convergence in appearance. Such beetles tend to be exceptionally convex and broad, have thick bristles, very coarse and irregular punctures, and stouter limbs. Sand burrowers have to push their way through an abrasive and fluid medium, rather than a firm medium which provides traction while the beetle is digging and pushing through it (HARRIS, 1970). It is interesting that the more specialised psammophilous beetles of New Zealand, while similar in appearance to certain sand-dwelling members of the same families elsewhere, are not at all closely related to them (except the Australian examples mentioned).

All genera mentioned above are confined to coastal sandy beaches, with the exception of *Pericoptus*, which has one coastal species, and another occasionally coastal but usually found in river sands, often many miles inland. Other insects of sandy beaches are much less specialised for psammophily, and mostly belong to genera which have other species in different habitats inland, e.g. *Mimopeus* (Tenebrionidae), *Anabarhynchus* (Therevidae). Three species of *Prodontria* (Scarabaeidae) are psammophilous, two being coastal, and the most specialised being found only in sand hills of fluviatile origin at Cromwell, Central Otago.

There are few insects specialised for life on rocky coasts. The lichen feeders of southern and subantarctic coasts are briefly discussed elsewhere by DUGDALE (q.v.). The supralittoral fauna of rocky coasts appears to be considerably richer in the far south and subantarctic than in the north. A number of species of Coelopid flies breed in decaying seaweed on most types of shore. The large coastal earwig *Anisolabis littorea* occurs commonly on rocky shores, and less frequently amongst driftwood on sandy beaches.

Ground-dwelling and Soil Insects

A striking feature of the New Zealand insect fauna, which contrasts strongly with faunas of oceanic islands, is the considerable number and variety of large flightless ground-surface insects. The number of these which exhibit Paleoaustral distributions is also significant. It seems that they would have required land connections, or at least very much smaller ocean gaps than there are now, to achieve their present distributions.

In prehuman times, terrestrial mammals were absent from New Zealand, and the ecological niches which they would normally occupy on a continent were partly filled by ground-dwelling birds and insects. The habits and life of giant wetas of the genus *Deinacrida* (Henicidae) are similar to those of some small rodents. Significantly, these large wetas, which were probably once widespread in the North Island lowlands rapidly became extinct throughout most of their ranges in the early days of European settlement, probably coincident with the establishment and spread of European rats (especially *Rattus rattus*). *Deinacrida rugosa* is now confined to small islands in Cook Strait, while *D. heteracantha*, once common in forests of the northern North Island, persists near Te Kuiti and on Little Barrier Island (WATT, 1964). Other species of *Deinacrida* are confined to rat-free islands, or have a montane or alpine distribution in areas where rats are infrequent or absent. *Deinacrida* wetas feed on the ground or on vegetation at night. Their powerful spiny hind legs, which are raised defensively over the succulent, vulnerable abdomen when they are molested, are insufficient protection from a determined rat, although they would probably deter most birds.

Deinacrida heteracantha is New Zealand's largest and heaviest insect, adult females measuring up to 85 mm × 32 mm (excluding the ovipositor). Unlike the more southern species, it tends to be arboreal, perhaps explaining why it still persists on at least one part of the mainland, while *D. rugosa* does not.

The devastating effect of European rats on large ground-dwelling insects was demonstrated recently on Big South Cape Island (southwest of Stewart Island). This island was visited by entomologists in 1955, who found a large flightless weevil *Hadramphus stilbocarpae* associated with *Stilbocarpa lyalli*, to be quite common, as was a large Lucanid *Dorcus helmsi*. Some time before March 1964, *Rattus rattus* was introduced to the island, probably via a stern line from a fishing boat sheltering in a cove. In November 1968 and February 1969, entomological expeditions visited Big South Cape Island, and made an extensive and intensive search for *Hadramphus*, but found only a few remains of adults and no larvae, indicating that the species is probably extinct on the island (KUSCHEL, 1971). *Dorcus helmsi* and some other previously common large flightless insects are now quite rare.

To what extent the extinction of *Hadramphus stilbocarpae* was due to actual predation by rats, and to what extent it is attributable to reduction of the host-plant, is unknown. At the height of the population explosion of rats, *Stilbocarpa lyalli* was almost completely eaten out, but it has since recovered well.

Several large flightless weevils, wetas, Lucanids and Elaterids are apparently confined now to islands free of rats. Suitable habitats for these species are present on the mainland, where presumably they occurred in prehuman times. The effects of the Polynesian rat *(Rattus exulans)* on large flightless insects is not known, except that it has generally been said to be almost exclusively vegetarian (WODZICKI, 1950).

Extensive speciation and complex patterns of distribution and geographical variation are characteristic of the large flightless ground-dwelling insects. Distributions tend to be much more extensive and numbers of species fewer in the North Island than in the more mountainous South Island. The environment in the latter island is much more diverse, and geographical and ecological barriers are more numerous and more effective. They were even more effective during Pleistocene glaciations.

Fig. 6. Argosarchus horridus, a wingless Phasmid, on *Nothofagus menziesii*. Photo: B. S. Eykel, DSIR.

In some flightless Carabidae (e.g. *Megadromus capito*) and Tenebrionidae (e.g. *Mimopeus opaculus*), isolated populations occur on mountains or islands in or off the north of the North Island, but on the mainland lowlands the species occur only much further south. These isolated northern populations appear to be relicts of more northerly ranges of the species during the last glaciation.

Flightlessness in New Zealand Lucanidae and some other insects has been discussed by Holloway (1963). Amongst insects which are normally fully winged, there are flightless New Zealand species of Phasmatodea (100% flightless), Acrididae, Dermaptera, Blattodea, Plecoptera, and members of several families of Hemiptera, of Lepidoptera, of Diptera, of Hymenoptera and of Coleoptera. In prehuman times, with limited vertebrate predators, the need for wings was not as great as in continental areas. Development of wings requires a fair amount of metabolic energy and materials at a critical and vulnerable time in the insect's life cycle. When temperatures are generally low (as they were several times during the Pleistocene in New Zealand), there is no energy to spare, and selection will strongly favour reduction of wings if they are not essential.

New Zealand has a very rich fauna of leaf litter insects. Compared with warmer climates, the litter itself tends to be deep and persistent, providing an ideal habitat for many insects. A striking feature is the virtual absence of termites and the relative scarcity of ants in litter and soil. Termites in particular are responsible for the rapid breakdown of litter in the tropics, while many tropical ants are rapacious predators.

The most numerous soil and litter insects are usually Collembola. Some feed directly on decaying leaves, while others feed on fungi. They are important contributors to the processes of litter breakdown and soil formation. Little has been published so far on the ecology of New Zealand Collembola, except in pastures (cf. ADAMS, 1971).

Beetles form a major constituent of the litter fauna in New Zealand, and many species are largely confined to litter (e.g. many Colydiidae, Tenebrionidae, Scydmaenidae, Pselaphidae, Staphylinidae, Curculionidae, some Corylophidae and larvae of Cantharidae). Each type of litter appears to have its characteristic assemblage of species, but although much information has accumulated, it has yet to be analysed. Litter beetles are mainly saprophagous or fungivorous forms and predators.

The most important soil beetles in New Zealand are Melolonthine Scarabaeidae, including some 'grassgrub' species, particularly *Costelytra zealandica*, whose larvae feed on the roots of pasture grasses. Distribution of species is related to soil type, vegetation cover, slope, aspect etc. Some species are characteristic of tussock grasslands, others of lowland scrub, others of various forest types, others of vegetated coastal dunes, others of river flats, others of subalpine scrub and others of various alpine habitats (GIVEN, 1952). Each of the largest genera (*Odontria*, *Costleya*, *Pyronota*) include alpine species. The genus *Prodontria* contains only flightless species which have very limited distributions in the south of the South Island, Stewart Island and the Snares. The loss of wings and consequent restriction of dispersibility has resulted in limited distributions and apparently more speciation in this genus than in the related fully-winged *Odontria* in the same area. Larvae of some Melolonthines feed mainly on plant roots, but others seem to feed mainly on dead plant material. Soil-inhabiting Elateridae (some of which are predators) are considerably less numerous as species, and usually as individuals, in most New Zealand soils than are Melolonthinae.

Carabidae include numerous important predators whose larvae are mainly soil-inhabiting, although adults tend to hide during the day under logs and stones or in surface litter, whence they emerge at night. The larger flightless forms (*Mecodema*, *Megadromus*, *Holcaspis* etc.) have speciated very extensively.

Dipterous larvae form an important part of the soil and litter insect fauna, the majority being saprophagous or fungivorous (e.g. Tipulidae, Chironomidae, Mycetophilidae, Sciaridae, Phoridae, Syrphidae, Muscidae). Others are predators (e.g. Asilidae, Empididae, Therevidae). Other insects represented in soil and litter in New Zealand include various Diplura, Protura, Thysanura, Archaeognatha, Dermaptera, Orthoptera, Hemiptera, Psocoptera, Lepidoptera and Hymenoptera.

Insects of Dead Wood and Fungi

Some insects of soil and litter also inhabit logs and fallen branches, so there is not a sharp division between soil/litter and some dead wood habitats. Most insects associated with dead wood derive their nourishment mainly from saprophytic or parasitic fungi, rather than from the wood itself. Some actually go to the extent of cultivating fungi in gallery systems in (usually dead) wood (notably ambrosia beetles and weevils of the genus *Psepholax*).

When a tree dies (perhaps with the assistance of live-wood borers or defoliators), there begins a whole succession of invasions by wood-borers, saprophagous and fungivorous insects together with their predators and parasites. The most important xylophagous species are Cerambycidae and certain weevils, mainly of the subfamilies Cossoninae, Scolytinae and Cryptorhynchinae, especially *Psepholax*. These commence life in or under the bark, but some of them enter the xylem as they grow larger. Their gallery systems and frass soon become infected with fungi, and are entered by saprophagous and fungivorous insects such as some Colydiidae, Rhizophagidae, Zopheridae, Mycetophilidae, and certain predators, e.g. *Thoramus* (Elateridae), Cleridae, Trogossitidae, *Chaetosoma* (Chaetosomatidae). As the bark loosens, typical subcortical insects such as Aradidae, Cucujidae, Pyrochroidae, *Diagrypnodes* (Inopeplidae), Cerylonidae, *Pristoderus* (Colydiidae), certain Sciaridae, Stratiomyidae, and Chironomidae move under it. As the wood decays, further insects such as *Thelyphassa* (Oedemeridae), certain Tipulidae, Scatopsidae, Phoridae, *Syrphetodes* (Zopheridae), several Tenebrionidae (including *Zolodinus*), *Izatha* (Oecophoridae) and predators of several groups become established.

Dead twigs and small branches support a different fauna from larger branches and the trunk of the same tree. Logs and branches on the ground support a number of litter insects which are not present in standing dead trees. The latter provide a greater diversity of niches through greater moisture, heat and light gradients than are found in wood on the ground. Dead tree-fern rachides, dead leaves of the nikau palm *(Rhopalostylis sapida)*, dead flower stalks of *Aciphylla* and other plants (e.g. *Phormium*), dead flowers, decaying fruits and cones all have characteristic insects associated with them in New Zealand.

New Zealand has a rich flora of bracket fungi (Polyporaceae and Thelyphoraceae), plus members of a few other groups with fairly large fruiting bodies (e.g. *Hericium, Daldinia*). All of these have Coleoptera associated with them. The most important groups are Scaphidiidae (mainly on the exterior of fresh fruiting bodies), Ciidae, *Brouniphylax*

(Zopheridae) and Anobiidae (all boring in large dead bracket fungi; Ciidae also occur in small brackets such as *Trametes*, and encrusting forms such as *Poria*). *Euxestus* (Cerylonidae) breeds in *Hericium coralloides* (Hydnaceae). Few Diptera are associated with bracket fungi, but some (particularly Phoridae) breed in the fruiting bodies of Agaricales (Trichoceridae, Mycetophilidae and Sciaridae).

An important habitat for fungivorous insects and associated predators is provided by the sooty fungi which grow on the trunks and branches of various trees, especially *Nothofagus fusca*. These sooty fungi grow on honey dew produced by Coccids (*Ultracoelostoma assimile* on *N. fusca*.). The honey dew itself is collected by bees, wasps and ants. The fungus supports larvae of Diptera and Coleoptera (including Derodontidae, Colydiidae, Nitidulidae, Salpingidae), and various other fungivorous insects.

Cave and Ground-water Insects

There are extensive systems of limestone caves in several parts of New Zealand. These contain a limited, but very interesting insect fauna. Caves provide a uniform cool dark humid environment, but fairly limited food. The basis for the food web is vegetable matter and animals entering, falling into or washed into the cave from outside. New Zealand caves differ from many overseas in the absence of bat colonies (cf. MAY, 1963), so that this important source of food is lacking here.

Cave inhabitants can be divided into: 1. *Troglobites*, which are obligate cavernicoles, living exclusively underground, and exhibiting characteristic modifications such as depigmentation, reduction or loss of eyes, loss of wings, and elongation of mouthparts, limbs, sensory appendages and setae. In New Zealand caves, species of Carabid beetles of the genera *Duvaliomimus*, *Neanops*, *Scototrechus*, *Erebotrechus* (Trechini) and *Pholeodytes* (Harpalini) are true troglobites. *In Duvaliomimus* some species are trogloxenes, living in damp, densely vegetated gullies, others are troglophiles, and yet others are true troglobites. There are also some Collembola which apparently are troglobites (MAY, 1963). Only in Australia and New Zealand are there troglobite Harpalini, but Trechini are characteristic cave-dwellers in many parts of the world. 2. *Troglophiles* are facultative cavernicoles, which live and reproduce in caves but also exist successfully outside caves in dark, moist habitats in forests, and are preadapted to the cave environment by their tolerance of high humidity. Insects of this type in New Zealand caves include Collembola, cave wetas (Rhaphidophoridae), Psychodidae, *Arachnocampa luminosa* (Mycetophilidae: the celebrated New Zealand Glow-worm), Chironomidae, some Carabidae, certain Catopidae. 3. *Trogloxenes* are creatures of the outside which do not breed in caves, and only enter them temporarily or accidentally.

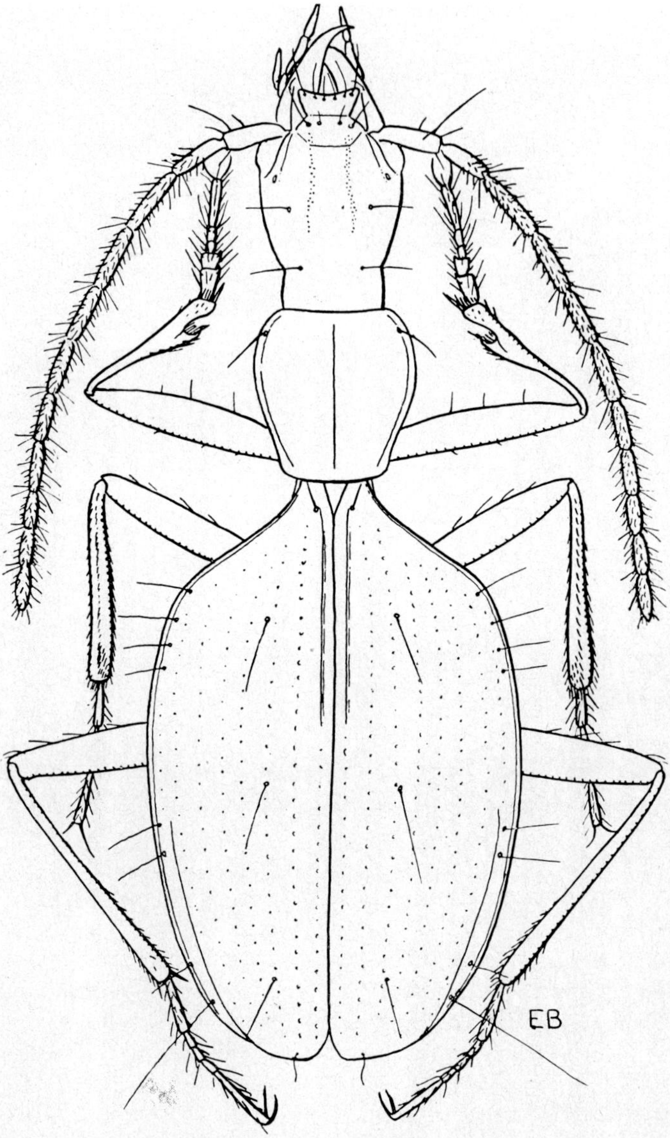

Fig. 7. Duvaliomimus lamberti, a troglobitic Trechine Carabid from marble caves near Nelson (after BRITTON, 1960).

The New Zealand Glow-worm is a tourist attraction in the Waitomo limestone caves, and is present in many other caves, artificial tunnels, and on damp, cool overhanging banks, logs etc. in forest. Larvae, pupae and adults of both sexes are luminous. The larva builds itself a hollow tubular nest of mucus and silk, which is suspended from the ceiling by a

Fig. 8. Larva of the Glow-worm, *Arachnocampa luminosa*, and its snare of pendulous silk 'fishing lines', Waitomo Cave. Photo: J. Pybus, DSIR (after Richards, 1960).

number of fine silk threads. From the nest the larva lets down up to 70 silk strings bearing sticky droplets. These 'fishing lines', up to 50 cm in length, are remarkable for their great tensile strength. The main food of the larva is Chironomid midges, which breed in the mud on the floor of caves. Adults are attracted by the lights of the glow-worm larvae, and fly upwards to become ensnared in the long, sticky fishing-lines. A tug on a fishing line causes the larva to move to the top of it and haul it up; it then consumes the midge. Other insects consumed less frequently include other Diptera, mayflies, stoneflies and caddis flies. In *Arachnocampa* adults, the light assists the sexes to find each other (RICHARDS, 1960).

Inhabitants of interstitial underground waters exhibit similar modifications (e.g. depigmentation, anophthalmy) to those found in cavernicolous forms. Until very recently no ground-water ('phreatic') insects were known in New Zealand, although several specialised Crustacea had been found in ground waters in Canterbury. Three species of phreatic Dytiscidae have recently been discovered in filters attached to pumps on the top of enclosed bore-holes near Nelson and in South Canterbury. The beetles are small, lack eyes, are lightly-pigmented, parallel-sided, with a distinct 'waist' between prothorax and elytra and are sparsely covered with long, very fine hairs. In these respects they resemble Dytiscidae and a Noterid of ground waters in Europe and Japan. They have the diagnostic characters of Hydroporinae, Bidessini, like their European and Japanese counterparts (*Siettitia* spp. and *Morimotoa phreatica* – see VANDEL, 1965), but have almost certainly evolved their phreatic adaptations independently to the Palaearctic forms. The largest New Zealand species appears to have diverged furthest from the typical Dytiscid shape in being more convex, with a more strongly marked 'waist', than in any of the other phreatic species.

Alpine Insects

New Zealand is a very mountainous country, with about 5,000,000 hectares (12,500,000 acres) mostly in the South Island, of alpine biotope (i.e. country above the real or hypothetical treeline), compared with only about 12,000 hectares (30,000 acres) in mainland Australia (DUGDALE, pers. comm.).

Above the limit of vegetation in New Zealand, there are relatively few permanent insect inhabitants, notably certain Rhaphidophoridae of the genus *Pharmacus*, Collembola and a few small Carabidae. It is not uncommon during summer to find lowland insects at these altitudes, where they are apparently carried by updraughts, together with light plant debris. These provide a source of food for the insects inhabiting this apparently barren wasteland.

As one descends and the vegetation cover increases, the insect fauna becomes more diverse, and can be very rich, especially in alpine scrub communities. Phytophagous alpine insects are numerous, and some (e.g. grasshoppers) may have a significant effect on the vegetation cover under some circumstances.

Many of the species of insects inhabiting the alpine zone live also at lower altitudes in suitable habitats (e.g. screes, scrub, river flats). There is also a significant exclusively alpine insect fauna. Alpine species tend to be restricted in distribution (sometimes to a single mountain range), and to have fairly specialised habitat requirements. Alpine insects have not yet been adequately studied on many mountains, and it is likely that some still await discovery, while the known distribution of others is likely to be extended considerably.

The larger flightless insects (Henicidae, Rhaphidophoridae, Carabidae, Tenebrionidae, Curculionidae etc.) include alpine species which have very restricted distributions. In the more extensively distributed species, populations from each mountain range, or even each individual mountain, are recognisably distinct from each other. In these cases the species presumably had a reasonably continuous distribution in the area at lower altitudes during the last glaciation, and became isolated following it.

A number of Scarab beetles of the subfamily Melolonthinae are alpine species, the best known of which are the large diurnal beetles of the genus *Costleya*. A few *Odontria* species also are exclusively alpine, and in Central Otago there are flightless alpine *Prodontria* and *Scythrodes*.

Some groups of flying insects have over 50% of their New Zealand species confined to the alpine zone, but these belong to relatively few genera. They include the important dipterous family Tachinidae (65% alpine species), the alpine members of which parasitise caterpillars, weevils and Melolonthinae (cf. DUGDALE, 1969). Amongst Lepidoptera, about 50% of New Zealand Geometridae are alpine, and about 60% of Tortricidae. The very large alpine Tortricine fauna consists entirely of Archipini, and is rich in species rather than genera (DUGDALE, 1966). Alpine Lepidoptera in New Zealand include a large number of brightly-coloured day-flying species, especially of these two families. There are only 12 established species of butterflies (two of which are alpine) in New Zealand, and the niches which they occupy elsewhere have frequently been filled by day-flying moths (DUGDALE, pers. comm.).

Other insect groups with a number of exclusively alpine species include Cicadidae (DUGDALE, 1972), Lygaeidae (EYLES, pers. comm.), cockroaches, Tipulidae (JOHNS, 1969), Acrididae (BIGELOW, 1967), Staphylinidae, Helodidae, Melyridae, Chrysomelidae (Coleoptera).

Despite the apparent youth of the alpine biotope in New Zealand, there are several genera of insects confined to the alpine zone. These genera do not seem to be very isolated taxonomically from their nearest relatives, and presumably they could have evolved since the beginning of the Kaikoura orogeny (middle Miocene), especially in view of the tremendous selection pressures and geographical isolation that must have affected mountain insects during Pleistocene glaciations.

Moss Insects

Some phytophagous moss insects are discussed elsewhere in this volume by DUGDALE (q.v.). In addition to phytophagous insects, moss supports a rich fauna of predators, parasites, detritus feeders and scavengers.

The Peloridiidae are amongst the most interesting moss-inhabiting insects. There are seven described species in New Zealand, and others in eastern Australia, Lord Howe Island and Chile. EVANS (1967) believes that the Peloridiidae, which are an ancient and primitive group, could only have achieved their present distribution by direct land connections. Certainly it is most unlikely that they, or many other insects restricted to very humid, protected habitats, could successfully cross the ocean gaps existing between these countries today.

In New Zealand 'water weevils' of the Subfamily Erirhininae include numerous species which inhabit more or less damp moss (KUSCHEL, pers. comm.). In all other Erirhininae (confined to continents) the weevils live in water, and their larvae live in tissues of aquatic plants. The adults have a coating of scales cemented together. A reservoir of air is trapped under the very convex elytra, enabling them to remain submerged for fairly long periods.

The family Helodidae is another group of beetles in which many New Zealand members have unusual habits. Larvae of some species live in small bodies of water, even in the axils of leaf bases of *Collospermum* and similar plants in wet rainforest, but others live in damp moss (especially on or near seepages), and yet others live in moist forest soils.

Certain insects inhabiting moss also are found in litter. These are general detritus-feeders, fungivores and predators, including diverse flies, beetles, Collembola, cockroaches and bugs.

Insects Associated with Vertebrates

In the absence of terrestrial mammals, New Zealand lacks native coprophagous insects, except for those associated with bird roosting and nesting sites. Nests contain also litter inhabitants, fungivorous and saprophagous insects, carrion feeders and parasites.

Cadavers are attacked by a succession of insects. Calliphorids, Muscids, Sarcophagids and Phorids are the primary necrophagous insects, followed by various other Diptera (Drosophilidae, Sepsidae, Syrphidae etc.), *Monopis* (Tineidae) and Coleoptera (Histeridae, Leiodidae, Dermestidae, Staphylinidae, *Necrophilus* etc.), some of which are predators on dipterous maggots.

The New Zealand flea fauna (SMIT, 1965), is sparse, although several undescribed native species are known. The main naturally occurring genus *Parapsyllus* includes fleas associated with various sea birds and the

Kea *(Nestor notabilis)*. Mallophaga have been collected from numerous New Zealand birds (especially sea birds and Anatidae), but the only comprehensive published account is limited to the Kermadec Islands fauna (WATT, 1972).

Biting Diptera, especially Culicidae and Simuliidae, can make life unpleasant for mammals during summer in the wetter parts of the country. These insects must have fed on birds in prehuman times. Some mosquitoes transmit an influenza-like virus of avian origin on the west coast of the South Island (PILLAI, 1966). The New Zealand mosquito fauna is reviewed very comprehensively by BELKIN (1968), and the Simuliidae by DUMBLETON (1972). The Tabanidae of New Zealand comprise 16 known species with Austral affinities (MACKERRAS, 1957), none of which are blood-suckers.

Mammalian parasites such as botflies, Anoplura, *Melophagus* and *Cimex* were introduced accidentally with their hosts. There are a few native Hippoboscidae which feed on birds.

Insects form an important part of the diet of some native birds. Flycatchers (Muscicapidae) including Fantails (*Rhipidura* spp.) take small Diptera and other insects on the wing. Other birds which feed regularly on flying insects include Meliphagidae (Tuis and Bellbirds), Silvereyes *(Zosterops lateralis)*, various warblers (Sylviidae) and New Zealand Wrens (Acanthisittidae). The Morepork *(Ninox novaeseelandiae)* includes large insects, especially moths, wetas and beetles, in its diet. The Kiwi *(Apteryx)* takes soil-inhabiting insect larvae by thrusting its beak into the soil, as well as consuming adults which are on the surface at night (WATT, 1971).

The Tuatara *(Sphenodon)* and lizards include many insects, particularly ground-dwelling forms, in their diet. Tree-dwelling geckoes consume many flying insects, especially Diptera.

Endoparasitic Insects

There is only one known Strepsipteran, *Elenchus maorianus*, in New Zealand, which parasitises a Cicadellid (GOURLAY, 1953). Colydiids of the genus *Bothrideres* parasitise immature Buprestidae and Cerambycidae. Several species of Rhipiphoridae are known, all of which apparently parasitise wood-boring Cerambycid larvae.

The most important group of parasitic Diptera in New Zealand, the Tachinidae, include numerous flies (Voriinae and Goniinae) which parasitise Lepidopterous larvae, others (Tachininae) which parasitise larvae of Scarabaeidae, Cerambycidae, Curculionidae, Lucanidae and Oedemeridae, and others (Phasiinae) which probably parasitise Hemiptera (DUGDALE, 1969). Coleoptera are unusual hosts for Tachininae, which normally parasitise Lepidoptera elsewhere in the world.

The vast majority of New Zealand Hymenoptera are parasitic, and there are very few phytophagous species, of which only two *Xiphydria* are endemic (VALENTINE, 1970). The superfamilies Scolioidea, Vespoidea, Megalyroidea, Chrysidoidea and Trigonaloidea are absent altogether, while Sphecoidea and Apoidea are very poorly represented. The absence of Chrysidoidea is easily explained, as they parasitise Vespoids and Sphecoids. Other parasitic groups which are lacking have no apparently suitable hosts in New Zealand. Suitable host-plants for most secondarily phytophagous Hymenoptera are absent, but the groups to which these latter belong are all poorly represented here, and the absence of endemic secondary phytophages such as Cynipinae, Eurytomidae and Torymidae may be due as much to this as to lack of suitable host-plants.

The known representatives of *Coccophagus* (Aphelinidae) in New Zealand parasitise Pseudococcidae, but the hosts for almost all other species of this genus elsewhere are Coccidae (VALENTINE, pers. comm.). In most parasitic Hymenoptera the host relations of the male are usually the same as those in the female. However, in certain Aphelinidae (e.g. *Bardylis*) the females are endoparasites on Diaspidid scales, but males develop on the female larvae of their own species, or on other Aphelinid or Encyrtid primary parasites in hosts other than the normal host of the female (VALENTINE, 1964).

Acknowledgements

I am indebted to several of my colleagues who provided information for this chapter, and I am very grateful to Drs. G. KUSCHEL, A. C. EYLES, G. W. RAMSAY and Messrs. J. S. DUGDALE and E. W. VALENTINE for reading and criticising a draft manuscript. I retain full responsibility for any errors of fact or opinion, however.

REFERENCES

ADAMS, E. C. G. 1971. Ecological studies of microarthropods in a New Zealand pasture soil with special reference to the Collembola (I). *Pedobiologia* 11: 321–337.

BELKIN, J. N. 1968. Mosquito studies (Diptera: Culicidae). VII. The Culicidae of New Zealand. *Contr. Am. ent. Inst.* 3(1): 1–180.

BIGELOW, R. S. 1964. Note on the Black Field Cricket in New Zealand. *N.Z. Ent.* 3(3): 9–10.

BIGELOW, R. S. 1967. The grasshoppers (Acrididae) of New Zealand. Their taxonomy and distribution. *University of Canterbury Publications* 9: 1–111, 40 pls.

BRUNDIN, L. 1966. Transantarctic relationships and their significance as evidenced by Chironomid Midges. *Kungel Svenska Vetensk. Handl.* 11(1): 1–472, 30 pl.

BRUNDIN, L. 1972. Phylogenetics and biogeography. *Syst. Zool.* 21: 69–79, 6 figs.

Darlington, P. J. 1965. Biogeography of the Southern end of the world. Harvard University Press. 236 pp. 38 figs.

Darlington, P. J. 1970. A practical criticism of Hennig-Brundin 'Phylogenetic Systematics' and Antarctic biogeography. *Syst. Zool.* 19: 1–18.

Dugdale, J. S. 1966. A revision of the New Zealand Schoenotenini and Cnephasiini (Lepidoptera: Tortricidae: Tortricinae). *N.Z. J. Sci.* 9: 731–775, 9 figs.

Dugdale, J. S. 1969. A classification of the New Zealand genera of Tachinidae. *N.Z. J. Sci.* 12: 606–646, 57 figs.

Dugdale, J. S. 1972. Genera of New Zealand Cicadidae (Homoptera). *N.Z. Jl. Sci.* 14(4): 856–882, 32 figs.

Dumbleton, L. J. 1967. Winter dormancy in the New Zealand biota and its paleoclimatic implications. *N.Z. J. Bot.* 5: 211–222.

Dumbleton, L. J. 1971. The biting midge *Styloconops myersi* (Tonnoir) (Diptera: Ceratopogonidae), description of male and redescription of female. *N.Z. Jl. Sci.* 14(2): 270–275, 6 figs.

Evans, J. W. 1967. The biogeographical significance of the Peloridiidae (Homoptera, Coelorrhyncha) and a new species from Lord Howe Island. *Proc. roy. Soc. Qd.* 79: 17–24, 5 figs.

Fleming, C. A. 1963. The nomenclature of biogeographic elements in the New Zealand biota. *Trans. Roy. Soc. N.Z. Gen.* 1, 2: 13–22.

Given, B. B. 1952. A revision of the Melolonthinae of New Zealand. Part 1: The adult beetles. *N.Z. Dep. Sci. Industr. Res. Bull.* 102: 1–137, 21 pls.

Gourlay, E. S. 1953. The Strepsiptera, an insect order new to New Zealand. *N.Z. Ent.* 1, 3: 3–8.

Gressitt, J. L. *et al.* 1961. Problems in the zoogeography of Pacific and Antarctic insects (with appendices). *Pacific Insects Mon.* 2: 1–127, 46 figs.

Harris, A. C. 1970. Coastal beetles of the Wanganui-Manawatu area *Bulletin of Natural Sciences, Victoria University of Wellington* 1: 45–87, 27 figs.

Hennig, W. 1960. Die Dipteren – Fauna von Neuseeland als systematisches und tiergeographisches Problem (with summ. in English & Russian). *Beitr. Ent.* 10, 3/4. 221–329, 27 figs. (English translation: 1966: *Pacif. Ins. Monog.* 9: 1–81).

Holloway, B. A. 1963. Wing development and evolution of New Zealand Lucanidae (Insecta: Coleoptera). *Trans. Roy. Soc. N.Z. Zool.* 3, 11: 99–116; 35 figs.

Holzapfel, E. P., Tsuda, D. M. & Harrell, J. C. 1970. Trapping of air-borne insects in the Antarctic area (Part 3). *Pacif. Ins.* 12(1): 133–156.

Illies, J. 1965. Phylogeny and zoogeography of the Plecoptera. *Ann. Rev. Ent.* 10: 117–140, 8 figs.

Johns, P. M. 1969. The mountain invertebrate fauna, pp. 391–399, in Knox, G. A. (Ed.), The natural history of Canterbury. Reed, Wellington. 620 pp., 133 pls., text figs.

Kloet, G. S. & Hincks, W. D. 1945. A check list of British insects. Stockport, 483 pp.

Kuschel, G. 1959. Nemonychidae, Belidae y Oxycoronidae de la fauna Chilene, con algunas consideraciones biogeograficas. *Inv. Zool. Chilenas* 5: 229–271, 96 figs.

Kuschel, G. 1971. Entomology of the Aucklands and other islands South of New Zealand: Coleoptera: Curculionidae. *Pacif. Ins. Monogr.* 27: 225–259, 24 figs.

McAlpine, J. F. & Martin, J. E. H. 1966. Systematics of Sciadoceridae and relatives with descriptions of two new genera and species from Canadian Amber and erection of family Ironomyidae (Diptera: Phoroidea). *Canad. Ent.* 98: 527–544, 24 figs.

MacKerras, I. M. 1957. Tabanidae (Diptera) of New Zealand. *Trans. Roy. Soc. N.Z.* 84(3): 581–610, 6 figs.

MacKerras, I. M. 1970 (Ed.). The insects of Australia. Melbourne University Press. 1029 pp., 9 col. pls., many text figs.

Manson, D. C. M. & Ward, A. 1968. Interceptions of insects, mites and other animals entering New Zealand, 1955–65. *N.Z. Dept. of Agriculture*, 407 pp.

MAY, B. M. 1963. New Zealand cave fauna II – The limestone caves between Port Waikato and Piopio Districts. *Trans. Roy. Soc. N.Z.* 3, 19: 181–204, 7 figs.
MILLER, D. 1956. Bibliography of New Zealand entomology. *N.Z. Dept. Sci. Industr. Res. Bull.* 120.
MILLER, D. 1970. Biological Control of weeds in New Zealand 1927–48. *N.Z. Dep. Sci. Industr. Res. Inf. Ser.* No. 74: 1–104, 31 figs.
MILLER, D. 1971. Common insects in New Zealand. A. H. & A. W. Reed, Wellington, 178 pp., 9 pls., 456 figs.
MILLIGAN, R. H. 1970. Overseas Wood- and bark-boring insects intercepted at New Zealand ports. *N.Z. Forest Service Technical Paper* No. 57, 80 pp.
MOSELY, M. E. & KIMMINS, D. E. 1953. The Trichoptera (Caddis-flies) of Australia and New Zealand. British Museum 550 pp., 353 pls.
RICHARDS, A. M. 1960. Observations on the New Zealand Glow-worm *Arachnocampa luminosa* (Skuse, 1890). *Trans. Roy. Soc. N.Z.* 88 (3): 559–574, pls. 27–38.
SMIT, F. G. A. M. 1965. Siphonaptera of New Zealand. *Trans. roy. Soc. N.Z. Zool.* 7: 1–50, 16 figs.
VALENTINE, E. W. 1964. Differential host relations of the sexes in some New Zealand parasitic Hymenoptera. *N.Z. Ent.* 3, 3: 6–8.
VALENTINE, E. W. 1967. A list of the hosts of entomophagous insects of New Zealand. *N.Z. Jl. Sci.* 10: 1100–1209.
VALENTINE, E. W. 1970. A list of the phytophagous Hymenoptera in New Zealand. *N.Z. Ent.* 4(4): 52–62, 4 figs.
VANDEL, A. 1965. Biospeleology: the biology of cavernicolous animals. Pergammon Press, Oxford 524 pp., 80 figs.
WATT, J. C. 1964. The rediscovery of a giant weta, *Deinacrida heteracantha*, on the North Island mainland. *N.Z. Ent.* 3, 2: 9–13, 1 fig.
WATT, J. C. 1970 (Ed.). Symposium: The present status of Taxonomic Entomology in New Zealand. *N.Z. Ent.* 4(3): 24–61.
WATT, J. C. 1971. The North Island Kiwi: a predator on pasture insects. *N.Z. Ent.* 5(1): 25–27.
WATT, J. C. 1972. Ectoparasitic insects on birds and mammals of the Kermadec Islands. *Notornis* 18(4): 227–244, 16 figs.
WODZICKI, K. A. 1950. Introduced mammals of New Zealand. *N.Z. Dep. Sci. Industr. Res. Bull.* 98: 1–255, 250 figs.

XIV. THE FRESHWATER INSECTS

by

I. D. McLELLAN

Introduction

The state of entomological research in a country depends very much upon its culture and economy. Because of the former factor research on aquatic insects in New Zealand only started to gain momentum during the present century. Our reliance upon agriculture has meant that the bulk of entomological research has been turned in that direction and consequently aquatic entomology has been carried out by a small number of workers either amateur or in universities and museums. The impact of so few workers is not great and in most aquatic orders much systematic work has to be accomplished. This lack of basic knowledge has had its affect upon both zoogeographical and ecological research. The influence of the state of a taxon's systematics on its zoogeography can be illustrated by looking at the stages that the systematics of N.Z. Plecoptera have passed through. 1. Species assigned to Northern genera and families. 2. Species assigned to new genera but still in Northern families. 3. New genera now in new tribes or sub-families still in Northern families. 4. All now in Southern families. Obviously these stages have not been completely passed through in other orders and the result gives a false impression of their relationships. This is particularly so where a considerable length of time has elapsed since work was done on certain taxa and also where poor taxonomic criteria are still in use.

Most ecological work consisted of natural history notes appended to systematic works but there are a few purely ecological works and these are either theses or population studies as part of fisheries research.

In the following accounts the major aquatic orders and also the main aquatic Diptera families have been dealt with. Some taxa which are not usually found in freshwater elsewhere are given some space while others have been omitted either because they inhabit only still waters or because useful information is not available.

EPHEMEROPTERA

4 families, 9 genera, 26 species. Siphlonuridae: 5 genera, 9 species. Siphlaenigmatidae: 1 genus, 1 species. Leptophlebiidae: 2 genera, 14 species. Ephemeridae: 1 genus, 2 species.

There are several undescribed genera and a number of undescribed species.

Table 1. Distribution of Austral genera of Siphlonuridae.

Subfamily	New Zealand	South America	Australia
Rallidentinae	Rallidens		
Siphlonurinae	Nesameletus	Metamonius	Ameletoides
Oniscigastrinae	Oniscigaster	Siphlonella	Tasmanophlebia
Amelotopsinae	Ameletopsis	Chiloporter Chaquihua	Mirawara
Coloburiscinae	Coloburiscus	Murphyella	Coloburiscoides

Zoogeography

All N.Z. species are endemic as are the genera in Siphlonuridae and Ephemeridae. Siphlonuridae also contain the endemic monotypic subfamily Rallidentinae and there is also the endemic monotypic family Siphlaenigmatidae which appears to be derived from Siphlonuridae. Also endemic is the subfamily Ichthybotinae in Ephemeridae. In Leptophlebiidae the genus *Zephlebia* is endemic and probably all N.Z. Leptophlebiidae genera may eventually be found to be endemic.

Our mayfly fauna follows the usual pattern found in other aquatic orders by having a disjunctive Austral distribution of closely related groups. Part of this can be seen for Siphlonuridae in Table 1. There is, however, the enigma of *Ichthybotus* which stands as the sole endemic N.Z. taxon without a sister group in S. America or Australia.

The situation in Leptophlebiidae is not as clear cut as in Siphlonuridae but it is obvious that the Austral components of the family have closer affinities to each other than to those of other regions.

In N.Z. all known genera are represented at low elevations but only *Oniscigaster*, *Nesameletus* and *Deleatidium* are common at elevations much above 700 m in South Island, 1000 m in North Island. A few species occur only from 650–2000 m. It is doubtful whether any species (except perhaps *Rallidens mcfarlanei*) is restricted to a single locality or even province. There are, however, sometimes slight differences between North and South Island forms of the same species.

Ecology

Literature on ecology of N.Z. mayflies is extremely sparse, consisting mainly of notes in predominantly systematic works. The only reasonably detailed study in this field is that on the nymph of *Coloburiscus humeralis* which has brought to light some unusual facts about this siphlonurid. Because of the inability of its fixed forked gills to absorb sufficient oxygen in slow streams the animal is restricted to rapid well-oxygenated waters. Its mode of obtaining sustenance also relies on water currents. Fine

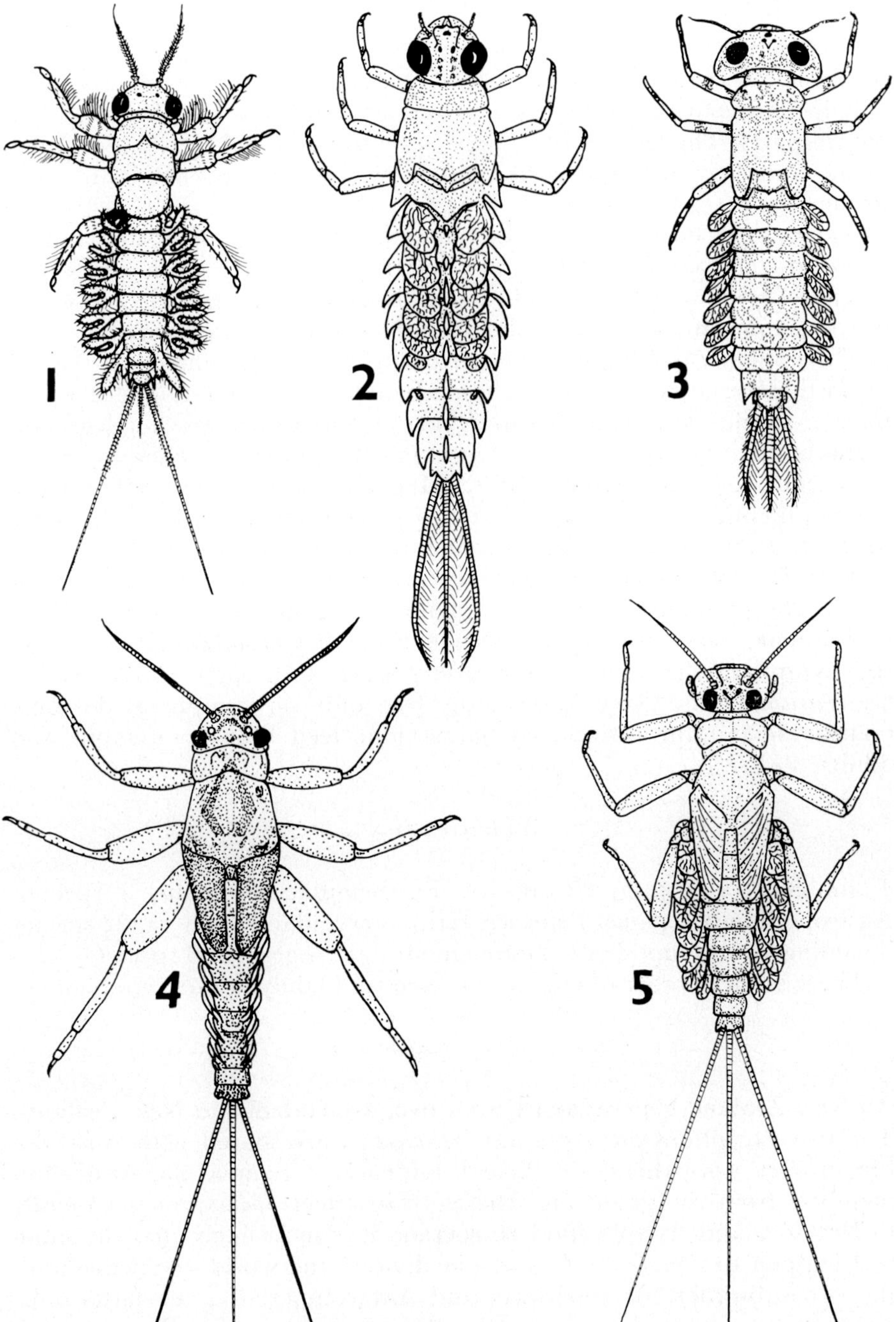

Fig. 1. Some Ephemeroptera larvae. 1: *Coloburiscus humeralis* (Siphlonuridae). 2: *Oniscigaster distans* (Siphlonuridae). 3: *Amelotopsis perscitus* (Siphlonuridae). 4: *Zephlebia* species (Leptophlebiidae). 5: *Deleatidium myzobranchia.*

particles (diatoms and algae mainly) are swept onto the long fringes of hairs on the front and middle legs and are combed off by the brush-like mouth-parts. The eggs of this mayfly are also adapted to rapid currents being equipped with spring-like adhesive filaments to anchor the eggs to the substrate.

As far as can be ascertained the remaining nymphs of our fauna appear to be grazers on algae and other water plants apart from *Ameletopsis perscitus* (Siphlonuridae) which is carnivorous and has been observed preying upon small leptophlebiid nymphs and other water insects.

The ephemerid *Ichthybotus hudsoni* occupies the same niche as that of its northern relatives. The nymphs are found in sand and mud of streams, have a similar gill structure to *Ephemera* and utilize the forelegs and mandibular tusks for digging. Several ephemeropteran niches appear to be vacant, e.g. lakes and ponds. Members of the genus *Deleatidium* (Leptophlebiidae) have atypical niches which are more or less equivalent to those of certain absent families such as Heptageniidae.

New Zealand mayfly nymphs are preyed upon by a number of invertebrate predators such as Trichoptera larvae of the family Rhyacophilidae, the larva of the Dobson Fly *(Archichauliodes diversus)* and the nymph of the stonefly *Stenoperla prasina*. Fish such as Eleotridae, *Cheimarrichthys* and *Anguilla* generally feed only on nymphs as do some Galaxiidae but the introduced Salmonidae feed both on nymphs and adults.

Plecoptera

4 families, 14 genera, 32 species. Eustheniidae: 1 genus, 1 species. Austroperlidae: 1 genus, 1 species. Gripopterygidae: 6 genera, 22 species including Antarctoperlinae. Notonemouridae: 6 genera, 8 species.

There are a number of undescribed species mainly in Gripopterygidae.

Zoogeography

All New Zealand genera, apart from two, are endemic to New Zealand. The two exceptions *Stenoperla* and *Spaniocerca* are shared with Australia but in *Spaniocerca* there is sufficient evidence to remove the Australian members from the genus and therefore in effect *Spaniocerca* is endemic to New Zealand. With a modern revision it is most likely that the same will happen to *Stenoperla*. Zwick has divided the world's stoneflies into the two sub-orders Arctoperlaria and Antarctoperlaria, the latter sub-order being restricted to the southern part of the world. This southern sub-order contains the families Eustheniidae, Austroperlidae and Gripopterygidae. Although the remaining family Notonemouridae is included in the Arctoperlaria the family has a southern distribution only, which is at variance with that suggested by the sub-ordinal name.

The relationships of our sole eustheniid, *Stenoperla prasina*, presents no problems. Australia has a diverse Archiperlarian fauna and as the genus (or as it may be proved a closely related genus) is also present there, the odds are in favour of a recent Australian origin.

According to Illies, *Austroperla cyrene*, our only austroperlid, forms with the Chilean fauna, an evolutionary offshoot from the remainder of the Austroperlidae. This would suggest a South American origin but I hesitate to accept this for a number of reasons. Smith & Hallam (1970) put the dispersal of Gondwanaland about Upper Cretaceous-Tertiary times. This means that the ancestor of *A. cyrene* would have been in the New Zealand region from that time and have been subjected to division of its gene pool by geological events during the next 100 million years. Considering this and the wide range of biotopes inhabited, one would think that a number of genera, or species, would have arisen. (This speciation has occurred in the Antarctoperlinae which almost certainly came to New Zealand in this way.) Apart from a small degree of wing shortening in specimens from alpine areas this widespread species shows very little variation which is compatible with variation found in recent invaders. From this reasoning it is more likely that *A. cyrene* is one of the numerous recent travellers by wind drift from Australia.

To understand the position of the Autarctoperlinae it is necessary to refer to Zwick (1969) who on sound evidence set this subfamily in limbo between the rest of Gripopterygidae and Austroperlidae. The Antarctoperlinae are found only in New Zealand and South America therefore dispersal must have been through the Antarctic. Our ancestral stock was most likely small in species because the same general morphological features, differing to a degree from the South American fauna, are found in all New Zealand species.

The remainder of the Gripopterygidae present a different problem. All New Zealand gripopterygids lack tibial spurs (a characteristic shared with Antarctoperlinae) and all belong to the subfamily Leptoperlinae. All Australian leptoperlinids possess tibial spurs as do those of South America except in the monotypic genus *Notoperlopsis* which bears a similarity to our *Megaleptoperla*. The Australian leptoperlinids also differ in a number of other ways from those of New Zealand. It appears that the New Zealand gripopterygids are of long established stock with a closer relationship to those of South America than to those of Australia. Like the Antarctoperlinae this presupposes dispersal through Antarctica.

The family Notonemouridae has, as intimated earlier, a possible northern origin. It is best looked at from its two elements, the *Spaniocerca* and the *Notonemoura* groups of Zwick. The *Spaniocerca* group (named from the New Zealand genus) is found in Madagascar, South Africa, Australia, New Zealand and South America. The South Africa-Madagascar genera are less closely related to the remainder, but the Australian genera are more closely akin to the New Zealand ones than those of

South America so it is possible that they have arrived in New Zealand via Australia either by wind drift or indirectly through the Inner Melanesian Arc. (Illies used the idea of a route from Asia through Malaysia and Indonesia to Australia and New Zealand then to South America for the whole of Notonemouridae with the South African and Madagascan fauna as an offshoot from Asia.) It is also within reason to assume that the group was present in Gondwanaland before break up. As most exponents of Continental Drift put the breakaway of India and then Africa before that of the rest of Gondwanaland the African-Madagascan fauna would, in isolation, have had more time to develop in their peculiar line.

The *Notonemoura* group has representatives in Australia, New Zealand and South America. As there are great basic similarities between the genera of these three regions it is likely that the group had its origin in Gondwanaland. These similarities are so marked that certain genera can be easily grouped on a morphological basis e.g. *Neofulla* (South America), *Notonemoura* (New Zealand), *Varioperla* (Australia). One thing is certain, they differ greatly from the *Spaniocerca* group in that they have certain features which are decidedly apomorphic in comparison with the *Spaniocerca* group.

Excluding the Subantarctic islands which have their own endemic species the situation in New Zealand is as follows. The representatives of Austroperlidae and Eustheniidae are found throughout New Zealand and therefore this would put their arrival in New Zealand prior to the formation of Cook Strait, i.e., prior to the end of the Pleistocene and probably not long before the Pleistocene because of lack of speciation. Within the Gripopterygidae and Antarctoperlinae there is an almost complete coverage of the country by some species, dependent on availability of habitat, whereas other species are restricted in range. For instance, in *Zelandoperla*, (Gripopterygidae) *Z. decorata*, *Z. fenestrata* and *Z. agnetis* are found throughout the country but *Z. denticulata* is restricted to the southwest of the South Island and in this area the four species may be collected from certain streams. *Zelandobius* (Antarctoperlinae) also shows this trend. The species *Z. furcillatus* and *Z. confusus* are widespread but so far *Z. illiesi* is restricted to heavily wooded streams in the northwest of the South Island and other species are restricted to the southeast and southwest. *Aucklandobius* (Gripopterygidae) exhibits the same trend with a greater degree of endemism in the south. This applies to a lesser extent with the Notonemouridae but nevertheless the south again has its own forms. The most decisive fact to appear is that the North Island has no endemic species. In addition, the North Island lacks the apterous forms which are so common in the alpine areas of the South Island. One example of extreme brachyptery in *Aucklandobius fulvescens* has been found in the Tararua Ranges in the North Island in a species which, in the south, has never been found in that condition, even at comparable altitudes. One of the alpine apterous forms in the South Island appears to have a discontinuous distribution in that up to the present it has been found only in Fiordland in the southwest and on the mountains in the northwest. These two areas are considered to have been refugia during the Pleistocene glaciations.

Auckland and Campbell Islands are the only Subantarctic islands having Plecoptera. Six gripopterygids are present on these islands which are endemic to the islands but the genera to which they belong (*Apteryoperla* and *Aucklandobius*) are also found on the mainland of New Zealand. These stoneflies have followed the usual trend of gripopterygids in high altitudes in that all but one species are apterous and most have terrestrial larvae. These species probably originated from New Zealand stock and they show quite marked adaptive radiation from two ancestors.

Ecology

In alpine regions and in the Subantarctic islands wingless stoneflies are common. These are found in all the species of the gripopterygid genera *Apteryoperla*, *Vesicaperla* and in some species of *Aucklandobius* (Gripopterygidae) and *Zelandobius* (Antarctoperlinae). There is a tendency for the larvae, in the first three abovementioned genera, to seek out terrestrial habitats. Some *Apteryoperla* plus *Vesicaperla* and the apterous *Aucklandobius* live in the humid atmosphere of thick alpine vegetation and *Vesicaperla* larvae have the odd habit of congregating in skeletons. Other *Apteryoperla* live under stones and in seepages. The larvae of winged *Aucklandobius* species living at low altitudes on the main New Zealand islands spend their first instars in water before moving out under stones on the flood plains of streams. This is made possible by the univoltine life cycle of these species which is as follows. The adults emerge in spring and early summer and from then until autumn no larvae are found so it is thought that the eggs remain dormant until the temperature falls to the required level. During autumn early instar larvae are found in the water but later when moist conditions and lower temperatures occur they move out into their terrestrial habitat for the remainder of their larval existence.

The larvae of the remaining gripopterygids and of *Zelandobius* inhabit a variety of biotopes in running water and we have obvious sets of associated species in certain biotopes. For example, *Zelandoperla agnetis* and *Zelandobius confusus* are associates in stony streams which carry little suspended material in flood but they are replaced by *Zelandoperla decorata* and *Zelandobius furcillatus* where the amount of suspended material is greater. The only gripopterygid which inhabits vegetation in running water is *Megaleptoperla diminuta* – a strange contrast to its generic partner *M. grandis* which lives among the boulders and stones of swift flowing streams. In the *Zelandobius* most species live in stony streams but *Z. illiesi* lives in small forest streams where dead terrestrial vegetation covers the

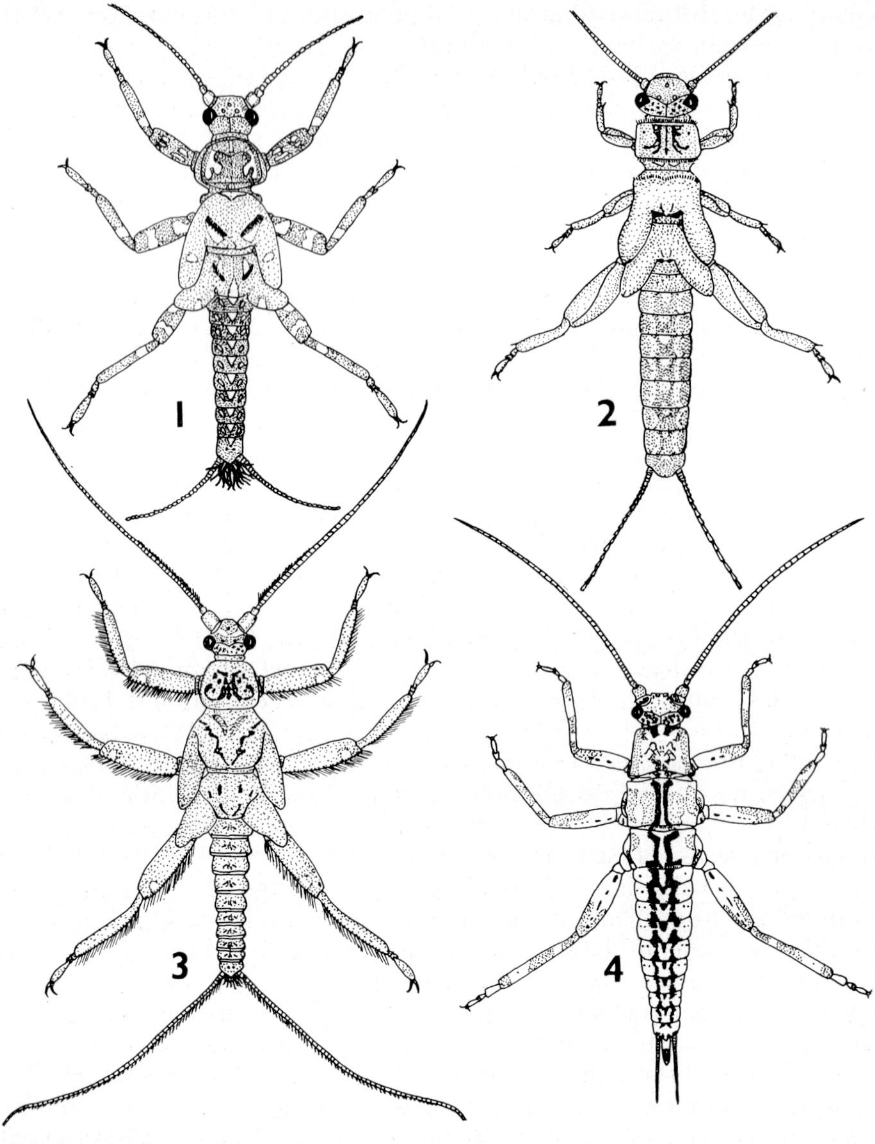

Fig. 2. Some Plecoptera. 1: Larva of *Aucklandobius trivacuatus* (Gripopterygidae). 2: Larva of *Notonemoura alisteri* (Notonemouridae). 3: Larva of *Zelandoperla decorata* (Gripopterygidae). 4: ♀ of *Vesicaperla substirpes* (a wingless gripopterygid with an ovipositor).

substrate of sand, mud and stones. Here the larvae are found mainly on the fronds of a tree fern *(Cyathea smithii)*.

The anal gill rosette is probably the most outstanding feature of gripopterygid and *Zelandobius* larvae. *Megaleptoperla* larvae retract and extend the rosette rhythmically and increase the beat as the level of dissolved oxygen drops. *Zelandobius* larvae waggle the abdomen from side to side with the rosette extended and again a drop in oxygen content results in an increased speed of waggle. The other gripopterygids have no similar respiratory behaviour but are able to retract the rosette under unfavourable conditions.

In Notonemouridae all of the *Notonemoura* group inhabit hygropetric (madicole) biotopes – either in the splash zones of steep streams or in pure seepages from between strata. The adults of this group are much more alert than other New Zealand stoneflies and use their long hind legs to leap away from danger.

In the *Spaniocerca* group the larvae of *Spaniocerca zelandica* inhabit many types of running water, from sea level to about 1,000 metres. These habitats include lowland and mountain rivers and streams, in forest or grassland, and with stony, sandy or muddy substrate. Generally the larvae are found under stones, on logs, in leaf litter or moss. *Spaniocercoides hudsoni* is a little more selective but still with a wide habitat range. These consist of alpine seepages, alpine bogs of the type inhabited by *Europetala carovei* (Odonata), weedy streams and forest strems containing dead branches, twigs and leaves.

Austroperla cyrene (Austroperlidae) inhabits pockets of dead leaves, crevices in logs and aquatic vegetation in many different types of running water. Because of its ability to exist in such a variety of conditions it is our most widespread stonefly.

Stenoperla prasina (Eusteniidae) exists also in a wide range of biotopes ranging from alpine rivulets to large lowland rivers. It is important that these biotopes are well oxygenated and not too warm.

The only predatory stonefly in New Zealand is *Stenoperla prasina* which preys upon mayfly, stonefly, simuliid and blepharocerid larvae. Detritus feeders are common and one notable example, *Austroperla cyrene*, reduces dead leaves to a network of veins and can even exist on a diet of twigs. Most gripopterygid larvae are either grazers of algae or eaters of dead leaves. Some of the terrestrial larvae strip the outer layers of plant stems ingesting any algae and fungi growing thereon. The Notonemouridae follow a similar pattern. The *Notonemoura* group graze on algae but are not averse to preying on other stonefly larvae and possibly other invertebrates if the opportunity arises. The *Spaniocerca* group is not well known but the type of biotope inhabited suggests a diet of dead vegetable material such as leaves and possibly algae and fungi on its surfaces.

Odonata

The small New Zealand fauna consists of eleven species. In the suborder Zygoptera there are two species in Coenagriidae and one in Sympecmatidae. The suborder Anisoptera contains one species in Petaluridae, two in Aeshnidae, four in Corduliidae and one in Libellulidae. Six species are endemic but with strong affinities with the Australian fauna. The remaining species are common to both Australia and New Zealand.

Possibly the most interesting dragonfly is *Uropetala carovei* the sole member of the small family Petaluridae. (The petalurids have a preference for alpine bogs and are restricted to Chile, Australia, New Zealand, North America and Japan.) The nymphs of this species construct burrows in alpine and subalpine spring fed bogs where there is a permanent flow of water with an oxygen content of about 6.23–7.25 ml/litre which is sufficient to prevent the formation of highly acid peat from the 65–80% organic matter present. These bogs which are of recent origin, are about one metre deep with a shingle or rock base.

The nymphs, using masks as shovels together with mandibles and the lateral lobes of the labium to cut vegetation, construct burrows which are about 35 cm deep. Because of cannabalistic tendencies only one inhabitant is found in each burrow.

Early instar nymphs feed on ciliates, flagellates, rotifers, nematodes, small crustaceans and microtrichopteran larvae. Late instar nymphs feed entirely on terrestrial arthropods which are taken on their nocturnal wanderings. This most unusual food for dragonfly nymphs consists of beetles, grasshoppers, cockroaches, ants and spiders.

Megaloptera

Only one megalopteron is aquatic in New Zealand. This is *Archichauliodes diversus* (Corydalidae) which is found in stony rivers and streams throughout the country. The genus has an Austral distribution in Australia, Chile and New Zealand.

Ecology

The larvae of this species are common and they rest on the undersides of rocks during the day then emerge to search for food at night. They are one of the most voracious invertebrate predators in our streams and subsist on a diet of most larval aquatic insects, particularly Ephemeroptera. The larval existence is probably over two years or more. When they are full grown they migrate to the stream edge during periods of high water in spring and early summer. As the water recedes and leaves them some distance from normal water level they construct pupal cells where they may remain in the active larval stage for several months. Pupation lasts from two to four weeks and the adults emerge from November to February. During flight adults occasionally dip onto the surface of water and rise again with little difficulty. This action plus a gliding flight high above the stream may be mistaken for egg laying behaviour, however, eggs are laid on objects directly above running water.

Mecoptera

New Zealand's sole mecopteron *Choristella philpotti* is a member of the family Nannochoristidae which has an Austral disjunctive distribution in Australia, New Zealand and South America. The larva of this insect is aquatic and is found in muddy backwater pools in small streams in South Island localities. The aquatic habit is also found in *Nannochorista* the Australian genus. Professor Pilgrim (University of Canterbury, Christchurch, New Zealand) has just completed investigating the biology of this species.

Trichoptera

14 families, 41 genera, 120 species. Rhynchopsychidae: 1 genus, 1 species. Sericostomatidae: 11 genera, 22 species. Helicopsychidae: 1 genus, 4 species. Philanisidae: 2 genera, 2 species. Pycnocentrellidae: 2 genera, 2 species. Helicophidae: 1 genus, 1 species. Philorheithridae: 1 genus, 2 species. Leptoceridae: 4 genera, 8 species. Hydropsychidae: 2 genera, 10 species. Polycentropodidae: 2 genera, 6 species. Psychomyidae: 2 genera, 2 species. Philopotamidae: 1 genus, 2 species, Rhyacophilidae: 9 genera, 54 species. Hydroptilidae: 2 genera, 4 species.

Zoogeography

Three elements constitute the New Zealand fauna. 1. Those with South-east Asian progenitors (these may be recent arrivals). 2. An element known only from both Australia and New Zealand plus certain subfamilies whose affinities and origins are difficult to determine. 3. Subfamilies or genera represented in Australia or New Zealand or both and in South America. This can be seen as each family in New Zealand is dealt with in detail.

Rhynchopsychidae: Found only in New Zealand and Chile it consists of two genera, *Kokiria* which is endemic to New Zealand and *Rynchopsyche* which is endemic to Chile.

Sericostomatidae: All eleven genera are endemic to New Zealand. The family is cosmopolitan but none of the genera occurring in the Australian-New Zealand area occur elsewhere. There is a possibility that eleven of the 26 species belong to Goeridae.

Helicopsychidae: The New Zealand species are assigned to the cosmopolitan genus *Helicopsyche.*

Philanisidae: A family of marine caddis is present only in Australia and New Zealand. *Philanisus plebeius* is common to Australia and New Zealand. *Chathamia brevipennis* (which should be transferred to the *Philanisus*) is found at the Chatham Islands.

Pycnocentrellidae: An endemic family containing the genera *Alloe-*

centrella and *Pycnocentrella* which were formerly included in Beraeidae.

Helicophidae: This family is restricted to Australia and New Zealand. The sole New Zealand genus *Zelolessica* is endemic.

Philorheithridae: A family restricted to Australia and New Zealand. The sole New Zealand *Philorheithrus* is endemic to New Zealand. Four genera are endemic to Australia.

Leptoceridae: The species of this cosmopolitan family are assigned to the following genera – *Triplectides* which is of worldwide distribution excluding Africa, the Palaearctic and Nearctic regions. Two species are assigned to this genus, *T. cephalotes* which is also found in Australia and *T. obsoleta* an endemic species which is apparently a sister species of *T. jaffueli* in Patagonia. *T. magna* is not found in New Zealand despite earlier reports. *Triplectidina* is an Australasian genus. It is possible that the New Zealand species should be in *Triplectides. Hudsonema* of Austral distribution contains two endemic species.

Hydropsychidae: Two genera of this cosmopolitan family are recorded from New Zealand; *Diplectrona* with two endemic species and *Hydropsyche* with nine endemic species. There is justification for removing the New Zealand species from the latter genus; they are clearly related to *Abacaria* and *Herbertorossia* from Fiji and New Guinea.

Polycentropodidae: The two genera of this cosmopolitan family have an almost world-wide distribution but the New Zealand species assigned to them are endemic.

Psychomyidae: The genus *Ecnomina* of this widespread family is shared with Australia but the sole New Zealand species in it is endemic. *Zelandoptila* which is endemic was recently transferred from Hydroptilidae.

Philopotamidae: Two genera of this world-wide family are present in New Zealand. *Dolophilodes* is endemic to Australasia but with two endemic New Zealand species. *Neobiosella* is an endemic monotypic genus.

Rhyacophilidae: All genera are endemic and belong to Hydrobiosinae which is in New Zealand, Australia and extends through South America to Mexico (this excludes the aberrant *Apsilochorema*)

Hydroptilidae: Two genera of this cosmopolitan family are represented in New Zealand. Cosmopolitan *Oxyethira* which has one endemic species and the endemic *Paroxyethira* with two species.

Three of the above families are thought to have evolved in Australia. These are Philansidae, Philorheithridae and Helicophidae. They possibly represent developments that are known to have occurred in the Australian Upper Permian and Triassic.

Distribution within New Zealand is not well known. There are some species with a restricted range of habitat such as *Pycnocentrella* which is confined to strongly flowing water usually in forest. There are, of course, large regions where the substrate of the streams is unsuitable as caddis habitat. The large papa (blue clay) area of the central North Island is such a region. Here fine clay particles prevent the growth of diatoms by either covering or shading them from the light and therefore disrupting possible food chains. The rivers of Canterbury frequently have a substrate of shifting sand and gravel which has a destructive affect on aquatic insects.

Ecology

As with the other aquatic orders the ecological work carried out on Trichoptera has been spasmodic and mainly concerned with life histories. There are, however, current investigations on such topics as population dynamics, ventilation mechanisms and trophic relationships.

The Trichoptera form considerable part of the first order consumers of our running waters. Sericostomatidae and Helicopsychidae make up the bulk of diatom browsers. The Hydropsychidae net drifting algae as well as other organic material. The Hydroptilidae feed on filamentous algae. Some Sericostomatidae larvae live in the crevices of submerged rotting wood eating either wood or fungi on it. Many of the algal browsers carry out their activities at night. The Hydrobiosinae (Rhyacophilidae) are the principal invertebrate carnivores of our aquatic fauna and possess chelate forelimbs to facilitate capture and holding of prey. *Neurochorema* preys on the pupae of *Hydropsyche* which are available for most of the year. Associated with this feeding on an almost inert prey it shows a reduction of the chelae of the forelimb and a narrow fore part enabling it to enter the confined space of its prey's pupal shelter (fig. 3). Species inhabiting torrents have campodeiform bodies and venters of abdominal segments modified into suction pads enabling them to move in swift currents after blepharocerid and simuliid larvae and pupae. Most of *Costachorema* and some of *Hydrobiosis* have these adaptations. Except for *Neurochorema*, which seems to be specific in its food intake, most larvae of this family take whatever they can capture. Mayfly and stonefly larvae are grist to the mill but chironomid and simuliid larvae make up the bulk of the diet. There is even a degree of cannibalism. The prey is held in the chelae and slowly cut up into chunks and swallowed. One section of the Sericostomatidae differ in body form, limbs and habitat from characteristic browsers. These are the species previously mentioned as possibly belonging to Goeridae. They inhabit semifluid organic mud and it is thought that they feed on chironomid larvae and annelids. *Hudsonema* (Leptoceridae) inhabits lakes, slow waters and stable streams with vegetation and moderate currents. The larvae of *H. aliena* swims in slow water and at times leaves the water. It eats floating dead insects or sets upon injured individuals which cannot escape. The other species work along the bottom in depressions where the current tends to be swifter. They often do this in groups and possibly pick up chironomids or eat detritus.

Some adaptations of Trichoptera have obvious uses but others have not. For example, the *Philorheithrus agilis* larva, an inhabitant of swift

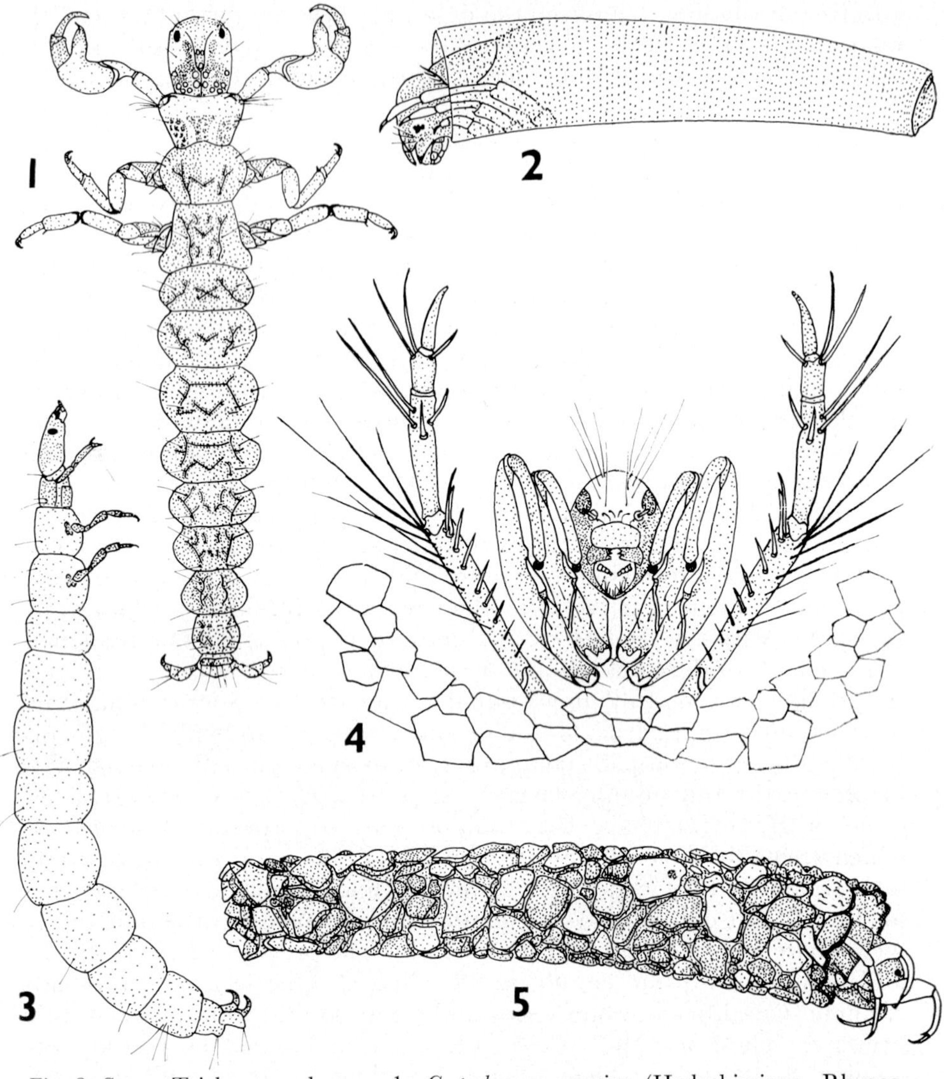

Fig. 3. Some Trichoptera larvae. 1: *Costachorema* species (Hydrobiosinae, Rhyacophilidae), a predator of swift streams. 2: *Olinga* species (Sericostomatidae), with horny case devoid of stones or vegetation. 3: *Neurochorema confusum* (Hydrobiosinae, Rhyacophilidae), a predator adapted for preying on *Hydropsyche* pupae. 4: *Kokiria miharo* (Rhynchopsychidae); note the peculiar adaptation of fore and mid limbs. 5: *Philorheithrus agilis* (Philorheithridae); an inhabitant of swift waters with a heavy case and with mid-leg tibia and tarsus fused.

streams, has a large heavy case to prevent being swept away but there is no obvious reason for fusion of tibia and tarsus of the middle limb (fig. 3). The larva of *Kokiria miharo* (fig. 3) has a puzzling adaptation. The front and middle legs have the tarsi missing and the claws reduced to thumbnail-like remnants which rest on a raised callous-like structure on each femur. Each hind leg has a row of spines and some long hairs. All this plus the wide gaping mouth and hairs on the head suggest a funnel-like net.

Predators on Trichoptera are many. The introduced Salmonidae prey upon many species of Trichoptera at all stages. In the author's locality *Salmo trutta* of the Buller River are often found full of larval cases of *Pycnocentria evecta.* At Lake Coleridge *Oecetis* larvae, which are found 30 metres below the surface, form an important part of the diet of the trout in the lake. Elsewhere *Pycnocentrodes, Triplectides obseleta* and *Olinga feredayi* larvae are commonly taken by trout.

Larval Trichoptera also form an important part of the food of those native fishes mentioned previously and frequently the larval cases are encountered in the gut of *Anguilla* species. Invertebrate predators are *Uropetala carovei* (Odonata) which preys on Hydroptilidae larvae, *Archichauliodes diversus* (Megaloptera) and of course the Hydrobiosinae (Rhyacophilidae).

Chironomidae

The New Zealand fauna consists of 63 species in 23 genera contained in the subfamilies Tanypodinae, Diamesinae, Orthocladiinae, Chironominae, Podonominae and Clunioninae.

Zoogeography

There is only one detailed account on the relationships of New Zealand chironomids and this deals only with the subfamilies Podonominae and Diamesinae. In these subfamilies the following conclusions have been drawn. 1. The sister group of a New Zealand group lives in South America or South America and Australia. 2. There is no direct phylogenetic connection between Australian groups and New Zealand groups. 3. South American connections to New Zealand appear to have been via West Antarctica and those to Australia via east Antarctica. 4. The New Zealand fauna shows a strong independence which is stressed by the strong relative plesiomorphy exhibited by several groups regardless of rank and by a greater degree of endemism than that of Australia.

In the previously mentioned subfamilies there are 3 endemic genera, 2 of which are the comparatively apomorphic *Maoridiamesa* and *Lobodiamesa.* The other is the peculiar *Zelandochlus* which exhibits a most interesting mosaic of plesiomorphic and apomorphic characteristics. Notable among the latter are double dististyles and a long outer spur on the hind tibia. The apomorphic podonomine genera (South American-Australian) *Rheochlus* and *Podonomopsis* and the apomorphic subfamily Aphroteniinae are absent.

Chironomus zelandicus, a pest of sewage treatment works, is the only New Zealand chironomid which has been extensively studied but as that part of its ecology concerns its lentic biotope it will not be dealt with here.

The podonomine midges are generally cold-tolerant, rheophil and polyoxybiontic and the larvae feed on algae, diatoms and algal detritus. Two New Zealand species which show not only this cold tolerance but also tolerance to mineral particles in suspension are *Zelandochlus latipalpis* and *Parochlus glacialis* which inhabit the upper reaches of glacial torrents. The diamesine chironomids are broadly and effectively adapted to cool running waters. Although only four species are found in the mountain streams of New Zealand these are as prominent quantitatively as those in temperate South America. Generally the diamesinae are typical algal feeders, living mainly on diatoms, but one exception is *Lobodiamesa* which feeds on small arthropods plus a smaller amount of diatoms.

TONNOIR has described the most unusual chironomid in New Zealand. This is *Dactylocladius commensalis* which is a commensal on the larva of the blepharocerid *Neocurupira hudsoni*. The larvae and pupae of the chironomid inhabit the ventral surface of the blepharocerid larvae (fig. 4).

BLEPHAROCERIDAE

The New Zealand fauna consists of 8 species in the genera *Neocurupira*, *Peritheates* and *Nothohoraia* which are contained in the subfamily Apistomyiinae.

Zoogeography

TILLYARD considered Apistomyiinae was of southern origin while TONNOIR considered it was of northern origin. Both HENNIG and DUMBLETON believed the various genera of Apistomyiinae were derived independently from Paltostominae indicated a northern origin. DUMBLETON alternatively suggested *Apistomyia* may have arisen in Australia from *Neocurupira* and then migrated north. The latest author CRAIG (1969) postulated an Asian origin for *Apistomyia*, considered that the Apistomyiinae genera have a monophyletic origin and that blepharocerids entered New Zealand from the north during the Cretaceous. CRAIG also considered that when New Zealand was divided into two islands in Oligocene, *Peritheates* evolved in the northern island and the more plesiomorphic *Neocurupira* evolved in the southern island. He assumed that during the Miocene, when New Zealand was again a single land mass, *Peritheates* migrated south to at least the limits of the present South Island, and that *Neocurupira*, having evolved in the cooler south, was prevented from migrating northwards. Speciation is postulated for *Peritheates* during the Pliocene when New Zealand was again two islands.

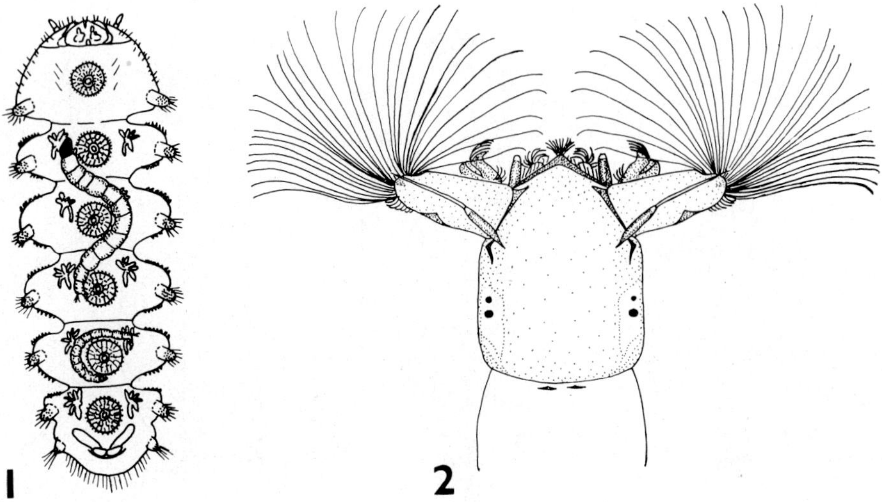

Fig. 4. 1: Larva of *Neocuripira hudsoni* (Blepharoceridae) with larval *Dactylocladius commensalis* (Chironomidae) in association. 2: Head of *Austrosimulium ungulatum* (Simuliidae).

No hypothesis is given for the present distribution of *Neocurupira* species apart from *N. chiltoni* which is found on Banks Peninsula formed from the remains of two volcanoes extinct from about the Pleistocene-Holocene and connected to the East Coast plains of the South Island. Here it is considered that because the plains did not contain suitable blepharocerid habitats wind drift in the prevailing westerly was responsible for colonisation of *N. tonnoiri* type blepharocerids from the Southern Alps.

The ancestral stock of the monotypic genus *Nothohoraia* (fig. 5) which appears to be closely related to *Horaia* of the Himalayas probably constituted a separate invasion from that of the other blepharocerids.

Ecology

The chironomid previously mentioned, *D. commensalis*, has now been found on other blepharocerids and also on the rocks within the habitat and is therefore a facultative commensal. Mermithid nematodes are found in blepharocerid larvae and pupae and it is thought that they may possibly castrate males. Ryacophilid (Trichoptera) larvae are one of the most important groups of predators. These lift small larvae off the substrate and devour them whole but tear large larvae into chunks of a suitable size for consumption. Epizooic algae of the same species on the rocks cover the larvae and pupae and are eaten off by ephemeropteran nymphs. Tolerances to water velocity varies considerably in New Zealand. The range favourable for *Neocurupira chiltoni* larvae in Purau

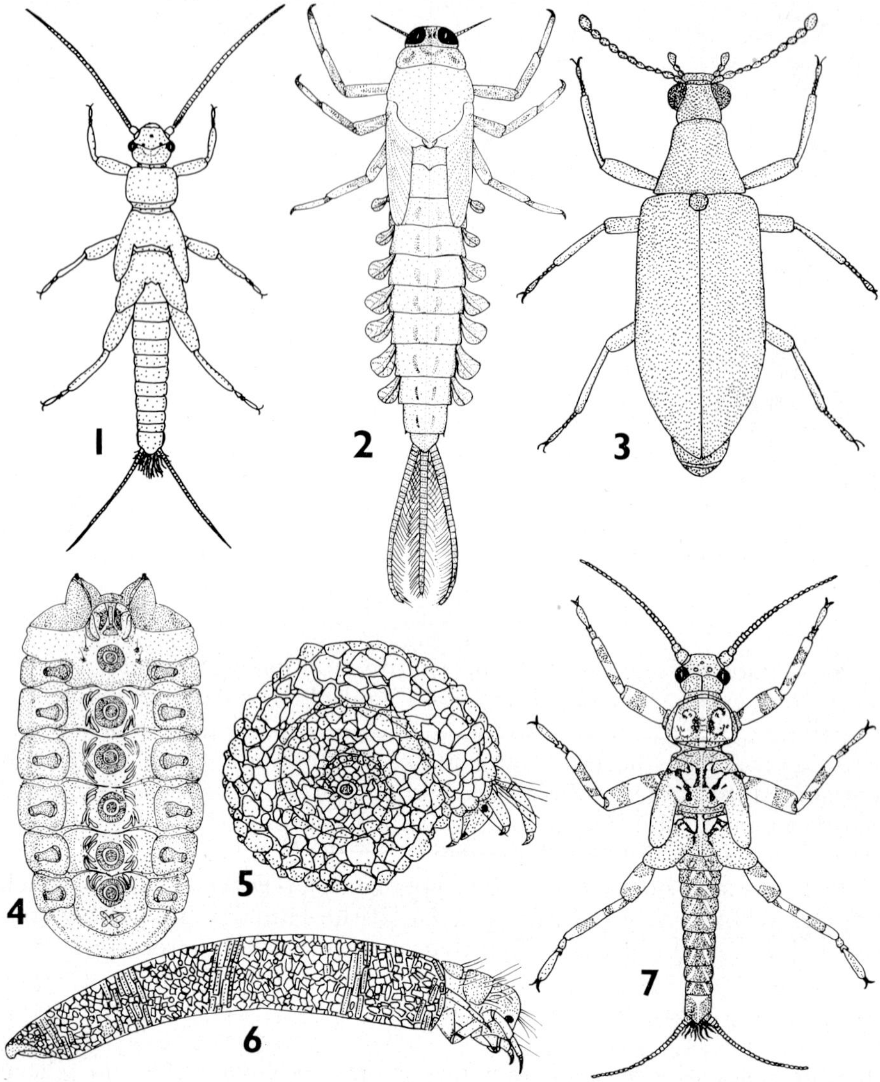

Fig. 5. Some of the fauna of a steep stony stream in lowland forest (Windy Point Creek, Buller Gorge, South Island). 1: Larva of *Zelandobius confusus* (Plecoptera, Antarctoperlinae). 2: Larva of *Nesameletus flavitinctus* (Ephemeroptera, Siphlonuridae). 3: Elmid species adult (Coleoptera). 4: *Nothohoraia micrognathia* larva (Blepharoceridae). 5: *Helicopsyche albescens* larva (Trichoptera, Helicopsychidae). 6: *Zelolessica chiera* larva (Trichoptera, Helicophidae). 7: *Aucklandobius fulvescens* larva (Plecoptera, Gripopterygidae).

Stream (Banks Peninsula) is 26–117 cm/sec. At Bealey Chasm (Southern Alps) larvae of other blepharocerids were not present if the water velocity was below 83 cm/sec and some larvae were found in a current of 420 cm/sec; the highest ever recorded for a blepharocerid. Water depth also has an affect in that no larvae were found in more than 30 cm of water. Apparently temperature has an affect on wing length because brachyptery in *N. campbelli* females was found to have an inverse relationship to the water temperature.

Simuliidae

Zoogeography

Only one genus *Austrosimulium* occurs in New Zealand and this is one of the most specialised. The same genus occurs in Australia and there is one species in southern South America. Five species groups are recognised in Australia and New Zealand. Two are confined to Australia, one to New Zealand, and one is common to both countries. The 12 species in New Zealand do not differ markedly from one another in the adult stage but there are divergences in the larvae, pupae and cocoons.

The New Zealand fauna has its strongest affinity with that of Australia. It is not possible to identify with certainty the source and route of dispersal of the common stock from which this fauna derives. The evidence at least does not strengthen the case for a Palaeantarctic origin which was once proposed for it.

The identity or close parallelism of Australian and New Zealand taxa in this family, separated since the Cretaceous, implies a very slow or closely parallel evolution in the isolated segments. This may well be an ancient condition in this taxon. It has been suggested for instance that the simuliid genera were segregated in the Jurassic. Given such characteristics the austral distribution of this taxon could well have resulted from dispersals from the north.

A comparison of simuliid with blepharocerid fauna suggests that the New Zealand one is derived from the Australian, whatever the source and route of entry of the latter.

The New Zealand *Austrosimulium* species are distributed as follows. Kermadec and Chatham Islands have no simuliids. Stewart Island has 3 species, one endemic. Auckland Islands has one endemic species, Campbell Island has one endemic species, North Island has 4 species which are not endemic to North Island, and South Island has the 4 North Island species which are widespread plus 3 species which are endemic to South Island plus 2 species which are shared with Stewart Island.

The present southern bias in the distribution of the species of *Austrosimulium* in New Zealand, even if it is superimposed on an older original bias, is not necessarily evidence of their southern origin. It would appear to be explicable in terms of greater and more recent speciation in the South Island, and more frequent and more widespread marine transgressions, and possibly vulcanicity also, in the North Island.

Ecology

DUMBLETON, 1972, has classified habitat preferences as follows:
A. Associated with cold snowfed streams; occasionally on boulders free of vegetation.
a. Small rivulets at high altitudes – 1550 m above bushline; stones angular *A. bicorne*
b. Larger streams, medium altitudes – 600–900 m; not predominantly glacier fed; below bushline, usually in open; stones rounded; larvae often inter-stitial *A. unicorne*
c. Large open rivers at lower altitudes – 300 m; stones rounded *A. albovelatum*
B. Associated with cold streams often within forests.
a. Low and at high altitudes in forest or in streams with high scrub covered banks; in peaty areas or on stones *A. vexans & A. campbellense*
b. Small streams within forest especially in steep areas – 300–900 m; may occur in tussock sheltered streams outside forests; on angular, sub-angular or rounded stones *A. ungulatum*
c. Small cold lowland streams with muddy substrate; in peaty or forested areas; on vegetation *A. stewartense*
d. Often associated with forest but sometimes in open; above or below bushline – 150–1000 m; on stones or leaves *A. multicorne*
C. Lowland streams and larger rivers; mostly open; usually with markedly warmer summer temperature.
a. Extending into colder summer temperatures; on stones usually rounded *A. laticorne*
b. Predominantly in warmer water; stones usually rounded *A. tillyardianum*
D. Mature streams predominantly in open lowland; slower and more constant flow permitting growth of emergent vegetation or having pendant vegetation *A. australense, A. longicorne*

THAUMALEIDAE

The two described species belong to *Austrothaumalea* a genus peculiar to Australia, New Zealand and southern South America. TONNOIR found Thaumaleidae only in South Island but a species does exist in North Island. Up to the present the New Zealand larvae and pupae have not been described although they have been collected. They inhabit the edges of small waterfalls and seepages on steep banks.

Coleoptera

The Coleoptera inhabiting running waters in New Zealand consist of the following. *Huxelhydrus syntheticus* (Dytiscidae) which lives both as larva and adult at the edges of shingle rivers and the side pools of main rivers. Other Dytiscidae of the genera *Rhantus*, *Antiporus*, *Liodessus* and *Copelatus* inhabit water weed and the backwaters of small streams. Species of *Orchymontia* (Hydraenidae) are river dwelling and apparently inhabit crevices in stones or the sand under stones in the headwaters of shingle rivers. Hydrophilidae species (which appear similar to Dytiscidae in their invasion of running water) are present but are not well known. Elmidae species, which are very common in stony rivers and streams, are also in this category.

Because of the many gaps in the knowledge of New Zealand aquatic beetles, little is known of their origin. However, the presence of the Dytiscidae genera *Homoeodytes* and *Antiporus* suggest an Australian influence in that family.

There are records of a gyrinid found in New Zealand in the 1870s but none have been recorded since. It is unlikely that such conspicuous beetles could be overlooked and it is considered that the reports are either erroneous or that wind borne specimens from Australia were collected.

Population Comparisons

The surveys of populations carried out as part of research into trout streams may afford some interesting comparisons with similar work done in the Northern Hemisphere. In the best known of these surveys, (Allan, 1951) insect larvae and pupae make up 93% of the total invertebrate fauna of the Horokiwi Stream. Of the three main aquatic orders Trichoptera accounts for 50.77% of the total, Ephemeroptera 14.28% and Plecoptera 0.29%. These figures give a picture of the relative abundance of the three main aquatic orders in small stony rivers in New Zealand. A survey of the Tiropahi River (a similar stony stream but running through forest) gave figures of 61% for Trichoptera, 31% for Ephemeroptera and 1.64% for Plecoptera. The notable thing about these figures is the extremely small percentage of Plecoptera netted. This is consistent in the surveys I have studied. However, in the streams concerned a skilled collector working by hand can amass large numbers of nymphs of a number of species at the same times of the year as the surveys have been carried out. It is possible that the collecting methods used in the surveys do not effectively collect Plecoptera. Species numbers are interesting too. Most rivers of the above type would contain about 6 gripopterygids, 1 austroperlid, 1 eustheniid and 2 or 3 notonemourids (the species composition also is fairly uniform for a great deal of the country). This total of 10 or 11 is lower than the 17 species recorded in Miron's survey of Pingaratzi Stream in the Eastern Carpathians or the 31 species recorded from the survey made by SHELDON & JEWETT of Sagehen Creek in the Sierra Nevada, California. Comparison of the number of genera (Tiropahi 8, Pingaratzi 6, Sagehen 15) shows that the number of species per genus is much lower in the Tiropahi than in the Northern Hemisphere streams. The relationship of genera to species in South America and Europe is South America 1:2 and Europe 1:11. The New Zealand situation is therefore similar to that found in South American Plecoptera.

REFERENCES

BRUNDIN, L. 1966. Transantarctic Relationships and their Significance as evidenced by Chironomid Midges (With a Monograph of the Subfamilies Podonominae and Aphroteniinae and the Austral Heptaginae). Kongliga Svenska Vetenskapsakademeins Handlingar. Uppsala & Stockholm. Series 4, 11(1): 1–472.

CRAIG, D. A. 1963. The Occurrence of Nematodes in the Family Blepharoceridae (Diptera). *New Zealand Entomologist.* 3(2): 25.

CRAIG, D. A. 1966. The Biology of some New Zealand Blepharoceridae. Ph.D. Thesis. Zoology Department, University of Canterbury, Christchurch.

CRAIG, D. A. 1969. A Taxonomic Revision of New Zealand Blepharoceridae and the Origin and Evolution of the Australasian Blepharoceridae (Diptera: Nematocera). *Transactions of the Royal Society of New Zealand, Biological Sciences.* 11(9): 101–151.

DUMBLETON, L. J. 1963a. New Zealand Blepharoceridae (Diptera, Nematocera). *New Zealand Journal of Science* 6(2): 234–58.

DUMBLETON, L. J. 1963b. Evolution in some Aquatic Nematocera (Diptera). *New Zealand Entomologist* 3(2): 34–41.

DUMBLETON, L. J. 1972. The Genus Austrosimulium. *New Zealand Journal of Science* 15(4).

FRASER, F. C. 1960. A Handbook of the Dragonflies of Australasia. Royal Zoological Society New South Wales, Handbook No. 14.

FREEMAN, P. 1959. A Study of N.Z. Chironomidae (Diptera-Nematocera). *Bulletin British Museum (Natural History)* 7: 9.

HAMILTON, A. 1940. The New Zealand dobson-fly *Archichauliodes diversus* (Walker): life-history and bionomics. *New Zealand Journal of Science & Technology* 22: 44–45.

HENNIG, W. 1960. Die Dipteren-Fauna von Neuseeland als systematisches und tiergeographisches Problem. *Beitrage zur Entomologie* 10(3 and 4): 222–329.

ILLIES, J. 1963. Plecoptera of the Auckland and Campbell Islands. *Records of the Dominion Museum, New Zealand* 4(19): 255–265.

ILLIES, J. 1964. Insects of Campbell Island. Plecoptera. Pacific Insects Monograph 7: 208–215.

ILLIES, J. 1965. Phylogeny and Zoogeography of the Plecoptera. *Annual Review of Entomolgy* 10: 117–140.

ILLIES, J. 1969. Revision der Plecopterenfamilie Austroperlidae. *Entomologisk Tidskrift* 9: 1–2; 19–51.

MCFARLANE, A. G. 1951. Caddisfly larvae (Trichoptera) of the family Rhyacophilidae.

Morphology and Description of Species. *Records of Canterbury Museum, Christchurch, New Zealand* 5: 267–289.
McLellan, I. D. 1966. Genitalia and Nymphs of Some New Zealand Gripopterygidae (Plecoptera). *Transactions of the Royal Society of New Zealand, Zoology* 8: 2: 5–22.
McLellan, I. D. 1967. New Gripopterygids (Plecoptera) of New Zealand. *Transactions of the Royal Society of New Zealand, Zoology* 9: 1: 1–15.
Mosely, M. E. & Kimmins, D. E. 1953. The Trichoptera of Australia and New Zealand. British Museum (Natural History) London 550 pp.
Penniket, J. G. 1961. Notes on New Zealand Ephemeroptera I. *New Zealand Entomologist* 2, 6: 1–11.
Penniket, J. G. 1962. Notes on New Zealand Ephemeroptera II & III. *Records of the Canterbury (N.Z.) Museum* Vol. VII No. V: 375–398.
Penniket, J. G. 1966. Notes on New Zealand Ephemeroptera IV. *Records of the Canterbury (N.Z.) Museum* 8(2): 163–175.
Phillips, J. S. 1930. A Revision of the New Zealand Ephemeroptera. *Transactions of the New Zealand Institute* 61: 271–390.
Robb, J. A. 1966. An Ecological Study on *Chironomus zelandicus*. Christchurch Drainage Board Publication 1–176.
Ross, H. H. 1967. The Evolution and Past Dispersal of the Trichoptera. *Annual Review of Entomology* 12: 169–206.
Smith, A. G. & Hallam, A. 1970. The Fit of the Southern Continents. *Nature* 225: 139–144.
Tillyard, R. J. 1917. The Biology of Dragonflies (Odonata or Paraneuroptera). Cambridge University Press, Cambridge.
Tillyard, R. J. 1922c. A revision of the New Zealand Blepharoceridae (order Diptera). *New Zealand Journal of Science & Technology* 5: 101–107.
Thomson, M. S. 1933. An Account of the Systematics, Anatomy and Bionomics of *Austroperla cyrena* Newman. Thesis Zoology Department, University of Canterbury, Christchurch, New Zealand.
Tonnoir, A. L. 1927. Thaumaleidae (Orphnephilidae). *Records of the Canterbury Museum, Christchurch, New Zealand* 3: 108–112.
Tonnoir, A. L. 1930. Notes on the Genus *Apistomyia* (Diptera) and description of a new species. *Proceedings of the Linnean Society of New South Wales, Australia* 1(2): 136–144.
Winterbourn, M. J. 1966. Studies on New Zealand Stoneflies 2. The Ecology and Life History of *Zelandoperla maculata* (Hare) and *Aucklandobius trivacuatus* (Tillyard) (Gripopterygidae). *New Zealand Journal of Science* 9: 312–323.
Wise, K. A. J. 1965. An Annotated List of the Aquatic and Semiaquatic Insects of New Zealand. *Pacific Insects, Bishop Museum, Hawaii* 7: 2: 192–216.
Wisely, B. 1962a. Studies on Ephemeroptera I. – *Coloburiscus humeralis* (Walker); early life history and nymph. *Transactions of the Royal Society of New Zealand Zoology* 1: 249–257.
Wisely, B. 1962b. Studies on Ephemeroptera II. – *Coloburiscus humeralis* (Walker); ecology and distribution of nymphs. *Transactions of the Royal Society of New Zealand Zoology* 2: 209–220.
Wolfe, L. S. 1953. A Study of the genus *Uropetala* Selys (Order Odonata) from New Zealand. *Transactions of the Royal Society of New Zealand* 80: 245–275.
Zwick, P. 1969. Das phylogenetische System der Plecopteren als Ergebnis vergleichend-anatomischer Untersuchungen. Kiel 1969, Kurzfassung.

XV. THE INSECTS IN RELATION TO PLANTS

by

J. S. DUGDALE

Abstract

Insect-host plant associations in New Zealand are characterised by a high degree (over 60% in some families) of host specificity, by a low incidence of monophages on the warmth-adapted plants, and a higher incidence on the cold-adapted plants. Fossil age and assumed provenance show no correlation with incidence of monophages on a plant genus. Many host associations agree with other Austral (as against Holarctic) situations, but many have arisen *de novo*, indicating the long isolation of New Zealand since the Cretaceous. The scanty information on pollinators suggests that pollinator-flower relationships are, by and large, non-specific.

Introduction

Assessment of specificity in the relation between an insect and a plant relies on repeated field observations. Some free-living defoliators wander by day, returning to the true host at night. Insects thought to be restricted to one host may, through repeated field studies, be shown to have a wide host range. Again, in New Zealand, an insect can have different hosts on each main island. The rather high rate of monophagy recorded in New Zealand may well be a result of the limited number of workers, of limited time, and the high speed at which indigenous communities have been 'europeanised' and modified beyond recall over an ever-increasing area.

Much information for this chapter has been gathered from unpublished sources, so it is fitting to quote G. M. THOMSON (1927, p. 124), whose comments on pollination can be taken to embrace the wider field: 'Everyone who will take the trouble to record the simplest observation made, may bring to light some fact of scientific importance, and it is by the accumulation of such little-recorded facts that much of our knowledge of natural history is built up.'

This chapter attempts to synthesise unpublished information, and records by HARRISON, 1959, HOY, 1962, HUDSON, 1928, 1939, 1950 (including records by CHAPPELL, FEREDAY, LEWIS, PURDIE), LAMB, 1960; WATT, 1916, 1920, 1921a, 1921b, 1923, 1924, WISE, 1959. Sections of the 'Biological Flora' that have appeared include annotated lists of known phytophages for each plant species (FRANKLIN, 1969, MACMILLAN, 1973, WARDLE, 1966, 1967, WARDLE *et al.*, 1971). For the purposes of this chapter, a phytophage is considered to be an insect that

Table 1. Plant Genera Supporting Monophage Insects in New Zealand.

Plant Family, Genus	No. of Plant Spp.	Phytophage categories and numbers of monophage genera involved	Floristic Element*	Fossil Record**
FERNS				
Aspleniaceae				
Asplenium	11	Defoliator (1); miner (1); sorus eater (1)	subcosmopolitan	?
Blechnaceae				
Blechnum	15	Defoliators (2); miner (1, Three Kings)	mainly austral	?
Cyatheaceae				
Cyathea	7	Defoliators (4); miner (1)	?Palaeoaustral	U. Cretaceous
Dennstaedtiaceae				
Hypolepis	5	Defoliators (2)	Tropical-subtropical	
Dicksoniaceae				
Dicksonia	3	Defoliators (3); gallmaker (1)	Tropical-subtropical	U. Cretaceous
Dryopteridaceae				
Polystichum	4	Defoliators (2) sori destroyer (1, shared with *Ctenitis*, *Rumohra*)	Austral-Polynesian	
Gleicheniaceae				
Gleichenia	5	Defoliator (1)		
Polypodiaceae				
Phymatodes	3	Defoliators (2); miner (1)		U. Eocene
Pyrrosia	1	miner (1), sori destroyer (1)		
Pteridaceae				
Histiopteris	1	Defoliators (2)	Pantropical	
CONIFERS				
Araucariaceae				
Agathis	1	Miner (1)	Palaeoaustral	U. Jurassic
Cupressaceae				
Libocedrus	2	Defoliator (1, also on introduced Cupressaceae); fruit eater (1)	Austral	Pliocene
Podocarpaceae				
Dacrydium	7	Defoliators (2, 1 shared occasionally with *Phyllocladus*) psyllid (1)	Palaeoaustral	U. Cretaceous/ U. Oligocene
Phyllocladus	3	Defoliators (4); flower eaters (2, shared with *Podocarpus*); barkbeetle (1); stem borer (1)	Malayo-Pacific	Eocene
Podocarpus	7	Defoliators (3); flower eaters (2, shared with *Phyllocladus*)	Palaeoaustral	Jurassic
MONOCOTS				
Agavaceae				

Cyperaceae				
Carex	73	Defoliators (2); fruit eater/leaf miner (1); stem borer (2); crown destroyer (1)	Cosmopolitan	?
Carpha	1	Defoliator (1)	Austral	
Cyperus	1	Miner/fruit eater (1); crown destroyer (1)	Cosmopolitan	?
Gahnia	6	Defoliators (2); stem borers (2), miner (1)	Malayo-Australian	?
Oreobolus	3	Crown destroyer (1)	Austral	
Gramineae				
Chionochloa	20	Defoliators (6); miners (2); fruit eaters (2); stem borer (1); crown destroyers (3)	NZ-Australian	
Cortaderia	4	Defoliators (2); miners (1); fruit eaters (1); crown destroyers (1)	NZ-Australian	
Poa	28	Defoliators (5); miners (2) (1 shared with *Hierochloe*)	Cosmopolitan	
Juncaceae				
Juncus	16	miner (1)	Cosmopolitan	L. Oligocene
Liliaceae				
Astelia	13	Defoliators (2); miners (5); stem borers (2); fruit eaters (1); crown destroyer (1)	Palaeoaustral	M. Oligocene
Collospermum	2	Defoliators (2, shared with *Astelia*); miners (3)	Malayo-Pacific	?
Palmae				
Rhopalostylis	2	sheath/fruit/inflorescence miner (1), sheath miner (1)	Endemic	U. Eocene
Pandanaceae				
Freycinetia	1	miner (1); gallmaker (1)	Qld-Pacific	Pliocene
Restionaceae				
Leptocarpus	1	Defoliator (1)	Austral	?
Typhaceae				
Typha	1	Fruit eaters (2)	Cosmopolitan	?
DICOTS				
Apocynaceae				
Parsonsia	2	Scolytid (1); miner (1)	Malayo-Pacific	
Araliaceae				
Pseudopanax	14	Defoliators (3); psyllid (1); miners (1); scolytid (1)	Austral	U. Eocene
Schefflera	1	Defoliators (2, 1 shared with *Pseudopanax*); psyllid (1); scolytid (1)	Malayo-Pacific	U. Eocene
Stilbocarpa	3	Rootstock borer (1); stem borer (1)	Endemic	
Caprifoliaceae				
Alseuosmia	8	psyllid (1)	Endemic	
Chenopodiaceae				
Atriplex etc	—	miner (1, + 1 introduced)		
Compositae				
Brachyglottis	1	Defoliators (3); stem borer (1); psyllid (1); miner (1)	Endemic	Pliocene
Cassinia	5	Defoliators (6); stem borer (2); psyllid (2); fruit eater (3)		
Celmisia	60	Defoliators (4); crown destroyers (4); fruit eaters (4); miners (1)		

Table 1. (cont.)

Plant Family, Genus	No. of Plant Spp.	Phytophage categories and numbers of monophage genera involved	Floristic Element*	Fossil Record**
Craspedia	6	fruit eater (1), defoliator (1)		
Erechtites	8	miner (1)		
Helichrysum	9	Defoliator (1); galler (1)		
Olearia	32	Defoliators (7, 1 shared with Brachyglottis); miners (3); fruit eaters (1); gallmakers (2); psyllid (2, 9 species)	Malayo Pacific	Pliocene
Pachystegia	1	Fruit eater (1)	Endemic	
Raoulia	20	Defoliators (4); gallmakers (2); fruit eaters (2)	Endemic	
Senecio	40	Fruit eaters (1); defoliators (2); miners (2); stem borer (1); psyllid (1)	Cosmopolitan	
Traversia	1	(miner (1)		
Convolvulaceae				
Calystegia *Convolvulus* *Dichondra* *Ipomoea*	10	Defoliator (1); miner (1); (cosmopolitan, family – monophages)	?subcosmopolitan	U. Miocene
Coriariaceae				
Coriaria	8	Defoliator (1); miner (1); stemborer (1)		
Cornaceae				
Corokia	3	miner/fruit borer/green-shoot borer (1)	Endemic	
Griselinia	2	miner (1); defoliator (1 – shared with *Pseudopanax*)	Austral	U. Miocene
Cunoniaceae				
Weinmannia	2	Gallmaker (psyllid) (1); callus inducer (1); miner (1); cambium borer (1)	Austral	Miocene
Elaeocarpaceae				
Aristotelia	2	Scolytid (1); flower feeders (2, + 1 also on *Elaeocarpus*); psyllid (1)	Neoaustral	L. Miocene
Elaeocarpus	2	Scolytid (1); flower feeder (1, shared with *Aristotelia*); miner (1)	Malayo-Pacific	
Epacridaceae				
Archeria	2	flower feeder (1)	Australian	
Cyathodes	8	Defoliators (3 + 1 shared with *Nothofagus*)	Australian	
Dracophyllum	35	Defoliators (6); shoot apex destroyers (2); flower eaters (1): flower stalk borers (1); gallers (1); psyllid (1); miner (1)		
Escalloniaceae				
Carpodetus	1	Galler (1)	New Guinea-NZ	?
Fagaceae				
Nothofagus	4	Defoliators (8, + 1 shared with *Cyathodes*); miners (2); gallers (2);	Palaeoaustral	U. Cretaceous

Rhabdothamnus	1	Galler (1)	Endemic	
Haloragaceae				
Haloragis	7	Defoliator (1); stem borer (1); miner (1)	Australian	Pleistocene
Myriophyllum	5	Defoliator (1)		
Icacinaceae				
Pennantia	2	Defoliator (1)	Pacific	
Loranthaceae				
Loranthoid genera (5)	7	Defoliators (2); miner (1); flower feeder (1, possibly same as miner)	Australian	U. Eocene
Malvaceae				
Hoheria	5	Defoliators (4); flower/fruit feeders (3); miner (2)	Endemic	?
Plagianthus	2		Australian	Eocene
Myoporaceae				
Myoporum	1	Galler (1), miner (1); stem borer (1)	Australian	?
Myrsinaceae				
Myrsine	9	Galler (1)	Malayo-Pacific	?
Myrtaceae				
Eugenia	1	Miner(1); psyllid (1)	Australian	?
Leptospermum	2	Psyllid (1); defoliators (3); stem borer (1, shared with *Nothofagus*)	Australian	?
Metrosideros	11	Scolytid (1); psyllid (1); miner (1); phloem miner (1); flower eater (2); galler (1)	Pacific	U. Cretaceous
Oleaceae				
Nestegis	4	Miner (2); defoliator (1); flower/fruit eater (2)	Malayo-Pacific	?
Onagraceae				
Epilobium	50	Defoliator (1)	Cosmopolitan	?
Fuchsia	4	Defoliator (1); stem borer (1); psyllid (1)	Cosmopolitan	U. Oligocene
Papilionaceae				
Carmichaelia	38	Defoliator (6); scolytid (1); psyllid (1); miner (1); galler (1); stem borer (1); flower/fruit feeders 1 (15 spp)	Endemic	?
Clianthus	1	Miner (1)	Australian	?
Sophora	3	Defoliator (3); miner (1); psyllid (1); scolytid (1); fruit eater (1)	Malayo-Pacific	?
Pittosporaceae				
Pittosporum	26	Defoliators (3); psyllid (1); miners (2); scolytid (1); stem borer (1); flower/fruit feeders (2)	Malayo-Pacific	?
Polygonaceae				
Muehlenbeckia	5	Defoliators (10); miners (1); psyllid (1); galler (1); flower/fruit feeder (1, 5 spp)	Neoaustral	L. Miocene
Primulaceae				
Samolus	1	Defoliator (1)		
Ranunculaceae				

Table 1. (cont.)

Plant Family, Genus	No. of Plant Spp.	Phytophage categories and numbers of monophage genera involved	Floristic Element*	Fossil Record**
Clematis	10	Miner (1); defoliator (1, shared with *Ranunculus*)	Cosmopolitan	?
Ranunculus	43	Miner (1); defoliators (2, + 1 shared with Clematis on *R. lyalli* only); flower feeder (1)	Cosmopolitan	?
Rhamnaceae				
Discaria	1	Defoliators (2); psyllid (1)		
Rosaceae				
Rubus	5	Defoliators (2); fruit eaters (1); galler (1)	Cosmopolitan	
Rubiaceae				
Coprosma	45	Defoliators (8); gallers (4); miners (4); stem borers (3); fruit feeders (1)	Malayo-Pacific	Oligocene
Nertera	6	miner (1)	Austral	
Sapindaceae				
Alectyron	2	Miner (1); fruit feeder (1)	Australian-Pacific	L. Miocene
Dodonaea	1	Psyllid (1)	Australian	?
Scrophulariaceae				
Euphrasia	15	Defoliator (1)	Cosmopolitan	?
Hebe	79	Defoliators (13); psyllid (1); miners (2); gallers (4); flower/fruit feeders (1); stem borers (3)	Neoaustral	Pleistocene
Ourisia	10	Defoliators (1); defoliators/flower-fruit feeders (1); galler (1)	Austral	?
Solanaceae				
Solanum	3	Stem/leaf/fruit borers (2)	Cosmopolitan	?
Thymelaceae				
Drapetes	5	Defoliator (1)	Austral	
Pimelea	15	Defoliators (4); stem borer (1)	Australian	?
Umbelliferae				
Aciphylla	40	Defoliators (2); crown destroyers (1); rootstock borer (1)	Australian	?
Anisotome	16	Defoliators (1); rootstock borer (1)	Australian	?
Coxella	1	Rootstock borer (1)	(Endemic)	
Urticacae				
Elatostemon	1	Defoliator (1, shared with *Urtica*)		
Urtica	6	Defoliators (5); miner (1)	Cosmopolitan	?
Verbenaceae				
Teucridium	1	miner (1)	Endemic	?
Violaceae				
Melicytus	4	Defoliators (3); miner (1); fruit feeder (1)	Endemic	?
Winteraceae				
Pseudowintera	2	Defoliator (1)	Endemic	M. Oligocene

demands living tissue of photosynthetic plants at some major stage of its life history. Sapsuckers, except for psyllids (for which information is relatively good) are not discussed. Scolytid beetles, which show high specificity are included.

It has long been recognised elsewhere that some insects feed on many, taxonomically diverse plants, others on a few such, while others are restricted to one host, or to a group of taxonomically related hosts, or to hosts with some common chemical characteristic. Biogeographical interest usually centres on the phytophages that are restricted to plants of one family (family-monophagous) genus (genus-monophagous) or species (species-monophagous). For brevity, they will be referred to as monophages in this chapter. HERING (1951) erected a large classification based on monophagy, oligophagy and polyphagy, and THORSTENSTEIN (1961) proposed another, simpler classification based on the chemical stimuli that induce feeding. He distinguishes between two associations (1) in which feeding is induced by chemical stimuli in nearly all plants, but not all plants are consumed because some have inhibitors – these either present at random among plant taxa or absent in taxonomically related groups; (2) in which feeding is induced by chemical stimuli that are extraordinary, and occur in some plants and not in others.

EHRLICH & RAVEN (1964), discussing the Papilionoidea, concluded that radiation of an insect group was partly consequent on overcoming some chemical barrier set up by the plant, with the result that the barrier becomes, as it were, a stimulus.

New Zealand has been isolated since Cretaceous times and the biota show affinity with those of Australia, the Pacific, South America and the Northern Hemisphere. The host associations developed over the past millions of years indicate a seemingly haphazard occurrence; unusual host relationships have developed, and certain plant species, particularly those of warmer areas in New Zealand, are singularly poor in monophages. The fall-off in monophage numbers from south to north, the taxonomic isolation of many northern or lowland monophages, and the fact that cold-tolerant plant genera are rich in monophages could suggest that most monophage associations are those that could withstand the Pleistocene glacial regimes. Some may have arisen in late Pliocene times.

All tracheophyte genera with one or more monophages associated are listed in Table 1. Of the 404 tracheophyte genera noted in CHEESEMAN (1926, Gramineae), ZOTOV (1961, Gramineae), ALLAN (1961) BARLOW (1965) and MOORE & EDGAR (1970), 110, or 27% are known to have monophages. All our conifer genera have monophages, at least 21% of the fern genera likewise, and 32% of all dicotyledonous (dicot) genera likewise. The low figure for monocotyledonous (monocot) genera (15%) possibly reflects the difficulties in identification of the host, rather than a true representation of the degree of monophage associations. Table 1 also lists the Floristic Element, or distribution outside New Zealand of plant genera, and, where known, the fossil record. It is immediately apparent that early appearance in the fossil record is no guarantee of high monophage numbers – cf *Agathis*, and *Knightia*, *Nothofagus* and *Hebe*. Origin seems of little importance – cf *Melicytus* (endemic), *Hymenanthera* (Pacific) and *Astelia* (Palaeoaustral), *Collospermum* (Pacific).

The plant genera with the greatest number of monophage genera – *Pittosporum* (10); *Astelia* (11); *Carmichaelia* (12); *Dracophyllum* (12); *Cassinia* (13); *Celmisia* (14); *Chionochloa* (14); *Muehlenbeckia* (14); *Olearia* (15) and the extremely well-endowed *Coprosma* (20), *Nothofagus* (22) and *Hebe* (24), cover all ages in the fossil record from Jurassic to Pleistocene, and cover a wide range of provenance, being endemic *(Carmichaelia)*, palaeoaustral *(Nothofagus, Astelia)*, neoaustral *(Hebe, Muehlenbeckia)*, Australian *(Cassinia, Celmisia, Chionochloa, Dracophyllum)*, or Pacific *(Coprosma, Pittosporum)*. All are widely distributed in New Zealand, all colonise young sites, many dominate communities and all have alpine representatives, except for *Nothofagus*, which however, forms the upper tree-line over much of the country.

Some genera with 8 or less monophage genera may have lost phytophages; one possible candidate for loss is *Metrosideros* (Pacific species have a gracillariid leaf miner, Clarke, 1971); possibly the malvaceous *Hoheria* and *Plagianthus* may have lost monophages, but information on exotic *Plagianthus* is lacking. It is noteworthy that apart from *Nothofagus*, no forest canopy tree genus is particularly rich in monophages, and that the genera *Agathis* (kauri), *Beilschmiedia* (tawa, taraire), *Dacrydium* (rimu), *Vitex* (puriri) and *Weinmannia* (kamahi, towai), so prevalent in many remaining forests, are conspicuously poor in monophages. Thus the original northern North Island forests (rimu dominant over tawa, or kauri dominant over taraire) are poorer in monophages in the canopy than the *Nothofagus* forests of the south. It follows also that a non-forest community containing *Carmichaelia*, *Dracophyllum*, *Hebe* and *Olearia* will also be rich in monophages (such areas persist as frost hollows in the areas west of Lake Taupo in the North I.).

Some Characteristics of Insect-Hostplant Associations in New Zealand

Associations involving warm-temperate plant genera

Plant genera confined to warm-temperate communities (Wardle, 1964, p. 353) are largely Pacific in origin and support few, or no monophages (Table 2). Of the 46 genera mentioned, 19 (43%) have monophages, and only *Collospermum*, *Nestegis* and *Sophora* have a moderate monophage fauna (and *Collospermum* shares three monophage genera with *Astelia*). Of the six endemic plant genera, four each have a specific monophage, and two have none. Some monophages are anomalous elements in the fauna, such as the two monophages on *Alectryon* (titoki) –

Table 2. Monophages on Plant Genera Characteristic of Warm Temperate Forests in New Zealand (cf Table 3).

Plant Genus and Family		Distribution	Fossil Record	No. Monophage Genera
Ackama	(1) Cunoniaceae	Australian	?	—
Agathis	(1) Araucariaceae	Pacific-Australian	U. Jurassic	1
Alectryon	(1) Sapindaceae	Australian-Pacific	L. Miocene	2
Alseuosmia	(8) Caprifoliaceae	Endemic	?	1
Arthropodium	(2) Liliaceae	Australian	?	—
Ascarina	(1) Chloranthaceae	Malayo-Pacific	U. Palaeocene	—
Avicennia	(1) Avicenniaceae	Tropical	Miocene	2
Beilschmiedia	(2) Lauraceae	Tropical	Miocene	—
Calystegia	(4) Convolvulaceae	Cosmopolitan	?	2
Clianthus	(1) Papilionaceae	Australian	?	1
Collospermum	(2) Liliaceae	Pacific	?L. Oligocene	5
Corokia	(3) Cornaceae	Endemic	?	1
Corynocarpus	(1) Corynocarpaceae	Pacific	?	—
Dodonea	(1) Sapindaceae	Australian	?	1
Dysoxylum	(1) Meliaceae	Tropical	Eocene	—
Elatostema	(1) Urticaceae	Tropical	?	2
Entelea	(1) Tiliaceae	Endemic	?	—
Eugenia	(1) Myrtaceae	Australian	?	2
Freycinetia	(1) Pandanaceae	Queensland-Pacific	Pliocene	2
Geniostoma	(1) Loganiaceae	Malayo-Australian	?	—
Hedycarya	(1) Monimiaceae	Pacific	?	—
Knightia	(1) Proteaceae	NZ-New Caledonia	U. Cretaceous	—
Laurelia	(1) Monimiaceae	Chile-NZ	M. Oligocene	—
Litsaea	(1) Lauraceae	Tropical	?	—
Lophomyrtus	(2) Myrtaceae	Endemic	?	—
Macropiper	(1) Piperaceae	Pacific	L. Miocene	—
Melicope	(2) Rutaceae	Malayo-Australian	?	—
Mida	(1) Santalaceae	Malayo-Australian	?	—
Myoporum	(1) Myoporaceae	Australian	?	3
Nestegis	(4) Oleaceae	Pacific	?	5
Paratrophis	(3) Moraceae	Pacific	?	—
Parsonsia	(2) Apocynaceae	Malayo-Australian	?M. Oligocene	1
Pennantia	(1) Icacinaceae	NZ-Australia	?	1
Peperomia	(1) Piperaceae	Australian-Pacific	?	—
Persoonia	(1) Proteaceae	Australian	?	—
Phebalium	(1) Rutaceae	Australian	?	—
Planchonella	(1) Sapotaceae	Malayo-Pacific	?	—
Pomaderris	(7) Rhamnaceae	Australian	?	—
Rhabdothamnus	(1) Gesneriaceae	Endemic	?	1
Rhopalostylis	(1) Palmae	NZ-Norfolk I.	U. Eocene	2
Ripogonum	(1) Smilacaceae	Australian	Miocene	—
Solanum	(2) Solanaceae	Australian	?	2
Sophora	(2) Papilionaceae	Widespread	?	7
Tetrapathaea	(1) Passifloraceae	Malayo-Pacific	?	—
Teucridium	(1) Verbenaceae	Endemic	?	1
Vitex	(1) Verbenaceae	Tropical	?	—

one is closely related to South African species on Rhamnaceae and Combretaceae (Vari, 1961) and the other is congeneric with Australian species for which hosts are unknown. Both monophages on *Solanum* are the sole representatives of their genera in New Zealand. The yponomeutid moth on *Rhopalostylis* (nikau) is an isolated element in New Zealand.

Despite the poverty of monophages, there is a horde of oligo- or polyphagous phytophages, particularly arboreal Lepidoptera. They are not confined to warm-temperature communities, but many of them show close Australian (e.g. the hepialid *Aenetus virescens*) or Pacific (e.g. the geometrid *Cleora scriptaria*) affinities.

Associations involving plant genera not restricted to warm-temperate communities

These plant genera span lowland and upland forest, or these and intermontane grassland and alpine scrub/grassland. Of the 52 genera in Table 3, 47 (92%) have monophages. The twelve plant genera richest in monophages are in 10 unrelated families, and neither provenance nor known fossil age appear to be significant, despite the fact that Lepidoptera (a predominantly phytophagous order) are known from Cretaceous times. Note that entries under 'Numbers of monophage genera' differ so markedly in Tables 2 and 3.

Forest plant genera (e.g. *Nothofagus*) can be as monophage-rich as non-forest genera (e.g. *Celmisia*), but those plant genera with forest and non-forest species tend to have less monophages in the site with the fewer species. Thus there are fewer monophage genera associated with the two forest-dwelling *Chionochloa* species than with the 18 non-forest species.

There is good evidence, particularly concerning *Nothofagus*, that there is faunal loss at low altitudes e.g. below 300 m at 41 °S. While 12 tychiine species are found above this altitude, i.e. in upland or in inland valleys and up to the tree line, only three species have been found to extend below this altitude coastally. Similarly, there can be loss of monophages, usually north of 38 °S, or replacement. Some northern monophages have restricted distributions, suggesting that they are remnants of a tropical fauna. An yponomeutid moth bores the stems of *Coprosma*, but is found only around the Auckland isthmus; *Haloragis* is mined by a momphid moth, in the same area. Both monophages are more closely related to Australian or Pacific species than to New Zealand species.

In the main, a tracheophyte in a genus spanning lowland, upland and alpine zones, whether forest or non-forest, is likely to support some or all of the following monophages: defoliators, miners, live-stem borers, gallmakers, flower or fruit eaters; it may also support psyllids and bark beetles. If the plant is a rosette herb, it is likely to support monophages boring in the rootstock or the rosette-centre.

None of the purely alpine genera e.g. *Haastia*, *Hectorella*, *Pygmaea*,

Table 3. Monophages on Plant Genera Spanning Warm and Cool Sites in New Zealand (cf Table 2).

Plant Genera and Family		Distribution	Fossil Record	No. Monophage genera
Aciphylla	Umbelliferae	Australian	?	4
Archeria	Epacridaceae	Australian	?	1
Aristotelia	Elaeocarpaceae	Austral	L. Miocene	4
Astelia	Liliaceae	Pacific	M. Oligocene	11
Brachyglottis	Compositae	Endemic	Pliocene	6
Carmichaelia	Papilionaceae	Endemic	?	12
Carpodetus	Escalloniaceae	NZ-N. Guinea	?	1
Cassinia	Compositae	Australian	?	13
Celmisia	Compositae	Australian	?	14
Chionochloa	Gramineae	Australian	?	14
Clematis	Ranunculaceae	Cosmopolitan	?	2
Coprosma	Rubiaceae	Pacific	Oligocene	20
Coriaria	Coriariaceae	Widespread	U. Miocene	3
Cyathodes	Epacridaceae	Australian	Pleistocene	3
Dacrydium	Podocarpaceae	Palaeoaustral	U. Cretaceous	3
Dracophyllum	Epacridaceae	Australian	U. Eocene	11
Elaeocarpus	Elaeocarpaceae	Tropical	?	3
Fuchsia	Onagraceae	Austral	U. Oligocene	3
Gahnia	Cyperaceae	Pacific	?	5
Gaultheria	Ericaceae	Cosmopolitan	?	—
Griselinia	Cornaceae	Endemic	U. Miocene	2
Hebe	Scrophulariaceae	Neoaustral	Pleistocene	24
Hoheria	Malvaceae	Endemic	Eocene	8
Plagianthus		Pacific		
Hymenanthera	Violaceae	Pacific		5
Melicytus		Endemic		
Ileostylis et al.	Loranthaceae	Australian	U. Eocene	3
Ixerba	Escalloniaceae	Endemic	U. Pliocene	—
Leptospermum	Myrtaceae	Australian	uncertain	5
Libocedrus	Cupressaceae	N. Caledonia	Pliocene	2
Metrosideros	Myrtaceae	Pacific	uncertain	7
Muehlenbeckia	Polygonaceae	Austral	L. Miocene	14
Myrsine	Myrsinaceae	widespread	L. Oligocene	1
Neomyrtus	Myrtaceae	Endemic	U. Pliocene	—
Nothofagus	Fagaceae	Palaeoaustral	U. Cretaceous	22
Olearia	Compositae	Australian	Pliocene	15
Phormium	Agavaceae	Endemic	L. Miocene	3
Phyllocladus	Podocarpaceae	Pacific	Eocene	8
Pimelea	Thymeleaceae	Australian	—	5
Pittosporum	Pittosporaceae	S. Hemisphere	?	10
Podocarpus	Podocarpaceae	Palaeoaustral	U. Jurassic	5
Pseudopanax	Araliaceae	Austral	?U. Eocene	6
Pseudowintera	Winteraceae	Endemic	M. Olyocene	1
Quintinia	Escalloniaceae	Australian	U. Pliocene	—
Rubus	Rosaceae	Cosmopolitan		4
Schefflera	Araliaceae	Tropical	U. Eocene	4
Urtica	Urticaceae	Cosmopolitan		6
Weinmannia	Cunoniaceae	S. Hemisphere	Miocene	4

support monophages. Some common non-forest genera e.g. *Chrysobactron*, *Epilobium*, *Gentiana*, *Myosotis*, also lack monophages (or support only one), and this fact warrants investigation.

DIFFERENCES BETWEEN NORTH AND SOUTH ISLAND, AND BETWEEN THESE AND OUTLYING ISLANDS

Where a monophage is found on both main islands, it cannot necessarily be assumed that the host is the same on both. The satyrine butterfly *Dodonidia* exists in two populations, one on forest snowgrasses, *Chionochloa conspicua* and *Ch. cheesemani* from eastern Nelson southwards, the other on sedges, *Gahnia* spp (usually *G. setifolia*) in northwest Nelson and over the North I. to North Auckland. Conversely, the two geometrids monophagous on the araliad *Pseudopanax*, although structurally distinct, with *Declana egregia* in the South I. and *D. atronivea* in the North I., both have identical host ranges.

Fern feeding monophages on *Cyathea* are, so far as is known, restricted to the North I. Thus the elegant, silver-washed, geometrid, *Pseudocoremia aristarcha* is absent from the South I. The flower feeding and fruit boring weevils associated with *Nestegis* are absent in the South I., but the gracillariid leafminer is present. Conversely, two geometrid moths and one weevil associated with *Dracophyllum* are absent in the North I., despite the fact that suitable – indeed identical – host species are present.

It is generally assumed that most of the fauna of the several groups of outlying islands – excluding Three Kings for the moment – is of New Zealand derivation. While most monophages retain their mainland habits e.g. *Dracophyllum* feeders retain their host association on Auckland and Campbell Is, *Coprosma* feeders likewise on Antipodes, Auckland, Campbell and Chatham, some are at variance. A weevil, *Stephanorhynchus* sp., bores the young stems of *Olearia lyalli* on Big South Cape I. (mainland *Stephanorhynchus* species bore in dead wood or seeds), and *S. purus* utilises the dead petioles of the giant thistle *Sonchus grandifolius* on Chatham Is.

On the Chatham Is also, a representative of a tortricid genus elsewhere feeding on herbs and low shrubs, defoliates tree ferns. Liverworts support the byrrhid genus *Epichorius* (FARRELL, 1974), and lichens the byrrhid genus *Synorthus* (WATT, 1971) on the Subantarctic Islands. Elsewhere, mosses are the hosts of byrrhids.

Thus island situations have led to new hosts, or to new niches on hosts, e.g. the *Coprosma* bark-eating tortricid on Chatham I. The alpine biotope, which was archipelagic during interglacials and is again today, with each area being separated by unsuitable lowland is analogous to an island situation. Where a genus has species in alpine and lowland situations, there may be a host difference. The weevil genus *Gromilus* has alpine and non-alpine species, the former browsing only on angiosperms, the latter only on ferns. The colourful diurnal geometrids ('replacing' butterflies in New Zealand) of the genus *Dasyuris* are typically herb-feeders, and are derived from herb-feeders *(Helastia)*; but one section of *Dasyuris* is restricted to shrubs.

DIFFERENCES BETWEEN NEW ZEALAND AND ELSEWHERE

Changes in host between the mainland and outlying biota have been mentioned above; likewise there are differences between the New Zealand situation and those elsewhere in the world. In some instances, other Austral faunae share similar host niches with our fauna, but are different from Holarctic situations, and in others, the New Zealand situation is unique.

Austral-wide situations are exemplified by the vast weevil subfamily Cryptorhynchinae, which in the Northern Hemisphere is not reported to be associated with monocot angiosperms, but which in the Austral Region has many groups associated with monocots, as borers in corms, rhizomes and tiller bases. What may be a unique New Zealand elaboration is the leafmining habit found in *Microcryptorhynchus* (Mrs. B. M. MAY, pers. comm.). Another weevil subfamily, Hylobiinae is elsewhere restricted to dead wood; our hylobiines are live-tissue feeders on Umbelliferae and the herbaceous araliad *Stilbocarpa*. Conversely, our plume moths in *Platyptilia* and related genera are largely restricted to Scrophulariaceae, as are most overseas species.

New Zealand has a rich flora of ferns, mosses, liverworts and lichens. It is not surprising that this flora supports a rich fauna. Ferns support dipterous, coleopterous and lepidopterous miners, borers, sorus-eaters and defoliators; liverworts support the ancient lepidopterous family Micropterigidae (Fig. 1), and mosses support, besides the usual groups (Scopariine pyralid moths, byrrhid beetles) some Chrysomelidae. The vast assemblage of moss-weevils (Erirhininae) in New Zealand differ from those overseas by not being aquatic. Lichens too are hosts to the extraordinary lichen-gouging anthribid beetles *(Lichenobius)* and to *Artystona* and *Pseudohelops* species, the only tenebrionid beetles known to browse lichens.

New Zealand lacks the arboreal lepidopterous families so familiar overseas. Their place is taken by arboreal hadenine Noctuidae, and ennomine, larentiine and oenochromatine Geometridae. We have no Cetoniine scarabaeids, and their place is taken by a small group of diurnal melolonthines *(Pyronota, Costleya)*. We lack hymenopterous gall makers, and most of our gallers produce either leaf galls (43%) or stem and/or bud galls (41%) (figures derived from LAMB, 1960). MANI (1964, pp. 18–19) records that in South America and India 55–60% of gallers are leaf gallers, and 27–35% are stem/bud gallers. We lack bruchid beetles, and our Papilionaceae fruits are attacked by tychiine weevils (on *Carmichaelia*) and stathmopodid moths (on *Sophora*).

1

2

3

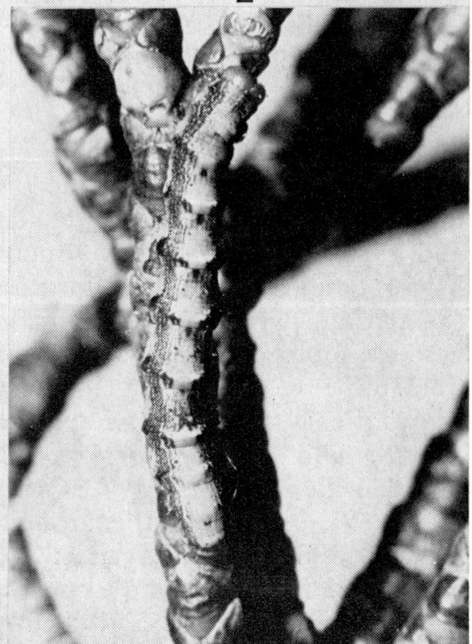

4

Fig. 1. Micropardalis sp., Micropterigidae on liverwort, Reefton, New Zealand.
Fig. 2. Chrysomeline chrysomelid larvae (Coleoptera); characteristic feeding scars (windows) on the montane forest-floor *Ourisia macrophylla* var. *lactea* (Scrophulariaceae).
Fig. 3. Declana atronivea (Walker) (Lepidoptera: Geometridae), an example of bizarre structure and specialised procrypsis, on *Pseudopanax arboreus* (Araliaceae).
Fig. 4. Pseudocoremia sp. indet. (Lepidoptera: Geometridae), an example of mimicry of the stem of the host plant, the whipcord *Hebe hectori* (Scrophulariaceae).

Miners are present in around 18% of the indigenous tracheophyte families in New Zealand, a poor representation compared with Europe, where HERING (1957, p. 156) reports that 85% of tracheophyte families have miners. We have no mining Chrysomelidae. Also, whereas in Europe there is little difference between Lepidoptera and Diptera in numbers of families attacked, in New Zealand, the Lepidoptera attack 18, and Diptera 10. Other anomalies lie in the choice of hosts; Araliaceae are rich in miners in New Zealand, poor in Europe (HERING, 1957, p. 184). In Europe and India, Onagraceae are hosts to all but a few Momphidae (Lepidoptera) (BRADLEY *et al.*, 1973), whereas in New Zealand, our two momphids mine *Muehlenbeckia* (Polygonaceae) and *Haloragis* (Haloragaceae) respectively.

Chrysomeline and galerucine chrysomelid beetles are lacking in the forest canopy, contrasting with Australia, where canopy plant genera (*Eucalyptus*, *Acacia*) have a rich chrysomeline fauna. Our forest chrysomelids with above-ground larvae are forest floor inhabitors, feeding on *Ourisia* (Fig. 2), mosses or ferns.

One other difference concerns dead tissue feeders rather than live tissue ones. There are over 120 species of Lepidoptera, i.e. about 10% of the indigenous fauna, mostly in Oecophoridae and Tortricidae, which feed on litter, or on dead and dying plant tissue. Australia has a similar proportion (Dr. I. F. B. COMMON, pers. comm.), but there is little evidence (FITTKAU & KLINGE, 1973) that such is the case in the tropics or in the Northern Hemisphere. The New Zealand situation differs from the Australian in that there are many Lepidoptera that attack dead wood (fallen branches, or twigs, or subcortical in dead trunks).

SOME CHARACTERISTIC PHYTOPHAGE TYPES IN NEW ZEALAND

Defoliators

The defoliating larvae of Lepidoptera and Coleoptera may be free-living or may construct a refuge, from which they emerge to browse. The free-living caterpillars of arboreal Geometridae and Noctuidae may mimic very closely some characteristic part of the host. The bizarre 4th and 5th instar larva of *Declana atronivea* (Fig. 3) mimics a shoot of *Pseudopanax* attacked by scale insects and coated with a white fungus. Earlier instars rest coiled on the leaf, resembling a bird dropping, or amongst the fruiting heads, where the dorsal tubercles on the second abdominal segment resemble the top of the fruit. *Declana feredayi* larvae, restricted to *Nothofagus*, rest with the body kinked so that they resemble the characteristic, gently zigzag shape of a beech twig. *Tatosoma fasciatum* larvae restricted to *Nothofagus* rest with the body bent so that the head capsule rests by the anal shield; the whole effect looks exactly like a beech leaf, particularly *Nothofagus menziesii*. Some noctuid caterpillars have a similar resting pose, but are reverse-countershaded, so that they rest upside-down.

While such elegant procrypsis argues for a long association between insect and host (and it is noteworthy that *Declana* and *Tatosoma* are South American, not Australian, in affinity, suggesting a Cretaceous origin), there are other, equally elegant examples in the genus *Pseudocoremia*, of Australian-Pacific affinity. A *Pseudocoremia* species on the whipcord *Hebe hectori* (Fig. 4), is characteristic of a colour pattern that has arisen also in other unrelated geometrid genera *(Pasiphila, Dasyuris)* using the same host. The whipcord habit is generally considered to be an alpine phenomenon and hence no older than very late Pliocene.

Of all the refuge-making defoliators, the larvae of chrysomelid beetles *Chalcolampra* are the most bizarre. The refuge tunnel may kill the host rosette or twig; the larva sits head-down in the refuge, blocking the entrance with a large, hard, tuberculate, anal plate. The larva browses on surrounding leafy or flowering parts of the host.

Flower, Fruit and Seed-eating Insects

For a country whose flora is noted for its inconspicuous flowers there is a surprisingly large assemblage of phytophages associated with these organs. The largest single insect group attacking flowers and fruit is undoubtedly the weevils, and includes the nemonychids whose agile larvae eat the male sporophylls of Podocarpaceae (Fig. 5). The angiosperm families Elaeocarpaceae, Epacridaceae, Papilionaceae, Pittosporaceae, have flowers in cymes, racemes or panicles, and there are weevils (invariably tychiines) associated with each family. In *Dracophyllum* (Epacridaceae) species with solitary flowers are not favoured. The absence of specific flower-eaters from such copiously flowering genera as *Carpodetus, Clematis, Ixerba, Phormium, Rubus* and *Sophora* is surprising. Fruit and seed eaters are more generally distributed through the flora. There is an extraordinary record of an apionine weevil in the fruit of the conifer *Libocedrus*. Compositae is a favoured family, and in the daisy genus *Celmisia*, unrelated taxa of four insect orders compete for sole rights to a capitulum. In general, where the fruit is a nut or a drupe, the kernel is attacked, and the flesh is untouched. Exceptions are attacks by oligophagous or polyphagous Tortricidae which eat the flesh but leave the seed. In berry fruit (e.g. *Astelia, Melicytus*) only the flesh is eaten whereas in capsule fruits (e.g. those of Scrophulariaceae) and in achenes, the seeds are eaten.

In Lepidoptera, seed eaters predominate in Carposinidae with some species restricted to Malvaceae, Myrtaceae, Epacridaceae and Rosaceae, and in Pterophoridae, with species on Scrophulariaceae (as elsewhere in the world), Compositae and Gentianaceae.

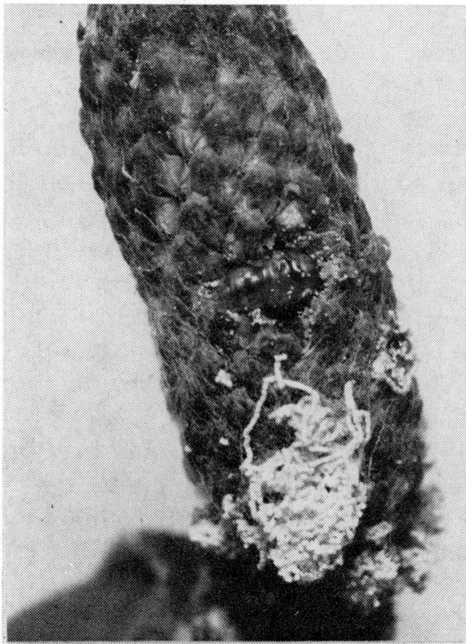

5

6

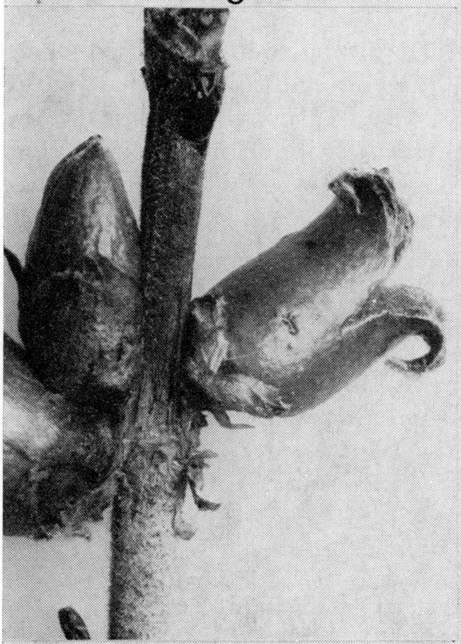

7

8

Fig. 5. Frass pellets and collapsed ♂ sporophylls; the work of the nemonychid weevil *Rhinorhynchus rufulus* on *Podocarpus dacrydioides* (Podocarpaceae).

Fig. 6. Apion sp., (Coleoptera: Curculionidae); stem gall on *Nothofagus menziesii* (Fagaceae).

Fig. 7. Cecidomyidae (Diptera); fruit galls (one empty) induced on *Coprosma propinqua* (Rubiaceae).

Fig. 8. Cecidomyidae (Diptera); trumpet gall in rosette centre on *Celmisia viscosa* (Compositae).

Gallmakers

Most of our leafgallers are erinea, caused by eriophyid mites, or dimple galls caused by coccids or psyllids. Although most of our galls are rather ordinary, the nail-head galls on *Phyllocladus* and midrib galls on *Hebe salicifolia* are rather spectacular. Our largest galls, in general, are fungal (e.g. those on *Nothofagus* caused by the fungus *Cyttaria*, RAWLINGS, 1956).

Stem galls (Fig. 6) and bud galls (Figs. 7, 8, 9) are frequent. Some *Celmisia* daisies are attacked by a cecidomyid causing an elegant trumpet gall (Fig. 8) up to 3 cm long. The psyllid, *Atmetocranium myersi* induces a stem gall on *Weinmannia* resembling cicada oviposition scars, up to 7 cm long. Only one insect-caused root gall is known, that of a cecidomyid on roots of *Metrosideros*. Galls on conifers are relatively common, being either leaf or bud galls induced by coccids or cecidomyids. A complex of gall-makers and successori and parasites remains to be worked out for the singularly phytophage-rich matai *(Podocarpus spicatus)*.

Gallmakers (particularly cecidomyids) are the only known monophages on *Carpodetus* (Escalloniaceae), *Rhabdothamnus* (Gesneriaceae) and *Myrsine* (Myrsinaceae).

Successori have not been closely studied, although consistent sequences have been noted in some instances. The dying gall tissue caused by the *Carpodetus* cecidomyid is taken over by a weevil, *Hoplocneme* (but is not the sole *Hoplocneme* breeding site); stem galls on *Hebe* induced by a eugnomine weevil are taken over, as the tissues degrade, by another weevil genus, *Oreocalus;* and the fusiform galls on the liane *Muhlenbeckia* caused by the moth *Morova subfasciata* are often taken over by a tortricid, *Trachybathra scoliastis.*

Callus tissue is induced by some phytophages, and is cropped by the larva. This mode of feeding is usually associated with a refuge within the stem of the plant (e.g. the ghost moth, *Aenetus;* moths such as '*Heliostibes*' *callispora*), accompanied by swelling of the stem or around the site of the workings. The carposinid *Carposina eriphylla* is often a successorus of *Aenetus* workings, and the refugia of *H. callispora* may be inhabited later by *Chalcodrya* larvae (Coleoptera: Chalcodryidae) or *Artystona* (Coleoptera: Tenebrionidae) or *Cryptolechia* (Lepidoptera: Oecophoridae). The last two taxa are lichen browsers.

A false gall, caused by *Oreocalus* weevils in *Hebe* is formed when the apices of the bud-leaves are cemented together. The larva develops inside the apex, feeding on the budleaf tissue continually being thrust towards it by basal elongation of the leaf tissues. The bud, whose axis is also elongating, bends at a right angle, and affected buds are a conspicuous feature of such *Hebe* species as *H. divaricata* (GOURLAY, 1964).

Miners

Miners are represented in the following insect orders and families:

Coleoptera:	Curculionidae (Tychiinae, Figs. 10, 11, Apioninae, Cryptorhynchinae)
Diptera:	Agromyzidae
	Cecidomyidae (mostly on *Coprosma*)
	Drosophilidae (? all introduced)
	Ephydridae
Lepidoptera:	Coleophoridae
	Elachistidae
	Gelechiidae (Fig. 12)
	Glyphipterygidae
	Gracillariidae (Fig. 13)
	Lyonetidae
	Momphidae
	Nepticulidae (Fig. 14)
	Opostegidae
	Tortricidae
	Yponomeutidae (Fig. 15)

Multiple use of a host is a characteristic in Europe (e.g. the multitude of diverse miners on *Betula*, HERING, 1957, pp. 74–79), but is not pronounced in New Zealand. *Nothofagus* has a nepticulid and a gracillariid, and *Coprosma* species in southern New Zealand may support a nepticulid, a weevil, a cecidomyid and a tortricid, and *Astelia* may support a glyphipterygid, a yponomeutid, and a weevil, but these are exceptions. In the great majority, a host species will have but one miner. Early instar larvae of some monophages, e.g. a plume moth on shrubby *Euphrasia*, mine leaves, but are ectophytic for the rest of their larval development. Other monophages may adopt a mining habit in the absence of other suitable sites on the host (Fig. 15).

Anomalies such as the absence of momphine miners on New Zealand Onagraceae (and their presence on the unrelated plant families Polygonaceae and Haloragaceae), and the development of a group of miners on Araliaceae have been mentioned. The Araliad miners, first studied by WATT (1920), have diverged into stem-mining and leafmining species, with at least one species restricted not only to a site (e.g. *Parectopa panacitorsens*, on the underside of the lamina) but also to a host species (e.g. *Parectopa panacivagans*, on *Pseudopanax crassifolius*). Two species compete for leaves of *Pseudopanax arboreus*. All have readily distinguishable mines. Another, undescribed, species of *Parectopa* mines the undersides of leaves of *Pittosporum eugenioides* (Fig. 13).

More extensive recording and rearing needs to be done, particularly with those miners on the adventive flora. Another study is the investigation of host range in genus-specific miners e.g. *Phytomyza clematadis;* all mines found on *Clematis foetida* were incomplete, with dead larvae.

Live stem Borers and Shoot Apex Destroyers

Among the live stem borers are insects which are ambivalent in their requirements; a cryptorhynchine weevil on *Cassinia*, and the lamiine

9

10

11

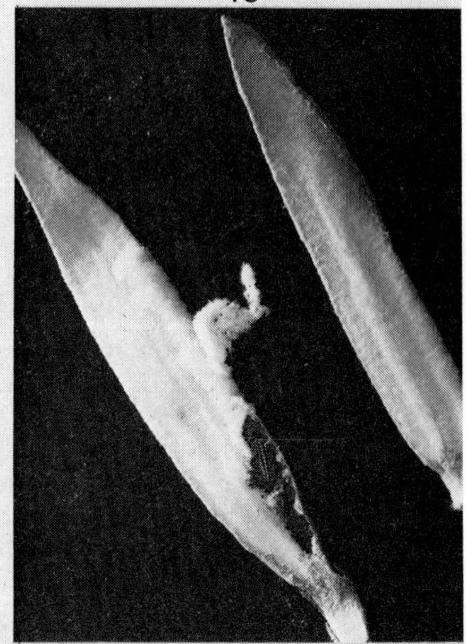

12

Fig. 9. Cecidomyidae (Diptera); longitudinal section through a big bud gall containing a pupa on *Hebe pauciramosa* var. *masonae* (Scrophulariaceae).
Fig. 10. Aneuma fasciatum (Coleoptera: Curculionidae); leafmine with larva and frass trail in *Pittosporum eugenioides* (Pittosporaceae).
Fig. 11. Phorostichus lineatus Broun (Coleoptera: Curculionidae), in lateral vascular bundles of the lily *Astelia fragrans* (Liliaceae).
Fig. 12. Hierodoris sp. (Lepidoptera: Gelechiidae) workings under tomentum on underside of leaf of *Celmisia spectabilis* (Compositae); unattacked leaf on upper right.

13

14

15

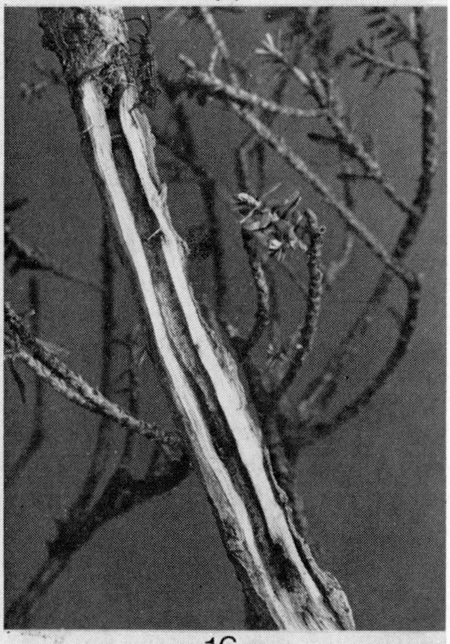

16

Fig. 13. 'Parectopa' sp. (Lepidoptera: Gracillariidae); variation in mine shapes (simplest at bottom, most complex at top) in leaves of *Pittosporum eugenioides* (Pittosporaceae).

Fig. 14. Nepticula sp. (Lepidoptera: Nepticulidae) mine in leaf of rangiora, *Brachyglottis repanda* (Compositae).

Fig. 15. Doxophyrtis hydrocosma Meyrick (Lepidoptera: Yponomeutidae); leaf base mining habit adopted in the absence of host fruit, on *Rhopalostylis sapida* (Palmaceae).

Fig. 16. Brounopsis hudsoni Blair (Coleoptera; Cerambycidae), adult beside longitudinal section of terminal part of larval workings and exit hole on branch of *Cassinia vauvilliersii* (Compositae).

cerambycid *Hybolasius cristus* on *Coprosma* may develop in live or recently killed wood. In several beetle genera (e.g. *Navamorpha*, *Oreocalus*) some species attack live tissue while others feed only on dead tissue.

The live stem borers (mostly cerambycine cerambycids) are a conspicuous element both in damage and, often, in appearance (Fig. 16). Damage can be spectacular, as in the case of a cerambycine on tawhero *(Weinmannia sylvicola)*; the larva can fell a branch, or the upper trunk, 5 cm in diameter. Repeated attack by the ubiquitous, polyphagous cerambycine *Oemona hirta* can weaken understorey trees such as mahoe *(Melicytus)*. In northwest Nelson and in Otago, *Hebe* shrubs attain an open-centred appearance, as central stems are weakened and killed by repeated attacks at ground level by the lamiine cerambycid *Mesolamia*. Stems of manuka *(Leptospermum)* and beeches *(Nothofagus)* are characteristically disfigured by the large cerambycine *Ochrocydus*.

Live stem borers are poorly represented in Lepidoptera (the callus-inducers have been mentioned before). Carposinidae and a few Gelechiidae bore live stems, usually in shrubby Compositae and Malvaceae (Carposinidae) or Papilionaceae (Gelechiidae: *Anisoplaca*). The Yponomeutidae include a number of species of larvae which attack several sites on the host, such as the *Orthenches* on *Corokia buddleoides* which bores stems, mines leaves and eats the fruit; similarly *Doxophyrtis* on nikau palm mines petioles, bores in the inflorescence and eats out the fruits. *Tanaoctena* is a yponomeutid genus common to New Zealand, Queensland and Rapa I.; the New Zealand species bores green shoots of *Coprosma* (CROWHURST, 1967).

Where the plant is a sub-shrub or rosette herb, the growing apex or crown may be a favoured site, and damage can be very striking. Many crown destroyers have habits that appear to be defensive or insulating. The large noctuid moth *Graphania nullifera* on speargrasses *(Aciphylla)* lacks these (Fig. 17), the vicious spines of the host plant possibly being protection enough. The larva of a large aterpine weevil on *Celmisia spectabilis* (Fig. 18) has characteristics shared by other crown destroyers on *Celmisia*. The central base of the rosette is destroyed, and the workings are concealed by a large mass of frass, tomentum and other rubbish, often over 2.5 cm in depth. Tortricid moths, *Gelophaula* species, do the same, covering the entrance with a stout, tough, cap of frass and silk.

Shoot apex damage, whether on rosette fruticose or arboreal plants leads to the production of multiple side shoots. In some plants, notably *Celmisia* and *Aciphylla* such multiple shoots tend to remain weak, and the whole clump degrades and dies. In others, multiple shoots may well be advantageous in a windy climate such as the New Zealand one. One plant group that appears to suffer irrecoverably from shoot apex damage is *Chionochloa*, the snow grasses. The three monophages that affect the tiller bases of alpine and open country *Chionochloa* are a large yponomeutid moth, and two aterpine weevil groups, the *Anagotus lewisi* group, and

17

18

Fig. 17. Graphania nullifera (Walker) (Lepidoptera: Noctuidae); last instar larva beside the eaten-out centre of a rosette of *Aciphylla colensoi* (Umbelliferae).
Fig. 18. Anagotus sp. (Coleoptera: Curculionidae) larval damage to *Celmisia spectabilis* (Compositae). The larva bores beneath the pile of tomentum in the centre.

the species of *Lyperopais*. All three taxa bore into and kill the tiller base: damage is indicated by the fact that the central leaf or leaves are either wilting or are dead. The yponomeutid and *Lyperopais* larvae may account for over three tillers per larva; the *Anagotus* species are restricted to those *Chionochloa* species with a very stout tiller base (e.g. *Ch. flavescens Ch. rigida*, *Ch. rubra*) and usually destroy less than three tillers per larva. Progressive death of the tillers opens out or collapses the tussock, weakening its performance as ground cover.

New zealand phytophages as botanists

There are some phenomena which provoke future inquiry. The extensive shrub genus *Coprosma* is attacked by three well defined groups of geometrid species, e.g. *Austrocidaria* (Dugdale, 1971). These three groups (*Austrocidaria; 'Eucymatoge'; 'Asaphodes' parora* – *'Xanthorrhoe' stricta* complex) are unrelated to other larentiine groups, and all are solely on *Coprosma*. Only one other geometrid genus, *Pasiphila*, has species on *Coprosma* (e.g. *Pasiphila sandycias*). None of the widely polyphagous *Pseudocoremia*, *Cleora*, *Gellonia* or *Declana* species have been collected on *Coprosma*. This type of host relationship suggests that the *Coprosma* feeders have broken through a chemical barrier. Noteworthy also is the fact that *Coprosma* defoliators are parasitised by a tachinid, *Pales aurea*, itself only parasitising *Coprosma* geometrids. In Noctuidae and Tortricidae, only two species in each family are restricted to *Coprosma*, and also in each family, one other, a polyphagous species, sparingly attacks these hosts. There is a need for a biochemical evaluation here.

It is not surprising that, on the Auckland Is, the leafmining weevil *Notinus cordipennis* mines both *Coprosma* (a shrub) and *Nertera* (a creeping perennial). Both belong to Rubiaceae. It is more unusual however, that the geometrid *'Epirhanthis' alectoraria* defoliates not only *Pseudopanax* species (except *Ps. anomalus*), but also *Griselinia*. This tree genus is equivocal in position (Philipson, 1967), with araliad as well as cornacead affinities. Also interesting in this regard is the faunal relationship between Araliaceae and Pittosporaceae: species groups of plume moths *('Aciptilia')*, geometrids *('Epirhanthis')* and gracillariids (*'Parectopa' aellomacha* group) are common to both, although each plant genus has different species. Is the apparent relationship based on coincidental chemical stimuli, or on close common ancestry?

The adventive flora and fauna

Table 4 outlines the types of phytophages found on 21 adventive shrub and tree genera which now cover so much of New Zealand. The paucity of indigenous phytophages that have adapted to these now dominant, adventive communities in much of our lowland areas is striking. Those

Table 4. Some Introduced Plant Genera and Associated Phytophages.

(Note: the list is incomplete; most introduced plants lack their native phytophages)

Host	Source	Introduced Phytophage genus	Indigenous (adapted) Phytophage
Conifers:			
Chamaecyparis	Holarctic	*Phloeosinus* (Scolytidae)	oligophagous *Orthenches* (Yponomeutidae)
Cupressus	Holarctic	*Phloeosinus* (Scolytidae)	
Picea	Holarctic	*Elatobium* (Aphididae)	polyphagous and oligophagous Coleoptera, Homoptera, Lepidoptera, Orthoptera
Pinus	Holarctic	*Pineus* (Adelgidae); *Hylastes* (Scolytidae), *Sirex* (Siricidae)	
Pseudotsuga	Holarctic	*Megastigmus* (Torymidae)	
Dicotyledonous Angiosperms:			
Acacia	Australia	*Acrocercops* [2 spp] (Gracillariidae); *Anarsia* (Gelechiidae) *Dasypodia* [2 spp] (Noctuidae); *Saccolacmus* (Curculionidae) *Psylla* [7 spp] (Psyllidae)	"
Azalea	Holarctic	*Caloptilia* (Gracillariidae)	
Corylus	Holarctic	*Nepticula* (Nepticulidae)	
Crataegus	Holarctic	*Caliroa* (Tenthredinidae); *Megastigmus* (Torymidae)	
Eucalyptus	Australia	*Antheraea* (Saturniidae); *Ctenairytaina* (Psyllidae); *Eriococcus* (Eriococcidae); *Flockiella* (Eulophidae); *Gonipterus* (Curculionidae); *Paropsis* (Chrysomelidae); *Phoracantha* (Cerambycidae); *Rhicnopeltella* (Eulophidae); *Spilonota* (Tortricidae); *Xyleborus* (Scolytidae).	"
Hakea	Australia	*Psylla* (Psyllidae); *Stegommata* (Lyonetidae)	—
Malus	Holarctic	*Dasyneura* (Cecidomyidae); *Eriosoma* (Pemphigidae)	"
Prunus	Holarctic	*Caliroa* (Tenthredinidae)	"
Pyrus	Holarctic	*Caliroa* (Tenthredinidae); *Dasyneura* (Cecidomyidae); *Carpocapsa* (Tortricidae)	
Quercus	Holarctic	*Asterolecanium* (Asterolecaniidae); *Lithocolletis* (Gracillariidae)	"
Ribes	Holarctic	*Synanthedon* (Sessidae)	
Rubus	Holarctic	*Eutorna* (Oecophoridae, Australian); *Priophorus* (Tenthredinidae)	*Carposina adreptella*, polyphagous tortricids (Carposinidae) [from NZ *Rubus* spp]
Salix	Holarctic	*Pontania* (Tenthredinidae)	
Sarothamnus	Holarctic	none (cf *Ulex*)	
Ulex	Holarctic	*Apion* (Curculionidae); *Lampides* (Lycaenidae)	*Anisoplaca ptyoptera* (Gelechiidae) [from NZ *Carmichaelia*, *Clianthus*] *Oemona hirta* (Cerambycidae) [polyphagous]
Vitis	Holarctic	*Phalaenoides* (Agaristidae) [Australian, usually on *Hibbertia*]	

indigenous phytophages that have taken advantage are, with two exceptions, species that are family-oligophagous or polyphagous. Of the 19 species of Lepidoptera regularly found in *Pinus*, *Pseudotsuga* and exotic Cupressaceae, 14 are oligophagous or polyphagous on conifers and dicots, while the remaining five species are oligophagous on endemic conifers. In one geometrid genus, *Tatosoma*, none of the Podocarpaceae-restricted species have moved to the exotic conifers and it is the dicot-oligophage, *Tatosoma tipulatum*, that is occasionally recorded from exotic conifers.

Besides the few monophages introduced along with their hosts, the adventive Holarctic dicots rarely attract endemic monophages. The two exceptions are *Ulex* (Papilionaceae), now an occasional host for the *Carmichaelia* stem-borer, the gelechiid moth *Anisoplaca ptyoptera*, and the other, *Rubus* is now host to the endemic fruit and bud borer, *Carposina adreptella*, one of the so-called 'raspberry budworms'. This paucity contrasts with Australian plants (cf. *Eucalyptus* and *Pinus* on Table 4). Australian plant genera introduced by man support a large assemblage of Australian monophages, either introduced by man or which have a history of recurrent dispersal across the Tasman Sea. Opportunities for such dispersal arise between once and twice a month (TOMLINSON, 1973), with a drop in February, and a peak in September-October. Reasonably well documented evidence exists to suggest that although the central western shores of New Zealand are the most usual landfalls, there is occasionally a doubling back by airmasses passing to the south of New Zealand so that coastal Otago or South Canterbury receive Australian insects (TOMLINSON, 1973). Distribution of the butterflies *Zizeeria otis labradus* and the endemic *Z. otis oxleyi*, the latter restricted to eastern and intermontane South I. east of the Alpine Fault suggest such a phenomenon. *Z. o. labradus* could only establish once the western lowlands of both islands were 'europeanised' by the introduction of herbaceous legumes. Our only endemic herbaceous legume, *Swainsonia* has (or had) a distribution similar to that of *Z. otis oxleyi*. Similarly, another lycaenid, *Lampides boeticus* has established on western shores where gorse *(Ulex)* is common (HARDING, 1972, 1973).

POLLINATORS

While flower structure has been used by some workers (e.g. VAN DER PIJL, 1969) to characterise the type of pollinator, the system breaks down in island situations such as that in New Zealand. Precise information is largely lacking but in view of the interest by botanists in breeding systems in our flora, pollinators should receive more attention. Annotated lists have been published by THOMSON (1927) and HEINE (1938) but both authors stress the incompleteness of their work.

While few records have been kept over recent years (to my knowledge) most entomologists find that the commonest flower-haunters are diurnal

Coleoptera, Diptera, and our few, short-tongued, bees. The flower selection habits of the short-tongued bees are under study and appear to be complex. They are essentially pollengatherers. All the diverse flower-haunting weevils (ca 200 species) are pollen-eaters. Hover-flies, tabanid flies and butterflies are poorly represented; all our tabanids are nectar-feeders. Moths, both diurnal and nocturnal, abound but probably the bulk of pollination of plants without anemophilous flowers is effected by beetles of the families Helodidae, Melyridae, Mordellidae and Curculionidae, and by calyptrate Diptera (Muscidae and Tachinidae). The last three families have high numbers of species but the first three, though fewer in species, occur in often incredible numbers of individuals.

VAN DER PIJL (1969) described inflorescence types that 'suit' various insect groups. Most New Zealand inflorescences fall within his beetle-favouring category. Deep flowers such as those of *Phormium* and *Sophora* and brush-flowers (Myrtaceae) are largely bird-pollinated, although THOMSON (1927) records moths of many species (as does HUDSON, 1928) on flowers of Myrtaceae. However, VAN DER PIJL's types are not useful in the New Zealand context, since the long-flowered Loranthaceae are visited, not only by moths, but also by small bees. These small bees prise open the appressed corolla lobes and dive into the corolla tube, which is at least three times the length of the bee (Dr. E. J. GODLEY, pers. comm.). I have seen similar bees do the same to unexpanded flowers on racemes of *Hebe gracillima*.

The outstanding characteristic of most insects recorded in New Zealand as pollinators (THOMSON, 1927; HEINE, 1938), in the light of recent experience, is that they can be found on all flowers capable of insect-pollination. As yet, flower-specificity has not been convincingly demonstrated, although as interest redevelops, and observations become more frequent, such specificity could well emerge.

REFERENCES

ALLAN, H. H. 1961. Flora of New Zealand Vol. I. Govt. Printer, Wellington, New Zealand. 1085 pp.

BRADLEY, J. D., CARTER, D. J., SANKARAN, T. & NARAYANAN, E. 1973. A new species of *Mompha* Huebner (Lepidoptera: Momphidae) from Assam, N.E. India, a potential agent for biological control of *Ludwigia adscendens*. *Bulletin of Entomological Research* 63: 57–63.

BARLOW, B. A. 1966. A Revision of the Loranthaceae of Australia and New Zealand. *Australian Journal of Botany* 14: 421–99.

CHEESEMAN, T. F. 1925. Manual of the New Zealand Flora, 2nd edition. Wellington, New Zealand. 1163 pp.
CLARKE, J. F. G. 1971. Lepidoptera of Rapa I. Smithsonian Contributions to Zoology No. 56.
COUPER, R. A. 1953. Upper Mesozoic and Cainozoic spores and Pollen Grains from New Zealand. N.Z. Geological Survey Palaeontological Bulletin 22.
COUPER, 1960. New Zealand Mesozoic and Cainozoic Plant Microfossils. N.Z. Geological Survey Palaeontological Bulletin 32.
CROWHURST, P. S. 1967. Notes on *Tanaoctenota dubia* Philpott (Lepidoptera ?Hyponomeutidae). *The New Zealand Entomologist* 3(5): 18–19.
DUGDALE, J. S. 1971. Lepidoptera, excluding non-crambine Pyralidae, in Entomology of the Aucklands and other Islands South of New Zealand. *Pacific Insects Monograph* 27: 55–172.
EHRLICH, P. R. & RAVEN, P. H. 1964. Butterflies and Plants: a study in coevolution. *Evolution* 18: 586–608.
FARRELL, J. A. K. 1974. Observations on *Epichorius longulus* (Coleoptera: Byrrhidae) in rata forests of the Auckland Is. *N.Z. Journal of Zoology* 1: 161–169.
FITTKAU, E. J. & KLINGE, H. 1973. On biomass and trophic structure of the central Amazonian plain forest ecosystem. *Biotropica* 5(1): 2–14.
FLEMING, C. A. 1962. New Zealand Biogeography. A Palaeontologist's Approach. *Tuatara, Wellington,* 10: 53–108.
FRANKLIN, D. A. 1969. Biological Flora of New Zealand – 3. *Dacrydium cupressinum* Lamb. (Podocarpaceae) Rimu. *New Zealand Journal of Botany* 6: 493–513.
GOURLAY, E. S. 1964. Notes on New Zealand Insects and Records of Introduced Species. *The New Zealand Entomologist* 3(3): 45–51.
HARDING, J. W. 1973. Additional records of *Lampides boeticus* (Lepidoptera: Lycaenidae). *N.Z. Entomologist* 5: 272–273.
HARRISON, R. A. 1959. Acalyptrate Diptera of New Zealand. N.Z. Dept. Scientific and Industrial Research Bulletin 128.
HEINE, E. M. 1938. Observations on the pollination of New Zealand flowering plants. *Transactions of the Royal Society of New Zealand* 67: 133–148.
HERING, E. M. 1951. Biology of the Leaf Miners. Junk, The Hague 420 pp.
HERING, E. M. 1957. Bestimmungstabellen der Blattminen von Europa Vol. 3. Junk, The Hague 221 pp.
HOY, J. M. 1962. Eriococcidae of New Zealand. N.Z. Dept Scientific and Industrial Research, Bulletin 146.
HUDSON, G. V. 1928. The Butterflies and Moths of New Zealand. Ferguson and Osborn, Wellington.
HUDSON, G. V. 1939. Supplement to the Butterflies and Moths of New Zealand. Ferguson and Osborn, Wellington.
HUDSON, G. V. 1950. Fragments of Entomology. Ferguson and Osborn, Wellington.
LAMB, K. P. 1960. A Check List of New Zealand Plant Galls (Zoocecidia). Transactions of the Royal Society of New Zealand 88: 121–139.
MANI, M. S. 1964. Ecology of Plant Galls. Monographiae Biologicae, Vol. 12 W. Junk, The Hague, 434 pp.
MOORE, L. B. & EDGAR, E. 1971. Flora of New Zealand, Vol. II. Government Printer, Wellington.
PHILIPSON, W. R. 1967. *Griselinia* Forst. f. – anomaly or link? *New Zealand Journal of Botany* 5: 134–165.
PIJL, L. VAN DER, 1960. Ecological Aspects of Flower Evolution 1. Phyletic Evolution. *Evolution* 14: 403–416.
RAWLINGS, G. B. 1956. Australasian Cyttaraceae. *Transactions of the Royal Society of New Zealand* 84: 19–28.
THOMSON, G. M. 1927. The pollination of New Zealand flowers by birds and insects. *Transactions of the New Zealand Institute* 57: 106–125.

THORSTENSTEIN, A. J. 1961. Host selection in phytophagous insects. *Annual Review of Entomology* 5: 193–218.

TOMLINSON, A. I. 1973. Meteorological aspects of transTasman insect dispersal. *N.Z. Entomologist* 5: 253–268.

VARI, L. 1961. South African Lepidoptera, I. Lithocolletidae. *Transvaal Museum Memoir* 12: 238 pp.

WARDLE, P. 1964. Facets of the distribution of forest vegetation in New Zealand. *N.Z. Journal of Botany* 2: 352–366.

WARDLE, P. 1966. Biological Flora of New Zealand – 1. Weinmannia racemosa Linn. (Cunoniaceae), Kamahi. *New Zealand Journal of Botany* 4: 114–131.

WARDLE, P. 1967. Biological Flora of New Zealand – 2. *Nothofagus menziesii* (Hook. f.) Oerst. (Fagaceae). Silver Beach. *New Zealand Journal of Botany* 5: 276–302.

WARDLE, P. 1968. Evidence for an indigenous pre-Quaternary element in the mountain flora of New Zealand. *New Zealand Journal of Botany* 6: 120–125.

WARDLE, P. 1969. Biological Flora of New Zealand – 4. *Phyllocladus alpinis* Hook. f. (Podocarpaceae), Mountain Toa Toa, Celery Pine. *New Zealand Journal of Botany* 7: 76–85.

WARDLE, P., FIELD, T. R. O. & SPAIN, A. V. 1971. Biological Flora of New Zealand. 5. *Olearia colensoi* Hook. f. (Compositae) Leatherwood, Tupari. *New Zealand Journal of Botany* 9: 186–214.

WATT, J. C. 1971. Entomology of the Auckland and other Islands south of New Zealand: Coleoptera: Scarabaeidae, Byrrhidae, Ptinidae, Tenebrionidae. *Pacific Insects Monograph* 27: 193–224.

WATT, M. N. 1916. Contributions to the entomology of New Zealand: No. 8. *Parectopa citharoda* Meyr. (Order Lepidoptera). *Transactions of the New Zealand Institute* 48: 407–413.

WATT, M. N. 1920. The leafmining insects of New Zealand. *Transactions of the New Zealand Institute* 52: 439–466.

WATT, M. N. 1921a. A new tortricid moth. *New Zealand Journal of Science and Technology* 4: 257–258.

WATT, M. N. 1921b. The leafmining insects of New Zealand, Part II. The genus *Nepticula* (Lepidoptera). *Transactions of the New Zealand Institute* 53: 197–219.

WATT, M. N. 1923. The leafmining insects of New Zealand, Part III – species belonging to the genera *Agromyza* (Fallen) and *Phytomyza* (Fallen) (Diptera). *Transactions of the New Zealand Institute* 54: 465–489.

WATT, M. N. 1924a. The leafmining insects of New Zealand, Part IV – *Charixena iridoxa* Meyr., *Apatetris melanombra* Meyr., *Philocryptica polypodii* Watt. *Transactions of the New Zealand Institute* 55: 327–340.

WATT, M. N. 1924b. Part 5 – The genus *Nepticula* (Lepidoptera) and the *Agromyzidae* (Diptera) continued, and *Gracilaria selenitis* Meyr. (Lepidoptera). *Transactions of the New Zealand Institute* 55: 674–687.

WISE, K. A. J. 1962. *Parectopa leucocyma* (Meyrick) (Lepidoptera: Gracillariidae) rediscovered as a leafminer of kauri (*Agathis australis* Salisb.). *Transactions of the Royal Society of New Zealand (Zoology)* 1: No. 31. 373–375.

ZOTOV, V. D. 1963. Synopsis of the grass subfamily Arundinoideae in New Zealand. *N.Z. Journal of Botany* 1: 78–136.

XVI. ADAPTATION AND CHANGE IN MAORI CULTURE

by

R. C. GREEN

The Origins of the Maori

Ultimate Origins

The Asian origins of the Maori are lost in the mists of an antiquity still unexplored by modern archaeology. The probable source, the direction of movement, the general level of cultural development, and the approximate age are known, but the actual sites with their ancient occupation layers have yet to be discovered. In large part this is because there has been little excavation on which such knowledge would depend in the islands and mainland of Southeast Asia. For this reason only an outline can be presented of the ancestral languages, populations, and cultural complexes which eventually gave rise to those we now call Maori.

The Maori language belongs to the Austronesian family of some 500 languages centred in the Pacific. Oceanic representatives of this family are spread throughout the islands of Polynesia, Micronesia, and Melanesia including coastal areas of New Guinea. It also includes the indigenous languages of Taiwan, Borneo, Philippines, Celebes and Indonesia in Island Southeast Asia and the Malayan and Cham languages on the Southeast Asian mainland itself (Map 1). There is as well a remote outpost in Madagascar. The Austronesian family exhibits some of its greatest linguistic diversity and may have had its origin in the eastern area of Island Southeast Asia (Pawley 1971).

Austronesian economy also had its roots in an ancient Pacific cultural adaptation worked out in the same area (Map 2). One part of the ancestral economy was based on root crops like taro and yam. These were combined with tree crops like breadfruit, banana, and coconut. Nearly all have their source as domesticates in an Indo-Malayan centre that extends from the Asian mainland out into the Pacific Islands (Barrau 1965). The food resources also included the domestic dog, pig, and chicken plus the rat. The other part of the economy featured exploitation of the common Indo-Pacific marine resources of fish, shell-fish, sea mammals and turtles.

The technology of the Austronesian populations in the Pacific derives from a cultural pattern which made its initial expansion into the Pacific Island world some 5000 to 7000 years ago. One aspect of it included a maritime component of outrigger canoes, rafts and their accessories

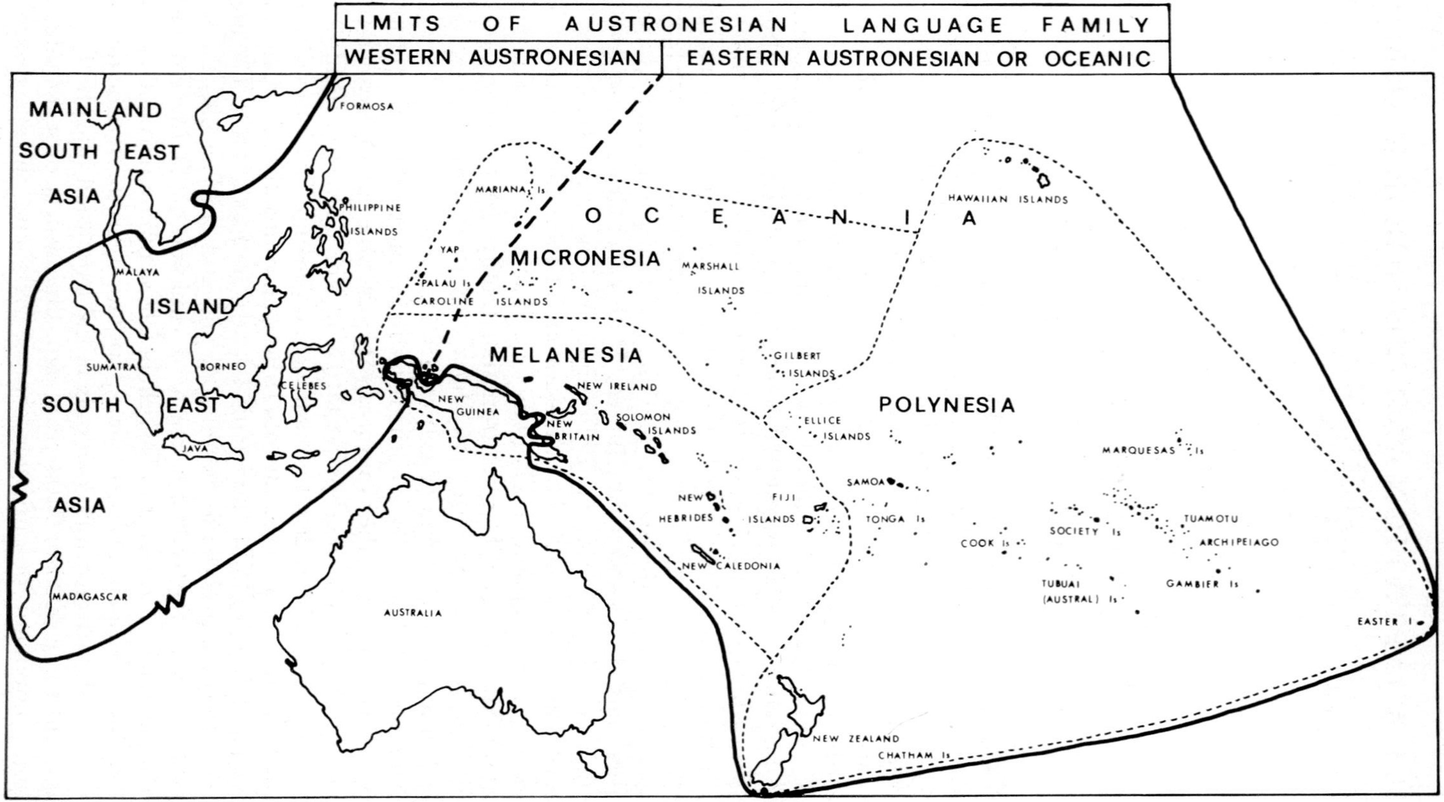

Map 1. Southeast Asian and Pacific Culture Areas in Relation to the Distribution of Austronesian Languages.

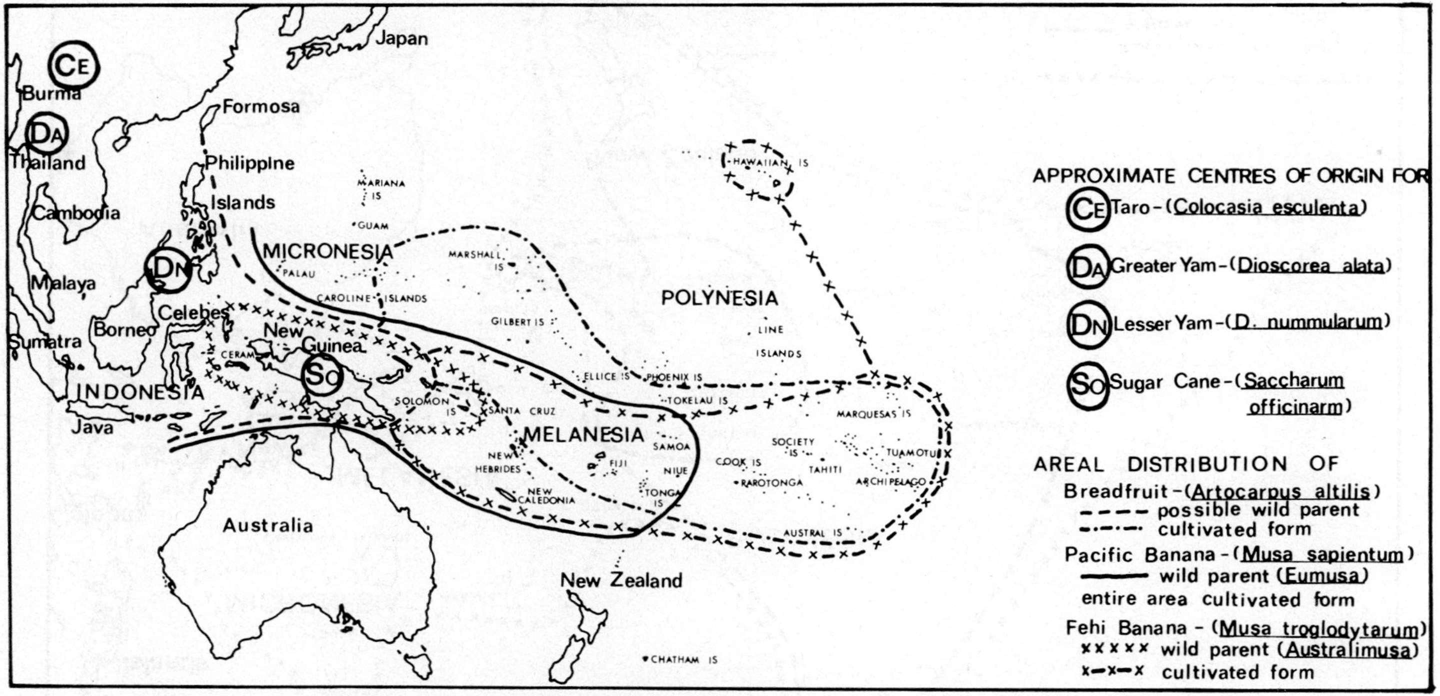

Map 2. Indo-Malaysian and Pacific Sources of some Common Pacific Food Plants (after BARRAU 1965).

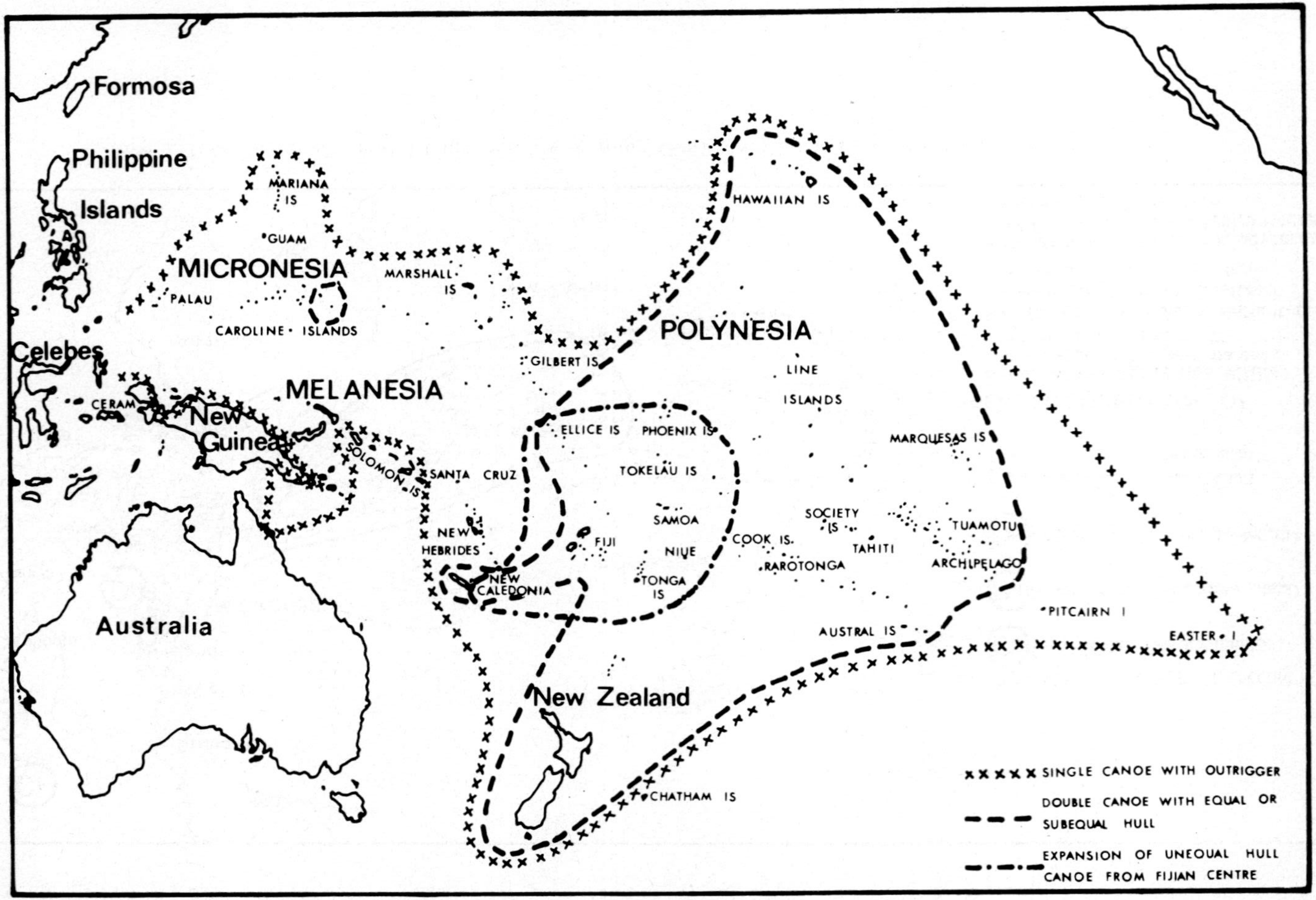

Map 3. Single canoe with outrigger and double canoe distribution in Oceania. The area at contact of double canoes with equal or subequal hulls and with masts forward of midships is indicated. This excludes Fiji, where the unequal and dissimilar hull form was encountered, a type which expanded in the 19th century to the area indicated (after HADDON & HORNELL 1938 III).

which permitted people to cross open stretches of sea, and so reach the major island groups of Oceania (Map 3). Another aspect related to items required in exploiting the maritime resources. Included were at least some of the widespread oceanic techniques of nets, traps, gorges, simple fishhooks, spears and harpoons (ANELL 1955). Still another aspect involved the working of wood or the clearing of forests. It is reflected by polished stone tools, especially the quadrangular stone adze, and by assemblages of stone flakes used in scraping, cutting, and drilling. A final aspect was a well developed art of pottery manufacture and use. Both pottery and polished stone adzes have ages in excess of 5,000 B.C. in Southeast Asia (GORMAN 1971) and are even older in Japan. No longer is their presence in the Pacific at this time a cause for surprise.

The racial affinities of the ancestors of the Austronesians are unknown. They appear to lie with the same ancestral stock which gave rise to the Mongolians, American Indians, and to the various aboriginal groups of Taiwan, Borneo, the Philippines, Indonesia and parts of Mainland Southeast Asia. They may be contrasted with those Australoid populations which settled Australia, New Guinea and Tasmania some 30,000 years earlier when they still formed a single super-continent in the Pacific. In fact, genetic hybridization between these two ancestral stocks at a later date in the western part of Melanesia, followed by gene flow to the east as far as Fiji, probably has produced the more Australoid looking Austronesian speakers of Melanesia. This, rather than a separate Melanesian migration may explain their contrast to Eastern Micronesians and Polynesians, who were not so affected.

OCEANIC ORIGINS

The Oceanic origins of Maori language and culture are based in more certain knowledge. Linguistically Maori belongs to an Oceanic Branch of the Austronesian family, a block of some 300 languages which constitutes the major grouping in the central Pacific (Map 1). The other major branch is Western Austronesian which forms a block of some 200 languages encompassing the entire area of Island Southeast Asia and two enclaves on the Mainland. All but the Eastern Micronesian and Polynesian subgroups among the numerous subdivisions of the Oceanic Branch of the Austronesian family are found in Melanesia. The region in this area with the greatest diversity is to the northwest in New Guinea, the Admiralties and New Britain. Among the subgroups of the Oceanic Branch, the one to which Maori belongs is Eastern Oceanic, languages of which are found in the Southeast Solomons, the New Hebrides and the Fiji-Polynesian areas (fig. 1).

It is not yet possible to trace the cultural origins of Oceanic groups in Melanesia to a time much before 1,000–1,200 B.C., although there are probably ancestral cultural complexes of greater age in this region from

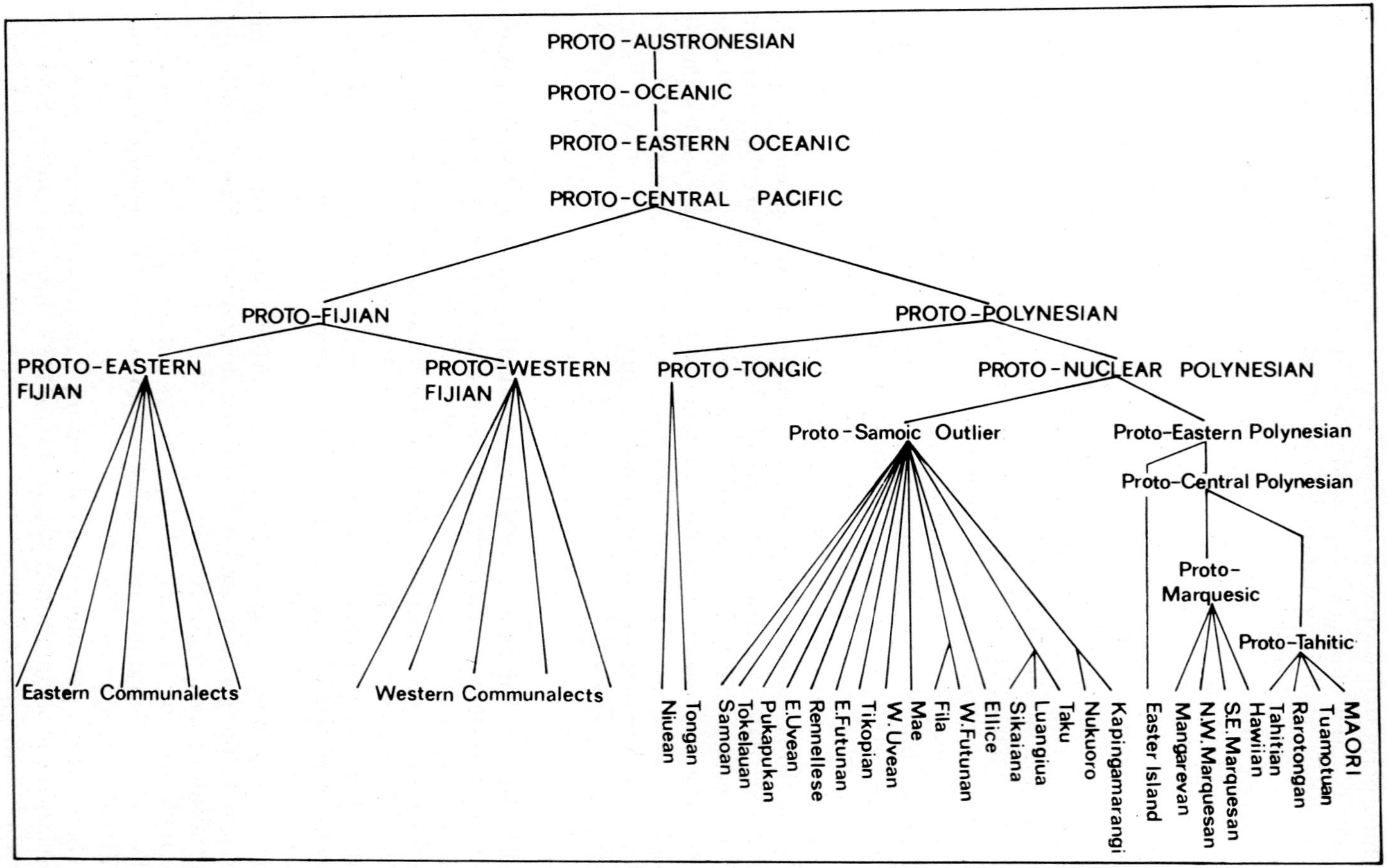

Fig. 1. The Position of Maori in the Central Pacific Subgroup of Eastern Oceanic Languages.

which some of the known complexes derive. The Oceanic cultural complex ancestral to the Polynesian pattern is termed Lapita after a site of this name in New Caledonia, where its distinctive pottery was found. The cultural content appears somewhat variable because not all items have been found in all sites, though all possess the distinctive pottery. It is almost certainly this cultural complex, which on reaching Tonga, slowly changed into the Polynesian pattern. The Lapita complex probably gave rise to other cultural traditions in Eastern Melanesia as well, particularly in Fiji and the New Hebrides.

Lapita sites in Melanesia are often found on small islands lying off the coasts of larger islands from New Britain and New Ireland through Santa Cruz and Espiritu Santo to New Caledonia. Occasionally they are found in similar coastal environments on some of the larger islands such as Santa Cruz, Efate, New Caledonia, and Viti Levu. While these sites appear to represent the earliest settlements in Fiji and Tonga, this is probably not true for the rest of Melanesia.

Lapita sites attest to reasonably permanent settlements of fairly large size (1,100–14,000 m^2) containing ovens, fire pits and postholes of former structures. The associated food refuse indicates the intensive exploitation of a wealth of marine resources. Though the extent to which the people possessed and used domesticated plants and animals is not yet known, it is evident from the Polynesian data that both were present at the time that area was settled. The most common objects, however, are literally thousands of sherds of Lapita pottery, a low-fired friable sand or shell tempered earthenware. The tempers and possibly whole pots were sometimes traded over long distances, together with obsidian and chert. Pottery forms consist of deep dishes, large ovoid bowls, shouldered jars, and collared vessels, many of which have extensive decoration, mainly around the upper surface. A plain ware is also common. The decoration includes incised motifs but the most distinctive technique is a unique form of toothed or 'dentate' stamping, (Plate 1).

Lapita manufacturing techniques include slab, and perhaps ring and coil building in conjunction with paddle and anvil finishing. There is sometimes a red slip and occasionally a white calcareous infill of the decorative impressions to emphasize the motifs. Associated artifacts include stone adzes of quadrangular, lenticular or planoconvex cross-section, together with a range of shell adzes. Other items are flake tools of chert, obsidian and other stones; shell food scrapers, arm rings and necklace units; drilled shark teeth and sling stones.

Polynesian Origins

Theories of immediate Polynesian origins are many, but today most can be dismissed as not supported by the now rapidly accumulating evidence (Green 1967a; 1972a). This includes those theories advocating both

Plate 1. Human face mask on sherd of Lapita style pottery from main Reef Islands, Santa Cruz Group.

Micronesian and South American routes. Although Polynesians have had contact and derived particular items from both these areas, neither has served as a source for the distinctive Polynesian language, physical type, or cultural complex. The error has been in thinking there was a migration to Polynesia of a people physically identical with the Polynesians as we know them, and already possessing distinctive features of Polynesian language and culture (EMORY 1959). Today we seek proto-languages, earlier cultural assemblages, and ancestral populations which were not Polynesian, but from which the Polynesian language, the basic form of

Table 1. Proto-Polynesian Reconstructions Indicating Physical Nature of their Homeland.

PPN	Meaning	PPN	Meaning	PPN	Meaning
Inanimate Environment					
awa	channel	*lanu*	fresh water	*muriwai*	mouth of river
fatu	stone	*lase*	coral, lime	*rano*	lake, swamp
feo	coral	*loto*	lagoon, lake	*tahi*	sea
hakau	coral reef	*mato*	cliff	*waitafe*	river, stream
huʔa	high tide	*maʔunga*	mountain	*wasa*	open sea
kele	dirt, earth	*motu*	island	*ʔone*	sand
kilikili	gravel				
Land Fauna					
fonu	turtle	*kumaa*	rat, mouse	*ngata*	snake
kulii	dog	*moko*	lizard	*peka*	bat
Plus 17 common Pacific birds,				*puaka*	pig
and 17 common Pacific insects.					
Land Flora					
ango	turmeric	*ifi*	chestnut tree	*kulu*	breadfruit
asi	sandal wood	*kafika*	Malaya apple	*maele*	fern, sedge
fara	pandanus	*kape*	Alocasia sp.	*niu*	coconut
fau	hibiscus	*kawa*	kava	*renga*	turmeric
futi	banana	*kea*	breadfruit	*talo*	taro
futu	*Barringtonia asiatica*	*kofe*	bamboo	*tongo*	mangrove
Plus numerous other species.				*ʔufi*	yam
Sea Life					
At least 44 Indo-Pacific fish, including:					
kanahe	mullet	*sakulaa*	swordfish	*ʔatu*	bonito
malolo	flying fish	*toke*	sea eel	*ʔono*	barracuda
mangoo	shark	*tuna*	freshwater eel		
Other Sea Life					
feke	octopus, squid	*paka*	crab	*tifa*	pearl oyster
kina	sea urchin	*pipi*	shellfish sp.	*tio*	oyster sp.
kuku	mussel	*pule*	cowry shell	*tupa*	land crab
				ʔura	crayfish

their culture, and even their general physical type could derive and subsequently develop within Polynesia itself.

These sources all lie, as indicated above, in the immediately adjacent area of Eastern Melanesia and especially in Fiji. Here the dissolution of a proto-language, Central Pacific, to which both Polynesian and Fijian belong (PAWLEY 1970) gave rise between 1500 B.C. and 200 B.C. to proto-Fijian on the one hand and proto-Polynesian on the other (fig. 1). Reconstruction of the proto-Polynesian lexical vocabulary indicates a physical environment which most closely corresponds to a central Pacific island group (Table 1). Geographical and linguistic arguments from this data favour Tonga as the most probable homeland (PAWLEY & K. GREEN 1971).

This hypothesis is supported by the archaeological evidence from Polynesia which shows Tonga to have been the first island group settled over 3,000 years ago by people possessing a culture identical with Lapita assemblages of this age in Fiji. Between 1000 B.C. and 200 B.C. Lapita assemblages in Tonga and Samoa exhibit changes in the pottery, stone adzes, and shell artifacts which are in the Polynesian direction (GROUBE 1971). By the 1st century A.D. when widespread settlement in Samoa is documented archaeologically (GREEN & DAVIDSON 1969), the ancestral

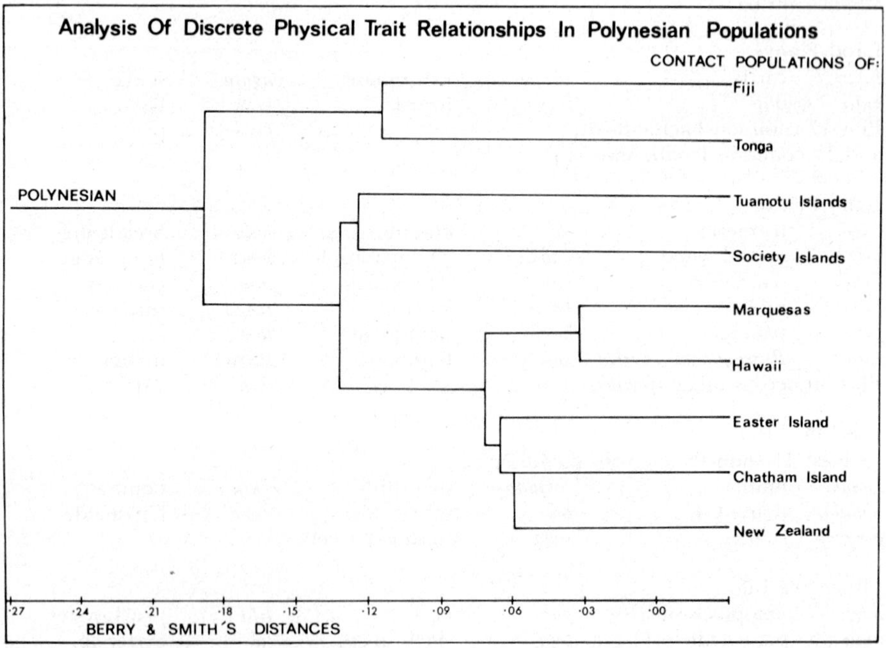

Fig. 2. The Position of the Maori Among Polynesian Populations on the Basis of Single Trait Relationships (after PIETRUSEWSKY 1970).

Polynesian cultural pattern had evolved. It appears to have differentiated out of the earlier assemblages of Tonga and Samoa, at the time when people with Lapita culture established themselves on the Oceanic high islands of West Polynesia.

By this time the dissolution of the Polynesian languages was also underway, one branch, to which Maori belongs, becoming Eastern Polynesian and the other Samoic-Outlier. The linguistic differentiation which led to Maori and other present day languages of Polynesia has been charted (fig. 1). A similar dissolution has been graphed for the breakup of the Polynesian population itself into its present representatives (fig. 2).

Settlement of East Polynesia from West Polynesia apparently took place in the first few centuries A.D. Only in the Marquesas, however, have occupation sites of this age been encountered which allow us to document the process. Still, by tracing sequences in Hawaii, the Society Islands, the Marquesas, and New Zealand back to their earliest cultural assemblages, it is possible to show they all diverge from very similar types of adzes, fishhooks, ornaments, burials, etc. (fig. 3). In addition the earliest assemblages in the Marquesas, dating between 300 and 600 A.D. contain adzes of preponderantly West Polynesian types, as well as pottery and octopus lures similar to those in early contexts in Samoa (GREEN 1968, SINOTO 1970).

The content of the ancestral East Polynesian cultural complex is variable, and as good a case can be argued for its origin in the Marquesas as has formerly been argued for its origin in the Society Islands (EMORY & SINOTO 1964; BELLWOOD 1970). While the archaeological evidence is more than sufficient to establish the cultural origins of New Zealand's earliest assemblages as East Polynesian, a conclusion which SKINNER (1921, 1923, 1924) demonstrated by the comparative method more than 50 years ago, the precise homeland has yet to be convincingly proven. The Marquesas, the Southern Cooks, and the Society Islands all have some claims, though the vital settlement period assemblages on which to decide the issue are only known in the Marquesas and New Zealand.

EAST POLYNESIAN ADAPTATION

While interesting, the determination of the actual homeland for the Maori or the precise route by which they travelled is not of primary importance in analyzing the ecological and cultural adaptations undergone by those East Polynesians who initially colonized New Zealand. The broad origins framework sketched above is quite sufficient to specify some of the basic cultural and environmental orientations which these settlers possessed when they arrived. It is this background which provides an understanding of the changes in a tropical East Polynesian culture that were consequent upon its transfer to a quite different ecological

Plate 2. Mr. TAMATI REEDY and Mrs TILLY REEDY wearing dogskin cloaks at the display of Maori clothing to HRH Princess ALEXANDRA at the Auckland Museum 21.4.71. (Photo C. A. SCHOLLUM).

situation in New Zealand. This aspect is often neglected in accounts of the early phases of Maori prehistory.

As one proceeds from Melanesia to East Polynesia, the adequacy of the indigenous natural resources to support a population rapidly diminishes. So does the size of the islands. Also the distances of open sea between islands increases and major water gaps are frequently encountered between one group and the next. An increasing ability to cross stretches of open sea on one-way voyages and make a landfall was required. Thus in the passage from the larger continental islands of Melanesia to the smaller volcanic high islands and coral atolls of East Polynesia, the maritime technology and economy became more elaborate. For example the distribution of the double canoe (Map 3) suggests that with the exception of probable borrowings in the central Caroline Islands and in New Caledonia, the vessel is a Polynesian innovation based on the widespread single outrigger canoe (HADDON & HORNELL 1938, III). This vessel would have permitted ancestors of the Polynesians to successfully establish founder populations of sufficient size, and with adequate provisions of plants, animals and equipment, on the far flung islands of Polynesia.

Similarly from the Melanesian islands to East Polynesia there is an increasing attenuation of the varied New Guinea flora and fauna as one moves east. It involves many wild plants and trees, marsupial animals, and even birds of economic importance to man. In fact, settlement of Polynesia required that colonizers be accustomed to bringing with them on their one-way voyages of discovery (SHARP 1963) the domestic plants and animals on which they were largely to rely. It also required that their culture be fully capable of exploiting the products of the sea, for at times they would be almost completely dependent on these resources if they were to establish themselves in some parts of the Polynesian island world. Not surprisingly, a general survey of man's use of the maritime environments in the Pacific (REINMAN 1967) shows that through time and to the East, especially in Polynesia, there was a gradually increasing interest in the exploitation of the aquatic environment. A consequence was an increasingly specialized technological response to its requirements (REINMAN 1970: 48). Items illustrating this include the use throughout Polynesia of the trolling lure (fig. 3) for taking fish in surface waters offshore, while in deeper waters they involve the use of the one and two-piece fishhooks on a still line. The latter devices, uncommon or unknown in West Polynesia, early reached positions of importance in East Polynesia.

The larger continental islands of Melanesia provide a fair number of rock, soil, and plant types. They also have reasonably extensive riverine, esturine and marine resource zones and a varied range of coastal, fringing and barrier reef environments. The change to the typical volcanic high islands or atolls of Polynesia with their more limited land environments, lagoons and barrier reefs also had its effect. For example suitable rocks for adzes, flakes and other stone tools were largely confined to the fine grained Oceanic olivine basalts, and occasional deposits of volcanic glass on some islands. Many of the large shell species of Melanesia were no longer available and items such as the very large giant clams used for adzes and ornaments, or the *trochus* shells for arm rings, drop from the Polynesian inventory. Pearlshell, however, becomes far more important.

Among the food plants the yam loses its importance as primary crop beyond Tonga, and breadfruit and taro become dominant food plants. Moreover, as the breadfruit is seasonal, the storage of this crop in large and small circular and rectangular pits, from Tonga and Samoa to Tahiti and the Marquesas, forms a part of an economic technology not developed in Melanesia (YEN 1971a: 9–10; LAW 1970: 114).

On present though limited evidence the sweet potato was introduced into East Polynesia after the settlement of that area around the 1st to 3rd century A.D. and the time before the 8th to 9th centuries A.D. when Hawaii, Easter Island and New Zealand were colonized. The source was South America and the agent human, but whether of Polynesian, South American or other identity is unknown (YEN 1971a, 1971b). Like the yam, the sweet potato seems not to have played a major role in the horticulture of central East Polynesia. However, on transfer to the areas of Hawaii, Easter Island and New Zealand, where the other tropical Polynesian food crops become marginal, or drop from the roster, it achieved varying degrees of dominance. Confirmation of its addition to the East Polynesian horticultural complex before the time of settlement of New Zealand is therefore a vital question on which better evidence is required. Without the sweet potato, accounts of early Maori prehistory might well be constructed quite differently.

In the shift from land forms of the large continental Pacific islands to smaller Oceanic high islands and atolls, contrasts between inland and coastal communities assume less relevance. The preferred location of population on most Oceanic high islands is along an ecotome, where with minimum effort it is possible for extended family groups to exploit both the agriculturally productive soils of inland valleys, and the resources of the coastal strip and adjacent sea (GREEN in GREEN *et al.* 1967a: 216–17; BELLWOOD 1971a: 156–162). The result is that nucleated villages are seldom a feature of Polynesian settlement with a few late exceptions. Rather, the pattern ranges from dispersed households to loosely clustered though largely unplanned settlements which are organized in relation to the local kingroup ownership of land. Only on the communal or political level, where the residence of a principal chief or a complex of religious structures provided a focus, does any tendancy towards nucleation appear (GREEN 1970a: 22).

Multiple origins or cultural isolation

The possibility of settlement from multiple sources has long been an issue in reconstructions of New Zealand prehistory. The prevailing view of two settlements usually held that a later group commemorated in numerous canoe traditions was responsible for the 'Coming of the Maori'. This interpretation has lost favour for three reasons. First, the hypothesis has proved to be grounded in an unwarranted manipulation of authentic Maori tradition and the unjustified use of corrupt sources (Simmons 1969b). Second, it has attributed a navigational ability and a pattern of deliberate two-way voyaging to the Maori that is not well founded (Sharp 1958, 1963). Third, it has attracted only negative rather than substantive evidence from the archaeological record (Golson & Gathercole 1962: 272–74; Green 1963: 7; Groube 1968: 142–43). Thus, at present there are few grounds on which to infer significant contact at a later date or to postulate that sources outside East Polynesia played an important role in New Zealand prehistory.

The possibility of one or more early settlements from East Polynesia is less easily resolved. Hawaii and New Zealand provided suitable conditions for effective secondary settlement in the early stages of their prehistory because of the large size of their land areas and the relative lateness of their initial occupation, factors which increased possibilities for multiple discovery and settlement from an already almost completely occupied West and Central Polynesia. The requirements were that discovery and decisive secondary settlement must occur before the population of the primary group was sufficiently great in relation to the available land area. This would allow the second group room to become established and their influence and their presence documented in the linguistic, physical and archaeological record (Green 1966: 29). Data on which to base the claim for two early settlements in New Zealand from separate sources in East Polynesia are few, and the proposition speculative, though it can not be entirely discounted (Green 1966: 31–33; 1967b: 81, 89).

Golson (1959) in the most thorough recent review of 'Culture Change in Prehistoric New Zealand' favoured a single origin. This was followed by development in isolation of the distinctive varieties of Maori culture within a continuum of change thought to have proceeded along several lines. The view of Duff (1956) and Green (1963), who both postulated an agricultural transfer after initial settlement, (Green favouring its introduction at an earlier stage than Duff) have not received much support from the archaeological record. Following criticisms by Golson (1965) and Groube (1967) and closer examination of the issues by Law (1970) and Yen (1971a), Green (1970b: 50) has abandoned his former position, and now also favours the introduction of food plants with the initial settlers. The question of unitary versus multiple origins from separate sources in East Polynesia he leaves open.

The reader may wonder at the concern indicated above by various investigators over the issue of origins and outside contacts. It is vital, however, in evaluating the degree to which change in the cultural

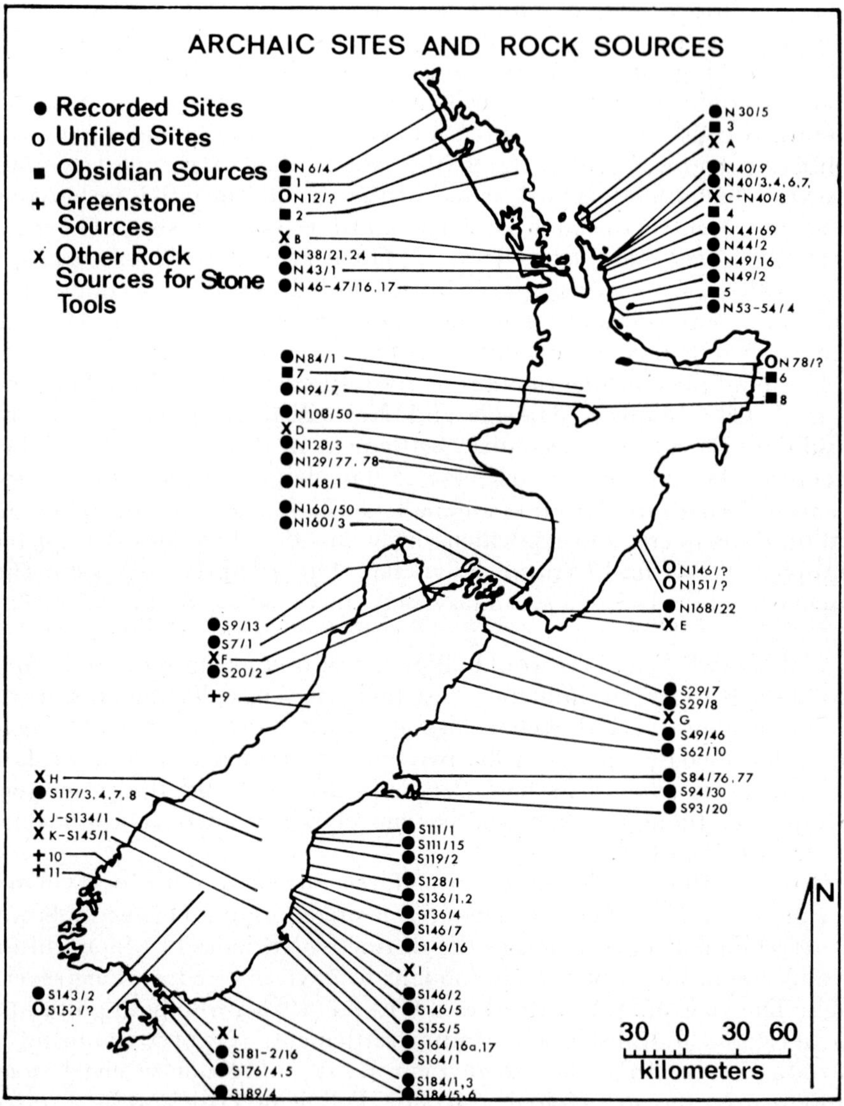

Map 4. New Zealand Archaic Sites and Rock Sources (see Table II for site names and details).

Table 2. Common Names for Archaic Sites and Rock Sources (see Map 4 for locations).

Archaic Sites			
North Island		South Island	
N 6/4	– Houhora	S 7/1	– Heaphy River
N 12/?	– Moturua garden	S 9/13	– Anapai
N 30/5	– Harataonga	S 20/2	– Tahuanui
N 38/21	– Pig Bay	S 29/7	– Wairau Bar
N 38/24	– Sunde	S 29/8	– Grassmere Spit
N 40/3	– Opito Midden	S 49/46	– Kaikoura
N 40/4	– Fisher's Midden	S 62/10	– Hurunui Mouth
N 40/6	– Black's Midden	S 84/76	– Redcliff's Sand Dunes
N 40/7	– Skipper's Ridge	S 84/77	– Moa-bone Point Cave
N 40/9	– Sarah's Gully	S 93/20	– Rakaia River
N 44/2	– Tairua North	S 94/30	– Tumbledown Bay
N 44/69	– Hot Water Beach	S 111/1	– Dashing Rocks
N 46-47/16	– Bramley	S 111/15	– Normanby No. 1
N 46-47/17	– University	S 119/2	– Pareora Mouth
N 49/2	– Whangamata Wharf	S 117/3	– Woolshed Flat
N 49/16	– Whitipirorua	S 117/4	– Ahuriri
N 53–54/4	– Whiritoa	S 117/7	– Junction Point
N 78/?	– Paerata Ridge	S 117/8	– Gooseneck Bend
N 84/1	– Tokoroa	S 128/1	– Waitaki Mouth
N 94/7	– Whakamoenga Cave	S 136/1	– Tai Rua South
N 108/50	– ?	S 136/2	– Ototara Glen
N 128/3	– Kaupokonui	S 136/4	– Awamoa
N 128/77	– Ohawe	S 143/?	– Hawksburn
N 128/78	– Te Rangatapu	S 146/2	– Waimataitai Lagoon
N 146/?	– Black Head	S 146/5	– Shag Point
N 148/1	– Foxton	S 146/7	– Waianakarua River
N 151/?	– Porangahua	S 146/16	– Hampden Beach
N 160/3	– Makara	S 152/?	– Upper Waikaia
N 160/50	– Paremata	S 155/5	– Shag River Mouth
N 168/22	– Palliser Bay	S 164/1	– Little Papanui
		S 164/17	– Kaikai's Beach
Obsidian Sources		S 164/16a	– Murdering Beach
1	– Kaeo	S 176/5	– Tihaka
2	– Huruiki	S 176/4	– Wakapatu
3	– Great Barrier	S 181–2/16	– Tiwai Point
4	– Whitianga	S 184/1	– Pounawea
5	– Mayor Island	S 184/3	– False Island
6	– Whakarewarewa	S 184/5	– Papatowai
7	– Maraetai	S 184/6	– King's Rock
8	– Taupo	S 189/4	– Old Neck
Other Rock Sources		Greenstone Sources	
A	– Tryphena andesitic basalt	9	– Arahura & Taramakai Rivers Nephrite
		10	– Anita Bay – Bowenite
		11	– Wakatipu Mts – Nephrite

Table 2 (continued)

Archaic Sites			
North Island		South Island	
Other Rock Sources		Other Rock Sources	
B	– Motutapu greywacke	F	– D'Urville baked argillites and serpentines
C	– Tahanga andesitic basalt	G	– Amuri Bluff cherts
D	– Taranaki indurated sandstone/greywacke	H	– Gray's Hills – quartzite
		I	– Moeraki porcellanite
E	– Mungaroa chert	J	– Oturehua quartzite
		K	– Nenthorn quartzite
		L	– Orepuki-Riverton Bluff actinolitic tuffs

sequence should be ascribed to innovation and continuing adaptation in isolation. It is also necessary in assessing the impact the New Zealand environment had on the economy and culture of the early inhabitants.

Whatever its origin, a voyage from East Polynesia to New Zealand involved a distance over open sea of not less than 2500 km. Following SHARP'S (1956, 1963) examination of the navigational ability of Polynesians, it has become evident that repeated two-way voyaging between island groups seldom involved distances over 480–500 km. Thus it is likely that an early landfall or landfalls in New Zealand resulting from those one-way voyages of discovery also effected its successful settlement. The position of subsequent isolation implied in this view is supported, as noted above, by the lack of evidence for significant later contact. It is also supported by the frequencies of canoe types and voyaging patterns which obtained in New Zealand in the 18th century A.D. By then the dominance of sea-going single outrigger and double sailing canoes characteristically found in East Polynesia had altered in response to the local conditions in New Zealand. Single outrigger canoes were only infrequently attested, while double canoes were of limited occurrence, though slightly more common in the South than the North Island. Emphasis had shifted to several types of single canoes without outrigger, including the famous highly decorated war canoe, to dugouts, and to simple and shaped rafts. Frequencies of each were determined by local requirements for different combinations of sea, sea and river, and lake and river transport to which types were most suited (BEST 1925; BUCK 1949: 201–04; BATHGATE 1969). While sails were sometimes employed, the most common mode of propulsion was the paddle.

When these changes occurred is unknown but it appears likely they followed closely the sequence of settlement of different localities and environmental situations. One might infer from the necessity for sailing to New Zealand in ocean-going outriggers or double canoes, and the rapid spread of the early inhabitants from one end of the country to the other (Map 4), that the colonizers initially maintained their East Polynesian preferences in vessels while in the process of exploring and settling the country. The seasonal nature of many early sites, their open beach locations, the evidence for early and widespread trade in materials like obsidians from a few quarries in the North Island, or baked argillite in the South Island, also suggest a highly mobile coastal and sea-oriented people employing such craft. The high frequencies of trolling lures and one-piece fishhooks in the earliest Archaic assemblages implies a continued reliance on offshore fishing, and, as in East Polynesia, would have required numbers of sea-worthy fishing vessels as part of the technique of exploiting these maritime resources. However, the lack of islands near enough to maintain a pattern of two-way voyaging, increasing warfare, changing methods for economic exploitation of aquatic environments and new patterns of social interaction between groups would increasingly have led to changes in the types of vessels employed. By the 18th century the pattern of Maori voyaging testifies to adaptation in isolation and as well to the limited role New Zealand's inhabitants would have played as donors to cultures in the rest of Polynesia. This lack of contact on the part of the Maori helped to maintain their isolation as receivers.

Archaic economic adaptation

Having established a case for the relative isolation of New Zealand, after its decisive primary settlement by what was probably a single voyage from East Polynesia, it is possible to outline some of the economic adaptations which occurred. As Yen has noted 'Polynesian colonization of New Zealand may be looked on as one endpoint in a migration that represents the area of most rigorous climate reached. The islands are presumably marginal for the unassisted horticulture of tropical species, and it is not too much to suppose they were so at initial contact' (Yen 1971a: 2). Also the North and South Islands were almost completely covered in heavy bush. The dominant trees belonged either to a podocarp or mixed podocarp-broadleaf subtropical rainforest or to a *Nothofagus* subarctic beach forest. Minor exceptions included areas of swamps, sand-dunes, and shingle riverbeds. A major exception was the eastern interior of the North Island where a succession of volcanic ash showers had destroyed the heavy vegetation leaving them in more open scrubland (Cumberland 1961: 139–40 and fig. 1; Simmons 1969a: 22–25). Adapted to life in the forest and forest fringe communities were six genera and over twenty species of flightless birds now collectively known as *moa*, this being the East Coast Maori term for them which was adopted by Europeans (SIMMONS 1968; HANSEN 1970). As well there was a wide range of other bush, water, and seashore birds, of which at least 10 species, plus all the moa, later became extinct. In the sea there were the usual resources of fish and shellfish, along with an unaccustomed abundance of sea mammals, especially seals. There was also an unaccustomed abundance of mudflat species of shellfish, of freshwater eels and crayfish, and of species of esturine and freshwater fish. Finally there was a wide variety of indigenous edible plants, many of them unknown elsewhere in Polynesia. In combination these factors all militated against the immediate re-establishment of the East Polynesian subsistence pattern in the new homeland. They favoured instead a segregation of its components and their realignment in ways which were more closely adapted to the new environment (YEN 1971a: 2). As a result, agriculture, on the one hand, was initially relegated to a minor role. On the other hand, greater reliance was placed on the somewhat richer naturally occurring land biota and maritime resources. This meant that hunting and gathering from the beginning achieved a prominence not possible in East Polynesia. In short the maritime portion of the East Polynesian economy, already highly developed, was retained and emphasized, a formerly rather unimportant hunting component was elaborated, and a gathering aspect was greatly expanded.

The results are reflected in the earliest Archaic sites. These contain the bones of most genera and a representative sample of the species of moa, numerous bones of bush and sea birds plus waterfowl. Usually included are a number of now extinct species. Bones of more important animals in the diet include seals and dogs along with occasional sea lions and porpoises. As well there are quantities of fishbones, especially of snapper in the North Island. These sites also contain non-concentrated deposits of shellfish, the earliest layers of sites on the northeast coast of New Zealand often featuring rocky-shore rather than sand beach or mudflat species where both were easily available. Perhaps this reflects an initial use of shellfish most similar in type and habitat to ones taken in East Polynesia, but it quickly gave way to exploitation of the abundant, readily gathered, and not so easily depleted supplies of sand beach or mudflat shellfish widespread throughout New Zealand. Thereafter variations in shellfish gathering depended on site location and the types of associated economic activities undertaken in conjunction with it (DAVIDSON 1964). The one identifiable trend has been a development of shellfishing from an unspecialized economic activity carried out in conjunction with many other economic pursuits, as in Polynesia, to a pattern of more concentrated seasonal collecting (and dumping) as a separate and important economic activity in its own right. It is worth noting that after leaving the extensive shellfish midden deposits associated with the prehistoric sequences of New Caledonia, Fiji, and Tonga, one does not strike with any frequency in Polynesia the pattern of numerous refuse dumps that are the result of intensive shellfish gathering until again reaching another continental island with extensive intertidal environments like New Zealand.

While it was once thought that the earliest inhabitants of New Zealand failed to introduce domesticated plants from the tropics, it was evident from the onset that they did manage to bring the dog and the rat, though not the pig or chicken. The dog, moreover, was an important item of food for the initial inhabitants, as well as being a possible source for clothing as among the later Maori (ALLO 1970).

As indicated above there are now sound theoretical reasons and a growing body of indirect evidence to indicate that the transfer of a limited number of East Polynesian cultigens to New Zealand was accomplished at the time of discovery. The food plants in question are: sweet potato or *kumara* (*Ipomoea batatas* (L.) Lam.); taro (*Colocasia antiquorum* (L.) Schott), yam (*Dioscorea* sp. probably *D. alata*), the introduced cabbage tree or *ti pore* (*Cordyline terminalis* (L.) Kunth), and the gourd (*Lagenaria siceraria* (Molina) Standl.). To these may be added the paper mulberry *(Brousonettia papyrifera)* used to make barkcloth. Because of the marginal climatic situation for tropical plants in New Zealand, the coconut, breadfruit, banana, and pandanus would not have survived if they were introduced.

The above list can be roughly ranked in order of increasing distribution in New Zealand, with all introduced cultigens present in the northern part of the North Island, especially along the eastern coast. The introduced cabbage tree, in contrast to the native form much favoured as food in the South Island, seems to have been confined to Northland. The paper mulberry and yam were observed growing in coastal areas as far as the East Cape and the taro as far south as the northern end of the South Island. The sweet potato and gourd extended down the South Island's east coast to just south of the Banks Peninsula. (LEWTHWAITE 1965: 17–18, LAW 1969). Because the settlements of the first inhabitants were widespread throughout New Zealand, these food plants even if successfully introduced in the most favourable parts of the North Island at the time of discovery, could have played only a minor and quite variable role in the economy of many of the more southern sites. The range would have been from none-at-all in the lower South Island and Chathams to one of slowly altering importance as one moves north. For this reason there could be no single or uniform early economy but only a series of economies, in each case closely adapted to the local environmental situation.

Variation in the early economies can be demonstrated in several other ways. For example there is the occurrence of the moa itself, which initially led to the first settlers in New Zealand being called Moahunters. It is now recognized that this is an inappropriate economic designation and it has been replaced by a culturally defined term-Archaic (see below).

Plate 3. Mr SYD JACKSON wearing a *korowai* cloak and holding a *maro* at the display of Maori clothing to HRH Princess ALEXANDRA at the Auckland Museum, 21.4.71. (Photo C. A. SCHOLLUM).

The shift to a cultural definition, based on recurrent assemblages of early artifact types derived from East Polynesia has permitted a closer assessment of the importance of the moa in early sites. The result is a growing body of data indicating that the industrial occurrence of moa, and its economic contribution to the diet varied considerably from site to site, including complete exclusion. It appears that in no site where the food resources have been analyzed, even in a preliminary fashion, has the moa been a mainstay in the diet, usually fish, seals, the dog or even shellfish being more important food resources in terms of meat weight. Moreover the importance of moa and bushbirds in progressively later layers of the same site, or in later Archaic sites decreases (SIMMONS 1968). Finally nothing like sufficient numbers of moa seem to have been killed and eaten by man to directly have brought about their extinction. Rather it is man's indirect role as an active agent in firing and forest clearance and as a new predator accompanied by the dog and rat that is important.

The decrease to extinction of moa and other birds may be explained by a combination of factors, and probably occurred earlier in the North than in the South Island. For the moa the factors seem to have included inherited genetic tendencies and their direct exploitation by man, but the most telling was probably an increasing restriction of the forest, especially those stands of podocarp and broadleaf species with berries which provided a majority of their food (SIMMONS 1968; 1969a: 24–25; H. LEACH 1969: 33). The reasons for the forest reduction are very controversial but were probably a result of cultural interference in conjunction with various natural processes.

In the South Island from the time of settlement in the 11th century A.D. or before to somewhere in the 15th century, the succession of plant cover involved increasing destruction and withdrawal of the podocarp forest from many of the coastal areas and much of the inland plains. Generally tussock grassland, often associated with *manuka (Leptospermum scoparium)* and under suitable conditions bracken fern, took its place. It was probably this restriction of forest that most seriously affected survival of the moa, reducing the populations to levels where man eventually accomplished its extinction. While the trend to grassland may have been a naturally occurring post-Pleistocene succession, perhaps assisted by further minor climatic changes, it was certainly accelerated by man, particularly in his use of fire (CUMBERLAND 1961: 141–42, 1962; MOLLOY 1969). The effect, however, was to render the inland plains of reduced economic importance to man (AMBROSE 1968), and to further restrict the potential role that the hunting of forest birds, including the moa, could play in the economy. Later Archaic sites in the South Island, as LOCKERBIE (1959), DUFF (1963), SIMMONS (1969a) and HIGHAM (1970) have all indicated, exhibit a variety of seasonal settlements, most of them coastal. Each was occupied at different times for a wide range of hunting, fishing, shellfishing, sealing, eeling, and gathering activities. The result was a quite varied range of Archaic phase economies, none of which can be characterized as moa hunting. Even where climatically possible, agricultural pursuits seem to have contributed little to these economies.

In the North Island the trend in the post-Pleistocene succession of plant cover was from podocarp to mixed podocarp-broadleaf and broadleaf forests in the lower lying and coastal areas. To some extent this natural change in forest type may perhaps have limited the size and distribution of populations of North Island moa and certain bush birds, although the archaeological record from 11th and 12th century sites indicates an initial situation remarkably similar to the South Island.

Also similar, though seemingly occurring with somewhat greater rapidity in the North Island, is the reduction in the 13th and 14th century sites of moa species and certain bush birds. While a reduction in forest is implied, the mechanism appears more subtle. It is far more likely that in the North Island the forest was progressively cleared from coastal areas for agricultural production and on occasion perhaps simply to increase supplies of fernroot. Unlike many parts of the South Island, after clearance of primary forest in the North Island, an area with soils suitable for initial agricultural use could usually be expected to produce extensive stands of bracken fern *(Pteridium esculentum)* in the following years. The underground stem (rhizome) of this crop, sometimes neglected in accounts of Maori economy, was often of great value as a food source as were any of the agricultural products (K. SHAWCROSS 1967). Where land was left fallow for sufficiently long it would have returned to forest. Usually, however, this did not happen, for unlike much of the South Island, once land in the agriculturally suitable parts of the North Island was cleared of heavy forest it was a valuable resource, which after a sufficient fallow could be returned to crops with far less labour (VAYDA 1960: 111–114). As MOLLOY (1969: 63) suggests, fire in the hands of man seems to have replaced climate as a major controlling factor of the vegetation over 50 to 75% of New Zealand during the last 1,000 years. For the South Island the 13th to 15th century A.D. change in the vegetation cover somewhat restricted forest hunting and with man assisted in the extinction of some of the avifauna including eventually the moa. In the North Island, forest clearance had much the same effect on the avifauna and on its exploitation by man, but the benefits derived were of far more importance to a changing economy in which fernroot and agricultural products played an increasingly important, though not necessarily dominant role. Again as in the South Island, the result was a series of Archaic economies which varied according to local differences in the environmental resources available for exploitation.

The evidence for agriculture in the 12th, 13th, 14th and mid-15th centuries is slowly accumulating in the North Island. Some is in the form of garden soils, some in the form of boundary walls around garden plots, while other parts consist of storage pits to overwinter the sweet potato for replanting or for its later consumption (GROUBE 1967, 1968: 142–43; LAW 1969; GREEN 1972b; B. F. & H. LEACH 1971). The occurrence of this evidence at a point in the prehistoric sequence when forest hunting and exploitation of sea mammals is diminishing, and the primary forest is being removed from coastal area, suggests that in contrast to the South Island, in some regions of the North Island later Archaic economies were evolving in the direction of economies characteristic of the 18th century Maori practices in that area.

ARCHAIC CULTURAL ADAPTATION

The early period in New Zealand prehistory is often referred to as the Archaic phase of New Zealand Eastern Polynesia Culture (GOLSON 1959). Many of the artifact types which characterize it were first recognized by SKINNER (1921) under the term Southern Culture (Map 5), because they persisted much later in the southern part of New Zealand and the Chatham Islands whereas they had been replaced by new items farther north. Evidence for the earlier chronological status of these recurrent assemblages throughout New Zealand was assembled by DUFF (1947, 1956) on the basis of their consistent association with the moa, his hypothesis being most effectively demonstrated by excavations at S29/7 on the end of Wairau Bar (Map 4). The characteristic artifact types encounered in these cultural assemblages, recognized from their repeated occurrence in sites which were palaeotologically, stratigraphically and chronometrically old, were listed by GOLSON (1959) when formally defining the cultural content of the Archaic phase. Typical items which he assigned to this phase are illustrated in figure 3b. A few will be discussed here as a means of describing various adaptive processes which have occurred in the transfer of specific items from East Polynesia to New Zealand.

The Polynesian techniques of trolling offshore for surface feeding fish, along with typical forms of the lure employed for this purpose, were transferred to New Zealand virtually intact. In fact at least one dorsal-ventrally perforated flat lure shank made in pearlshell was brought here from East Polynesia (GREEN 1967b), though it was not long before lures of similar form were reproduced in local New Zealand stone, bone and shell (fig. 3b). The most common Polynesian lure shanks, with bilaterally drilled holes in triangular sectioned heads (fig. 3a), were typically rendered in stone in New Zealand, although they retained an identical shape to the pearlshell prototypes (fig. 3b). As well oval to round sectioned lure shanks with grooves on the head end for the line attachment were manufactured in bone and stone. Unlike the first two forms, the last form lacks good Polynesian antecedents and may be a New Zealand innovation.

Plate 4. Miss KAREN RIKA wearing a *korowai* cloak with decorative sewn bands, at the display of Maori clothing to HRH Princess ALEXANDRA at the Auckland Museum, 21.4.71. (Photo C. A. SCHOLLUM).

Points for the lures, formerly in pearlshell, were in New Zealand usually made of bone.

Lures of these three forms seem to have persisted with some changes in frequency but few formal modifications for most of the Archaic phase. On the East Cape of the North Island a slightly modified version with eyes on each side survived to the time of European contact. Yet from the beginning Archaic lures had not only to be rendered in new materials, but also adapted to taking quite different fish from the bonito commonly caught on them in Polynesia. In the North Island and the northern part of the South Island they were used primarily for the common surface feeding sea trout, *Arripus trutta*, called by the Maori *kahawai*, and only secondarily for the less frequent barracouta. In the southern part of South Island the barracouta, *Thyrsites atun*, was a common surface feeder taken in the November to June season by means of lures. The sharp teeth, large size (up to five feet) and more voracious nature of the barracouta seems to have led to the southern lures being larger in size, more solid in make, and possessing firmer point lashings. It also led to the evolution in later Archaic levels of a distinctive 'barracouta lure' consisting of a plain bone point which was inserted in a hole fashioned in a wooden shank, rather than being lashed to the shank as in the Polynesian form of lure (HJARNØ 1967). In the southern South Island Archaic this form slowly replaced the typical Polynesian type of lure. It also appears likely that the change in fish being caught may in time have resulted in the New Zealand lures employing a single attachment to the line at the head of the shank, rather than usual Polynesian practice of taking the line back to the point with a secondary attachment at the head end of the shank (CROSBY 1966: 129–131).

The East Polynesian technique of line fishing employing pearlshell fishhooks was also transferred virtually intact to New Zealand, where it was used to take a wide range of sub-surface and bottom feeding fish, many of them new. In the process of the conversion from pearlshell to moa bone as the principal material of manufacture, the same hook shapes and features of line attachment and points were retained. However, because the grain at the base of the curved part of the shank in bone hooks normally paralleled lines of stress when in use, they easily and often broke at this point. As a consequence bone hooks in New Zealand usually had much thicker bases than their pearlshell counterparts, and two-piece simple fishhooks became more common.

The examples provided by these items of fishing gear furnish models common to many categories. Most forms were replicated in new materials, often with little typological change, even where new flora or fauna were being exploited. In time, however, some local innovations occurred, usually based on the Polynesian models. In some categories substitution in materials was accompanied by innovation in certain formal characteristics of the object, but there is no suggestion that the items themselves

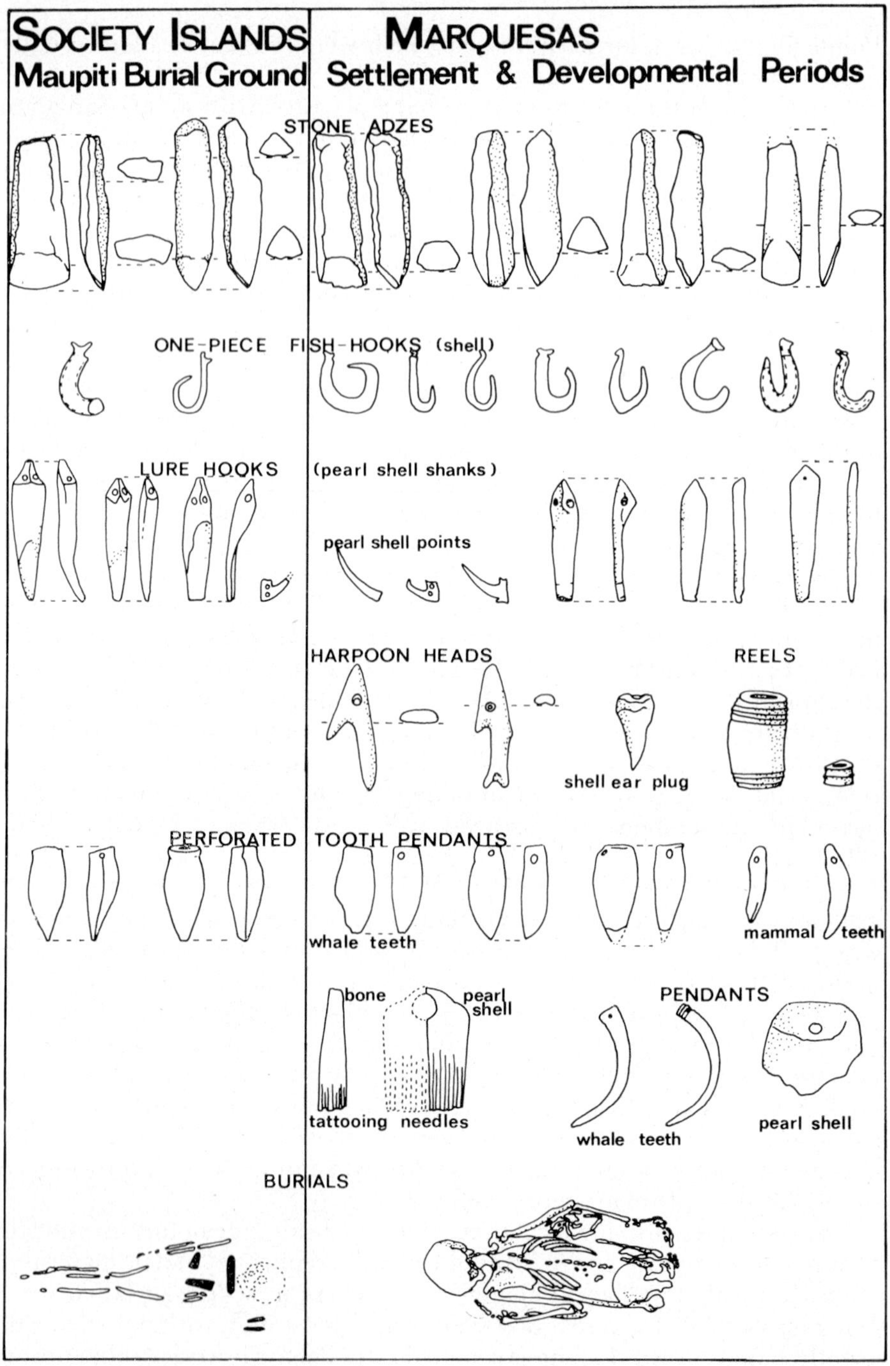

Fig. 3a. Early East Polynesian Cultural Assemblages From: the Society Islands Maupiti Burial Ground and the Marquesas Settlement and Developmental Periods.

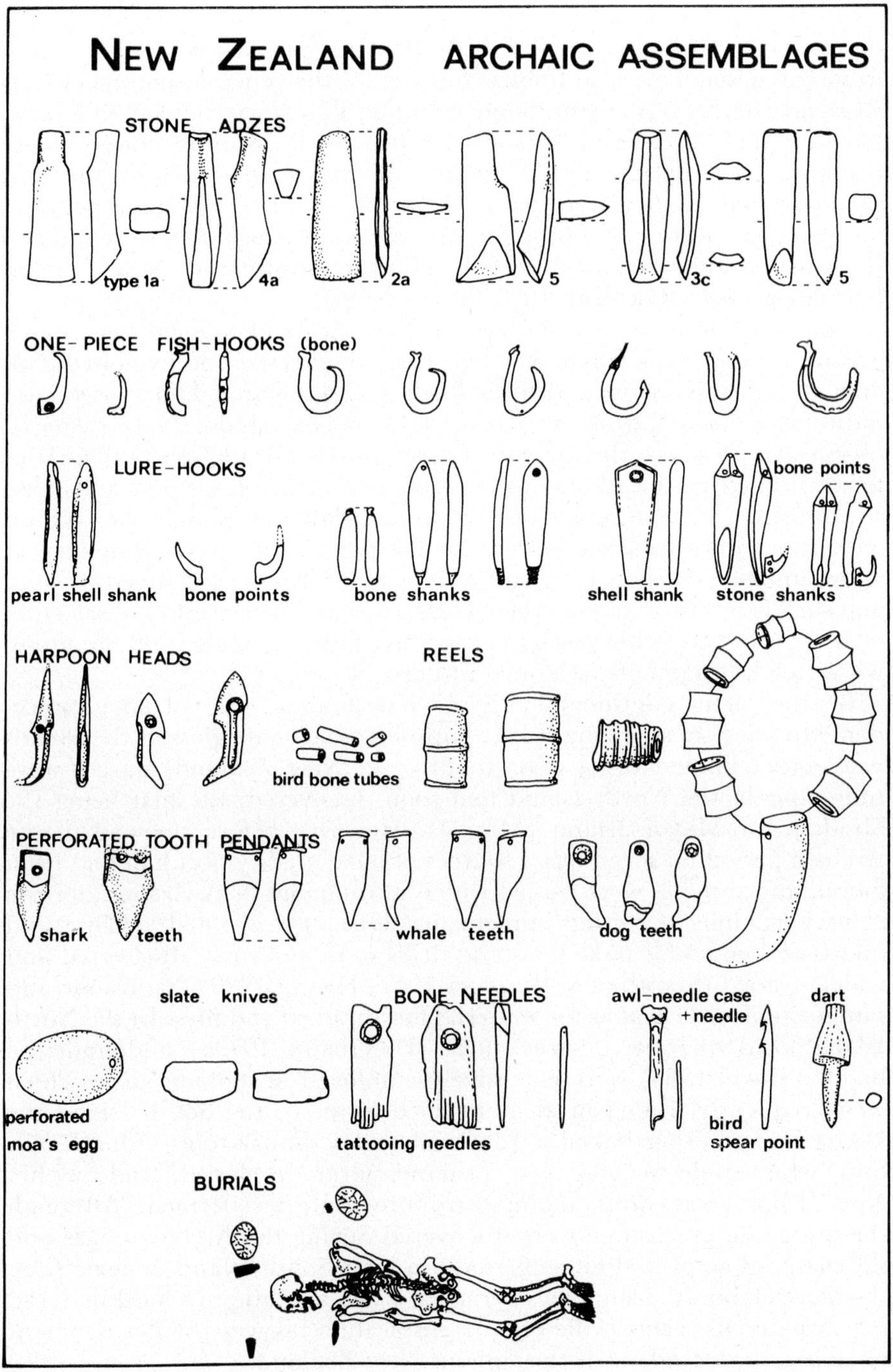

Fig. 3b. The New Zealand Archaic.

or the raw materials to make them were ever traded to New Zealand from Polynesia. This is additional support for the general isolation of New Zealand after its settlement. Some common Polynesian objects, of course, like the shell breadfruit scrapers or pearlshell coconut graters, were dropped from the inventory, but for climatic reasons so were the food plants on which they were used. Other objects, like the stone pounder for taro paste were never brought, apparently because they were innovations which appeared in East Polynesia after transfer to New Zealand had taken place (GROUBE 1968; SINOTO 1970).

The stone and bone resources in New Zealand, which were much greater than in East Polynesia, were early discovered and exploited. Not only did the living moa furnish food, but also large bones useful for industrial purposes. Moreover industrial moa bone could also be gathered from swamps where these creatures were often trapped naturally. After being limited in the islands to the bones of men, pigs, dogs, and some sea birds, the very large size and readily available supplies of moa, plus a wide range of other bones except for the pig, meant that a wide variety of ornaments as well as tools and fishing gear formerly fashioned in ivory and shell could be made in bone. The results are seen in a broad selection of awls, abraders, birdspears, harpoons, needles, reels, and imitation whale teeth ornaments fashioned in bone.

Because of its continental origin, New Zealand offered an unprecedented variety of stone materials. Supplies of volcanic glass in the islands were often limited or of poor quality; in New Zealand sources were numerous in the North Island and soon discovered, the first being the obsidian on Mayor Island (Map 4). However before the end of the Archaic period all seven other sources of flake quality obsidian had been found, and supplies were traded widely throughout New Zealand for use in various kinds of cutting implements (GREEN *et al* 1967b). Chert and quartzite sources for flake tools and drills were also early discovered and widely traded (SIMMONS & WRIGHT 1967; KEYES 1970). Sandstone and pumice replaced coral as the material for abraders and files. In the North Island local quarries of greywacke (DAVIDSON 1970a) and andesitic basalt (SHAW 1963) were extensively employed in making adzes which were frequently traded in an unfinished form. In the South Island the D'Urville and other baked argillites played a similar role (DUFF 1956). Still, while trade beyond New Zealand never developed, trade within New Zealand was common almost from the time of settlement. Although the sources of greenstone were discovered during the Archaic phase and the material employed for some tools in later South Island Archaic sites, the more elaborate technology required when working this hard material into ornaments seems to be of later inspiration, as was the development of an extensive trade in it throughout New Zealand.

In contrast to the later period, Archaic sites are overwhelmingly coastal in their distribution (Map 4). A few inland camps in rock shelters, caves or the open are known in both the North and South Islands but most of these exhibit very brief occupations, probably of a seasonal nature. It is on the coast, often in the dunes behind sand beaches or on bars and spits, that sites containing evidence for a number of activities are encountered. Usually reflected are the results of various economic endeavours like fishing and hunting, domestic concerns like food preparation and cooking, industrial practices like the manufacture of tools and ornaments, and occasionally even religious rites in the form of burials. As well there are the remains of structures in the form of postholes, pits, fireplaces, and ovens. It appears that while many Archaic sites were specialized around particular and perhaps short term seasonal activities, others were more permanent home bases repeatedly reoccupied over longer periods as their larger size, multiple layers, and depths of deposit attest. In short the far larger land mass of New Zealand, an initially small population and an economy in which agriculture played a restricted or nil role, all promoted a quite different pattern of settlement from Polynesia. Not only would the small population have restricted group size and the number of social units that could be formed, it would have also affected the range of settlement types possible. While the introduced plants in northern New Zealand may have required fixed home bases from the beginning, the need for mobility in exploiting the resources of the environment generally militated against permanent settlements or the establishment of domestic structures designed for continuous occupation. It is perhaps no surprise that the remains of houses have been so difficult to find when a tradition of permanent residence, as in tropical Polynesia, was not maintained.

In this respect it should be noted that New Zealand also remained cut off from all later developments of the elaborate *marae* or religious structures of East Polynesia. The Polynesian-wide ancestral pattern of upright position stones on an open *marae* may be reflected in the few religious complexes or *tu ahu* of the Maori of which we have knowledge. No religious structures at all have been recovered from certain Archaic contexts.

Archaic phases and periods

The approximate age of the Archaic phase in both the North and South Islands has been determined by radiocarbon dating. In the North Island associations with pumice ash showers and the thickness of hydration rims on obsidian flakes have also been employed as an independent means of relative dating. The radiocarbon age results have been summarized by W. Shawcross (1969) who concludes 'If attention is concentrated on the zone in which the majority of dates are clustered, it will be found that almost all of the sites which may be identified as early on artifactual and palaeontological grounds were first occupied during the twelfth century A.D. With the recent availability of determinations for the Mt.

Plate 5. Mr. Rauru Kirikiri wearing a *kiwi* feather cloak, at the display of Maori clothing to HRH Princess Alexandra, at the Auckland Museum, 21.4.71. (Photo C. A. Schollum).

Plate 6. Mr. ALBERT PRYOR, wearing a *korowai* cloak at the display of Maori clothing to HRH Princess ALEXANDRA, Auckland Museum, 24.1.71. (Photo C. A. SCHOLLUM).

Camel site, Houhora [N 6/4] in the far north, it is now possible to recognize a pattern of a dozen or so early settlements strung throughout the whole length of the country – a length of a thousand or so miles.' Reasonably well supported dates in the 11th century A.D. are known for North Island sites like Tairua, N 44/2, (GREEN 1967b: 83) and for South Island ones like Kaikai's Beach, N 164/17 (HJARNØ 1967: 42) and OTUREHUA, S 134/1 (F. LEACH 1969: 72). The New Zealand-wide coastal distribution of sites dating to 12th century A.D., plus a few sites in the 11th century which are also geographically well separated, suggests that settlement of New Zealand began before 1,000 A.D. How much before, it is difficult to say as two 9th century A.D. dates, a northern one for the Moturua Island garden (GROUBE 1968) and a southern one for the Redcliff's Sand Dune (TROTTER 1967) have not been supported by subsequent investigation. However, most would agree with GROUBE (1968: 145) that a date for settlement of New Zealand can be set with some assurance several centuries before 1,000 A.D.

In the South Island an Archaic tradition in artifacts persisted throughout the area until the 15th century A.D. and lasted even later in the southern part of the island and on the west coast (LOCKERBIE 1959; WILKES & SCARLETT 1967). LOCKERBIE (1959) was the first to divide the southern Archaic into three periods largely on the basis of economic changes, although he also took into account correlated changes in artifact frequencies. HJARNØ (1967) has since identified early and late periods in Archaic phase fishing gear, and SIMMONS (1967) has shown that parallel changes occur in adzes as well as the rocks in which they are manufactured. Similar changes were noted in bird spears and ornaments. F. LEACH'S (1969) analysis of the flake tool tradition has demonstrated that blade making, an early and widespread southern South Island practice, exhibited an accelerating decline which reached its peak in the 14th century A.D. In the Canterbury-Marlborough area DUFF (1963) has employed a Residual phase, following his Moa-hunter, to characterize the later part of GOLSON'S Archaic in that area.

In the North Island GREEN (1963) proposed Settlement, Developmental, and Experimental phases for assemblages of New Zealand Eastern Polynesian culture in the Auckland Province. Distinctions between phases were based on a number of criteria including a series of smallscale successive changes in ecological orientation, economy, settlement type. The divisions have not been much employed or developed by subsequent workers who have remained content with criticisms of the evolutionary basis of its theoretical structure (GOLSON 1965; GROUBE 1967). The distinctions made there between earlier and later Archaic sites appear valid, although the criteria for an Experimental phase distinct from a Developmental one has not proved viable. The important new artifacts which can be associated with later Archaic phase sites in the North Island are largely structural in form and consist of pits for food storage, garden soils, and garden boundary walls. The early North Island Archaic presents a slowly changing cultural continuum very similar to that in the South Island. From the 14th century A.D. on, however, the North Island Archaic exhibits changes different from those in the South Island. Presumably in the 15th and 16th centuries these changes quickened in the direction of the later Maori culture. It is precisely this period of change which is most poorly documented archaeologically. As a result the initial steps are virtually completed by the 17th century when the archaeological record again becomes more satisfactory.

New Zealand Maori Culture

Background

The later development of Maori culture in New Zealand is marked by a great deal of regional diversity in linguistic, economic, technological, social, political and artistic spheres. Skinner (1921) early emphasized this point when he used such criteria to divide New Zealand into eight culture areas (Map 5). The diversity has been reinforced by later studies of Maori art styles (McEwen 1966), artifact distributions (Crosby 1966; Simmons 1969a, fig. 1) and economic endeavours (Davidson 1964; H. Leach 1969). It is also evident in the archaeological record of this period in the Auckland Province (Green 1963) and has often been commented on with respect to items like pits and defensive earthworks on fortified sites, not to mention the range of site types themselves. Many of these differences have historical rather than ecological processes as a major part of their explanation.

One ecological factor, however, had overwhelming significance. This was the potential of the northern part of the North Island to support an annual population increase of not more than 1 to 1.5% (W. Shawcross 1969; Groube 1970) by further developing certain of those economic and technological practices already present in that region's Archaic phase. The greater potential in food resources of this region resulted from favourable combinations of warmer climate, more fertile soils, and extensive inter-tidal harbours, factors which were either greatly reduced or lacking in regions farther south (W. Shawcross 1970, esp. fig. 1; Gorbey 1970). As a result, the evolution of a distinctive culture called Maori by Duff (1956) or a new phase termed Classic Maori by Golson (1959) occurred in that part of the North Island where general cultural development in New Zealand also reached its climax. Moreover, because this climax was achieved in isolation, it was unlike any encountered elsewhere in Polynesia. As a consequence it is not easily explained by influences from outside New Zealand.

The broad differences in climate, soils, and food resources in New Zealand allowed two geographers, Cumberland (1949) and Lewth-

Plate 7. An old Maori woman from the Te Teko area – she is one of the few women with a moko (traditional tatoo). (Photo C. A. Schollum).

Plate 8. A Maori warrior 'Challenging' the guests of honour on their arrival at the Te Tokanganui-a-noho meeting house during the centenary celebrations, 1972. (Photo C. A. Schollum).

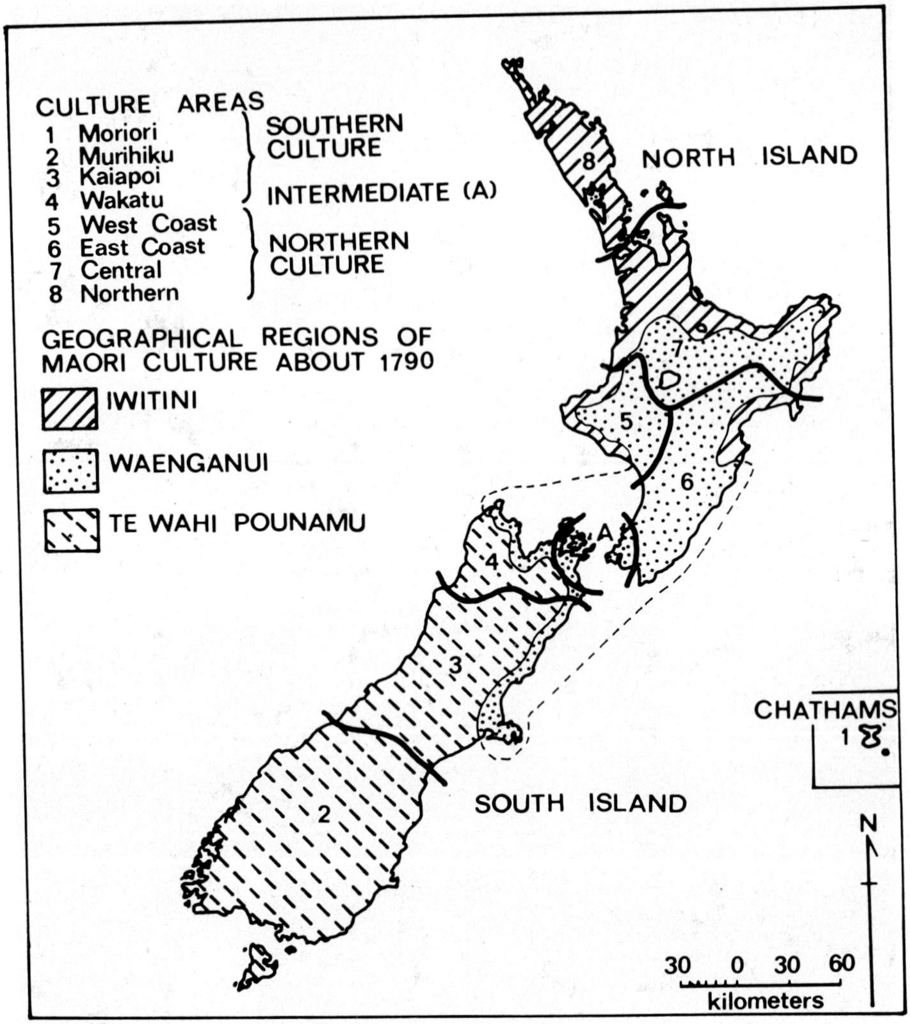

Map 5. Cultural and Geographical Diversity in New Zealand Maori Culture in the 18th Century (see text for details).

WAITE (1949), to formulate three regions which possessed a close relationship to the relative density of late 18th century Maori population and its related cultural development (Map 5). Closer examination of their work (GORBEY 1970) has shown that while it is possible to use density and proximity of fortified sites to arrive at quite reasonable estimates of relative population distribution, deduction of population numbers from fortified sites is extremely difficult. In this respect the estimates of the geographers are over generous and their figure of 250,000 (LEWTHWAITE 1950, CUMBERLAND 1965) is more than double those of the archaeologists. Among the archaeologists, GROUBE (1970) has used the numbers of *pa* or fortified sites to arrive at a population estimate of 100,000, while W. SHAWCROSS (1970) has employed the resources of harbour oriented and interior economies to calculate the bulk of New Zealand's prehistoric population at approximately 106,200 $\pm$ 22,300. These figures are in substantial agreement with the first historical estimate by the Forsters, often attributed to Captain James Cook, of no more than 100,000 and the results of a close historical study of all pre-censal records by POOL (1964: Ch. 2) which indicated a pre-European population of 115,000.

GORBEY's (1970) detailed study of the distribution of over 3,100 *pa* sites has verified that the region of Iwitini as defined by the geographers is a valid indication of the extent of the most populous part of prehistoric New Zealand during the late 18th century. The region may well have contained somewhere in the neighbourhood of 80% of the population estimated by CUMBERLAND (1949) because its food resources were varied and abundant, because it provided access to multiple environments within a restricted locality and because its soils and climate were most favourable to sweet potato and fernroot production. Waenganui was a region with a colder climate, less fertile soils, a more rugged landscape, fewer resources, lower agricultural productivity, and restricted access to the coast over much of the region except in the South Island. All contributed to a smaller population which Cumberland estimates at some 15% of the total. The third area, Te Wahi Pounamu, was considered the least favourable, except as a source for greenstone, and its population was estimated as no more than 3 or 4% of the total (CUMBERLAND 1965).

MAORI ECONOMIC ADAPTATION

In a recent survey of the literature on Maori economy H. LEACH (1969: 20) found that the concept of a single staple vegetable food had little relevance. Rather, the viewpoint adopted is that of COLENSO, an acute naturalist and a close observer of mid-19th century Maori use of the native fauna and flora:

'Their means of obtaining subsistence were as varied as the things themselves. They were not . . . deficient in food; although . . . what they had and used was not to be obtained without a large amount of daily labour. At the same time, there must have been a great difference in the food of the natives of the Northern and of the Southern and Stewart Island;' (COLENSO 1868: 345).

Still in view of the great variation encountered in the Archaic economy discussed above, it is perhaps not too surprising that similar variations also appear to have held for the Maori.

The internal and regional variation in the early historic economies of the Maori of the East and West coast portions of Te Wahi Pounamu have been explored in detail. In current taxonomies all would be classed as hunter-gatherers as would their Archaic predecessors in the area. Yet just as the southern Archaic economies had changed with the altering environment, so the historic economies possessed important regional differences closely related to the now altered distribution of key resources in the various micro-environments. Thus 'the inhabitants of South Canterbury consumed quantities of fresh-water fish and cabbage tree roots; the Otago natives lived primarily on fernroot, birds and seafish; the Westland Maori ate mostly freshwater fish and preserved birds and their diet included less vegetable matter than in the east' (H. LEACH 1969: 75).

Moving north to Waenganui and Iwitini, other food sources, not present farther south are encountered. The differential distributions of the various domestic food plants, already described for the Archaic phase above, can also be applied to the Maori. Interwoven with them are still other distributions of indigenous plants like fern or important berry trees like the *karaka (Corynocarpus laevigatus)* which occurred only in coastal areas and *hinau (Elaeocarpus dentatus)* which was not found south of the sparse forests of mid-Canterbury (H. LEACH 1969).

Unfortunately only in a few parts of the North Island have detailed regional examinations of Maori economy been carried out. For the Waenganui region an example is BEST's (1902) account of the late 19th century Tuhoe tribe of the inland Urewera mountains, where the climate and topography prohibited all domestic food plants, restricted fernroot production, and limited fish to a small freshwater species. In this region berries, especially of the *hinau*, birds, rats, and greens were prominant items in the diet. At the opposite extreme are the early historic accounts of the Maori economy of the Bay of Islands, one of the most favourable areas in Iwitini (K. SHAWCROSS 1966; 1967: Table I; KENNEDY 1969). Coastal communities in this locality were fairly stable, permanently occupied fishing villages, which possessed extensive gardens nearby where most of the known food plants were or could be grown. Despite this, fernroot seems to have been a major food resource of the region, more important than the sweet potato (K. SHAWCROSS 1967). Forest foods on the other hand played an insignificant role, berries of the *karaka* being favoured and those of the *hinau* neglected.

In summary typical characterizations of Maori prehistoric economic systems which assign them to a single general agricultural category are misleading. Nor is the situation much improved by stressing the contrast between a category of agriculturalists in northern New Zealand and hunters and gatherers in the southern part. Rather it appears Maori economy, like that of the Archaic, exhibited an environmentally predictable regional diversity. The economies ranged from hunting and gathering with techniques for the preservation and storage of natural food products to semi-agricultural with the preservation and storage of both natural and garden products (H. LEACH 1969).

Artifact assemblages of the Maori are frequently distinguished by a few commonly associated portable artifacts, particularly certain adze and fishhook types (GOLSON 1959). They are also identified by the addition of new items not present in the Archaic phase. Among those recovered in archaeological contexts are greenstone ornaments in a variety of styles and stone and bone fighting clubs of the *patu* class. As well there are a large number of wooden objects such as combs, weapons, and carvings, which have been recovered only under special conditions of preservation, usually in swamps and lakes (W. SHAWCROSS 1964; BELLWOOD 1971b).

In many cases the origin of or explanation for the adoption of these items is not well understood. Some of the items replaced have Archaic prototypes; others are seemingly unique innovations. Still others involve shifts in the statistical proportions with which different types occur. For example an unspecialized plain quadrangular adze, found in the Archaic, was the typical form in most Maori cultural assemblages (GOLSON 1959, fig. 14). It has a distribution very similar to that for sweet potato crops, for pits in which to store the tubers, and for defensive earthworks. A likely interpretation is that many of these adzes were employed in clearing and working gardens and for digging pits and ditches rather than solely as woodworking tools (DAVIDSON 1971a: 32). In the same way the numbers of adzes and chisels fashioned in the more durable greenstone may reflect the increasing importance of Maori woodcarving. These suggestions, however, have yet to receive empirical verification from studies indicating different kinds of damage which might be expected on their cutting edges.

Projected against those changes in Archaic trolling lures already discussed, the *kahawai* lure of the Maori, employing a wooden shank with a *haliotis* shell inlay, may be singled out as a further and probably later northern New Zealand innovation comparable to the barracouta lure farther south. Also in many sites where fishing remained an important activity, there is a noticeable decline in the numbers of one-piece fishhooks. This has led to the suggestion that net fishing had become the favoured technique (HJARNØ 1967; W. SHAWCROSS 1967).

The emergence of weapons as an important class of artifacts is almost certainly correlated with the growing importance of warfare. The growth of this activity is also reflected in the large number of fortified sites which in large part belong to the Maori cultural stage. Given warfare as a culturally inherited Polynesian pattern for settling differences, it is doubtful that its practice in the Archaic was unknown despite the limited amount of direct evidence. Its emergence as a major activity, however, has as part of its explanation the increasing population of northern part of New Zealand, the closer proximity in which small groups lived, and the numerous opportunities this presented for frictions to arise in the course of normal social interchanges. Another factor may lie within the greater competition for and added pressure on food resources.

Groube (1970) has shown that it was not a simple case of economic pressure on scarce agricultural land or diminished access to other natural resources because such pressure had not reached sufficiently critical proportions to have induced intensive warfare when the evidence for it first appears. He argues therefore that the crucial factor may have been the speed with which the Maori in the most favourable areas of Iwitini could convert rain forest into land suitable for agriculture and fernroot production. While the total supply of land and other resources was sufficient for the large and growing population and many suitable areas were never fully utilized (Gorbey 1970), once the easy environments were occupied, it was necessary to clear and pioneer localities which were ever more difficult. In a growing population, warfare forced the excess from prime lands into exile. Their survival required clearing new lands and pioneering the more marginal environments. Simmons (1969a) and Groube (1970) see this process as setting in train a series of internal migrations from Northland into areas further south, particularly Taranaki

Fig. 4. Reconstruction of period III fighting stages at Otakanini, South Kaipara Harbour. (Courtesy P. Bellwood 1971).

and the Bay of Plenty. These movements they believe are what is reflected in the so-called 'Maori fleet' migrations, the 'fleet' itself being an invalid European concept derived from some twelve different ancestral canoe traditions which are found among many North Island Maori tribes. Their hypothesis is controversial, but has some merit as a contrast to the interpretation of these traditions as representing the coming of the Maori to New Zealand as part of a later migration.

Whatever the origin of fortified sites, they are a dominant structural artifact of Maori culture. The earliest known defensive earthworks and palisades are 14th century in date (Bellwood 1971b); the majority of excavated sites in the Waikato, Tauranga, Kaipara (fig. 4) and Auckland Isthmus area date to the 15th century A.D. or after and are usually associated with small assemblages of portable artifacts belonging to Maori culture.

While classified under a single Maori term, *pa*, such sites served a variety of functions and reflect several different kinds of settlement. It is important to stress that they were not all permanently occupied fortified villages as so many older works affirm. Certainly in the ecologically very favourable environment of the coastal Bay of Islands continuously inhabitated fortified and unfortified villages with fluctuating populations had evolved. They provide a contrast both with many earlier Archaic sites and with the typical dispersed settlement pattern noted above for many other Polynesian island groups at the time of contact. Other regularly inhabited fortified villages called swamp *pa* occur in and around the freshwater lakes of inland Waikato, where it appears the sites were utilised throughout the year, although the population in residence may have fluctuated in number. In other areas settlement was of an even more flexible type centered around a *pa* or an unfortified home base, or possibly a number of *pa*, which were inhabited seasonally (the defenses and structures being permanent) together with a number of hamlets and similar seasonal settlements employed for various economic purposes. In some areas the largest *pa* may have served as citadels into which kin-related occupants from a number of adjacent *pa* could withdraw at times of political stress (Groube 1964, 1965; Gorbey 1970).

Pits for the storage of sweet potato have been encountered on most fortified sites that have been recorded or excavated. As well, undefended pit complexes are well known, and where satisfactorily recorded, are often more numerous than the fortified sites (Davidson 1970a, b, 1971b). Some of these undefended pit complexes also possessed sufficient evidence in the form of postholes of small huts, ovens, stone tools and domestic rubbish to qualify as undefended settlements or hamlets (Davidson 1970b, 1970c, Leahy 1970). It has also become apparent from excavations that *pa* sites during some periods were in fact not settlements, but defended food stores with little evidence of occupation (Ambrose 1967; Gorbey 1970). The distribution of prehistoric storage pits has been shown to correlate closely with those areas where sweet potato agriculture was possible, especially in the critical area of the South Island (Groube 1965, Law 1969). Not only was their adaptation to the storage of sweet potato in New Zealand an important environmentally based innovation, their distribution was also closely related to biogeographic factors. So it appears were some of the types such as the raised rimmed pit of the central and southern east coast of the North Island (Groube 1965).

Maori culture phases and periods

If the definition of phases or periods within the Archaic phase seemed beset with difficulties, those for Maori are even more so. A distinction can be drawn in many parts of Iwitini between a Classic Maori phase and an Early European Maori phase along the lines suggested by Green (1963, 1970b). The latter would date from sometime after 1810 to the 1840's. It resulted from the impact of continuous contact and trade with Europeans on the still dominant Maori culture. An intrusive Classic Maori culture from the north dating to the late 18th century A.D. can also be recognized in the Otago region (Hjarnø 1967; Simmons 1969). However, satisfactory distinctions between earlier and later stages of Classic Maori culture have still to be developed, although the outlines are evident in local sequences on Motutapu (Davidson 1970a, 1970b, 1971c) and Kauri Point (Green 1963, Shawcross 1964, Ambrose 1967). Beyond this all one can say is that elements of Classic Maori culture are in evidence in numbers of excavated sites by the 17th century A.D., and some elements may well be of earlier age in parts of Iwitini.

Summary

The Asian origins of the Maori ultimately become lost in the mists of an antiquity still unexplored by modern archaeology. Only in general can the source of the languages, populations and cultural complexes ancestral to the Maori be traced to Islands and Mainland of Southeast Asia. However, the Oceanic origins are better known. Linguistically Maori can be traced to one branch of the Oceanic block of Austronesian languages centered in Melanesia. Archaeologically their culture can be traced to the Lapita complex of the same area.

In the last decade the proto-languages, earlier cultural assemblages and ancestral populations of the Polynesians have been demonstrated to lie in the neighbouring island groups of eastern Melanesia, especially Fiji. People from this area, after arrival in Tonga, moved out into the islands of West Polynesia where they formed their own distinctive languages, physical type and culture, all of which we now call Polynesian. The Maori, it has long been known, derive from those Polynesians who first settled East Polynesia. This is well supported by recent linguistic, archaeological and physical anthropological studies.

Given this background it is feasible to sketch the general cultural orientation which the initial colonizers of New Zealand possessed when they first arrived in New Zealand. From the earliest New Zealand evidence it is then possible to set out the numerous cultural adaptations which their transfer to the non-tropical environment of New Zealand required.

It is also necessary to examine the degree to which further changes in New Zealand took place in isolation. In particular this requires examination of a long held belief that the 'Coming of the Maori' represents a second migration to New Zealand, a migration which in the popular literature is often referred to as the 'fleet'. Inquiry into this question reveals that neither the concept of the 'fleet' nor evidence of significant later contact with Polynesians or peoples of other areas can be established for any point in the New Zealand prehistoric sequence until the coming of the European. Further changes in Maori culture after the period of settlement are therefore interpreted as adaptive innovations in isolation. Many of these are influenced by the quite different environmental circumstances encountered in New Zealand in contrast to tropical Polynesia.

Within a context of general isolation changes in the economy and artifactual content of the two major prehistoric cultures recognized in New Zealand are examined. The earlier culture, New Zealand Eastern Polynesia, is often called the Archaic phase, though the term Moa-hunter is still in use. Its cultural content, while adapted to New Zealand conditions, is recognizably of East Polynesian origin. Its economy, however, is varied and a response to the diversity of New Zealand's environment. The later culture, Maori, is a northern North Island development in the ecologically most favourable area of New Zealand for population growth. It is based on adequate and expandable food resources, particularly sweet potato and fernroot. The development of this culture to a prehistoric climax, usually called Classic Maori, appears to coincide with an increase in warfare. It also correlates with changes in the cultural content and economy of the population, particularly in the North Island of New Zealand. Some of these changes have biogeographic factors as a part of their explanation and they are described. Others have various historical explanations which are not considered in the context of a book on biogeography and ecology.

Conclusion

This review of adaptation and change in the culture of the prehistoric inhabitants of New Zealand has as its theme the response in isolation of a Polynesian culture to a non-tropical environment. These issues have been

Plate 9. Centennial Marae Commitee, Te Tokanga-Nui-A-Noho Meeting House, Te Kuiti, 1972. Photo C.A. SCHOLLUM

examined at the expense of other and equally interesting functional and historical explanations. In a book largely concerned with biogeography and ecology this seems a suitable viewpoint to have taken, but the reader is warned that it leaves many other facets of change in New Zealand prehistory completely unexplored.

The argument is developed that only by tracing the origin of the Maori to their Polynesian and Oceanic predecessors, is it possible to specify those changes in their culture which colonization of New Zealand entailed. While some of these changes were immediate and fundamental, change did not cease after the period of initial adaptation. Rather the diversity of micro-environments in New Zealand promoted further developments and varied responses. For the most part these took place in isolation, so that outside influences have been assigned a minimal role once New Zealand was settled.

Because of substantial differences in biogeographic factors like climate, soils, and potential food resources, only one region of New Zealand, Iwitini, was able to support a rapidly increasing population. As a result a cultural climax in the form of Classic and Early European Maori culture was reached in the northern part of the North Island of New Zealand which was in many ways markedly different from its Archaic or Polynesia antecedents or any contemporary Polynesian society in the tropical Pacific. Its genesis lay in the genius of the Maori people to innovate successfully in ways that were both culturally and environmentally adaptive. Because of this ability they remain a vigorous and important element of New Zealand's multi-racial society today, (Plate 9).

REFERENCES

Allo, Jan L. 1970. The Maori Dog. Unpublished M. A. Thesis, Dept. of Anthropology, University of Auckland, Auckland.

Ambrose, W. 1967. Excavations at Kauri Point. Paper presented to the N.Z. Archaeological Assoc. Conference, New Plymouth.

Ambrose, W. 1968. The unimportance of the inland plains in South Island prehistory. *Mankind* 6 (11): 585–593.

Anell, B. 1955. Contribution to the history of fishing in the Southern Seas. Studia Ethnographica Upsaliensia 9. Uppsala.

Barrau, J. 1965. Histoire et préhistoire horticoles de l'Océanie tropicale. *J. Soc. Océanistes* 21: 55–78.

Bathgate, M. 1969. Maori River and Ocean going craft in Southern New Zealand. *J. Polynesian Soc.* 78 (3): 344–377.

Bellwood, P. 1970. Dispersal Centers in East Polynesia, with Special Reference to the Society and Marquesas Islands in Green & Kelly (eds.), 1970: 93–104.

Bellwood, P. 1971a. Varieties of Ecological Adaptation in the Southern Cook Islands. *Archaeology and Physical Anthropology in Oceania* 6 (2): 145–169.

Bellwood, P. 1971b. Fortifications and Economy in Prehistoric New Zealand. *Proc. of the Prehistoric Soc. for 1971* 37: 56–95.

Best, E. 1902. Food products of Tuhoeland: Being Notes on the food-supplies of a non-agricultural tribe of the Natives of N.Z. . . . *Trans. Proc. N.Z. Inst.* 35: 45–111.

Best, E. 1925. The Maori Canoe. Dominion Mus. Bull. 7 Wellington, Gov. Printer.

Buck, P. (Te Rangi Hiroa) 1949. The Coming of the Maori, Wellington, Whitcombe and Tombs Ltd.

Colenso, W. 1868. On the Maori Races of New Zealand. *Trans. Proc. N.Z. Inst.* 1: 339–424.

Crosby, Eleanor B. V. 1966. Maori Fishing Gear. Unpublished M. A. Thesis. Dept. of Anthropology, Univ. of Auckland, Auckland.

Cumberland, K. B. 1949. Aotearoa Maori: New Zealand about 1780. *Geographical Review* 39: 401–424.

Cumberland, K. B. 1961. Man in Nature in New Zealand. *N.Z. Geographer* 17: 137–154.

Cumberland, K. B. 1962. 'Climatic Change' or Cultural Interference? New Zealand in Moa-Hunter Times, in Land and Livelihood. Murray McCaskill (ed.), Christchurch, N.Z. Geographical Soc. pp. 88–142.

Cumberland, K. B. 1965. The Maori. New Zealand Topical Geographies 1b. Christchurch, Whitcombe and Tombs.

Davidson, Janet M. 1964. The Physical Analysis of Refuse in New Zealand Archaeological Sites, Unpublished M. A. Thesis, Dept. of Anthropology, Univ. of Auckland, Auckland.

Davidson, Janet M. 1970a. Survey of Archaeological Sites on Motutapu Island, New Zealand. *Rec. Auckland Inst. Mus.* 7: 1–12.

Davidson, Janet M. 1970b. Excavation of an 'undefended' site N38/37, on Motutapu Island, New Zealand. *Rec. Auckland Inst. Mus.* 7: 31–60.

Davidson, Janet M. 1970c. Salvage excavations at Hamlin's Hill, N42/137, Auckland, New Zealand. *Rec. Auckland Inst. Mus.* 7: 105–122.

Davidson, Janet M. 1971a. Early Man in New Zealand. *New Zealand's Heritage* 1 (2): 29–32.

Davidson, Janet M. 1971b. Te Paki Archaeological Survey. Dept. of Lands and Survey (mimeographed, with map and appendix of sites) Auckland.

Davidson, Janet M. 1971c. Field work on Motutapu Island, Hauraki Gulf. *N.Z. Archaeological Assoc. Newsletter* 14 (3): 80–81.

Duff, R. S. 1947. The Evolution of Native Culture in New Zealand; Moa-hunters, Morioris, Maoris. *Mankind* 3 (10): 281–291, (11): 313–322.

Duff, R. S. 1956. The Moa-hunter Period of Maori Culture, (2nd ed.) Wellington, Gov. Printer.

Duff, R. S. 1963. Aspects of the Cultural Succession in Canterbury-Marlborough, with wider reference to the New Zealand area. *Trans. Roy. Soc. N.Z., General Series* 1 (4): 27–37.

Emory, K. P. 1959. Origin of the Hawaiians. *J. Polynesian Soc.* 68 (1): 29–35.

Emory, K. P. & Y. H. Sinoto 1964. Eastern Polynesian burials at Maupiti. *J. Polynesian Soc.* 73 (2): 143–160.

Freeman, J. D. & W. R. Geddes (eds.), 1959. Anthropology in the South Seas. New Plymouth, Thomas Avery & Sons Ltd.

Golson, J. 1959. Culture Change in prehistoric New Zealand, in Freeman & Geddes (eds.), 1959: 29–74.

Golson, J. 1965. Some considerations of the role of theory in New Zealand archaeology. *N.Z. Archaeological Assoc. Newsletter* 8 (2): 72–92.

Golson, J. & P. W. Gathercole, 1962. The Last Decade in New Zealand Archaeology. *Antiquity* 36: 168–174 Part I, 271–278 Part II.

GORBEY, K. C. 1970. Pa distribution in New Zealand. Unpublished M. A. Thesis. Dept. of Anthropology, Univ. of Auckland, Auckland.
GORMAN, C. F. 1971. The Hoabinhian and after: subsistence patterns in Southeast Asia during the late Pleistocene and early Recent periods. *World Archaeology* 2 (3): 300–320.
GREEN, R. C. 1963. A Review of the prehistoric sequence in the Auckland Province. N.Z. Archaeological Association No. 2. Auckland, Univ. of Auckland Bindery.
GREEN, R. C. 1966. Linguistic subgrouping within Polynesia: the implications for prehistoric settlement. *J. Polynesian Soc.* 75 (1): 6–33.
GREEN, R. C. 1967a. The immediate origins of the Polynesians, in Polynesian Culture History. G. A. HIGHLANDS, R. W. FORCE, A. HOWARD, M. KELLY, Y. H. SINOTO (eds.), B. P. Bishop Mus. Spec. Pub. 56, Honolulu pp. 215–40.
GREEN, R. C. 1967b. Sources of New Zealand's East Polynesian Culture: the evidence of a pearl shell lure shank. *Archaeology and Physical Anthropology in Oceania* 2 (2): 81–90.
GREEN, R. C. 1968. West Polynesian prehistory, in YAWATA & SINOTO (eds.), 1968: 99–109.
GREEN, R. C. 1970a. Settlement Pattern Archaeology in Polynesia, in GREEN & KELLY (eds.), 1970: 13–32.
GREEN, R. C. 1970b. A Review of the prehistoric sequence in the Auckland Province. (2nd ed.) Dunedin, University Book Shop Ltd.
GREEN, R. C. 1972a. Aspects of the Neolithic in Oceania: Polynesia, in Early Chinese Art and its Possible Influence in the Pacific Basin N. BARNARD (ed.), New York, Intercultural Arts Press.
GREEN, R. C. 1972b. Moa-hunters, Agriculture and Changing Analogies in New Zealand Prehistory. *N.Z. Archaeological Assoc. Newsletter*, 15 (1): 16–39.
GREEN, R. C., KAYE GREEN, R. A. RAPPAPORT, ANN RAPPAPORT & JANET M. DAVIDSON, 1967a. Archaeology on the Island of Mo'orea, French Polynesia. Anthropology Papers Am. Mus. Nat. Hist. 51 (2) New York.
GREEN, R. C., R. R. BROOKS, R. D. REEVES, 1967b. Characterization of New Zealand Obsidians by Emission Spectroscopy. *N.Z. J. of Sci.* 10 (3): 675–682.
GREEN, R. C. & JANET M. DAVIDSON (eds.), 1969. Archaeology in Western Samoa, Vol. 1. *Auckland Inst. Mus. Bull.* 6, Auckland.
GREEN, R. C. & MARION KELLY (eds.), 1970. Studies in Oceanic Culture History, Vol. 1. Pacific Anthro. Rec. No. 11. B. P. Bishop Mus., Honolulu.
GREEN, R. C. & MARION KELLY (eds.), 1971. Studies in Oceanic Culture History, Vol. 2. Pacific Anthro. Rec. No. 12. B. P. Bishop Mus., Honolulu.
GROUBE, L. M. 1964. Settlement pattern in prehistoric New Zealand. Unpublished M. A. Thesis, Dept. of Anthropology, Univ. of Auckland, Auckland.
GROUBE, L. M. 1965. Settlement patterns in New Zealand prehistory. Anthropology Dept., Univ. of Otago, Occasional Papers in Archaeology: 1, Dunedin (incomplete manuscript).
GROUBE, L. M. 1967. Models in prehistory: a consideration of the New Zealand evidence. *Archaeology and Physical Anthropology in Oceania* 2 (1): 1–27.
GROUBE, L. M. 1968. Research in New Zealand since 1956, in YAWATA & SINOTO (eds.), 1968: 141–149.
GROUBE, L. M. 1970. The Origin and Development of Earthwork Fortifications in the Pacific, in GREEN & KELLY (eds.), 1970: 133–164.
GROUBE, L. M. 1971. Tonga, Lapita Pottery, and Polynesian Origins. *J. Polynesian Soc.* 80 (3): 278–316.
HADDON, A. C. & J. HORNELL 1938. Canoes of Oceania, Vol. III: Definitions of Terms, General Survey, and Conclusions. B. P. Bishop Mus. Spec. Pub. 29, Honolulu.
HANSEN, T. 1970. The Early Development of prehistoric Archaeology in New Zealand. Unpublished M. A. Thesis. Dept. of Anthropology, Univ. of Auckland, Auckland.
HIGHAM, C. F. W. 1970. The Role of economic prehistory in the interpretation of the settlement of Oceania, in GREEN & KELLY (eds.), 1970: 165–174.

HJARNØ, J. 1967. Maori Fishhooks in Southern New Zealand. *Rec. Otago Mus., Anthropology* no. 3, Dunedin.
KENNEDY, JEAN 1969. Settlement in the South East Bay of Islands; 1772. A study in text-aided field archaeology. Anthropology Dept., Univ. of Otago, Studies in Prehistoric Anthropology, vol. 3, Dunedin.
KEYES, I. W. 1970. Wairarapa Chert Sources in New Zealand Prehistory *N.Z. Archaeological Assoc. Newsletter* 13 (3): 128–134.
KEYES, J. W. 1971. Use of Inland Taranaki Rock Sources for Artifacts. *N.Z. Archaeological Assoc. Newsletter* 14 (4): 157–160.
LAW, R. G. 1969. Pits and Kumara Agriculture in the South Island. *J. Polynesian Soc.* 78 (2): 223–251.
LAW, R. G. 1970. The Introduction of Kumara into New Zealand. *Archaeology and Physical Anthropology in Oceania* 5 (2): 114–127.
LEACH, B. F. 1969. The Concept of Similarity in Prehistoric Studies. Anthropology Dept., Univ. of Otago, Studies in Prehistoric Anthropology, vol. 1, Dunedin.
LEACH, HELEN 1969. Subsistence Patterns in Prehistoric New Zealand. Anthropology Dept., Univ. of Otago, Studies in Prehistoric Anthropology, vol. 2, Dunedin.
LEACH, B. F. & HELEN LEACH 1971. Radiocarbon Dates for the Wairarapa – II. *N.Z. Archaeological Assoc. Newsletter* 14 (4): 201.
LEAHY, ANN 1970. Excavations at Site N38/20, Motutapu Island. *Rec. Auckland Inst. Mus.* 7: 61–82.
LEWTHWAITE, G. R. 1949. Human Geography of Aotearoa about 1790. Unpublished M. A. Thesis. Dept. of Geography, Univ. of Auckland, Auckland.
LEWTHWAITE, G. R. 1965. Life and Landscape in Ancient New Zealand, Ch. 1 in Land and Society in New Zealand. R. F. WALTERS (ed.), Wellington, A. H. & A. W. Reed, pp. 13–23.
LOCKERBIE, L. 1959. From Moa-hunter to Classic Maori in Southern New Zealand, in FREEMAN & GEDDES (eds.), 1959: 75–110.
MCEWEN, J. M. 1966. Maori Art, in An Encyclopaedia of New Zealand. A. H. MCLINTOCK (ed.), Wellington, Gov. Printer, pp. 408–429.
MALLOY, B. P. J. 1969. Evidence for Post-Glacial Climate change in New Zealand. *Journal of Hydrology New Zealand* 8 (2): 56–67.
PAWLEY, A. 1970. Grammatical Reconstruction and Change in Polynesian and Fiji, in Pacific Linguistic Studies in Honour of Arthur Capell. S. A. WURM & D. C. LAYCOCK (eds.), Sydney, A. H. & A. W. Reed, pp. 301–367.
PAWLEY, A. 1971 Austronesian Languages, Working Papers in Anthropology, No. 15. Dept. of Anthropology, Univ. of Auckland, Auckland.
PAWLEY, A. & KAYE GREEN, 1971. Lexical Evidence for the Proto-Polynesian Homeland. *Te Reo* 14: 1–35.
PIETRUSEWSKY, M. 1970. An Osteological View of Indigenous Populations in Oceania in GREEN & KELLY (eds.), 1970: 1-12.
POOL, D. I. 1964. The Maori Population of New Zealand. Unpublished PhD dissertation. Australian National Univ., Canberra.
REINMAN, F. M. 1967. Fishing: an aspect of Oceanic Economy. *Fieldiana: Anthropology* 56 (2): 95–208.
REINMAN, F. M. 1970. Fishhook variability: implications for the history and distribution of fishing fear in Oceania, in GREEN & KELLY (eds.) 1970: 47–59.
SHARP, C. A. 1956. Ancient Voyagers in the Pacific. Polynesian Soc. Mem 32. Wellington.
SHARP, C. A. 1958. Maori Genealogies and the 'Fleet', *J. Polynesian Soc.* 67 (1): 12–13.
SHARP, C. A. 1963. Ancient Voyagers in Polynesia. Hamilton, Paul's Book Arcade.
SHAW, ELIZABETH 1963. Maori Quarry, Tahanga Hill, Opito. *N.Z. Archaeological Assoc. Newsletter* 6 (1): 34–36.
SHAWCROSS, F. W. 1964. An archaeological assemblage of Maori Combs. *J. Polynesian Soc.* 73 (4) 382–398.

Shawcross, F. W. 1967. An Investigation of Prehistoric Diet and Economy on a coastal site at Galatea Bay, New Zealand. *Proc. Prehistoric Soc.* 33: 107–31.

Shawcross, F. W. 1969. Archaeology with a short, isolated timescale: New Zealand. *World Archaeology* 1 (2): 184–199.

Shawcross, F. W. 1970. Ethnographic Economics and the study of Populations in Prehistoric New Zealand: viewed through Archaeology. *Mankind* 7 (4): 279–91.

Shawcross, K. A. 1966. Maoris of the Bay of Islands, 1769–1840. A study in changing Maori Attitudes towards Europeans. Unpublished M. A. Thesis, Dept. of History, Univ. of Auckland, Auckland.

Shawcross, K. A. 1967. Fern-Root, and the total scheme of 18th century Maori food production in agricultural areas. *J. Polynesian Soc.* 76 (3): 330–352.

Simmons, D. R. 1967. Little Papanui and Otago Prehistory. *Rec. Otago Mus., Anthropology*, No. 4, Dunedin.

Simmons, D. R. 1968. Man, Moa and the Forest. *Trans. Roy. Soc. N.Z., General Series* 2 (7): 115–177.

Simmons, D. R. 1969a. Economic Change in New Zealand Prehistory. *J. Polynesian Soc.* 78 (1): 3–34.

Simmons, D. R. 1969b. A New Zealand Myth, Kupe, Toi and the 'Fleet'. *N.Z. J. Hist.* 3 (1): 14–31.

Simmons, D. R. & J. B. Wright 1967. Use of the Polarising Microscope for classifying quartzite artifacts from South Island Sites. *Trans Roy. Soc. N.Z., General Series* 2 (4): 71–78.

Sinoto, Y. H. 1970. An archaeologically based assessment of the Marquesas as a dispersal center in East Polynesia, in Green & Kelly (eds.), 1970: 105–132.

Skinner, H. D. 1921. Culture Areas in New Zealand. *J. Polynesian Soc.* 30: 71–78.

Skinner, H. D. 1923. The Morioris of the Chatham Islands. Mem. B. P. Bishop Mus. vol. 9, no. 1, Honolulu.

Skinner, H. D. 1924. Origin and Relationships of Maori Material Culture and Decorative Art. *J. Polynesian Soc.* 33: 229–43.

Trotter, M. M. 1967. Investigations of a moa-hunter site at Redcliffs, Sumner. *Rec. Canterbury Mus.* 8 (3): 247–250.

Vayda, A. P. 1960. Maori Warfare. Polynesian Soc. Maori Monographs No. 2. Wellington.

Wilkes, O. R. & R. J. Scarlett 1967. Excavation of a Moa-hunter site at the Mouth of the Heaphy River. *Rec. Canterbury Mus.* 8 (3): 177–208.

Yawata, I. & Y. H. Sinoto (eds.), 1968. Prehistoric Culture in Oceania: A Symposium. Honolulu, Bishop Museum Press.

Yen, D. E. 1971a. The Development of Agriculture in Oceania, in Green & Kelly (eds.), 1971: 1–12.

Yen, D. E. 1971b. Construction of the Hypothesis for Distribution of the Sweet Potato, in Man Across the Sea. C. L. Riley, J. C. Kelley, C. W. Pennington, R. L. Rands (eds.), Austin, University of Texas Press. pp. 328–342.

XVII. THE INFLUENCE OF MAN ON THE BIOTA

by

J. T. SALMON

Introduction

When New Zealand was discovered by European man it was inhabited by Polynesian peoples who had occupied the country for at least 600 years. As a land mass the New Zealand islands had been geologically isolated from the rest of the world over the entire tertiary geological era. During this time New Zealand was surrounded by extensive seas and only occasional migrants, such as wandering birds, marine animals and occasional plants reached this country from outside areas. Globally the tertiary era was the age of mammals and flowering plants but during this period of isolation few of these species from the outside world became established in New Zealand. Those, however, that did reach New Zealand were able to evolve and spread without challenge by new arrivals. Many plants developed new characteristics, or, new species especially suited to the New Zealand environment were evolved. This environment embraced zones that varied from warm humid regions through exposed wind swept plateaus to alpine fell-fields, all of which became densely populated with a luxuriant plant cover. This contained some genera which gave rise to large numbers of species, often with peculiar adapted forms and structures, as for instance in *Hebe* and *Coprosma*. Plant communities evolved unique to New Zealand and outstanding amongst these were the kauri forests of the North Island, the beech forests of the South Island, the dense shrubby undergrowth and epiphytes of the podocarp forests, the cushion plant formations of the high alpine regions, the subalpine and alpine tussock lands and herb fields, and the ferns. During this period of evolution of unique plant communities land mammals were excluded and the vegetation of New Zealand emerged from this period of isolation unadapted to compete with browsing animals.

According to Captain Cook and the early naturalists who visited New Zealand, the animal life consisted mainly of birds and insects, the tuatara and a few other species of lizards, two species of bats, the Maori dog and a species of rat *(Rattus exulans)*.

The first feral animals to be introduced to the country were pigs, liberated by Captain Cook, and possibly also some goats. According to Cook and the early naturalists the New Zealand forests carried a heavy population of native birds. This is highlighted in Cook's and other naturalists' descriptions of what is referred to as the 'dawn chorus' of the New Zealand forest birds.

Recent research seems to indicate that the first impact of man on these primitive biota was made by the peoples who migrated here from Polynesia in the north. The Maori used fire extensively to clear land and possibly also as a method of hunting down the flightless birds called moas, which were used as food. It now seems fairly certain that the various species of moa in New Zealand were exterminated by hunting them for food and in this process large areas of country were burned of forest and reverted to tussock. This theory is supported by several ecologists at the present time and by carbon dating of moa remains and logs which have been unearthed in various parts of the country. However, when European man first started to come to New Zealand the biota were still largely in their original primitive state. The land dominated by luxuriant widespread forests; the forests prolific with birds. On the coasts were many seal colonies which the Europeans very quickly exploited until many were almost exterminated.

The kauri forests of the northern part of the North Island formed one of the most extensive and valuable natural resources of New Zealand. This was quickly appreciated and a very lucrative export of kauri timber developed by the middle of the nineteenth century and continued until these vast forests were almost completely destroyed. Not only was the timber used but the gum was also harvested and used as an extremely valuable ingredient in the paint and varnish industry. It could be tapped from the trees but a more valuable form of gum was found by digging in the ground. Subterranean gum had qualities which produced finer forms of varnish. The gum digging industry followed the felling of the kauri forests and persisted for a long time until synthetic gums largely replaced natural gums in the paint and varnish industry. The forests, to the early settlers, were almost an enemy and had to be cleared in order that crops could be grown. The greatest ecological change which has been brought about in New Zealand resulted from the clearing of the mixed podocarp forests and the conversion of forested lands to farmland, mostly pasture, for the grazing of sheep and cattle.

In its original state, as discovered by European man it is estimated that 68% of the land surface of New Zealand was covered by forests. Today this percentage has been reduced to about 14% and most of the remaining forest is locked up in scenic reserves, national parks and crown lands preserved as watersheds. With the destruction of the forests the native birds retreated and their populations dwindled. Their places were taken by introduced european thrushes, sparrows, starlings and blackbirds, which with other introductions, such as the chaffinch are the common birds of farming country in New Zealand today.

New Zealand landholders in the early days were compelled by law to improve their land and many did so simply by burning the forest to clear the land. Much of this burned over land subsequently reverted to scrub and fern.

Fig. 1. Unharmed indigenous New Zealand forest, West Coast, South Island. Photo: N.Z. Forest Service by J. H. Johns, A.R.P.S.

At the close of the Great War much of this second rate or marginal land was again cleared by Government order and turned into farms for returned soldiers. Marginal lands of this nature in New Zealand have never been successfully farmed until the advent of aerial topdressing. It is still a moot point as to whether such lands can be kept in production economically by ordinary farming methods. As a result of much of this indiscriminate forest clearance which was done without regard to land classification or land form, the pattern of watersheds and water control was changed radically in many parts of the country. Flooding and

Fig. 2 and 3. 36 years' usage by people of a scenic reserve forest near Wellington. Fig. 2 the stream banks as they were, thick with shrubby undergrowth in March, 1933. Fig. 3. the same place in February, 1969, the shrubs mostly destroyed or replaced by sedges. These processes of destruction are slow enough to pass unnoticed from year to year but their cumulative effects pose a major problem in scenic reserve conservation. Photos: J. T. Salmon, F.R.P.S.

erosion has followed as a natural consequence until today flood control by building of stopbanks, dams and other forms of engineering works costs vast sums of money throughout the country. As more land is opened up for farming flooding becomes more severe and more frequent. As high country land became more intensively farmed, particularly following the turn of the century, accelerated erosion increased in severity until today the instability of some parts of New Zealand, notably in the foothills of the mountains bordering South Island's Canterbury Plains, in the Whangamomona County and the Gisborne district of the North Island has no equal anywhere else in the world.

With the passing of the Soil Conservation and Rivers Control Act in 1941, the first steps were taken by Government to curb the menace of accelerated erosion in New Zealand. After 27 years operation this Act was consolidated and amended by the Water and Soil Conservation Act of 1968. Under the old Act Catchment Boards were set up throughout the country and a great deal of excellent work has been done in combating the menace of erosion, but much still remains to be accomplished.

Over recent years the Forest Service has assisted with afforestation schemes, for instance in the Gisborne district, primarily to control severe erosion and slipping. Much of the accelerated erosion in New Zealand has resulted from over-grazing by sheep and this is particularly evident in the South Island. With the clamour for more and better quality land quite early in the development of the country attempts were made to drain swamps. The drainage of wetlands has become, over the years, a major feature of New Zealand land development. These drainage schemes have completely changed the ecology of large areas seriously affecting the populations of water birds from one end of the country to the other. Today, with few of these areas left, public opinion is very much against further drainage of wetlands in New Zealand.

With the development of hydro electric power large engineering works have been undertaken on the Waikato River in the North Island, on the Waitaki and Clutha Rivers in the South Island, and attempts have been made to alter the levels of lakes, notably at Waikaremoana in the North Island and at Lakes Hawea and Manapouri in the South Island. Almost all the other lakes in New Zealand have also been investigated for electric power generation. The damming of rivers for this purpose has brought about great changes in water tables and the general ecology of the valley regions so affected.

Second only to the development of pasture lands in New Zealand have been the development of pine forests from the introduced species *Pinus radiata*. In both the North and South Islands very extensive areas of man-made forests are now coming into production and supporting a booming forestry industry. The ecological change in the country brought about by the growth of these pine forests is as yet little understood but the introduction and growth of a monoculture on such a vast scale as is practised in New Zealand cannot help but have a profound effect in the long term ecology of the area.

In common with most other advanced countries in the world New Zealand has suffered heavily from the effect of pesticides and herbicides, particularly organo-chloride insecticides which have been used very extensively all over the country. The long term effects of the use of these chemical substances on the native insects and on the general ecology of soils is little understood and has not been extensively investigated. In common with many other countries, however, New Zealand recently has banned the use of DDT in farming. Much of the success of the New Zealand farming industries has stemmed from the use of artificial fertilisers. When these substances, particularly superphosphate, began to be applied to marginal lands and later to ordinary farmlands by aerial topdressing methods it was not long before consequential changes in the ecology of waterways began to be noticed by biologists. Today it is quite common knowledge that the ecology of streams, rivers and lakes is being seriously affected by the leeching of superphosphate and other agricultural chemicals from the lands on to which they are applied, into the waterways. Eutrophication is becoming widespread and very few streams and rivers in the farming areas of New Zealand are now free from pollution by agricultural chemicals. In addition to this pollution from sewage coming from domestic sources and farms is common. It has become so serious that measures are now being taken in many parts of the country to install sewage treatment plants for towns and cities while many farms are being equipped with recycling sprays to return the washings from cow bales, piggeries and other farm sources back to the land and to prevent it from draining into streams and rivers.

Following on the conversion of forest land to pasture, native insects such as the grass grub which is the larvae of a cockchafer beetle, *Costelytra zealandica*, have become widespread pests of pastures. As important as the grass grub is the caterpillar of several species of Hepialid moths known now as *Wiseana* but previously as *Porina*. With the establishment of a monoculture of grass pasture these insects responded to the ecological change which was favourable to them and developed huge subterranean populations. These feed upon the roots of the pasture grasses bringing about extensive damage to the pastures often resulting in large patches of bare ground. They cause tremendous economic losses. These insects have been resonably well controlled by the use of DDT and other related compounds. The damage caused to other soil inhabiting organisms by the use of pesticides of this nature to control these pasture pests is as yet little investigated nor understood in New Zealand.

However, it is quite clear, biologically speaking, that man's impact on the primitive biota in New Zealand through conversion of forest to pasture and of great areas to introduced pines has resulted in far reaching consequential ecological changes. These changes have not yet reached a

Fig. 4. The effects of browsing animals on New Zealand forest – an animal proof enclosure with young seedlings thriving, while outside the fence undergrowth is completely destroyed by noxious animals. Photo: N.Z. Forest Service by E. W. J. Wilson.

new state of equilibrium and are probably unlikely to do so for many years yet. Apart from the changes which have been directly brought about by forest clearance and the introduction of new kinds of plant species for cropping, the other most striking impact of man on the New Zealand biota has resulted from the introduction into the country of many forms of domestic and feral animals. Domestic animals were mostly brought here for farming purposes and as household pets. Feral animals were introduced purposely and liberated for sporting or fur production purposes. Some of the domestic species have also escaped and gone wild – sheep and goats for instance – and these wild browsing animals have brought about drastic changes in the natural ecology of the regions into which they were purposely introduced.

Sheep, cattle and horses were introduced early for farming purposes. Dogs and cats were brought for domestic purposes. Species of deer, species of birds, wallabies and rabbits were brought for sporting purposes, although rabbits were possibly brought more for food. The Australian opossum was brought to establish a fur industry. A number of species were brought to control others which turned out to be pests. More escaped through the avenues presented by commerce while others just simply appeared, presumably arriving in this country more or less as stowaways on early trading ships.

The wild pig, *Sus scrofa* L., was one of the first animals to be purposely introduced to New Zealand. It is known that Captain Cook liberated some of these animals in both the North and South Islands, near Cape Kidnappers and Queen Charlotte Sound respectively, but pigs were also liberated in New Zealand by early settlers and other Europeans who visited these shores as traders and explorers in the early eighteenth century. Liberated pigs quickly adapted and became feral, though it is recorded that many Maoris kept pigs in a semi-domesticated state near their pas and transported them from place to place. The wild pig is now firmly established throughout New Zealand and some of the outlying islands, such as Mayor Island, Chatham, Pitt and the Auckland Islands. They were liberated and flourished on the Poor Knights Islands but were exterminated from there in 1936 by the Department of Internal Affairs.

Goats were also liberated in New Zealand by Captain Cook at Queen Charlotte Sound on 2 June, 1773. Goats have also been liberated from time to time in other parts of the country and on some of the outlying islands, such as the Three Kings, Great Barrier and Auckland Islands. Goats were also brought into the country to check the spread of blackberry or other weeds and to help consume scrub to facilitate bush clearing for sheep grazing. It is certain that many goats escaped and became feral and that from these the serious goat infestations of Taranaki, Marlborough and Hawke's Bay have undoubtedly developed.

The opossum, which is now widespread in New Zealand, belongs to the marsupial family *Phalangeridae* and was brought here from Australia. It belongs to the species *Trichosurus vulpecula* Kerr. The animal was introduced ostensibly to establish a fur industry. The first liberations were made in 1858 in the South Island near Riverton. Other liberations were made in Canterbury in 1865 and 1875 and in 1890 they were liberated on the Auckland Islands, on Stewart Island, in the West Coast Sounds and in the Nelson district. During 1894 and 1895 opossum were liberated in the Otago district, and in 1905 in Westland near Kumara. The first liberations in the North Island were between 1869 and 1876 in the vicinity of Auckland and in 1870 at Kawau Island. In 1890 they were liberated near Gisborne; in 1892 near Paraparaumu and in 1893 on Kapiti Island. Between 1893 and 1915 they were liberated from Wellington through Taranaki to Rotorua and in 1920 on the Coromandel Peninsula. They have multiplied at an ever increasing rate and now assume epidemic proportions in different parts of the country. When this occurs drastic means of control have to be implemented in order to reduce their number.

Between 1867 and 1905 some eight species of deer were liberated in New Zealand, the first of these being *Cervus axis* which was liberated

Fig. 5. Red Deer herd, Upper Dart Valley. Photo: N.Z. Forest Service by J. H. Johns, A.R.P.S.

near Palmerston South in Otago in 1867 and later in 1893 on Kapiti Island, in 1907 in the Tongariro National Park and in 1908 at Dusky Sound. Fallow deer, *Dama dama*, were first liberated in 1864 near Nelson and between then and 1900 were liberated at many places in different parts of both islands. Japanese deer were liberated near Oamaru in 1885 and on the Kaimanawa Ranges in the North Island in 1905. Moose were liberated in 1900, near Hokitika, and in 1910 in Dusky Sound. Samba deer, *Cervus unicolor*, were liberated in the North Island in the Rangitikei Valley in 1875 and later in 1876 in the Waikato, the last liberation of this species being in 1920 near Rotorua on Mt Tarawera. Virginia deer were first released in 1901 at the Takaka Valley in Nelson and later in 1905 both on Stewart Island and the Rees Valley at the head of Lake Wakatipu. Wapiti were released in the first instance in New Zealand in 1870 on Kawau Island but they did not survive. They were later released in 1905 at George Sound and it is from this introduction that the present day Wapiti herd has grown. The Himalayan Thar was liberated in the Mt Cook area in 1904 and again in 1909. Chamois was also liberated in the Mt Cook area in 1889 and again in 1907 and 1913. By the end of World War II herds of these animals had grown considerably in numbers and had spread extensively throughout the Southern Alps mountain region. Wallabies from Australia were also introduced into New Zealand about 1870 on Kawau Island and in 1903 at Mt Nimrod near Timaru.

Hedgehogs were first brought into the country in 1885 and became well established. The first rabbits were brought to New Zealand shortly before 1838 and quickly became firmly established.

Hares were first introduced in 1867, near Waihola in Otago and were introduced on a number of occasions between 1868 and 1905 at different places throughout both islands. Weasels, stoats and ferrets were brought into the country between 1885 and 1905 and quickly established themselves. In addition to these, domestic animals have escaped and become feral, notably cats which are wild throughout the country and also at the bird sanctuary on Little Barrier Island. Dogs have also gone wild, as also have horses, sheep and cattle. Rats and mice became established in the country almost concurrently with European settlement and are now widespread. In addition to these animals which have established themselves successfully, a number of other animals were brought to New Zealand and liberated but failed to establish themselves. Amongst these were such things as bandicoots, kangaroos, racoons, squirrels, bharals, gnu, camels and zebras.

Many species of birds have been introduced to New Zealand. These include the European blackbirds, thrushes, sparrows, rooks, yellow hammers, chaffinches, the budgerigar and several species of pheasants for sporting purposes. Weasels and allied animals, rodents, cats and dogs that have gone wild have not so far caused serious upset to the natural environment in New Zealand. But rabbits, opossums, wallabies, deer, thar and chamois have caused serious damage to forests and alpine meadows and have brought about extensive changes in the balance of native vegetation. There is a tendency for unpalatable species to increase and for species of plants palatable to these animals to diminish, almost in some cases to the point of disappearance. The rabbit caused widespread destruction to pasture land in many parts of New Zealand. Some idea of the rate of increase in the rabbit population can be gleaned from the export of skins. In 1873 33,000 rabbit skins were exported. Four years later in 1877 nearly 1 million were exported and by 1882 over 9 million rabbit skins were exported. By 1925 the annual export of rabbit skins had reached just over 20 million. The economic loss to pastures and crops in New Zealand, as a result of the vast increase in rabbits is incalculable. The carrying capacity of sheep stations was seriously affected and as an example of this Castle Rock Station in Southland with an area of 70,000 acres carried 50,000 sheep before it was invaded with rabbits. After the rabbit invasion the stock of sheep was progressively reduced to stabilize at 20,000. The Rabbit Act of 1928 set up Rabbit Boards throughout the country whose purpose was to endeavour to control rabbits. These Boards used trapping, poisoning and fumigation, digging out of nests and shooting as means of control but the rabbit as a pest was never brought under control until aerial dropping of poisoned baits was extensively used. Attempts were also made to introduce the disease myxomytosis, but without success. Poisoning from air dropped baits has, however, greatly reduced the rabbit population throughout New Zealand and for the first time brought this animal under control. The rabbit successfully colonized all parts of the country suitable to it and it has been shown that the value of rabbits in skins and carcases did not even equate the losses in grazing capacity alone which resulted from heavy rabbit infestation of pasture and cropping land. To this must be added the cost of rabbit control and the hidden costs in rabbit induced soil erosion in hilly country. When all this is taken into account it is abundantly clear that the introduction of rabbits into New Zealand was disastrous from every point of view.

Pigs have spread throughout the forest areas of New Zealand and are extensively hunted for sport. The chief damage arises from their rooting and churning up of the forest litter and subsoil. They are not regarded as seriously interfering with the basic ecology of the country and never appear to have multiplied to anything like the extent of the browsing animals brought into New Zealand.

Goats, which were both liberated and escaped from farms, caused widespread damage to scrub and forest in lowland and subalpine regions. They have been used successfully to combat blackberry and other weeds but have multiplied to form extensive herds. By their ability to scramble in steep places and get out on to the limbs of trees they have become a serious menace to the remaining forests of New Zealand on both the mainland and outlying islands. Shortly after the close of the second world war goats were exterminated from Great Island of the Three Kings group and extensive and interesting studies have been made of the regeneration and recovery of the forests of this island.

The most serious menace to the forests of New Zealand, however, is the introduced Australian opossum. First brought here as the basis for a fur industry, this animal has multiplied beyond all expectations and spread throughout the forests of both islands of New Zealand. It feeds by eating the young shoots and fresh leaves of many of the native trees forming the forest canopy and also of many of the shrubs and ferns that form the undergrowth and lower canopy of our forests. On the west coast of the South Island, on Mt Egmont and in the vicinity of Wellington ratas are being extensively damaged by the browsing of these animals and large mature trees are killed in 2–3 years as a result of heavy infestations of opossum. These animals also attack pohutukawa, invading urban areas where these trees grow. The fuschia is rapidly disappearing

Fig. 6. A rata tree completely defoliated and killed by opossum, Kokotahi Valley, Westland, South Island. Photo: N.Z. Forest Service by J. H. Johns, A.R.P.S.

from the New Zealand bush and the once beautiful grove of fuschia trees in the crater of Mt Tauhara near Taupo has been exterminated by opossum. Five finger trees in the undergrowth of forests are extensively damaged by the browsing of opossum when they have the appearance after a severe attack of having been burned. These trees often die as a result. There are few of our indigenous shrubs and trees that are immune to the attack of opossum and this pest is now invading and attacking the man-made forests of introduced pines. As the opossum attacks the canopy and the undergrowth of the indigenous forests, these forests are opened up, as it were, and made less impenetrable for deer which follow the opossum and inflict further serious damage on the forests. Deer browse on the undergrowth and the lower branches of trees. Herds of these animals moving through the forests completely eating out the undergrowth, leave the forest floor bare. Only totara seedlings which are tough and spikey are left. Both deer and opossum browse on the seedlings of beech and other forest trees so that recovery of the forest is negligible.

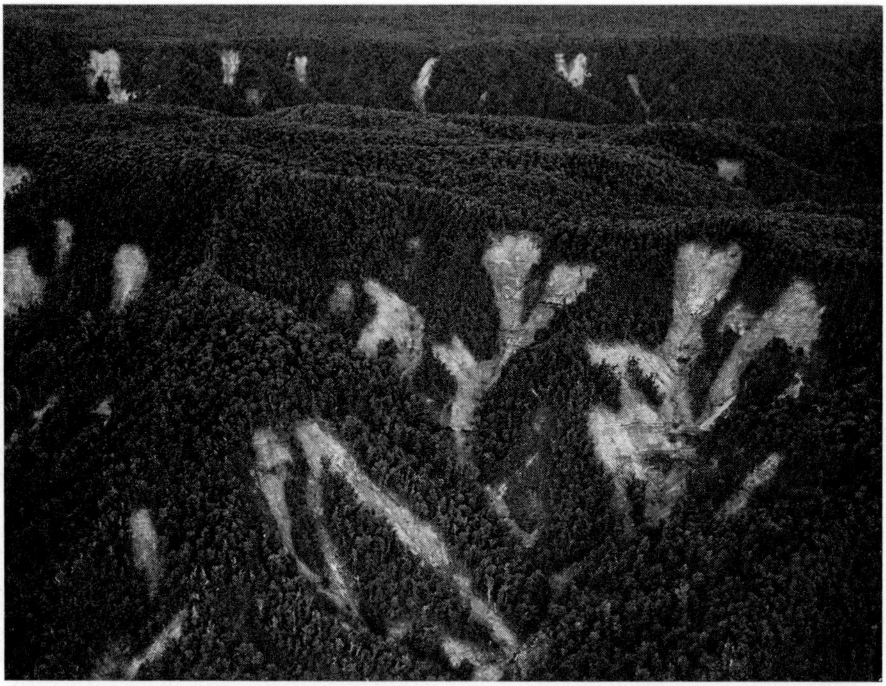

Fig. 7. Noxious animal induced erosion, Urewera Country, North Island – the combined effects of opossum, red and Japanese deer. Photo: N.Z. Forest Service by J. H. JOHNS, A.R.P.S.

After severe attack of this nature the forest floor usually becomes covered with hard fern and, at this stage, it seems, a forest is unable to replace its dominant trees and, with regeneration of most species reduced to a minimum, is doomed. Most of the species of deer introduced into New Zealand will leave the bush-line and advance out into the sub-alpine and alpine meadows of the mountains during the summer, browsing on tussock and other small shrubs. In the higher more inaccessible regions of the South Island mountains populations of thar and chamois have become widely distributed. These animals have taken a heavy toll of the alpine and sub-alpine vegetation in many parts of the main mountain chains of the South Island. Their numbers had reached such proportions after the close of World War II that Government hunters were appointed whose sole purpose was the culling of the herds of deer, thar and chamois. The populations of these animals were reduced to a level at which the vegetation in the mountain regions could start to recover. In areas where effective control of animals has been established it is remarkable how quickly the alpine vegetation reappears and how luxuriant it can again become. It seems evident, however, that the balance of species is being

Fig. 8. Sheep induced erosion through overstocking and overgrazing of land, Mt Fitzwilliam, Harper-Avoca Catchment. Photo: N.Z. Forest Service by J. H. JOHNS, A.R.P.S.

changed in both the alpine vegetation and in the forest undergrowth browsed by animals and the forests are unlikely to be the same ecologically as they were before the infestation of browsing animals reached serious proportions. There seems to be a danger, too, that certain species once quite common will become rare as a result of the pressure of browsing animals, particularly the opossum. Where a forest is seriously damaged by these animals and the undergrowth removed it appears that the micro-climate of the lower storey of the forest is dramatically changed. This change weakens the resistance of the main forest species so that in storms a collapse can occur followed by slipping and incipient erosion which, if conditions prevent healing, leads to serious accelerated erosion. This kind of thing is evident in many parts of the alpine and forest regions of New Zealand. Accelerated erosion, however, does not result entirely from the depradations of these feral introduced animals. It also arises in much of our hill country from poorly managed saw milling and, particularly in the South Island, from over-grazing by sheep. For example the almost complete destruction of the vegetation on parts of the Torlesse Range and in the Craigieburn Mountains of Canterbury, as well as other parts of North and South Canterbury, has resulted almost entirely

Fig. 9. Silt and gravel filling the Waipaoa River bed, North Island, as a result of forest clearing in the high back country. Photo: N.Z. Forest Service by J. H. JOHNS, A.R.P.S.

from the effect of over-grazing by sheep assisted in some places by rabbits. The destruction in parts of the Torlesse Range and the Craigieburns is almost complete. The author himself can recall areas which today are moving shingle slides but which thirty years ago were in heavy tussock.

Fig. 10. Thar herd on ridge in Rangitata River catchment, Canterbury, South Island. Photo: N.Z. Forest Service by J. H. JOHNS, A.R.P.S.

Molesworth Station in Marlborough was a classical example of this sort of thing, but today, after nearly three decades of careful management and conservation is clear evidence of what scientific land management and conservation can accomplish.

The damaging effect of introduced plants and animals upon the natural ecology of the New Zealand countryside became obvious quite early in the settlement of this country. After establishment of the provincial councils in the early 1840's laws were soon passed to control nuisances caused by introductions into the colony. In 1844, for instance, laws were enacted to control dogs wandering at large in towns. The first Rabbit Nuisance Act appeared in the statutes in 1876 and this was followed in 1882 by the Small Bird Nuisances Act. Sir WALTER BULLER campaigned vigorously against the introduction of ferrets, stoats and similar animals, but without success. As early as 1861 the colonial Parliament passed legislation to provide for the protection of selected introduced animals, birds and fish brought into the country for sporting purposes. This act also included hares and deer. It was not until 1907 that legislation was passed in New Zealand giving protection to New Zealand's native animals and birds. This first act gave protection to the tuatara and some thirty-six species of native birds. This type of protection was further extended by the Animals and Game Protection Act of 1921 which extended the mantle of conservation to native bats, frogs and a further 132 species of native birds. It also afforded protection to the Australian opossum, the introduced magpie and other species of introduced animals. However, the mounting damage to native forests and alpine regions caused by introduced browsing animals together with the subsequent increased erosion in forest and high country soils became of such magnitude during the 1920's that a conference was called in 1930 presided over by the then Minister of Internal Affairs. This conference recommended that Government remove the protection previously given to certain of these introduced animals. This was done and organised culling parties began to operate on a large scale. As a result of these activities a considerable reduction in the number of browsing animals was brought about and this policy has been pursued ever since. During some of these campaigns against rabbits, deer and opossum the poison 1080 was used as a means of control. This caused considerable public disquiet and a further conference on wildlife control was called by the Minister of Forests in 1958. As a result of this conference control measures on introduced animals were intensified and Government in 1963 appointed the Lands and Agricultural Committee of the House of Representatives to further inquire into the whole question of noxious animal control. The report of this committee was issued in 1965 and recommended no relaxation in the relevant provisions of either the Noxious Animals Act or of the National Parks Act. The committee also recommended increased financial aid for the control of noxious animals in National Parks. Opossum, deer, goats and other introduced animals in New Zealand were declared noxious by the Noxious Animals Act of 1956. Although there has been considerable opposition to this Act by certain sporting sectors of the community the Act still stands. It is administered today by the Forest Service and the campaign against these noxious animals in our forests and alpine regions continues. The Ecology Division of the Department of Scientific and Industrial Research is conducting research into the opossum with a view to obtaining better control of this animal. The amount of research authorised by Government in connection with noxious animal control is pitiful in relation to the enormity of the problem. Although extermination of these animals is the official objective considerable divergence of opinion exists in scientific circles as to whether extermination would ever be possible. Many biologists consider that the best that can be achieved would be a measure of control. It is unlikely, however, that this dilemma will ever be solved so long as the amount of money and effort made available for research into noxious animals remains at such a low level.

The changes in the composition of our forests and alpine vegetation due to the impact of these animals are still proceeding and it is impossible to predict at present what the final ecological equilibrium is likely to be. Very few areas of unaltered indigenous forest remain in

New Zealand and it seems highly probable that the New Zealand indigenous forests of the future will be of a very different composition from that of the forests encountered by our early settlers. Under present circumstances and in the light of our existing ecological knowledge of this situation the current policy of extermination of introduced feral animals seems the only sensible policy to pursue. Until New Zealanders as a society can make up their minds whether the current policy of total extermination or the alternative policy of control should be followed the policy of extermination is the only policy likely to achieve any measure of control. The author believes that extermination is possible given a sufficient effort in research to develop methodology. Until this is done the policy of extermination will at least give us a measure of control even though it is based largely on opinion and has very little factual knowledge to support it. Research which has been carried out on this problem has been shared by Government agencies, University staffs and students and dedicated individuals. Much more effort is needed before wildlife management in New Zealand can be placed on a secure footing. Even if control is decided upon the management of wildlife in a controlled programme is a scientific business which must be based on sound ecological knowledge and this can only be obtained from long periods of painstaking research.

So long as the protagonists of total extermination or control by management continue to argue in the absence of scientific data provided by research there can be no true solution to New Zealand's major wildlife problems. Therefore, until a decision on which of these two policies to follow is made wildlife control and forest and range management in New Zealand must continue to be largely a hit and miss affair. To arrive at a sensible solution to the dilemma the value of the introduced browsing animals as a recreational and sporting facility for hunters must be evaluated against the damage which they are doing to the total environment of New Zealand.

This means not just the damage which they do to bush and alpine vegetation but all the consequences in erosion and damage to agricultural and pastoral lands which follows from the diminishing effectiveness of the damaged vegetation to control and regulate the flow of water from watersheds to the sea.

In an attempt possibly to settle this question Government set up a commission to enquire into the organisation of wildlife management and research in New Zealand. This commission issued its report in November 1968. It is known as the Hunn report after the chairman of the commission. Although this report recommended far reaching changes in wildlife management and research it has failed to secure any agreement for its implementation between the opposing factors on wildlife control in New Zealand. No effective action has so far been taken by Government on any of its recommendations. It seems clear, therefore, that for the next decade, at least, official policy on noxious animals in New Zealand will 'muddle along' in much the same way as it has done in the past. The problem will grow in intensity and seems unlikely to be solved until New Zealand is faced with dire catastrophe arising from the impact of these many introduced animals on the basic ecology of these islands.

GENERA INDEX

(a solidus, '/', between two numbers indicates position of fold-outs).

C

D

G

H

N

O

T

SUBJECT INDEX